TREATISE ON ANALYSIS

Volume IV

This is Volume 10-IV in
PURE AND APPLIED MATHEMATICS
A series of Monographs and Textbooks
Editors: SAMUEL EILENBERG AND HYMAN BASS
A list of recent titles in this series appears at the end of this volume.

Volume 10

TREATISE ON ANALYSIS

10-I. Chapters I–XI (Foundations of Modern Analysis, enlarged and corrected printing, 1969)
10-II. Chapters XII–XV, 1970
10-III. Chapters XVI–XVII, 1972
10-IV. Chapters XVIII–XX, 1974

TREATISE ON

ANALYSIS

J. DIEUDONNÉ

Nice, France

Volume IV

Translated by

I. G. Macdonald

University of Manchester
Manchester, England

ACADEMIC PRESS **New York and London** **1974**

ACADEMIC PRESS, INC.
111 Fifth Avenue, New York, New York 10003

United Kingdom Edition published by
ACADEMIC PRESS, INC. (LONDON) LTD.
24/28 Oval Road, London NW1

Library of Congress Cataloging in Publication Data

Dieudonné, Jean Alexandre, Date
Foundations of modern analysis.

(Pure and applied mathematics; a series of monographs and textbooks, 10)
Vols. 2- have title: Treatise on analysis.
Vol. 1, "an enlarged and corrected printing" of the author's Foundations of modern analysis, published in 1960.
Includes bibliographies.
1. Mathematical analysis. I. Title. II. Title: Treatise on analysis. III. Series.
QA3.P8 vol. 10 1969 510'.8s [515] 73-10084
ISBN 0-12-215504-1 (v. 4)

PRINTED IN THE UNITED STATES OF AMERICA

"Treatise on Analysis," Volume IV

First published in the French Language under the title "Elements d'Analyse," tome 4 and copyrighted in 1971 by Gauthier-Villars, Éditeur, Paris, France.

SCHEMATIC PLAN OF THE WORK

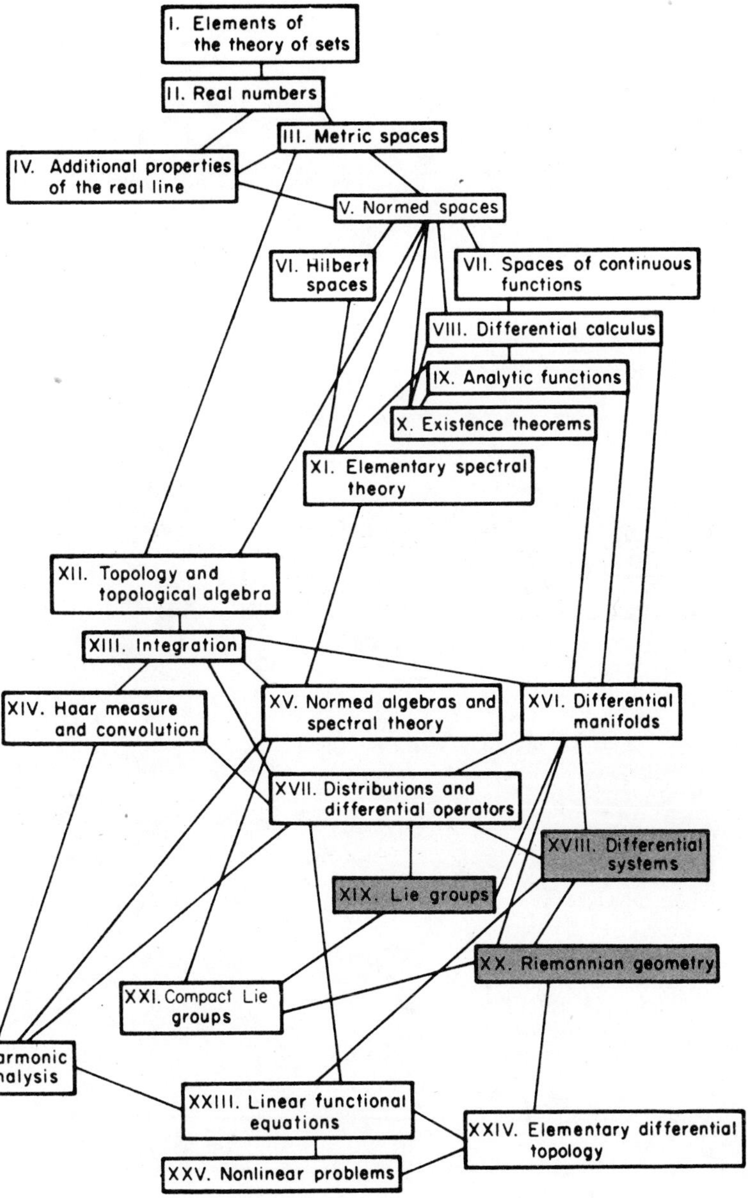

CONTENTS

NOTATION

In the following definitions, the first number indicates the chapter in which the notation occurs and the second number indicates the section within that chapter.

E	unit vector field on $\mathbf{R}$, such that $\tau_t(E(t)) = 1$: 18.1
$v'(t)$	velocity vector (or derivative) $T(v) \cdot E(t)$ of a mapping v of an interval of $\mathbf{R}$ into a manifold, at a point $t \in \mathbf{R}$: 18.1
$\mathrm{J}(x_0)$, $t^-(x_0)$, $t^+(x_0)$	interval of definition of the maximal integral curve v of X such that $v(0) = x_0$, and its endpoints: 18.2
F_X	flow of a vector field X: 18.2
$\mathrm{dom}(\mathrm{F}_X)$	domain of the flow of X: 18.2
v'', $v^{(r)}$	successive derivatives of a mapping v of an interval of $\mathbf{R}$ into a manifold: 18.3
Ω_Z, Ω	domain of definition of the exponential of a spray Z over M: 18.4
$\exp_Z$, $\exp$	exponential mapping defined by a spray Z: 18.4
$\tilde{X}$	canonical lifting to T(M) of a vector field X on M: 18.6, Problem 3
E_1, E_2	canonical vector fields on $\mathbf{R}^2$: 18.7
$f'_t(t, \xi)$, $f'_\xi(t, \xi)$	partial derivatives of a mapping of an open subset of $\mathbf{R}^2$ into a manifold: 18.7

∇_t, ∇_ξ	covariant derivatives in the directions of the vector fields E_1, E_2: 18.7
$\mathscr{E}_0$	**R**-algebra of C^∞ real-valued functions on M: 18.9
$\mathscr{E}_p$	space of C^∞ real differential p-forms on M: 18.9
$\mathscr{A}$	direct sum of the $\mathscr{E}_p$: 18.9
$s_0(x)$, $(s_1x, \mathbf{u}_1)$, ..., $s_q(x, \mathbf{u}_1, \ldots, \mathbf{u}_q)$	integers defining an increasing sequence of integral elements: 18.10
$\mathrm{rk}_x(\alpha)$	rank of $\alpha(x)$, where α is a differential 2-form: 18.16, Problem 2
$\gamma(s)f$	f a section of a bundle E with base B, s an element of a Lie group acting equivariantly on E and B: 19.1
$\gamma(s)P$	P a differential operator from a vector bundle E to a vector bundle F, s an element of a Lie group acting equivariantly on E (resp. F) and B: 19.1
$\gamma(s)X$	X a vector field on M, s an element of a Lie group acting on M: 19.1
$\delta(s)f$, $\delta(s)P$	actions on the sections defined by a right action: 19.2
$\mathrm{Int}(s)$	inner automorphism $x \mapsto sxs^{-1}$: 19.2
$\mathrm{Ad}(s)$	automorphism $\mathbf{h}_e \mapsto s \cdot \mathbf{h}_e \cdot s^{-1}$ of the tangent space at the identity element: 19.2
$\mathfrak{G}$	infinitesimal algebra of a Lie group G: 19.3 and 19.17
$\mathfrak{g}$	Lie algebra of left-invariant vector fields on a Lie group G: 19.3 and 19.17
$\mathfrak{G}_e$	algebra of distributions with support contained in $\{e\}$: 19.3 and 19.17
$\mathfrak{g}_e$, Lie(G)	Lie algebra of a Lie group G: 19.3 and 19.17
$X_\mathbf{u}$	left-invariant vector field, equal to $\mathbf{u}$ at the point e: 19.3
$f_*(P)$	image of a field of distributions $P \in \mathfrak{G}$ under a homomorphism f: G → G′: 19.3
f_*	derived homomorphism $f_* : \mathfrak{G} \to \mathfrak{G}'$ of a Lie group homomorphism f: G → G′: 19.3
Lie(f), f_*	homomorphism Lie(G) → Lie(G′) derived from a Lie group homomorphism f: G → G′: 19.3
P_M	differential operator on M associated with a left-invariant differential operator P on G, where G acts on M on the right: 19.3
$Z_{\mathbf{u},\mathrm{M}}$, $Z_\mathbf{u}$	Killing field on M corresponding to $\mathbf{u} \in$ Lie(G), where G acts on M on the right: 19.3
$\mathfrak{gl}(\mathrm{E})$, $\mathfrak{gl}(n, \mathbf{R})$, $\mathfrak{gl}(n, \mathbf{C})$, $\mathfrak{gl}(n, \mathbf{H})$	Lie algebras of general linear groups: 19.4

$o_m(\mathrm{U})$	set of functions f such that $f(\mathbf{x})/\Vert\mathbf{x}\Vert^{m+1}$ is bounded in U: 19.5
$\tilde{f} = \sum_{\alpha} a_\alpha \mathrm{X}^\alpha$	formal Taylor series of f: 19.5
$f \sim \sum_{\alpha} a_\alpha \mathbf{x}^\alpha$	Taylor series of f: 19.5
Δ_α	distribution with support $\{e\}$ whose local expression for the chart φ is $\mathrm{D}^\alpha\varepsilon_0/\alpha!$: 19.5
Z_α	left-invariant differential operator taking the value Δ_α at the point e: 19.5
ε_i	multi-index $(\delta_{ij})_{1 \leq j \leq n}$: 19.5
X_i	invariant vector field Z_{ε_i}: 19.5
X_α	if $\alpha = (\alpha_1, \ldots, \alpha_n)$, $X_\alpha = X_1^{\alpha_1} X_2^{\alpha_2} \cdots X_n^{\alpha_n}$: 19.6
$\mathrm{U}(\mathfrak{g})$	enveloping algebra of the Lie algebra $\mathfrak{g}$: 19.6
$\mathfrak{sl}(n, \mathbf{R})$, $\mathfrak{sl}(n, \mathbf{C})$	Lie algebras of special linear groups: 19.7
$\exp_{\mathrm{G}}$, exp	exponential mapping of Lie(G) into G: 19.8
$\log_{\mathrm{G}}$, log	inverse of the exponential mapping: 19.8
$\mathscr{N}(\mathfrak{m})$, $\mathfrak{N}(\mathfrak{m})$	normalizers in G and $\mathfrak{g}_e$ of a vector subspace $\mathfrak{m}$ of $\mathfrak{g}_e$: 19.11
$\mathscr{Z}(\mathfrak{m})$, $\mathfrak{Z}(\mathfrak{m})$	centralizers in G and $\mathfrak{g}_e$ of a vector subspace $\mathfrak{m}$ of $\mathfrak{g}_e$: 19.11
(H, K)	commutator group of H and K: 19.12
$\mathscr{D}(\mathrm{G})$	derived group of a group G: 19.12
$\mathfrak{D}(\mathfrak{g})$	derived ideal of a Lie algebra $\mathfrak{g}$: 19.12
$\mathscr{D}^n(\mathrm{G})$	nth derived group of a group G: 19.12
$\mathfrak{C}^p(\mathfrak{g})$, $\mathfrak{C}_p(\mathfrak{g})$	ideals of the descending and ascending central series of a Lie algebra $\mathfrak{g}$: 19.12, Problem 3
Aut(G)	automorphism group of a Lie group G: 19.13
$\mathrm{Aut}(\mathfrak{g}_e)$	automorphism group of the Lie algebra $\mathfrak{g}_e$: 19.13
Int(G)	group of inner automorphisms of a Lie group G: 19.13
Ad(G)	image of a Lie group G in $\mathrm{Aut}(\mathfrak{g}_e)$ under the mapping $s \mapsto \mathrm{Ad}(s)$: 19.13
$\mathscr{A}(\mathrm{G})$	automorphism group of a group G: 19.14
$\mathrm{N} \times_\sigma \mathrm{L}$	semidirect product relative to a homomorphism σ of L into $\mathscr{A}(\mathrm{N})$: 19.14
$\mathfrak{n}_e \times_\varphi \mathfrak{l}_e$	semidirect product of Lie subalgebras $\mathfrak{n}_e$, $\mathfrak{l}_e$ of $\mathfrak{g}_e$, relative to a homomorphism $\varphi: \mathfrak{l}_e \to \mathrm{Der}(\mathfrak{n}_e)$: 19.14
$d_x f$, df	(left) differential of a mapping f of a manifold M into a Lie group G: 19.15
$f^{-1}\,df$	logarithmic differential of a mapping $f: \mathrm{M} \to \mathrm{A}$ of a manifold M into a finite-dimensional algebra A: 19.15

$\boldsymbol{\omega}$, $\boldsymbol{\omega}_G$	canonical differential form on a Lie group G: 19.16
$G_{\|\mathbf{R}}$, $\mathfrak{g}_{e\|\mathbf{R}}$	real Lie group and Lie algebra obtained by restriction of scalars to **R** from a complex Lie group G and its Lie algebra $\mathfrak{g}_e$: 19.17
$\mathfrak{g}_{(\mathbf{C})}$	complexification of a real Lie aglebra $\mathfrak{g}$: 19.17
Isom(B × F, E)	frame bundle of a vector bundle E with base B: 20.1
Isom(F, E_b)	set of isomorphisms of the vector space F onto the fiber E_b of E: 20.1
R(M)	frame bundle of a differential manifold M: 20.1
A(n, **R**)	affine group of $\mathbf{R}^n$: 20.1, Problem 1
Aff($\mathbf{R}^n$, E_b)	set of affine-linear bijections of $\mathbf{R}^n$ onto the fiber E_b of E: 20.1, Problem 1
$\boldsymbol{P}$	principal connection in a principal bundle: 20.2
$\boldsymbol{P}_b$	value of a principal connection at a point b of the base: 20.2
$\mathbf{t}_{r_b}$	mapping $\mathbf{u} \mapsto r_b \cdot \mathbf{u}$ of the Lie algebra $\mathfrak{g}_e$ into the space of vertical tangent vectors G_{r_b} at the point r_b of the principal bundle R with group G: 20.2
$\boldsymbol{\omega}$	differential 1-form with values in $\mathfrak{g}_e$ of a principal connection $\boldsymbol{P}$: 20.2
$D\boldsymbol{\alpha}$, $D_{\boldsymbol{P}}\boldsymbol{\alpha}$	covariant exterior differential of a vector-valued differential q-form $\boldsymbol{\alpha}$ on a principal bundle endowed with a principal connection $\boldsymbol{P}$: 20.3
$\boldsymbol{\Omega}$	curvature form of a principal connection: 20.3
$\boldsymbol{\omega} \wedge_\rho \boldsymbol{\alpha}$	exterior product of the connection form $\boldsymbol{\omega}$ with a vector-valued q-form $\boldsymbol{\alpha}$ on the principal bundle R, with values in V, relative to an action ρ of G on V: 20.3, Problem 1
rel($\mathbf{h}_x$)	horizontal lifting at the point $r_x \in R_x$ of a tangent vector $\mathbf{h}_x \in T_x(M)$: 20.5
$\boldsymbol{\sigma}$	canonical form on the frame bundle R(M): 20.6
$\boldsymbol{\Theta}$	torsion form of a principal connection in R(M): 20.6
ω_{ij}	canonical components of the connection form of a principal connection in R(M): 20.6
σ_i	canonical components of the canonical form $\boldsymbol{\sigma}$ on R(M): 20.6
$\boldsymbol{R}$, $(\mathbf{e}_j)$	moving frame: 20.6
$\boldsymbol{\omega}^{(\boldsymbol{R})}$, $\boldsymbol{\sigma}^{(\boldsymbol{R})}$	connection form and canonical form on M, corresponding to the moving frame $\boldsymbol{R}$: 20.6
$\omega_{ij}^{(\boldsymbol{R})}$, $\sigma_i^{(\boldsymbol{R})}$, ω_{ij}, σ_i (by abuse of notation)	canonical components of $\boldsymbol{\omega}^{(\boldsymbol{R})}$, $\boldsymbol{\sigma}^{(\boldsymbol{R})}$: 20.6

$\sum_i \sigma_i \mathbf{e}_i, \sum_j \omega_{ji} \mathbf{e}_j$	abuse of notation for $\sum_i \sigma_i \otimes \mathbf{e}_i$ and $\sum_j \omega_{ji} \otimes \mathbf{e}_j$: 20.6
$\boldsymbol{\Omega}^{(\mathbf{R})}, \boldsymbol{\Theta}^{(\mathbf{R})}$	curvature and torsion forms of a linear connection on M, corresponding to a moving frame $\mathbf{R}$: 20.6
$\Omega_{ij}^{(\mathbf{R})}, \Theta_i^{(\mathbf{R})}, \Omega_{ij}, \Theta_i$ (by abuse of notation)	canonical components of $\boldsymbol{\Omega}^{(\mathbf{R})}, \boldsymbol{\Theta}^{(\mathbf{R})}$: 20.6
$H_{\mathbf{a}}$	horizontal vector field on R(M): 20.6, Problem 1
$\mathbf{B}_q^p(\mathrm{M})$	$\left(\bigwedge^q \mathrm{T(M)}^*\right) \otimes \left(\bigwedge^p \mathrm{T(M)}\right)$: 20.6, Problem 2
$\mathbf{B}(\mathrm{M})$	direct sum of the $\mathbf{B}_q^p(\mathrm{M})$: 20.6, Problem 2
$\mathscr{B}_q^p(\mathrm{M}), \mathscr{B}(\mathrm{M})$	$\mathscr{E}(\mathrm{M})$-module of C^∞-sections of $\mathbf{B}_q^p(\mathrm{M})$ (resp. $\mathbf{B}(\mathrm{M})$): 20.6, Problem 2
$\boldsymbol{d}$	differential operator in $\mathscr{B}(\mathrm{M})$: 20.6, Problem 2
$\boldsymbol{dU}$	matrix $(\boldsymbol{du}_{ij})$ for a matrix $\boldsymbol{U} = (\boldsymbol{u}_{ij})$: 20.6, Problem 2
$\tilde{X}$	canonical lifting to R(M) of a vector field X on M: 20.6, Problem 5
$\boldsymbol{A}_X$	operator $Y \mapsto \theta_X \cdot Y - \nabla_X \cdot Y$: 20.6, Problem 6
$\mathfrak{a}(\boldsymbol{P})$	Lie algebra of infinitesimal automorphisms of a principal connection $\boldsymbol{P}$ in R(M): 20.6, Problem 6
$\mathfrak{a}(n, \mathbf{R})$	Lie algebra of the affine group $\mathbf{A}(n, \mathbf{R})$: 20.6, Problem 20
$\mathrm{S_G(M)}$	G-structure on a differential manifold M: 20.7
$\tilde{\mathrm{H}}$	canonical image of H in $\mathbf{GL}(\mathfrak{g}_e/\mathfrak{h}_e)$ (G a Lie group, H a Lie subgroup of G): 20.7
$\boldsymbol{g}$	pseudo-Riemannian or Riemannian metric tensor: 20.8
$(\mathbf{h}_x \mid \mathbf{k}_x)_{\boldsymbol{g}}, \Vert\mathbf{h}_x\Vert_{\boldsymbol{g}}$	scalar product (resp. norm) in $\mathrm{T}_x(\mathrm{M})$ for a pseudo-Riemannian (resp. Riemannian) metric tensor $\boldsymbol{g}$ on M: 20.8
$(\mathbf{h}_x \mid \mathbf{k}_x), (X \mid Y), \Vert\mathbf{h}_x\Vert$	abuses of notation for scalar product and norm relative to a pseudo-Riemannian or Riemannian metric tensor: 20.8
g_{ij}	components of a pseudo-Riemannian metric tensor $\boldsymbol{g}$, relative to a moving frame: 20.8
G_x	isomorphism of $\mathrm{T}_x(\mathrm{M})$ onto $\mathrm{T}_x(\mathrm{M})^*$ defined by a pseudo-Riemannian metric tensor: 20.8
G	M-isomorphism $\mathbf{h}_x \mapsto G_x \cdot \mathbf{h}_x$ of T(M) onto T(M)*: 20.8
G_j^i	isomorphism of lowering the ith contravariant index to the jth place: 20.8
grad(f)	gradient of a real-valued function on a pseudo-Riemannian manifold: 20.8

$\mathbf{g}_r$, $\mathbf{g}_r^*$	symmetric covariant tensor fields on $\bigwedge^r \mathrm{T(M)}$ and $\bigwedge^r \mathrm{T(M)}^*$, corresponding to a pseudo-Riemannian metric tensor $\mathbf{g}$ on M: 20.8
$\lVert \mathbf{h}_1 \wedge \mathbf{h}_2 \wedge \cdots \wedge \mathbf{h}_r \rVert$	r-dimensional area of a tangent r-vector: 20.8
$v_{\mathbf{g}}$, $\mathrm{vol}_{\mathbf{g}}$, v, vol	Riemannian volume: 20.8
$\upsilon_{\mathbf{g}}$, υ	canonical volume form an on oriented Riemannian manifold: 20.8
$*\alpha$	adjoint of a differential r-form α on a Riemannian manifold: 20.8
α_{M}, α	differential 1-form ${}^t\mathrm{G}(\kappa_{\mathrm{M}})$ on T(M), for a pseudo-Riemannian manifold M: 20.8, Problem 3
$\mathrm{E}(\mathbf{h})$	$(\mathbf{h} \mid \mathbf{h})$ for a pseudo-Riemannian manifold M and $\mathbf{h} \in \mathrm{T(M)}$: 20.8, Problem 3
$S(\mathbf{h}_x)$	geodesic field $\mathbf{C}_x(\mathbf{h}_x, \mathbf{h}_x)$: 20.9, Problem 2
$\mathbf{g}_{\mathrm{T}}$	pseudo-Riemannian metric tensor on T(M) induced by a pseudo-Riemannian metric tensor $\mathbf{g}$ on M: 20.9, Problem 3
i(M)	Lie algebra of infinitesimal isometries of a pseudo-Riemannian manifold M: 20.9, Problem 7
$\mathfrak{pgl}(n+1, \mathbf{R})$	Lie algebra of the projective group $\mathbf{PGL}(n+1, \mathbf{R})$: 20.9, Problem 12
$\boldsymbol{K}$	Riemann–Christoffel tensor of a pseudo-Riemannian manifold: 20.10
K_{hijk}	components of the Riemann–Christoffel tensor relative to a moving frame: 20.10
$\boldsymbol{r}'(X, Y)$	M-morphism $Z \mapsto (\boldsymbol{r}(Z \wedge X)) \cdot Y$ of T(M) into itself: 20.10
$\boldsymbol{r}''(X)$	$\boldsymbol{r}'(X, X)$: 20.10
$\boldsymbol{K}'$	Ricci tensor, defined by $\langle \boldsymbol{K}', X \otimes Y \rangle = \mathrm{Tr}(\boldsymbol{r}'(X, Y))$: 20.10
K'_{jk}	components of the Ricci tensor relative to a moving frame: 20.10
$\mathrm{Ric}(\mathbf{h}_x)$	Ricci curvature in the direction of $\mathbf{h}_x \in \mathrm{T}_x(\mathrm{M})$: 20.10
$\tilde{r}_x$	endomorphism of $\mathrm{T}_x(\mathrm{M})$ defined by $(\tilde{r}_x(\mathbf{h}_x) \mid \mathbf{k}_x) = \langle \boldsymbol{K}'(x), \mathbf{h}_x \otimes \mathbf{k}_x \rangle$: 20.10
$\mathrm{S}(x)$	scalar curvature at x: 20.10
$\mathrm{K}(x)$	Gaussian curvature of a surface: 20.10
$\mathrm{div}(X)$	divergence of a vector field X: 20.10, Problem 3
$\mathfrak{so}(n+1)$	Lie algebra of $\mathbf{SO}(n+1)$: 20.11
$\mathbf{Y}_n$	hyperbolic n-space: 20.11

$Q_{n-1}(\mathbf{C})$	complex quadric of dimension $n-1$: 20.11, Problem 5
$\mathbf{g}'$	first fundamental form of a submanifold of a Riemannian manifold: 20.12
$\boldsymbol{I}_\alpha$, $\boldsymbol{I}$	second fundamental forms, vector-valued second fundamental form of a submanifold of a Riemannian manifold: 20.12
$K(\mathbf{n}_x)$	total curvature of a submanifold of $\mathbf{R}^N$ in the normal direction $\mathbf{n}_x$: 20.12, Problem 4
$\kappa(M)$	integral curvature of a submanifold of $\mathbf{R}^N$: 20.12, Problem 4
ds^2	abuse of notation for $\mathbf{g}$: 20.13
$\boldsymbol{I}$	second fundamental form of a hypersurface: 20.14
$\rho_j(x)$	principal curvatures of a hypersurface at a point x: 20.14
$H(x)$, $K(x)$	mean curvature and total curvature of a hypersurface at a point x: 20.14
$[\mathbf{u}_1, \ldots, \mathbf{u}_p, \omega_1, \ldots, \omega_{n-p}]$	scalar differential $(n-p)$-form, constructed from p vector-valued functions and $n-p$ vector-valued differential 1-forms, with values in $\mathbf{R}^n$: 20.14, Problem 9
$L(\gamma)$	length of a piecewise-C^1 path: 20.16
$d(x, y)$	Riemannian distance from x to y: 20.16
$L(\xi)$, $E(\xi)$	length and energy of a path $C_\xi : t \mapsto f(t, \xi)$: 20.20
$I(\mathbf{w}_1, \mathbf{w}_2)$	index form of two liftings of v: 20.20
$A(\mathbf{h}_x, \mathbf{k}_x)$	sectional curvature: 20.21
$x \otimes y$	tensor product of two vectors: A.20.1
$E \otimes_K F$	tensor product of two K-vector spaces: A.20.1
X_i	indeterminates, elements of $K^{\mathbf{N}^n}$: A.21.2
$\sum_\alpha a_\alpha X^\alpha$	formal power series with coefficients in K: A.21.2
$K[[X_1, \ldots, X_n]]$	algebra of formal power series in n indeterminates over a field K: A.21.2
$f(g_1, \ldots, g_n)$	formal power series obtained by substitution of formal power series g_j without constant terms in a formal power series f: A.21.3

CHAPTER XVIII

DIFFERENTIAL CALCULUS ON A DIFFERENTIAL MANIFOLD II. ELEMENTARY GLOBAL THEORY OF FIRST- AND SECOND-ORDER DIFFERENTIAL EQUATIONS. ELEMENTARY LOCAL THEORY OF DIFFERENTIAL SYSTEMS

Once the concept of a tangent vector to a differential manifold M has been established, it is easy to generalize the notion of a solution of a first-order differential equation (10.4.2) to functions defined on an interval of the real line, with values in M: the derivative $u'(t)$ is replaced by the image of the unit tangent vector to **R** at the point t, and the right-hand side of the equation must therefore be a tangent vector to M at the point $u(t)$. Locally, by means of a chart, we can reduce such an equation to a differential equation in the usual sense (10.4.1). However, since we now have an intrinsic formulation of the notion of a differential equation, problems of a *global* nature present themselves: the existence of a maximal integral curve and its behavior as a function of the "initial condition" which defines it or of "parameters" on which the equation depends. Here we shall indicate only the first rudiments of this extremely difficult subject, which involves subtle considerations of topology and integration theory (the reader will see some examples in Chapter XXV, and in [21] and [51]).

The notion of a differential equation of the second order on a manifold M is less obvious, because it requires the concept of the "second derivative" of a function defined on an interval of **R** with values in M. Since the values of the "first derivative" are tangent vectors to M, it is to be regarded as a function with values *in the tangent bundle* T(M), and a second-order equation on M is therefore a first-order equation *on the manifold* T(M). The local and global questions which we shall consider for such equations (Sections 18.3

to 18.7) are largely concerned with a special type of second-order equation, namely, those which give rise to *geodesics* of connections, and which will be studied in more detail in Chapter XX in the context of Riemannian manifolds.

The curves defined by a differential equation are characterized geometrically by the requirement that at each point they should touch a given line in the tangent space to the manifold at that point. Replacing the line by a vector subspace of arbitrary dimension, we have the general notion of a "Pfaffian system" on a manifold, which is the intrinsic counterpart of the "partial differential equations" of classical mathematics. Here, the global theory is practically nonexistent, except for completely integrable systems (see [69]) and the linear partial differential equations (and allied types), which we shall encounter in chapters XXIII and XXV. In this chapter, we shall be concerned exclusively with *local* problems of existence and uniqueness, so that it would be possible to work in $\mathbf{R}^N$ throughout. But the language of manifolds and especially the exterior differential calculus are extremely useful even in this local situation, as E. Cartan showed, in order to comprehend the intrinsic nature of the problems independently of any adventitious coordinate system. We have hardly skimmed the surface of the immense work of E. Cartan in this field, to which we urge the reader to refer ([54] and [59]).

1. FIRST-ORDER DIFFERENTIAL EQUATIONS ON A DIFFERENTIAL MANIFOLD

(18.1.1) Given an open subset U′ of $\mathbf{R}^n$, a (real) *autonomous system of differential equations* in U is a vector differential equation

(18.1.1.1) $$\mathrm{D}x = \mathbf{f}(x),$$

where $\mathbf{f}: \mathrm{U} \to \mathbf{R}^n$ is a continuous mapping. A solution of this equation is therefore a continuously differentiable mapping u of an *open* interval $\mathrm{I} \subset \mathbf{R}$ into U such that $\mathrm{D}u(t) = \mathbf{f}(u(t))$ for all $t \in \mathrm{I}$ (10.4). An autonomous system of differential equations is therefore a vector differential equation in which the variable t "does not appear" on the right-hand side. If u is a solution of (18.1.1.1) defined on $\mathrm{I} \subset \mathbf{R}$, it is clear that for each $a \in \mathbf{R}$, the function $t \mapsto u(t + a)$ is also a solution of (18.1.1.1), defined on the interval $\mathrm{I} + (-a)$.

If we identify the tangent bundle T(U) with $\mathrm{U} \times \mathbf{R}^n$ by means of the canonical trivialization (16.15.5), then the mapping $x \mapsto (x, \mathbf{f}(x))$ is the local expression of a continuous *vector field* X on U, defined by $X(x) = \tau_x^{-1}(\mathbf{f}(x))$ (16.5.2). Let E denote the vector field on $\mathbf{R}$ (called the *unit vector field*) defined by the condition that $\tau_t(E(t))$ is the vector $1 \in \mathbf{R}$. Then, if u is a C^1 mapping of I into U, we may write

$$Du(t) = \tau_{u(t)}(T_t(u) \cdot E(t))$$

and then the relation $Du(t) = \mathbf{f}(u(t))$ is equivalent to

(18.1.1.2) $$T(u) \cdot E(t) = X(u(t))$$

for all $t \in I$.

(18.1.2) The form (18.1.1.2) in which we have expressed that u is a solution of (18.1.1.1) no longer involves the trivialization of T(U), and therefore allows us to generalize the notion of an autonomous system of differential equations to an arbitrary differential manifold M. Given a vector field X of class C^r ($r \geqq 0$) on M, the (first-order) *differential equation on* M *defined by* X is the relation

(18.1.2.1) $$T(u) \circ E = X \circ u$$

and a C^1 mapping v of an *open* interval $I \subset \mathbf{R}$ into M is said to be a *solution* of this equation if for each $t \in I$ we have

(18.1.2.2) $$T(v) \cdot E(t) = X(v(t)).$$

For such a mapping v, we shall write†

(18.1.2.3) $$v'(t) = T(v) \cdot E(t) \in T_{v(t)}(M)$$

for all $t \in I$ ($v'(t)$ is the *derivative* or *velocity vector* of v at the point t), and the equation (18.1.2.2) now takes the form

(18.1.2.4) $$v'(t) = X(v(t))$$

for all $t \in I$.

† We are therefore now assigning two different meanings to the symbols v' and Dv when v is a mapping of an interval I into a *vector space* $\mathbf{R}^n$ (although in (8.1) we regarded these two symbols as synonymous). In this particular case we have

$$Dv(t) = \tau_{v(t)}(v'(t)).$$

If f is any real-valued function of class C^1 on M, it follows immediately from the definitions ((16.5.4) and (17.14.1)) that, by virtue of (18.1.2.4), we have

(18.1.2.5)
$$\frac{d}{dt}(f(v(t)) = \theta_{X(v(t))} \cdot f$$

for all $t \in \mathrm{I}$.

A solution of (18.1.2.1) is also called an *integral curve of the vector field* X. A differential equation (18.1.2.1) is also called a *dynamical system* on M.

If (U, φ, n) is a chart on M and if $z \mapsto (z, \mathbf{f}(z))$ is the corresponding local expression of X (so that

$$(z, \mathbf{f}(z)) = \mathrm{T}(\varphi) \cdot X(\varphi^{-1}(z))),$$

the function $(\varphi \circ v) | v^{-1}(\mathrm{U})$ is a solution in $v^{-1}(\mathrm{U})$ of the vector differential equation (18.1.1.1), which is called the *local expression* of the equation (18.1.2.1) corresponding to the chart (U, φ, n).

Remarks

(18.1.3) Consider an arbitrary vector differential equation

(18.1.3.1)
$$\mathrm{D}x = \mathbf{f}(x, t),$$

where $\mathbf{f}$ is defined and continuous on an open subset H of $\mathbf{R}^n \times \mathbf{R}$ and takes values in $\mathbf{R}^n$. This equation may be reduced to an autonomous system as follows. Consider the continuous mapping $z \mapsto \mathbf{F}(z)$ of H into $\mathbf{R}^{n+1}$ defined by $\mathbf{F}(x, t) = (\mathbf{f}(x, t), 1)$, and the autonomous system

(18.1.3.2)
$$z' = \mathbf{F}(z).$$

If $u = (v, \varphi)$ is a solution of this equation defined on an open interval $\mathrm{I} \subset \mathbf{R}$, where $v(t) \in \mathbf{R}^n$ and $\varphi(t) \in \mathbf{R}$, then we have $\varphi'(t) = 1$, so that $\varphi(t) = t + \alpha$, and $v'(t) = \mathbf{f}(v(t), t + \alpha)$ for all $t \in \mathrm{I}$. Putting $w(t) = v(t - \alpha)$, we have $w'(t) = \mathbf{f}(w(t), t)$ in $\mathrm{I} + \alpha$, and w is therefore a solution of (18.1.3.1) in this interval. Conversely, if v is a solution of (18.1.3.1) in I, then it is clear that $t \mapsto (v(t), t)$ is a solution of (18.1.3.2) in I. Hence the problem of solving (18.1.3.1) is equivalent to that of solving (18.1.3.2). The counterpart of (18.1.3.1) on a differential manifold M is therefore a differential equation defined by a vector field Y on $\mathrm{M} \times \mathbf{R}$ such that $Y(x, t) = (X(x, t), E(t))$, when $\mathrm{T}(\mathrm{M} \times \mathbf{R})$ is identified with $\mathrm{T}(\mathrm{M}) \times \mathrm{T}(\mathbf{R})$.

2. FLOW OF A VECTOR FIELD

(18.2.1) *Let X be a vector field of class C^r ($r \geqq 1$) on M. If v_1 and v_2 are two integral curves of X, defined respectively on open intervals I_1, I_2 in $\mathbf{R}$, and such that $v_1(t_0) = v_2(t_0)$ for some point $t_0 \in I_1 \cap I_2$, then v_1 and v_2 coincide in the interval $I_1 \cap I_2$.*

It is enough to show that the set A of points $t \in I_1 \cap I_2$ such that $v_1(t) = v_2(t)$ is both open and closed in $I_1 \cap I_2$ (3.19.1), since by hypothesis A is not empty. Now A is closed, because v_1 and v_2 are continuous (3.15.1). On the other hand, if $t_1 \in A$, there exists a neighborhood $J \subset I_1 \cap I_2$ of t_1, and a chart (U, φ, n) of M at the point $v_1(t_1) = v_2(t_1)$ for which $\varphi \circ (v_1 | J)$ and $\varphi \circ (v_2 | J)$ are solutions of the same equation (18.1.1.1), where $\mathbf{f}$ is of class C^r with $r \geqq 1$, and these two solutions take the same value at the point t_1. It follows, therefore, from (10.5.2) that v_1 and v_2 coincide in J and the proof is complete.

(18.2.2) Let X be a vector field of class C^r ($r \geqq 1$) on M. Then for each $x_0 \in M$, there exists an open neighborhood J of 0 in $\mathbf{R}$ and an integral curve v of X defined in J and such that $v(0) = x_0$. For by considering a chart (U, φ, n) of M at the point x_0 such that the differential equation (18.1.2.1) has a local expression of the form (18.1.1.1), with $\mathbf{f}$ of class C^r ($r \geqq 1$), the existence of a solution of this equation, taking the value $\varphi(x_0)$ at the point 0 and defined in a neighborhood J of 0 in $\mathbf{R}$, is guaranteed by (10.4.5), which proves the assertion. Now let $J(x_0)$ be the open interval which is the *union* of all the open intervals J containing 0 and in which there exists an integral curve of X which is equal to x_0 at $t = 0$. Since, by virtue of (18.2.1), any two of these functions agree in the intersection of their intervals of definition, it follows that there exists a unique function v defined on $J(x_0)$, such that each of the integral curves is a restriction of v, and it is clear that v itself is an integral curve of X. This function v is said to be the *maximal integral curve* of X such that $v(0) = x_0$ (or with *origin* x_0). We denote by $t^-(x_0)$ and $t^+(x_0)$ the left- and right-hand endpoints of the interval $J(x_0)$, so that $t^-(x_0) < 0 < t^+(x_0)$; it can happen that $t^-(x_0) = -\infty$ or $t^+(x_0) = +\infty$. Also, we denote by $F_X(x_0, t)$ the value at $t \in J(x_0)$ of the maximal integral curve of x which takes the value x_0 at $t = 0$. The function F_X is defined on the set $\operatorname{dom}(F_X) \subset M \times \mathbf{R}$, consisting of the points (x, t) such that $t^-(x) < t < t^+(x)$ (the union of the sets $\{x\} \times J(x)$). The function F_X is called the *flow* of the vector field X, and $\operatorname{dom}(F_X)$ its *domain*.

(18.2.3) With the notation of (18.2.2), for each point $t_0 \in J(x)$, we have

(18.2.3.1) $$J(F_X(x, t_0)) = J(x) + (-t_0)$$

and, for each $t \in J(x) + (-t_0)$,

(18.2.3.2) $$F_X(F_X(x, t_0), t) = F_X(x, t + t_0).$$

For it is clear that the function $t \mapsto F_X(x, t_0 + t)$ is an integral curve of X defined on $J(x) + (-t_0)$ and taking the value $F_X(x, t_0)$ at $t = 0$; and this function cannot be the restriction of an integral curve v defined on a strictly larger interval J' and taking the same value at $t = 0$, otherwise the function $t \mapsto v(t - t_0)$ would be an integral curve defined on an interval strictly larger than $J(x)$ and taking at t_0 the same value as the function $t \mapsto F_X(x, t)$; which, in view of (18.2.1), would contradict the definition of $J(x)$. The formula (18.2.3.2) is a consequence of this argument and the definition of F_X.

(18.2.4) *Let* M *be a differential manifold,* N *a closed submanifold of* M, *and* X *a vector field on* M *of class* C^r ($r \geqq 1$). *Suppose that* $X(x) \in T_x(N)$ *for all* $x \in N$ (in other words, that the field X is *tangent to* N *at all points of* N). *Then every integral curve of* X *which meets* N *is contained in* N.

Let $t \mapsto v(t)$ be an integral curve of X defined in I, and suppose that $v(t_0) \in N$ for some $t_0 \in I$. If the *open* set (3.11.4) of points $t \in I$, such that $t > t_0$ and $v(t) \notin N$ is not empty, then it will have a greatest lower bound $t_1 \geqq t_0$, and we shall have $v(t_1) \in N$. But if Y is the restriction to N of the vector field X, there exists an integral curve w of Y, defined on an open interval $J \subset I$ containing t_1 and such that $w(t_1) = v(t_1)$ (18.2.2), and it is clear that w is also an integral curve of X. But then we have $w(t) = v(t)$ for all $t \in J$, by (18.2.1), hence $v(t) \in N$ for all $t \in J$, contradicting the definition of t_1. Hence the set of points $t \in I$ such that $t > t_0$ and $v(t) \notin N$ is empty; likewise is the set of $t \in I$, such that $t < t_0$ and $v(t) \notin N$.

(18.2.5) *Let* X *be a vector field of class* C^r *on* M, *with* $r \geqq 1$ (*possibly* $r = \infty$). *Then* $\mathrm{dom}(F_X)$ *is an open subset of* $M \times \mathbf{R}$ *and* F_X *is a* C^r *mapping of* $\mathrm{dom}(F_X)$ *into* M.

Let $x_0 \in M$ and let $t_0 \in J(x_0)$. Then we have to show that there exists an open interval $]a, b[\subset J(x_0)$ *containing* the points 0 and t_0, and a neighborhood V of x_0 in M, such that for each $x \in V$, the interval $]a, b[$ is *contained in* $J(x)$ and that F_X is of class C^r in $V \times]a, b[$. Clearly, we may assume that $t_0 \geqq 0$.

Let $[c, d]$ be a compact interval contained in $J(x_0) =]t^-(x_0), t^+(x_0)[$, such that $c < 0 \leqq t_0 < d$. Let L be the compact subset of M which is the image of $[c, d]$ under $F_X(x_0, .)$, and let W be a relatively compact open neighborhood of L in M. Then by (16.25.1) there exists an embedding of

W in $\mathbf{R}^N$, for sufficiently large N, and we may therefore assume that $M = W \subset \mathbf{R}^N$. By (16.12.11) and (16.4.3) there exists an extension of X to a vector field Y of class C^r, defined on a neighborhood U of M in $\mathbf{R}^N$. Since the tangent bundle T(U) may be canonically identified with $U \times \mathbf{R}^N$, we may write $Y(y) = (y, \mathbf{f}(y))$ for all $y \in U$, where $\mathbf{f}$ is a mapping of class C^r of U into $\mathbf{R}^N$. Since M is *locally closed* in $\mathbf{R}^N$ (16.8.3) we may, by restricting U if necessary, assume that M is *closed* in U (12.2.3); and, replacing U again by a smaller open set containing L, that $D\mathbf{f}$ is bounded in U, say $\|D\mathbf{f}(y)\| \leqq k$ for all $y \in U$. Let $\delta > 0$ be chosen sufficiently small that, for each $t \in]c, d[$, the open ball with center $F_X(x_0, t)$, and radius δ is contained in U (3.17.11), and then choose $\varepsilon > 0$ such that $\varepsilon e^{k(d-c)} < \delta$. Then it follows from (10.5.6) that for each point $z \in U$ such that $\|z - x_0\| \leqq \varepsilon$, there exists a solution $t \mapsto v(z, t)$ of the differential equation $Dy = \mathbf{f}(y)$, *defined on the interval* $]c, d[$ and such that $v(z, 0) = z$ and $\|v(z, t) - F_X(x_0, t)\| \leqq \varepsilon$ for all $t \in]c, d[$. By (18.2.4) applied to U and M, for each $x \in M$ satisfying $\|x - x_0\| \leqq \varepsilon$, the function $t \mapsto v(x, t)$ is an integral curve of X in the interval $]c, d[$ that takes the value x at $t = 0$. Consequently, we have $]c, d[\subset J(x)$ and $v(x, t) = F_X(x, t)$ for all $t \in]c, d[$. Moreover, it follows from (10.7.4) that by replacing the interval $]c, d[$, if necessary, by a smaller interval $]a, b[$ such that $a < 0 \leqq t_0 < b$, we may suppose that the function $(x, t) \mapsto v(x, t)$ is of class C^r in the product of a neighborhood of of L in M and the interval $]a, b[$. This completes the proof.

This proposition leads directly to the following corollaries:

(18.2.6) *For each* $t \in \mathbf{R}$, *the set of points* $x \in M$ *such that* $(x, t) \in \text{dom}(F_X)$ *is open in* M.

This follows from (3.20.12).

(18.2.7) *The function* $x \mapsto t^+(x)$ *is lower semicontinuous, and the function* $x \mapsto t^-(x)$ *is upper semicontinuous on* M.

For the set of points x such that $t^+(x) > \alpha > 0$ is equal to the set of points x such that $(x, \alpha) \in \text{dom}(F_X)$, hence is open by (18.2.6). The first assertion therefore follows from (12.7.2) and the second assertion is proved similarly.

(18.2.8) *Let* U *be an open set in* M *and* a *a real number* > 0 *such that*

$$U \times]-a, a[\subset \text{dom}(F_X).$$

Then, for each $t \in]-a, a[$, *the mapping* $x \mapsto F_X(x, t)$ *is a homeomorphism* (*of class* C^r) *of* U *onto an open subset* U_t *of* M, *and the mapping* (*of class* C^r) $x \mapsto F_X(x, -t)$ *is the inverse homeomorphism.*

This follows from (18.2.5) and (18.2.3.2).

(18.2.9) For a C^∞ vector field X, the numbers $t^+(x)$ and $t^-(x)$ may be finite: take, for example, $M = \mathbf{R}$ and $X(x) = (x, x^2)$. When this is the case, the "global" analog of (10.5.5) is the following proposition:

(18.2.10) *Let X be a vector field on* M *of class* C^r $(r \geqq 1)$, *and x a point of* M *such that* $t^+(x) < +\infty$. *Then, for each compact subset* K *of* M, *there exists* $\varepsilon > 0$ *such that, for each* $t > t^+(x) - \varepsilon$, *the point* $F_X(x, t)$ *does not lie in* K. (In other words, the integral curve "ends outside" *every compact subset* of M.)

The proof is by contradiction. If the assertion is false, there will exist an increasing sequence (t_n) of real numbers strictly less than $t^+(x)$, with $t^+(x)$ as limit and such that $F_X(x, t_n) \in K$ for all n. Passing to a subsequence, we may assume that the sequence of points $F_X(x, t_n)$ converges to a point $z \in K$. By virtue of (18.2.5), there exists an open neighborhood U of z in M and a real number $a > 0$ such that $t^+(y) > a$ for all $y \in U$. Now, if n is sufficiently large, we have $t^+(x) < t_n + a$ and $F_X(x, t_n) \in U$, and therefore

$$t^+(F_X(x, t_n)) > a;$$

but by (18.2.3.1), $t^+(F_X(x, t_n)) = t^+(x) - t_n$, whence $t^+(x) > t_n + a$, a contradiction.

There is an analogous result for t^-, the statement of which we shall leave to the reader. In particular:

(18.2.11) *Let X be a vector field of class* C^r $(r \geqq 1)$ *on* M, *with compact support* (in particular, this condition will be automatically satisfied if the manifold M is *compact*). *Then* $J(x) = \mathbf{R}$ *for all* $x \in M$.

Let K be the support of X. For each $x \notin K$ we have $J(x) = \mathbf{R}$, and the integral curve $t \mapsto F_X(x, t)$ is the constant function $t \mapsto x$. Hence, if $x \in K$, the function $F_X(x, t)$ takes no values outside K, and therefore by (18.2.10) we have $J(x) = \mathbf{R}$ in this case also.

If X is a vector field *of class* C^∞ with compact support K, then for *each* $t \in \mathbf{R}$ we have a *diffeomorphism*

(18.2.11.1) $$h_t : x \mapsto F_X(x, t)$$

of M *onto* M, such that

(18.2.11.2) $$h_{t+t'} = h_t \circ h_{t'} = h_{t'} \circ h_t$$

for all t, $t' \in \mathbf{R}$ and such that $h_0 = 1_M$. This follows from (18.2.8) (taking $U = M$ and $]-a, a[= \mathbf{R}$) and (18.2.3.2). The h_t form a group, called the *one-parameter group of diffeomorphisms of* M *defined by* X. Notice that if $x \notin K$, we have $h_t(x) = x$ for all $t \in \mathbf{R}$.

Remarks

(18.2.12) If M is a real-analytic manifold and X is an analytic vector field on M, then it follows from the proof of (18.2.5) and from (10.7.5) that the flow F_X is *analytic* in the open set $\operatorname{dom}(F_X)$.

(18.2.13) If v is a C^∞ solution of (18.1.1.1) in I and if $X(v(t)) \neq 0$ at a point $t \in I$, then v is an *immersion* at t. But it can happen that v is an injective immersion of I in M but not an embedding (16.9.9.3).

(18.2.14) Suppose that the vector field X is of class C^1, and put $g_t(x) = F_X(x, t)$. Then $g_{-t}(g_t(x)) = x$ for all sufficiently small $t \in \mathbf{R}$ (18.2.3.2). If Y is any C^1 vector field on M, put

(18.2.14.1) $$Y_t(x) = T_{g_t(x)}(g_{-t}) \cdot Y(g_t(x)),$$

which is a tangent vector at x and is defined for all sufficiently small $t \in \mathbf{R}$. With this notation, we have the following interpretation of the *Lie bracket* $[X, Y]$:

(18.2.14.2) $$[X, Y](x) = \frac{d}{dt} Y_t(x)\big|_{t=0}$$

in the vector space $T_x(M)$, endowed with its canonical topology (12.13.2). To prove this, we may assume that M is an open subset of $\mathbf{R}^n$; then the fields X, Y can be written in the form $y \mapsto (y, \mathbf{G}(y))$ and $y \mapsto (y, \mathbf{H}(y))$, where $\mathbf{G}$ and $\mathbf{H}$ are C^1 mappings of M into $\mathbf{R}^n$. Hence, for a fixed x, we have

$$\frac{d}{dt} g_t(x) = \mathbf{G}(g_t(x)).$$

Consequently, for all sufficiently small t, we may write

(18.2.14.3) $$g_t(x) = x + t\mathbf{G}(x) + t\mathbf{h}(t),$$

where $\mathbf{h}(t)$ tends to 0 with t (8.6.2). On the other hand, if $Dg_t(y)$ denotes the derivative at y of the function $z \mapsto g_t(z)$, then $t \mapsto Dg_t(x)$ is the solution of the linear differential equation

$$U' = D\mathbf{G}(g_t(x)) \circ U,$$

which reduces to the unit matrix I at $t = 0$ (10.7.3). Hence

(18.2.14.4) $$\mathrm{D}g_t(x) = I + t\mathrm{D}\mathbf{G}(x) + tW(t),$$

where the matrix $W(t)$ tends to 0 with t. It follows that the right-hand side of (18.2.14.2) is of the form $(x, \mathbf{V}(x))$, where

$$\mathbf{V}(x) = \lim_{t\to 0,\, t\neq 0} \frac{1}{t}(\mathrm{D}g_{-t}(g_t(x))\cdot\mathbf{H}(g_t(x)) - \mathbf{H}(x)).$$

Now, for (t, y) close to $(0, x)$, we have $g_{-t}(g_t(y)) = y$ and therefore

$$\mathrm{D}g_{-t}(g_t(x)) \circ \mathrm{D}g_t(x) = I$$

by differentiating. Hence

$$\mathbf{V}(x) = \lim_{t\to 0,\, t\neq 0} \frac{1}{t}\mathrm{D}g_{-t}(g_t(x))\cdot(\mathbf{H}(g_t(x)) - \mathrm{D}g_t(x)\cdot\mathbf{H}(x)).$$

But, by virtue of (18.2.14.3) and (18.2.14.4), we have

(18.2.14.5) $$\mathbf{H}(g_t(x)) = \mathbf{H}(x) + t\mathrm{D}\mathbf{H}(x)\cdot\mathbf{G}(x) + to_1(t),$$

(18.2.14.6) $$\mathrm{D}g_t(x)\cdot\mathbf{H}(x) = \mathbf{H}(x) + t\mathrm{D}\mathbf{G}(x)\cdot\mathbf{H}(x) + to_2(t),$$

where $o_1(t)$ and $o_2(t)$ tend to 0 with t. Since $\mathrm{D}g_{-t}(g_t(x))$, the inverse of $\mathrm{D}g_t(x)$, tends to I as $t \to 0$, we obtain

$$\mathbf{V}(x) = \mathrm{D}\mathbf{H}(x)\cdot\mathbf{G}(x) - \mathrm{D}\mathbf{G}(x)\cdot\mathbf{H}(x),$$

which proves our assertion (17.14.3.2).

More generally, if $\mathbf{Z}$ is any C^1 *tensor field* of type (r, s) on M, and if we put

(18.2.14.7) $$\mathbf{Z}_t(x) = \mathbf{T}_s^r(\mathrm{T}_{g_t(x)}(g_{-t}))\cdot\mathbf{Z}(g_t(x)),$$

then we have the formula

(18.2.14.8) $$(\theta_X\cdot\mathbf{Z})(x) = \frac{d}{dt}\mathbf{Z}_t(x)\big|_{t=0}$$

in the vector space $(\mathbf{T}_s^r(\mathrm{M}))_x$ endowed with its canonical topology. This follows immediately from the uniqueness statement in (17.14.6), since the right-hand side of (18.2.14.8) evidently satisfies the conditions of (17.14.6) by virtue of (8.1.4).

PROBLEMS

1. Let F be a closed set in $\mathbf{R}^n$ and a a frontier point of F. A vector $\mathbf{u} \neq 0$ is said to be an *outward normal* to F at the point a if there exists a point $b = a + \rho\mathbf{u}$, with $\rho > 0$, such that the (Euclidean) open ball with center b and radius ρ is contained in the complement of F. A vector $\mathbf{v} \in \mathrm{T}_a(\mathbf{R}^n)$ is said to be *tangent* to F if $(\boldsymbol{\tau}_a(\mathbf{v}) \mid \mathbf{u}) = 0$ for all outward normals $\mathbf{u}$ to F at a. A vector field X defined on an open neighborhood U of F is said to be *tangent to* F *along* F if, for each frontier point a of F, the vector $X(a)$ is tangent to F.

(a) Let $t \mapsto x(t)$ be a C^1 curve, i.e., a C^1 mapping of an open interval $\mathrm{I} \subset \mathbf{R}$ into $\mathbf{R}^n$. For each $t \in \mathrm{I}$ let $\delta(t)$ denote the (Euclidean) distance $d(x(t), \mathrm{F})$, and let y be a point of F such that $\|x(t) - y\| = d(x(t), \mathrm{F})$. Show that, if $x(t) \notin \mathrm{F}$ and if $\mathbf{u}$ is a unit vector proportional to $x - y$, then

$$\liminf_{h \to 0,\, h \neq 0} \frac{\delta(t+h) - \delta(t)}{|h|} \geqq -|(x'(t) \mid \mathbf{u})|.$$

(If (h_n) is a sequence of real numbers converging to 0 and if y_n is a point of F such that $d(x(t+h_n), y_n) = d(x(t+h_n), \mathrm{F})$, observe that $\delta(t) \leqq \|x(t) - y_n\|$.)

(b) Let X be a *Lipschitzian* vector field defined on an open neighborhood U of F (we identify $\mathrm{T}_x(\mathbf{R}^n)$ with $\mathbf{R}^n$ by means of $\boldsymbol{\tau}_x$). Suppose that X is tangent to F along F. Show that there exists a constant $k > 0$ such that, with the notation of (a),

$$\liminf_{h \to 0,\, h \neq 0} \frac{\delta(t+h) - \delta(t)}{|h|} \geqq -k\,\delta(t)$$

for each integral curve $t \mapsto x(t)$ of the vector field X. (Use (a), and the definition of a tangent vector to F at the point y.)

(c) Show that every integral curve of X which meets F is contained in F. (Argue by contradiction, and suppose that an integral curve $t \mapsto x(t)$ satisfies $x(t_0) \in \mathrm{F}$ and $x(t) \notin \mathrm{F}$ for $t_0 < t < t_1$. Use (b) and the following lemma: if f is a continuous real-valued function defined on an open interval of $\mathbf{R}$, which satisfies the inequality

$$\liminf_{h \to 0,\, h \neq 0} \frac{f(t+h) - f(t)}{|h|} \geqq -k,$$

where $k > 0$, at each point of the interval, then f is Lipschitzian for the constant k. To prove this lemma, assume, e.g., that $f(s_1) - f(s_2) = c(s_1 - s_2)$, with $s_1 < s_2$ and $-c > k$, and consider the function $t \mapsto f(t) - c(t - s_1)$ for $s_1 \leqq t \leqq s_2$.)

2. Let X_1, X_2 be two C^∞ vector fields on an open set $\mathrm{U} \subset \mathbf{R}^n$.

(a) Let λ_1, λ_2 be two real-valued functions of class C^∞ on U, and let

$$Z(x) = \lambda_1(x)X_1(x) + \lambda_2(x)X_2(x).$$

Suppose that an integral curve $t \mapsto x(t)$ of the vector field Z is defined for $0 \leqq t \leqq 1$ and that $x(0) = x_0$. For each positive integer n, consider the continuous function $t \mapsto z_n(t)$

defined on [0, 1] by the following conditions:

$$z_n(0) = x_0,$$

$$z_n'(t) = \lambda_1\left(\frac{k}{n}\right) X_1(z_n(t)) \quad \text{for } \frac{k}{n} \leqq t \leqq \frac{2k+1}{2n},$$

$$z_n'(t) = \lambda_2\left(\frac{k}{n}\right) X_2(z_n(t)) \quad \text{for } \frac{2k+1}{2n} \leqq t \leqq \frac{k+1}{n}$$

($0 \leqq k \leqq n$). Show that $z_n \to x$ uniformly on [0, 1]. (Consider the function y_n which is affine-linear on each interval $[k/n, (k+1)/n]$ and such that $y_n(k/n) = z_n(k/n)$ for $0 \leqq k \leqq n$, and use (**10.5.1**).)

(b) Let $Z = [X_1, X_2]$. Suppose that an integral curve $t \mapsto x(t)$ of the vector field Z is defined for $0 \leqq t \leqq 1$ and that $x(0) = x_0$. For each positive integer n, consider the continuous function $t \mapsto z_n(t)$ defined on [0, 1] by the following conditions:

$$z_n(0) = x_0,$$

$$z_n'(t) = X_2(z_n(t)) \quad \text{for } \frac{k}{n} \leqq t \leqq \frac{4k+1}{4n},$$

$$z_n'(t) = X_1(z_n(t)) \quad \text{for } \frac{4k+1}{4n} \leqq t \leqq \frac{2k+1}{2n},$$

$$z_n'(t) = -X_2(z_n(t)) \quad \text{for } \frac{2k+1}{2n} \leqq t \leqq \frac{4k+3}{4n},$$

$$z_n'(t) = -X_1(z_n(t)) \quad \text{for } \frac{4k+3}{4n} \leqq t \leqq \frac{k+1}{n}$$

($0 \leqq k \leqq n$). Show that $z_n \to x$ uniformly on [0, 1]. (Use the same method as in (a).)

3. Let $X_1, \ldots, X_r$ be C^∞ vector fields on an open set $U \subset \mathbf{R}^n$. Let F be a closed set contained in U, and suppose that each X_j is tangent to F along F (Problem 1). Let $\mathscr{H}$ by the smallest sub-$\mathscr{E}(U; \mathbf{R})$-module of $\mathscr{T}_1^0(U)$ which contains the X_j and is a Lie algebra for the bracket operation $[Y, Z]$. Show that every vector field $Z \in \mathscr{H}$ is tangent to F along F. (Use Problems 1 and 2.)

4. With the notation of (**18.2.14**), assume that M is a real-analytic manifold and that the vector field X is analytic. For each analytic tensor field $\mathbf{Z}$ on M, show that $\mathbf{Z}_t$ (defined for t sufficiently small) is given in a sufficiently small neighborhood of $x \in M$ by the *Lie series*

$$\mathbf{Z}_t = \mathbf{Z} + \sum_{n=1}^{\infty} \frac{t^n}{n!} (\theta_X \cdot \mathbf{Z}).$$

(Calculate the derivative of $t \mapsto \mathbf{Z}_t$.) (Cf. Section **18.14**, Problem 14 for a counterexample for C^∞ vector fields.)

5. Let X be a C^1 vector field on a differential manifold M. A point $x \in M$ is said to be a *critical point* of X if $X(x) = 0$. At the point $\mathbf{0}_x \in T_x(M) = T(M)_x$, the tangent space $T_{\mathbf{0}_x}(T(M))$ is the direct sum of the subspace $T_{\mathbf{0}_x}(\mathbf{0}_{T(M)})$ tangent to the zero section of

T(M), and the subspace $T_{0_x}(T_x(M))$ of vertical vectors at the point $\mathbf{0}_x$. Let p_2 be the canonical projection of $T_{0_x}(T(M))$ onto $T_{0_x}(T_x(M))$, and let ρ_x be the composite linear mapping $\tau_{0_x} \circ p_2 : T_{0_x}(T(M)) \to T_x(M)$. If x is a critical point of X, then $\text{Hess}_x(X) = \rho_x \circ T_x(X)$ is an endomorphism of $T_x(M)$, called the *Hessian* of X at the critical point x.

(a) Show that, at a critical point x of X, we have

$$T_x(F_X(.\,, t)) = \exp(t \cdot \text{Hess}_x(X)),$$

the exponential being taken in the Banach algebra $\text{End}(T_x(M))$. (Use (**18.2.3.2**) to calculate the derivative of $t \mapsto T_x(F_X(.\,, t))$ by first calculating its value at $t = 0$.)
(b) A critical point x of X is said to be *nondegenerate* if the endomorphism $\text{Hess}_x(X)$ is invertible, or equivalently if $X : M \to T(M)$ is transversal at the point x over the submanifold $\mathbf{0}_{T(M)}$ of T(M) (Section **16.8**, Problem 9). Show that the nondegenerate critical points of X are isolated (observe that they are the points of $X^{-1}(\mathbf{0}_{T(M)})$).
(c) Suppose that M is *compact*. Show that the set of vector fields of class C^r on M, all of whose critical points are nondegenerate, is open in the space of all vector fields of class C^r, relative to the C^r-topology (Section **17.2**, Problem 1).

6. Let $M = \mathbf{R}^2$ and consider the vector field $X(\xi^1, \xi^2) = ((\xi^1, \xi^2), (\sin \xi^2, \cos^2 \xi^2))$. Show that there exists no homeomorphism of $\mathbf{R}^2$ onto itself which transforms the images of the integral curves of X into parallel lines.

7. Let X be a C^∞ vector field on a differential manifold M of dimension n. Suppose that, for some $x_0 \in M$, the image Γ of the integral curve $t \mapsto F_X(x_0, t)$ is *closed* in M and is not reduced to a point. Then $X(x) \neq 0$ at all points $x \in \Gamma$.

(a) Let $x \in \Gamma$ and let $c = (U, \varphi, n)$ be a chart of M at the point x, such that $\varphi(U) = I \times V$, where V is open in $\mathbf{R}^{n-1}$ and I is an open interval in $\mathbf{R}$, and such that X is equal in U to the vector field X_1 associated with the chart c (Section **17.14**, Problem 4). Then the image $\varphi(\Gamma \cap U)$ is of the form $I \times E$, where E is a *closed* subset of V. The point x is said to be *transversally isolated* in Γ if the projection of $\varphi(x)$ on V is an *isolated* point of E. Show that this condition is independent of the choice of chart c having the above properties.
(b) Show that if $t \mapsto F_X(x_0, t)$ is not injective, then the interval $J(x_0)$ of definition of this mapping is the whole of $\mathbf{R}$. The set H of numbers $t \in \mathbf{R}$ such that $F_X(x_0, t) = x_0$ is a discrete subgroup $\neq 0$ of $\mathbf{R}$, and Γ is compact. The mapping $t \mapsto F_X(x, t)$ is *periodic* for each $x \in \Gamma$, and H is the group of its periods.
(c) With the notation of (a), show that the set E is at most denumerable and contains at least one isolated point (use Baire's theorem). On the other hand, the set of transversally isolated points of Γ has as its inverse image under the mapping $t \mapsto F_X(x_0, t)$ a set which is both open and closed in $J(x_0)$. Deduce that *all* points of Γ are transversally isolated, and hence that Γ is a *submanifold* of M, which is diffeomorphic to $\mathbf{R}$ if $F_X(x_0, .)$ in injective, and to $\mathbf{T}$ otherwise.

8. Let X be the vector field on $\mathbf{R}^2$ defined by

$$X(\xi^1, \xi^2) = ((\xi^1, \xi^2), (\xi^2, -f(\xi^1)\xi^2 - g(\xi^1))),$$

where f and g are *odd* continuous functions such that $f(t) > 0$ and $g(t) > 0$ for $t > 0$. Suppose, moreover, that there exists $k > 0$ such that $f(t) < kg(t)$ for $0 < t < T$.

(a) Let

$$G(t) = \int_0^t g(s)\, ds,$$

$$F(\xi^1, \xi^2) = G(\xi^1) + \tfrac{1}{2}(\xi^2)^2.$$

Show that, if $t \mapsto u(t) = (u^1(t), u^2(t))$ is an integral curve of X, then we have

$$\frac{d}{dt} F(u^1(t), u^2(t)) < 0$$

for all t sucb that $u^1(t) > 0$. Deduce that, if $a > 0$ and $0 < b < (2G(a))^{1/2}$ and if $x_0 = (0, b)$, the integral curve $t \mapsto F_X(x_0, t) = (u^1(t), u^2(t))$ is such that $u^2(t_0) = 0$, $u^1(t_0) < a$ and $G(u^1(t_0)) < \frac{1}{2}b^2$ for some $t_0 > 0$. Furthermore, u^1 is strictly increasing and u^2 strictly decreasing, in the interval $[0, t_0]$.

(b) Show that, if $x_1 = (c, 0)$ is such that $0 < c < T$ and $(2G(c))^{1/2} < 1/k$, the integral curve $t \mapsto F_X(x_1, t) = (v^1(t), v^2(t))$ is such that $v^1(t_1) = 0$, $-1/k < v^2(t_1) < 0$ for some $t_1 > 0$, and v^1 and v^2 are strictly decreasing in the interval $[0, t_1]$.

(c) Deduce from (a) and (b) that, for all $x_0 = (0, b)$ satisfying

$$0 < b < \inf((2G(T))^{1/2}, 1/k),$$

the integral curve $t \mapsto F_X(x_0, t)$ is *periodic*.

9. Let X be a vector field of class $C^r (r \geq 1)$ on a differential manifold M, such that $J(x) = \mathbf{R}$ for all $x \in M$. Then the group $\mathbf{R}$ acts on M by $(t, x) \mapsto F_X(x, t)$, and the *orbits* of this action are the images of the integral curves $t \mapsto F_X(x, t)$.

(a) For each real number $\alpha > 0$, let $\mathscr{P}_\alpha(X)$ be the set of orbits corresponding to integral curves $t \mapsto F_X(x, t)$ which are constant or of period $\leq \alpha$. Show that the union $P_\alpha(X)$ of the orbits belonging to $\mathscr{P}_\alpha(X)$ is closed in M. (Let (x_k) be a sequence of points of $P_\alpha(X)$ converging to a point $x \in M$; observe that for each k there exists a number t_k such that $\frac{1}{2}\alpha \leq t_k \leq \alpha$ and $F_X(x_k, t_k) = x_k$, and use the compactness of the interval $[\frac{1}{2}\alpha, \alpha]$.)

(b) Suppose that M is *compact* and that X has only *nondegenerate* critical points (Problem 5). Show that there exists $\beta > 0$ such that every strictly positive period of a periodic integral curve $t \mapsto F_X(x, t)$ is $\geq \beta$. (Argue by contradiction, and assume that there exists a sequence (t_k) of real numbers >0 and tending to 0, and a sequence (x_k) of points of M such that $F_X(x_k, t_k) = x_k$; we may also assume that the sequence (x_k) converges to a point $x \in M$ and, by virtue of (a), that either x is a critical point or else $t \mapsto F_X(x, t)$ is periodic with period >0. Show that the latter alternative is impossible, by using Section **17.14**, Problem 4. On the other hand, if x is a critical point, use the fact that for each integer $m > 0$ the endomorphism $T_x(F_X(\cdot, mt_k))$ has an eigenvalue equal to 1, and proceed as in (a) to show that there exists $t > 0$ such that the endomorphism $T_x(F_X(\cdot, t))$ has an eigenvalue equal to 1; then use Problem 5(a) to obtain a contradiction.)

10. (a) Let x be a vector field of class C^r $(r \geq 1)$ on a differential manifold M, such that $J(x) = \mathbf{R}$ for each $x \in M$. Let $t \mapsto F_X(x, t)$ be an integral curve of X which admits a period $b > 0$. Consider the mapping $\Phi_X : M \times \mathbf{R} \to M \times M \times \mathbf{R}$ defined by $\Phi_X(x, t) = (x, F_X(x, t), t)$. Show that Φ_X is transversal (Section **16.8**, Problem 9) over the submanifold $\Delta \times \mathbf{R}$ of $M \times M \times \mathbf{R}$ (where Δ is the diagonal of $M \times M$) at the point

(x, b) if and only if the eigenvalue 1 of the endomorphism $T_x(F_X(., b))$ occurs with multiplicity 1.

(b) Let α be a real number >0. Let $\mathscr{G}_\alpha^{(r)}$ denote the set of vector fields X of class C^r on M ($r \geqq 1$) such that $J(x) = \mathbf{R}$ for all $x \in M$ and such that, for each period $b \in]0, \alpha]$ of an integral curve $t \mapsto F_X(x, t)$ of X, the eigenvalue 1 of $T_x(F_X(., b))$ occurs with multiplicity 1. Show that, if $X \in \mathscr{G}_\alpha^{(r)}$ and if the integral curve $t \mapsto F_X(x, t)$ belongs to $P_\alpha(X)$ (Problem 9), then there exists a neighborhood U of the corresponding orbit which does not meet any other orbit corresponding to an integral curve $t \mapsto F_X(y, t)$ which is either constant or admits a period $<\alpha$. (Argue by contradiction, by supposing that there exists a sequence (t_k) in the interval $[\frac{1}{2}\alpha, \alpha]$ and a sequence of points (x_k) in M belonging to distinct orbits, such that $F_X(x_k, t_k) = x_k$ and such that the sequence $((x_k, t_k))$ has a limit (x, t) satisfying $F_X(x, t) = x$. Using the transversality of Φ_X at the point (x, t), show that there exists a neighborhood V of (x, t) in $M \times \mathbf{R}$ such that $V \cap \Phi_X^{-1}(\Delta \times \mathbf{R})$ is the image of a neighborhood I of 0 in $\mathbf{R}$ under the mapping $s \mapsto (x, t+s)$ if x is a critical point, and under the mapping $s \mapsto (F_X(x, s), t)$ if $s \mapsto F_X(x, s)$ is periodic; obtain a contradiction by remarking that $(x_k, t_k) \in \Phi_X^{-1}(\Delta \times \mathbf{R})$.)

(c) Suppose that M is *compact*. Show that if $X \in \mathscr{G}_\alpha^{(r)}$, the set $P_\alpha(x)$ (Problem 9) is *finite*. Furthermore, there exists a number $\varepsilon > 0$ such that no integral curve $t \mapsto F_X(x, t)$ admits a period $b \in]\alpha, \alpha + \varepsilon[$. (Argue by contradiction as in (b).)

(d) Show that if M is compact, the set $\mathscr{G}_\alpha^{(r)}$ is open in $\Gamma^{(r)}(M, \mathbf{T}_0^1(M))$ relative to the C^r-topology.

11. Let M and N be two differential manifolds. Let $f: M \times \mathbf{R} \to N$ be a C^∞ mapping such that, for each $t \in \mathbf{R}$, $x \mapsto f(x, t)$ is an *embedding* (16.8.4) of M in N. Let $F(x, t) = (f(x, t), t)$; then F is an embedding of $M \times \mathbf{R}$ in $N \times \mathbf{R}$.

(a) Let I be the unit interval $[0, 1]$ in $\mathbf{R}$, and suppose that for each $t \leqq 0$ (resp. each $t \geqq 1$) the mapping $x \mapsto f(x, t)$ is an embedding f_0 (resp. f_1) independent of t. Suppose, moreover, that there exists a compact subset K of M such that the mapping $t \mapsto f(x, t)$ is constant for each $x \in M - K$. Let W be a compact neighborhood of $f(M_0 \times I)$ in N. Show that there exists a C^∞ vector field Z on $N \times \mathbf{R}$ such that

(i) $Z(F(x, u)) = T_u(F(x, .)) \cdot E(u)$ for each $(x, u) \in M \times I$;
(ii) $Z(y, t) = (\mathbf{0}_y, E(y))$ for all $(y, t) \in W \times I$;
(iii) the projection of $Z(y, t)$ on $T_t(\mathbf{R})$ is $E(t)$ for each $(y, t) \in N \times \mathbf{R}$. (Use (16.12.11).).

(b) Deduce from (a) that there exists a C^∞ mapping $g: N \times \mathbf{R} \to N$ such that $g(y, t) = y$ for each $y \notin W$, and such that for each $t \in \mathbf{R}$ the mapping $x \mapsto g(x, t)$ is a *diffeomorphism* of N onto N and $g(f(x, 0), t) = f(x, t)$ for each $(x, t) \in M \times \mathbf{R}$. (Let $u \mapsto G(y, t, u)$ be the integral curve of the vector field Z such that $G(y, t, 0) = (y, t)$, and observe that $G(y, t, u) = (h(y, t, u), t + u)$.)

12. Let X be a C^∞ vector field on M. The notation is as in (18.2.14).

(a) Let α be a C^∞ differential p-form on M, let $h: N \to M$ be a C^∞ mapping of a differential manifold N of dimension p into M, and let U be a relatively compact open set in N. For sufficiently small $u \in \mathbf{R}$, put

$$I(u) = \int_U {}^t(g_u \circ h)(\alpha).$$

Show that

$$\frac{dI}{du} = \int_U {}^t(g_u \circ h)(\theta_X \cdot \alpha).$$

(Use (18.2.14.8).)
(b) Deduce from (a) that, in order that $I(u)$ should be independent of u in a neighborhood of 0 in $\mathbf{R}$, for all choices of N, h, and U, it is necessary and sufficient that $\theta_X \cdot \alpha = 0$. If this condition is satisfied, the form α is said to be *invariant under the vector field* X, or X-*invariant*. (Take N = U equal to the domain of a chart c of M for which X is the vector field X_1 associated with c (Section 17.14, Problem 4).)
(c) A C^∞ real-valued function f is said to be a *first integral* of X if f is constant on each integral curve of X. Show that this is the case if and only if $\theta_X \cdot f = 0$ (i.e., if and only if f is an X-invariant 0-form).
(d) If M has dimension n and if v is an X-invariant n-form then, (with the same notation) ${}^t g_u(v) = 0$ for all sufficiently small u. Deduce that, for u small, the Lebesque measure μ_v associated with v (16.24.1) is invariant under the diffeomorphism g_u of M (Cf. Section 13.4, Problem 8).
(e) The set of C^∞ exterior differential forms on M which are X-invariant is a *subalgebra* of the $\mathscr{E}_0$-algebra $\mathscr{A}$ and is stable under exterior differentiation.

13. With the hypotheses and notation of Problem 12, an X-invariant form α is said to be an *absolute integral invariant* of X if in addition $i_X \cdot \alpha = 0$, or equivalently, if simultaneously $i_X \cdot \alpha = 0$ and $i_X \cdot d\alpha = 0$ (17.15.3.4). A first integral is an absolute integral invariant of order 0.

(a) Let α be a C^∞ differential p-form on M, and let N be a differential manifold of dimension p. Let h be a C^∞ mapping of $\mathbf{R} \times \mathrm{N}$ into M such that, for each $y \in \mathrm{N}$, $t \mapsto h(t, y)$ is an integral curve of the field of 1-directions on M defined by the vector field X (assumed to be everywhere nonzero) (18.8.5). In order that α should be an absolute integral invariant of X, it is necessary and sufficient that, for each choice of N and h satisfying the conditions above and for each relatively compact open subset U of N, the integral

$$I(u) = \int_U {}^t h(u, .)(\alpha)$$

should be independent of u in a neighborhood of 0.
(b) The set of C^∞ differential forms on M which are absolute integral invariants of X is a subalgebra of $\mathscr{A}$, stable under exterior differentiation, and is equal to the set of absolute integral invariants of fX, where f is any C^∞ real-valued function on M.

14. With the hypotheses and notation of Problem 12, a C^∞ differential p-form α on M is said to be a *relative integral invariant* of X if $d\alpha$ is an absolute integral invariant of X, or equivalently if $i_X \cdot d\alpha = 0$, or equivalently again if $X(x)$ is a Cauchy characteristic vector of the differential system formed by the single equation $d\alpha = 0$ (18.16.1).

Suppose that M is of dimension n, that $X(x) \neq 0$ for all x, and that there exists a relative integral invariant α of X which is an $(n-2)$-form such that $d\alpha(x) \neq 0$ for all $x \in \mathrm{M}$. Show that there exists no *compact connected* submanifold N of dimension $(n-1)$ which is *transversal* (Section 16.8, Problem 9) at each of its points to the integral curve of X passing through that point. (If $j: \mathrm{N} \to \mathrm{M}$ is the canonical injection, consider the $(n-1)$-form ${}^t j(d\alpha)$ on N, and use the formula (17.15.5.1).)

15. Let M, N be two differential manifolds, I an open neighborhood of 0 in **R**, and $\varphi: \mathrm{I} \times \mathrm{N} \to \mathrm{M}$ a C^∞ mapping. For each $(s, z) \in \mathrm{I} \times \mathrm{N}$, let $\varphi_s'(s, z) \in \mathrm{T}_{\varphi(s, z)}(\mathrm{M})$ denote the derivative at s (18.1.2.3) of the mapping $s \mapsto \varphi(s, z)$ of I into M. For each p-covector $\mathbf{u}^* \in \bigwedge^p (\mathrm{T}_{\varphi(s, z)}(\mathrm{M})^*)$, $i(\varphi_s'(s, z)) \cdot \mathbf{u}^*$ is a $(p-1)$-covector in $\bigwedge^{p-1} (\mathrm{T}_{\varphi(s,z)}(\mathrm{M})^*)$ (A.15.4). Now define a $(p-1)$-covector $\mathbf{v}^* \in \bigwedge^{p-1} (\mathrm{T}_z(\mathrm{N})^*)$ by the requirement that, for any $p-1$ vectors $\mathbf{h}_1, \ldots, \mathbf{h}_{p-1}$ in $\mathrm{T}_z(\mathrm{N})$,

$$\begin{aligned} \langle \mathbf{v}^*, \mathbf{h}_1 \wedge \mathbf{h}_2 \wedge \cdots \wedge \mathbf{h}_{p-1} \rangle &= \langle i(\varphi_s'(s, z)) \cdot \mathbf{u}^*, \mathbf{k}_1 \wedge \mathbf{k}_2 \wedge \cdots \wedge \mathbf{k}_{p-1} \rangle \\ &= \langle \mathbf{u}^*, \varphi_s'(s, z) \wedge \mathbf{k}_1 \wedge \mathbf{k}_2 \wedge \cdots \wedge \mathbf{k}_{p-1} \rangle, \end{aligned}$$

where $\mathbf{k}_j = \mathrm{T}_z\, \varphi(s, .) \cdot \mathbf{h}_j$ for $1 \leq j \leq p-1$. If now α is a C^∞ p-form on M, we define for each $z \in \mathrm{N}$ a $(p-1)$-covector in $\bigwedge^{p-1} (\mathrm{T}_z(\mathrm{N})^*)$ by replacing in the definition of $\mathbf{v}^*$ the p-covector $\mathbf{u}^*$ by $\alpha(\varphi(s, z))$. Show that this defines, for each $s \in \mathrm{I}$, a differential $(p-1)$-form of class C^∞. We denote this $(p-1)$-form by ${}^t\varphi(s, .)(i_{\varphi_s'(s,.)} \cdot \alpha)$.

On the other hand, ${}^t\varphi(s, .)(\alpha)$ is a differential p-form on N; hence its value ${}^t\varphi(s, .)(\alpha)(z)$ at a point z is a p-covector belonging to $\bigwedge^p (\mathrm{T}_z(\mathrm{N})^*)$, which is a function of s, and has a derivative at $s = 0$. We thus define a differential p-form on N, which we denote by

$$\frac{d}{ds}({}^t\varphi(s, .)(\alpha))|_{s=0}.$$

With this notation, prove the formula

$$\frac{d}{dt}({}^t\varphi(s, .)(\alpha))|_{s=0} = {}^t\varphi(0, .)(i_{\varphi_s'(0, .)} \cdot d\alpha) + d({}^t\varphi(0, .)(i_{\varphi_s'(0, .)} \cdot \alpha)).$$

(Show that it is enough to prove the formula in the cases where α is a function f or a differential df on M.)

3. SECOND-ORDER DIFFERENTIAL EQUATIONS ON A MANIFOLD

(18.3.1) Let v be a mapping of class C^1 of an open set $\mathrm{I} \subset \mathbf{R}$ into a differential manifold M. We have seen (18.1.2) that for each $t \in \mathrm{I}$, the vector

$$v'(t) \in \mathrm{T}_{v(t)}(\mathrm{M})$$

is defined as the image of $E(t)$ under $\mathrm{T}(v)$. The mapping $t \mapsto v'(t)$ is therefore a continuous mapping of I into the *differential manifold* T(M).

If (U, φ, n) is a chart of M at the point $v(t)$, we have

$$v'(t) = \mathrm{T}(\varphi^{-1}) \cdot ((\varphi \circ v)(t), \mathrm{D}(\varphi \circ v)(t)).$$

If v is *of class* C^2, this local expression shows that v' is a mapping *of class* C^1 of I into T(M) and is a *lifting* of v; hence we may define the vector $v''(t)$, which belongs to the tangent space $\mathrm{T}_{v'(t)}(\mathrm{T}(\mathrm{M}))$ to the manifold T(M), and thus v''

is a continuous mapping of I into the differential manifold T(T(M)). In this way we can define successively the higher derivatives of v; the derivative $v^{(r)}$ is defined if v is of class C^r, and is a continuous mapping of I into $T^r(M)$, where $T^r(M)$ is the manifold defined inductively by the conditions $T^1(M) = T(M)$ and $T^r(M) = T(T^{r-1}(M))$ for $r > 1$.

(18.3.2) In order to define an (autonomous) *second-order* differential equation on M, we must therefore start with a vector field Z of class C^r ($r \geqq 0$) *on the tangent bundle* T(M), and consider mappings v of class C^2 of an open interval $I \subset \mathbf{R}$ into M which are such that the mapping $t \mapsto v'(t)$ of I into T(M) is an integral curve of the vector field Z, or in other words satisfies the equation

(18.3.2.1) $$v''(t) = Z(v'(t))$$

for $t \in I$. However, this is possible only if the vector field Z satisfies a supplementary condition. For v' has to be a *lifting* of v; that is to say, $v(t) = o_M(v'(t))$; differentiating this relation, we obtain $T(v) \cdot E(t) = T(o_M) \cdot (T(v') \cdot E(t))$, which may be written as $v'(t) = T(o_M) \cdot v''(t)$, and by virtue of (18.3.2.1), this gives

$$v'(t) = T(o_M) \cdot Z(v'(t)).$$

Since we wish to have solutions satisfying arbitrary initial conditions, $v'(t)$ must be able to take all values in T(M). Hence we must impose on the vector field Z the condition

(18.3.2.2) $$T(o_M) \cdot Z(\mathbf{h}_x) = \mathbf{h}_x$$

for all $\mathbf{h}_x \in T(M)$. A vector field Z on T(M) satisfying this condition is called a *vector field defining a second-order differential equation*. In particular, (18.3.2.2) implies that $Z(\mathbf{h}_x) \neq \mathbf{0}_{\mathbf{h}_x}$ if $\mathbf{h}_x \neq \mathbf{0}_x$. An (autonomous) second-order equation on M is therefore by definition a differential equation of the first order *on* T(M), defined by a vector field Z satisfying (18.3.2.2), and a *solution* of such an equation is a mapping v of an open interval $I \subset \mathbf{R}$ into M, of class C^2, satisfying (18.3.2.1) for all $t \in I$. In terms of a local chart (U, φ, n) of M, the tangent bundle T(M) is identified locally with $\varphi(U) \times \mathbf{R}^n$, and a vector field Z satisfying (18.3.2.2) has a local expression

(18.3.2.3) $$(x, \mathbf{y}) \mapsto ((x, \mathbf{y}), (\mathbf{y}, \mathbf{f}(x, \mathbf{y}))),$$

where $\mathbf{f} : \varphi(U) \times \mathbf{R}^n \to \mathbf{R}^n$ is a mapping of class C^r; if v is a solution of (18.3.2.1), the function $u = \varphi \circ v : \mathrm{I} \to \varphi(\mathrm{U})$ satisfies the second-order vector differential equation

(18.3.2.4) $$\mathrm{D}^2 u(t) = \mathbf{f}(u(t), \mathrm{D}u(t)).$$

It comes to the same thing to say that a function v of class C^2 on an open interval $\mathrm{J} \subset \mathbf{R}$ is a solution of the second-order equation defined by Z, or to say that $v = o_{\mathrm{M}} \circ \mathbf{w}$, where $\mathbf{w}$ is an integral curve of the vector field Z, defined in J. For we have $v'(t) = \mathrm{T}(o_{\mathrm{M}}) \cdot \mathbf{w}'(t) = \mathrm{T}(o_{\mathrm{M}}) \cdot Z(\mathbf{w}(t)) = \mathbf{w}(t)$ by (18.3.2.2). A *maximal* solution of the second-order equation defined by Z is a solution which cannot be extended to a solution on a strictly larger interval; or, equivalently, a solution of the form $o_{\mathrm{M}} \circ \mathbf{w}$, where $\mathbf{w}$ is a *maximal* integral curve of the vector field Z.

Differential equations of higher orders on M are defined analogously.

(18.3.3) The results of (18.2) can of course be applied to second-order differential equations since what is involved is a particular case of integral curves of vector fields on T(M). Particular interest attaches to the set of solutions of a second-order differential equation for which $v(0)$ is a *given* point $a \in \mathrm{M}$ and $v'(0)$ takes *all* possible values in the fibre $\mathrm{T}_a(\mathrm{M})$. From (18.2) we deduce:

(18.3.4) *Let Z be a vector field of class C^r $(r \geqq 1)$ on* T(M), *defining a second-order differential equation.*

(i) *For each $a \in \mathrm{M}$, there exists a real number $\alpha > 0$ and a neighborhood* U *of $\mathbf{0}_a$ in* T(M) *such that, for each point $\mathbf{h}_x \in \mathrm{U}$, there is a solution $t \mapsto \gamma(t, \mathbf{h}_x)$ of the second-order equation defined by the vector field Z which is defined in the open interval $]-\alpha, \alpha[\subset \mathbf{R}$ and is such that $\gamma(0, \mathbf{h}_x) = x = o_{\mathrm{M}}(\mathbf{h}_x)$ and*

$$\gamma'(0, \mathbf{h}_x) = \mathbf{h}_x.$$

(ii) *We can moreover choose α and* U *such that, for each point $x \in \mathrm{U} \cap \mathrm{M}$ (where* M *is identified with the zero section of* T(M)) *and each $t_0 \neq 0$ in $]-\alpha, \alpha[$, the mapping $\mathbf{h}_x \mapsto \gamma(t_0, \mathbf{h}_x)$ is a homeomorphism of $\mathrm{U} \cap \mathrm{T}_x(\mathrm{M})$ onto an open neighborhood V_x of x in* M, *this homeomorphism and its inverse both being of class C^r. If d is a distance which defines the topology of* M, *we may assume also that (for fixed t_0) there exists $\rho > 0$ such that V_x contains the open ball with center x and radius ρ, for each $x \in \mathrm{U} \cap \mathrm{M}$.*

(i) We apply (18.2.5) to the integral curves of the vector field Z: the number α and the neighborhood U are chosen so that $\mathrm{U} \times]-\alpha, \alpha[$ is contained in the open set $\mathrm{dom}(\mathrm{F}_Z)$.

(ii) Since the question is local as regards T(M), we may assume that M is an open set in $\mathbf{R}^n$ and that the vector field Z has the local expression

$$(x, \mathbf{y}) \mapsto ((x, \mathbf{y}), (\mathbf{y}, \mathbf{f}(x, \mathbf{y}))),$$

where $\mathbf{f}$ is continuously differentiable in a neighborhood of $(a, 0)$ in $\mathrm{M} \times \mathbf{R}^n$. Changing notation, let us write $(x_0, \mathbf{y}_0)$ in place of $\mathbf{h}_x$ and

$$(u(t, x_0, \mathbf{y}_0), \mathbf{v}(t, x_0, \mathbf{y}_0))$$

in place of $\gamma(t, \mathbf{h}_x)$, so that the vector-valued function

$$t \mapsto (u(t, x_0, \mathbf{y}_0), \mathbf{v}(t, x_0, \mathbf{y}_0))$$

is the solution of the system of two vector differential equations

(18.3.4.1) $$x' = \mathbf{y}, \qquad \mathbf{y}' = \mathbf{f}(x, \mathbf{y}),$$

such that $u(0, x_0, \mathbf{y}_0) = x_0$, $\mathbf{v}(0, x_0, \mathbf{y}_0) = \mathbf{y}_0$, where u and $\mathbf{v}$ are of class C^r in $]-\alpha, \alpha[\times \mathrm{U}$. Put

$$A(t, x_0, \mathbf{y}_0) = \mathrm{D}_1\mathbf{f}(u(t, x_0, \mathbf{y}_0) \ \mathbf{v}(t, x_0, \mathbf{y}_0)),$$
$$B(t, x_0, \mathbf{y}_0) = \mathrm{D}_2\mathbf{f}(u(t, x_0, \mathbf{y}_0), \mathbf{v}(t, x_0, \mathbf{y}_0)).$$

It follows from (10.7.3) that, for $(t, x_0, \mathbf{y}_0)$ in a neighborhood of $(0, a, 0)$ in $\mathbf{R} \times \mathbf{R}^{2n}$, the functions (with values in $\mathscr{L}(\mathbf{R}^n)$)

$$Z(t, x_0, \mathbf{y}_0) = \mathrm{D}_3\,\mathbf{g}(t, x_0, \mathbf{y}_0),$$
$$W(t, x_0, \mathbf{y}_0) = \mathrm{D}_3\,\mathbf{h}(t, x_0, \mathbf{y}_0),$$

where

$$\mathbf{g}(t, x_0, \mathbf{y}_0) = (u(t, x_0, \mathbf{y}_0) - x_0) - t\mathbf{y}_0,$$
$$\mathbf{h}(t, x_0, \mathbf{y}_0) = \mathbf{v}(t, x_0, \mathbf{y}_0) - \mathbf{y}_0$$

form (as functions of t) the solution of the linear differential system

$$\begin{cases} Z' = W, \\ W' = A \circ Z + B \circ W + (tA + B), \end{cases}$$

such that $Z(0, x_0, \mathbf{y}_0) = 0$ and $W(0, x_0, \mathbf{y}_0) = 0$. Application of Gronwall's lemma (10.5.1) to this system shows that we may choose α_0 and

$$\mathrm{U} = \{(x_0, \mathbf{y}_0) : \|x_0 - a\| < r, \|\mathbf{y}_0\| < r\}$$

such that, for all $|t| < \alpha_0$ and $(x_0, \mathbf{y}_0) \in \mathrm{U}$,

(18.3.4.2) $$\|W(t, x_0, \mathbf{y}_0)\| \leq b|t|, \qquad \|Z(t, x_0, \mathbf{y}_0)\| \leq b|t|^2,$$

where $b > 0$ is a constant. Now consider the mapping

(18.3.4.3) $$\mathbf{y}_0 \mapsto u(t_0, x_0, \mathbf{y}_0) - x_0 = t_0\mathbf{y}_0 + \mathbf{g}(t_0, x_0, \mathbf{y}_0)$$

for some $t_0 \neq 0$ in $]-\alpha, \alpha[$, where $\alpha < \alpha_0$. By virtue of (18.3.4.2) and the mean-value theorem, we have

(18.3.4.4)
$$\|\mathbf{g}(t_0, x_0, \mathbf{y}_1) - \mathbf{g}(t_0, x_0, \mathbf{y}_2)\| \leqq b\alpha \|t_0(\mathbf{y}_1 - \mathbf{y}_2)\| \leqq \tfrac{1}{2}\|t_0(\mathbf{y}_1 - \mathbf{y}_2)\|$$

for $\|\mathbf{y}_1\| < r$ and $\|\mathbf{y}_2\| < r$, provided that $\alpha < \alpha_0$ has been chosen sufficiently small so that $b\alpha < \frac{1}{2}$. Hence we see that (18.3.4.3) is an *injective* continuous mapping of the compact ball $B' : \|\mathbf{y}_0\| \leqq \frac{1}{2}r$ into $\mathbf{R}^n$, hence is a *homeomorphism* of B' onto its image (3.17.12). It remains to show that this image contains a ball with center 0 and radius *independent of* x_0 (for $\|x_0 - a\| < r$). For this, we shall show that (10.1.1) can be applied to the function

$$(\mathbf{X}, \mathbf{Y}) \mapsto \mathbf{X} - \mathbf{g}(t_0, x_0, t_0^{-1}\mathbf{Y})$$

defined in the product of the balls

$$B : \|\mathbf{X}\| < \tfrac{1}{16} r |t_0| \qquad \text{and} \qquad B'' : \|\mathbf{Y}\| < \tfrac{1}{2} r |t_0|.$$

Now, by virtue of (18.3.4.1), there exists a constant c such that

$$\|u(t, x_0, 0) - x_0\| \leqq c|t|^2$$

for $|t| < \alpha$ and $\|x_0 - a\| < r$, hence

(18.3.4.5)
$$\|\mathbf{g}(t_0, x_0, 0)\| \leqq c|t_0|^2 \leqq c\alpha|t_0|.$$

Take $\alpha < \alpha_0$ sufficiently small so that $c\alpha < \frac{1}{16}r$. Then, for $\mathbf{Y}_1$ and $\mathbf{Y}_2$ in B'', we have

$$\|\mathbf{g}(t_0, x_0, t_0^{-1}\mathbf{Y}_1) - \mathbf{g}(t_0, x_0, t_0^{-1}\mathbf{Y}_2)\| \leqq \tfrac{1}{2}\|\mathbf{Y}_1 - \mathbf{Y}_2\|$$

by virtue of (18.3.4.4), and for $\mathbf{X} \in B$,

$$\|\mathbf{X} - \mathbf{g}(t_0, x_0, 0)\| \leqq \tfrac{1}{8} r |t_0| < \tfrac{1}{4} r |t_0|.$$

Hence there exists a continuous mapping $\mathbf{X} \mapsto \mathbf{F}(\mathbf{X})$ of B into B'' such that

$$\mathbf{X} = \mathbf{F}(\mathbf{X}) + \mathbf{g}(t_0, x_0, t_0^{-1}\mathbf{F}(\mathbf{X}))$$

and this proves that the image of B' under the mapping (18.3.4.3) contains B. Hence we obtain the assertion (ii) of (18.3.4) by taking $\rho = \frac{1}{16} r |t_0|$, and observing that in a metrizable compact space all distances are uniformly equivalent (3.16.5).

4. SPRAYS AND ISOCHRONOUS SECOND-ORDER EQUATIONS

(18.4.1) In the theory of second-order differential equations, one is often interested less in the *solutions* $t \mapsto v(t)$ (which are *unending paths* in M (16.27)) than in the *images* of these solutions. These images are called the *trajectories* of the equation or of the vector field Z that defines the equation. The image under v of a *compact* interval $[\alpha, \beta]$ contained in the open interval of definition of v is called an *arc of trajectory*, and $v(\alpha)$ and $v(\beta)$ are called respectively the *origin* and *endpoint* of the arc. The tangent vectors $v'(\alpha)$ and $v'(\beta)$ at the origin and endpoint of the arc are well-defined. If v is a solution defined in an interval $]-\alpha, \alpha[$, then for each positive real number c the function $w : t \mapsto v(ct)$, defined on $]-c^{-1}\alpha, c^{-1}\alpha[$, has the *same image* as v, but is *not*, in general, a solution of the same second-order equation. In the case where M is an open set in $\mathbf{R}^n$, v is a solution of (18.3.2.4), and we have

$$w'(t) = cv'(ct), \qquad w''(t) = c^2 v''(ct),$$

so that w is a solution of the vector differential equation

$$w''(t) = c^2 \mathbf{f}(w(t), c^{-1}w'(t))$$

for $-c^{-1}\alpha < t < c^{-1}\alpha$. We are therefore led, in this case, to consider mappings $\mathbf{f}$ satisfying the condition

(18.4.1.1) $$\mathbf{f}(x, c\mathbf{y}) = c^2 \mathbf{f}(x, \mathbf{y})$$

for *all* $c \in \mathbf{R}$. To express this condition in an intrinsic form for an arbitrary differential manifold M, we introduce the mapping $m_c : \mathbf{h}_x \mapsto c \cdot \mathbf{h}_x$ of T(M) into T(M): for a vector field Z on T(M) defining a second-order differential equation, the condition corresponding to (18.4.1.1) is

(18.4.1.2) $$Z(m_c(\mathbf{h}_x)) = c \cdot (\mathrm{T}(m_c) \cdot Z(\mathbf{h}_x)).$$

The vector field Z is said to be *isochronous*, or to be a *spray* over M, if it satisfies this relation for all $\mathbf{h}_x \in \mathrm{T(M)}$ and all $c \in \mathbf{R}$. The corresponding differential equation is also said to be *isochronous* (if the variable t represents time, the equation "does not depend on the *unit of measure* of time"). We remark that if Z is an isochronous field, then $Z(\mathbf{0}_x) = \mathbf{0}_{\mathbf{0}_x}$ for all $x \in \mathrm{M}$ and $Z(\mathbf{h}_x) \neq \mathbf{0}_{\mathbf{h}_x}$ whenever $\mathbf{h}_x \neq \mathbf{0}_x$; if v is a solution of the corresponding equation, then *either* v is constant *or* $v'(t) \neq \mathbf{0}_{v(t)}$ throughout *every* open interval I of definition of v (and if Z is of class C^∞, then v is an *immersion* of I in M). If v is a solution defined in $]-\alpha, \alpha[$, then the function $t \mapsto v(-t)$

is also a solution defined in the same interval. Since for each $c \in \mathbf{R}$ the function $t \mapsto v(t + c)$ is also a solution, it follows that an arc of trajectory with origin a and endpoint b in M is also an arc of trajectory with origin b and endpoint a, and hence a and b are also called the *endpoints* of the arc.

(18.4.2) *Let Z be a spray over* M *of class* C^r $(r \geq 0)$. *With the notation of* (18.2.3), *for each real number* $c \neq 0$ *and each vector* $\mathbf{h}_x \in T(M)$, *we have*

(18.4.2.1) $$J(c\mathbf{h}_x) = c^{-1}J(\mathbf{h}_x),$$

(18.4.2.2) $$F_Z(c\mathbf{h}_x, t) = c \cdot F_Z(\mathbf{h}_x, ct)$$

for all $t \in J(c\mathbf{h}_x)$.

This follows directly from the definitions.

The *trajectories*, which are the images of the solutions $t \mapsto o_M(F_Z(c\mathbf{h}_x, t))$ (where $t \in J(c\mathbf{h}_x)$), are therefore *independent of the choice of* $c \neq 0$ for $\mathbf{h}_x \neq 0$; they are called the *maximal trajectories* of the second-order equation defined by Z (or of the vector field Z) passing through the point x and *tangent at this point to the direction defined by the vector* $\mathbf{h}_x$ (or any of the vectors $c\mathbf{h}_x$, $c \neq 0$).

(18.4.3) In what follows, we shall denote by Ω_Z, $\Omega_Z(M)$, or simply Ω, the set of points $\mathbf{h}_x \in T(M)$ such that the open interval $J(\mathbf{h}_x)$ contains the closed unit interval $[0, 1]$. In the notation of (18.2.3), this is equivalent to $t^+(\mathbf{h}_x) > 1$. Since t^+ is a lower semicontinuous function on T(M) (18.2.7), the set Ω is *open* in T(M), and it follows from (18.4.2.1) that the relation $\mathbf{h}_x \in \Omega$ implies $c\mathbf{h}_x \in \Omega$ for all $c \in [0, 1]$ (in other words, each of the sets $\Omega \cap T_x(M)$ is *star-shaped* in the fiber $T_x(M)$; but it is not necessarily symmetrical with respect to $\mathbf{0}_x$ (18.4.9)). The mapping $\mathbf{h}_x \mapsto o_M(F_Z(\mathbf{h}_x, 1))$ is called the *exponential mapping defined by Z* and is written $\mathbf{h}_x \mapsto \exp_Z(\mathbf{h}_x)$, or simply $\mathbf{h}_x \mapsto \exp(\mathbf{h}_x)$; its value at $\mathbf{h}_x$ is the value at $t = 1$ of the solution v of the second-order differential equation defined by Z which satisfies $v(0) = x$, $v'(0) = \mathbf{h}_x \in \Omega$.

(18.4.4) *For each* $\mathbf{h}_x \in T(M)$, *the function* $v : t \mapsto \exp(t\mathbf{h}_x)$, *defined in the interval* $J(\mathbf{h}_x)$, *is the maximal solution of the second-order differential equation defined by Z, which satisfies the initial conditions* $v(0) = x$, $v'(0) = \mathbf{h}_x$. *For each* $t \neq 0$ *in* $J(\mathbf{h}_x)$, *we have*

(18.4.4.1) $$v'(t) = t^{-1}F_Z(t\mathbf{h}_x, 1).$$

For it follows from (18.4.2.2) that the function which takes the value $\mathbf{h}_x$ at $t = 0$ and the value $t^{-1}F_Z(t\mathbf{h}_x, 1)$ at $t \neq 0$ in $J(\mathbf{h}_x)$ is equal to $F_Z(\mathbf{h}_x, t)$ for all $t \in J(\mathbf{h}_x)$.

We recall also (18.2.3.1) that for $t_0 \in J(\mathbf{h}_x)$ we have

(18.4.4.2) $$J(v'(t_0)) = J(\mathbf{h}_x) + (-t_0).$$

(18.4.5) *Let Z be a spray over M of class C^r, where $r \geqq 1$. Then the exponential map is a C^r mapping of Ω_Z into M. For each $x \in M$, the tangent linear mapping $T_{\mathbf{0}_x}(\exp)$ at the point $\mathbf{0}_x$ of the zero section $\mathbf{0}_M$ of T(M), when restricted to the tangent space $T_{\mathbf{0}_x}(T_x(M))$ of vertical vectors (identified with $T_x(M)$), is the identity mapping, and when restricted to the tangent space $T_{\mathbf{0}_x}(\mathbf{0}_M)$ to the zero section is the canonical bijection of this space onto $T_x(M)$.*

The first assertion follows from (18.2.5), and the others follow from the facts that the restriction of exp to the zero section of T(M) is by definition the canonical bijection $\mathbf{0}_M \to M$ (the restriction of the projection

$$o_M : T(M) \to M),$$

and that, considering the linear mapping $r : t \mapsto t\mathbf{h}_x$ of $\mathbf{R}$ into the vector space $T_x(M)$, we have $\mathbf{h}_x = \tau_{t\mathbf{h}_x}(r'(t))$ and in particular $\mathbf{h}_x = r'(0)$, so that $T_{\mathbf{0}_x}(\exp) \cdot \mathbf{h}_x$ (being equal to $v'(0)$, where $v(t) = \exp(t\mathbf{h}_x)$) is equal to $\mathbf{h}_x$ (18.4.4).

Hence we obtain, by virtue of (10.2.5):

(18.4.6) *For each $a \in M$ and each neighborhood U_0 of $\mathbf{0}_a$ in T(M), there exists an open neighborhood $U \subset U_0$ of $\mathbf{0}_a$ in T(M) such that the mapping*

$$\mathbf{h}_x \mapsto (x, \exp(\mathbf{h}_x))$$

is a homeomorphism of class C^r, with inverse of class C^r, of U onto an open neighborhood of (a, a) in $M \times M$, and such that the mapping $\mathbf{h}_a \mapsto \exp(\mathbf{h}_a)$ is a homeomorphism of $U \cap T_a(M)$ onto an open neighborhood V of a in M.

(18.4.7) *Let a be a point of M and let U be a neighborhood of $\mathbf{0}_a$ in T(M) such that $U \cap T_a(M)$ is star-shaped and contained in Ω, and such that the mapping $\mathbf{h}_a \mapsto \exp(\mathbf{h}_a)$ is a homeomorphism of $U \cap T_a(M)$ onto an open neighborhood V of a in M. Then, for each $y \neq a$ in V, the arc of trajectory which is the image of [0, 1] under the mapping $t \mapsto \exp(t\mathbf{h}_a)$, where $\mathbf{h}_a$ is the unique solution of $\exp(\mathbf{h}_a) = y$ in $U \cap T_a(M)$, is the unique arc of trajectory with origin a and endpoint y which is contained in V.*

An arc of trajectory L with origin a and endpoint y is necessarily the image of an interval $[0, \alpha]$ under a mapping $t \mapsto \exp(t\mathbf{h}'_a)$, where $\mathbf{h}'_a \in T_a(M)$. We shall show that if $L \subset V$, then $t\mathbf{h}'_a \in U$ for $0 \leqq t \leqq \alpha$ and therefore $\alpha\mathbf{h}'_a = \mathbf{h}_a$, which will establish the uniqueness of the arc in question. Now, it is clear that $t\mathbf{h}'_a \in U$ for all sufficiently small t; if the closed set of $t \in [0, \alpha]$ such that $t\mathbf{h}'_a \notin U$ were not empty, it would have a least element $\beta > 0$. But, by hypothesis, we have $z = \exp(\beta\mathbf{h}'_a)$; if $\psi : V \to U \cap T_a(M)$ is the inverse of exp, then $t\mathbf{h}'_a = \psi(\exp(t\mathbf{h}'_a))$ tends to $\psi(z) \in U$ as $t \to \beta$, contrary to the hypothesis that $\beta\mathbf{h}'_a \notin U$.

It should nevertheless be remarked that in general there will exist other arcs of trajectories with origin a and endpoint a point $y \in V$, but *not contained wholly in* V. In other words, the exponential mapping *is not necessarily injective* on the whole of Ω. Likewise, we shall meet, in Chapter XIX, examples where it is not surjective; and finally, even if M is compact, the open set Ω *is not necessarily the whole of* T(M).

Examples

(18.4.8) Take M to be the open interval $\mathbf{R}_+^* =]0, +\infty[$ of $\mathbf{R}$, and identify the tangent bundle T(M) with $M \times \mathbf{R}$. Consider the spray

$$Z(x, y) = ((x, y), (y, y^2/x))$$

corresponding to the differential equation $x'' = x'^2/x$ in M. Here we have $\Omega = T(M)$, and for each $x \in M$ the exponential mapping restricted to $T_x(M) = \{x\} \times \mathbf{R}$ is the diffeomorphism $(x, y) \mapsto xe^y$ of $T_x(M)$ onto M. This example will be generalized in Chapter XIX, and will justify the nomenclature of exponential mapping.

(18.4.9) Let $M = \mathbf{R}$ and identify T(M) with $\mathbf{R} \times \mathbf{R}$. Consider the spray $Z(x, y) = ((x, y), (y, y^2))$, corresponding to the differential equation $x'' = x'^2$ in $\mathbf{R}$. It is easily seen that in this case the set Ω is the set of points (x, y) with $y < 1$, and that the exponential mapping, restricted to $\Omega \cap T_x(\mathbf{R})$, is the mapping $(x, y) \mapsto x - \log(1 - y)$, which is a diffeomorphism of $\Omega \cap T_x(\mathbf{R})$ onto $\mathbf{R}$.

(18.4.10) In the preceding example, the vector field Z was translation-invariant. Consider then the quotient group $\mathbf{T} = \mathbf{R}/\mathbf{Z}$, for which $\mathbf{R}$ is a covering, and let $\pi : \mathbf{R} \to \mathbf{T}$ denote the canonical homomorphism. There exists a unique vector field Z_0 on the tangent bundle $T(\mathbf{T})$ for which Z is a lifting (relative to the morphism $T(\pi) : T(\mathbf{R}) \to T(\mathbf{T})$); Z_0 is isochronous, and if v is any solution of the differential equation defined by Z, then $t \mapsto \pi(v(t))$ is a solution of the differential equation defined by Z_0. We see therefore that although $\mathbf{T}$ is compact, Ω_{Z_0} is not the whole of $\mathbf{T}$.

5. CONVEXITY PROPERTIES OF ISOCHRONOUS DIFFERENTIAL EQUATIONS

(18.5.1) Let M be a differential manifold and let Z be a spray of class C^r ($r \geqq 1$) over M. The manifold M is said to be *convex relative to* Z if there exists a C^1 mapping $\mathbf{s}$ of $M \times M$ *into the open set* Ω_Z in T(M) satisfying the following conditions:

(18.5.1.1) $$o_M(\mathbf{s}(x_1, x_2)) = x_1, \qquad \exp(\mathbf{s}(x_1, x_2)) = x_2 .$$

In other words, given any two points x_1, x_2 of M, the mapping

(18.5.1.2) $$t \mapsto \exp(t\mathbf{s}(x_1, x_2))$$

of the interval [0, 1] into M is a path with origin x_1 and endpoint x_2 and is the restriction to [0, 1] of a solution of the differential equation defined by Z (18.4.4). A manifold which is convex relative to a spray is therefore necessarily *connected* (3.19.3). An open set U in M is said to be *convex relative to* Z if it is convex relative to the restriction of Z to T(U), or equivalently, if there exists a C^1 mapping $\mathbf{s} : U \times U \to \Omega_Z$ satisfying (18.5.1.1) and in addition such that

(18.5.1.3) $$\exp(t\mathbf{s}(x_1, x_2)) \in U$$

for all $x_1, x_2 \in U$ and all $t \in [0, 1]$.

(18.5.2) *Suppose that the open set* $U \subset M$ *is convex relative to* Z. *Then, with the notation of* (18.5.1), *for each* $x \in U$, *the image of* U *under the mapping* $y \mapsto \mathbf{s}(x, y)$ *is the star-shaped open set* $T_x(M) \cap \Omega_{Z_U}(U)$, *where* Z_U *denotes the restriction of the vector field* Z *to* T(U). *The mapping* $y \mapsto \mathbf{s}(x, y)$ *is a homeomorphism of* U *onto this open set, and the inverse homeomorphism is* $\mathbf{h}_x \mapsto \exp(\mathbf{h}_x)$; *both homeomorphisms are of class* C^r. *In particular, for each* $y \in U$, *the image of the interval* [0, 1] *under the mapping* $t \mapsto \exp(t\mathbf{s}(x, y))$ *is the only arc of trajectory with origin x and endpoint y contained in* U.

Since only the vector field Z_U features in the statement of the proposition, we may assume that $U = M$. The relation $\exp(\mathbf{s}(x, y)) = y$ shows that the composition of the tangent linear mappings of $\mathbf{h}_x \mapsto \exp(\mathbf{h}_x)$ and $y \mapsto \mathbf{s}(x, y)$ is the identity, and therefore each of them is *bijective* (since the tangent spaces $T_{\mathbf{h}_x}(T_x(M))$ and $T_y(M)$ have the same dimension (16.1.5)). It follows that the image A_x of M under the mapping $y \mapsto \mathbf{s}(x, y)$ is a nonempty *open* subset of

$T_x(M)$ (16.7); and since the mapping $y \mapsto \mathbf{s}(x, y)$ is injective, it is a homeomorphism of M onto A_x, and clearly the inverse homeomorphism is

$$\mathbf{h}_x \mapsto \exp(\mathbf{h}_x).$$

Now the latter is of class C^r (18.2.5), and the bijectivity of the tangent mappings implies that $y \mapsto \mathbf{s}(x, y)$ is also of class C^r (10.2.5). It remains to show that $A_x = \Omega_Z \cap T_x(M)$. Now A_x is the set of all $\mathbf{h}_x \in \Omega_Z \cap T_x(M)$ such that $\mathbf{h}_x = \mathbf{s}(x, \exp(\mathbf{h}_x))$, and hence is *closed* in $\Omega_Z \cap T_x(M)$. But we have already seen that A_x is also open in $\Omega_Z \cap T_x(M)$, and this space is connected because it is star-shaped; hence the result. The final assertion of the proposition is a consequence of the fact that every arc of trajectory with origin x is the image of $[0, 1]$ under $t \mapsto \exp(t\mathbf{h}_x)$, where $\mathbf{h}_x \in \Omega_Z \cap T_x(M)$ (18.4.4), and the mapping $\mathbf{h}_x \mapsto \exp(\mathbf{h}_x)$ is bijective, as we have just seen.

It should be noted that the intersection of two open sets which are convex relative to Z is not necessarily connected (nor, *a fortiori*, convex relative to Z) (Section 18.6, Problem 8).

(18.5.3) *Let* M *be a differential manifold,* Z *a spray over* M, *and let* Q *be a real-valued function of class* C^2 *defined in an open neighborhood* V *of the diagonal of* M × M, *satisfying the following conditions:*

(i) $Q(a, a) = 0$ *for all* $a \in M$.

(ii) *For each* $a \in M$, *the function* $Q_a = Q(a, .) : x \mapsto Q(a, x)$ *has zero differential at the point* $x = a$.

(iii) *For each* $a \in M$, *the Hessian* $\text{Hess}_a(Q_a)$ (16.5.11) *is a positive definite quadratic form.*

Under these conditions, for each point $a \in M$, *there exists a neighborhood* W *of* a *and a real number* $r > 0$ *such that, for each point* $b \in W$ *and each* $\rho \in]0, r[$, *the connected component* C_b *of* b *in the open set defined by the inequality* $Q(b, x) < \rho$ *is* convex *relative to* Z.

The question is clearly local with respect to M, and we may therefore take M to be an open set in $\mathbf{R}^n$, the vector field Z being given by

$$(x, \mathbf{y}) \mapsto ((x, \mathbf{y}), (\mathbf{y}, \mathbf{f}(x, \mathbf{y}))),$$

where $\mathbf{f}(x, c\mathbf{y}) = c^2\mathbf{f}(x, \mathbf{y})$. Consider the expression

(18.5.3.1) $$H(x_1, x_2, \mathbf{u}) = D^2Q_{x_1}(x_2) \cdot (\mathbf{u}, \mathbf{u}) + DQ_{x_1}(x_2) \cdot \mathbf{f}(x_2, \mathbf{u})$$

for $(x_1, x_2, \mathbf{u}) \in V \times S_{n-1}$. When $x_1 = x_2 = a$, we have $DQ_a(a) = 0$ by hypothesis, and the hypotheses also imply that the function

$$\mathbf{u} \mapsto D^2 Q_a(a) \cdot (\mathbf{u}, \mathbf{u})$$

is strictly positive on S_{n-1}. Hence ((3.17.10) and (3.17.11)) there exists a compact neighborhood W_0 of a in M and a real number $\alpha > 0$ such that, for all x_1, x_2 in W_0 and $\mathbf{u} \in S_{n-1}$, we have $H(x_1, x_2, \mathbf{u}) \geqq \alpha$. For each solution $t \mapsto v(t)$ of the vector differential equation $D^2 x = \mathbf{f}(x, Dx)$ such that $v(t) \in W_0$, and each point $b \in W_0$, we have therefore

$$\text{(18.5.3.2)} \quad \frac{d^2}{dt^2}(Q_b(v(t))) = D^2 Q_b(v(t)) \cdot (v'(t), v'(t)) + DQ_b(v(t)) \cdot \mathbf{f}(v(t), v'(t))$$
$$= H(b, v(t), v'(t)) \geqq \alpha \|v'(t)\|^2.$$

Let U be an open neighborhood of $\mathbf{0}_a$ in T(M), contained in Ω_Z and such that $o_M(U) \subset W_0$ and $\exp(U) \subset W_0$ (18.4.6); we may also suppose that, for each $x \in o_M(U)$, the set $U \cap T_x(M)$ is *star-shaped* with respect to $\mathbf{0}_x$; in other words, that $t\mathbf{h}_x \in U$ for all $\mathbf{h}_x \in U$ and $t \in [0, 1]$. By virtue of (18.4.7), there exists a neighborhood $W_1 \subset W_0$ of a in M and a C^1 mapping $\mathbf{s}$ of $W_1 \times W_1$ into Ω_Z such that

$$o_M(\mathbf{s}(x_1, x_2)) = x_1, \qquad \exp(\mathbf{s}(x_1, x_2)) = x_2$$

for all x_1, x_2 in W_1, and such that $\mathbf{s}(a, a) = \mathbf{0}_a$. Now let W′ be a compact neighborhood of a such that

$$W' \times W' \subset (W_1 \times W_1) \cap \mathbf{s}^{-1}(U),$$

and let L be the frontier of W′. We shall show that

$$\text{(18.5.3.3)} \qquad Q_a(z) > 0 \qquad \text{for all} \quad z \in L.$$

Indeed, we have $\mathbf{s}(a, z) \in U$, hence $t \cdot \mathbf{s}(a, z) \in U$ for $0 \leqq t \leqq 1$, and therefore the function $v(t) = \exp(t \cdot \mathbf{s}(a, z))$ is defined in an open interval I containing [0, 1]. By the definition of L, we have $z \neq a$, hence $\mathbf{s}(a, z) \neq \mathbf{0}_a$ and therefore $v'(0) \neq 0$; and because $\mathbf{f}(x, 0) = 0$, we have also $v'(t) \neq 0$ for all $t \in I$, since otherwise v would be constant in I. It follows then from (18.5.3.2) that the function $t \mapsto Q_a(v(t))$ has its second derivative everywhere >0 in I. Since $Q_a(v(0)) = Q_a(a) = 0$ and $D(Q_a \circ v)(0) = DQ_a(a) \cdot v'(0) = 0$, we conclude (8.14.2) that $Q_a(v(t)) > 0$ for $0 < t \leqq 1$, and in particular, that $Q_a(z) = Q_a(v(1)) > 0$.

Since L is compact, there exists $r > 0$ and a neighborhood $W'' \subset W'$ of a such that, for all $b \in W''$,

$$\text{(18.5.3.4)} \qquad Q_b(c) \geqq r > 0 \qquad \text{for all} \quad c \in L.$$

Now let ρ be such that $0 < \rho < r$, and let W be the open neighborhood of a consisting of the points $x \in W''$ such that $Q(x, x) < \rho$. For each $b \in W$, let C_b be the connected component of b in the open set defined by the inequality $Q(b, x) < \rho$. It follows from (18.5.3.4) that C_b cannot intersect the frontier L of W′, and hence (3.19.9) we have $C_b \subset W'$. Now let x_1, x_2 be two points of C_b. Since $C_b \subset W'$, we have $\mathbf{s}(x_1, x_2) \in U$ and therefore $t \cdot \mathbf{s}(x_1, x_2) \in U$ for $0 \leqq t \leqq 1$. The function $t \mapsto \varphi(t) = Q_b(\exp(t \cdot \mathbf{s}(x_1, x_2)))$ is therefore defined in the interval [0, 1], and its second derivative $\varphi''(t)$ is $\geqq 0$ for all $t \in [0, 1]$ by virtue of (18.5.3.2). Hence we have the inequality

(18.5.3.5) $$\varphi(t) \leqq (1-t)\varphi(0) + t\varphi(1) \leqq \sup(\varphi(0), \varphi(1))$$

for $0 \leqq t \leqq 1$. (For the function $\psi(t) = \varphi(t) - ((1-t)\varphi(0) + t\varphi(1))$ has second derivative $\psi''(t) = \varphi''(t) \geqq 0$ in [0, 1], and $\psi(0) = \psi(1) = 0$; consequently the derivative $\psi'(t)$ is increasing, and if we had $\psi(\alpha) > 0$ for some $\alpha \in]0, 1[$, we should have $\psi'(\beta) > 0$ for some $\beta \in [0, \alpha]$, whence $\psi'(t) > 0$ for $t \in [\beta, 1]$ and so $\psi(1) > \psi(\beta) > 0$, a contradiction.) From (18.5.3.5) we deduce that

(18.5.3.6) $$Q_b(\exp(t \cdot \mathbf{s}(x_1, x_2))) \leqq \sup(Q_b(x_1), Q_b(x_2))$$

for all $t \in [0, 1]$, by the definition of the mapping $\mathbf{s}$. This implies that

$$\exp(t \cdot \mathbf{s}(x_1, x_2)) \in C_b$$

for $0 \leqq t \leqq 1$, by (3.19.7) and the definition of a connected component, and proves that C_b is convex relative to Z. Q.E.D.

Proposition (18.5.3) has the following corollary:

(18.5.4) *Under the hypotheses of* (18.5.3), *if* K *is any compact subset of* M, *there exists a finite covering of* K *by open sets* C_i $(1 \leqq i \leqq m)$ *which, together with all their finite intersections, are convex relative to Z.*

Let d be a distance defining the topology of M. For each $a \in K$, it follows from (18.3.4(ii)) and (18.4.7) that there exists $r_a > 0$ such that, for each point x satisfying $d(a, x) < \frac{1}{2}r_a$ and each x' satisfying $d(x, x') < \frac{1}{2}r_a$, there is only one arc of trajectory with origin x and endpoint x' contained in the open ball with center x and radius $\frac{1}{2}r_a$. Cover K by a finite number of open balls B_k $(1 \leqq k \leqq m)$ with centers a_k and radii $\frac{1}{2}r_{a_k}$. Then there exists $r' > 0$ such that $r' < \frac{1}{2}r_{a_k}$ for $1 \leqq k \leqq m$, and such that for each $x \in K$ the open ball B(x) with center x and radius r' is contained in one of the balls B_k (3.16.6). By definition, for each point $x' \in B(x)$, there is only one arc of trajectory with origin x and endpoint x' contained in $B(x)$. This being so, it follows from (18.5.3) that for each $b \in K$ there exists an open set C_b containing b which is convex relative to Z and of diameter $< r'$; moreover, if E is the intersection of a finite number of these open sets C_{b_k} $(1 \leqq k \leqq p)$, then for each k and each pair of points x, x' of E, there exists one and only one arc of trajectory L_k with origin x and endpoint x' contained in C_{b_k} (18.5.2); but since all these arcs L_k $(1 \leqq k \leqq p)$ have diameter $< r'$, they are all the *same*, and therefore E is convex (18.5.1.3). Finally, K can be covered by a finite number of the open sets C_{b_k}.

We shall see in (20.17.5) that there exist functions Q with the properties of (18.5.3) on any differential manifold M.

6. GEODESICS OF A CONNECTION

(18.6.1) Let $\mathbf{C}$ be a linear connection on a differential manifold M (17.18.1). For each $x \in M$ and each $\mathbf{h}_x \in T_x(M)$, the vector

$$G(\mathbf{h}_x) = \mathbf{C}_x(\mathbf{h}_x, \mathbf{h}_x) \in T_{\mathbf{h}_x}(T(M))$$

is such that $T(o_M) \cdot G(\mathbf{h}_x) = \mathbf{h}_x$; furthermore, for each $c \in \mathbf{R}$, we have

$$G(c \cdot \mathbf{h}_x) = \mathbf{C}_x(c \cdot \mathbf{h}_x, c \cdot \mathbf{h}_x) = c \cdot \mathbf{C}_x(\mathbf{h}_x, c \cdot \mathbf{h}_x)$$

by virtue of (17.16.3.3), and

$$T(m_c) \cdot G(\mathbf{h}_x) = \mathbf{C}_x(\mathbf{h}_x, c \cdot \mathbf{h}_x)$$

by (17.16.3.5). It follows from these formulas that $\mathbf{h}_x \mapsto G(\mathbf{h}_x)$ is a *spray* over M. This spray is called the *geodesic spray* of the linear connection $\mathbf{C}$; the *solutions* of the second-order differential equation on M defined by G are called the *geodesics* of the connection $\mathbf{C}$; and the *trajectories* (resp. *arcs of trajectories*) of this equation are called the *geodesic trajectories* (resp. *geodesic arcs*) of $\mathbf{C}$.

The differential equation of the geodesics of $\mathbf{C}$ is therefore

(18.6.1.1) $$v''(t) = \mathbf{C}_{v(t)}(v'(t), v'(t)),$$

which, by virtue of (17.17.2.1), can also be written in the form

(18.6.1.2) $$\nabla_E \cdot v' = \mathbf{0}$$

with the notation of (18.1.1).

Relative to a chart (U, φ, n) of M for which the local expression of the connection **C** is given by (17.16.4.1), this equation is equivalent to the following system of (*nonlinear*) second-order differential equations:

$$\frac{d^2u^i}{dt^2} + \sum_{h,l} \Gamma^i_{hl}(u) \frac{du^h}{dt}\frac{du^l}{dt} = 0 \qquad (1 \leqq i \leqq n), \tag{18.6.1.3}$$

where $u = \varphi \circ v$ has components $u^i = \varphi^i \circ v$ $(1 \leqq i \leqq n)$.

(18.6.2) (i) *In order that two connections* **C** *and* **C′** *on* M *should have the same geodesic spray, it is necessary and sufficient that the bilinear morphism* $B : T(M) \oplus T(M) \to T(M)$, *the difference of* **C** *and* **C′** (17.16.6), *should be antisymmetric.*

(ii) *Given any connection* **C** *on* M, *there exists a unique connection* **C′** *on* M *which has zero torsion* (17.20.6) *and the same geodesic spray as* **C**.

From (17.16.6.3) we have

$$B_x(\mathbf{h}_x, \mathbf{h}_x) = \tau_{\mathbf{h}_x}(\mathbf{C}_x(\mathbf{h}_x, \mathbf{h}_x) - \mathbf{C}'_x(\mathbf{h}_x, \mathbf{h}_x))$$

and to say that B is antisymmetric means that $B_x(\mathbf{h}_x, \mathbf{h}_x) = 0$ for all $\mathbf{h}_x \in T(M)$, whence (i) follows. Furthermore, if $\boldsymbol{t}$ and $\boldsymbol{t}'$ are the torsions of **C** and **C′**, respectively, then (17.20.7.2) we have $\boldsymbol{t}' - \boldsymbol{t} = 2B$ if B is antisymmetric; hence the only connection **C′** having the same geodesic spray as **C** and zero torsion is obtained by taking $B = -\frac{1}{2}\boldsymbol{t}$.

(18.6.3) The geodesics of a connection **C** can be interpreted in a different way by introducing the notion of *parallel transport* relative to **C**. Consider a *path* $v : [\alpha, \beta] \to M$ of class C^1 in M (resp. an *unending path* $v : I \to M$ of class C^1, where I is an *open* interval in **R**). Then a *parallel transport along the path* v (resp. *along the unending path* v) is by definition a C^1 mapping $\mathbf{w}$ of $[\alpha, \beta]$ (resp. I) into T(M) such that

$$o_M(\mathbf{w}(t)) = v(t) \quad \text{and} \quad \nabla_E \cdot \mathbf{w} = \mathbf{0} \quad \text{in} \quad [\alpha, \beta] \quad \text{(resp. in I)}, \tag{18.6.3.1}$$

which by definition (17.17.2.1) is equivalent to

$$o_M(\mathbf{w}(t)) = v(t) \qquad \text{and} \qquad \mathbf{w}'(t) = \mathbf{C}_{v(t)}(v'(t), \mathbf{w}(t)). \tag{18.6.3.2}$$

(18.6.4) *For each path* v *of class* C^1 *in* M, *defined on* $[\alpha, \beta]$, *and each tangent vector* $\mathbf{h}_{v(\alpha)} \in T_{v(\alpha)}(M)$, *there exists a unique parallel transport* $\mathbf{w}$ *along the path* v, *such that* $\mathbf{w}(\alpha) = \mathbf{h}_{v(\alpha)}$.

Let U be a relatively compact open neighborhood in M of the compact set $L = v([\alpha, \beta])$. By virtue of (16.25.1) there exists an embedding of U in $\mathbf{R}^N$ for N sufficiently large, and we may assume therefore that $M = U \subset \mathbf{R}^N$. Then by (17.18.5) there exists an open set V in $\mathbf{R}^N$ such that M is a closed submanifold of V, and a connection on V which extends the given connection on M. In view of (18.2.4), we are reduced to proving (18.6.4) in the case where M is an open set in $\mathbf{R}^n$. But then, in the notation of (17.16.2), we may write $\mathbf{w}(t) = (v(t), \mathbf{u}(t))$, where $\mathbf{u}(t) \in \mathbf{R}^n$ and $\mathbf{u}$ is a solution of the vector differential equation

(18.6.4.1) $$\mathbf{u}'(t) + \Gamma_{v(t)}(v'(t), \mathbf{u}(t)) = 0$$

(the local expression of (18.6.3.2)) which is *homogeneous linear*, by the definition of a connection. The result therefore follows from (10.6.3).

The mapping $\mathbf{w}$ satisfying the conditions of (18.6.4) is called the *parallel transport of the vector* $\mathbf{h}_{v(\alpha)}$ *along the path* v, and $\mathbf{w}(t)$ is said to be obtained by *parallel transport of the vector* $\mathbf{h}_{v(\alpha)}$ *from* $v(\alpha)$ *to* $v(t)$ *along* v.

The same method proves:

(18.6.5) *If* $(\mathbf{e}_j)_{1 \leqq j \leqq n}$ *is a basis of the tangent space* $T_{v(\alpha)}(M)$ *and if, for each index* j, $\mathbf{w}_j$ *is the parallel transport of the vector* $\mathbf{e}_j$ *along the path* v, *then for each* $t \in [\alpha, \beta]$ *the* $\mathbf{w}_j(t)$ $(1 \leqq j \leqq n)$ *form a basis of* $T_{v(t)}(M)$.

This is a consequence of (10.8.4). We may therefore say that the $\mathbf{w}_j$ form a *basis* of the vector space of parallel transports along v.

(18.6.6) If now we compare equations (18.6.3.2) and (18.6.1.2), we see that every *geodesic* of the connection $\mathbf{C}$ may be defined as a C^2 mapping v of an open interval $I \subset \mathbf{R}$ into M such that $v' : I \to T(M)$ is a *parallel transport* along the unending path v, and that for any two points $s < t$ in I the tangent vector $v'(t)$ is obtained by parallel transport of the tangent vector $v'(s)$ along v.

Remarks

(18.6.7) Let $f: M_1 \to M$ be a *local diffeomorphism* and let $\mathbf{C}_1$ be the connection on M_1 which is the inverse image of $\mathbf{C}$ under f (17.18.6). Then it follows immediately from the definitions that, if v_1 is an unending path in M_1, and $\mathbf{w}_1$ a parallel transport along v_1 (relative to $\mathbf{C}_1$), then $f \circ \mathbf{w}_1$ is a parallel transport along $f \circ v_1$. If v_1 is a geodesic relative to $\mathbf{C}_1$, then $f \circ v_1$ is a geodesic relative to $\mathbf{C}$.

(18.6.8) More generally, we may define a *tensor parallel transport* of type (r, s) along v as a *lifting* $\mathbf{Z}$ of class C^1 of v to $\mathbf{T}_s^r(M)$, such that $\nabla_E \cdot \mathbf{Z} = 0$ (17.18.3). With the notation of (18.6.5), it follows from (17.18.2) that if, for each $t \in [\alpha, \beta]$, $(\mathbf{w}_j^*(t))$ $(1 \leqq j \leqq n)$ is the basis of $T_{v(t)}(M)^*$ dual to the basis $(\mathbf{w}_j(t))$ of $T_{v(t)}(M)$, then the liftings

$$\mathbf{w}_{j_1}^* \otimes \mathbf{w}_{j_2}^* \otimes \cdots \otimes \mathbf{w}_{j_s}^* \otimes \mathbf{w}_{i_1} \otimes \mathbf{w}_{i_2} \otimes \cdots \otimes \mathbf{w}_{i_r}$$

form a *basis* of the space of tensor parallel transports of type (r, s) along v.

PROBLEMS

1. Show that, if j is the canonical involution of $T(T(M))$ (Section 16.20, Problem 2(b)), then $j(v'') = v''$ for every C^2 mapping $v : I \to M$ (where I is an open interval in $\mathbf{R}$).

2. Let E be a vector bundle with base B and projection π. A real-valued function f on E is said to be *homogeneous of degree* r (where r is an integer $\geqq 0$) if the restriction of f to each fiber E_b is a *homogeneous function of degree* r, i.e., if $f(c \cdot \mathbf{u}_b) = c^r f(\mathbf{u}_b)$ for all $\mathbf{u}_b \in E_b$ and all $c \in \mathbf{R}$. A homogeneous function of degree 0 is therefore a function of the form $g \circ \pi$, where g is a real-valued function on B.

   A vector field Z on E is said to be *homogeneous of degree* r if

$$Z(c \cdot \mathbf{u}_b) = c^{r-1}(T(m_c) \cdot Z(\mathbf{u}_b))$$

for all $\mathbf{u}_b \in E_b$ and all $c \in \mathbf{R}$.

(a) When B is an open set in $\mathbf{R}^n$, and $E = B \times \mathbf{R}^n$, in order that

$$Z(x, \mathbf{y}) = (\mathbf{g}(x, \mathbf{y}), \mathbf{f}(x, \mathbf{y}))$$

should be homogeneous of degree r, it is necessary and sufficient that

$$\mathbf{g}(x, c\mathbf{y}) = c^{r-1}\mathbf{g}(x, \mathbf{y}) \qquad \text{and} \qquad \mathbf{f}(x, c\mathbf{y}) = c^r \mathbf{f}(x, \mathbf{y}).$$

(b) If Z is a vector field on E, homogeneous of degree r and of class C^0, and if f is a real-valued function on E, homogeneous of degree s and of class C^1, then $\theta_Z \cdot f$ is a homogeneous function of degree $r + s - 1$. Conversely, let Z be a C^0 vector field on E such that (i) for every C^1 homogeneous function f of degree 0 on E, the function $\theta_Z \cdot f$ is homogeneous of degree $r - 1$; (ii) for each C^1 section $\mathbf{s}^*$ of the dual bundle E^*, if we define f by

$$f(\mathbf{u}_b) = \langle \mathbf{s}^*(b), \mathbf{u}_b \rangle$$

(so that f is of class C^1 and homogeneous of degree 1 on E), then $\theta_Z \cdot f$ is homogeneous of degree r. Under these conditions, show that Z is homogeneous of degree r. A homogeneous vector field of degree 0 consists of vertical vectors in T(E).
(c) If Z_1, Z_2 are C^1 homogeneous vector fields on E, of degrees r_1, r_2, respectively, then $[Z_1, Z_2]$ is homogeneous of degree $r_1 + r_2 - 1$.
(d) Let $\boldsymbol{g}$ be a B-morphism of $E^{\otimes r}$ into E. With the notation of Section 16.19, Problem 11, show that the mapping

$$\mathbf{u}_b \mapsto \lambda(\mathbf{u}_b, \boldsymbol{g}(\mathbf{u}_b \otimes \mathbf{u}_b \otimes \cdots \otimes \mathbf{u}_b))$$

is a field of vertical tangent vectors on E, homogeneous of degree r. This vector field is denoted by $\boldsymbol{g}^{\mathrm{V}}$.

(e) For each $\mathbf{u}_b \in \mathrm{E}_b$, let

$$H(\mathbf{u}_b) = \tau_{\mathbf{u}_b}^{-1}(\mathbf{u}_b) \in \mathrm{T}_{\mathbf{u}_b}(\mathrm{E}_b);$$

H is a C^∞ field of vertical vectors on E, homogeneous of degree 1, and is sometimes called the *Liouville field*. In order that a real-valued function f of class C^1 on E should be homogeneous of degree r, it is necessary and sufficient that $\theta_H \cdot f = rf$ (*Euler's identity*). For a vector field Z of class C^1 on E to be homogeneous of degree r, it is necessary and sufficient that $[H, Z] = (r-1)Z$.

(f) Suppose that $\mathrm{E} = \mathrm{T(B)}$. If $\delta : \mathbf{h}_x \mapsto (\mathbf{h}_x, \mathbf{h}_x)$ is the diagonal B-morphism of T(B) into $\mathrm{T(B)} \times_{\mathrm{B}} \mathrm{T(B)}$, then we have $H = \lambda \circ \delta$. With the notation and definitions of Section **17.19**, Problems 2, 3 and 4, show that

$$H \barwedge \boldsymbol{J} = \boldsymbol{J} \barwedge H = 0,$$
$$[\boldsymbol{J}, H] = -[H, \boldsymbol{J}] = \boldsymbol{J},$$
$$[i_{\boldsymbol{J}}, \theta_H] = i_{\boldsymbol{J}}, \quad [i_H, d_{\boldsymbol{J}}] = i_{\boldsymbol{J}}, \quad [d_{\boldsymbol{J}}, \theta_H] = d_{\boldsymbol{J}}, \quad [i_{\boldsymbol{J}}, i_H] = 0.$$

3. Let X be a C^1 vector field on M. Show that there exists a unique vector field $\tilde{X}$ on T(M) such that

$$\theta_{\mathbf{h}_x} \cdot (\theta_X \cdot f) = (\theta_{\tilde{X}} \cdot \tilde{d}f)(\mathbf{h}_x)$$

for every function f of class C^2 on M (the notation is that of Section **16.20**, Problem 2). $\tilde{X}$ is a homogeneous vector field of degree 1, and we have $\tilde{X} = j \circ \mathrm{T}(X)$ and $\mathrm{T}(o_{\mathrm{M}}) \circ \tilde{X} = X \circ o_{\mathrm{M}}$. The field $\tilde{X}$ is called the *canonical lifting* of X.

4. Let G be the geodesic spray of a linear connection $\mathbf{C}$ on M. Let $\mathbf{g}$ be a covariant tensor field of order r and class C^1 on M, and for each $\mathbf{h}_x \in \mathrm{T(M)}$ put

$$\check{\mathbf{g}}(\mathbf{h}_x) = \langle \mathbf{g}(x), \mathbf{h}_x \otimes \mathbf{h}_x \otimes \cdots \otimes \mathbf{h}_x \rangle$$

show that

$$(\theta_G \cdot \check{\mathbf{g}})(\mathbf{h}_x) = \langle \nabla \mathbf{g}, \mathbf{h}_x \otimes \mathbf{h}_x \otimes \cdots \otimes \mathbf{h}_x \rangle$$

(where there are $r+1$ factors $\mathbf{h}_x$ in the tensor product). (Consider the particular case where $\mathbf{g} = \omega_1 \otimes \omega_2 \otimes \cdots \otimes \omega_r$, the ω_j being differential 1-forms on M.)

5. Let M be a differential manifold and $\boldsymbol{g}$ an M-morphism of $\mathbf{T}_0^r(\mathrm{M}) = (\mathrm{T(M)})^{\otimes r}$ into T(M). Let G be the geodesic spray of a torsion-free connection $\mathbf{C}$ on M. In Problem 2(d) we defined the field $\boldsymbol{g}^{\mathrm{V}}$ of vertical vectors on T(M), which is homogeneous of degree r. Define also

$$\boldsymbol{g}^{\mathrm{H}}(\mathbf{h}_x) = \mathbf{C}_x(\mathbf{h}_x, \boldsymbol{g}(\mathbf{h}_x \otimes \mathbf{h}_x \otimes \cdots \otimes \mathbf{h}_x))$$

for all $\mathbf{h}_x \in \mathrm{T(M)}$; then $\boldsymbol{g}^{\mathrm{H}}$ is a homogeneous vector field on T(M), of degree $r+1$. Finally, let $\boldsymbol{r} \cdot \boldsymbol{g}$ denote the M-morphism of $\mathbf{T}_0^{r+2}(\mathrm{M})$ into T(M) defined by

$$(\boldsymbol{r} \cdot \boldsymbol{g})(\mathbf{k}_1 \otimes \mathbf{k}_2 \otimes \cdots \otimes \mathbf{k}_{r+2}) = (\boldsymbol{r} \cdot (\boldsymbol{g} \cdot (\mathbf{k}_1 \otimes \cdots \otimes \mathbf{k}_r) \wedge \mathbf{k}_{r+1})) \cdot \mathbf{k}_{r+2}.$$

Show that

$$[G, \boldsymbol{g}^{\mathrm{V}}] = (\nabla \boldsymbol{g})^{\mathrm{V}} - \boldsymbol{g}^{\mathrm{H}},$$
$$[G, \boldsymbol{g}^{\mathrm{H}}] = (\nabla \boldsymbol{g})^{\mathrm{H}} + (\boldsymbol{r} \cdot \boldsymbol{g})^{\mathrm{V}}.$$

6. Let X be a C^1 vector field on a differential manifold M, and let $g_t(x) = F_X(x, t)$, in the notation of (18.2.2). Let **C** be a linear connection on M and let **Z** be a C^1 tensor field on M. For each $t \in J(x)$, let $\mathbf{Z}_t$ be the parallel transport along the curve $s \mapsto g_s(x)$ which takes the value $\mathbf{Z}(g_t(x))$ at $s = t$. Show that

$$(\nabla_X \cdot \mathbf{Z})(x) = \frac{d}{dt} \mathbf{Z}_t(x)|_{t=0}.$$

7. Let M be a differential manifold, **C** a linear connection on M; let a be a point of M and let U be an open neighborhood of $\mathbf{0}_a$ in $T_a(M)$ such that the restriction of $\exp_a$ to U is a diffeomorphism of U onto an open neighborhood V of a in M (18.4.6). For each vector $\mathbf{h} \in U$, let $Z_{\mathbf{h}}$ denote the C^∞ vector field on V for which $Z_{\mathbf{h}}(\exp \mathbf{u})$ is the tangent vector obtained by parallel transport of $\mathbf{h}$ from a to $\exp(\mathbf{u})$ along the path $t \mapsto \exp_a(t\mathbf{u})$. For each real-valued function f of class C^1 on M, we have

$$(\theta_{Z_{\mathbf{h}}} \cdot f)(\exp_a t\mathbf{h}) = \frac{d}{dt}(f(\exp_a t\mathbf{h}))$$

(cf. (18.1.2.5)).

Deduce that, if M is a real-analytic manifold and if f is analytic in a neighborhood of a, then in a neighborhood of $t = 0$ we have

$$f(\exp_a t\mathbf{h}) = \sum_{n=0}^{\infty} \frac{t^n}{n!} (\theta^n_{Z_{\mathbf{h}}} \cdot f)(a).$$

8. On $T(\mathbf{R}^2)$ (identified with $\mathbf{R}^2 \times \mathbf{R}^2$) consider the spray $Z(x, y) = ((x, y), (y, 0))$, the trajectories of which are the lines in $\mathbf{R}^2$. Let M be the cylinder $\mathbf{R}^2/(\mathbf{Z} \times \{0\})$, of which $\mathbf{R}^2$ is a covering, and let Z_0 be the unique spray over M which lifts to Z. Give an example of two open sets in M which are convex relative to Z_0 but whose intersection is not connected.

9. Let G be a spray of class C^∞ over a differential manifold M, and let B be an anti-symmetric bilinear M-morphism of $T(M) \oplus T(M)$ into $T(M)$. Show that there exists on M a unique linear connection **C** with torsion equal to B, and geodesic spray G. (If $\mathbf{C}'$ is a linear connection on M with torsion equal to B (17.20.7), show that there exists a symmetric bilinear M-morphism $A : T(M) \oplus T(M) \to T(M)$ such that $G(\mathbf{h}_x) - \mathbf{C}'_x(\mathbf{h}_x, \mathbf{h}_x) = \lambda(\mathbf{h}_x, A(\mathbf{h}_x, \mathbf{h}_x))$ (in the notation of Section 16.19, Problem 11).

7. ONE-PARAMETER FAMILIES OF GEODESICS AND JACOBI FIELDS

(18.7.1) Let E_1 and E_2 denote the vector fields (called *canonical*) on $\mathbf{R}^2$ defined by

$$\tau_{(t_1, t_2)}(E_1(t_1, t_2)) = (1, 0), \qquad \tau_{(t_1, t_2)}(E_2(t_1, t_2)) = (0, 1).$$

Given two open intervals I, J in $\mathbf{R}$, a mapping $f : I \times J \to M$ of class C^r ($r \geqq 1$), with values in a differential manifold M, is often called a *one-parameter family of curves*. (Consider for each $\xi \in J$ the mapping $t \mapsto f(t, \xi)$

of I into M; the function f may be regarded as describing the "variation" of this family of "curves.") When there is no risk of confusion, we shall write, by abuse of notation (generalizing the notation of (18.1.2.3))

(18.7.1.1)
$$f'_t(t, \xi) = \mathrm{T}(f) \cdot E_1(t, \xi),$$
$$f'_\xi(t, \xi) = \mathrm{T}(f) \cdot E_2(t, \xi),$$

so that the mappings $(t, \xi) \mapsto f'_t(t, \xi)$ and $(t, \xi) \mapsto f'_\xi(t, \xi)$ are C^{r-1} mappings of $\mathrm{I} \times \mathrm{J}$ into T(M) which *lift f*.

(18.7.2) *Let* **C** *be a connection on* M, *and for each lifting* $\mathbf{w} : \mathrm{I} \times \mathrm{J} \to \mathrm{T}(\mathrm{M})$ *of f, of class* C^r $(r \geqq 3)$, *write (by abuse of notation)*

$$\nabla_t \cdot \mathbf{w} = \nabla_{\dot{E}_1} \cdot \mathbf{w}, \qquad \nabla_\xi \cdot \mathbf{w} = \nabla_{E_2} \cdot \mathbf{w}$$

(17.17.3). *Then we have*

(18.7.2.1)
$$\nabla_t \cdot (\nabla_\xi \cdot \mathbf{w}) - \nabla_\xi \cdot (\nabla_t \cdot \mathbf{w}) = (\boldsymbol{r} \cdot (f'_t \wedge f'_\xi)) \cdot \mathbf{w},$$

where $\boldsymbol{r}$ *is the curvature of* **C**.

This is a particular case of (17.20.4.1) since $[E_1, E_2] = 0$.

(18.7.3) With the notation of (18.7.1), and assuming M endowed with a connection **C**, the mapping f is said to be a *one-parameter family of geodesics* of M if f is of class C^r with $r \geqq 3$ and if, for each $\xi \in \mathrm{J}$, the mapping $t \mapsto f(t, \xi)$ is a geodesic.

(18.7.4) *If the connection* **C** *is torsion-free and if f is a one-parameter family of geodesics in* M, *then*

(18.7.4.1)
$$\nabla_t \cdot (\nabla_t \cdot f'_\xi) = (\boldsymbol{r} \cdot (f'_t \wedge f'_\xi)) \cdot f'_t$$

for all $(t, \xi) \in \mathrm{I} \times \mathrm{J}$.

Since $[E_1, E_2] = 0$, the formula (17.20.6.4) gives us

(18.7.4.2)
$$\nabla_t \cdot f'_\xi = \nabla_\xi \cdot f'_t$$

since the torsion is zero. Using (18.7.2.1), we obtain

$$\nabla_t \cdot (\nabla_t \cdot f'_\xi) = \nabla_t \cdot (\nabla_\xi \cdot f'_t) = \nabla_\xi \cdot (\nabla_t \cdot f'_t) + (\boldsymbol{r} \cdot (f'_t \wedge f'_\xi)) \cdot f'_t.$$

But since f is a family of geodesics, we have $\nabla_t \cdot f'_t = \boldsymbol{0}$ (18.6.1.2), whence the result.

(18.7.5) Throughout the remainder of this section we shall assume that the connection **C** is *torsion-free*.

Under the hypotheses of (18.7.4), fix a value $\alpha \in \mathrm{J}$ and put $v(t) = f(t, \alpha)$, so that v is a geodesic. If we then put $\mathbf{w}(t) = f'_\xi(t, \alpha)$, the mapping $\mathbf{w}$ of I into T(M) is of class C^2 and satisfies the *homogeneous linear* equation

(18.7.5.1) $$\nabla_t \cdot (\nabla_t \cdot \mathbf{w}) = (\mathbf{r} \cdot (v' \wedge \mathbf{w})) \cdot v'.$$

For each geodesic v of **C**, defined on the interval I, a C^2 mapping $\mathbf{w}$ of I into T(M) which satisfies (18.7.5.1) is called a *Jacobi field along the geodesic* v. If $v_1 = v \circ \varphi$ is a geodesic obtained from v by an affine change of parameter $\varphi : t \mapsto \lambda t + \mu$, then it is immediately verified that $\mathbf{w} \circ \varphi$ is a Jacobi field along v_1, for each Jacobi field $\mathbf{w}$ along v.

(18.7.6) *For each nonconstant geodesic v of* **C**, *defined on an open interval* $\mathrm{I} \subset \mathbf{R}$, *each point* $\alpha \in \mathrm{I}$ *and each pair of vectors* $\mathbf{h}, \mathbf{k} \in \mathrm{T}_{v(\alpha)}(\mathrm{M})$, *there exists a unique Jacobi field* $\mathbf{w}$ *along* v *which is defined on* I *and satisfies*

$$\mathbf{w}(\alpha) = \mathbf{h}, \qquad (\nabla_t \cdot \mathbf{w})(\alpha) = \mathbf{k}.$$

Let $(\mathbf{e}_i)_{1 \leq i \leq n}$ be a basis of the vector space $\mathrm{T}_{v(\alpha)}(\mathrm{M})$, and let $\mathbf{u}_i$ denote the parallel transport of $\mathbf{e}_i$ along the path v (18.6.4). Then for each $t \in \mathrm{I}$ the vectors $\mathbf{u}_i(t) \in \mathrm{T}_{v(t)}(\mathrm{M})$ form a *basis* of $\mathrm{T}_{v(t)}(\mathrm{M})$ (18.6.5), and every C^2 mapping $\mathbf{w} : \mathrm{I} \to \mathrm{T}(\mathrm{M})$ which lifts v can be written uniquely in the form $\mathbf{w} = \sum_{i=1}^{n} w_i \mathbf{u}_i$, where the w_i are real-valued functions of class C^2 defined on I, by virtue of Cramer's rule. By definition, we have $\nabla_t \cdot \mathbf{u}_i = \mathbf{0}$ (18.6.3.1), and therefore it follows from the rule (17.17.2.5) for calculating covariant derivatives that

$$\nabla_t \cdot \mathbf{w} = \sum_{i=0}^{n} w'_i \mathbf{u}_i, \qquad \nabla_t \cdot (\nabla_t \cdot \mathbf{w}) = \sum_{i=0}^{n} w''_i \mathbf{u}_i.$$

Now we may assume that the basis $(\mathbf{e}_i)$ has been chosen so that $\mathbf{e}_n = v'(\alpha)$, since v is not constant. Since v is a geodesic, it follows from (18.6.6) that $\mathbf{u}_n(t) = v'(t)$ for all $t \in \mathrm{I}$. If we put

$$(\mathbf{r} \cdot (\mathbf{u}_j(t) \wedge \mathbf{u}_k(t))) \cdot \mathbf{u}_i(t) = \sum_{l=1}^{n} r^l_{ijk}(t) \cdot \mathbf{u}_l(t),$$

then the equation (18.7.5.1) for $\mathbf{w}$ is equivalent to the system of n second-order differential equations

(18.7.6.1) $$w''_i(t) = \sum_{j=1}^{n} r^i_{nnj} w_j(t).$$

We have here a system of homogeneous linear equations of the second order with C^∞ coefficients; hence the result follows from (10.6.3).

Furthermore ((10.7.4) and (10.8.4)):

(18.7.7) *Suppose that* M *is connected. Then the Jacobi fields along a non-constant geodesic of* **C** *defined on* I *are of class* C^∞ *and form a vector space of dimension* 2 dim(M). *Those which vanish at a given point* $\alpha \in I$ *form a vector space of dimension equal to* dim(M).

(18.7.8) We have derived the notion of a Jacobi field by considering a one-parameter family of geodesics, and the Jacobi field derived from this family along one of the geodesics appears as the "derivative with respect to the parameter." In fact, we shall see that *every* Jacobi field which vanishes at one point at least can be obtained in this way, and in a canonical fashion (cf. Problem 3).

(18.7.9) *Let a be a point of* M, *let* $\mathbf{h}_a$ *be a nonzero vector in* $T_a(M)$, *and let* v *be the geodesic* $t \mapsto \exp(t\mathbf{h}_a)$ *of* **C**, *defined in an open interval* I *of* **R** *containing the point* 0. *For each open interval* I_1 *of* **R** *containing* 0 *and such that* $\bar{I}_1 \subset I$, *and each vector* $\mathbf{k}_a \in T_a(M)$, *there exists an open interval* $J \subset \mathbf{R}$ *containing* 0 *such that, for each* $\xi \in J$, *the geodesic* $t \mapsto \exp(t(\mathbf{h}_a + \xi\mathbf{k}_a)) = f(t, \xi)$ *is defined on* I_1. *The Jacobi field* $t \mapsto f'_\xi(t, 0)$ *is then the unique Jacobi field* $\mathbf{w}$ *along* v *in* I_1 *such that* $\mathbf{w}(0) = \mathbf{0}_a$ *and* $(\nabla_t \cdot \mathbf{w})(0) = \mathbf{k}_a$.

The existence of J follows immediately from (18.2.5), and it is clear that f is a one-parameter family of geodesics such that $f(0, \xi) = a$. Consequently, if we put $\mathbf{w}(t) = f'_\xi(t, 0)$, then we have $\mathbf{w}(0) = \mathbf{0}_a$. On the other hand, if we put

$$\mathbf{g}(t, \xi) = t(\mathbf{h}_a + \xi\mathbf{k}_a) \in T_a(M)$$

and if we denote by $\exp_a$ the restriction of exp to $\Omega \cap T_a(M)$, then

$$f'_\xi(t, \xi) = T_{\mathbf{g}(t,\xi)}(\exp_a) \cdot \boldsymbol{\tau}^{-1}_{\mathbf{g}(t,\xi)}(t\mathbf{k}_a)$$

and therefore, for $\xi = 0$,

(18.7.9.1) $$\mathbf{w}(t) = T_{t\mathbf{h}_a}(\exp_a) \cdot \boldsymbol{\tau}^{-1}_{t\mathbf{h}_a}(t\mathbf{k}_a) = t(T_{t\mathbf{h}_a}(\exp_a) \cdot \boldsymbol{\tau}^{-1}_{t\mathbf{h}_a}(\mathbf{k}_a)).$$

It remains to verify that $(\nabla_t \cdot \mathbf{w})(0) = \mathbf{k}_a$. Using the formula (17.17.2.5), this reduces to the fact that $T_{\mathbf{0}_a}(\exp_a)$ is the identity mapping (18.4.5).

(18.7.10) The properties of Jacobi fields which vanish at a point a of a geodesic $v : t \mapsto \exp(t\mathbf{h}_a)$ are closely related to the properties of the restriction $\exp_a$ of the exponential mapping at a point $\mathbf{h}_a \in \Omega \cap T_a(M)$:

(18.7.11) *For each point* $\mathbf{h}_a \in \Omega \cap T_a(M)$, *the rank* (16.5.3) *of the mapping* $\exp_a$ *at the point* $\mathbf{h}_a$ *is equal to* $\dim_a(M) - s$, *where s is the dimension of the vector space of Jacobi fields along the geodesic* $v_{\mathbf{h}_a} : t \mapsto \exp(t\mathbf{h}_a)$ *which vanish at* $t = 0$ *and* $t = 1$.

For it follows from (18.7.9.1) that as $\mathbf{w}$ runs through the n-dimensional vector space J of Jacobi fields along $v_{\mathbf{h}_a}$ which vanish at $t = 0$, the mapping $\mathbf{w} \mapsto \mathbf{w}(1)$ is of rank s, because the mapping $\mathbf{w} \mapsto (\nabla_t \cdot \mathbf{w})(0)$ is an isomorphism of J onto $T_a(M)$, by virtue of (18.7.6).

(18.7.12) The points $\mathbf{h}_a \in T_a(M) \cap \Omega$ at which $T_{\mathbf{h}_a}(\exp_a)$ is not invertible are called the *singular points* of $\exp_a$, and the images $\exp(\mathbf{h}_a)$ of these points are called the *conjugates* of a on the geodesic $v_{\mathbf{h}_a}$. The set of nonsingular points of $\exp_a$ is open in $T_a(M)$ and contains the origin $\mathbf{0}_a$ (18.4.5), and the restriction of $\exp_a$ to this open set is a *local diffeomorphism* onto an open set in M. But it should be observed that, for each $\mathbf{h}_a \in \Omega \cap T_a(M)$, there may exist *no conjugate* of a on $v_{\mathbf{h}_a}$ (in other words, all the points of $\Omega \cap T_a(M)$ may be nonsingular) even though the mapping $\exp_a$ is not injective.

Example

(18.7.13) In general, if M is a *covering* of M_0, with projection $\pi : M \to M_0$, and if $\mathbf{C}_0$ is a connection on M_0 and $\mathbf{C}$ its *inverse image* on M (17.18.6), then the geodesics for $\mathbf{C}_0$ are the mappings $\pi \circ v$, where v is a geodesic for $\mathbf{C}$ (16.28.2), and every Jacobi field along $\pi \circ v$ is of the form $T(\pi) \circ \mathbf{w}$, where $\mathbf{w}$ is a Jacobi field along v. If no pair of points is conjugate on v, the same is true for $\pi \circ v$.

Take, for example, $M = \mathbf{R}$ and $M_0 = \mathbf{T}$, and identify T(M) with $M \times \mathbf{R} = \mathbf{R}^2$. Let $\mathbf{C}$ be the connection on M defined by

$$\mathbf{C}_x((x, \mathbf{y}), (x, \mathbf{z})) = ((x, \mathbf{y}), (\mathbf{y}, 0));$$

then the geodesic field is

$$Z(x, \mathbf{y}) = ((x, \mathbf{y}), (\mathbf{y}, 0))$$

and the geodesics are therefore the solutions of the differential equation $D^2 x = 0$. Since $\mathbf{C}$ is translation-invariant, it is the inverse image of a connection $\mathbf{C}_0$ on $\mathbf{T}$. It is clear that for $\mathbf{C}$ we have $\Omega = T(M)$, and the exponential mapping is $(x, \mathbf{y}) \mapsto x + \mathbf{y}$, so that $\exp_a$ is a diffeomorphism for each $a \in M$. Consequently the exponential for $\mathbf{C}_0$ is also a local diffeomorphism, but evidently $\exp_{a_0}$ is not injective, for any $a_0 \in M_0$.

(18.7.14) Conversely, it can happen that $\exp_a$ admits singular points, but is injective on $T_a(M) \cap \Omega$: in other words, a point conjugate to a on a geodesic v is not necessarily a point through which there passes a second geodesic passing through a with a direction distinct from that of v, but is a point b at which certain geodesics passing through a, with directions "infinitesimally close" to that of v, have a distance from b which is "infinitesimally small to a higher order."

(18.7.15) *Let v be a geodesic for* **C**, *defined on an open interval* $I \subset \mathbf{R}$, *and let* $[\alpha, \beta]$ *be a compact interval contained in* I *such that* $\alpha \neq \beta$ *and* $a = v(\alpha)$ *and* $b = v(\beta)$ *are not conjugate points on* v. *Then, given any two tangent vectors* $\mathbf{h} \in T_{v(\alpha)}(M)$ *and* $\mathbf{k} \in T_{v(\beta)}(M)$, *there exists a unique Jacobi field* $\mathbf{w}$ *along* v *such that* $\mathbf{w}(\alpha) = \mathbf{h}$ *and* $\mathbf{w}(\beta) = \mathbf{k}$.

Clearly there is no loss of generality in assuming that $\alpha = 0$ and $\beta = 1$. Then we have $v(t) = \exp(t\mathbf{h}')$ for some $\mathbf{h}' \in T_{v(0)}(M)$, and since by hypothesis the rank of $\exp_{v(0)}$ is equal to n at the point $\mathbf{h}'$, there exists, by virtue of (18.7.9.1), a Jacobi field $\mathbf{w}_1$ along v such that $\mathbf{w}_1(0) = \mathbf{0}_a$ and $\mathbf{w}_1(1) = \mathbf{k}$. Likewise, we have $v(1 - t) = \exp((1 - t)\mathbf{k}')$ for some $\mathbf{k}' \in T_{v(1)}(M)$, and since the rank of $\exp_{v(1)}$ is equal to n at the point $\mathbf{k}'$, there exists a Jacobi field $\mathbf{w}_2$ along v such that $\mathbf{w}_2(1) = \mathbf{0}_b$ and $\mathbf{w}_2(0) = \mathbf{h}$. The Jacobi field $\mathbf{w}_1 + \mathbf{w}_2$ then has the desired properties, and is the only one because no Jacobi field $\neq 0$ can vanish simultaneously at the points 0 and 1.

PROBLEMS

1. Let M be a differential manifold, X a C^1 vector field on M, and $\tilde{X}$ the canonical lifting of X to T(M) (Section 18.6, Problem 3).

   (a) Let $v : I \to M$ be an integral curve of X and let $\mathbf{w}$ be a *lifting* of v to T(M), of class C^1. Then we have

   $$T(o_M) \cdot (\mathbf{w}'(t) - \tilde{X}(\mathbf{w}(t))) = 0$$

   for all $t \in I$. Hence there exists a unique vector $\mathbf{u}(t) \in T_{v(t)}(M)$ such that

   $$\lambda(\mathbf{w}(t), \mathbf{u}(t)) = \mathbf{w}'(t) - \tilde{X}(\mathbf{w}(t))$$

   for each $t \in I$, in the notation of Section 16.19, Problem 11. The mapping $t \mapsto \mathbf{u}(t)$ (which is a lifting of v to T(M)) is called the *Lie derivative of* $\mathbf{w}$ *with respect to* X, and is denoted by $\theta_X \cdot \mathbf{w}$. For each C^1 function $\varphi : I \to \mathbf{R}$, we have

   $$\theta_X(\varphi\mathbf{w}) = \varphi'\mathbf{w} + \varphi(\theta_X \cdot \mathbf{w}).$$

   For each C^1 vector field Y on M, we have

   $$\theta_X \cdot (Y \circ v) = [X, Y] \circ v.$$

(b) A C^1 lifting $\mathbf{w}$ of v to T(M) is said to be *X-invariant* if $\theta_X \circ \mathbf{w} = 0$. For each $t_0 \in \mathrm{I}$ and each tangent vector $\mathbf{h} \in \mathrm{T}_{v(t_0)}(\mathrm{M})$, there exists a unique X-invariant lifting $\mathbf{w}$ of v such that $\mathbf{w}(t_0) = \mathbf{h}$.
(c) If J is a neighborhood of 0 in **R** and if $f : \mathrm{I} \times \mathrm{J} \to \mathrm{M}$ is a C^1 mapping such that $t \mapsto f(t, \xi)$ is an integral curve of X for each $\xi \in \mathrm{J}$, then $t \mapsto f'_\xi(t, 0)$ is an X-invariant lifting of $t \mapsto f(t, 0)$ to T(M).

2. Let M be a differential manifold, **C** a torsion-free linear connection on M, and G the geodesic spray of **C**.

(a) Let v be a nonconstant geodesic of **C**. In order that $\mathbf{w}$ should be a Jacobi field along v, it is necessary and sufficient that $\mathbf{w} = \mathrm{T}(o_\mathrm{M}) \circ \mathbf{z}$, where $\mathbf{z}$ is a G-invariant lifting of v' to T(T(M)) (Problem 1).
(b) Let X be a C^1 vector field on M whose canonical lifting $\tilde{X}$ (Section **18.6**, Problem 3) satisfies the condition $[\tilde{X}, G] = 0$. Then, for each nonconstant geodesic v of **C**, the field $X \circ v$ is a Jacobi field along v.
(c) Let v be a nonconstant geodesic of **C** and let $\mathbf{w}$ be a C^1 lifting of v to T(M). Put

$$\mathbf{w}^\mathrm{V} = \lambda(v', \mathbf{w}), \qquad \mathbf{w}^\mathrm{H} = \mathbf{C}(v', \mathbf{w});$$

these are liftings of v' to T(T(M)). Show that every C^1 lifting $\mathbf{z}$ of v' to T(T(M)) can be expressed uniquely in the form $\mathbf{z} = \mathbf{w}_1^\mathrm{V} + \mathbf{w}_2^\mathrm{H}$, where $\mathbf{w}_1$ and $\mathbf{w}_2$ are two liftings of v to T(M). Moreover, for each lifting $\mathbf{w}$ of v to T(M), of class C^1, we have

$$\theta_G \cdot \mathbf{w}^\mathrm{V} = (\nabla_t \cdot \mathbf{w})^\mathrm{V} - \mathbf{w}^\mathrm{H},$$
$$\theta_G \cdot \mathbf{w}^\mathrm{H} = (\nabla_t \cdot \mathbf{w})^\mathrm{H} + ((\mathbf{r} \cdot (\mathbf{w} \wedge v')) \cdot v')^\mathrm{V}.$$

3. Let M be a differential manifold, **C** a linear connection on M, and let H be the Liouville field on T(M) (Section **18.6**, Problem 2(e)).

(a) Consider the C^∞ mapping $\mathbf{c} = \mathrm{T}(\exp) \circ H : \Omega \to \mathrm{T}(\mathrm{M})$, which is a lifting of the mapping $\exp : \Omega \to \mathrm{M}$. Show that $\theta_H \cdot \mathbf{c} = \mathbf{c}$.
(b) Let L be a C^2 vector field on T(M), homogeneous of degree 1 (Section **18.6**, Problem 2). Let $\mathbf{s} = \mathrm{T}(\exp) \circ L$, which is a C^2 mapping of Ω into T(M). Show that

$$\nabla_H \cdot (\nabla_H \cdot \mathbf{s}) = (\mathbf{r} \cdot (\mathbf{c} \wedge \mathbf{s})) \cdot \mathbf{c} + \nabla_H \cdot \mathbf{s}.$$

(c) Let a be a point of M, let $\mathbf{h}_a$ be a nonzero vector in $\mathrm{T}_a(\mathrm{M})$, and put $\mathbf{u}(t) = t\mathbf{h}_a$ for $t \in \mathbf{R}$. A lifting $\mathbf{z}$ of $\mathbf{u}$ to T(T(M)) is said to be *homogeneous of degree* 1 if $\mathbf{z}(ct) = \mathrm{T}(m_c) \cdot \mathbf{z}(t)$ for all $c \in \mathbf{R}$ and all $t \in \mathbf{R}$. A necessary and sufficient condition for this is that

$$\mathbf{z}(t) = \lambda(\mathbf{u}(t), \mathbf{h}'_a) + \mathbf{C}_a(\mathbf{u}(t), \mathbf{h}''_a)$$

for two vectors $\mathbf{h}'_a$, $\mathbf{h}''_a$ in $\mathrm{T}_a(\mathrm{M})$. An equivalent condition is that there should exist a vector field L on T(M), homogeneous of degree 1, and such that $\mathbf{z} = L \circ \mathbf{u}$.
(d) Let $\mathbf{z}$ be a lifting of $\mathbf{u}$ to T(T(M)) which is homogeneous of degree 1. Show that $\mathbf{w} = \mathrm{T}(\exp) \circ \mathbf{z}$ is a Jacobi field along the geodesic $t \mapsto \exp(t\mathbf{h}_a)$ of **C** and that

$$\mathbf{z}(t) = \lambda(t\mathbf{h}_a, t(\nabla_{E(0)} \cdot \mathbf{w})) + \mathbf{C}_a(t\mathbf{h}_a, \mathbf{w}(0)).$$

In particular, the Jacobi fields along the geodesic $t \mapsto \exp(t\mathbf{h}_a)$ such that $(\nabla_t \cdot \mathbf{w})(0) = \nabla_{\mathbf{E}0} \cdot \mathbf{w} = 0$ are the fields of the form $t \mapsto \mathbf{C}_a(t\mathbf{h}_a, \mathbf{k}_a)$ for some $\mathbf{k}_a \in \mathrm{T}_a(\mathrm{M})$.

4. Let M be a real-analytic manifold, **C** an analytic linear connection on M; let a be a point of M and let U be a neighborhood of $\mathbf{0}_a$ in $T_a(M)$ such that the restriction of $\exp_a$ to U is a diffeomorphism of U onto a neighborhood of a in M. Given two vectors **h**, **k** in $T_a(M)$, put $f(t, \xi) = \exp_a(t(\mathbf{h} + \xi\mathbf{k}))$ and $\mathbf{w}(t) = f'_\xi(t, 0)$, defined for sufficiently small t. Let $\exp_1(z)$ denote the entire function $(\exp(z) - 1)/z$. With the notation of Section 18.6, Problem 7, show that

$$\mathbf{w}(t) = (\exp_1(-t\theta_Z) \cdot (tZ_{\mathbf{k}}))(\exp_a(t\mathbf{h})).$$

(Imitate the proof of (19.16.5), replacing $\mathfrak{g}_e$ by End(E) and G by **GL**(E), where E is a Banach space; apply this to the case where E is the Banach space of restrictions of C^∞ vector fields to a compact neighborhood of a, and consider the linear mapping $\theta_{Z_{\mathbf{h}}} : Y \to [Z_{\mathbf{h}}, Y]$ of this space into itself.) Deduce that

$$T_{t\mathbf{h}}(\exp_a) \cdot \mathbf{k} = (\exp_1(-t\theta_{Z_{\mathbf{h}}}) \cdot Z_{\mathbf{k}})(\exp_a(t\mathbf{h})).$$

5. With the notation and hypotheses of (18.7.9) show that, for each $t \in I_1$, $\mathbf{w}(t)$ is obtained by *parallel transport* along v of the vector

$$t\mathbf{k}_a + \tfrac{1}{6}((\mathbf{r}(a) \cdot (\mathbf{h}_a \wedge \mathbf{k}_a)) \cdot \mathbf{h}_a)t^3 + o_3(t) \in T_a(M),$$

where $o_3(t)$ is such that $\lim_{t \to 0} o_3(t)/t^3 = 0$. (Use (18.7.6.1).)

8. FIELDS OF p-DIRECTIONS, PFAFFIAN SYSTEMS, AND SYSTEMS OF PARTIAL DIFFERENTIAL EQUATIONS

Let M be a differential manifold. A *p-direction* (or a *tangent p-direction*) at a point $x \in M$ is a vector subspace of $T_x(M)$ of dimension p, that is to say (16.11.8), it is an element of the *Grassmannian* $\mathbf{G}_p(T_x(M))$. A *field of p-directions* on M is a mapping which assigns to each $x \in M$ a p-direction $L_x \in \mathbf{G}_p(T_x(M))$. If we identify a sufficiently small neighborhood V of a point $x_0 \in M$ with an open set in $\mathbf{R}^n$ by means of a chart, then for each point $x \in V$ the tangent space $T_x(M)$ may be identified with $\{x\} \times \mathbf{R}^n$, and $\mathbf{G}_p(T_x(M))$ with $\{x\} \times \mathbf{G}_p(\mathbf{R}^n)$, and a field of p-directions is therefore a mapping $x \mapsto (x, L'_x)$, where $L'_x \in \mathbf{G}_p(\mathbf{R}^n)$. The field is said to be *of class* C^r (r an integer ≥ 0, or $+\infty$) if the mapping $x \mapsto L'_x$ of V into the differential manifold $\mathbf{G}_p(\mathbf{R}^n)$ (16.11.8) is of class C^r.

This definition is independent of the choice of chart, by virtue of the following proposition:

(18.8.1) *In order that a field $x \mapsto L_x$ of p-directions on* M *should be of class* C^r *it is necessary and sufficient that, for each $x_0 \in M$, there should exist an open neighborhood* U *of x_0 in* M *and p vector fields X_j $(1 \leq j \leq p)$ of class C^r on* U *such that, for each $x \in U$, the vectors $X_j(x)$ form a basis of the subspace L_x of* $T_x(M)$.

(For $r = \infty$, therefore, an equivalent condition is that the *union* L of the subspaces L_x for all $x \in M$ be a *vector subbundle* (16.17.1) of T(M).)

By choosing a suitable basis $((x_0, \mathbf{e}_j))_{1 \leq j \leq n}$ of $T_{x_0}(M)$, we may assume that L_{x_0} is the subspace spanned by the first p basis vectors $(x_0, \mathbf{e}_1), \ldots, (x_0, \mathbf{e}_p)$. By identifying as before a neighborhood U of x_0 with an open subset of $\mathbf{R}^n$, we may assume (since L'_x is a continuous function of x) that for each $x \in U$ the subspace L'_x of $\mathbf{R}^n$ is spanned by p vectors $\mathbf{e}_1(x), \ldots, \mathbf{e}_p(x)$ such that the projection of $\mathbf{e}_j(x)$ on $\mathbf{R}^p$ ($\mathbf{R}^n$ being identified with $\mathbf{R}^p \times \mathbf{R}^{n-p}$, and $\mathbf{R}^p$ with L'_{x_0}) is $\mathbf{e}_j$ for $1 \leq j \leq p$; this follows from the description of the atlas of a Grassmannian given in (16.11.10), and linear algebra (A.4.5). By the definition of the differential structure of $\mathbf{G}_p(\mathbf{R}^n)$ (16.11.10), to say that $x \mapsto L_x$ is of class C^r then signifies that each of the mappings $x \mapsto \mathbf{e}_j(x)$ of U into $\mathbf{R}^n$ is of class C^r, and it is now clear that the p mappings $x \mapsto (x, \mathbf{e}_j(x))$ are *vector fields* satisfying the required conditions.

Conversely, if there exist p vector fields $X_j(x) = (x, \mathbf{v}_j(x))$ satisfying these conditions, we may assume that (with the same notation) $\mathbf{v}_j(x_0) = \mathbf{e}_j$. If $\mathbf{u}_j(x)$ is the projection of $\mathbf{v}_j(x)$ on $\mathbf{R}^p$, the continuity of the X_j implies the continuity of the $\mathbf{u}_j$, and therefore we may assume that the $\mathbf{u}_j(x)$ form a *basis* of $\mathbf{R}^p$ for all $x \in U$. If $P(x)$ is the $p \times p$ matrix whose columns are the vectors $\mathbf{u}_j(x)$ (relative to the basis $(\mathbf{e}_j)_{1 \leq j \leq p}$) and if $Q(x)$ is the $n \times n$ matrix

$$\begin{pmatrix} P(x) & 0 \\ 0 & I_{n-p} \end{pmatrix},$$

then the vectors $\mathbf{e}_j(x) = P(x)^{-1} \cdot \mathbf{v}_j(x)$ $(1 \leq j \leq p)$ form a basis of L'_x such that the projection of $\mathbf{e}_j(x)$ on $\mathbf{R}^p$ is $\mathbf{e}_j$ for $1 \leq j \leq p$. If the X_j are of class C^r, then so are the $\mathbf{u}_j$ and Q^{-1}, hence also the mappings $x \mapsto \mathbf{e}_j(x)$. Consequently, $x \mapsto L'_x$ is of class C^r, and the proof is complete.

Remarks

(18.8.2) If M is a pure manifold of dimension n, there exists a unique field of n-directions on M, for which $L_x = T_x(M)$ for all $x \in M$. On the other hand, there does not necessarily exist any C^0 field of 1-*directions* on M, as we shall see in Chapter XXIV. Moreover, if $x \mapsto L_x$ is a C^r field of p-directions on M, there need not exist C^r vector fields X_j *defined on all of* M such that the vectors $X_j(x)$ are a basis of L_x for each $x \in M$, even in the case where M is an open set in $\mathbf{R}^n$ and $p = 1$ (Problem 2).

We remark also that it follows from the proof of (18.8.1) that when the conditions of the proposition are satisfied, the set of vector fields X of class C^r on U such that $X(x) \in L_x$ for all $x \in U$ is a *free module* over the ring $\mathscr{E}_{\mathbf{R}}^{(r)}(U)$ of real-valued functions of class C^r on U, and has a basis the X_j $(1 \leq j \leq p)$.

(18.8.3) Let $x \mapsto L_x$ be a field of p-directions on M, of class C^r ($r \geqq 0$), and for each $x \in M$ let L_x^0 be the *annihilator* of L_x in the dual $T_x(M)^*$ of $T_x(M)$, so that $\dim(L_x^0) = n - p$ (A.9.5).

(18.8.4) *With the notation of* (18.8.3), *each point* $x_0 \in M$ *admits an open neighborhood* U *such that there exist* $n - p$ *differential* 1-*forms* ω_j $(1 \leqq j \leqq n - p)$ *of class* C^r *on* U *with the property that, for each* $x \in U$, *the* $\omega_j(x)$ *form a basis of* L_x^0.

Conversely, if ω_j $(1 \leqq j \leqq n - p)$ *are differential* 1-*forms of class* C^r *on* U *such that, for each* $x \in U$, *the covectors* $\omega_j(x)$ *are linearly independent, and if* L_x *is the p-dimensional subspace of* $T_x(M)$ *defined by the* $n - p$ *equations* $\langle \mathbf{h}_x, \omega_j(x) \rangle = 0$ $(1 \leqq j \leqq n - p)$, *then* $x \mapsto L_x$ *is a* C^r *field of p-directions on* U.

To prove the first assertion, let

$$\mathbf{e}_j(x) = \mathbf{e}_j + \sum_{k=1}^{n-p} \lambda_{kj}(x)\mathbf{e}_{p+k} \qquad (1 \leqq j \leqq p),$$

with the notation of the proof of (18.8.1), and take

(18.8.4.1) $$\omega_j(x) = \left(x, \mathbf{e}_{p+j}^* - \sum_{k=1}^{p} \lambda_{kj}(x)\mathbf{e}_k^*\right) \qquad (1 \leqq j \leqq n - p),$$

where $(\mathbf{e}_i^*)_{1 \leqq i \leqq n}$ is the basis dual to the basis $(\mathbf{e}_i)_{1 \leqq i \leqq n}$ of $\mathbf{R}^n$.

Conversely, if the ω_j are such that the covectors $\omega_j(x)$ are linearly independent at each point $x \in U$, then in a neighborhood of each point x_1 of U we may assume (after a permutation of the basis vectors of $\mathbf{R}^n$) that the $\omega_j(x)$ are given by (18.8.4.1) by virtue of (A.4.5) and the continuity of ω_j; then Cramer's rule shows that the λ_{kj} are of class C^r, and the vectors

$$\mathbf{e}_j(x) = \mathbf{e}_j + \sum_{k=1}^{n-p} \lambda_{jk}(x)\mathbf{e}_{p+k}$$

form a basis of L'_x. Hence the result, by virtue of (18.8.1).

It follows, as in (18.8.2), that the set of C^r differential 1-forms ω on U, such that $\omega(x) \in L_x^0$ for each $x \in U$, is a *free module* over the ring $\mathscr{E}_{\mathbf{R}}^{(r)}(U)$, having the ω_j $(1 \leqq j \leqq n - p)$ as a basis.

(18.8.5) Let $x \mapsto L_x$ be a C^r field of p-directions on a differential manifold M. For each $q = 1, \ldots, p$, an *integral manifold of dimension* q *of the field* $x \mapsto L_x$ is defined to be a pair consisting of a pure differential manifold N of dimension q and a mapping $j : N \to M$ of class C^r and (maximum) rank q at each point of N, such that for each point $z \in N$ the image of $T_z(N)$ under

$T(j)$ is *contained in* $L_{j(z)}$. If $r \geqq 1$, it follows from (16.7.8) that each $z \in N$ has an open neighborhood V in N such that the restriction of j to V is *injective*; and if j is of class C^∞, then j is an *immersion* of N in M. Consequently, as far as the *local* study of the integral manifolds of the field $x \mapsto L_x$ is concerned, we may always assume that N is an open subset of $\mathbf{R}^n$.

(18.8.6) When $p = 1$, the notion of an integral manifold (necessarily 1-dimensional) of a field of 1-directions $x \mapsto L_x$ on M coincides *locally* with the notion of an *integral curve* of a vector field *which is everywhere nonzero.* To see this, we may assume that N is an open interval $I \subset \mathbf{R}$, and if $j : I \to M$ is an integral manifold, we may also assume that I is sufficiently small so that $j(I)$ is contained in an open set U of M on which there exists a vector field X for which, at each $x \in U$, the vector $X(x)$ is nonzero and forms a basis of the line L_x (18.8.1). By hypothesis, for each $t \in I$, we may write $j'(t) = \rho(t)X(j(t))$, where $\rho(t)$ is a nonzero scalar and the mapping $t \mapsto \rho(t)$ is continuous in I. Since ρ cannot change sign in I, the mapping $\varphi : t \mapsto \int_{t_0}^{t} \rho(u)\,du$ is a C^1 homeomorphism of I onto an open interval $I' \subset \mathbf{R}$ (4.2.2), and the inverse mapping φ^{-1} is also a C^1 homeomorphism; the mapping $f = j \circ \varphi^{-1}$ is an integral curve of the vector field X defined on I'. The converse is clear.

(18.8.7) Throughout the remainder of this chapter (except for (18.14.6)) we shall limit our considerations to the case of fields of p-directions *of class* C^∞ defined *on an open set* M in $\mathbf{R}^n$, and we shall be concerned only with the *local* existence and uniqueness of integral manifolds of this field. In view of the remarks in (18.8.5), we may therefore always assume (by replacing M by a smaller open set if necessary) that the integral manifolds are *closed submanifolds* of M.

By virtue of (18.8.4), we may assume that there exist $n - p$ differential 1-forms ω_j $(1 \leqq j \leqq n - p)$ of class C^∞ on M such that the covectors $\omega_j(x)$ form a *basis* of L_x^0 for each point $x \in M$. A submanifold N of M, of dimension $q \leqq p$, is an integral manifold of the field $x \mapsto L_x$ if and only if *the restrictions to* N *of the* $n - p$ *forms* ω_j *are zero*, and this property is expressed by saying that N is an *integral manifold of the Pfaffian system*

$$\omega_j = 0 \qquad (1 \leqq j \leqq n - p). \tag{18.8.7.1}$$

It should be observed that this definition still makes sense when the $\omega_j(x)$ are not everywhere linearly independent.

Conversely, if we are given $n - p$ differential 1-forms ω_j of class C^∞ on M, such that the $\omega_j(x)$ are linearly independent at each point $x \in M$, then the subspace L_x of the tangent space $T_x(M)$ on which all the covectors $\omega_j(x)$ vanish is of dimension p at each $x \in M$, and $x \mapsto L_x$ is a C^∞ field of p-directions on M. Every integral manifold of the Pfaffian system (18.8.7.1) is an integral manifold of this field.

(18.8.8) If x^i $(1 \leqq i \leqq n)$ are the canonical coordinates of a point $x \in \mathbf{R}^n$, then we may write

(18.8.8.1)
$$\omega_j(x) = \sum_{i=1}^{n} a_{ji}(x)\, dx^i \qquad (1 \leqq j \leqq n - p)$$

for all $x \in$ M, where the a_{ji} are C^∞ functions on M. Let N be a q-dimensional submanifold of M, and let $h : \mathrm{N} \to \mathrm{M}$ be the canonical injection. Replacing M by an open subset, we may assume that there exists a chart φ of N in $\mathbf{R}^q$. Let $z^i = \mathrm{pr}_i \circ h \circ \varphi^{-1}$ $(1 \leqq i \leqq n)$; then, in order that N should be an integral manifold of the Pfaffian system (18.8.7.1), it is necessary and sufficient that, in the open set $\varphi(\mathrm{N}) \subset \mathbf{R}^q$, the n functions $(u^1, \ldots, u^q) \mapsto z^i(u^1, \ldots, u^q)$ should satisfy the system of $q(n - p)$ partial differential equations

(18.8.8.2)
$$\sum_{i=1}^{n} a_{ji}(z^1, \ldots, z^n) \frac{\partial z^i}{\partial u^k} = 0 \qquad (1 \leqq k \leqq q, \quad 1 \leqq j \leqq n - p).$$

(18.8.9) Conversely, let us show how the determination of the solutions of a *system of first-order partial differential equations* can usually be reduced to the determination of the *integral manifolds of a Pfaffian system.* Let q and n be two positive integers, let $\mathrm{N} = q + n + qn$, and consider, on an open set $\mathrm{U} \subset \mathbf{R}^{\mathrm{N}}$, a certain number of functions F_h $(1 \leqq h \leqq r)$ of class C^∞. A *solution* of the system of first-order partial differential equations

(18.8.9.1)
$$\mathrm{F}_h(x^1, \ldots, x^q, z^1, \ldots, z^n, p^{11}, \ldots, p^{1q}, \ldots, p^{n1}, \ldots, p^{nq}) = 0 \qquad (1 \leqq h \leqq r)$$

is *by definition* a system of n functions

$$(x^1, \ldots, x^q) \mapsto v^i(x^1, \ldots, x^q) \qquad (1 \leqq i \leqq n)$$

of class C^1 on an open set $\mathrm{V} \subset \mathbf{R}^q$ such that, for each point

$$x = (x^1, \ldots, x^q) \in \mathrm{V},$$

the point of $\mathbf{R}^{\mathrm{N}}$ with coordinates

$$x^1, \ldots, x^q, v^1(x), \ldots, v^n(x), \mathrm{D}_1 v^1(x), \ldots, \mathrm{D}_q v^n(x)$$

belongs to U and satisfies the equations (18.8.9.1).

Assume that the set of points of U satisfying (18.8.9.1) is a *submanifold* M of U. Then it is immediately clear that, if we consider the n differential 1-forms

(18.8.9.2) $$\varpi_j = dz^j - \sum_{i=1}^{q} p^{ij}\, dx^j \qquad (1 \leqq j \leqq n)$$

on U, and if we denote by ω_j the 1-form *induced* by ϖ_j on the submanifold M, then the submanifold of $\mathbf{R}^q \times \mathbf{R}^n$ defined by the relations

$$x \in \mathrm{V}, \qquad z^j = v^j(x^1, \ldots, x^q) \qquad (1 \leqq j \leqq n)$$

is the *projection* on $\mathbf{R}^q \times \mathbf{R}^n$ of an n-dimensional integral manifold of the Pfaffian system

(18.8.9.3) $$\omega_j = 0 \qquad (1 \leqq j \leqq n)$$

on M. Conversely, however, the projection on $\mathbf{R}^q \times \mathbf{R}^n$ of an integral manifold of the system (18.8.9.3) cannot always be defined in a neighborhood of a point by equations of the form $z^j = v^j(x^1, \ldots, x^q)$ $(1 \leqq j \leqq n)$, where the v^j form a system of solutions of (18.8.9.1).

(18.8.10) Finally, given a system of equations which contain, besides the variables x^i $(1 \leqq i \leqq q)$ and the unknown functions z^j $(1 \leqq j \leqq n)$, the partial derivatives of each z^j up to order m_j $(1 \leqq j \leqq n)$, it is easy to reduce this system to one of type (18.8.9.1) containing only first-order partial derivatives. The method is the same as in the case of systems of ordinary differential equations, which is the case $q = 1$ (cf. (10.6.6)): we take as new unknowns the derivatives of each z^j *of order* $< m_j$ (because of the symmetry of partial derivatives (8.12.4), it is not necessary to take *all* these derivatives). For example, suppose that we have two variables x and y, one unknown function z, and a single second-order equation

$$\mathrm{F}\left(x, y, z, \frac{\partial z}{\partial x}, \frac{\partial z}{\partial y}, \frac{\partial^2 z}{\partial x^2}, \frac{\partial^2 z}{\partial x\, \partial y}, \frac{\partial^2 z}{\partial y^2}\right) = 0,$$

where F is of class C^∞ on an open set $\mathrm{U} \subset \mathbf{R}^8$. Introduce two new unknown functions p, q with the additional equations

$$p - \frac{\partial z}{\partial x} = 0, \qquad q - \frac{\partial z}{\partial y} = 0;$$

then the original equation is reduced to the system of type (18.8.9.1) consisting of these two equations and the equation

$$F\left(x, y, z, p, q, \frac{\partial p}{\partial x}, \frac{\partial p}{\partial y}, \frac{\partial q}{\partial y}\right) = 0.$$

If the equation $F(x, y, z, p, q, r, s, t) = 0$ defines a *submanifold* M of U, then the solution of this system of equation reduces to finding the solutions of the Pfaffian system (18.8.9.3) *on* M in which the left-hand sides are the 1-forms induced on M by the three differential forms

$$dz - p\,dx - q\,dy, \qquad dp - r\,dx - s\,dy, \qquad dq - s\,dx - t\,dy.$$

PROBLEMS

1. Let M be a connected differential manifold which is a submanifold of $\mathbf{R}^N$, so that T(M) may be identified with a submanifold of $T(\mathbf{R}^N) = \mathbf{R}^N \times \mathbf{R}^N$. Let $x \mapsto L_x$ be a C^∞ field of 1-directions on M. For each $x \in M$, consider the two points on the line $L_x \subset T_x(M)$ whose Euclidean distance from x is equal to 1. The set M′ of these points is a two-sheeted covering of M, whose projection on M is the restriction of o_M. Show that a necessary and sufficient condition that there should exist a C^∞ vector field X on M such that $X(x) \neq \mathbf{0}_x$ and $X(x) \in L_x$ for all $x \in M$ is that M′ is not connected. In particular, such a vector field will exist if the fundamental group $\pi_1(M)$ contains no subgroups of index 2, for example, if M is simply connected.

2. (a) On $M = \mathbf{R}^2 - \{0\}$, let $x \mapsto L_x$ be the C^∞ field of 1-directions such that, for $x = (r\cos\theta, r\sin\theta)$ with $r > 0$, L_x is the line with gradient $\tan(\frac{1}{2}\theta)$. Show that there exists no continuous vector field X on M such that $X(x) \in L_x$ and $X(x) \neq 0$ for all $x \in M$.
(b) Let $p : \mathbf{R}^2 \to \mathbf{T}^2$ be the canonical mapping. For each point $z = (\xi, \eta) \in \mathbf{R}^2$, let $L'_z \subset T_z(\mathbf{R}^2)$ be the line with gradient $\tan \frac{1}{2}\pi(1 - 2\xi)$. Show that, for each point $x \in \mathbf{T}^2$, there exists a unique line $L_x \subset T_x(\mathbf{T}^2)$ such that $T(p)L'_z = L_x$ (where $x = p(z)$) and that $x \mapsto L_x$ is a C^∞ field of 1-directions on $\mathbf{T}^2$ with the property that there exists no continuous vector field X on $\mathbf{T}^2$ such that $X(x) \in L_x$ and $X(x) \neq 0$ for all $x \in \mathbf{T}^2$.

3. An *analytic* field of p-directions on a real-analytic manifold is defined exactly as in (18.8.1). Let M, M′ be two real-analytic manifolds and let $x \mapsto L_x$ (resp. $x' \mapsto L'_{x'}$) be an analytic field of p-directions on M (resp. of p'-directions on M′). Let $f : M \to M'$ be an analytic mapping. If M is connected, and if there exists a nonempty open subset U of M such that $T(f)(L_x) \subset L'_{f(x)}$ for all $x \in U$, show that this relation holds for all $x \in M$. (Use the principle of analytic continuation.)

9. DIFFERENTIAL SYSTEMS

(18.9.1) Let M be a pure differential manifold of dimension n. We shall denote by $\mathscr{E}_p$ the space of real C^∞ differential p-forms on M (which was denoted by $\mathscr{E}_{p,\mathbf{R}}(M)$ in (17.6.1)), and by $\mathscr{E}_0$ the **R**-algebra of real-valued C^∞ functions on M (which was denoted by $\mathscr{E}(M; \mathbf{R})$ in (16.15.8)). Also we shall put

$$\mathscr{A} = \mathscr{E}_0 \oplus \mathscr{E}_1 \oplus \cdots \oplus \mathscr{E}_n.$$

We recall that, if M is an *open set in* $\mathbf{R}^n$, then $\mathscr{A}$ is the *exterior algebra* of the free $\mathscr{E}_0$-module $\mathscr{E}_1$ (A.13.5), and each $\mathscr{E}_p$ is a free $\mathscr{E}_0$-module with a basis consisting of the p-forms $dx^{i_1} \wedge dx^{i_2} \wedge \cdots \wedge dx^{i_p}$ $(i_1 < i_2 < \cdots < i_p)$.

Consider now a Pfaffian system (18.8.7.1) on an open set M of $\mathbf{R}^n$. If N is an integral manifold of this system, it is clear that the restrictions to N of the *exterior differentials* $d\omega_j$ are also zero (17.15.3.2). Hence the same is true of the restrictions to N of the differential p-forms $(1 \leqq p \leqq n)$ which belong to the *graded ideal* $\mathfrak{a}$ of $\mathscr{A}$ (A.18.1) *generated by the* ω_j *and the* $d\omega_j$. For if a differential p-form α on M is such that its restriction to N is zero, then the same is true for each $(p+q)$-form $\alpha \wedge \beta$ ((16.20.15.4) and (16.20.9.5)); hence the set of elements $\alpha = \alpha_1 + \alpha_2 + \cdots + \alpha_n \in \mathscr{A}$ (where $\alpha_p \in \mathscr{E}_p$) such that the restriction of each α_p to N is zero is a *graded ideal* $\mathfrak{b}$ of $\mathscr{A}$ which contains the ω_j and the $d\omega_j$, and hence contains $\mathfrak{a}$ (in general $\mathfrak{b} \neq \mathfrak{a}$).

Let $\mathfrak{a}_p = \mathfrak{a} \cap \mathscr{E}_p$, the set of p-forms contained in $\mathfrak{a}$, so that we have $\mathfrak{a} = \mathfrak{a}_0 \oplus \mathfrak{a}_1 \oplus \cdots \oplus \mathfrak{a}_n$. Then $\mathfrak{a}_1$ is just the $\mathscr{E}_0$-module generated by the ω_j. We shall show that *the relation* $\alpha_p \in \mathfrak{a}_p$ *implies* $d\alpha_p \in \mathfrak{a}_{p+1}$. For this, we remark that the set $\mathfrak{c}$ of elements $\gamma_1 + \gamma_2 + \cdots + \gamma_n \in \mathscr{A}$ (with $\gamma_p \in \mathscr{E}_p$) such that $d\gamma_p \in \mathfrak{a}_{p+1}$ for all p is a *graded ideal* of $\mathscr{A}$, as follows from the formula giving the exterior differential of an exterior product (17.15.2.1); since $\mathfrak{c}$ contains the ω_j, it contains $\mathfrak{a}$.

(18.9.2) These results suggest the following generalization of the notion of a Pfaffian system. For brevity, let us define a *differential ideal* to be any graded ideal $\mathfrak{a}$ of the algebra $\mathscr{A}$ such that, for each p-form $\alpha \in \mathfrak{a}$, we have also $d\alpha \in \mathfrak{a}$. We remark that $\mathfrak{a} = \mathfrak{a}_0 \oplus \mathfrak{a}_1 \oplus \cdots \oplus \mathfrak{a}_n$ (where $\mathfrak{a}_p = \mathfrak{a} \cap \mathscr{E}_p$) but that here we may have $\mathfrak{a}_0 \neq \{0\}$. Such an ideal is said to define a *differential system* on M; a submanifold N of M is said to be an *integral manifold* of the system if the restrictions to N of *all* the differential forms belonging to the different homogeneous components of the idea $\mathfrak{a}$ are zero.

A Pfaffian system corresponds therefore to a particular type of differential system, in which the differential ideal is *generated by* 1-*forms*.

PROBLEM

Given any differential system on an open set $U \subset \mathbf{R}^n$, and a p-dimensional integral manifold N of this system, show that there exists a nonempty open set $V \subset U$ such that $N \cap V$ is the graph of a solution of a system of first-order partial differential equations in $n - p$ unknown functions of p variables. Consequently N is also (in general) an integral manifold of a Pfaffian system.

10. INTEGRAL ELEMENTS OF A DIFFERENTIAL SYSTEM

We shall make repeated use of the following lemma, which generalizes the remark at the beginning of (10.3).

(18.10.1) *Let* X *be a topological space, and let* $(f_\lambda)_{\lambda \in L}$ *be a family of continuous mappings of* X *into the dual* E* *of a finite-dimensional real vector space* E. *Then the rank of the system of linear forms* $(f_\lambda(x))_{\lambda \in L}$ *is a lower semicontinuous function of* x (12.7), *and the set of points* $x \in X$ *at which this rank is locally constant is a dense open set in* X.

If $(\mathbf{e}_i)_{1 \leqq i \leqq r}$ is a basis of E, and $(\mathbf{e}_i^*)$ the dual basis of E*, then we may write

$$f_\lambda(x) = \sum_{i=1}^{r} c_{i\lambda}(x)\mathbf{e}_i^*,$$

where the $c_{i\lambda}$ are continuous real-valued functions on X. If the system of linear forms $f_\lambda(x)$ has rank p, then there exist p indices λ_k $(1 \leqq k \leqq p)$ such that at least one of the $p \times p$ minors of the matrix $(c_{i\lambda_k}(x))_{1 \leqq i \leqq r, 1 \leqq k \leqq p}$ is nonzero; the corresponding minor of the matrix $(c_{i\lambda_k}(x'))$ is then also $\neq 0$ for all x' in a suitably small neighborhood of x, and therefore for x' in this neighborhood the rank of the system $(f_\lambda(x'))$ is $\geqq p$. This proves the first part of the lemma. Next, for each open set V in X, the maximum of the values taken by the rank of the system $(f_\lambda(x))$ for $x \in V$ is a finite number q. If $x_0 \in V$ is such that the rank of $(f_\lambda(x_0))$ is equal to q, then it is equal to q at all points of a neighborhood W of x_0 contained in V, by lower semicontinuity. This proves the second part of the lemma.

(18.10.2) With the notation of (18.9.1), let $\mathfrak{a}$ be a differential ideal of $\mathscr{A}$, and let $\mathfrak{a}_p$ denote its homogeneous component of degree p. Recall that, at any point $x \in M$, the tangent space at x to M is identified canonically with $\{x\} \times \mathbf{R}^n$. For each $x \in M$ and each $p > 0$, a p-dimensional vector subspace

F_x of $\{x\} \times \mathbf{R}^n$ is said to be a *p-dimensional integral element* of the differential system defined by $\mathfrak{a}$ if: (i) all the functions $f \in \mathfrak{a}_0$ vanish at x; (ii) for each integer q such that $1 \leqq q \leqq p$, and each q-form $\varpi_q \in \mathfrak{a}_q$, the restriction to F_x of the q-covector $\varpi_q(x)$ is zero, or equivalently,

(18.10.2.1) $$\langle \mathbf{u}_1 \wedge \mathbf{u}_2 \wedge \cdots \wedge \mathbf{u}_q, \varpi_q(x) \rangle = 0$$

for any q vectors $\mathbf{u}_1, \ldots, \mathbf{u}_q$ in the space F_x.

From this definition it is immediately clear that if F_x is an integral element of dimension p, every *vector subspace* G_x of F_x, of dimension $q < p$, is an integral element of dimension q.

An integral element of a *Pfaffian system* (18.8.7.1) is by definition an integral element of the corresponding differential system (18.9.2).

It is clear that if N is an *integral manifold* (18.9.2) of the system defined by $\mathfrak{a}$, containing the point x and of dimension q at x, then the subspace $T_x(N)$ of $\{x\} \times \mathbf{R}^n$ tangent to N at x is an integral element of dimension q. The existence and uniqueness theorem for differential systems will establish *under certain conditions* a converse of this proposition.

(18.10.3) For this purpose we shall consider a point $x \in M$ at which all the functions $f \in \mathfrak{a}_0$ vanish, and then successively the integral elements of dimension 1 passing through x, and then for each of these the integral elements of dimension 2 containing them, and so on; and we shall seek to answer the following question: *given* an integral manifold N of dimension q, the tangent space $T_x(N)$ at a point $x \in N$ (which is an integral element of dimension q) and an integral element F_x of dimension $q+1$ *containing* $T_x(N)$, does there exist (at any rate locally) an integral manifold N′ of dimension $q+1$ *containing* N and such that $T_x(N') = F_x$?

More precisely, we shall first define a sequence of integers $\geqq 0$ as follows. For each point $x \in M$ (not necessarily a common zero of the functions $f \in \mathfrak{a}_0$), let s_0 be the rank of the system of linear forms $\varpi_1(x)$ on $\{x\} \times \mathbf{R}^n$, where ϖ_1 runs through the homogeneous component $\mathfrak{a}_1$ of $\mathfrak{a}$. Also let $M_1(x)$ denote the subspace of $\{x\} \times \mathbf{R}^n$ of dimension $n - s_0(x)$ on which all the forms $\varpi_1(x)$ vanish.

Now consider a fixed nonzero vector $\mathbf{u}_1 \in M_1(x)$, and the vectors $\mathbf{v} \in \{x\} \times \mathbf{R}^n$ such that

(18.10.3.1) $$\begin{cases} \langle \mathbf{v}, \varpi_1(x) \rangle = 0, \\ \langle \mathbf{u}_1 \wedge \mathbf{v}, \varpi_2(x) \rangle = 0 \end{cases}$$

for all $\varpi_1 \in \mathfrak{a}_1$ *and* $\varpi_2 \in \mathfrak{a}_2$. The rank of the system (18.10.3.1) of linear equations in $\mathbf{v}$ is at least $s_0(x)$, say $s_0(x) + s_1(x, \mathbf{u}_1)$. Let $M_2(x, \mathbf{u}_1)$ denote the subspace of all vectors $\mathbf{v} \in \{x\} \times \mathbf{R}^n$ which satisfy (18.10.3.1), so that the dimension of $M_2(x, \mathbf{u}_1)$ is $n - s_0(x) - s_1(x, \mathbf{u}_1)$. Clearly $\mathbf{u}_1 \in M_2(x, \mathbf{u}_1)$.

Proceeding now by induction, let $\mathbf{u}_1, \ldots, \mathbf{u}_q \in \{x\} \times \mathbf{R}^n$ be q *linearly independent* vectors such that, *for each* $r = 1, \ldots, q$ and each sequence of integers $(i_1, \ldots, i_r)$ satisfying $1 \leqq i_1 < \cdots < i_r \leqq q$, we have

$$\langle \mathbf{u}_{i_1} \wedge \mathbf{u}_{i_2} \wedge \cdots \wedge \mathbf{u}_{i_r}, \varpi_r(x) \rangle = 0 \tag{18.10.3.2}$$

for all $\varpi_r \in \mathfrak{a}_r$. Consider the vectors $\mathbf{v} \in \{x\} \times \mathbf{R}^n$ such that, for $r = 1, \ldots, q+1$, we have

$$\langle \mathbf{u}_{i_1} \wedge \mathbf{u}_{i_2} \wedge \cdots \wedge \mathbf{u}_{i_{r-1}} \wedge \mathbf{v}, \varpi_r(x) \rangle = 0 \tag{18.10.3.3}$$

for all sequences $(i_1, \ldots, i_{r-1})$ satisfying $1 \leqq i_1 < \cdots < i_{r-1} \leqq q$. (When $r = 1$, (18.10.3.3) is replaced by $\langle \mathbf{v}, \varpi_1(x) \rangle = 0$.) Denote by

$$s_0(x) + s_1(x, \mathbf{u}_1) + \cdots + s_q(x, \mathbf{u}_1, \ldots, \mathbf{u}_q)$$

the rank of this system of linear equations in $\mathbf{v}$, and by $M_{q+1}(x, \mathbf{u}_1, \ldots, \mathbf{u}_q)$ the subspace of $\{x\} \times \mathbf{R}^n$ defined by the equations (18.10.3.3). It is clear that this subspace contains $\mathbf{u}_1, \ldots, \mathbf{u}_q$, and therefore

$$s_0(x) + s_1(x, \mathbf{u}_1) + \cdots + s_q(x, \mathbf{u}_1, \ldots, \mathbf{u}_q) \leqq n - q. \tag{18.10.3.4}$$

If we have *strict* inequality in (18.10.3.4) then we can choose a vector $\mathbf{u}_{q+1} \in M_{q+1}(x, \mathbf{u}_1, \ldots, \mathbf{u}_q)$ linearly independent of $\mathbf{u}_1, \ldots, \mathbf{u}_q$, and the induction can continue.

It follows immediately from these definitions that if x is a common zero of all the functions $f \in \mathfrak{a}_0$, and if I_q denotes the q-dimensional vector subspace of $\{x\} \times \mathbf{R}^n$ spanned by the vectors $\mathbf{u}_1, \ldots, \mathbf{u}_q$, then the I_r for $1 \leqq r \leqq q$ form an increasing sequence of *integral elements*.

(18.10.4) From now on we shall assume that the set M_0 of points $x \in M$ at which all $f \in \mathfrak{a}_0$ vanish is a *submanifold* of M, of dimension $\rho \geqq 1$, and that $\mathfrak{a}_0$ is generated as an $\mathscr{E}_0$-module by $n - \rho$ functions f_k $(1 \leqq k \leqq n - \rho)$ such that, at each $x \in M_0$, the rank of the system of differential 1-forms df_k is $n - \rho$ (16.8.9). Now let M_1 be the set of pairs $(x, \mathbf{u}_1) \in M \times \mathbf{R}^n$ such that $x \in M_0$, $\mathbf{u}_1 \in M_1(x)$, and $\mathbf{u}_1 \neq 0$. Proceeding by induction, suppose that M_r has been defined, and define M_{r+1} to be the set of points

$$((x, \mathbf{u}_1, \ldots, \mathbf{u}_r), \mathbf{u}_{r+1}) \in M_r \times \mathbf{R}^n$$

such that

$$\mathbf{u}_{r+1} \in M_{r+1}(x, \mathbf{u}_1, \ldots, \mathbf{u}_r) \qquad \text{and} \qquad \mathbf{u}_1 \wedge \mathbf{u}_2 \wedge \cdots \wedge \mathbf{u}_{r+1} \neq 0.$$

The projection of M_{r+1} on M_r is the set of points $(x, \mathbf{u}_1, \dots, \mathbf{u}_r)$ such that

$$s_0(x) + s_1(x, \mathbf{u}_1) + \cdots + s_r(x, \mathbf{u}_1, \dots, \mathbf{u}_r) < n - r.$$

Since the df_k belong to $\mathfrak{a}_1$, we have $s_0(x) \geqq n - \rho$ and therefore the set M_{r+1} will certainly be *empty* when $r \geqq \rho$.

By virtue of (18.10.1), the set V_0 of points $x \in M_0$ such that the rank $s_0(x)$ is *locally constant* at x is a dense open set *in* M_0. By induction again, suppose that the open subset V_r of M_r has been defined. Then the set U_{r+1} of points $(x, \mathbf{u}_1, \dots, \mathbf{u}_{r+1}) \in M_{r+1}$ such that

$$s_0(x) + \cdots + s_{r+1}(x, \mathbf{u}_1, \dots, \mathbf{u}_{r+1})$$

is *locally constant* at this point is a dense *open* set in $M_{r+1} \subset M_r \times \mathbf{R}^n$, and therefore $V_{r+1} = U_{r+1} \cap \mathrm{pr}_1^{-1}(V_r)$ is also *open* in M_{r+1}.

An integral element of dimension r contained in $\{x\} \times \mathbf{R}^n$ is said to be *regular* if it is spanned by r vectors $\mathbf{u}_1, \dots, \mathbf{u}_r$ such that $(x, \mathbf{u}_1, \dots, \mathbf{u}_r) \in V_r$. We have then, by definition, $(x, \mathbf{u}_1, \dots, \mathbf{u}_s) \in V_s$ for $s < r$, and therefore the integral element of dimension s spanned by $\mathbf{u}_1, \dots, \mathbf{u}_s$ is also *regular.* Moreover, the definition shows that, for $(x', \mathbf{u}_1', \dots, \mathbf{u}_r')$ *sufficiently near to* $(x, \mathbf{u}_1, \dots, \mathbf{u}_r)$ *in* M_r, the integral element spanned by $\mathbf{u}_1', \dots, \mathbf{u}_r'$ is also regular, and we have

$$s_q(x, \mathbf{u}_1', \dots, \mathbf{u}_q') = s_q(x, \mathbf{u}_1, \dots, \mathbf{u}_q)$$

for $1 \leqq q \leqq r$. Observe that if $(x, \mathbf{u}_1, \dots, \mathbf{u}_{r-1}) \in V_{r-1}$ and if

(18.10.4.1)

$$s_0(x) + s_1(x, \mathbf{u}_1) + \cdots + s_{r-1}(x, \mathbf{u}_1, \dots, \mathbf{u}_{r-1}) + s_r(x, \mathbf{u}_1, \dots, \mathbf{u}_r) = n - r,$$

the integral element generated by $(x, \mathbf{u}_1, \dots, \mathbf{u}_r)$ is *regular* by definition, and there exists *no* integral element of dimension $r + 1$ containing $(x, \mathbf{u}_1, \dots, \mathbf{u}_r)$.

An integral element which is not regular is said to be *singular.*

(18.10.4.2) The vectors $\mathbf{u}_1, \dots, \mathbf{u}_r$ will certainly generate a regular integral element if $x, \mathbf{u}_1, \dots, \mathbf{u}_r$ are successively chosen so that the values of

$$s_0(x), s_1(x, \mathbf{u}_1), \dots, s_r(x, \mathbf{u}_1, \dots, \mathbf{u}_r)$$

are *as large as possible* (18.10.1).

(18.10.5) So far we have not made use of the fact that $\mathfrak{a}$ is an *ideal* in the exterior algebra $\mathscr{A}$. This has the following supplementary consequences:

(18.10.6) (i) *In order that a subspace* F_x *of* $\{x\} \times \mathbf{R}^n$ (*where* $x \in M_0$) *should be an integral element of dimension p, it is necessary and sufficient that, for each* $\varpi_p \in \mathfrak{a}_p$, *the restriction to* F_x *of the p-covector* $\varpi_p(x)$ *should be zero.*

(ii) *If* F_x *is an integral element of dimension p, the sum*

$$s_0(x) + s_1(x, \mathbf{u}_1) + \cdots + s_p(x, \mathbf{u}_1, \ldots, \mathbf{u}_p) \tag{18.10.6.1}$$

and the subspace $M_{p+1}(x, \mathbf{u}_1, \ldots, \mathbf{u}_p)$ *are independent of the basis* $\mathbf{u}_1, \ldots, \mathbf{u}_p$ *of* F_x.

Assertion (ii) is a consequence of (i), because the equations (18.10.3.3) for $r < p + 1$ are consequences of the equations (18.10.3.3) for $r = p + 1$ by virtue of (i). Hence it is enough to prove (i), and for this it is enough to show that if each $\varpi_p \in \mathfrak{a}_p$ is such that the restriction of $\varpi_p(x)$ to F_x is zero, then the same is true for all $\varpi_q \in \mathfrak{a}_q$, $q < p$.

Let $\mathbf{u}_1, \ldots, \mathbf{u}_q$ be a sequence of q linearly independent vectors in F_x; then we have to show that $\langle \mathbf{u}_1 \wedge \cdots \wedge \mathbf{u}_q, \varpi_q(x)\rangle = 0$. Extend the sequence $(\mathbf{u}_j)_{1 \leqq j \leqq q}$ to a basis $(\mathbf{u}_j)_{1 \leqq j \leqq p}$ of F_x; let $(\mathbf{u}_j^*)$ be the dual basis of F_x^*, and consider the $(p-q)$-covector $\mathbf{u}_{q+1}^* \wedge \cdots \wedge \mathbf{u}_p^* = \mathbf{z}^*$, say. There exists a C^∞ differential $(p-q)$-form ϖ_{p-q} on M such that $\varpi_{p-q}(x) = \mathbf{z}^*$ (16.4.3). The p-form $\varpi_p = \varpi_q \wedge \varpi_{p-q}$ belongs to the ideal $\mathfrak{a}$, and $\varpi_p(x) = \varpi_q(x) \wedge \mathbf{z}^*$. Now, by hypothesis we have

$$\langle \mathbf{u}_1 \wedge \mathbf{u}_2 \wedge \cdots \wedge \mathbf{u}_p, \varpi_p(x)\rangle = 0,$$

and it follows immediately from the choice of $\mathbf{z}^*$ and the rules of calculation in the exterior algebra (A.14.2.1) that the relation above is equivalent to

$$\langle \mathbf{u}_1 \wedge \cdots \wedge \mathbf{u}_q, \varpi_q(x)\rangle = 0.$$

Example

(18.10.7) Consider the case in which the differential system corresponds to a Pfaffian system on an open set M in $\mathbf{R}^4$, consisting of a single equation

$$\omega = dx^4 - A(x^1, x^2, x^3)\, dx^1 - B(x^1, x^2, x^3)\, dx^2 - C(x^1, x^2, x^3)\, dx^3 = 0,$$

where A, B, C are of class C^∞. The corresponding ideal $\mathfrak{a}$ is generated by ω and the differential 2-form

$$\begin{aligned} d\omega = {} & A'(x^1, x^2, x^3)\, dx^2 \wedge dx^3 + B'(x^1, x^2, x^3)\, dx^3 \wedge dx^1 \\ & + C'(x^1, x^2, x^3)\, dx^1 \wedge dx^2, \end{aligned}$$

where

$$A' = \frac{\partial B}{\partial x^3} - \frac{\partial C}{\partial x^2}, \qquad B' = \frac{\partial C}{\partial x^1} - \frac{\partial A}{\partial x^3}, \qquad C' = \frac{\partial A}{\partial x^2} - \frac{\partial B}{\partial x^1}.$$

It is immediately verified that we have $s_0(x) = 1$ for all $x \in M$, and that $M_1(x)$ is the hyperplane H_x with equation

$$u^4 - A(x)u^1 - B(x)u^2 - C(x)u^3 = 0.$$

Every vector $\mathbf{v} \in \mathbf{R}^3$ is therefore the projection of a unique vector $\mathbf{u} \in H_x$. In order that two nonproportional vectors $\mathbf{u}_1, \mathbf{u}_2 \in H_x$ should define a 2-dimensional integral element, it is necessary and sufficient that their projections $\mathbf{v}_1, \mathbf{v}_2$ on $\mathbf{R}^3$ should satisfy the relation $\mathbf{t}(x) \wedge \mathbf{v}_1 \wedge \mathbf{v}_2 = 0$, where

$$\mathbf{t}(x) = (A'(x), B'(x), C'(x)).$$

If $\mathbf{t}(x) \neq 0$, then $s_1(x, \mathbf{u}_1) = 1$ for each vector $\mathbf{u}_1 \in H_x$ such that $\mathbf{t}(x) \wedge \mathbf{v}_1 \neq 0$, and therefore $M_2(x, \mathbf{u}_1)$ is a plane whose projection on $\mathbf{R}^3$ is the plane containing $\mathbf{t}(x)$ and $\mathbf{v}_1$; there is no integral element of dimension >2 containing $\mathbf{u}_1$. If $\mathbf{t}(x)$ and $\mathbf{v}_1$ are proportional, then $s_1(x, \mathbf{u}_1) = 0$ and $M_2(x, \mathbf{u}_1)$ is equal to H_x; but if we choose $\mathbf{u}_2 \in H_x$ not proportional to $\mathbf{u}_1$, we see again that there is no integral element of dimension >2 containing $\mathbf{u}_1$ and $\mathbf{u}_2$. If on the other hand $\mathbf{t}(x) = 0$, then H_x is an integral element of dimension 3. In order that s_1 should be locally constant near the point $(x, \mathbf{u}_1)$, it is therefore necessary and sufficient that *either* $\mathbf{t}(x) \neq 0$ and $\mathbf{v}_1$ be not proportional to $\mathbf{t}(x)$, *or* $\mathbf{t}(x') = 0$ for all x' sufficiently close to x. Consequently, the singular integral elements of dimension 1 are the lines in $\{x\} \times \mathbf{R}^4$ spanned by a vector $\mathbf{u}_1 \in H_x$, such that either x is a frontier point of the closed set $\mathbf{t}^{-1}(0)$, or else $\mathbf{t}(x) \neq 0$ and the projection $\mathbf{v}_1$ of $\mathbf{u}_1$ is proportional to $\mathbf{t}(x)$. The singular integral elements of dimension 2 are the planes contained in H_x, where x is a frontier point of $\mathbf{t}^{-1}(0)$. We see therefore that a *regular* integral element of dimension 2 may contain *singular* integral elements of dimension 1.

(18.10.8) Suppose that the differential ideal $\mathfrak{a}$ is generated by a set S of q-forms, where $q \leqq r$. Then the condition (i) of (18.10.6) is met by requiring that *for each* $\varpi_q \in S$ *the restriction of* $\varpi_q(x)$ *to* F_x *is zero.* For, by hypothesis, every element of $\mathfrak{a}_p$ is a sum of exterior products $\alpha_1 \wedge \cdots \wedge \alpha_s$, where α_j is a p_j-form, with $p_1 + p_2 + \cdots + p_s = p$, and at least one of the forms belongs to S. Now, in order that the restriction to F_x of a q-covector should be identically zero, it is sufficient that the same should be true of the exterior product of this q-covector by an arbitrary covector (A.14.2.1). This remark will be particularly useful in the case where the differential system is a *Pfaffian system* (18.8.7.1); we may then take S to consist of the ω_j and the $d\omega_j$.

PROBLEMS

1. Assume that the differential ideal $\mathfrak{a}$ is *generated by* $\mathfrak{a}_1 \oplus \mathfrak{a}_2$ (so that $\mathfrak{a}_0 = \{0\}$).

(a) With the notation of (18.10.4), let F_x be an integral element of $\mathfrak{a}$, of dimension $r + 2$, generated by $r + 2$ vectors $\mathbf{u}_1, \dots, \mathbf{u}_{r+2}$, and suppose that the integral element generated by $\mathbf{u}_1, \ldots, \mathbf{u}_{r+1}$ is *regular*. Let G_x be the subspace of F_x generated by $\mathbf{u}_1, \ldots, \mathbf{u}_{r-1}$, and let H_x be a supplement of G_x in $\{x\} \times \mathbf{R}^n$, containing $\mathbf{u}_r$, $\mathbf{u}_{r+1}$, and $\mathbf{u}_{r+2}$. Then $M_r(x, \mathbf{u}_1, \ldots, \mathbf{u}_{r-1})$, is the direct sum of G_x and $V_x = H_x \cap M_r(x, \mathbf{u}_1, \ldots, \mathbf{u}_{r-1})$, of dimension

$$m = n - r + 1 - (s_0(x) + s_1(x, \mathbf{u}_1) + \cdots + s_{r-1}(x, \mathbf{u}_1, \ldots, \mathbf{u}_{r-1})).$$

Let $(\mathbf{w}_1, \ldots, \mathbf{w}_m)$ be a basis of V_x such that $\mathbf{w}_1 = \mathbf{u}_r$, $\mathbf{w}_2 = \mathbf{u}_{r+1}$, $\mathbf{w}_3 = \mathbf{u}_{r+2}$. For $j = 1, 2, 3$, let W_j denote the $(m-2)$-dimensional subspace of V_x generated by $\mathbf{w}_j$ and $\mathbf{w}_4, \ldots, \mathbf{w}_m$. Every integral element F'_x of dimension $r+2$ containing G_x and sufficiently close to F_x is determined by its intersections $\mathbf{R}\mathbf{v}_j$ with $W_j (j = 1, 2, 3)$, where the $\mathbf{v}_j$ are subject to the relations $\langle \mathbf{v}_j \wedge \mathbf{v}_k, \varpi_2(x) \rangle = 0$ for all forms $\varpi_2 \in \mathfrak{a}_2$ and $j \neq k$. Let f be the mapping

$$(\mathbf{v}_1, \mathbf{v}_2, \mathbf{v}_3) \mapsto (\mathbf{v}_1 \wedge \mathbf{v}_2, \mathbf{v}_2 \wedge \mathbf{v}_3, \mathbf{v}_3 \wedge \mathbf{v}_1)$$

of $W_1 \times W_2 \times W_3$ into $\left(\bigwedge^2 V_x\right)^3$. Show that f is an *immersion* in a neighborhood of $(\mathbf{w}_1, \mathbf{w}_2, \mathbf{w}_3)$. Next, let L be the vector subspace of $\bigwedge^2 V_x$ on which all the linear forms $\varpi_2(x)$ $(\varpi_2 \in \mathfrak{a}_2)$ vanish. Then L is of codimension $s_r = s_r(x, \mathbf{u}_1, \ldots, \mathbf{u}_r)$ and $f^{-1}(L^3)$ is the set of all triples $(\mathbf{v}_1, \mathbf{v}_2, \mathbf{v}_3) \in W_1 \times W_2 \times W_3$ such that $\langle \mathbf{v}_j \wedge \mathbf{v}_k, \varpi_2(x) \rangle = 0$ for all $\varpi_2 \in \mathfrak{a}_2$ and $j \neq k$. Show that the intersection of $f^{-1}(L^3)$ with a small neighborhood of $(\mathbf{w}_1, \mathbf{w}_2, \mathbf{w}_3)$ is a submanifold of $W_1 \times W_2 \times W_3$ of dimension

$$3(m-2) - (2s_r + s_{r+1}),$$

where $s_{r+1} = s_{r+1}(x, \mathbf{u}_1, \ldots, \mathbf{u}_{r+1})$; and deduce finally that $s_{r+1} \leqq s_r$.

(b) Let E_x be an integral element of $\mathfrak{a}$, of dimension r and generated by r vectors $\mathbf{u}_1, \ldots, \mathbf{u}_r$; suppose also that E_x is *regular* and such that

$$s_0(x) + s_1(x, \mathbf{u}_1) + \cdots + s_r(x, \mathbf{u}_1, \ldots, \mathbf{u}_r) = n - r,$$

so that there is *no* integral element of dimension $r+1$ containing E_x. Let G_x be the subspace E_x generated by $\mathbf{u}_1, \ldots, \mathbf{u}_{r-2}$, and let H_x be a supplement of G_x in $\{x\} \times \mathbf{R}^n$, containing $\mathbf{u}_{r-1}$ and $\mathbf{u}_r$. Then $M_{r-1}(x, \mathbf{u}_1, \ldots, \mathbf{u}_{r-2})$ is the direct sum of G_x and $V_x = H_x \cap M_{r-1}(x, \mathbf{u}_1, \ldots, \mathbf{u}_{r-2})$, of dimension

$$m = n - r + 2 - (s_0(x) + s_1(x, \mathbf{u}_1) + \cdots + s_{r-2}(x, \mathbf{u}_1, \ldots, \mathbf{u}_{r-2})).$$

Let $(\mathbf{w}_1, \ldots, \mathbf{w}_m)$ be a basis of V_x such that $\mathbf{w}_1 = \mathbf{u}_{r-1}$ and $\mathbf{w}_2 = \mathbf{u}_r$, and such that $\mathbf{w}_1, \mathbf{w}_2, \ldots, \mathbf{w}_{m-s_{r-1}}$ (where $s_{r-1} = s_{r-1}(x, \mathbf{u}_1, \ldots, \mathbf{u}_{r-1})$) form a basis of the subspace W_x of V_x on which vanish all the linear forms $\mathbf{v} \mapsto \langle \mathbf{v} \wedge \mathbf{w}_1, \varpi_2(x) \rangle$ for $\varpi_2 \in \mathfrak{a}_2$. Suppose now that s_{r-1} is the largest value taken by $s_{r-1}(x, \mathbf{u}_1, \ldots, \mathbf{u}_{r-2}, \mathbf{v})$ as $\mathbf{v}$ runs through V_x, so that $s_{r-1}(x, \mathbf{u}_1, \ldots, \mathbf{u}_{r-2}, \mathbf{u}_r) \leqq s_{r-1}$. Show that the intersection of W_x and $M_r(x, \mathbf{u}_1, \ldots, \mathbf{u}_{r-2}, \mathbf{u}_r)$ is the line $\mathbf{R}\mathbf{u}_{r-1}$ (for otherwise there would exist an integral element of dimension $r+1$ containing E_x), and finally deduce that $s_r(x, \mathbf{u}_1, \ldots, \mathbf{u}_r) \leqq s_{r-1}$.

2. Suppose that $\mathfrak{a}_0 = \{0\}$ and that $M = P \times Q$, where P (resp. Q) is an open subset of $\mathbf{R}^p$ (resp. $\mathbf{R}^{n-p}$). For each vector $\mathbf{u} \in \{x\} \times \mathbf{R}^n$, where $x = (y, z)$, let $\mathbf{u}''$ denote its projection on $\{z\} \times \mathbf{R}^{n-p}$ (identified with $T_z(Q)$). Given a p-dimensional integral element $(x, \mathbf{u}_1, \ldots, \mathbf{u}_p)$ of $\mathfrak{a}$ such that the integral element $(x, \mathbf{u}_1, \ldots, \mathbf{u}_{p-1})$ is *regular*, the differential system defined by $\mathfrak{a}$ is said to be *in involution relative to* P (or relative to the first p coordinates) in a neighborhood of $(x, \mathbf{u}_1, \ldots, \mathbf{u}_p)$ if the projections $\mathbf{u}'_j$ of the $\mathbf{u}_j (1 \leqq j \leqq p)$ on $\{y\} \times \mathbf{R}^p$ (identified with $T_y(P)$) are linearly independent. An equivalent condition is that the projections of

$$M_1(x), \quad M_2(x, \mathbf{u}_1), \ldots, \quad M_p(x, \mathbf{u}_1, \ldots, \mathbf{u}_{p-1})$$

on $\{y\} \times \mathbf{R}^p$ should be *surjective*. For each $q < p$, let

$$s_0''(x) + s_1''(x, \mathbf{u}_1) + \cdots + s_q''(x, \mathbf{u}_1, \ldots, \mathbf{u}_q)$$

denote the rank of the system of linear equations in $\mathbf{v}'' \in \{z\} \times \mathbf{R}^{n-p}$

$$\langle \mathbf{u}_{i_1} \wedge \cdots \wedge \mathbf{u}_{i_{r-1}} \wedge \mathbf{v}'', \varpi_r(x) \rangle = 0,$$

where $1 \leqq r \leqq q$ and $i_1 < i_2 < \cdots < i_{r-1}$, and ϖ_r runs through $\mathfrak{a}_r$. In order that the system defined by $\mathfrak{a}$ should be in involution relative to P in a neighborhood of $(x, \mathbf{u}_1, \ldots, \mathbf{u}_p)$, it is necessary and sufficient that

$$s_q''(x, \mathbf{u}_1, \ldots, \mathbf{u}_q) = s_q(x, \mathbf{u}_1, \ldots, \mathbf{u}_q)$$

for $0 \leqq q \leqq p - 1$, or equivalently that

$$\begin{aligned} &p s_0(x) + (p-1) s_1(x, \mathbf{u}_1) + \cdots + s_{p-1}(x, \mathbf{u}_1, \ldots, \mathbf{u}_{p-1}) \\ &\quad = p s_0''(x) + (p-1) s_1''(x, \mathbf{u}_1) + \cdots + s_{p-1}''(x, \mathbf{u}_1, \ldots, \mathbf{u}_{p-1}). \end{aligned}$$

The system defined by $\mathfrak{a}$ is said to be *in involution relative to* P *at the point* x if there exists an integral element $(x, \mathbf{u}_1, \ldots, \mathbf{u}_p)$ of dimension p with these properties; in that case, the system is in involution relative to P in a neighborhood of x.

3. (a) With the notation of Problem 2, suppose that $p = 2$ and that $\mathfrak{a}_2$ is generated by m differential 2-forms $\varpi_2^{(j)}$ $(1 \leqq j \leqq m)$ such that the 2-covectors $\varpi_2^{(j)}(y)$ $(1 \leqq j \leqq m)$ are linearly independent for all y sufficiently close to x. Suppose, moreover that $s_0''(x) = s_0(x)$ and that the projection of $M_1(x)$ on P (identified with $\mathbf{Re}_1 \oplus \mathbf{Re}_2$) is surjective. Show that, if $s_1''(x, \mathbf{u}_1) = m$ for some $\mathbf{u}_1 \in M_1(x)$, then the system defined by $\mathfrak{a}$ is in involution relative to P at the point x. Prove also that the converse is true when the restrictions to $\{z\} \times \mathbf{R}^{n-2}$ of all the 2-covectors $\varpi_2(x)$, as ϖ_2 runs through $\mathfrak{a}_2$, are zero.
(b) Consider for $n = 6$ the differential system

$$dx^1 \wedge dx^3 + dx^2 \wedge dx^4 = 0, \qquad dx^2 \wedge dx^3 = 0, \qquad dx^2 \wedge dx^5 = 0, \qquad dx^5 \wedge dx^6 = 0.$$

Show that it is in involution relative to P.
(c) Consider for $n = 5$ the differential system

$$dx^1 \wedge dx^3 = 0, \qquad dx^2 \wedge dx^3 = 0, \qquad dx^3 \wedge dx^4 = 0, \qquad dx^3 \wedge dx^5 = 0.$$

Show that it is not in involution relative to P.
(d) Consider for $n = 5$ the differential system

$$dx^3 \wedge dx^4 = 0, \qquad dx^4 \wedge dx^5 = 0, \qquad dx^3 \wedge dx^5 = 0.$$

Show that it is in involution relative to P, although $s''(x, \mathbf{u}_1) < m$.

4. Let $\pi : M' \to M$ be a surjective submersion. For each differential system on M defined by a differential ideal $\mathfrak{a}$, let ${}^t\pi(\mathfrak{a})$ denote the differential ideal on M' generated by the forms ${}^t\pi(\alpha)$ for $\alpha \in \mathfrak{a}$. Describe how to obtain the integral elements of the system defined by ${}^t\pi(\mathfrak{a})$. If $M = P \times Q$ and $M = P \times Q'$, and $\pi = 1_P \times \sigma$, where $\sigma : Q' \to Q$ is a surjective submersion, then the differential system defined by ${}^t\pi(\mathfrak{a})$ is in involution relative to P at a point $x' \in M'$ if and only if the differential system defined by $\mathfrak{a}$ is in involution relative to P at the point $\pi(x')$.

11. FORMULATION OF THE PROBLEM OF INTEGRATION

(18.11.1) With the assumptions and notation of (18.10.4), we shall consider, at a point $x_0 \in M_0$, a *regular* integral element F_{x_0} of dimension p which is assumed to be the *tangent space* at x_0 to an *integral manifold* $N_0 \subset M_0$ of dimension p of the differential system defined by the ideal $\mathfrak{a}$. Let $(\mathbf{u}_{10}, \ldots, \mathbf{u}_{p0})$ be a basis of F_{x_0} such that the point $(x_0, \mathbf{u}_{10}, \ldots, \mathbf{u}_{p0})$ belongs to the open set $V_p \subset M_p$ defined in (18.10.4), and write $s_q = s_q(x_0, \mathbf{u}_{10}, \ldots, \mathbf{u}_{q0})$ for $0 \leqq q \leqq p$. Then there exists a neighborhood W of $(x_0, \mathbf{u}_{10}, \ldots, \mathbf{u}_{p0})$ in V_p such that $s_q(x, \mathbf{u}_1, \ldots, \mathbf{u}_q) = s_q$ for all $(x, \mathbf{u}_1, \ldots, \mathbf{u}_p) \in W$ and $0 \leqq q \leqq p$. By making an affine-linear transformation, we may assume that $x_0 = 0$ and that $\mathbf{u}_{10}, \ldots, \mathbf{u}_{p0}$ are the first p vectors $\mathbf{e}_1, \ldots, \mathbf{e}_p$ of the canonical basis of $\mathbf{R}^n$. It follows (16.8.3.2) that, by replacing M if necessary by an open neighborhood of x_0, we may suppose that N_0 is the set of points satisfying the equations

(18.11.1.1) $$x^{p+k} = f_k(x^1, \ldots, x^p) \qquad (1 \leqq k \leqq n-p),$$

where the functions f_k are of class C^∞ and the Jacobian matrix $(D_j f_k(0))$ is *zero*.

For each point $x \in N_0$ close to 0, the tangent space at x to N_0 may therefore be identified with the subspace of $\{x\} \times \mathbf{R}^n$ generated by the vectors

$$\mathbf{u}_j = \mathbf{e}_j + \sum_{k=1}^{n-p} \frac{\partial f_k}{\partial x^j} \mathbf{e}_{p+k},$$

which are arbitrarily close to $\mathbf{e}_j$ $(1 \leqq j \leqq p)$. Since this subspace is a p-dimensional integral element by hypothesis, the point $(x, \mathbf{u}_1, \ldots, \mathbf{u}_p)$ belongs to the set W defined above, provided that x is sufficiently close to x_0 in N_0.

(18.11.2) Let $r = s_0 + s_1 + \cdots + s_p$. We shall *assume* that $r < n - p$. Then the subspace $G_{x_0} = M_{p+1}(x_0, \mathbf{u}_{10}, \ldots, \mathbf{u}_{p0})$ of $\{x_0\} \times \mathbf{R}^n$, which contains F_{x_0}, is of dimension $n - r > p$. Let $\mathbf{u}_{p+1,0}$ be a vector of G_{x_0}, *not contained in* F_{x_0}, and let $F'_{x_0} = F_{x_0} \oplus \mathbf{R}\mathbf{u}_{p+1,0}$. Then the integration problem which we wish to examine consists of determining:

(i) whether there exists, in a sufficiently small neighborhood of x_0, an *integral manifold of dimension* $p+1$ of the differential system, which *contains* N_0 and whose tangent space at the point x_0 is F'_{x_0};

(ii) if the answer to (i) is affirmative, what is the "indeterminacy" of these solutions.

To make the second question more precise, let H_{x_0} be a vector space supplement of G_{x_0} in $\{x_0\} \times \mathbf{R}^n$ (so that $\dim(H_{x_0}) = r$). Let P_0 be a submanifold of M, of dimension $p+r+1$, which contains N_0 and has $F'_{x_0} \oplus H_{x_0}$ as tangent space at x_0, but otherwise arbitrary. The problem is then whether there exists a *unique* integral manifold satisfying the condition in (i) above and subjected to the supplementary condition of being *contained in* P_0 (Cauchy's problem).

We may assume that $\mathbf{u}_{p+1,0} = \mathbf{e}_{p+1}$ and that G_{x_0} is spanned by the vectors $\mathbf{e}_1, \ldots, \mathbf{e}_{n-r}$, and H_{x_0} by $\mathbf{e}_{n-r+1}, \ldots, \mathbf{e}_n$. Applying (16.8.3.2) again, we may suppose that P_0 is defined by a system of equations

(18.11.2.1)
$$x^{p+1+k} = g_k(x^1, \ldots, x^{p+1}, x^{n-r+1}, \ldots, x^n) \qquad (1 \leqq k \leqq n-r-p+1),$$

where the g_k are C^∞ functions subject *only* to the conditions

$$g_k(x^1, \ldots, x^p, f_1(x^1, \ldots, x^p), f_{n-r-p+1}(x^1, \ldots, x^p), \ldots, f_{n-p}(x^1, \ldots, x^p)) = f_{k+1}(x^1, \ldots, x^p)$$

and $D_j g_k(0) = 0$, for $1 \leqq k \leqq n-r-p-1$ and

$$j = 1, \ldots, p+1, n-r+1, \ldots, n.$$

The Jacobian of the n functions

$$x^j \qquad (1 \leqq j \leqq p+1, \quad n-r+1 \leqq j \leqq n),$$
$$x^{p+1+k} - g_k(x^1, \ldots, x^{p+1}, x^{n-r+1}, \ldots, x^n) \qquad (1 \leqq k \leqq n-r-p-1)$$

is equal to 1 at the origin and therefore, by the implicit function theorem, there exists a diffeomorphism of a neighborhood of 0 onto a neighborhood of 0 such that the image of P_0 is the *linear subspace* defined by $x^{p+1+k} = 0$ for $1 \leqq k \leqq n-r-p-1$. Having effected this diffeomorphism, it follows that we may assume that the manifold P_0 is given by the equations

(18.11.2.2)
$$x^{p+1+k} = 0 \qquad (1 \leqq k \leqq n-r-p-1),$$

and since this manifold must contain N_0, the equations of N_0 are of the form

(18.11.2.3)
$$\begin{cases} x^{p+1} = f_1(x^1, \ldots, x^p), & \\ x^{p+1+k} = 0 & (1 \leqq k \leqq n-r-p-1), \\ x^{p+k} = f_k(x^1, \ldots, x^p) & (n-r-p+1 \leqq k \leqq n-p). \end{cases}$$

The Jacobian of the $r + p + 1$ functions

$$\begin{aligned} &x^j && (1 \leqq j \leqq p),\\ &x^{p+1} - f_1(x^1, \ldots, x^p), &&\\ &x^{p+k} - f_k(x^1, \ldots, x^p) && (n - r - p + 1 \leqq k \leqq n - p) \end{aligned}$$

is equal to 1, and therefore, by the same reasoning as before, we may assume that a second diffeomorphism (leaving $x^{p+2}, \ldots, x^{n-r}$ invariant) has been performed, in such a way that the manifold N_0 is given by the equations

(18.11.2.4) $$x^{p+k} = 0 \qquad (1 \leqq k \leqq n - p).$$

(18.11.3) It follows from (16.8.3.2) that if there exists an integral manifold of dimension $p + 1$ *contained in* P_0, *containing* N_0 and having F'_{x_0} as *tangent space at the point* x_0, then it will be given by equations of the form

(18.11.3.1) $$\begin{cases} x^{p+1+k} = 0 & (1 \leqq k \leqq n - r - p - 1)\\ x^{n-r+k} = v_k(x^1, \ldots, x^p, x^{p+1}) & (1 \leqq k \leqq r), \end{cases}$$

with the "initial conditions" for the C^∞ functions v_k:

(18.11.3.2) $$v_k(x^1, \ldots, x^p, 0) = 0 \qquad (1 \leqq k \leqq r).$$

In view of (18.10.6), the conditions that the v_k must satisfy (in a neighborhood of 0), in addition to (18.11.3.2), are that *for each* $(p + 1)$*-form*

(18.11.3.3) $$\varpi_{p+1} = \sum A_{i_1 i_2 \cdots i_{p+1}} dx^{i_1} \wedge dx^{i_2} \wedge \cdots \wedge dx^{i_{p+1}}$$

belonging to $\mathfrak{a}_{p+1}$, where the $A_{i_1 \ldots i_{p+1}}$ are C^∞ functions on M, the coefficient of $dx^1 \wedge \cdots \wedge dx^{p+1}$ must be zero when we replace:

the $A_{i_1 \cdots i_{p+1}}(x)$ by

$$\begin{aligned} &B_{i_1 \cdots i_{p+1}}(x^1, \ldots, x^{p+1})\\ &\quad = A_{i_1 \cdots i_{p+1}}(x^1, \ldots, x^{p+1}, 0, \ldots, 0, v_1(x^1, \ldots, x^{p+1}), \ldots, v_r(x^1, \ldots, x^{p+1})); \end{aligned}$$

the dx_i by

$$0 \qquad \text{for} \quad p + 2 \leqq j \leqq n - r;$$

and the dx^{n-r+k} by

$$\sum_{j=1}^{p+1} \frac{\partial v_k}{\partial x^j} dx^i \qquad \text{for} \quad 1 \leqq k \leqq r.$$

We obtain therefore for *each* $\varpi_{p+1} \in \mathfrak{a}_{p+1}$ a *partial differential equation* of the form

(18.11.3.4)

$$\sum_{j=1}^{p+1} C_j\left(x^1, \ldots, x^{p+1}, v_1, \ldots, v_r, \frac{\partial v_1}{\partial x^1}, \ldots, \frac{\partial v_1}{\partial x^p}, \ldots, \frac{\partial v_r}{\partial x^1}, \ldots, \frac{\partial v_r}{\partial x^p}\right) \frac{\partial v^j}{\partial x^{p+1}}$$
$$= D\left(x^1, \ldots, x^{p+1}, v_1, \ldots, v_r, \frac{\partial v_1}{\partial x^1}, \ldots, \frac{\partial v_1}{\partial x^p}, \ldots, \frac{\partial v_r}{\partial x^1}, \ldots, \frac{\partial v_r}{\partial x^p}\right),$$

where the expressions

$$C_j(x^1, \ldots, x^{p+1}, w_1, \ldots, w_r, w_{11}, \ldots, w_{1p}, \ldots, w_{r1}, \ldots, w_{rp})$$

and

$$D(x^1, \ldots, x^{p+1}, w_1, \ldots, w_r, w_{11}, \ldots, w_{1p}, \ldots, w_{r1}, \ldots, w_{rp})$$

are *polynomials* in the *rp* variables w_{ij} which are *linear* in each w_{ij} and whose coefficients are C^∞ *functions* in the variables $x^1, \ldots, x^{p+1}, w_1, \ldots, w_r$.

(18.11.4) In view of (18.10.6), the system of linear equations in $\mathbf{y}$:

(18.11.4.1) $$\langle \mathbf{u}_{10} \wedge \cdots \wedge \mathbf{u}_{p0} \wedge \mathbf{y}, \varpi_{p+1}(x_0) \rangle = 0,$$

where ϖ_{p+1} runs through $\mathfrak{a}_{p+1}$, has rank r, and the set of solutions $\mathbf{y}$ is G_{x_0}. This implies that the left-hand sides of the equations (18.11.4.1) are, as linear forms in $\mathbf{y}$, linear combinations of the coordinate forms $y^{n-r+1}, \ldots, y^n$, and that there exist among them r *linearly independent* forms corresponding to $(p+1)$-forms $\varpi_{p+1}^{(\alpha)} \in \mathfrak{a}_{p+1}$ $(1 \leqq \alpha \leqq r)$. In conformity with (18.11.3.3), let us put

(18.11.4.2) $$\varpi_{p+1}^{(\alpha)} = \sum A_{i_1 \cdots i_{p+1}}^{(\alpha)} \, dx^{i_1} \wedge \cdots \wedge dx^{i_{p+1}}.$$

Then it follows from above that the determinant

$$\det(A_{1,2,\ldots,p,n-r+k}^{(\alpha)}(0)) \qquad (1 \leqq \alpha \leqq r, \quad 1 \leqq k \leqq r)$$

does not vanish. Now, to each of the forms $\varpi_{p+1}^{(\alpha)}$ there corresponds an equation of the type (18.11.3.4) for the unknown functions v_k; in other words, we have for these r functions a *system of r partial differential equations*

(18.11.4.3)

$$\sum_{j=1}^{r} C_j^{(\alpha)}\left(x^1, \ldots, x^{p+1}, v_1, \ldots, v_r, \frac{\partial v_1}{\partial x^1}, \ldots, \frac{\partial v_1}{\partial x^p}, \ldots, \frac{\partial v_r}{\partial x^1}, \ldots, \frac{\partial v_r}{\partial x^p}\right) \frac{\partial v_j}{\partial x^{p+1}}$$
$$= D^{(\alpha)}\left(x^1, \ldots, x^{p+1}, v_1, \ldots, v_r, \frac{\partial v_1}{\partial x^1}, \ldots, \frac{\partial v_r}{\partial x^p}\right) \qquad (1 \leqq \alpha \leqq r).$$

We shall show that

(18.11.4.4) $$\det(C_j^{(\alpha)}(0, \ldots, 0)) \neq 0.$$

Indeed, by virtue of the definitions in (18.11.3), in order to obtain

$$C_j^{(\alpha)}(x^1, \ldots, x^{p+1}, v_1, \ldots, v_r, 0, \ldots, 0).$$

we have to replace the dx^j in $\varpi_{p+1}^{(\alpha)}$ by 0 for $p+2 \leqq j \leqq n-r$, and the dx^{n-r+k} by

$$\frac{\partial v_k}{\partial x^{p+1}} dx^{p+1} \quad \text{for} \quad 1 \leqq k \leqq r;$$

this produces a $(p+1)$-form which is a multiple of $dx^1 \wedge dx^2 \wedge \cdots \wedge dx^{p+1}$, the scalar coefficient being

$$\sum_{j=1}^{r} A_{1,2,\ldots,p,n-r+j}^{(\alpha)}(x^1, \ldots, x^{p+1}, 0, \ldots, 0, v_1, \ldots, v_r) \frac{\partial v_j}{\partial x^{p+1}}.$$

It follows that

$$C_j^{(\alpha)}(x^1, \ldots, x^{p+1}, v_1, \ldots, v_r, 0, \ldots, 0) \\ = A_{1,2,\ldots,p,n-r+j}^{(\alpha)}(x^1, \ldots, x^{p+1}, 0, \ldots, 0, v_1, \ldots, v_r),$$

which, in view of the remarks made earlier, proves (18.11.4.4).

The system of equations (18.11.4.3) may therefore, by virtue of Cramer's formulas, be written in the form

(18.11.4.5)
$$\frac{\partial v_j}{\partial x^{p+1}} = H_j\left(x^1, \ldots, x^{p+1}, v_1, \ldots, v_r, \frac{\partial v_1}{\partial x^1}, \ldots, \frac{\partial v_r}{\partial x^p}\right) \qquad (1 \leqq j \leqq r),$$

wherein the functions

$$H_j(x_1, \ldots, x^{p+1}, w_1, \ldots, w_r, w_{11}, \ldots, w_{rp})$$

are *rational functions* in $w_{11}, \ldots, w_{rp}$ with coefficients which are C^∞ *functions* in $x^1, \ldots, x^{p+1}, w_1, \ldots, w_r$; and these rational functions are defined when

$$x^1 = \cdots = x^{p+1} = 0, \qquad w_1 = \cdots = w_r = 0, \qquad w_{11} = \cdots = w_{rp} = 0,$$

because their denominator is (18.11.4.4).

(18.11.5) *Contrary to what one might expect by analogy with the existence theorem* (10.4.5) *for differential equations*, it is possible to give examples of systems of partial differential equations of the form (18.11.4.5) in which the

H_j are C^∞ functions on $\mathbf{R}^N$ (where $N = p + 1 + r + rp$), *affine-linear* with respect to the w_{jk}, and which admit *no* solution $(v_1, \ldots, v_r)$ of class C^1 in any neighborhood of the origin in $\mathbf{R}^{p+1}$ (see the problem below).

However, we shall see in (18.12) that, when the H_j are *analytic* in a neighborhood of the origin, the existence of a solution of (18.11.4.5) can be established by means of a method (called the "method of majorants") which uses in an essential way the properties of power series.

We shall also show (in 18.14, 18.17, and Chapter XXIII) that for *some* types of systems of partial differential equations in which the H_j are *only of class* C^∞, the existence of a solution can be established.

PROBLEM

Consider the following system of two partial differential equations in $\mathbf{R}^3$, in two unknown real functions v_1, v_2:

$$(*)\qquad \begin{cases} \dfrac{\partial v_1}{\partial x^1} = \dfrac{\partial v_2}{\partial x^2} - 2x^2 \dfrac{\partial v_1}{\partial x^3} - 2x^1 \dfrac{\partial v_2}{\partial x^3} - f'(x^3), \\[2ex] \dfrac{\partial v_2}{\partial x^1} = -\dfrac{\partial v_1}{\partial x^2} + 2x^1 \dfrac{\partial v_1}{\partial x^3} - 2x^2 \dfrac{\partial v_2}{\partial x^3}, \end{cases}$$

where $f : \mathbf{R} \to \mathbf{R}$ is of class C^∞. Putting $u = v_1 + iv_2$, this system is equivalent to the single equation

$$-\frac{\partial u}{\partial x^1} - i\frac{\partial u}{\partial x^2} + 2i(x^1 + ix^2)\frac{\partial u}{\partial x^3} = f'(x^3).$$

Suppose that this equation admits a complex solution $u(x^1, x^2, x^3)$ of class C^1 in a neighborhood V of 0 in $\mathbf{R}^3$. Let

$$w(y, x^3, \theta) = u(\sqrt{y}\cos\theta, \sqrt{y}\sin\theta, x^3)$$

for $0 \leqq y \leqq r$, $|x^3| < r$, $0 \leqq \theta \leqq 2\pi$, and r sufficiently small; also let

$$U(y, x^3) = \int_0^{2\pi} \sqrt{y}\, e^{i\theta} w(y, x^3, \theta)\, d\theta,$$

which is continuous for $0 \leqq y < r$ and $|x^3| < r$. Show that, for $y > 0$,

$$\frac{\partial U}{\partial y} = \frac{1}{2}\int_0^{2\pi} \left(\frac{\partial u}{\partial x^1} + i\frac{\partial u}{\partial x^2}\right) d\theta,$$

and deduce that, if $V(x^3, y) = U(y, x^3) + i\pi f(x^3)$, then we have

$$\frac{\partial V}{\partial x^3} + i\frac{\partial V}{\partial y} = 0$$

for $0 < y < r$ and $|x^3| < r$. Deduce that there exists a function $g(z)$, *holomorphic* in $|z| < r$, and such that $V(x^3, y) = g(x^3 + iy)$ for $|x^3+iy| < r$ and $y \geqq 0$.† Show that this leads to a *contradiction* when the function f is not analytic in some neighborhood of 0 (*H. Lewy's example*).

12. THE CAUCHY–KOWALEWSKA THEOREM

(18.12.1) (Cauchy–Kowalewska theorem) *Given a system of partial differential equations* (18.11.4.5) *in which the* H_j *are* analytic *in a neighborhood of* 0 *in* $\mathbf{R}^{p+1+r+rp}$, *there exists a neighborhood* V_0 *of* 0 *in* $\mathbf{R}^{p+1}$ *with the property that, for each connected open neighborhood* $V \subset V_0$ *of* 0, *there is a unique solution* $(v_1, \ldots, v_r)$ *of* (18.11.4.5) *consisting of functions* analytic *in* V *and such that* $v_j(x^1, \ldots, x^p, 0) = 0$ $(1 \leqq j \leqq r)$ *in* $V \cap \mathbf{R}^p$.

The hypotheses imply that there exists an open ball B with center 0 and radius R_0 in $\mathbf{R}^{p+1+r+rp}$ such that we have

(18.12.1.1)
$$H_j(x^1, \ldots, x^{p+1}, u_1, \ldots, u_r, w_{11}, \ldots, w_{rp}) = \sum_{\alpha,\beta,\gamma} c_{\alpha\beta\gamma}^{(j)} x^\alpha u^\beta w^\gamma \qquad (1 \leqq j \leqq r)$$

for $|x^i| < R_0$, $|u_j| < R_0$, and $|w_{jk}| < R_0$, where $1 \leqq i \leqq p+1$, $1 \leqq j \leqq r$, and $1 \leqq k \leqq p$; as usual, we have put

$$x^\alpha = (x^1)^{\alpha_1} \cdots (x^{p+1})^{\alpha_{p+1}}, \qquad u^\beta = u_1^{\beta_1} \cdots u_r^{\beta_r}, \qquad w^\gamma = \prod_{j=1}^{r} \prod_{k=1}^{p} w_{jk}^{\gamma_{jk}},$$

and the series on the right-hand side of (18.12.1.1) are convergent in B.

Assume for the moment that there exists a real number $\rho > 0$ and a system of power series

(18.12.1.2)
$$v_j(x^1, \ldots, x^{p+1}) = \sum_\lambda a_{j\lambda} x^\lambda \qquad (1 \leqq j \leqq r)$$

convergent for $|x^i| < \rho$ $(1 \leqq i \leqq p+1)$ and such that, for these values of the x^i, the point

$$\left(x^1, \ldots, x^{p+1}, v_1(x), \ldots, v_r(x), \frac{\partial v_1}{\partial x^1}, \ldots, \frac{\partial v_1}{\partial x^p}, \ldots, \frac{\partial v_r}{\partial x^1}, \ldots, \frac{\partial v_r}{\partial x^p}\right)$$

belongs to B, the equations (18.11.4.5) are satisfied, and

$$v_j(x^1, \ldots, x^p, 0) = 0 \qquad (1 \leqq j \leqq r).$$

† Cf. J. Dieudonné, "Calcul Infinitésimal," pp. 260–261, Hermann, Paris, 1968.

Then for $|x^i| < \rho$ $(1 \leqq i \leqq p+1)$, we have

$$\text{(18.12.1.3)} \qquad \frac{\partial v_j}{\partial x^k} = \sum_{\lambda} \lambda_k a_{j\lambda} x^{\lambda - \varepsilon_k} \qquad (1 \leqq j \leqq r, \quad 1 \leqq k \leqq p+1),$$

where ε_k is the multi-index $(\delta_{kl})_{1 \leqq l \leqq p+1}$ (Kronecker delta) and the series on the right-hand side of (18.12.1.3) are convergent. Moreover, the constant terms of the series (18.12.1.2) and (18.12.1.3) for $k \leqq p$ are *zero* by hypothesis. It follows, therefore, from the substitution theorem for power series (9.2.2) that there exists a positive real number $\rho' < \rho$ such that for $|x^i| < \rho'$ $(1 \leqq i \leqq p+1)$, the series (18.12.1.2) may be substituted for the u_j $(1 \leqq j \leqq r)$ and the series (18.12.1.3) for the w_{jk} $(1 \leqq j \leqq r,\ 1 \leqq k \leqq p)$ in the power series H_j. Let us now express that each of the power series (18.12.1.3) for $k = p+1$ and $1 \leqq j \leqq r$ is equal to the power series obtained from H_j by these substitutions; then, for $1 \leqq j \leqq r$ and each multi-index λ (9.1.6), we have

(18.12.1.4)

$$\lambda_{p+1} a_{j\lambda} = \sum_{\alpha, \beta, \gamma, (\mu_{mh}), (\nu_{nhk})} c_{\alpha\beta\gamma}^{(j)} \left(\prod_{h=1}^{r} \prod_{m=0}^{\beta_h} a_{h, \mu_{mh}} \right) \left(\prod_{h=1}^{r} \prod_{k=1}^{p} \prod_{n=0}^{\gamma_{hk}} (\nu_{nhk})_k a_{h, \nu_{nhk}} \right),$$

where μ_{mh} and ν_{nhk} are *multi-indices* (so that $(\nu_{nhk})_k$ is the kth component of ν_{nhk}) and for *given* λ the summation is over the *finite* set of multi-indices satisfying the relation

$$\text{(18.12.1.5)} \qquad \lambda - \varepsilon_{p+1} = \alpha + \sum_{h=1}^{r} \sum_{m=0}^{\beta_h} \mu_{mh} + \sum_{h=1}^{r} \sum_{k=1}^{p} \sum_{n=0}^{\gamma_{hk}} (\nu_{nhk} - \varepsilon_k)$$

(which expresses the equality of the coefficients of $x^{\lambda - \varepsilon_{p+1}}$ in the two series of the jth equation (18.11.4.5), by considering products of terms in the same way as in (9.2)).

In particular, looking at the components of index $p+1$ in (18.12.1.5), we obtain

$$\text{(18.12.1.6)} \qquad \lambda_{p+1} - 1 = \alpha_{p+1} + \sum_{h=1}^{r} \sum_{m=0}^{\beta_h} (\mu_{mh})_{p+1} + \sum_{h=1}^{r} \sum_{k=1}^{p} \sum_{n=0}^{\gamma_{hk}} (\nu_{nhk})_{p+1}.$$

From this we shall deduce that the $a_{j\lambda}$ are *uniquely determined* by the relations (18.12.1.4). When $\lambda_{p+1} = 0$, all the $a_{j\lambda}$ are *zero*, by virtue of the hypothesis $v_j(x^1, \ldots, x^p, 0) = 0$ (9.1.6). Now argue by induction on λ_{p+1}: by virtue of (18.12.1.6), the multi-indices μ_{mh} and ν_{nhk} which feature on the right-hand side of (18.12.1.4) are such that $(\mu_{mh})_{p+1} < \lambda_{p+1}$ and $(\nu_{nhk})_{p+1} < \lambda_{p+1}$, for a given multi-index λ. Hence, when $\lambda_{p+1} \geqq 1$, if the $a_{h\sigma}$ for $\sigma_{p+1} < \lambda_{p+1}$ are known, the relations (18.12.1.4) determine the $a_{j\lambda}$ *uniquely*. This already establishes the uniqueness assertion in (18.12.1), in view of (9.1.6) and the principle of analytic continuation (9.4.2).

It remains to show that, conversely, if we *define* by induction on λ_{p+1} the real numbers $a_{j\lambda}$ so that $a_{j\lambda} = 0$ for $\lambda_{p+1} = 0$ and so that $a_{j\lambda}$ is given by (18.12.1.4) when $\lambda_{p+1} \geqq 1$, then the power series (18.2.1.2) with these numbers as coefficients will *converge* in some neighborhood of 0. For if this is established, then since the constant terms of the series (18.12.1.2) and their derivatives (18.12.1.3) for $1 \leqq k \leqq p$ are *zero*, these series may be substituted for the u_j and the w_{jk}, respectively, in the series (18.12.1.1) for $|x^i| < \rho''$, where ρ'' is a sufficiently small positive real number (9.2.2), and the analytic functions v_j defined by (18.12.1.2) for $|x^i| < \rho''$ $(1 \leqq i \leqq p+1)$ will then satisfy the required conditions, by virtue of the relations (18.12.1.4) satisfied by the $a_{j\lambda}$.

To establish this convergence we shall use Cauchy's "method of majorants." The idea is to find power series

(18.12.1.7) $$G_j(x^1, \ldots, x^{p+1}, u_1, \ldots, u_r, w_{11}, \ldots, w_{rp}) = \sum_{\alpha, \beta, \gamma} C^{(j)}_{\alpha\beta\gamma} x^\alpha u^\beta w^\gamma$$

convergent in some neighborhood of 0 and having the following two properties:

(i) for all values of the indices, $C^{(j)}_{\alpha\beta\gamma}$ is a real number $\geqq 0$ and

(18.12.1.8) $$|c^{(j)}_{\alpha\beta\gamma}| \leqq C^{(j)}_{\alpha\beta\gamma};$$

(ii) the system of partial differential equations

(18.12.1.9)
$$\frac{\partial v_j}{\partial x^{p+1}} = G_j\left(x^1, \ldots, x^{p+1}, v_1, \ldots, v_r, \frac{\partial v_1}{\partial x^1}, \ldots, \frac{\partial v_r}{\partial x^p}\right) \qquad (1 \leqq j \leqq r)$$

admits a solution consisting of functions analytic in a neighborhood of the origin,

(18.12.1.10) $$v_j(x^1, \ldots, x^{p+1}) = \sum_\lambda A_{j\lambda} x^\lambda \qquad (1 \leqq j \leqq r),$$

where the $A_{j\lambda}$ are *real numbers* $\geqq 0$.

Suppose that the $C^{(j)}_{\alpha\beta\gamma}$ and the $A_{j\lambda}$ have been found to satisfy these conditions. Then the first part of the proof shows that

(18.12.1.11)

$$\lambda_{p+1}A_{j\lambda} = \sum_{\alpha, \beta, \gamma, (\mu_{mh}), (\nu_{nhk})} C^{(j)}_{\alpha\beta\gamma}\left(\prod_{h=1}^{r}\prod_{m=0}^{\beta_h} A_{h, \mu_{mh}}\right)\left(\prod_{h=1}^{r}\prod_{k=1}^{p}\prod_{n=0}^{\gamma_{hk}} (\nu_{nhk})_k A_{h, \nu_{nhk}}\right).$$

It is enough to show that for all choices of the indices, these relations imply

(18.12.1.12) $$|a_{j\lambda}| \leqq A_{j\lambda}.$$

But this is obvious when $\lambda_{p+1} = 0$, because then $a_{j\lambda} = 0$; and for $\lambda_{p+1} > 0$ it follows immediately by induction on λ_{p+1}, from comparison of the relations (18.12.1.4) and (18.12.1.11), and the inequalities (18.12.1.8).

It remains, therefore, to construct power series G_j with the properties (i), (ii) listed above. We shall first show that, as a matter of convenience, we may assume that the constant terms $c_j = c^{(j)}_{000}$ of the series (18.12.1.1) are all *zero*. For if v_j $(1 \leqq j \leqq r)$ are solutions of the system of equations (18.11.4.5) which reduce to 0 when $x^{p+1} = 0$, then the functions

$$\tilde{v}_j(x^1, \ldots, x^{p+1}) = v_j(x^1, \ldots, x^{p+1}) - c_j x^{p+1}$$

are solutions, vanishing for $x^{p+1} = 0$, of the analogous system of equations obtained from (18.11.4.5) by replacing each H_j by the function

$$\tilde{H}_j(x^1, \ldots, x^{p+1}, u_1, \ldots, u_r, w_{11}, \ldots, w_{rp})$$
$$= H_j(x^1, \ldots, x^{p+1}, u_1 + c_1 x^{p+1}, \ldots, u_r + c_r x^{p+1}, w_{11}, \ldots, w_{rp}) - c_j.$$

Since the series $\tilde{H}_j$ have zero constant terms, we have achieved the desired reduction.

Next we remark that the power series on the right-hand side of (18.12.1.1) converge also for *complex* values of the x^i, u_j, and w_{jk} less than R_0 in absolute value (9.1.2). Denote the sums of these series again by H_j, so that each H_j is now an analytic function of the *complex* variables x^i, u_j, and w_{jk}. Let R be a positive real number $< R_0$, and let M be an upper bound for the absolute values $|H_j|$ for all complex values of the variables x^i, u_j, and w_{jk} of absolute value $\leqq R$. Then from Cauchy's inequalities (9.9.5) we have

(18.12.1.13) $$|c^{(j)}_{\alpha\beta\gamma}| \leqq M/R^{|\alpha| + |\beta| + |\gamma|}.$$

It follows now that we may take all the functions G_j to be the *same* function:

(18.12.1.14)

$$G(x^1, \ldots, x^{p+1}, u_1, \ldots, u_r, w_{11}, \ldots, w_{rp})$$
$$= -M + M\left(1 - \frac{x^1 + \cdots + x^p + \theta x^{p+1}}{R}\right)^{-1} \prod_{j=1}^{r}\left(1 - \frac{u_j}{R}\right)^{-1} \prod_{j=1}^{r}\prod_{k=1}^{p}\left(1 - \frac{w_{jk}}{R}\right)^{-1},$$

where θ is a real number > 1, whose value we shall fix in a moment. For it is enough to remark that in a power

$$(x^1 + \cdots + x^p + \theta x^{p+1})^N$$

the coefficients of the various monomials x^α (where $|\alpha| = N$) are all $\geqq 1$, by reason of the choice of θ.

We look now for a solution of the corresponding system of equations (18.12.1.9) that is of the form

$$v_j(x^1, \ldots, x^{p+1}) = Y(x^1 + x^2 + \cdots + x^p + \theta x^{p+1}) \qquad (1 \leqq j \leqq r),$$

where $Y(x)$ is an analytic function of the single variable x, such that $T(0) = 0$. It is immediately seen that if we put

(18.12.1.15) $$F(x, u, w) = -M + M\left(1 - \frac{x}{R}\right)^{-1}\left(1 - \frac{u}{R}\right)^{-r}\left(1 - \frac{w}{R}\right)^{-rp}$$
$$= \sum C_{mnq}\, x^m u^n w^q \qquad (m + n + q > 0),$$

the function Y must be a solution of the differential equation

(18.12.1.16) $$(\theta - C_{001})Y' = F(x, Y, Y') - C_{001}Y',$$

and so we are reduced to showing that there exists a solution

(18.12.1.17) $$Y(x) = \sum_{n=1}^{\infty} A_n x^n$$

of (18.12.1.16) which converges in a neighborhood of 0 and has all its coefficients $A_n \geqq 0$. Now, *if we choose* θ *such that* $\theta > C_{001}$, the relation $A_n \geqq 0$ is a consequence of the existence of a solution (18.12.1.17) of (18.12.1.16) such that $Y'(0) = 0$. For, just as at the beginning of the proof, the A_n satisfy the relations

(18.12.1.18) $$(\theta - C_{001})nA_n = \sum_{\alpha, \beta, \gamma, (\mu_s), (\nu_t)} C_{\alpha\beta\gamma}\left(\prod_{s=0}^{\beta} A_{\mu_s}\right)\left(\prod_{t=0}^{\gamma} \nu_t A_{\nu_t}\right),$$

where now α, β, γ are integers $\geqq 0$ such that $\alpha + \beta + \gamma > 0$, the triple (0, 0, 1) being *excluded*, and $\mu_s \geqq 0$, $v_t \geqq 1$ are integers satisfying

$$\alpha + \sum_{s=0}^{\beta} \mu_s + \sum_{t=0}^{\gamma} (v_t - 1) = n - 1. \tag{18.12.1.19}$$

When $n = 1$, this relation is satisfied only by $\alpha = \beta = 0$, $\gamma \geqq 2$ and otherwise arbitrary, and $v_t = 1$ for $0 \leqq t \leqq \gamma$; hence for A_1 we have the equation

$$(\theta - C_{001})A_1 = \sum_{\gamma=2}^{\infty} C_{00\gamma} A_1^{\gamma}, \tag{18.12.1.20}$$

which admits $A_1 = 0$ as solution. Suppose now that $n \geqq 2$ and that $A_m \geqq 0$ for $m \leqq n - 1$, and observe that the relation $A_1 = 0$ implies that the only nonzero terms on the right-hand side of (18.12.1.18) correspond to the case $v_t \geqq 2$ when $\gamma \neq 0$, and $\mu_s \geqq 2$ when $\beta \neq 0$. Now it follows from (18.12.1.19) that if $\alpha \neq 0$, we have $\mu_s \leqq n - 2$ and $v_t \leqq n - 1$; if $\alpha = 0$ and $\beta \neq 0$, we have $2 \leqq \mu_s \leqq n - 1$ for all s, hence $v_t \leqq n - 2$ for all t; and if $\alpha = \beta = 0$, we must have $\gamma \geqq 2$, hence $v_t \leqq n - 1$ for all t. It follows that the right-hand side of (18.12.1.18), for $n \geqq 2$, is a *polynomial in* $A_2, \ldots, A_{n-1}$ *with coefficients* $\geqq 0$, and the inductive hypothesis (together with the relation $\theta - C_{001} > 0$) therefore implies that $A_n \geqq 0$.

We have, therefore, finally to show that with this choice of θ, the differential equation (18.12.1.16) admits a solution $Y(x)$ which is analytic in a neighborhood of 0, and such that $Y(0) = Y'(0) = 0$. This equation is not of the usual type (10.4.1), but may be reduced to it as follows: if we put

$$\Phi(x, u, w) = F(x, u, w) - \theta w,$$

we have $\Phi(0, 0, 0) = 0$ and

$$\frac{\partial \Phi}{\partial w}(0, 0, 0) = C_{001} - \theta \neq 0;$$

hence, by the implicit function theorem (10.2.4), there exists a function $\Psi(x, u)$ analytic in a neighborhood of (0, 0), such that $\Psi(0, 0) = 0$ and $\Phi(x, u, \Psi(x, u)) = 0$ identically. It is clear that every solution of the differential equation

$$Y' = \Psi(x, Y) \tag{18.12.1.21}$$

will also be a solution of (18.12.1.16). But now the existence theorem (10.5.3) applies to (18.12.1.21), and a solution $Y(x)$ of this equation satisfying

$Y(0) = 0$ will automatically also satisfy the condition $Y'(0) = 0$. The proof of the Cauchy–Kowalewska theorem is now complete.

(18.12.2) It is clear that we may replace **R** by **C** throughout in the statement of (18.12.1): the proof is unchanged (except, of course, that the extension of the H_j to $\mathbf{C}^{p+1+r+rp}$ is now superfluous).

(18.12.3) Consider now a system of partial differential equations depending on "parameters":

(18.12.3.1)

$$\frac{\partial v_j}{\partial x^{p+1}} = H_j\left(x^1, \ldots, x^{p+1}, v_1, \ldots, v_r, \frac{\partial v_1}{\partial x^1}, \ldots, \frac{\partial v_r}{\partial x^p}, z_1, \ldots, z_q\right) \quad (1 \leqq j \leqq r),$$

where the functions H_j are analytic in a neighborhood of 0 in

$$\mathbf{R}^{p+1+r+rp+q} \quad (\text{resp. } \mathbf{C}^{p+1+r+rp+q}).$$

Then there exists a neighborhood T of 0 in $\mathbf{R}^q$ (resp. $\mathbf{C}^q$) and a connected open neighborhood V of 0 in $\mathbf{R}^{p+1}$ (resp. $\mathbf{C}^{p+1}$) such that in $V \times T$ there exists a unique system of *analytic* functions $v_j(x^1, \ldots, x^{p+1}, z_1, \ldots, z_q)$ satisfying (18.12.3.1) and such that

$$v_j(x^1, \ldots, x^p, 0, z_1, \ldots, z_q) = 0$$

in $(V \cap \mathbf{R}^p) \times T$ (resp. $(V \cap \mathbf{C}^p) \times T$). The proof follows exactly the same pattern as before: the right-hand side of (18.12.1.1) has to be replaced by

$$\sum_{\alpha, \beta, \gamma, \delta} c^{(j)}_{\alpha\beta\gamma\delta} x^\alpha u^\beta w^\gamma z^\delta$$

and the right-hand side of (18.12.1.2) by

$$\sum_{\lambda, \mu} a_{j\lambda\mu} x^\lambda z^\mu.$$

We leave it to the reader to write down the relations corresponding to (18.12.1.4). In the expression of the "majorant" function (18.12.1.14), the second term on the right-hand side must be multiplied by

$$\prod_{l=1}^{q} \left(1 - \frac{z_l}{R}\right)^{-1}.$$

The function (18.12.1.15) is then replaced by a function

$$F(x, u, w, z) = \sum C_{mns}(z) x^m u^n w^s,$$

analytic in a neighborhood of 0, where the $C_{mns}(z)$ are power series in $z_1, \ldots, z_q$ which all converge in the same polydisk T and have *coefficients* $\geqq 0$. By a homothety on z, we may assume that the point $(1, 1, \ldots, 1)$ belongs to T. If we put

$$C_{001}(z) = \sum_{\delta} C_{001\delta} z^{\delta},$$

the coefficients $C_{001\delta}$ (which by construction are $\geqq 0$) are *bounded* by a number N independent of δ (9.9.5); we then choose the number θ such that $\theta > N$, and complete the proof as before.

(18.12.4) For systems of equations (18.11.4.5) in which the functions H_j are analytic, there is *no* general result analogous to (10.5.6) expressing that, for a "small" variation of the H_j, the solutions taking the same initial values are also subjected only to "small" variations (cf. Problem 5).

PROBLEMS

1. (a) Let c be a real number $\geqq 1$. Show that there exists a real number $C > 0$, depending only on c, such that for each pair of integers $m \geqq 0$, $n \geqq 0$ we have

$$\binom{m}{\mu}\binom{n}{\nu}\Gamma(c\mu + \nu + 1)\Gamma(c(m - \mu) + n - \nu + 1) \leqq C\Gamma(cm + n + 1)$$

for $0 \leqq \mu \leqq m$ and $0 \leqq \nu \leqq n$. (First establish the inequality for $c = 1$ and then show that when $c > 1$ and $\mu > 1/(c - 1)$, $m - \mu > 1/(c - 1)$, there exists a constant C_0, depending only on c, such that

$$\frac{\Gamma(c\mu + \nu + 1)}{(\mu + \nu)!} \cdot \frac{\Gamma(c(m - \mu) + n - \nu + 1)}{(m - \mu + n - \nu)!} \leqq C_0 \frac{\Gamma(cm + n + 1)}{(m + n)!}.$$

For this purpose, use the fact that

$$\Gamma(a)\Gamma(b) \leqq C_1\Gamma(a + b - 1)$$

whenever a or b is sufficiently large, by considering the integral expression for Euler's beta function $\mathbf{B}(a, b)$.)

(b) Deduce from (a) that there exists a constant C', depending only on c, such that

$$\sum_{4 \leqq c\mu + \nu \leqq cm + n - 4} \binom{m}{\mu}\binom{n}{\nu}\Gamma(c\mu + \nu - 3)\Gamma(c(m - \mu) + n - \nu - 3) \leqq C'\Gamma(cm + n - 3).$$

(Observe that the sum is unchanged by replacing μ by $m - \mu$ and ν by $n - \nu$, and hence that we may assume that $c\mu + \nu \geqq \frac{1}{2}(cm + n)$. Then majorize the sum by replacing $\Gamma(c(m - \mu) + n - \nu - 3)$ by $\Gamma(c(m - \mu) + n - \nu - 1)$, and remark that the sum $\sum_{p, q}(p + q)^{-4}$ is finite (p, q each running through the set of positive integers).)

(c) Let $w(x, t)$ be a real-valued function of class C^{∞} in a neighborhood of $(0, 0)$ in $\mathbf{R}^2$.

Assume that there exist real numbers $c \geqq 1$, $M > 0$, $N > 0$ and an integer $n \geqq 1$ such that, for all integers $m \geqq 0$,

$$\left| \frac{\partial^{m+k}}{\partial x^m \partial t^k} w(0, 0) \right| \leqq \begin{cases} MN^{m+2k-1}\Gamma(cm + k - 3) & \text{for } 1 \leqq k \leqq n, \\ 0 & \text{for } k = 0. \end{cases}$$

Show that there exists a real number $K > 0$, depending only on c (and not on M or N) such that, for *all* integers $j \geqq 1$,

$$\left| \frac{\partial^{m+k}}{\partial x^m \, \partial t^k} w^j(0, 0) \right| \leqq \left(\frac{KM}{N} \right)^j MN^{m+2k-1}\Gamma(cm + k - 3)$$

for $1 \leqq k \leqq n$. (Use (b) and Leibniz's formula, and induction on j.)

2. Let c_1, c_2, c_3 be three numbers in the interval $[1, +\infty[$. A function $f(x, t, y)$ of class C^∞ in an open subset D of $\mathbf{R}^3$ is said to be of *type* (c_1, c_2, c_3) if for each compact subset L of D, there exist $M > 0$ and $N > 0$ such that

$$\left| \frac{\partial^{i+j+k}}{\partial x^i \, \partial t^j \, \partial y^k} f(x, t, y) \right| \leqq MN^{i+j+k}\Gamma(c_1 i + 1)\Gamma(c_2 j + 1)\Gamma(c_3 k + 1)$$

for all $(x, t, y) \in L$ and all integers $i, j, k \geqq 0$. An equivalent condition is

$$\left| \frac{\partial^{i+j+k}}{\partial x^i \, \partial t^j \, \partial y^k} f(x, t, y) \right| \leqq MN^{i+j+k}\Gamma(c_1 i + c_2 j + c_3 k + 1)$$

or that

$$\left| \frac{\partial^{i+j+k}}{\partial x^i \, \partial t^j \, \partial y^k} f(x, t, y) \right| \leqq MN_1^i N_2^j N_3^k \, \Gamma(c_1 i - a_1)\Gamma(c_2 j - a_2)\Gamma(c_3 k - a_3)$$

for three numbers a_1, a_2, a_3 (with the convention that $\Gamma(u + 1)$ is replaced by 0 when $u \leqq 0$). If $c_1 = c_2 = c_3 = 1$, then f is *analytic* in D (Section 9.9, Problem 7).

Consider the partial differential equation

$$(*) \qquad \frac{\partial^p u}{\partial t^p} = f\left(x, t, \frac{\partial^{q+r} u}{\partial x^q \, \partial t^r} \right),$$

where $r < p$ and $q + r \leqq p$, the function f is of class C^∞ in a neighborhood of 0 in $\mathbf{R}^3$, and f is of *type* $(c, 1, 1)$, where $1 \leqq c \leqq (q - r)/q$. In a neighborhood of 0 in $\mathbf{R}^3$ we may therefore write

$$f(x, t, y) = \sum_{j=0}^{\infty} a_j(x, t) y^j,$$

where the series converges for all (x, y, t) sufficiently near 0, and the $a_j(x, t)$ are of type $(c, 1)$, so that

$$a_j(x, t) = \sum_{h=0}^{\infty} b_{jh}(x) t^h$$

for (x, t) sufficiently near 0, the series being convergent and the b_{jh} of type c.

(a) Show, by induction on $k \geqq p$, that a sequence (u_k) of functions of class C^∞ in a neighborhood of 0 in **R** may be determined such that

$$\frac{\partial^p}{\partial T^p}\left(\sum_{k=p}^{\infty} u_k(x)T^k\right) = \sum_{j=0}^{\infty}\left(\sum_{h=0}^{\infty} b_{jh}(x)T^h\right)\left(\frac{\partial^{q+r}}{\partial x^q\,\partial T^r}\left(\sum_{k=0}^{\infty} u_k(x)T^k\right)\right)^j$$

is an identity between formal power series in T (A.21.2).
(b) Show by induction on $k \geqq p$ that there exist two constants $M > 0$, $N > 0$ such that, for all $k \geqq p$ and all m such that $m + 2k \geqq q + 2r + 1$, we have

$$\left| k!\,\frac{\partial^m}{\partial x^m} u_k(x)\right| \leqq MN^{m+2k-q-2r-1}\Gamma(c(m-q)+k-r-3)$$

for all x in a neighborhood of 0. (Start with the inequalities

$$\left|\frac{\partial^{m+n}}{\partial x^m\,\partial t^n} a_j(x,t)\right| \leqq AB^{m+2n+j}\Gamma(cm+n-3)$$

and proceed by induction, using Problem 1(c) and the inequalities

$$m + 2(k-p) \leqq m + 2k - q - 2r - 1,$$
$$cm + (k-p) - 3 \leqq c(m-q) + k - r - 3.)$$

(c) Deduce from (b) that the equation (∗) has a unique solution

$$v(x,t) = \sum_{k=0}^{\infty} u_k(x)t^k$$

of class C^∞ in a neighborhood of (0, 0), analytic in t in this neighborhood and such that

$$\frac{\partial^k}{\partial t^k} v(x,0) = 0 \qquad \text{for} \quad 0 \leqq k \leqq p-1$$

in a neighborhood of 0; furthermore, this function is of type $(c, 1)$.

3. Generalize the results of Problem 2 by replacing the equation (∗) by a system of equations

$$\frac{\partial^{p_i} u_i}{\partial t^{p_i}} = f_i\left(x_1, \ldots, x_n, t, u_1, \ldots, u_m, \left(\frac{\partial x^{q_{ikj}+r_{ikj}}}{\partial x_i^{q_{ikj}}\,\partial t^{r_{ikj}}} u_k\right)_{1 \leqq k \leqq m,\ 1 \leqq j \leqq s}\right)$$

$(1 \leqq i \leqq m)$, where the functions $f_i(x_1, \ldots, x_n, t, u_1, \ldots, u_m, (w_{ikj})_{1 \leqq k \leqq m,\ 1 \leqq j \leqq s})$ are of class C^∞ and of type $(c, 1, 1)$ with respect to the three vectors $\mathbf{x} = (x_1, \ldots, x_n) \in \mathbf{R}^n$, $t \in \mathbf{R}$, and $\mathbf{u} = (u_1, \ldots, u_m, (w_{ikj})) \in \mathbf{R}^{m+ms}$; the initial conditions are replaced by

$$\frac{\partial^k}{\partial t^k} v_i(x_1, \ldots, x_n, 0) = 0 \qquad \text{for} \quad 0 \leqq k \leqq p_i - 1,$$

and the p_i are assumed to satisfy $p_i \geqq q_{ikj} + r_{ikj}$ and $p_i > r_{ikj}$ for all (i, k, j). Hence deduce another proof of the Cauchy–Kowalewska theorem.

4. Consider the partial differential equation (the heat equation)

$$\frac{\partial^2 u}{\partial x^2} = \frac{\partial u}{\partial y}.$$

Suppose that $v(x, y) = \sum_{k=0}^{\infty} v_k(y)x^k$ is a C^∞ solution in a neighborhood of (0, 0) and is analytic in x.

(a) Show that there exists a nonempty open interval $I \subset \mathbf{R}$ in which the v_k are defined, and a number $M > 0$, such that $|v_k(y)| \leqq M^k$ for all $k \geqq 0$ and *all* $y \in I$. (Observe that in a neighborhood V of 0 in **R**, the radius of convergence of the series $\sum_{k=0}^{\infty} v_k(y)x^k$ is >0 by hypothesis, for all $y \in V$. Then use Problem 9 of Section 12.7, and (12.16.2).)

(b) Show that the functions v_k are of class C^∞ in a neighborhood of 0, and that

$$v_k'(y) = (k+2)(k+1)v_{k+2}(y).$$

Deduce that there exists a neighborhood J of 0 in **R** such that the function u is of *type* (2, 1) (Problem 2) in $J \times I$. Hence the condition $c \leqq (p-r)/q$ in Problem 2 cannot be improved.

5. Consider the system of partial differential equations $(*)$ in the problem of Section 18.11, and replace f' by a function g *analytic* in a neighborhood of 0. Let $(v_1^{(g)}, v_2^{(g)})$ denote the unique analytic solution of this system of equations such that $v_1(0, x^2, x^3) = v_2(0, x^2, x^3) = 0$. Show that it is not possible that there should exist a compact neighborhood I of 0 in **R**, a neighborhood V of 0 in $\mathbf{R}^3$, an integer $k \geqq 0$ and a number $A > 0$ such that, for each function g analytic in a neighborhood of I, the functions $v_1^{(g)}, v_2^{(g)}$ are defined in V and satisfy in this neighborhood the inequalities

$$\left.\begin{array}{l} |v_j^{(g)}(x^1, x^2, x^3)| \\ |D_i v_j^{(g)}(x^1, x^2, x^3)| \end{array}\right\} \leqq A \cdot \sup_{t \in I}(|g(t)|, |g'(t)|, \ldots, |g^{(k)}(t)|)$$

for $i = 1, 2, 3$ and $j = 1, 2$ (cf. (14.11.3)).

13. THE CARTAN–KÄHLER THEOREM

(18.13.1) We shall take up again the problem posed in Section (18.11), but with the following additional hypotheses: M is a *real-analytic* manifold (in fact, it will again be an open subset of $\mathbf{R}^n$) and the differential forms in the ideal $\mathfrak{a}$ defining the differential system are *analytic*; furthermore, the only manifolds we shall consider will be *analytic* submanifolds of M. Under these conditions, the functions f_k (18.11.1.1), the functions g_k in (18.11.2.1) and the preliminary diffeomorphisms effected in (18.11.2) and (18.11.3) are *analytic*, and so are the functions C_j and D in (18.11.3.4). Hence we obtain, after the reductions performed in (18.11), a system of equations (18.11.4.5) in which the right-hand sides are *analytic*, and we seek *analytic* solutions.

(18.13.2) The Cauchy–Kowalewska theorem (18.12.1) therefore applies, and shows that (restricting M if necessary) the system of equations (18.11.4.3) admits a *unique* (analytic) *solution* $(v_1, \ldots, v_r)$ satisfying the initial conditions (18.11.3.2). In order to resolve the problem posed in (18.11.2), it is necessary to show that the v_j satisfy not only the equations (18.11.4.3) but *all* the equations (18.11.3.4) corresponding to *all* $(p+1)$-forms $\varpi_{p+1} \in \mathfrak{a}_{p+1}$. For this purpose, we shall need to make use of the hypothesis of *regularity* of the integral element F_{x_0} from which we started (18.11.1).

(18.13.3) We recall (18.10.4) that we are assuming that the subset M_0 of M on which all the functions $f \in \mathfrak{a}_0$ vanish is a submanifold of M. In applications it is most often the case that $M_0 \neq M$ (cf. (18.17) and (18.18)); but we can always reduce to the case $M_0 = M$ (i.e., $\mathfrak{a}_0 = \{0\}$), because in a neighborhood of the point x_0 we may assume that M_0 is defined by equations $x^{\rho+k} = g_k(x^1, \ldots, x^\rho)$ $(1 \leqq k \leqq n - \rho)$, where the g_k are analytic, and then as in (18.11.2) we reduce to the case where all the g_k are zero, i.e., $M_0 = M \cap \mathbf{R}^\rho$. But then we may operate in the space $\mathbf{R}^\rho$, by considering the differential forms induced on $\mathbf{R}^\rho$ by the forms belonging to the ideal $\mathfrak{a}$. Hence, we shall assume from now on that $M_0 = M$.

(18.13.4) Let q be any integer between 0 and p, and let $r_q = s_0 + \cdots + s_q$. Let $\varpi_{q+1}^{(\beta)}$ $(1 \leqq \beta \leqq r_q)$ be $(q+1)$-forms belonging to $\mathfrak{a}_{q+1}$ with the property that the r_q linear forms

(18.13.4.1) $$\mathbf{w} \mapsto \langle \mathbf{u}_{10} \wedge \cdots \wedge \mathbf{u}_{q0} \wedge \mathbf{w}, \varpi_{q+1}^{(\beta)}(x_0) \rangle$$

are linearly independent. By definition (18.10.3) and by (18.10.6) there exists such a system of $(q+1)$-forms, and *the hypothesis of regularity* implies that, for $(x, \mathbf{u}_1, \ldots, \mathbf{u}_q)$ *sufficiently close to* $(x_0, \mathbf{u}_{10}, \ldots, \mathbf{u}_{q0})$ *in* M_q, the r_q linear forms

(18.13.4.2) $$\mathbf{w} \mapsto \langle \mathbf{u}_1 \wedge \cdots \wedge \mathbf{u}_q \wedge \mathbf{w}, \varpi_{q+1}^{(\beta)}(x) \rangle$$

are still *linearly independent*, and that for *any* $(q+1)$-form, $\varpi_{q+1} \in \mathfrak{a}_{q+1}$, there exist r_q scalars $A_{q+1}^{(\beta)}(x, \mathbf{u}_1, \ldots, \mathbf{u}_g)$ such that

(18.13.4.3) $$\langle \mathbf{u}_1 \wedge \cdots \wedge \mathbf{u}_q \wedge \mathbf{w}, \varpi_{q+1}(x) \rangle = \sum_{\beta=1}^{r_q} A_{q+1}^{(\beta)}(x, \mathbf{u}_1, \ldots, \mathbf{u}_q) \langle \mathbf{u}_1 \wedge \cdots \wedge \mathbf{u}_q \wedge \mathbf{w}, \varpi_{q+1}^{(\beta)}(x) \rangle$$

for all vectors $\mathbf{w} \in \mathbf{R}^n$ and all $(x, \mathbf{u}_1, \ldots, \mathbf{u}_q) \in M_q$ sufficiently close to

$$(x_0, \mathbf{u}_{10}, \ldots, \mathbf{u}_{q0}).$$

(18.13.5) For $q \leqq p$, put

$$d_{q+1} = (q+1)s_0 + qs_1 + \cdots + 2s_{q-1} + s_q = r_0 + r_1 + \cdots + r_q.$$

We shall show that there exist d_{q+1} $(q+1)$-forms $\varpi_{q+1}^{(\beta)} \in \mathfrak{a}_{q+1}$ such that, in a neighborhood of the point $(x_0, \mathbf{u}_{10}, \ldots, \mathbf{u}_{q+1,0})$ in $\mathrm{M} \times \mathbf{R}^{(q+1)n}$, the (analytic) mapping

(18.13.5.1)
$$(x, \mathbf{u}_1, \ldots, \mathbf{u}_{q+1}) \mapsto (\langle \mathbf{u}_1 \wedge \cdots \wedge \mathbf{u}_{q+1}, \varpi_{q+1}^{(\beta)}(x)\rangle) \qquad (1 \leqq \beta \leqq d_{q+1})$$

of this neighborhood into $\mathbf{R}^{d_{q+1}}$ has a (total) derivative of *constant rank* d_{q+1}. This will imply that, in a neighborhood of the point

$$(x_0, \mathbf{u}_{10}, \ldots, \mathbf{u}_{q+1,0}),$$

the set M_{q+1} coincides with an *analytic submanifold* of dimension

$$(q+2)n - d_{q+1}$$

in $\mathrm{M} \times \mathbf{R}^{(q+1)n}$ (16.8.8).

The proof is by induction on q. For $q = 0$, we have $d_1 = s_0 = r_0$ and by (18.13.4) the mapping (18.13.5.1), which in this case is

$$(x, \mathbf{u}_1) \mapsto (\langle \mathbf{u}_1, \varpi_1^{(\beta)}(x)\rangle),$$

is, for all x sufficiently near x_0, linear in $\mathbf{u}_1$ and of rank equal to s_0, from which our assertion follows (8.1.3). Suppose now that the assertion has been proved for all integers $< q$. Then there are d_q differential q-forms

$$\varpi_q^{(\beta)} \in \mathfrak{a}_q \qquad (1 \leqq \beta \leqq d_q)$$

such that the Jacobian matrix A of the d_q functions

$$\langle \mathbf{u}_1 \wedge \mathbf{u}_2 \wedge \cdots \wedge \mathbf{u}_q, \varpi_q^{(\beta)}(x)\rangle$$

in $(q+1)n$ real variables has, at the point $(x_0, \mathbf{u}_{10}, \ldots, \mathbf{u}_{q0})$, rank equal to d_q, the number of rows of A. Now let ω be an analytic differential 1-form on M such that $\langle \omega(x_0), \mathbf{u}_{k0}\rangle = 0$ for $1 \leqq k \leqq q$ and $\langle \omega(x_0), \mathbf{u}_{q+1,0}\rangle \neq 0$ (for example, a differential form $\sum_{i=1}^{n} c_i\, dx^i$, where the c_i are constant on M). To each q-form $\varpi_q^{(\beta)}$ we shall associate the $(q+1)$-form

(18.13.5.2)
$$\varpi_{q+1}^{(\beta)} = \varpi_q^{(\beta)} \wedge \omega \in \mathfrak{a}_{q+1} \qquad (1 \leqq \beta \leqq d_q).$$

Let

(18.13.5.3)

$$\mathrm{F}_\beta(x, \mathbf{u}_1, \ldots, \mathbf{u}_q, \mathbf{u}_{q+1}) = \langle \mathbf{u}_1 \wedge \mathbf{u}_2 \wedge \cdots \wedge \mathbf{u}_{q+1}, \varpi_{q+1}^{(\beta)}(x) \rangle \qquad (1 \leqq \beta \leqq d_q).$$

Then it follows from the rule for calculating the derivative of a multilinear mapping (8.1.4) and the rules of calculation in exterior algebra (A.14.2.1) that, by reason of our choice of ω, at the point $(x_0, \mathbf{u}_{10}, \ldots, \mathbf{u}_{q+1,0})$ the Jacobian matrix of the d_q functions F_β is of the form

(18.13.5.4) $$(cA \quad 0),$$

where $c = \langle \mathbf{u}_{q+1,0}, \omega(x_0) \rangle \neq 0$; the fact that the last n columns (corresponding to differentiating with respect to the coordinates of $\mathbf{u}_{q+1}$) are zero comes from the fact that $\langle \mathbf{u}_{10} \wedge \cdots \wedge \mathbf{u}_{q0}, \varpi_q^{(\beta)}(x_0) \rangle = 0$ for all β. For each vector $\mathbf{w}$ sufficiently close to $\mathbf{u}_{q+1,0}$, the d_q equations

(18.13.5.5) $$\mathrm{F}_\beta(x, \mathbf{u}_1, \ldots, \mathbf{u}_q, \mathbf{w}) = 0 \qquad (1 \leqq \beta \leqq d_q)$$

therefore define, in a neighborhood of the point

$$(x_0, \mathbf{u}_{10}, \ldots, \mathbf{u}_{q0}) \in \mathrm{M} \times \mathbf{R}^{qn},$$

a submanifold of dimension $(q+1)n - d_q$ which *contains* M_q, and therefore coincides with M_q by virtue of the inductive hypothesis since it has the same dimension. Let us now introduce r_q differential forms

$$\varpi_{q+1}^{(\beta)} \in \mathfrak{a}_{q+1} \qquad (d_q + 1 \leqq \beta \leqq d_q + r_q = d_{q+1})$$

such that the r_q linear forms (18.13.4.1) are linearly independent. The matrix B (of r_q rows and n columns) of these forms is then of rank r_q. Hence it follows from above that, in a neighborhood of the point

$$(x_0, \mathbf{u}_{10}, \ldots, \mathbf{u}_{q+1,0}),$$

the set M_{q+1} is defined by the d_{q+1} equations

(18.13.5.6) $$\begin{cases} \mathrm{F}_\beta(x, \mathbf{u}_1, \ldots, \mathbf{u}_q, \mathbf{u}_{q+1}) = 0 & (1 \leqq \beta \leqq d_q), \\ \langle \mathbf{u}_1 \wedge \cdots \wedge \mathbf{u}_{q+1}, \varpi_q^{(\beta)}(x) \rangle = 0 & (d_q + 1 \leqq \beta \leqq d_{q+1}), \end{cases}$$

and that at the point $(x_0, \mathbf{u}_{10}, \ldots, \mathbf{u}_{q+1,0})$ the Jacobian matrix of the left-hand sides of these equations is of the form

$$\begin{pmatrix} cA & 0 \\ U & B \end{pmatrix}$$

and therefore is of rank d_{q+1} (A.7.4.1).

(18.13.6) Now let $q = p$. Notice first that for a given form $\varpi_{p+1} \in \mathfrak{a}_{p+1}$, the scalars $A_{p+1}^{(\beta)}(x, \mathbf{u}_1, \ldots, \mathbf{u}_p)$ $(d_p + 1 \leqq \beta \leqq d_{p+1})$ which feature in (18.13.4.3) may be calculated explicitly by Cramer's rule, and are therefore *defined and analytic* in a neighborhood V of $(x_0, \mathbf{u}_{10}, \ldots, \mathbf{u}_{p0})$ in $M \times \mathbf{R}^{pn}$, and not merely in $V \cap M_p$; but the identities (18.13.4.3) hold *only in* $V \cap M_p$. By virtue of (18.13.5), this implies (16.8.9) the existence of scalar-valued functions $B_{p+1}^{(\beta)}(x, \mathbf{u}_1, \ldots, \mathbf{u}_p)$ $(1 \leqq \beta \leqq d_p)$, *analytic* in V, and such that *in* $V \times \mathbf{R}^n$ we have

(18.13.6.1)

$$\begin{aligned}\langle \mathbf{u}_1 \wedge \mathbf{u}_2 \wedge \cdots \wedge \mathbf{u}_p \wedge \mathbf{w}, \varpi_{p+1}(x)\rangle &= \sum_{\beta=1}^{d_p} B_{p+1}^{(\beta)}(x, \mathbf{u}_1, \ldots, \mathbf{u}_p) F_\beta(x, \mathbf{u}_1, \ldots, \mathbf{u}_p, \mathbf{w}) \\ &\quad + \sum_{\beta=d_p+1}^{d_{p+1}} A_{p+1}^{(\beta)}(x, \mathbf{u}_1, \ldots, \mathbf{u}_p)\langle \mathbf{u}_1 \wedge \cdots \wedge \mathbf{u}_p \wedge \mathbf{w}, \varpi_{p+1}^{(\beta)}(x)\rangle.\end{aligned}$$

It has now to be shown that when we replace, on the right-hand side of (18.13.6.1),

x by the point of M:

(18.13.6.2) $$(x^1, \ldots, x^{p+1}, 0, \ldots, 0, v_1(x^1, \ldots, x^{p+1}), \ldots, v_r(x^1, \ldots, x^{p+1}));$$

$\mathbf{u}_k$ $(1 \leqq k \leqq p)$ by the vector in $\mathbf{R}^n$:

$$\left(\delta_{k1}, \ldots, \delta_{k,p+1}, 0, \ldots, 0, \frac{\partial v_1}{\partial x^k}, \ldots, \frac{\partial v_r}{\partial x^k}\right);$$

and $\mathbf{w}$ by the vector

$$\left(0, \ldots, 0, 1, 0, \ldots, 0, \frac{\partial v_1}{\partial x^{p+1}}, \ldots, \frac{\partial v_r}{\partial x^{p+1}}\right);$$

the result of these substitutions is 0.

By definition, these substitutions produce 0 in each of the functions

$$\langle \mathbf{u}_1 \wedge \cdots \wedge \mathbf{u}_p \wedge \mathbf{w}, \varpi_{p+1}^{(\beta)}(x)\rangle \qquad (d_p + 1 \leqq \beta \leqq d_{p+1}),$$

and so it is enough to show that they also produce 0 in each of the

$$F_\beta \qquad (1 \leqq \beta \leqq d_p).$$

For this purpose we observe that, by reason of the reductions performed in (18.11), we may simply take $\omega = dx^{p+1}$ for the form ω of (18.13.5). Now, by substituting for x the point (18.13.6.2) in each of the p-forms $\varpi_p^{(\beta)}$ $(1 \leqq \beta \leqq d_p)$, we obtain analytic p-forms

(18.13.6.3) $$\sum_{k=1}^{p+1}(-1)^k w_k^{(\beta)}(x^1, \ldots, x^{p+1})\, dx^1 \wedge \cdots \wedge \widehat{dx^k} \wedge \cdots \wedge dx^{p+1}$$

and with the above choice of ω, it is clear that our substitutions in F_β produce the function

(18.13.6.4) $$w_{p+1}^{(\beta)}(x^1, \ldots, x^{p+1}) \qquad (1 \leqq \beta \leqq d_p),$$

which we therefore have to prove *vanishes identically* in some neighborhood of 0 in $\mathbf{R}^{p+1}$.

We shall write S_β in place of $w_{p+1}^{(\beta)}$. Since the $(p+1)$-forms $\varpi_p^{(\beta)} \wedge dx^k$ for $1 \leqq k \leqq p$ also belong to $\mathfrak{a}_{p+1}$, there exist analytic functions $G_{k\beta\gamma}(x^1, \ldots, x^{p+1})$ such that we have, in a neighborhood of 0 in $\mathbf{R}^{p+1}$,

(18.13.6.5) $$w_k^{(\beta)} = \sum_{\gamma=1}^{d_p} G_{k\beta\gamma} S_\gamma \qquad (1 \leqq k \leqq p, \quad 1 \leqq \beta \leqq d_p).$$

(18.13.7) We have not yet used the fact that $\mathfrak{a}$ is a *differential* ideal (18.9.2). This implies that the forms $d\varpi_p^{(\beta)}$ $(1 \leqq \beta \leqq d_p)$ belong to $\mathfrak{a}_{p+1}$; hence we may apply to each of these forms the same reasoning that was used for the form ϖ_{p+1}, and in this way we obtain, after substituting for x the point (18.13.6.2) in $d\varpi_p^{(\beta)}$, having regard to the relations (18.13.6.5), a $(p+1)$-form

(18.13.7.1) $$\left(\sum_{\gamma=1}^{d_p} L_{\beta\gamma} S_\gamma\right) dx^1 \wedge \cdots \wedge dx^{p+1},$$

where the $L_{\beta\gamma}$ are analytic in a neighborhood of 0.

On the other hand, the formula for change of variables in an exterior differential (17.15.3.2) shows, by virtue of (18.13.6.3), that this $(p+1)$-form is also equal to

(18.13.7.2) $$\left(\frac{\partial w_1^{(\beta)}}{\partial x^1} + \frac{\partial w_2^{(\beta)}}{\partial x^2} + \cdots + \frac{\partial w_{p+1}^{(\beta)}}{\partial x^{p+1}}\right) dx^1 \wedge \cdots \wedge dx^{p+1}.$$

Replacing the $w_k^{(\beta)}$ in this expression by their values given by (18.13.6.5), we obtain finally a system of partial differential equations for the S_β of the form

(18.13.7.3) $$\frac{\partial S_\beta}{\partial x^{p+1}} = \sum_{\gamma=1}^{d_p}\left(L_{\beta\gamma} S_\gamma + \sum_{k=1}^{p} C_{k\beta\gamma}\frac{\partial S_\gamma}{\partial x^k}\right) \qquad (1 \leqq \beta \leqq d_p),$$

in which the right-hand sides are *linear and homogeneous* in the S_γ and the $\partial S_\gamma/\partial x^k$, and the $C_{k\beta\gamma}$ are analytic in a neighborhood of 0 in $\mathbf{R}^{p+1}$.

Finally, we make use of the fact that the submanifold N_0 of dimension p has been assumed to be an *integral manifold* (18.11.1). This implies that the restrictions to N_0 of the forms $\varpi_p^{(\beta)}$ must be *zero*, i.e., that we must have

(18.13.7.4) $$S_\beta(x^1, \ldots, x^p, 0) = 0 \qquad (1 \leqq \beta \leqq d_p)$$

in a neighborhood of 0 in $\mathbf{R}^p$.

We may now apply to the system (18.13.7.3) the Cauchy–Kowalewska theorem (18.12.1). Since clearly the relations (18.13.7.3) and (18.13.7.4) are satisfied when the functions S_β are replaced by 0, the *uniqueness* assertion of the Cauchy–Kowalewska theorem shows that the S_β are identically zero in a neighborhood of 0 in $\mathbf{R}^{p+1}$.

We have therefore established the following theorem:

(18.13.8) (Cartan–Kähler theorem) *Let* $\mathfrak{a}$ *be a differential ideal on a real* (resp. *complex*) *analytic manifold* M *of dimension* n, *generated by analytic differential forms. Let* $N_0 \subset M$ *be an analytic integral submanifold of dimension* p *of the differential system defined by* $\mathfrak{a}$, *and suppose that at a point* $x_0 \in N_0$ *the tangent space* $T_{x_0}(N_0) = F_{x_0}$ *is a* regular *integral element, and that*

$$s_0 + s_1 + \cdots + s_p < n - p.$$

Then there exists an open neighborhood U *of* x_0 *in* M *and* (*at least*) *one analytic integral submanifold* N_0' *of dimension* $p+1$ *of* $M \cap U$, *containing* $N_0 \cap U$.

The "indeterminacy" of the submanifold N_0' in (18.13.8) is specified by the manifold P_0 which must contain it (18.11.2), by reason of the uniqueness assertion of the Cauchy–Kowalewska theroem: one may therefore say that the solutions N_0' "depend on $n - r - p - 1$ arbitrary functions of $r + p + 1$ variables."

The Cartan–Kähler theorem has the following corollary:

(18.13.9) *For each integral element* L_{x_0} *of dimension* $p + 1$, *containing a regular integral element* F_{x_0} *of dimension* p, *there exists at least one analytic integral manifold* N *of the differential system defined by* $\mathfrak{a}$, *containing the point* x_0 *and having* L_{x_0} *as tangent space at* x_0.

With the notation of (18.11.1), let $F_{x_0}^{(q)}$ $(1 \leqq q \leqq p - 1)$ denote the subspace of F_{x_0} spanned by $\mathbf{u}_{10}, \ldots, \mathbf{u}_{q0}$, which is by definition a *regular* integral element. Then we have only to apply the Cartan–Kähler theorem by induction on q, provided that we first establish the existence of an integral manifold

C_0 of dimension 1 which contains the point x_0 and is such that $T_{x_0}(C_0) = F_{x_0}^{(1)}$. Now, by hypothesis, there exists in some neighborhood Z of x_0 in M_0 a system of s_0 differential 1-forms $\omega^{(\alpha)}$ $(1 \leq \alpha \leq s_0)$ on Z which form a *basis* of the module (over the ring of analytic functions on Z) of differential 1-forms on Z which are restrictions of 1-forms in $\mathfrak{a}_1$. By virtue of (18.8.4), these forms define a field of $(n - s_0)$*-directions* on Z. Consequently (18.8.1), there exists an analytic vector field X_1 on Z such that $X_1(x_0)$ is a nonzero vector in the line $F_{x_0}^{(1)}$. We may then take for C_0 the maximal integral curve of this vector field passing through x_0 (18.2.2).

PROBLEMS

1. (a) Let M be a *connected* open subset of $\mathbf{R}^n$ (resp. $\mathbf{C}^n$) and let $\mathfrak{a}$ be a differential ideal, generated by *analytic* differential forms, and such that $\mathfrak{a}_0 = \{0\}$. Then there exists a dense open subset U of M such that the inverse image U_r of U in M_r (in the notation of (18.10.4)) is contained in the open set V_r and is dense in M_r; furthermore, $s_r(x, \mathbf{u}_1, \ldots, \mathbf{u}_r)$ is a *constant* s_r in U_r. Likewise, with the hypotheses and notation of Section 18.10, Problem 2, we may assume (by modifying U if necessary) that the $s_r''(x, \mathbf{u}_1, \ldots, \mathbf{u}_r)$ are *constants* s_r'' in U_r. The condition for the differential system defined by $\mathfrak{a}$ to be in involution relative to P in a neighborhood of a point is then $s_q'' = s_q$ for $0 \leq q \leq p - 1$. The system is then said simply to be *in involution* (relative to P).

   When this is so, for each point $z = (x, y) \in U \cap (P \times Q)$ and each 1-dimensional integral element in $T_z(M)$ whose projection on $T_x(P)$ is not zero, there exists a p-dimensional integral manifold V of the given differential system, which is the graph of an analytic mapping f of a neighborhood of $x \in P$ into Q, and is such that the given 1-dimensional integral element is contained in $T_z(V)$.

   (b) Consider the Pfaffian system which is equivalent (18.8.9) to the system of partial differential equations (18.11.4.5). Show that this system is in involution relative to the first $p + 1$ coordinates.

2. Under the general hypotheses of Problem 1, suppose that $U = M$, that $\mathfrak{a}$ is generated by $\mathfrak{a}_0$ and $\mathfrak{a}_1$, and that $M = P \times Q$ (Section 18.10, Problem 2), where P (resp. Q) is open in $\mathbf{R}^p$ (resp. $\mathbf{R}^{n-p}$). Furthermore, suppose that the set M_0 defined in (18.10.4) is an analytic submanifold of $P \times Q$, whose projection onto P is a surjective submersion. Finally, suppose that there exists at least one analytic mapping $f : P \to Q$ whose graph is a p-dimensional integral manifold of the given differential system.

   Show that the numbers s_j'' (Problem 1) may be calculated by the following procedure. Let R be the integral domain of analytic functions on M, let K_0 be the field of fractions of R, let K be a field containing K_0, and let F^* be the K-vector space obtained by extension to K of the ring of scalars R of the module of differential 1-forms on M. Then F^* is a vector space of dimension n, and the forms dx^i $(1 \leq i \leq n)$ restricted to M constitute a basis of F^*. Let F_P^* be the p-dimensional subspace of F^* spanned by $dx^1, \ldots, dx^p$, and let S^* be the subspace $\mathfrak{a}_1 \otimes_{\mathbf{R}} K$ of F^*. We way assume that S^* has a K-basis consisting of differential 1-forms

$$\theta_k = dx^{p+k} - \sum_{i=1}^{p} a_{ki}\,\omega_i \qquad (1 \leq k \leq r),$$

where the a_{ki} are elements of K and the ω_i form a basis of F_P^*. Finally, let T^* be the subspace of F^* spanned by $dx^{p+r+1}, \dots, dx^n$, so that F^* is the direct sum of the three subspaces F_P^*, S^* and T^*. Let F be the *dual* of F^*, so that F is the direct sum of the duals F_P, S, T of F_P^*, S^*, T^*, respectively. A *distinguished integral field* (relative to P) is by definition a mapping $X \mapsto \Phi(X)$ of F_P into T which is K-*linear* and whose graph G (which is a p-dimensional vector subspace of F) is such that the restrictions to G of the forms $d\omega \in \mathfrak{a}_2$ (regarded as 2-covectors over F) are all zero. Let $f: P \to Q$ be an analytic mapping, and define $\tilde{f}: F_P \to T$ by the condition

$$\langle \tilde{f}(X), dx^{p+r+h} \rangle = \langle X, {}^t f(dx^{p+r+h}) \rangle \qquad (1 \leqq h \leqq n-p-r).$$

Then, in order that the graph G of f should be an integral manifold of dimension p of the differential system defined by $\mathfrak{a}$, it is necessary and sufficient that $\tilde{f}$ should be a distinguished integral field relative to P.

We may write

$$d\theta_k = \sum_{i,h} a_{kih}\omega_i \wedge dx^{p+r+h} + \sum_{i,l} b_{kil}\omega_i \wedge \theta_l + \sum_{i<j} c_{ijk}\omega_i \wedge \omega_j,$$

where the coefficients $a_{kih} = \partial a_{ki}/\partial x^{p+r+h}$, b_{kil} and c_{ijk} belong to K. In order that Φ should be a distinguished integral field relative to P, it is necessary and sufficient that

$$\sum_{i,h} a_{kih}(\langle X, \omega_i\rangle\langle\Phi(X'), dx^{p+r+h}\rangle - \langle X', \omega_i\rangle\langle\Phi(X), dx^{p+r+h}\rangle) + \sum_{i<j} c_{ijk}\langle\omega_i \wedge \omega_j, X \wedge X'\rangle = 0 \qquad (1 \leqq k \leqq r)$$

for all X, X' in F_P.

For each $X \in F_P$ and $Y \in T$, put

$$(*) \qquad \delta(X \otimes Y) = \sum_{k,i,h} a_{kih}\langle X, \omega_i\rangle\langle Y, dx^{p+r+h}\rangle \mathbf{v}_k,$$

where $(\mathbf{v}_k)$ is the basis of S dual to (θ_k). This defines a K-linear mapping δ of $F_P \otimes_K T$ into S which is *independent* of the choice of the basis (ω_i) of F_P^*. If $\Phi_0 = \tilde{f}$, where $f: P \to Q$ is such that the graph of f is an integral manifold of the given system, then in order that $\Phi = \Phi_0 + g$ should be a distinguished integral field, it is necessary and sufficient that

$$(**) \qquad \delta(X \otimes g(X')) = \delta(X' \otimes g(X))$$

for all X, X' in F_P.

Now, consider a basis $\mathbf{U} = (\mathbf{u}_1, \dots, \mathbf{u}_p)$ of F_P over K, and let (ω_i) denote the dual basis of F_P^*. Let $F_h(\mathbf{u}_1, \dots, \mathbf{u}_h)$ denote the vector subspace of F_P, of dimension $h \leqq p$, spanned by $\mathbf{u}_1, \dots, \mathbf{u}_h$. Let $I_h(\mathbf{u}_1, \dots, \mathbf{u}_h)$ denote the set of K-linear mappings g_h of $F_h(\mathbf{u}_1, \dots, \mathbf{u}_h)$ into T such that, for $i < j \leqq h$, we have

$$(***) \qquad \delta(\mathbf{u}_i \otimes g_h(\mathbf{u}_j)) = \delta(\mathbf{u}_j \otimes g_h(\mathbf{u}_i)).$$

Then $I_1(\mathbf{u}_1)$ may be identified with T, and $V = I_p(\mathbf{u}_1, \dots, \mathbf{u}_p)$ may be identified with *the set of all distinguished integral fields*. The restriction mapping

$$\rho_{h+1} : g_{h+1} \mapsto g_{h+1} | F_h(\mathbf{u}_1, \dots, \mathbf{u}_h)$$

is a K-linear mapping of $I_{h+1}(\mathbf{u}_1, \dots, \mathbf{u}_{h+1})$ into $I_h(\mathbf{u}_1, \dots, \mathbf{u}_h)$. If $N_h(\mathbf{u}_1, \dots, \mathbf{u}_{h+1}) = N_h$ is its kernel and if $\tau_h(\mathbf{u}_1, \dots, \mathbf{u}_{h+1}) = \dim(N_h)$, then for $1 \leqq h \leqq p-1$, we have

$$\dim I_{h+1}(\mathbf{u}_1, \dots, \mathbf{u}_{h+1}) \leqq \tau_h(\mathbf{u}_1, \dots, \mathbf{u}_{h+1}) + \dim I_h(\mathbf{u}_1, \dots, \mathbf{u}_h),$$

with equality only when ρ_{h+1} is *surjective*. Let τ_h denote the least value of

$$\tau_h(\mathbf{u}_1, \dots, \mathbf{u}_{h+1})$$

as the basis $\mathbf{U}$ is allowed to vary arbitrarily. Show that there exists a basis $\mathbf{U}_0 = (\mathbf{u}_{10}, \dots, \mathbf{u}_{p0})$ such that $\tau_h = \tau_h(\mathbf{u}_{10}, \dots, \mathbf{u}_{h0})$ for $h = 1, \dots, p-1$. Moreover, if $\mathbf{u}_1, \dots, \mathbf{u}_h$ are such that $\tau_k = \tau_k(\mathbf{u}_1, \dots, \mathbf{u}_k)$ for $1 \leqq k \leqq h < p-1$, then the set of $\mathbf{u}_{h+1} \in \mathrm{F_P}$ such that $\mathbf{u}_1, \dots, \mathbf{u}_h, \mathbf{u}_{h+1}$ are linearly independent and

$$\tau_{h+1}(\mathbf{u}_1, \dots, \mathbf{u}_h, \mathbf{u}_{h+1}) = \tau_{h+1}$$

is the (nonempty) complement of an *algebraic variety* in the vector space $\mathrm{F_P}$. (Use the fact that K is an infinite field.)

In the notation of Problem 1, show that $s_0'' = r$ and

$$\tau_h = n - p - (s_0'' + s_1'' + \cdots + s_h'').$$

Deduce that the Pfaffian system under consideration is *in involution* relative to P (Problem 1) if and only if

$$\dim \mathrm{V} = \tau_0 + \tau_1 + \cdots + \tau_{p-1},$$

where $\tau_0 = \dim \mathrm{T} = n - p - r$; or equivalently, if and only if all the linear mappings ρ_{h+1} $(1 \leqq h \leqq p-1)$ are *surjective*.

3. With the hypotheses and notation of Problem 2, consider the open set

$$\mathrm{M}^{(1)} = \mathrm{M} \times \mathbf{R}^{p(n-p-r)}.$$

Let x^i $(1 \leqq i \leqq n)$ denote the coordinates of a point of M, and

$$z^{hi} \qquad (1 \leqq i \leqq p,\ 1 \leqq h \leqq n-p-r)$$

the coordinates of a point of $\mathbf{R}^{p(n-p-r)}$. Let $\mathfrak{b}$ be the differential ideal on $\mathrm{M}^{(1)}$ generated by $\mathfrak{b}_0$ and $\mathfrak{b}_1$, where $\mathfrak{b}_0$ is generated by $\mathfrak{a}_0$ and the $\frac{1}{2}rp(p-1)$ functions

$$\mathrm{H}_{ijk} = \sum_h (a_{kih} z^{hj} - a_{kjh} z^{hi}) + c_{ijk},$$

and $\mathfrak{b}_1$ is generated by the 1-forms θ_k $(1 \leqq k \leqq r)$, the $d\mathrm{H}_{ijk}$ and

$$\theta_h^{(1)} = dx^{p+r+h} - \sum_{i=1}^{p} z^{hi} \omega_i \qquad (1 \leqq h \leqq n-p-r).$$

Let $f: \mathrm{P} \to \mathrm{Q}$ be an analytic mapping such that the graph of f is an integral manifold of the Pfaffian system defined by $\mathfrak{a}$. If f^{p+r+h} are the components of f relative to the last $n-p-r$ vectors of the canonical basis of $\mathbf{R}^n$, define scalar-valued functions f^{hi} analytic in a dense open subset $\mathrm{P}^{(1)}$ of P by the formulas

$$df^{p+r+h} = \sum_i f^{hi} \omega_i.$$

Then the graph of the mapping

$$(f, (f^{hi})_{1 \leqq i \leqq p,\ 1 \leqq h \leqq n-p-r})$$

of $\mathrm{P}^{(1)}$ into $\mathrm{Q}^{(1)} = \mathrm{Q} \times \mathbf{R}^{p(n-p-r)}$ is an integral manifold of the differential system defined by $\mathfrak{b}$.

The differential system defined by $\mathfrak{b}$ is called the *first-order prolongation* of the Pfaffian system defined by $\mathfrak{a}$. Replacing $\mathrm{P}^{(1)}$ if necessary by a dense open subset, we may assume also that the set $\mathrm{M}_0^{(1)}$ defined in (18.10.4) for the ideal $\mathfrak{b}$ is an analytic submanifold of $\mathrm{P}^{(1)} \times \mathrm{O}^{(1)}$ and that its projection onto $\mathrm{P}^{(1)}$ is a surjective submersion.

The *m*th *order prolongation* of the system defined by $\mathfrak{a}$ is then defined by induction as the first-order prolongation of the $(m-1)$th order prolongation.

Show that the study of the numbers s_j'' corresponding to the prolongations (of all orders) of the given system reduces to that of the following algebraic situation:

Consider a field K of characteristic 0 (for example, the field of rational functions in infinitely many indeterminates over K_0), three vector spaces F_P, S, T and a linear mapping $\delta : F_P \otimes_K T \to S$. Define F_h, I_h, τ_h, ρ_{h+1}, N_h, and $V = I_p$ as in Problem 2. The quadruple (F_P, S, T, δ) is called a *Cartan quadruple*. Define the *first-order prolongation* $(F_P, S^{(1)}, T^{(1)}, \delta^{(1)})$ to be the Cartan quadruple in which $S^{(1)} = T$, $T^{(1)} = V$, and $\delta^{(1)} : F_P \otimes V \to T$ is defined by $\delta^{(1)}(u \otimes g) = g(u)$. The *m*th order prolongation is then defined by induction to be the Cartan quadruple $(F_P, S^{(m)}, T^{(m)}, \delta^{(m)})$, which is the first-order prolongation of the $(m-1)$th order prolongation. Let $I_h^{(m)}$, $\tau_h^{(m)}$, ρ_{h+1}^m, $N_h^{(m)}$, and $V^{(m)}$ denote the objects defined for the *m*th order prolongation as above for the Cartan quadrupole (F_P, S, T, δ).

4. The hypotheses and notation are as in Problems 2 and 3.

(a) Show that $T^{(m)} = V^{(m-1)}$ may be identified with the vector space of *symmetric* *m*-linear mappings of F_P^m into T such that

$$\delta(\mathbf{u}_0 \otimes g^{(m)}(\mathbf{u}_1, \mathbf{u}_2, \ldots, \mathbf{u}_m)) = \delta(\mathbf{u}_1 \otimes g^{(m)}(\mathbf{u}_0, \mathbf{u}_2, \ldots, \mathbf{u}_m))$$

for all choices of $\mathbf{u}_0, \mathbf{u}_1, \ldots, \mathbf{u}_m \in F_P$. The mapping

$$\delta^{(m)} : F_P \otimes_K T^{(m)} \to S^{(m)} = T^{(m-1)}$$

is defined by

$$(\delta^{(m)}(\mathbf{u} \otimes g^{(m)}))(\mathbf{u}_1, \ldots, \mathbf{u}_{m-1}) = g^{(m)}(\mathbf{u}_1, \ldots, \mathbf{u}_{m-1}, \mathbf{u}).$$

The space $I_h^{(m)}(\mathbf{u}_1, \ldots, \mathbf{u}_h)$ may be identified with the space of $(m+1)$-linear mappings $g_h^{(m+1)}$ of $F_P^m \times F_h(\mathbf{u}_1, \ldots, \mathbf{u}_h)$ into T which satisfy the conditions:

(i) $g_h^{(m+1)}(\mathbf{u}_{\sigma(i_1)}, \ldots, \mathbf{u}_{\sigma(i_{m+1})}) = g_h^{(m+1)}(\mathbf{u}_{i_1}, \ldots, \mathbf{u}_{i_{m+1}})$,

where $i_1, \ldots, i_m$ are any integers between 1 and *p*, and $i_{m+1} \leqq h$, and σ is any permutation of $\{1, 2, \ldots, p\}$ such that $\sigma(i_{m+1}) \leqq h$;

(ii) $\delta(\mathbf{u}_i \otimes g_h^{(m+1)}(\mathbf{u}_{k_1}, \ldots, \mathbf{u}_{k_m}, \mathbf{u}_j)) = \delta(\mathbf{u}_j \otimes g_h^{(m+1)}(\mathbf{u}_{k_1}, \ldots, \mathbf{u}_{k_m}, \mathbf{u}_i))$

for all indices $k_l \leqq p$ and $i \leqq h$, $j \leqq h$.

The kernel $N_h^{(m)}$ of the homomorphism $\rho_{h+1}^{(m)}$ is canonically isomorphic to the space $Q_h^{(m)} \subset T^{(m)}$ of mappings $g^{(m)}$ such that

$$g^{(m)}(\mathbf{v}_1, \mathbf{v}_2, \ldots, \mathbf{v}_m) = 0$$

whenever at least one of the vectors $\mathbf{v}_j$ belongs to $F_h(\mathbf{u}_1, \ldots, \mathbf{u}_h)$. The isomorphism transforms $g_{h+1}^{(m+1)} \in N_h^{(m)}$ into the mapping

$$(\mathbf{v}_1, \ldots, \mathbf{v}_m) \mapsto g_{h+1}^{(m+1)}(\mathbf{v}_1, \ldots, \mathbf{v}_m, \mathbf{u}_{h+1}).$$

Define $Q_0^{(m)}$ to be $T^{(m)}$.

(b) For each $\mathbf{u} \in F_P$, let $\mu^{(m)}(\mathbf{u})$ be the linear mapping

$$g^{(m)} \mapsto \delta^{(m)}(\mathbf{u} \otimes g^{(m)})$$

of $T^{(m)}$ into $T^{(m-1)}$. The restriction $\lambda^{(m)}(\mathbf{u})$ of $\mu^{(m)}(\mathbf{u})$ to $Q_h^{(m)}$ maps $Q_h^{(m)}$ into $Q_h^{(m-1)}$, and the kernel of $\lambda^{(m)}(\mathbf{u}_{h+1})$ is $Q_{h+1}^{(m)}$.

(c) Deduce from (a), (b) and Problem 2 that the quadruple $(F_P, S^{(m)}, T^{(m)}, \delta^{(m)})$ is *involutory* relative to P if and only if $T^{(m+1)}$ is isomorphic to the direct sum of the $Q_h^{(m)}$ (for a suitable choice of the basis **U**). Deduce that an equivalent condition is that each of the mappings $\lambda^{(m+1)}(\mathbf{u}_{h+1})$ $(1 \leqq k \leqq p-1)$ is *surjective*.

5. If $\dim(T) = q$, the space L_m of symmetric m-linear mappings of F_P^m into T may be identified (after choosing bases of F_P and T) with the dual of the subspace $A_{m,1}$ of the space $A = K[Z_1, \ldots, Z_p, Z_1', \ldots, Z_q']$ of polynomials in $p+q$ indeterminates, consisting of the polynomials which are homogeneous of degree m in the Z_i and homogeneous of degree 1 in the Z_j'. In this identification, F_P is identified with the subspace $A_{1,0}$ of A, and $Z_1, \ldots, Z_p$ with the elements $\mathbf{u}_1, \ldots, \mathbf{u}_p$ of the chosen basis of F_P.

(a) The mapping $\delta^{(m)}(\mathbf{u})$ $(m \geqq 1)$ is defined everywhere on L_m. If U is the element of $A_{1,0}$ corresponding to $\mathbf{u}$, show that the transposed mapping ${}^t\delta^{(m)}(\mathbf{u})$ may be identified with the multiplication mapping $W_{m-1,1} \mapsto UW_{m-1,1}$, which is a linear mapping of $A_{m-1,1}$ into $A_{m,1}$. Let B_{mh} be the annihilator of $Q_h^{(m)}$ in $A_{m,1}$; then we have

$$UB_{m-1,h} \subset B_{mh},$$

and in order that $\lambda^{(m)}(\mathbf{u})$ should be surjective it is necessary and sufficient that the relation $UW \in B_{mh}$ for some $W \in L_{m-1}$ should imply $W \in B_{m-1,h}$.
(b) Let $\mathfrak{b}_h$ be the ideal of A generated by the B_{mh} for $m \geqq 1$ and fixed h. Then

$$\mathfrak{b}_h \cap A_{m,1} = B_{mh}.$$

(c) Prove the following algebraic lemma: let $\mathfrak{c}$ be an ideal of A and let

$$\mathfrak{c} = \mathfrak{q}_1 \cap \mathfrak{q}_2 \cap \cdots \cap \mathfrak{q}_r$$

be a reduced primary decomposition of $\mathfrak{c}$,† where each primary ideal $\mathfrak{q}_j$ corresponds to a prime ideal $\mathfrak{p}_j$; also let m be an integer such that $\mathfrak{p}_j^{m-1} \subset \mathfrak{q}_j$ for $1 \leqq j \leqq r$. Then there exists an element $U \in A_{1,0}$ such that the relation $UW \in \mathfrak{c} \cap A_{m,1}$ implies $W \in \mathfrak{c} \cap A_{m-1,1}$. (Consider each index j, and distinguish two cases according to whether $\mathfrak{p}_j \supset A_{1,0}$ or not; in the first case, we have $\mathfrak{q}_j \supset A_{m-1,0}$. Deduce that if U does not belong to any of the intersections $\mathfrak{p}_j \cap A_{1,0} \neq A_{1,0}$, then U has the desired property.)
(d) Deduce from (a) and (c) that, under the hypotheses of Problem 2, there exists an integer m_0 such that the mth order extensions of the given Pfaffian system are *involutory* relative to P for all $m \geqq m_0$ (*Cartan–Kuranishi prolongation theorem*). (Choose the elements $\mathbf{u}_1, \ldots, \mathbf{u}_p$ of the basis of F_P successively so that the $\lambda^{(m)}(\mathbf{u}_{h+1})$ are surjective and the $\tau_h^{(m)}(\mathbf{u}_1, \ldots, \mathbf{u}_h)$ take their minimum values; at each stage we have to choose $\mathbf{u}_{h+1}$ outside a certain algebraic subvariety of F_P, distinct from F_P, by (c) and Problem 2.)

14. COMPLETELY INTEGRABLE PFAFFIAN SYSTEMS

(18.14.1) We shall now take up again the problem of the existence of integral manifolds of a *field of p-directions* (18.8.5), which locally is equivalent to that of the existence of integral manifolds of a *Pfaffian system* (18.8.7.1)

(18.14.1.1) $$\omega_j = 0 \qquad (1 \leqq j \leqq n-p),$$

† See, for example, N. Bourbaki, "Commutative Algebra," Chapter IV, Section 2. Addison-Wesley, Reading, Massachusetts, 1972.

where the ω_j are C^∞ differential 1-forms such that the $\omega_j(x)$ are *linearly independent* at each point $x \in M$. These integral manifolds are also the integral manifolds of the differential system defined by the differential ideal $\mathfrak{a}$ generated by the ω_j (18.9.1).

(18.14.2) In particular, we propose to investigate under what conditions there exists, in a neighborhood of *each* point of M, an integral manifold of *maximum* dimension p passing through the point. When this is the case, the Pfaffian system (18.14.1.1) (or the field of p-directions $x \mapsto L_x$) is said to be *completely integrable.*

We remark first of all that the hypothesis of *linear independence* of the $\omega_j(x)$ at each point implies that $s_0(x) = n - p$ at each $x \in M$, in the notation of (18.10.3). Secondly, if L_x is a p-dimensional integral element, we must have $\langle \mathbf{u}_1 \wedge \mathbf{u}_2, d\omega_j(x)\rangle = 0$ for all vectors $\mathbf{u}_1$, $\mathbf{u}_2$ in L_x, and by definition (18.10.3) this implies that $s_1(x, \mathbf{u}) = 0$ for all vectors $\mathbf{u} \neq 0$ in L_x. It follows that we have also $s_q(x, \mathbf{u}_1, \ldots, \mathbf{u}_q) = 0$ for all sequences of linearly independent vectors $\mathbf{u}_1, \ldots, \mathbf{u}_q$ in L_x, for $2 \leqq q \leqq p - 1$: for each q-form $\varpi_q \in \mathfrak{a}_q$ is a linear combination of exterior products of forms, each of which contains at least one of the ω_j or $d\omega_j$, and the assertion is therefore a consequence of the rules of calculation in exterior algebra (A.14.2.1). Hence the condition imposed on the Pfaffian system (18.14.1.1) already implies that *every* vector subspace L_x is an integral element containing a *regular integral element* of dimension $p - 1$.

We could therefore apply the Cartan–Kähler theorem, provided that the manifold M and the differential forms ω_j are assumed to be *analytic*; but, as we shall see, it is possible to improve on this, and to obtain an existence and uniqueness theorem under the weaker hypotheses that M is merely a differential manifold and the ω_j are *of class* C^∞:

(18.14.3) *In order that the Pfaffian system* (18.14.1.1) *should be completely integrable, it is necessary and sufficient that for each* $x \in M$ *there should exist an open neighborhood* W *of* x *and* C^∞ *differential* 1-*forms* α_{jk} *defined in* W, *such that in* W *we have*

$$d\omega_j = \sum_{k=1}^{n-p} \omega_k \wedge \alpha_{jk} \qquad (1 \leqq j \leqq n - p). \tag{18.14.3.1}$$

When this condition is satisfied, for each $x_0 \in M$ *there exists an open neighborhood* W *of* x_0 *in* M, *a product* $U \times V$ *of a connected open set* $U \subset \mathbf{R}^p$ *and an open set* $V \subset \mathbf{R}^{n-p}$, *and a diffeomorphism* φ *of* $U \times V$ *onto* W *such that, in* $U \times V$, *the p-dimensional connected integral manifolds of the Pfaffian system*

${}^t\varphi(\omega_j) = 0$ $(1 \leqq j \leqq n-p)$ are exactly the submanifolds of the form $\mathrm{T} \times \{y_0\}$, where T is a connected open subset of U, and y_0 a point of V.

Since the question is local, we may assume that M is an open subset of $\mathbf{R}^n$. Linear algebra shows that for each point $x_0 \in \mathrm{M}$ there exists an open neighborhood W_0 of x_0 in M and p indices i_k $(1 \leqq k \leqq p)$ such that the ω_j and the dx^{i_k} form a basis of the $\mathscr{E}_0$-module of C^∞ differential 1-forms on W_0 (A.4.5). By performing a linear transformation on $\mathbf{R}^n$, we may assume that $i_k = k$ $(1 \leqq k \leqq p)$. Hence, for $p+1 \leqq j \leqq n$, we may write

(18.14.3.2)
$$dx^j = \sum_{k=1}^{n-p} \mathrm{A}_{jk}\,\omega_k + \sum_{h=1}^{p} \mathrm{B}_{jh}\,dx^h,$$

where the A_{jk} and B_{jh}, being uniquely determined by Cramer's formulas, are functions of class C^∞ on W_0, and the determinant $\det(\mathrm{A}_{jk})$ is $\neq 0$ in W_0.

On the other hand, the exterior products,

$$\begin{array}{ll} \omega_j \wedge dx^h & (1 \leqq j \leqq n-p,\quad 1 \leqq h \leqq p), \\ \omega_j \wedge \omega_k & (1 \leqq j < k \leqq n-p), \\ dx^h \wedge dx^l & (1 \leqq h < l \leqq p), \end{array}$$

form a basis of the $\mathscr{E}_0$-module of C^∞ differential 2-forms on W_0, and hence we may write

(18.14.3.3)
$$d\omega_j = \sum_{k,h} \mathrm{C}_{jkh}\,\omega_k \wedge dx^h + \sum_{i,k} \mathrm{D}_{jik}\,\omega_i \wedge \omega_k + \sum_{h,l} \mathrm{E}_{jhl}\,dx^h \wedge dx^l \quad (1 \leqq j \leqq n-p),$$

where the coefficients are again uniquely determined by Cramer's formulas, and are therefore C^∞ functions on W_0. For each $x \in \mathrm{V}$, the space L_x is the set of vectors $\mathbf{u}$ such that $\langle \mathbf{u}, \omega_j(x) \rangle = 0$ for $1 \leqq j \leqq n-p$, and therefore there exists a basis $(\mathbf{u}_h(x))_{1 \leqq h \leqq p}$ of L_x such that $\langle \mathbf{u}_h(x), dx^l \rangle = \delta_{hl}$ (Kronecker delta). Expressing that the restriction of $d\omega_j(x)$ to L_x is zero, we obtain the relations $\mathrm{E}_{jhl}(x) = 0$ for $x \in \mathrm{W}_0$, whence (18.14.3.1) follows immediately; the converse is obvious.

We now remark that the projection of L_x onto the subspace of $\mathbf{R}^n$ spanned by the first p vectors of the canonical basis is surjective. Hence (16.8.3.2) if there exists a p-dimensional integral manifold of the system (18.14.1.1) passing through x_0, then it is defined in a neighborhood of x_0 by $n-p$ equations

(18.14.3.4)
$$x^{p+k} = v_k(x^1, \ldots, x^p) \qquad (1 \leqq k \leqq n-p),$$

where the v_k are of class C^∞. Substituting into (18.14.3.2) and bearing in mind that the restrictions of the ω_j to this integral manifold must be zero, we obtain the system of partial differential equations

(18.14.3.5) $$\frac{\partial v_k}{\partial x^h} = B_{kh}(x^1, \dots, x^p, v_1(x^1, \dots, x^p), \dots, v_{n-p}(x^1, \dots, x^p))$$
$$(1 \leqq k \leqq n-p, \quad 1 \leqq h \leqq p)$$

and conversely, if the v_k satisfy this system of equations and are such that

$$v_k(x_0^1, \dots, x_0^p) = x_0^{p+k} \qquad (1 \leqq k \leqq n-p),$$

then it follows from (18.14.3.2) and the nonvanishing of $\det(A_{jk}(x))$ in W_0 that the restrictions of the ω_j to the manifold defined by (18.14.3.4) are zero.

This being so, let us show that the system (18.14.3.5) is *completely integrable in the sense of* (10.9.6). For this it is enough to take the exterior derivatives of the 1-forms in the equations (18.14.3.2); we obtain

(18.14.3.6) $$\sum_{k=1}^{n-p} (dA_{jk} \wedge \omega_k + A_{jk} \wedge d\omega_k) + \sum_{h=1}^{p} dB_{jh} \wedge dx^h = 0.$$

Replace dB_{jh} in this relation by

$$\sum_{l=1}^{n} \frac{\partial B_{jh}}{\partial x^l} dx^l,$$

and then each of $dx^{p+1}, \dots, dx^n$ by its expression given by (18.14.3.2). Expressing that the coefficient of $dx^h \wedge dx^l$ $(1 \leqq h < l \leqq p)$ in the 2-form so obtained is zero, we arrive at the conditions

(18.14.3.7) $$\frac{\partial B_{jh}}{\partial x^l} + \sum_{s=1}^{p} \frac{\partial B_{jh}}{\partial x^s} B_{sl} = \frac{\partial B_{jl}}{\partial x^h} + \sum_{s=1}^{p} \frac{\partial B_{jl}}{\partial x^s} B_{sh}$$
$$(1 \leqq j \leqq n-p, \quad 1 \leqq h < l \leqq p),$$

which are precisely the conditions of complete integrability (10.9.6).

We may therefore apply the result of (10.9.5). Identifying $\mathbf{R}^n$ with

$$\mathbf{R}^p \times \mathbf{R}^{n-p}$$

and putting $x_0 = (y_0, z_0)$, there exists an open ball U with center y_0 in $\mathbf{R}^p$, two open balls $V \subset V_0$ with center z_0 in $\mathbf{R}^{n-p}$ such that $U \times V_0 \subset M$, and a C^∞ mapping $(y, y_1, z) \mapsto (y, u(y, y_1, z))$ of $U \times U \times V_0$ into M such that:

(i) for all $(y_1, z) \in U \times V_0$, the function

$$y \mapsto u(y, y_1, z) = (v_1(y), \dots, v_{n-p}(y))$$

is the unique solution of (18.14.3.5) in U such that $u(y_1, y_1, z) = z$;

(ii) for each $(y, y_1, z) \in U \times U \times V$, we have $u(y, y_1, z) \in V_0$, and $u(y, y_1, z)$ is the unique point $z_1 \in V_0$ such that

$$u(y_1, y, z_1) = z.$$

It follows that

$$\varphi : (y, z) \mapsto (y, y(u, y_0, z))$$

is a *diffeomorphism* of $U \times V$ onto an open subset of $U \times V_0$, and the inverse diffeomorphism is $(y, z) \mapsto (y, u(y_0, y, z))$ on $\varphi(U \times V) = W$. The connected integral manifolds of the Pfaffian system ${}^t\varphi(\omega_j) = 0$ in $U \times V$ are obviously inverse images under φ of the connected integral manifolds of (18.14.1.1) contained in $\varphi(U \times V)$; in a neighborhood of each point of $U \times V$ they are therefore of the form $y \mapsto u(y_0, y, u(y, y_0, z)) = z$, hence the v_j are *constants*. The system of partial differential equations (18.14.3.5) corresponding to this Pfaffian system is therefore such that the right-hand sides of the equations *vanish identically* in $U \times V$, and the result therefore follows from (8.6.1). We may remark that the only connected integral manifolds which are *closed in* $U \times V$ are the sets $U \times \{z_0\}$.

(18.14.4) We have seen (18.8.1) that a C^∞ field of p-directions $x \mapsto L_x$ on M may be defined in a neighborhood U of any point of M by giving p vector fields X_j of class C^∞ such that, for each point $x \in U$, the vectors $X_j(x)$ form a basis of the vector space L_x. The condition of complete integrability of the field of p-directions $x \mapsto L_x$ may then be expressed as follows:

(18.14.5) *In order that the field of p-directions $x \mapsto L_x$ should be completely integrable in* U, *it is necessary and sufficient that, for each pair of indices i, j such that $1 \leqq i < j \leqq p$, we have*

(18.14.5.1) $$[X_i, X_j] = \sum_{k=1}^{p} c_{ijk} X_k,$$

where the c_{ijk} are C^∞ functions on U.

Suppose first that $x \mapsto L_x$ is completely integrable. In the notation of (18.14.3) we have, for $h = 1, \ldots, n-p$ (17.15.3.6),

$$\langle d\omega_h, X_i \wedge X_j \rangle = \theta_{X_i} \cdot \langle \omega_h, X_j \rangle - \theta_{X_j} \cdot \langle \omega_h, X_i \rangle - \langle \omega_h, [X_i, X_j] \rangle$$

and by definition $\langle \omega_h, X_i \rangle = 0$ for each index i. Since for each $x \in U$ the 2-covector $d\omega_h(x)$ is zero on restriction to L_x, we obtain $\langle \omega_h(x), [X_i, X_j](x) \rangle = 0$ for $1 \leqq h \leqq n-p$, whence we conclude that the vector $[X_i, X_j](x)$

belongs to L_x and is therefore uniquely expressible in the form

$$[X_i, X_j](x) = \sum_{k=1}^{p} c_{ijk}(x) X_k(x);$$

and the fact that the functions c_{ijk} are of class C^∞ follows from Cramer's formulas. The argument may clearly be reversed to show that, if the relations (18.14.5.1) are satisfied, then the restriction of $d\omega_h(x)$ to L_x is zero for each index h, and then the proof of (18.14.3) shows that the field $x \mapsto L_x$ is completely integrable in U.

(18.14.6) It is trivial that, for $p = 1$, every field of 1-directions is completely integrable, and we have seen that locally the notion of an integral manifold of such a field coincides with the notion of an integral curve of a vector field (18.8.6). We shall now show that, for $p > 1$, we can define, for a *completely integrable* field of p-directions on M, a notion which generalizes that of a maximal integral curve (18.2.2).

For this purpose we shall define a new *topology* $\mathscr{T}$ on the manifold M which is finer than the given topology $\mathscr{T}_0$. Let $\mathfrak{B}$ denote the set of *integral manifolds* N of dimension p of the field $x \mapsto L_x$ which are *connected submanifolds* of M, and are such that the closure $\bar{N}$ of N in M is *compact*. We assert that $\mathfrak{B}$ is a *basis* (12.2) of a topology $\mathscr{T}$ on M. Let N and P be two sets belonging to $\mathfrak{B}$, and let $x_0 \in N \cap P$. There exists a fundamental system of open neighborhoods U of x_0 in M such that in U there is only one connected closed submanifold of dimension p which is an integral manifold of the field $x \mapsto L_x$ and contains x_0 (18.14.3), and this manifold is the largest connected submanifold of dimension p in U which is an integral manifold. Moreover, we may suppose that U is sufficiently small so that $U \cap N$ is closed in U and connected; then the connected component of x_0 in $U \cap P$ is an integral manifold of dimension p and contains x_0, hence is contained in $U \cap N$, and therefore in $N \cap P$, which proves the assertion. This proof shows also that $\mathscr{T}$ is finer than the topology $\mathscr{T}_0$, and that $\mathscr{T}$ and $\mathscr{T}_0$ both induce the *same* topology *on* N. Furthermore, it shows also that if $\mathfrak{B}'$ is the set of $N' \in \mathfrak{B}$ such that $\bar{N}' \subset N$ for some $N \in \mathfrak{B}$, then $\mathfrak{B}'$ is another basis for the topology $\mathscr{T}$, and the topology induced on $\bar{N}'$ by $\mathscr{T}$ is *metrizable* and *compact*. Since a manifold is locally connected the space M endowed with the topology $\mathscr{T}$ is locally connected. We shall see that a connected component C of M (for the topology $\mathscr{T}$), which is therefore open in M (for the topology $\mathscr{T}$) (3.19.5) is a *denumerable* union of sets belonging to $\mathfrak{B}'$.

For this we remark that, by virtue of (18.14.3) and (12.6.1), there exists an at most denumerable locally finite open covering (W_α) of M consisting of relatively compact sets of the form $\varphi_\alpha(U_\alpha \times V_\alpha)$, where U_α is a connected open set in $\mathbf{R}^p$, V_α is open in $\mathbf{R}^{n-p}$ and φ_α is a diffeomorphism such that the only connected p-dimensional integral manifolds of (18.14.1.1) which are closed in W_α are of the form $\varphi_\alpha(U_\alpha \times \{z_\alpha\})$. By restricting the U_α, we may assume that these integral manifolds belong to $\mathfrak{B}'$. Now consider the relation $R(x, x')$ on M defined as follows:

"There exists a finite sequence $(\alpha_k)_{0 \leqq k \leqq m}$ of indices and a finite sequence of integral manifolds $N_{\alpha_k} \subset W_{\alpha_k}$ of dimension p, connected and closed in W_{α_k}, such that $x \in N_{\alpha_0}$, $x' \in N_{\alpha_m}$, and

$$N_{\alpha_k} \cap N_{\alpha_{k+1}} \neq \emptyset \qquad \text{for} \quad 0 \leqq k \leqq m-1."$$

Since a point of W_α belongs to a unique connected closed p-dimensional integral manifold in W_α, it is clear that $R(x, x')$ is an *equivalence relation* on M and that the equivalence classes are *open and closed* sets for the topology $\mathscr{T}$, and are *connected* (3.19.4). Hence they are precisely the connected components of M for the topology $\mathscr{T}$.

Let C be a connected component and let $x \in C$. Define inductively a sequence (C_m) of open sets (for the topology $\mathscr{T}$) as follows: C_1 is the union of the integral manifolds $N_\alpha \subset W_\alpha$ of dimension p, connected and closed in W_α, which contain x; and for $m > 1$, C_m is the union of C_{m-1} and the integral manifolds $N_\alpha \subset W_\alpha$ of dimension p, connected and closed in W_α, which meet C_{m-1}. By induction on m, it is immediately seen that each C_m is the union of a *finite* number of sets of $\mathfrak{B}'$, because it is relatively compact and therefore intersects at most finitely many of the open sets W_α. Now C is the union of the C_m; for if $x' \in C$, then (with the preceding notation) we have $N_{\alpha_0} \subset C_1$ and (by induction) $N_{\alpha_k} \subset C_{k+1}$.

We can now apply (12.4.7), since each space $\bar{N}_\alpha$ is metrizable and separable for the topology $\mathscr{T}$; it follows therefore that C is *metrizable* and *separable* for the topology $\mathscr{T}$. Moreover, if N and P belong to $\mathfrak{B}$, then the submanifold structures on N and P induce the same manifold structure on $N \cap P$; hence (16.2.5) we can define on each connected component C of M for the topology $\mathscr{T}$ a structure of a *differential manifold* of dimension p. The canonical injection $j : C \to M$ is clearly an *immersion*, but C *is not necessarily a submanifold of* M (18.2.13). The integral manifold of the Pfaffian system (18.14.1.1) formed by C and the immersion j (18.8.5) is *maximal* in the following sense: if a connected manifold N *of dimension p* and an injective immersion $h : N \to M$ constitute an integral manifold (18.8.5), the fact that h is an immersion implies immediately that h is continuous relative to the topology $\mathscr{T}$ on M; hence, as N is connected, $h(N)$ is *contained* in a connected component C of M (for the topology $\mathscr{T}$) and the mapping h factorizes as $h : N \xrightarrow{u} C \xrightarrow{j} M$, where u is a *diffeomorphism* of N onto an open subset of C (16.8.8).

Furthermore, there is the following result:

(18.14.7) *Let* P *be a differential manifold,* $f: \mathrm{P} \to \mathrm{M}$ *a* C^∞ *mapping* (M being endowed with its original structure of differential manifold, with the topology $\mathscr{T}_0$), *and suppose that* $f(\mathrm{P}) \subset \mathrm{C}$, *where* C *is a maximal integral submanifold of the Pfaffian system* (18.14.1.1). *Then f is still a* C^∞ *mapping of* P *into* C, *when* C *is endowed with the structure of differential manifold defined in* (18.14.6), *with the topology* $\mathscr{T}$.

The question is local with respect to P. Let $z_0 \in \mathrm{P}$ and let $x_0 = f(z_0) \in \mathrm{C}$. There exists a chart $c = (\mathrm{U}, \varphi, n)$ of M at the point x_0 such that U is relatively compact in M; $\varphi(x_0)$ is the origin in $\mathbf{R}^n$; $\varphi(\mathrm{U}) = \mathrm{V} \times \mathrm{W}$, where V (resp. W) is a connected open neighborhood of 0 in $\mathbf{R}^p$ (resp. $\mathbf{R}^{n-p}$); and such that every p-dimensional connected integral submanifold of the system (18.14.1.1) contained in U is of the form $\varphi^{-1}(\mathrm{Y} \times \{w\})$, where Y is a connected open subset of V, and w a point of W (18.14.3). By virtue of the definition of the structure of differential manifold of C, it is enough to show that there exists a neighborhood Z of z_0 in P such that $\varphi(f(\mathrm{Z})) \subset \mathrm{V} \times \{0\}$ (16.8.3.4). For this, we remark that $\mathrm{U} \cap \mathrm{C}$ is open in C for the topology $\mathscr{T}$, hence the set of its connected components is *at most denumerable*; these components are open (for $\mathscr{T}$) (3.19.5) and pairwise disjoint. Since every $\mathscr{T}$-open subset of $\mathrm{C} \cap \mathrm{U}$ is an integral submanifold of the system (18.14.1.1), it follows that the connected components of $\mathrm{U} \cap \mathrm{C}$ are of the form $\varphi^{-1}(\mathrm{Y}_m \times \{w_m\})$, where Y_m is open in V and $w_m \in \mathrm{W}$, and the integer m runs through an at most denumerable set. Let $c' = (\mathrm{U}', \psi, p)$ be a chart of P at the point z_0, such that $f(\mathrm{U}') \subset \mathrm{U}$ and such that $\psi(\mathrm{U}')$ is a ball with center $0 = \psi(z_0)$ in $\mathbf{R}^p$. Suppose that there exists a point $z \in \mathrm{U}'$ such that $\varphi(f(z)) \notin \mathrm{V} \times \{0\}$. We shall show that this will produce a contradiction. Since $f(z) \in \mathrm{U} \cap \mathrm{C}$ by hypothesis, we have

$$\varphi(f(z)) \in \mathrm{Y}_m \times \{w_m\}$$

for some m such that $w_m \neq 0$. Now we have $\psi(z) \neq 0$ in $\mathbf{R}^p$: consider then the composite mapping $t \mapsto \varphi(f(\psi^{-1}(t\psi(z)))) = u(t)$ of the interval [0, 1] into $\varphi(\mathrm{U})$. By our choice of the chart U′, we have $\psi^{-1}(t\psi(z)) \in \mathrm{U}'$, hence

$$u(t) \in \varphi(\mathrm{U} \cap \mathrm{C})$$

for $0 \leqq t \leqq 1$, and u is clearly continuous. Now let $j \in [p+1, n]$ be an index such that $\mathrm{pr}_j(w_m) \neq 0$; then the continuous mapping $t \mapsto \mathrm{pr}_j(u(t))$ of [0, 1] into $\mathbf{R}$ takes the value 0 at $t = 0$ and the value $\mathrm{pr}_j(w_m) \neq 0$ at $t = 1$; hence its image is an *interval* in $\mathbf{R}$ not reduced to a point ((3.19.1) and (3.19.7)). But for $0 \leqq t \leqq 1$ the function $t \mapsto \mathrm{pr}_j(u(t))$ takes only the values $\mathrm{pr}_j(w_k)$, which form an *at most denumerable* set: a contradiction by virtue of (2.2.17) and (4.1.7).

Remark

(18.14.8) Consider a Pfaffian system (18.14.1.1), *not necessarily completely integrable*. If two integral manifolds N_1, N_2 *of dimension p* of this system have a point x in common, then there exists a neighborhood U of x in M such that $N_1 \cap U = N_2 \cap U$. For we may reduce to the case where the Pfaffian system is equivalent to the system of partial differential equations (18.14.3.5), and then the assertion follows from the uniqueness theorem for differential equations, by the same reasoning which established the sufficiency of the condition of complete integrability in (10.9.4).

PROBLEMS

1. Consider a Pfaffian system (18.14.1.1) in which the $\omega_j(x)$ are linearly independent at each point $x \in M$. Show that this system is completely integrable if and only if

$$\omega_1 \wedge \omega_2 \wedge \cdots \wedge \omega_{n-p} \wedge d\omega_j = 0$$

for each index j.

2. Consider a Pfaffian system $\omega_i = 0$ $(1 \leq i \leq n)$ in an open subset U of $\mathbf{R}^n$, where $\omega_i = \sum_{j=1}^{n} a_{ij}\,dx^j$, the a_{ij} being functions of class C^∞ satisfying the following conditions: (i) $a_{ij} + a_{ji} = 0$; (ii) $D_k a_{ij} + D_i a_{jk} + D_j a_{ki} = 0$ for all i, j, k; (iii) at each point $x \in U$, the rank of the system of covectors $\omega_i(x)$ is p (independent of x). Show that the Pfaffian system is completely integrable. (We may restrict ourselves to the case $p < n$. Let $(u_i)_{1 \leq i \leq n}$, $(v_i)_{1 \leq i \leq n}$ be two systems of C^∞ functions satisfying the relations $\sum_j a_{ij} u_j = 0$, $\sum_j a_{ij} v_j = 0$ in U. Show that

$$\sum_{j,k} \left(\frac{\partial a_{ij}}{\partial x^k} - \frac{\partial a_{ik}}{\partial x^j} \right) u_j v_k = 0 \qquad (1 \leq i \leq n).)$$

3. Consider a completely integrable Pfaffian system (18.14.1.1) such that the $\omega_j(x)$ are linearly independent at each point x, and suppose that the partial system formed by the first $n - r$ $(< n - p)$ equations $\omega_j = 0$ $(1 \leq j \leq n - r)$ is also completely integrable. Show that in the statement of (18.14.3) we may require that $V = V_1 \times V_2$, where V_1 is open in $\mathbf{R}^{r-p}$ and V_2 is open in $\mathbf{R}^{n-r}$, such that the connected r-dimensional integral manifolds of the partial Pfaffian system ${}^t\varphi(\omega_j) = 0$ $(1 \leq j \leq n - r)$ are the manifolds of the form $S \times \{z_0\}$, where S is a connected open subset of $U \times V_1$, and z_0 is a point of V_2.

4. Let $x \mapsto L_x$ be a completely integrable field of p-directions on a connected differential manifold M of dimension n. For brevity's sake a maximal integral manifold (18.14.6) of this system will be called a *leaf*. A *distinguished open set* in M is by definition a set of the form $\varphi(\frac{1}{2}I^n)$, where $I =]-1, 1[$ and φ is a diffeomorphism of I^n onto an open

set in M, such that the only closed connected p-dimensional integral manifolds in $\varphi(I^n)$ are of the form $\varphi(I^p \times \{z\})$ for $z \in I^{n-p}$. The manifolds $\varphi(\frac{1}{2}I^p \times \{z\})$, where $z \in \frac{1}{2}I^{n-p}$, will be called the *plaques* contained in the distinguished open set $\varphi(\frac{1}{2}I^n)$. An open set contained in a distinguished open set $U = \varphi(\frac{1}{2}I^n)$ is said to be *saturated* for U if it contains the plaque (in U) of each of its points, i.e., if it is of the form $\varphi(\frac{1}{2}I^p \times V)$, where V is open in $\frac{1}{2}I^{n-p}$.

(a) Let U be a distinguished open set, $P \subset U$ a plaque of U and let W be an open subset of M containing P. For each point $x \in P$, there exists a distinguished open set $U' \subset U \cap W$ containing x and such that $U' \cap P$ is a plaque of U'. If moreover $U \cap W$ is saturated for U, then U' may be chosen to be saturated for U.
(b) Let U, U' be two distinguished open sets such that $U \subset U'$. If P is a plaque of U, then there exists a distinguished open set $V \subset U$ which is saturated for U, contains P and is such that for each plaque P' of U', the set $P' \cap V$ is either empty or is a plaque of V.
(c) Let U, U' be two distinguished open sets, such that $U \cup U'$ is contained in a distinguished open set U''. If P (resp. P') is a plaque of U (resp. U') such that $P \cap P' \neq \varnothing$, then there exists a distinguished open set $V' \subset U'$, saturated for U', such that P' is the only plaque of V' which meets P. (Use (b).)

5. Let X be a separable metrizable topological space, (A_n) an at most denumerable locally finite open covering of X, and (B_n) an open covering of X such that $\bar{B}_n \subset A_n$ for all n (12.6.2). For each pair of integers p, q and each point $x \in A_p \cap A_q$, let $V_{pq}(x)$ be a neighborhood of x contained in $A_p \cap A_q$. For each point $x \in X$ there exist only finitely many pairs (p, q) such that $x \in A_p \cap A_q$; let $V(x)$ be a neighborhood of x contained in all the $V_{pq}(x)$, in all the B_n containing x and in all the $\complement B_n$ containing x. Show that, if $V(x)$ and $V(y)$ intersect, then there exists an integer n such that $V(x) \cup V(y) \subset A_n$.

6. The hypotheses and notation are as in Problem 4.

(a) Let C be a leaf and x, y two points of C. Show that there exists a finite sequence $(U_i)_{1 \leq i \leq n}$ of distinguished open sets such that for each i there exists a plaque P_i of U_i contained in C, such that $x \in P_1$, $y \in P_n$ and $P_i \cap P_{i+1} \neq \varnothing$ for $1 \leq i \leq n-1$. We may also require that $U_i \cup U_{i+1}$ is contained in a distinguished open set (Problem 5). Show that for each distinguished open neighborhood W of y in M, there exists a distinguished open neighborhood V of x in M such that, for each $z \in V$, the leaf C' containing z meets W; furthermore, if P is a plaque of V contained in C and such that $x \notin P$, then there exists a plaque Q of W contained in C and such that $y \notin Q$. (Use Problem 4(c).)
(b) Deduce from (a) that if a point $x \in M$ lies in the closure of the union of a family (C_α) of leaves, then the same is true of every point y of the leaf C containing x. This leaf C is said to be *adherent* to $\bigcup_\alpha C_\alpha$.

(c) Show that the union of the leaves which intersect an open set in M is open in M. (Use (a).)

7. The hypotheses and notation are as in Problem 4.

(a) Let C be a leaf for which there exists a distinguished open set U such that $C \cap U$ is dense in U. If S is the (open) union of the leaves which meet U, then $C \cap S$ is dense in S. The leaf C is said to be *locally dense*.

(b) Show that, if C is contained in the closure of a leaf C′ and is locally dense, then C′ is locally dense and $\bar{C} = \bar{C}'$.
(c) If M is connected and if all the leaves are locally dense, then all the leaves are dense in M.

8. The hypotheses and notation are as in Problem 4.

(a) A leaf C is said to be *proper* if C is a submanifold of M. Show that C is proper if and only if there exists a distinguished open set U such that $U \cap C$ is a plaque of U.
(b) Let C be a proper leaf and (C'_α) a family of leaves in the closure of C. Show that if C is adherent to the union of the C'_α, then $C'_\alpha = C$ for all α. (Use (a).)
(c) Let C be a proper leaf. Show that the union of the leaves $\neq C$ that lie in the closure of C is *closed* in M.
(d) Suppose that all the leaves are proper and locally dense (Problem 7). Show that if a leaf C is relatively compact in M, then its closure $\bar{C}$ contains a compact leaf. (Let $C(x)$ be the leaf containing a point $x \in M$. Follow the reasoning of Section 12.10, Problem 6, by considering for a point $y \in \bar{C}$ the least upper bound $\lambda(y)$ of the distances from $t \in \overline{C(y)}$ to the closure of a leaf $\neq C(t)$ and contained in $C(t)$, by using (c).)

9. In order that a relatively compact leaf C should be closed (and therefore compact), it is necessary and sufficient that, for each distinguished open set U, the intersection $C \cap U$ should be the union of a finite number of plaques of U. (Use the fact that if $U = \varphi(\frac{1}{2}I^n)$ (Problem 4), then $\bar{U} \subset \varphi(I^n)$ and the integral manifolds in $\varphi(I^n)$ are known.)

10. Consider the manifold $\mathbf{R}^{n-1} \times \mathbf{T}^2$, whose universal covering is $\mathbf{R}^{n-1} \times \mathbf{R}^2$. Let ξ^j $(1 \leq j \leq n-1)$ denote the coordinates in $\mathbf{R}^{n-1}$ and (θ, φ) the coordinates in $\mathbf{R}^2$. Consider the two differential forms on $\mathbf{R}^{n-1} \times \mathbf{R}^2$

$$\alpha_1 = d\theta, \qquad \alpha_2 = ((1 - \sin\theta)^2 + (\xi^1)^2)\, d\varphi + \sin\theta\, d\xi^1.$$

These forms are the inverse images (under the canonical projection) of differential forms β_1, β_2 on $\mathbf{R}^{n-1} \times \mathbf{T}^2$. Let M denote the n-dimensional submanifold $\mathbf{S}_{n-2} \times \mathbf{T}^2$ of $\mathbf{R}^{n-1} \times \mathbf{T}^2$, and let ω_1, ω_2 denote the forms induced on M by β_1, β_2. Show that the Pfaffian system $\omega_1 = 0$, $\omega_2 = 0$ on M is completely integrable. For this system, every leaf is proper (Problem 8), there is one compact leaf diffeomorphic to $\mathbf{S}_1 \times \mathbf{S}_{n-3}$, and infinitely many compact leaves diffeomorphic to $\mathbf{S}_{n-2}$.

11. Let M be the 3-dimensional sphere in $\mathbf{R}^4$ defined by the equation $\sum_{j=1}^{4} (\xi^j)^2 = 2$. Let $f: \mathbf{R} \to \mathbf{R}$ be a C^∞ function such that $f(t) = 0$ for $t \leq 1$ and $f(t) > 0$ for $t > 1$ (16.4.1). Consider the differential form ω on M induced by the differential form on $\mathbf{R}^4$,

$$\xi^1\, d\xi^1 + \xi^2\, d\xi^2 + f((\xi^1)^2 + (\xi^2)^2 - 1)(\xi^1\, d\xi^2 - \xi^2\, d\xi^1)$$
$$- f((\xi^3)^2 + (\xi^4)^2 - 1)(\xi^3\, d\xi^4 - \xi^4\, d\xi^3),$$

of class C^∞. Show that the Pfaffian system $\omega = 0$ on M is completely integrable; all the leaves are proper and exactly one (contained in the closure of each of the others) is compact and diffeomorphic to $\mathbf{T}^2$ (*Reeb foliation*).

12. Let M be a differential manifold which is the product of a compact manifold S of dimension n and a Euclidean space $\mathbf{R}^m$; endow S with a distance compatible with its topology, and $\mathbf{R}^m$ with the Euclidean distance $\|z - z'\|$, and M with the distance derived from these by the procedure of (3.20). Let $x \mapsto L_x$ be a completely integrable C^∞ field of n-directions on M, such that $S \times \{0\}$ is an *integral manifold* of the field.

(a) Let $\gamma : [0, 1] \to S$ be a path in S, with origin a and endpoint b. Show that there exists a neighborhood B of 0 in $\mathbf{R}^m$ and a unique continuous function

$$(t, \mathbf{z}) \mapsto (\gamma(t), \mathbf{f}(t, \mathbf{z}))$$

from $[0, 1] \times B$ to M such that, for each $\mathbf{z} \in B$, the path $t \mapsto (\gamma(t), \mathbf{f}(t, \mathbf{z}))$ defined on $[0, 1]$ is contained in the leaf (Problem 4) passing through the point $(a, \mathbf{z})$. (Cover S by a finite number of open sets which are domains of charts, such that for each of these open sets U there exists a neighborhood V of 0 in $\mathbf{R}^m$ such that, for each point $(y, \mathbf{z}) \in U \times V$, the connected component of the intersection of $U \times \mathbf{R}^m$ and the leaf passing through $(y, \mathbf{z})$ project bijectively onto U. For this, use the integration method of (10.9.4) and the majoration of (10.5.6).) Show that $\mathbf{z} \mapsto \mathbf{f}(1, \mathbf{z})$ is of class C^∞ on B. (Use (10.7.3).)
(b) Let $\varphi : [0, 1] \times [\alpha, \beta] \to S$ be a homotopy of the path $\gamma = \varphi(., \alpha)$ onto $\gamma_1 = \varphi(., \beta)$ leaving the endpoints a, b fixed. For each $\xi \in [\alpha, \beta]$ there exists a neighborhood B_ξ of 0 in $\mathbf{R}^m$ and a function $(t, \mathbf{z}) \mapsto (\varphi(t, \xi), \mathbf{f}(t, \mathbf{z}, \xi))$ defined on $[0, 1] \times B_\xi$, having the properties stated in part (a). Show that we may assume that there exists a neighborhood W of 0 in $\mathbf{R}^m$ contained in all the B_ξ and such that $\mathbf{f}(1, \mathbf{z}, \xi)$ is independent of ξ for $\mathbf{z} \in W$. (Follow the proof of Cauchy's theorem (9.6.3).)
(c) Suppose that S is *simply-connected*. Show by using (b) that for each multi-index $\nu = (\nu_1, \dots, \nu_m) \in \mathbf{N}^m$ there exists a C^∞ function $\mathbf{g}_\nu$ defined on S with values in $\mathbf{R}^m$, such that at each point $b \in S$ the vector $\mathbf{g}_\nu(b)$ is the derivative of order ν of the function $\mathbf{z} \mapsto \mathbf{f}(1, \mathbf{z})$ (defined in (a) above) at the point 0.
(d) Suppose moreover that S is a *real-analytic manifold*. Then, with the notation of (a), the function $\mathbf{z} \mapsto \mathbf{f}(1, \mathbf{z})$ is analytic in a neighborhood of 0, and the functions $\mathbf{g}_\nu$ defined in (c) are analytic in S. Deduce that there exists a neighborhood V of 0 in $\mathbf{R}^m$ such that for each $\mathbf{z} \in V$ the leaf $F_\mathbf{z}$ through the point $(a, \mathbf{z})$ is compact and projects bijectively onto S. (Observe that there exists a neighborhood of 0 in $\mathbf{R}^m$ in which the Taylor series

$$\sum_\nu \frac{1}{\nu!} \mathbf{g}_\nu(b)\mathbf{z}^\nu$$

converge, for all $b \in S$.)
(e) Extend the result of (d) to the situation where S is a compact real-analytic manifold with *finite* fundamental group. (Consider the universal covering S′ of S, which is compact, and lift the field of n-directions $x \mapsto L_x$ to $S' \times \mathbf{R}^m$.) Compare with Problem 10.

13. (a) Let $X_1, \dots, X_p$ be C^∞ vector fields on a differential manifold M, and suppose that at some point $x_0 \in M$ the p tangent vectors $X_1(x_0), \dots, X_p(x_0)$ are linearly independent. With the notation of (18.2.2), show that there exists an open neighborhood U of 0 in $\mathbf{R}^p$ such that the mapping

$$(*) \qquad (t_1, \dots, t_p) \mapsto F_{t_1 X + \dots + t_p X_p}(x_0, 1)$$

is defined and of class C^∞ in U, and is an *embedding* of U in M. (Reduce to the case $M = \mathbf{R}^n$ and use (10.7.4).)
(b) With the same notation, let N be the image of U under the mapping (*). Show that if the vector fields $[X_i, X_j]$ $(1 \leq i < j \leq p)$ vanish *at all points of* N, then the fields

$X_1, \ldots, X_p$ are tangent to N at the points of N. (If X, Y are two linear combinations of the X_i with constant coefficients, then for all sufficiently small real numbers s, t, the point $u(s, t) = \mathrm{F}_{t(sX+Y)}(x_0, 1)$ belongs to N. Reduce to the case $\mathrm{M} = \mathbf{R}^n$ and form the differential equation satisfied by the function $t \mapsto u_s'(0, t)$ (in the notation of (18.7.1), i.e., the tangent vector at the point $s = 0$ to the curve $s \mapsto u(s, t)$), using (10.7.3).)

(c) Let M be a *real-analytic* manifold, $\mathscr{F}(\mathrm{M})$ the ring of real-analytic functions on M, and let $\mathfrak{L}$ be a submodule of the $\mathscr{F}(\mathrm{M})$-module of *analytic* vector fields on M. For each $x \in \mathrm{M}$, let $\nu(x)$ denote the dimension of the subspace of $\mathrm{T}_x(\mathrm{M})$ spanned by the tangent vectors $X(x)$ with $X \in \mathfrak{L}$. We shall not assume that $\nu(x)$ is constant. We define an *integral manifold* of $\mathfrak{L}$ to be an analytic manifold N together with an analytic immersion $j : \mathrm{N} \to \mathrm{M}$ such that $\nu(j(z)) = \dim_z \mathrm{N}$ for all $z \in \mathrm{N}$ and such that for each $X \in \mathfrak{L}$ the vector $X(j(z))$ belongs to the image of $\mathrm{T}_z(\mathrm{N})$ under $\mathrm{T}_z(j)$.

We shall assume that $\mathfrak{L}$ is a *Lie algebra* (in general infinite-dimensional over $\mathbf{R}$). Let $x_0 \in \mathrm{M}$ and put $p = \nu(x_0)$, so that there exist p vector fields $X_1, \ldots, X_p \in \mathfrak{L}$ such that the vectors $X_i(x_0)$ $(1 \leq i \leq p)$ are linearly independent. Also let $\mathfrak{L}_0$ be the set of all $X \in \mathfrak{L}$ such that $X(x_0) = 0$. Show that $\mathfrak{L}_0$ is a Lie subalgebra of $\mathfrak{L}$, and that the X_i can be chosen so that $[X_i, X_j] \in \mathfrak{L}_0$ for all pairs of indices i, j. With this choice of $X_1, \ldots, X_p$, form the p-dimensional submanifold N of M which is the image of U under the mapping (*) in the notation of part (a). Show that the vector fields belonging to $\mathfrak{L}_0$ vanish at each point of N (use Problem 4 of Section 18.2). Deduce, with the help of (b), that N is an *integral manifold* of $\mathfrak{L}$.

(d) With the hypotheses of (c), show that *maximal integral manifolds* of $\mathfrak{L}$ can be defined as in (18.14.6), and that they form a partition of M.

14. Let f be a C^∞ function on $\mathbf{R}$ such that f and all its derivatives vanish at the point 0, and such that $f(x) \neq 0$ for all $x \neq 0$ (16.4.1). On $\mathbf{R}^2$, consider the Lie algebra $\mathfrak{L}$ over the ring $\mathscr{E}(\mathbf{R}^2)$ of C^∞ functions, generated by the two vector fields

$$(t_1, t_2) \mapsto E_1(t_1, t_2) \quad \text{and} \quad (t_1, t_2) \mapsto f(t_1)E_2(t_1, t_2)$$

(in the notation of (18.7.1)). Show that for this Lie algebra the conclusion of Problem 13(c) does not hold.

15. SINGULAR INTEGRAL MANIFOLDS; CHARACTERISTIC MANIFOLDS

(18.15.1) Given an integral manifold N of a differential system on a differential manifold M (18.9.2), the tangent space $\mathrm{T}_x(\mathrm{N})$ to N at a point x may be a *regular* integral element, in which case the same is true of the tangent spaces to N at all points of some neighborhood of x in N (18.10.4); or it may be *singular* (18.10.4), but such that every neighborhood of x in N contains points at which the tangent space is regular; or finally it may happen that the tangent space $\mathrm{T}_{x'}(\mathrm{N})$ at *every* point x' of some neighborhood of x in N is a singular integral element. As far as the local theory is concerned we may assume, in this case, that $\mathrm{T}_x(\mathrm{N})$ is singular for *all* points $x \in \mathrm{N}$, and then N is said to be a *singular integral manifold.*

Example: Singular Integrals of a First-Order Scalar Differential Equation

(18.15.2) If we take $q = n = 1$ in the general definition (18.8.9.1) of a system of first-order partial differential equations, we have a *first-order scalar differential equation*

$$F(x, z, p) = 0 \tag{18.15.2.1}$$

(where F is a C^∞ function on an open set $U \subset \mathbf{R}^3$), and a *solution* of this equation is a function v, defined on an interval I of $\mathbf{R}$, such that for each $x \in I$ the point $(x, v(x), v'(x))$ belongs to U and satisfies the relation

$$F(x, v(x), v'(x)) = 0.$$

Scalar equations of type (10.4.1) are therefore particular cases of this general notion. A differential equation (18.15.2.1) is often written $F(x, z, z') = 0$.

In order to apply the general theory, we shall *assume* that the set M_0 of points of U satisfying (18.15.2.1) is a *submanifold* of dimension 2, at all points of which we have $dF \neq 0$. Then we have to consider the differential system on U defined by the differential ideal $\mathfrak{a}$ generated by the 0-form F and the 1-form $dz - p\,dx$, and the solutions of (18.15.2.1) will be the *projections* on $\mathbf{R}^2$ of the integral manifolds of dimension 1 of this differential system (provided that these projections are manifolds of dimension 1).

It is clear that in this case $\mathfrak{a}_1$ is generated by the two 1-forms

$$dz - p\,dx, \qquad dF = \frac{\partial F}{\partial x}dx + \frac{\partial F}{\partial z}dz + \frac{\partial F}{\partial p}dp; \tag{18.15.2.2}$$

hence (18.10.3) we have $s_0 = 2$, except at points where simultaneously

$$\frac{\partial F}{\partial p} = 0, \qquad \frac{\partial F}{\partial x} + p\frac{\partial F}{\partial z} = 0, \tag{18.15.2.3}$$

in which case $s_0 = 1$. In general, the set S of *singular* points (x_0, z_0, p_0) satisfying all three relations (18.15.2.1) and (18.15.2.3) will be empty or will consist entirely of *isolated* points: an example is the equation $x - zz' = 0$, where S consists of the points $(0, 0, \pm 1)$. In this example, through each point of S there passes exactly *one* integral manifold of the differential system defined by $\mathfrak{a}$.

However, it may also happen that there exists a *curve* C (i.e., a 1-dimensional manifold) contained in S. The hypothesis that the three partial derivatives of F do not simultaneously vanish at any point of M implies that C cannot be contained in any plane $x = x_0$. For the restriction of dF to C must be zero, which (in view of (18.15.2.3)) implies that the restriction of

$$\frac{\partial F}{\partial x} dx + \frac{\partial F}{\partial z} dz$$

to C must be zero; and we cannot have $\partial F/\partial z = 0$, otherwise it would follow from (18.15.2.3) that $\partial F/\partial x = 0$ also. Hence we may assume that locally C is given by $z = v(x)$, $p = w(x)$, and therefore we must have

$$\frac{\partial F}{\partial x} + \frac{\partial F}{\partial z} v'(x) = 0, \qquad \frac{\partial F}{\partial x} + w(x)\frac{\partial F}{\partial z} = 0,$$

whence $v'(x) = w(x)$, because $\partial F/\partial z \neq 0$. The curve C is therefore a *singular integral manifold* of the differential system, and its projection $z = v(x)$ on $\mathbf{R}^2$ is called a *singular integral* (or *singular solution*) of the differential equation $F(x, z, z') = 0$.

An example of such an integral is given by *Clairaut's equation*,

(18.15.2.4) $$z = xz' + g(z'),$$

where the derivative $g'(p)$ is not a constant. The second equation (18.15.2.3) is satisfied automatically in this case, and the set S is given by the equations

(18.15.2.5) $$x = -g'(p), \qquad z = g(p) - pg'(p).$$

The module $\mathfrak{a}_1$ is generated by $dz - p\,dx$ and $(x + g'(p))\,dp$. Through each point $(x_0, z_0, p_0) \in M_0$ not belonging to S, there passes a unique integral curve of the differential system, namely the line

(18.15.2.6) $$p = p_0, \qquad z = xp_0 + g(p_0);$$

whereas if $(x_0, z_0, p_0) \in S$ and if $g''(p_0) \neq 0$, there pass through this point *two* integral curves of the differential system, namely S itself and the line (18.15.2.6). The projection of S on $\mathbf{R}^2$ is the *envelope* of the projections of the lines (18.15.2.6): at each point it has the same tangent space as the line $z = xp_0 + g(p_0)$ passing through the point.

We shall not attempt a general investigation of the singular integral manifolds of a differential system (cf. [54] and [57]).

(18.15.3) Given a *regular* integral manifold of dimension p of a differential system, i.e., such that all the tangent spaces $T_x(N)$ are *regular* integral elements, it may happen that in each of these tangent spaces there exist *singular* integral elements (18.10.7) of dimension $q < p$. If there exists a q-dimensional integral submanifold P of N whose tangent space at *each* point is a singular integral element (so that P is a *singular* integral manifold of the differential system), then P is said to be a *characteristic* submanifold of N. For such a manifold, the uniqueness property of the Cartan–Kähler theorem (18.13.2) is in general no longer valid (cf. (18.17)).

16. CAUCHY CHARACTERISTICS

(18.16.1) Let M be a pure differential manifold of dimension n, and let $\mathfrak{a}$ be a differential ideal of the algebra $\mathscr{A}$ (18.9.2). We shall assume that $\mathfrak{a}_0 = \{0\}$, which is always possible in the local study of differential systems (18.13.3). For each point $x \in \mathrm{M}$, a tangent vector $\mathbf{h}_x \in \mathrm{T}_x(\mathrm{M})$ is a *Cauchy characteristic vector* of the ideal $\mathfrak{a}$ (or of the differential system defined by $\mathfrak{a}$) if, for each $p = 1, \ldots, n$ and each p-form $\varpi_p \in \mathfrak{a}_p$, the $(p-1)$-covector $i(\mathbf{h}_x) \cdot \varpi_p(x)$ (16.18.4) belongs to the vector space $\mathfrak{a}_{p-1}(x)$ spanned by the $(p-1)$-covectors $\varpi_{p-1}(x)$, where $\varpi_{p-1} \in \mathfrak{a}_{p-1}$.

(18.16.2) By reason of the hypothesis $\mathfrak{a}_0 = \{0\}$, for $p = 1$ this condition reduces to $\langle \mathbf{h}_x, \varpi_1(x) \rangle = 0$ for all 1-forms $\varpi_1 \in \mathfrak{a}_1$, or equivalently $\mathbf{h}_x \in \mathrm{M}_1(x)$ in the notation of (18.10.3). In other words, the lines containing a Cauchy characteristic vector $\mathbf{h}_x \neq 0$ are the *integral elements of dimension* 1 at the point x. More generally, let F_x be an integral element of dimension p, and let $(\mathbf{u}_j)_{1 \leqq j \leqq p}$ be a basis of this vector space. If a Cauchy characteristic vector $\mathbf{h}_x$ does not belong to F_x (i.e., if $\mathbf{u}_1 \wedge \mathbf{u}_2 \wedge \cdots \wedge \mathbf{u}_p \wedge \mathbf{h}_x \neq 0$), then we have

$$\langle \mathbf{u}_1 \wedge \cdots \wedge \mathbf{u}_p \wedge \mathbf{h}_x, \varpi_{p+1}(x) \rangle = \langle \mathbf{u}_1 \wedge \cdots \wedge \mathbf{u}_p, i(\mathbf{h}_x) \cdot \varpi_{p+1}(x) \rangle = 0$$

for all $(p+1)$-forms $\varpi_{p+1} \in \mathfrak{a}_{p+1}$, by virtue of the definition (18.16.1). Furthermore, the same argument shows that if $\mathbf{v}$ is a vector such that

$$\langle \mathbf{u}_1 \wedge \cdots \wedge \mathbf{u}_p \wedge \mathbf{v}, \varpi_{p+1}(x) \rangle = 0$$

for all $(p+1)$-forms $\varpi_{p+1} \in \mathfrak{a}_{p+1}$, then also

$$\langle \mathbf{u}_1 \wedge \cdots \wedge \mathbf{u}_p \wedge \mathbf{h}_x \wedge \mathbf{v}, \varpi_{p+2}(x) \rangle = 0$$

for all $\varpi_{p+2} \in \mathfrak{a}_{p+2}$; or, in other words, $s_{p+1}(x, \mathbf{u}_1, \ldots, \mathbf{u}_p, \mathbf{h}_x) = 0$ (in view of (18.10.6)). Consequently we also have $s_{q+1}(x, \mathbf{u}_{i_1}, \ldots, \mathbf{u}_{i_q}, \mathbf{h}_x) = 0$ for $1 \leqq q \leqq p$ and $i_1 < \cdots < i_q$.

Nonzero Cauchy characteristic vectors do not always exist (Problems 2 and 3). If $\mathbf{h}_x$ is a nonzero Cauchy characteristic vector that belongs to a *regular* integral element G_x of dimension r (18.10.4), then an integral element of dimension q such that $1 < q < r$, which is contained in G_x and *contains* $\mathbf{h}_x$, is necessarily *singular*, except when $s_q = 0$. This justifies the name "characteristic vector" (18.15.3).

(18.16.3) Let $\rho_p(x)$ denote the dimension of the subspace $\mathfrak{a}_p(x)$ of the space $\bigwedge^p (T_x(M))^*$ of p-covectors. We shall show that the set of points $x \in X$ at which ρ_p is *locally constant* is a *dense open set* in M. We may assume that M is an open subset of $\mathbf{R}^n$, and hence that $\bigwedge^p(T_x(M))$ is identified with $M \times \mathbf{R}^{\binom{n}{p}}$, and $\mathfrak{a}_p(x)$ with a vector space of linear forms on $\{x\} \times \mathbf{R}^{\binom{n}{p}}$. Let U be an open neighborhood of x, and let y be a point of U such that $\rho_p(y) = m_p$ is equal to the least upper bound of ρ_p in U; then there exist m_p differential p-forms $\varpi_p^{(\alpha)} \in \mathfrak{a}_p$ $(1 \leqq \alpha \leqq m_p)$ such that the $(y, \varpi_p^{(\alpha)}(y))$ form a basis of the vector space $\mathfrak{a}_p(y)$. It follows now from (18.10.1) that there exists a neighborhood $V \subset U$ of y such that, for each $z \in V$, the covectors $(z, \varpi_p^{(\alpha)}(z))$ for $1 \leqq \alpha \leqq m_p$ form a basis of $\mathfrak{a}_p(z)$. This proves our assertion.

We may moreover assume (A.4.5) that there exist $\binom{n}{p} - m_p$ differential p-forms $\theta_p^{(\beta)}$ $\left(1 \leqq \beta \leqq \binom{n}{p} - m_p\right)$, which we may take to be of the type

$$dx^{i_1} \wedge \cdots \wedge dx^{i_p},$$

such that for each $z \in V$ the $\binom{n}{p}$ p-covectors $\theta_p^{(\beta)}(z)$ and $\varpi_p^{(\alpha)}(z)$ form a basis of $\bigwedge^p(T_z(M))^*$.

(18.16.4) We shall deduce from the foregoing that if, for each $x \in M$, we denote by $\nu(x)$ the dimension of the *space* N_x *of Cauchy characteristic vectors* of the ideal $\mathfrak{a}$ at the point x, then ν is *locally constant* in a *dense open subset* of M. For by restricting to an open subset of M we may assume that the numbers $\rho_p(x) = m_p$ are *constant* in M, and also that for each p we have chosen p-forms $\varpi_p^{(\alpha)}$ and $\theta_p^{(\beta)}$ having the properties stated in (18.16.3). We have then to express that, for $1 \leqq p \leqq n$ and $1 \leqq \alpha \leqq m_p$, the $(p-1)$-covector $i(\mathbf{h}_x) \cdot \varpi_p^{(\alpha)}(x)$ belongs to $\mathfrak{a}_{p-1}(x)$, or equivalently that its $\binom{n}{p-1} - m_{p-1}$ components relative to the $\theta_{p-1}^{(\beta)}(x)$ are *zero*. But it is clear that each of these components is a linear form in the coordinates of $\mathbf{h}_x$ (relative to the canonical basis of $\{x\} \times \mathbf{R}^n$), whose coefficients are C^∞-functions of x. Hence the Cauchy characteristic vectors $\mathbf{h}_x$ are determined by a system of linear equations to which (18.10.1) applies, and this proves our assertion.

It follows from this and from (18.8.4) that there exists a *nowhere dense* closed subset F of M such that, in each connected component U_j of M − F, the function $x \mapsto N_x$ is a C^∞ *field of* v_j*-directions*, where v_j denotes the constant value of $v(x)$ for $x \in U_j$.

(18.16.5) *On every open subset of* M *in which* $v(x)$ *is constant, the field of directions* $x \mapsto N_x$ *(where* N_x *is the space of Cauchy characteristic vectors at* x *for the ideal* $\mathfrak{a}$*)) is completely integrable.*

In view of (18.14.5), it is enough to show that if X, Y are two vector fields (on an open subset of M) such that $X(x) \in N_x$ and $Y(x) \in N_x$ for all x, then also $[X, Y](x) \in N_x$. Now, for each p-form $\varpi \in \mathfrak{a}_p$, it follows from the formula (17.15.3.4)

$$\theta_X \cdot \varpi = i_X \cdot d\varpi + d(i_X \cdot \varpi),$$

and from the facts that $\mathfrak{a}$ is a differential ideal and $X(x) \in N_x$, that we have $i_X \cdot d\varpi \in \mathfrak{a}_p$ and $d(i_X \cdot \varpi) \in \mathfrak{a}_p$, and therefore $\theta_X \cdot \varpi \in \mathfrak{a}_p$. Next, if $X(x) \in N_x$ and $Y(x) \in N_p$ for all x, then for each p-form $\varpi \in \mathfrak{a}_p$ we have $\theta_X \cdot \varpi \in \mathfrak{a}_p$ and $i_Y \cdot (\theta_X \cdot \varpi) \in \mathfrak{a}_{p-1}$; also $i_X \cdot \varpi \in \mathfrak{a}_{p-1}$ and $\theta_X \cdot (i_Y \cdot \varpi) \in \mathfrak{a}_{p-1}$ from above. Hence $i_{[X,Y]} \cdot \varpi \in \mathfrak{a}_{p-1}$ by virtue of (17.14.10.2), and the proof is complete.

(18.16.6) With the same notation, if F_x is an integral element of dimension p and if $F_x \cap N_x = \{0\}$, then $F_x + N_x$ is an *integral element* of dimension $p + v$. For if $(\mathbf{u}_i)_{1 \leqq i \leqq p}$ is a basis of F_x and $(\mathbf{v}_j)_{1 \leqq j \leqq v}$ a basis of N_x, then for each $(p+v)$-form $\varpi_{p+v} \in \mathfrak{a}_{p+v}$ we have

$$\begin{aligned}&\langle \mathbf{u}_1 \wedge \cdots \wedge \mathbf{u}_p \wedge \mathbf{v}_1 \cdots \wedge \mathbf{v}_v, \varpi_{p+v}(x)\rangle \\ &\quad = \langle \mathbf{u}_1 \wedge \cdots \wedge \mathbf{u}_p, i(\mathbf{v}_1)i(\mathbf{v}_2) \cdots i(\mathbf{v}_v) \cdot \varpi_{p+v}(x)\rangle = 0,\end{aligned}$$

because, by the definition of Cauchy characteristic vectors, we have

$$i(\mathbf{v}_1) \cdots i(\mathbf{v}_v) \cdot \varpi_{p+v}(x) \in \mathfrak{a}_p(x).$$

(18.16.7) Corresponding to the above differential property there is an analogous property of integral manifolds which enables us to obtain, from a p-dimensional integral manifold of the differential system defined by $\mathfrak{a}$, a $(p+v)$-dimensional integral manifold when we know the integral manifolds (called *Cauchy characteristics*) of the field of directions $x \mapsto N_x$. For the sake of simplicity, we shall restrict ourselves to the case where the differential system is a *Pfaffian system*:

(18.16.7.1) $$\omega_j = 0 \quad (1 \leqq j \leqq n - r).$$

The fact that i_X is an antiderivation of $\mathscr{E}$ (A.18.5) shows that for a vector $\mathbf{h}_x \in T_x(M)$ to be a Cauchy characteristic vector it is necessary and sufficient that it satisfy the following two conditions:

(18.16.7.2) $\langle \mathbf{h}_x, \omega_j(x) \rangle = i(\mathbf{h}_x) \cdot \omega_j(x) = 0$ for all j;

(18.16.7.3) $i(\mathbf{h}_x) \cdot d\omega_j(x)$ is a linear combination of the $\omega_k(x)$, for all j.

We now restrict to an open set U in M for which the $\omega_j(x)$ are *linearly independent* at each point and the function $\nu(x) = \dim N_x$ is *constant*. By virtue of (18.16.5) and (18.14.3), and by replacing U by a smaller open set if necessary, we may assume that there exists a submersion π of U onto a differential manifold P such that the fibers $\pi^{-1}(z)$ for $z \in P$ are integral manifolds of dimension ν of the field of directions $x \mapsto N_x$. In this situation we have:

(18.16.8) *If* Q *is any integral manifold of the system* (18.16.7.1), *contained in* U *and such that* $N_x \cap T_x(Q) = \{0\}$ *for all* $x \in Q$, *then* $\pi^{-1}(\pi(Q))$ *is an integral manifold of* (18.16.7.1).

In more concrete terms, $\pi^{-1}(\pi(Q))$ is obtained by taking the *union* of the ν-dimensional integral manifolds of $x \mapsto N_x$ passing through the points of Q. Since the question is local, we may assume that $U = P \times W$, where P (resp. W) is an open subset of $\mathbf{R}^{n-\nu}$ (resp. $\mathbf{R}^\nu$) and π is the projection onto $\mathbf{R}^{n-\nu}$, and that $\mathbf{R}^\nu$ is spanned by the vectors $\mathbf{e}_{n-\nu+1}, \dots, \mathbf{e}_n$ of the canonical basis of $\mathbf{R}^n$. We shall show that, by replacing the ω_j by suitable linear combinations with coefficients in $\mathscr{E}_0$, we may suppose that the coefficients of the dx^k for $k > n - \nu$ in the ω_j are *zero*, and that the coefficients of the dx^h for $h \leqq n - \nu$ *do not depend on* $x^{n-\nu+1}, \dots, x^n$. It will then follow that the Pfaffian system (18.16.7.1.) may be regarded as a Pfaffian system *in* P, and that for each integral manifold S of this system the inverse image $\pi^{-1}(S) = S \times W$ is an integral manifold of the original system in U; this will establish the proposition.

By (18.16.7.2), we must have $i(\mathbf{e}_k) \cdot \omega_j(x) = 0$ for $k > n - \nu$ and all j, which shows already that the coefficient of dx^k in ω_j is zero. Since for each x the $n - r$ covectors $\omega_j(x)$ are linearly independent we may (by replacing P and W by smaller open sets, if necessary) replace the ω_j by linear combinations of these forms with coefficients in $\mathscr{E}_0$, such that for $1 \leqq j \leqq n - r$ we have

(18.16.8.1)
$$\omega_j(x) = dx^j + \sum_{k=n-r+1}^{n-\nu} a_{jk}(x)\, dx^k.$$

Consequently

$$d\omega_j(x) = \sum_{h=1}^{n} \sum_{k=n-r+1}^{n-v} \mathrm{D}_h a_{jk}(x)\, dx^h \wedge dx^k.$$

Expressing that $i(\mathbf{e}_k) \cdot d\omega_j(x)$ must be a linear combination of the forms $\omega_h(x)$ for $k > n - v$ (18.16.7.3), we see that since the covector $i(\mathbf{e}_k) \cdot d\omega_j(x)$ is a linear combination of the dx^i such that $n - r + 1 \leqq i \leqq n - v$ *alone*, it cannot be a linear combination of the $\omega_j(x)$ unless it is zero; hence we have

(18.16.8.2) $$\mathrm{D}_h a_{jk} = 0$$

for $h > n - v$, $1 \leqq j \leqq n - r$, and $n - r + 1 \leqq k \leqq n - v$. This completes the proof of (18.16.8).

Remark

(18.16.9) It should be carefully noted that it can happen that the coefficients of a Pfaffian system (18.16.7.1) are *analytic* but that not all the integral manifolds of the system are analytic (compare with (10.5.3)). We shall see an example in the next section.

PROBLEMS

1. Let α be a C^∞ differential p-form on M, and let $\mathfrak{a}$ be the differential ideal of $\mathscr{A}$ generated by α. With the notation of (18.16.4) show that if the dimension $\nu(x)$ of N_x is constant and equal to $n - q$ on an open subset U of M, then $q \geqq p$ and for each point $x_0 \in \mathrm{U}$ there exists a system of local coordinates $(y^j)_{1 \leqq j \leqq n}$ such that, in the domain of this chart, we have

$$\alpha = \sum_{1 \leqq i_1 < \cdots < i_p \leqq q} a_{i_1 \cdots i_p}(y^1, \ldots, y^q)\, dy^{i_1} \wedge \cdots \wedge dy^{i_p},$$

where the functions $a_{i_1 \cdots i_p}$ are of class C^∞. If $q = p$, we may choose the y^j such that

$$\alpha = dy^1 \wedge dy^2 \wedge \cdots \wedge dy^p.$$

2. Let α be a C^∞ differential 2-form on a differential manifold M of dimension n. For each $x \in \mathrm{M}$, $\alpha(x)$ may be identified with the alternating bilinear form

$$(\mathbf{h}_x, \mathbf{k}_x) \mapsto \langle \mathbf{h}_x \wedge \mathbf{k}_x, \alpha(x) \rangle$$

on $\mathrm{T}_x(\mathrm{M}) \times \mathrm{T}_x(\mathrm{M})$ (or also with the linear mapping

$$\mathbf{h}_x \mapsto \langle \mathbf{h}_x \wedge \cdot\,, \alpha(x) \rangle$$

of $\mathrm{T}_x(\mathrm{M})$ into $\mathrm{T}_x(\mathrm{M})^*$); the *rank* of $\alpha(x)$ is defined to be the rank of this linear mapping, and is denoted by $\mathrm{rk}_x(\alpha)$. It is an *even* number (A.16.1). The mapping $x \mapsto \mathrm{rk}_x(\alpha)$ is lower semicontinuous, and hence there exists a dense open subset of M on which the rank is locally constant. Show that if, in a neighborhood of a point $x_0 \in \mathrm{M}$, the rank of $\alpha(x)$ is constant and equal to $2r$, then there exist $2r$ differential 1-forms ω_j of class C^∞ on a sufficiently small neighborhood U of x_0 such that the $\omega_j(x)$ are linearly independent at each $x \in \mathrm{U}$ and such that

$$\alpha = \omega_1 \wedge \omega_2 + \omega_3 \wedge \omega_4 + \cdots + \omega_{2r-1} \wedge \omega_{2r}$$

in U. (Take a system of local coordinates and show that if (for example) the coefficient of $dx^1 \wedge dx^2$ in α does not vanish at the point x_0, then two linearly independent 1-forms ω_1, ω_2 can be found such that $\alpha - \omega_1 \wedge \omega_2$ contains only the dx^j with $j \geqq 3$.)

If $\mathfrak{a}$ is the differential ideal generated by α, then (with the notation of (**18.16.4**)) we have $\nu(x) = n - 2r$ for all $x \in \mathrm{U}$, and the field of $(n-2r)$-directions $x \mapsto \mathrm{N}_x$ is defined by the Pfaffian system $\omega_j = 0$ $(1 \leqq j \leqq 2r)$. (Use the fact that i_X is an antiderivation.) By using the theory of alternating bilinear forms, prove that the integral elements of the differential system defined by $\mathfrak{a}$ have dimension $\leqq n - r$; every integral element of dimension $< n - r$ is contained in an integral element of dimension $n - r$, and every integral element of dimension $n - r$ contains a regular integral element of dimension $n - r - 1$. If $2r = n$, every integral element is regular.

The rank $\mathrm{rk}_x(\alpha)$ is the largest integer k such that the exterior product

$$(\alpha(x))^{\wedge k} = \alpha(x) \wedge \alpha(x) \wedge \cdots \wedge \alpha(x)$$

(k factors) is $\neq 0$.

3. Let M be a differential manifold of dimension n, and let ω be a C^∞ differential 1-form on M such that $\omega(x) \neq 0$ for all $x \in \mathrm{M}$. At a point $x \in \mathrm{M}$, let $2r$ be the rank of $d\omega$ (Problem 2); the form ω is then said to be of *Pfaffian class* (or simply *class*) $2r$ if $\omega(x) \wedge (d\omega(x))^{\wedge r} = 0$, and of *class* $2r + 1$ if $\omega(x) \wedge (d\omega(x))^{\wedge r} \neq 0$. The class of ω at x is a lower semicontinuous function of $x \in \mathrm{M}$.

(a) Suppose that ω is of constant class $2r + 1$ in a neighborhood of a point $x_0 \in \mathrm{M}$. Show that there exists a C^∞ function f on an open neighborhood V of x_0, such that $f(x_0) = 0$ and such that $\omega_1 = (\omega \mid \mathrm{V}) - df$ does not vanish in V and is of constant class $2r$. (For each $x \in \mathrm{V}$, let N'_x denote the set of tangent vectors $\mathbf{h}_x \in \mathrm{T}_x(\mathrm{M})$ such that $i(\mathbf{h}_x) \cdot \omega(x) = i(\mathbf{h}_x) \cdot d\omega(x) = 0$. Show that $x \mapsto \mathrm{N}'_x$ is a completely integrable field of $(n - 2r - 1)$-directions; apply Problem 1 to the form $\omega \wedge (d\omega)^{\wedge r}$ and to the form $(d\omega)^{\wedge r}$, and use Section **18.14**, Problem 3.)
(b) Suppose that ω is of constant class $2r$ in a neighborhood of x_0. Show that there exists a C^∞ function g on an open neighborhood V of x_0, such that $g(x_0) = 0$ and such that $\omega_2 = (1 + g)(\omega \mid \mathrm{V})$ is of constant class $2r - 1$. (Same method; consider the forms $\omega \wedge (d\omega)^{\wedge(r-1)}$ and $(d\omega)^{\wedge r}$.)
(c) Deduce from (a) and (b) that if ω is of constant class $2r + 1$ on an open subset U of M, then at each point $x \in \mathrm{U}$ there exists a system of local coordinates $(y^j)_{1 \leqq j \leqq n}$ in a neighborhood $\mathrm{V} \subset \mathrm{U}$ of x, which vanish at the point x and are such that

$$\omega = dy^1 + y^2\, dy^3 + \cdots + y^{2r}\, dy^{2r+1}$$

in V. If ω is of constant class $2r$ in U, then likewise there exists a system of local coordinates $(y^j)_{1 \leqq j \leqq n}$ in a neighborhood $\mathrm{V} \subset \mathrm{U}$ of x, which vanish at x and are such that

$$\omega = (1 + y^1)\, dy^2 + y^3\, dy^4 + \cdots + y^{2r-1}\, dy^{2r}$$

in V (*Darboux's theorem*). (Argue by induction on the class of ω.)

Deduce that if the class of ω is constant and equal to $2r+1$ (resp. $2r$) in U, then the integral elements of the Pfaffian equation $\omega = 0$ have maximum dimension $n-r-1$ (resp. $n-r$). Every integral element of dimension $<n-r-1$ (resp. $<n-r$) is contained in an integral element of dimension $n-r-1$ (resp. $n-r$); every integral element of dimension $n-r-1$ (resp. $n-r$) is regular. Prove also (assuming *only* that the form ω is of class C^∞) that there always exist regular integral manifolds of dimension $n-r-1$ (resp. $n-r$). If $n = 2r+1$ (resp. $n = 2r$), every integral element is regular.

(d) Suppose that ω is of constant class $2r+1$. Show that, for each C^∞ function f which does not vanish on M, the class of $f\omega$ is $2r+1$ or $2r+2$ at each point. For each C^∞ function g on M, the class of $\omega + dg$ is $2r+1$ or $2r$ at each point.

(e) Suppose that ω is of constant class $2r$. Show that, for each C^∞ function f which does not vanish on M, the class of $f\omega$ is $2r$ or $2r-1$ at each point. For each C^∞ function g on M, the class of $\omega + dg$ is $2r$ or $2r+1$ at each point.

4. On $\mathbf{R}^n - \{0\}$ consider the differential form

$$\omega = F_1\, dx^1 + \cdots + F_n\, dx^n,$$

where the F_j $(1 \leq j \leq n)$ are *homogeneous* of degree g (positive or negative). Show that if $x^1F_1 + \cdots + x^nF_n$ is a constant $\neq 0$, then g must be -1, and the class of ω is odd; if on the other hand $x^1F_1 + \cdots + x^nF_n$ is identically zero, and $g \neq -1$, then the class of ω is even. Using Problem 3(e), show that if $x^1F_1 + \cdots + x^nF_n = F$ is $\neq 0$ on $\mathbf{R}^n - \{0\}$, and if the class of ω is an even number $2r$, then the class of $F^{-1}\omega$ is $2r-1$. If $g \neq -1$ and $F \neq 0$ on $\mathbf{R}^n - \{0\}$, and if the class of ω is an odd number $2r+1$, then the class of $\omega - (g+1)^{-1}\, dF$ is $2r$.

5. Let ω be a C^∞ differential 1-form on a differential manifold M of dimension n, such that $\omega(x) \neq 0$ for all $x \in M$ and such that $\omega \wedge d\omega = 0$. (The Pfaffian class of ω, which is equal to 1 or 2 at each point, is not assumed to be constant.) Show that, for each point $x_0 \in M$, there exists an open neighborhood V of x_0 and two C^∞ functions f, g on V such that $\omega = f\, dg$ in V. (Argue by induction on n, by reducing to the case where M is an open set in $\mathbf{R}^n$).

6. On $\mathbf{R}^2$ consider the differential 1-form

$$\omega = y^3(1-y)^2\, dx + (y^3 - 2(1-y)^2)\, dy,$$

the coordinates being x, y. We have $\omega \wedge d\omega = 0$ and $\omega(z) \neq 0$ for all $z \in \mathbf{R}^2$; show that nevertheless there cannot exist two C^∞-functions f, g defined on the whole space $\mathbf{R}^2$, such that $\omega = f\, dg$ (compare with Problem 5). (By considering the integral curves of $\omega = 0$, show that we must have

$$g(x, y) = h\left(x + \frac{1}{y^2} + \frac{1}{1-y}\right)$$

for $x \in \mathbf{R}$ and $y \in]0, 1[$, where h is of class C^∞ on $\mathbf{R}$. By observing that f is continuous and nonzero everywhere, show that the functions $t^{3/2}h'(t)$ and $t^2h'(t)$ must both tend to a finite nonzero limit as $t \to +\infty$.)

7. Determine the singular integral manifolds of each of the Pfaffian equations in $\mathbf{R}^5$:

$$x^5\,dx^1 + x^3\,dx^2 + x^1\,dx^4 + x^1\,dx^5 = 0,$$
$$x^1x^3\,dx^2 + x^1x^2\,dx^3 + (x^1 + x^3x^5)\,dx^4 + x^3x^4\,dx^5 = 0.$$

(Examine the set of points at which the Pfaffian form is not of maximum class.)

8. Let M be a differential manifold of dimension n, and let $\mathfrak{a}$ be a differential ideal in $\mathscr{A}$ generated by $\mathfrak{a}_1$. Suppose that for each $x \in \mathrm{M}$ the space $\mathfrak{a}_1(x)$, generated by the $\omega(x)$ for $\omega \in \mathfrak{a}_1$, is of constant dimension (which we denote by $\dim(\mathfrak{a}_1)$). With the notation of **(18.16.4)**, for each $x \in \mathrm{M}$ let $\gamma(\mathfrak{a}_1)(x)$ be the subspace of covectors at x which are orthogonal to N_x; we shall suppose that $n - \nu(x) = \dim(\gamma(\mathfrak{a}_1)(x))$ is constant in M (so that the Pfaffian system defined by $\gamma(\mathfrak{a}_1)$ is *completely integrable*) and we denote this constant dimension by $\dim(\gamma(\mathfrak{a}_1))$. Let $\mathfrak{a}_1^{(1)}$ denote the sub-$\mathscr{E}_0$-module of $\mathfrak{a}_1$ consisting of the 1-forms ω such that $d\omega \in \mathfrak{a}_1 \wedge \mathscr{E}_1$. The corresponding Pfaffian system is called the *first derivative* of the Pfaffian system defined by $\mathfrak{a}_1$. By induction we define the kth *derivative* $\mathfrak{a}_1^{(k)}$ to be $(\mathfrak{a}_1^{(k-1)})^{(1)}$. The least positive integer N such that $\mathfrak{a}_1^{(\mathrm{N}+1)} = \mathfrak{a}_1^{(\mathrm{N})}$ is called the *derived length* of $\mathfrak{a}_1$. We have

$$\mathfrak{a}_1^{(\mathrm{N})} \subset \mathfrak{a}_1^{(\mathrm{N}-1)} \subset \cdots \subset \mathfrak{a}_1^{(1)} \subset \mathfrak{a}_1 \subset \gamma(\mathfrak{a}_1).$$

We shall assume that the dimensions $\dim(\mathfrak{a}_1^{(k)}(x))$ are constant throughout M, and we shall denote them by $\dim(\mathfrak{a}_1^{(k)})$. The Pfaffian system defined by $\mathfrak{a}_1^{(\mathrm{N})}$ is therefore the *largest completely integrable system* contained in the given system.

(a) Let $p_0 = \dim(\gamma(\mathfrak{a}_1)) - \dim(\mathfrak{a}_1)$, $p_1 = \dim(\mathfrak{a}_1) - \dim(\mathfrak{a}_1^{(1)})$, and in general, $p_k = \dim(\mathfrak{a}_1^{(k-1)}) - \dim(\mathfrak{a}_1^{(k)})$. Show that

$$p_1 \leqq \tfrac{1}{2}p_0(p_0 - 1)$$

and that, in general,

$$p_{k+1} \leqq p_k(p_0 + p_1 + \cdots + p_{k-1}) + \tfrac{1}{2}p_k(p_k - 1).$$

(Consider the image of a supplement of $\mathfrak{a}_1^{(k+2)}$ in $\mathfrak{a}_1^{(k+1)}$ under exterior differentiation d, and use **(18.16.8)**.) In particular, either $p_0 = 0$ (in which case the Pfaffian system defined by $\mathfrak{a}_1$ is completely integrable) or else $p_0 \geqq 2$.

(b) With the notation of **(18.10.3)**, show that

$$s_0(x) = \dim(\mathfrak{a}_1), \qquad s_1(x, \mathbf{u}_1) \leqq \inf(p_0 - 1, p_1).$$

(Consider the image of a supplement of $\mathfrak{a}_1^{(1)}$ in $\mathfrak{a}_1$ under the mapping $\omega \mapsto i(\mathbf{u}_1) \cdot d\omega(x)$.)

(c) Let $\omega_1, \ldots, \omega_{p_1}$ be a basis of a supplement of $\mathfrak{a}_1^{(1)}$ in $\mathfrak{a}_1$. Show that, for each 1-form $\varpi \in \mathfrak{a}_1^{(1)}$,

$$d\varpi \wedge d(\omega_1 \wedge \cdots \wedge \omega_{p_1}) \in \mathfrak{a}^{(1)},$$

where $\mathfrak{a}^{(1)}$ is the ideal $\mathfrak{a}_1^{(1)} \cap \mathscr{A}$ generated by $\mathfrak{a}_1^{(1)}$.

9. With the notation and hypotheses of Problem 8, the least integer k such that

$$(d\omega)^{\wedge(k+1)} \in \mathfrak{a}_1 \wedge \mathscr{A}$$

for all $\omega \in \mathfrak{a}_1$ is called the *rank* of the Pfaffian system defined by $\mathfrak{a}_1$, and is denoted by $\rho(\mathfrak{a}_1)$. Equivalently, we have

$$d\omega_1 \wedge d\omega_2 \wedge \cdots \wedge d\omega_{k+1} \in \mathfrak{a}_1 \wedge \mathscr{A}$$

for all systems of $k+1$ forms $\omega_j \in \mathfrak{a}_1$, while

$$d\omega_1 \wedge \cdots \wedge d\omega_k \notin \mathfrak{a}_1 \wedge \mathscr{A}$$

for at least one choice of $\omega_1, \ldots, \omega_k$ in $\mathfrak{a}_1$ (and the ω_i may be chosen from a given basis of $\mathfrak{a}_1$).

(a) Show that

$$2\rho(\mathfrak{a}_1) \leqq p_0 \leqq (\dim(\mathfrak{a}_1)+1)\rho(\mathfrak{a}_1).$$

To establish the second inequality, let ω_j $(1 \leqq j \leqq \dim(\mathfrak{a}_1))$ be a basis of $\mathfrak{a}_1$, and write

$$d\omega_j = \sum_k \alpha_{jk} \wedge \omega_k + \sum_{h=1}^{s_j} \varpi_{2h,j} \wedge \varpi_{2h-1,j},$$

where the $\varpi_{2h,j}$ belong to $\gamma(\mathfrak{a}_1)$, and $s_j \leqq \rho(\mathfrak{a}_1)$. By writing down the conditions defining $\rho(\mathfrak{a}_1)$, deduce that each of the forms $\varpi_{2h,j}$ is a linear combination of $\varpi_{2h-1,j}$, the $\omega_k (1 \leqq k \leqq \dim(\mathfrak{a}_1))$, and $2\rho(\mathfrak{a}_1)$ fixed forms $\varpi_{l_i, k_i} (1 \leqq i \leqq \rho(\mathfrak{a}_1))$.)

(b) Prove that if $\rho(\mathfrak{a}_1) \geqq p_1 + 1$, then $\rho(\mathfrak{a}_1^{(1)}) \leqq p_1 - 1$. (Let $\zeta_1, \ldots, \zeta_{p_1}$ be a basis of a supplement of $\mathfrak{a}_1^{(1)}$ in $\mathfrak{a}_1$. For each 1-form $\omega \in \mathfrak{a}_1^{(1)}$, there exist 1-forms α_i such that

$$d\omega = \sum_{i=1}^{p_1} \alpha_i \wedge \zeta_i \qquad \operatorname{mod}(\mathfrak{a}_1^{(1)} \wedge \mathscr{A}).$$

Use Problem 8(c) to show that

$$\sum_{i=1}^{p_1} (-1)^{i+1} \alpha_i \wedge d\zeta_i \in \mathfrak{a}_1 \wedge \mathscr{A}$$

and deduce that

$$d\zeta_i \wedge \alpha_1 \wedge \cdots \wedge \alpha_{p_1} \in \mathfrak{a}_1 \wedge \mathscr{A}.$$

Deduce that, if $\rho(\mathfrak{a}_1) \geqq p_1 + 1$, then the classes mod $\mathfrak{a}_1$ of the α_i are linearly dependent (argue by contradiction), and hence show that $(d\omega)^{\wedge p_1} \in \mathfrak{a}_1^{(1)} \wedge \mathscr{A}$.)

10. The notation and hypotheses are as in Problem 8.

(a) Show that the hypothesis $p_1 = 1$ is *equivalent* to the hypothesis that the greatest value of $s_1(x, \mathbf{u}_1)$ (in the notation of (18.10.3)) is 1. (Begin by establishing the following lemma of exterior algebra: if E is a finite-dimensional vector space and if $\mathbf{z}_1^*$, $\mathbf{z}_2^*$ are two 2-covectors over E such that $i(\mathbf{x}) \cdot \mathbf{z}_1^*$ and $i(\mathbf{x}) \cdot \mathbf{z}_2^*$ are proportional for *all* $\mathbf{x} \in \mathrm{E}$, then $\mathbf{z}_1^*$ and $\mathbf{z}_2^*$ are proportional. Note however that three 2-covectors can be linearly independent, although their interior products by each vector $\mathbf{x} \in \mathrm{E}$ are linearly dependent.)

(b) Show that, if $p_1 = 1$, we have $p_0 = 2\rho(\mathfrak{a}_1)$. (Consider $d\zeta$ for a form ζ which generates a supplement of $\mathfrak{a}_1^{(1)}$ in $\mathfrak{a}_1$.) Every integral element then has dimension at most $n - \dim(\gamma(\mathfrak{a}_1)) + \rho(\mathfrak{a}_1)$, and there exist integral elements of this dimension, which are all regular.

If $p_0 > 2$ (or, equivalently, $\rho(\mathfrak{a}_1) > 1$), then we have $\mathrm{N} = 1$; in other words, the Pfaffian system defined by $\mathfrak{a}_1^{(1)}$ is *completely integrable* (Problem 9(b)). In a neighborhood of any point, we may by means of a diffeomorphism reduce the Pfaffian system defined by $\mathfrak{a}_1$ to the form

$$dx^j = 0 \qquad (1 \leqq j \leqq m),$$
$$dx^{m+1} - x^{m+2}\, dx^{m+3} - \cdots - x^{m+2p}\, dx^{m+2p+1} = 0.$$

(c) If $p_0 = 2$ and $p_1 = 1$, then $p_2 \leqq 2$. Show that if $p_2 = 1$ and $\dim(\mathfrak{a}_1^{(2)}) = m - 1$, then the Pfaffian system defined by $\mathfrak{a}_1$ may be reduced, in a neighborhood of any point, to the form

$$dx^j - a_j\, dx^m - b_j\, dx^{m+1} = 0 \qquad (1 \leqq j \leqq m+1),$$
$$dx^{m+1} - x^{m+2}\, dx^m = 0,$$

where the a_j and b_j are C^∞-functions of $x^1, \ldots, x^{m+1}$. (By applying (b) to $\mathfrak{a}_1^{(1)}$, observe that either this system is completely integrable, or else $\dim(\gamma(\mathfrak{a}_1^{(1)})) - \dim(\mathfrak{a}_1^{(1)}) = 2$.)
(d) Suppose that $p_0 = 2, p_1 = p_2 = \cdots = p_N = 1$. Show that in a neighborhood of any point the Pfaffian system defined by $\mathfrak{a}_1$ may be reduced by means of a diffeomorphism to the form

$$dx^j = 0 \qquad (1 \leqq j \leqq m),$$
$$dx^{m+2} - x^{m+3}\, dx^{m+1} = 0,$$
$$\vdots$$
$$dx^{m+N-1} - x^{m+N}\, dx^{m+1} = 0.$$

(Use induction.)
(e) Consider a Pfaffian system $\theta_j = 0$ $(1 \leqq j \leqq n-2)$ defined on an open subset M of $\mathbf{R}^n$, and such that $\dim(\mathfrak{a}_1) = n - 2$. Suppose that there exist n functions F_i $(1 \leqq i \leqq n)$ defined on an open subset U of $\mathbf{R}^r$, such that the mapping

$$(t, w_1, \ldots, w_{r-1}) \mapsto (F_1(t, w_1, \ldots, w_{r-1}), \ldots, F_n(t, w_1, \ldots, w_{r-1}))$$

is a submersion F of U into M, of rank n; suppose moreover that, for *all* C^∞ functions v defined on an interval $I \subset \mathbf{R}$ and such that, for all $t \in I$, the point

$$(t, v(t), Dv(t), \ldots, D^{r-2}v(t))$$

belongs to U, the curve

$$t \mapsto (F_1(t, v(t), \ldots, D^{r-2}v(t)), \ldots, F_n(t, v(t), \ldots, D^{r-2}v(t)))$$

is an integral curve of the system. Show that, if we put $\omega_j = {}^t F(\theta_j)$ and

$$\varpi_1 = dw_1 - w_2\, dt, \qquad \ldots, \qquad \varpi_{r-1} = dw_{r-1} - w_r\, dt, \qquad \varpi_r = dw_r,$$

then we may write

$$\omega_j = \sum_{k=1}^{r-1} a_{jk} \varpi_k$$

for $1 \leqq j \leqq n-2$. If this system is not completely integrable, we may assume that $a_{j,r-1} = 0$ for $1 \leqq j \leqq n-3$. Show that the Pfaffian system $\omega_j = 0$ $(1 \leqq j \leqq n-2)$ is then of the type described in (d), and that the same is true of the system

$$\theta_j = 0 \qquad (1 \leqq j \leqq n-2).$$

Consider the converse.

11. Consider the differential equation in two unknown functions

$$(*) \qquad \frac{dz}{dx} = G\left(x, y, z, \frac{dy}{dx}, \frac{d^2y}{dx^2}\right),$$

where G is defined in an open subset $V \subset \mathbf{R}^5$. Suppose that there exists a submersion of rank 3,

$$(t, w_1, \ldots, w_{r-1}) \mapsto (F_1(t, w_1, \ldots, w_{r-1}), F_2(t, w_1, \ldots, w_{r-1}), F_3(t, w_1, \ldots, w_{r-1}))$$

of an open set $U \subset \mathbf{R}^r$ into the projection of V on $\mathbf{R}^3$, such that for *each* C^∞-function v defined on an interval $I \subset \mathbf{R}$ and such that $(t, v(t), \ldots, D^{r-2}v(t)) \in U$ for all $t \in I$, the curve

$$x = F_1(t, v(t), \ldots, D^{r-2}v(t)), \qquad y = F_2(t, v(t), \ldots, D^{r-2}v(t)),$$
$$z = F_3(t, v(t), \ldots, D^{r-2}v(t))$$

is an integral curve of (*) in every interval where $dx/dt \neq 0$. Show that we must have

$$G\left(x, y, z, \frac{dy}{dx}, \frac{d^2y}{dx^2}\right) = A\left(x, y, z, \frac{dy}{dx}\right)\frac{d^2y}{dx^2} + B\left(x, y, z, \frac{dy}{dx}\right).$$

Consider the converse. (Reduce (*) to a Pfaffian system, and apply Problem 10(e).)
Show in particular that for the equation

$$\frac{dz}{dx} = y^{m+1}\frac{d^2y}{dx^2},$$

we may take as integral curves

$$x(t) = -2tv''(t) - v'(t),$$
$$y^{m+1}(t) = (m+1)^2 t^3 (v''(t))^2,$$
$$z(t) = (m-1)t^2 v''(t) - mtv'(t) + mv(t).$$

12. Let M_1, M_2 be two differential manifolds, $\mathfrak{a}^{(i)}$ a differential ideal on M_i $(i = 1, 2)$, and $\mathfrak{a}$ the differential ideal on $M = M_1 \times M_2$ generated by ${}^t\mathrm{pr}_1(\mathfrak{a}^{(1)})$ and ${}^t\mathrm{pr}_2(\mathfrak{a}^{(2)})$. In order that a connected submanifold V of M should be an integral manifold of maximum dimension of $\mathfrak{a}$, it is necessary and sufficient that it should be of the form $V_1 \times V_2$, where V_i is an integral submanifold of $\mathfrak{a}^{(i)}$ $(i = 1, 2)$; this manifold V is regular if and only if the V_i are regular. If moreover V_i contains no characteristic submanifold of dimension >0 for $i = 1, 2$, then the characteristic submanifolds of V are the submanifolds $\{x_1\} \times V_2$ and $V_1 \times \{x_2\}$, where $x_1 \in V_1$ and $x_2 \in V_2$.

13. Consider, in $\mathbf{R}^{4n+4}$, the Pfaffian system of two equations

$$dz - p_1\,dx_1 - \cdots - p_{n+1}\,dx_{n+1} = 0,$$
$$dp_{n+1} + u_1\,dp_1 + \cdots + u_n\,dp_n - v_1\,dx_1 - \cdots - v_{n+1}\,dx_{n+1} = 0.$$

In the notation of Problem 8, we have $p_0 = 4n + 2$, $p_1 = 2$, $N = 1$.† Show that the maximal integral manifolds are regular and of dimension $2n + 1$. In each manifold of dimension $3n + 3$ given by an equation $z = f(x_1, \ldots, x_{n+1})$, there exists a unique integral manifold V of dimension $2n + 1$. For each curve Γ with equations

$$u_j = g_j(x_{+1}) \qquad (1 \leq j \leq n),$$

there exists a unique *characteristic curve* in V whose projection is Γ (in a sufficiently small neighborhood of a point of V); and conversely every *characteristic submanifold* of V is obtained in this way (compare with Problems 12 and 14, and with **(18.18)**).

† For a more complete study of the case $p_1 = 2$, see [54, Vol. II_1, pp. 511–552].

14. The general hypotheses are those of Section **18.13**, Problem 2. Let

$$\varpi_h \qquad (1 \leq h \leq n-p-r)$$

be linear combinations of the θ_k and the dx^{p+r+h} which, together with the θ_k and the ω_i, form a basis of F^*, and write

$$d\theta_k = \sum_{i,\,h} a_{kih}\,\omega_i \wedge \varpi_h + \sum_{i,\,l} b_{kil}\,\omega_i \wedge \theta_l + \sum_{i<j} c_{ijk}\,\omega_i \wedge \omega_j.$$

Suppose, moreover, that the ω_i are chosen so that, for the dual basis $(\mathbf{u}_j)_{1 \leq j \leq p}$ of F_P, the calculation of the $s''_h(x, \mathbf{u}_1, \ldots, \mathbf{u}_h)$ gives the constant value s''_h defined in Section **18.13**, Problem 1.

This being so, *suppose that we have*

$$s''_2 = s''_3 = \cdots = s''_{p-1} = 0,$$

and put $s''_1 = s$.

(a) Show that the basis (ϖ_h) may be chosen so that

$$\begin{aligned} &a_{k1h} = 0 \qquad \text{for} \quad 1 \leq k \leq s \quad \text{and} \quad 1 \leq h \leq n-p-r \quad \text{and} \quad h \neq k; \\ &a_{k1h} = 1; \\ &a_{k1h} = 0 \qquad \text{for} \quad k \geq s+1 \quad \text{and all} \quad h. \end{aligned}$$

Deduce that the rank of the $n-p-r$ forms $\sum_{i,\,h} a_{kih}\,\omega_i \wedge \varpi_h$ is equal to s. (Observe that we must have $s''_1(x, \mathbf{u}_1 + t\mathbf{u}_j) \leq s$ for $t \in \mathbf{R}$ and all $j \geq 2$, and deduce that with the above choice of the ϖ_h, this implies that $a_{kih} = 0$ for all $k \geq s+1$.)

(b) Suppose that the characteristic equation

$$\chi(\lambda) = \det(\delta_{kh} + \lambda a_{k2h}) = 0$$

$(1 \leq k, h \leq s)$, where δ_{kh} is the Kronecker delta, has s *distinct* roots $c_k \in K_0$ $(1 \leq k \leq s)$; replacing P by a dense open subset, we may assume that the c_k are analytic in P. By changing the bases of the spaces spanned by the θ_k $(1 \leq k \leq s)$ and the ϖ_h $(1 \leq k \leq s)$, we may then assume that

$$(*) \qquad \sum_{i,\,h} a_{kih}\,\omega_i \wedge \varpi_h = (\omega_1 + c_k \omega_2) \wedge \varpi_k + \sum_h \sum_{i \geq 3} a_{kih}\,\omega_i \wedge \varpi_h.$$

(c) Suppose that the Pfaffian system under consideration is *in involution* relative to P. Show that we have then

$$a_{kih} = 0 \qquad \text{for} \quad i \geq 3 \quad \text{and} \quad h \neq k.$$

(Express that for each vector which projects onto $\mathbf{u}_1$ and annihilates the $\theta_k(x)$, there passes through this vector an integral element of order p, which may be defined as a subspace of $T_x(M)$ on which vanish the $\theta_k(x)$ and $n-p-r$ covectors of the form $\varpi_k(x) - \sum_i t_{ik}\,\omega_i(x)$. Then substitute the values $\sum_i t_{ik}\,\omega_i(x)$ for $\varpi_k(x)$ in the expressions for $d\theta_k(x)$, and use the equation $(*)$.)

Bearing in mind that we may assume that the coefficients c_{ijk} are zero in the expression for $d\theta_k$ (by reason of the existence of an integral manifold which is the graph of a mapping of P into Q, after replacing P by a smaller open set if necessary), show that $\mathfrak{a}_2$ is generated by the $\omega_i \wedge \theta_k$ and 2-forms of the type

$$(**) \qquad \beta_k = \left(\omega_1 + \sum_{i=2}^{p} c_{ki}\,\omega_i\right) \wedge \varpi_k \qquad (1 \leq k \leq s),$$

(d) Show that the singular integral elements passing through a point $x \in \mathrm{M}$ are those which are contained in one of the hyperplanes with equations $\langle \zeta_k(x), \mathbf{h}_x \rangle = 0$, where

$$\zeta_k = \omega_1 + \sum_{i=2}^{p} c_{ki}\,\omega_i$$

(these 1-forms are not in general linearly independent).
(e) For every system of real numbers t_k ($1 \leqq k \leqq s$), the subspace of $\mathrm{T}_x(\mathrm{M})$ annihilated by the $\theta_l(x)$ and the s covectors $\varpi_k(x) - t_k\zeta_k(x)$ ($1 \leqq k \leqq s$) is a regular integral element of maximum dimension. Is the converse true?

Deduce that, if we put

$$d\zeta_k = \sum_{i<j} \mathrm{A}_{ij}^{(k)}\omega_i \wedge \omega_j + \sum_{h=1}^{s} \mathrm{A}_{ih}^{\prime(k)}\omega_i \wedge \varpi_h + \sum_{l=1}^{r} \mathrm{B}_{il}^{(k)}\omega_i \wedge \theta_l,$$

the values of the differential 3-forms

$$\sum_{i<j} \mathrm{A}_{ij}^{(k)}\omega_i \wedge \omega_j \wedge \zeta_k, \qquad \sum_{h,i} \mathrm{A}_{ih}^{\prime(k)}\omega_i \wedge \zeta_h \wedge \zeta_k$$

at the point x vanish on each integral element of maximum dimension at x. (Express that the $d\beta_k$ vanish on such an integral element, and use the fact that the t_k are arbitrary.)
(f) Deduce from (e) that, for each k such that $1 \leqq k \leqq s$, there exists in each p-dimensional integral manifold V, which is the graph of a mapping of an open subset of P into Q, a 1-parameter family of *characteristic* submanifolds of dimension $p-1$, defined by the equation ${}^t j(\zeta_k) = 0$ (where $j : \mathrm{V} \to \mathrm{M}$ is the canonical injection) and that this equation is *completely integrable*.

15. The notation and hypotheses are those of Problems 8 and 9. Put $\dim(\mathfrak{a}_1) = q+1$ and $\rho(\mathfrak{a}_1) = m$. If $\theta_0, \theta_1, \ldots, \theta_q$ form a basis of $\mathfrak{a}_1$ over $\mathscr{E}_0$, we propose to find q functions $u_j \in \mathscr{E}_0$ ($1 \leqq j \leqq q$) such that, putting

$$\Theta = \theta_0 + u_1\theta_1 + \cdots + u_q\theta_q \in \mathscr{E}_1,$$

we have the equation

$$\text{(1)} \qquad \Theta \wedge (d\Theta)^{\wedge(m+1)} = 0$$

(in other words, Θ is of Pfaffian class $\leqq 2m+1$ (Problem 3)).

Let $p : \mathrm{M} \times \mathbf{R}^q \to \mathrm{M}$ be the projection, and consider the differential 1-form on $\mathrm{M} \times \mathbf{R}^q$ whose value at $z = (x, v^1, \ldots, v^q) \in \mathrm{M} \times \mathbf{R}^q$ is

$$\Omega(z) = {}^t p(\theta_0)(z) + v^1 \cdot {}^t p(\theta_1)(z) + \cdots + v^q \cdot {}^t p(\theta_q)(z).$$

The problem consists in finding an integral manifold of the differential system on $\mathrm{M} \times \mathbf{R}^q$ consisting of the single equation

$$\text{(2)} \qquad \Omega \wedge (d\Omega)^{\wedge(m+1)} = 0,$$

and this manifold must be the graph of a mapping of an open subset of M into $\mathbf{R}^q$. To simplify the notation, we shall write θ_j in place of ${}^t p(\theta_j)$.

(a) Let $\mathscr{F}_0$ (resp. $\mathscr{F}_1$) denote the $\mathbf{R}$-algebra of C^∞-functions (resp. the space of C^∞ differential 1-forms) on the manifold $\mathrm{M} \times \mathbf{R}^q$. Show that the hypothesis $\rho(\mathfrak{a}_1) = m$ implies that there exists a basis over $\mathscr{F}_0$ of $(n+q)$ differential 1-forms in $\mathscr{F}_1$, consisting of $\theta_0, \theta_1, \ldots, \theta_q$ and $n-1$ forms ω_i ($1 \leqq i \leqq n-1$), the first $n-q-1$ of which do not contain the differentials $dv^1, \ldots, dv^q$, and the last q of which are $dv^1, \ldots, dv^q$; and such that there exist well-determined 1-forms $\varpi_1, \ldots, \varpi_q$ and ζ satisfying

$$d\Omega = \omega_1 \wedge \omega_2 + \cdots + \omega_{2m-1} \wedge \omega_{2m} + \theta_1 \wedge \varpi_1 + \cdots + \theta_q \wedge \varpi_q + \zeta \wedge \Omega,$$

where each 1-form ϖ_j is such that $\varpi_j + dv^j$ does not contain the differentials $dv^1, \ldots, dv^q$.

(b) Show that every n-dimensional integral element of (2) at a point z may be defined as the set of vectors in $T_z(M \times \mathbf{R}^q)$ which annihilate q covectors of the form

$$\varpi_j(z) - \sum_{i=1}^{m} (a_{2i-1,j}\omega_{2i}(z) - a_{2i,j}\omega_{2i-1}(z)) - \sum_{k=1}^{q} b_{jk}\theta_k(z) - c_j\Omega(z),$$

where the constants a_{ij}, b_{jk}, and c_j are chosen so that the difference

$$d\Omega(z) - \sum_{i=1}^{m}\left(\left(\omega_{2i-1}(z) + \sum_{j=1}^{q} a_{2i-1,j}\theta_j(z)\right) \wedge \left(\omega_{2i}(z) + \sum_{j=1}^{q} a_{2i,j}\theta_j(z)\right)\right)$$

is the product of a 1-covector and $\Omega(z)$. Consider the converse.

Deduce that the sum

$$ns_0(z) + (n-1)s_1(z, \mathbf{u}_1) + \cdots + s_{n-1}(z, \mathbf{u}_1, \ldots, \mathbf{u}_{n-1})$$

(Section **18.10**, Problem 2) is equal to

$$2mq + \tfrac{1}{2}q(q+3).$$

(c) Take $\mathbf{u}_1, \ldots, \mathbf{u}_n$ to be the basis dual to the basis of $T_x(M)^*$ consisting of the covectors $\omega_i(x)$ $(1 \leqq i \leqq n-q-1)$ and $\theta_j(x)$ $(0 \leqq j \leqq q)$. Show that, when the $\mathbf{u}_k$ are arranged in a suitable order, we have

$$s_0''(z) = \cdots = s_{2m+1}''(z, \mathbf{u}_1, \ldots, \mathbf{u}_{2m-1}) = 0,$$

$$s_{2m+2}''(z, \mathbf{u}_1, \ldots, \mathbf{u}_{2m+2}) = \cdots = s_{2m+q+1}''(z, \mathbf{u}_1, \ldots, \mathbf{u}_{2m+q+1}) = 1,$$

$$s_{2m+q+2}''(z, \mathbf{u}_1, \ldots, \mathbf{u}_{2m+q+2}) = \cdots = s_{n-1}''(z, \mathbf{u}_1, \ldots, \mathbf{u}_{n-1}) = 0,$$

in the notation of Section **18.10**, Problem 2.

Deduce that the differential system defined by (2) is in involution relative to M, and hence that there exist $q+1$ forms $\Theta_0, \Theta_1, \ldots, \Theta_q$ forming a basis of $\mathfrak{a}_1$ over $\mathscr{E}_0$ (by restricting M if necessary), all of which are of class $\leqq 2m+1$.

16. Consider the differential system in $\mathbf{R}^5$ defined by the ideal generated by the forms

$$x^1\, dx^1 \wedge dx^2, \qquad x^1\, dx^3 \wedge dx^4, \qquad dx^3 \wedge dx^4 \wedge dx^5.$$

Show that the dimension of the space N_x of characteristic vectors is constant and equal to 1, but that the mapping $x \mapsto N_x$ is not a continuous field of directions.

17. (a) Give another proof of (**18.16.8**) by showing successively that the coefficients of ω_j do not depend on x^n, then x^{n-1}, and so on up to $x^{n-\nu-1}$. For this purpose, put

$$\omega_j(x) = \sum_{k=1}^{n} a_{jk}(x)\, dx^k$$

and

$$i(\mathbf{e}_n) \cdot d\omega_j(x) = \sum_{k=1}^{n-r} b_{jk}\, \omega_k(x),$$

which gives rise to the conditions

$$D_n a_{jk} + \sum_{h=1}^{n-r} b_{jh} a_{hk} = 0;$$

then consider the system of linear differential equations

$$D_n z_j + \sum_{h=1}^{n-r} b_{jh} z_h = 0 \qquad (1 \leqq j \leqq n-r)$$

(in which the x^j with $j \leqq n-1$ are regarded as constants) and a fundamental system of solutions of these equations, by means of which the a_{jk} may be expressed linearly with coefficients which do not depend on x^n.

(b) Generalize the result of (18.16.8) to an arbitrary differential system, using a method analogous to that of (a) above.

(c) Consider the differential system in $\mathbf{R}^3$ defined by the differential ideal generated by the three forms

$$(x^1)^2\, dx^2, \qquad (x^1)^2\, dx^3, \qquad dx^2 \wedge dx^3.$$

Show that, although $\nu(x)$ is constant throughout $\mathbf{R}^3$, the result of (18.16.8) is not applicable on the whole of $\mathbf{R}^3$, because the dimension of $\mathfrak{a}_1(x)$ is not constant.

18. Let ω be a C^∞ Pfaffian form on an open subset M of $\mathbf{R}^n$, and suppose that ω is of Pfaffian class $\geqq 3$ at each point of M, i.e., that $\omega(x) \wedge d\omega(x) \neq 0$ for all $x \in$ M. Suppose that $0 \in$ M and that in a neighborhood of this point we have

$$\omega(x) = dx^n + \sum_{i=1}^{n-1} a_i(x)\, dx^i$$

with $a_i(0) = 0$ for $1 \leqq i \leqq n-1$.

(a) Show that, for each neighborhood V of 0 in M and each real number $c > 0$, there exists a neighborhood $W \subset V$ of 0 in M and a number $\rho \in \,]0, c[$, such that each point of W is the endpoint of a path of the form

$$\gamma : t \mapsto (tx^1, \ldots, tx^{n-1}, \varphi(t)),$$

of class C^∞, defined on $[0, 1]$ and with values in V, contained in a 1-dimensional integral manifold of the Pfaffian equation $\omega = 0$, and whose origin is of the form $(0, \ldots, 0, x^n)$ with $|x_n| \leqq \rho$. (Use (10.7.4).)

(b) Show that, for each neighborhood V of 0 in M, there exists $c > 0$ such that, for all x^n satisfying $|x^n| \leqq c$, there exists a path

$$\gamma : t \mapsto (\lambda\varphi_1(t), \ldots, \lambda\varphi_{n-1}(t), \psi(\lambda, t))$$

of class C^∞ defined on $[0, 1]$, with values in V, contained in a 1-dimensional integral manifold of the Pfaffian equation $\omega = 0$, with origin and endpoint $(0, \ldots, 0, x^n)$. (First choose the φ_j $(1 \leqq j \leqq n-1)$ arbitrarily, subject to $\varphi_j(0) = \varphi_j^*(1) = 0$. We may then, by virtue of (10.7.4), allow λ to take all values in an open neighborhood of 0 in $\mathbf{R}$. By using the hypothesis on ω, show that the φ_j may be chosen so that $\lim_{\lambda \to 0} \lambda^{-2}\psi(\lambda, 1)$ is a number >0 (or <0); hence deduce the result.)

(c) Deduce from (a) and (b) that the C^∞ Pfaffian forms ω satisfying $\omega \wedge d\omega = 0$ are characterized by the following property: for each point $x_0 \in M$, there exists a neighborhood V of x_0 in M such that each neighborhood $W \subset V$ of x_0 in M contains a point x which cannot be the endpoint of a path with origin x_0 contained in V, which is the juxtaposition of a finite number of C^∞ paths, each contained in a 1-dimensional integral manifold of $\omega = 0$ (*Carathéodory's inaccessibility theorem*).

17. EXAMPLES: I. FIRST-ORDER PARTIAL DIFFERENTIAL EQUATIONS

We shall show how the general theory we have developed applies to the case of a *single* pth order partial differential equation in *one* unknown function z of n variables $x^1, \ldots, x^n$:

(18.17.1)
$$F(x^1, \ldots, x^n, z, (D^\alpha z)_{|\alpha| \leq p}) = 0,$$

F being of class C^∞ in an open subset of $\mathbf{R}^N$, where $N = n + 1 + \nu$ and ν is the number of multi-indices $\alpha = (\alpha_1, \ldots, \alpha_n)$ such that $|\alpha| \leq p$.

We shall study first the case $p = 1$, since first-order equations have properties which do not extend to arbitrary equations of higher order.

Consider therefore a first-order equation

(18.17.2)
$$F(x^1, \ldots, x^n, z, p^1, \ldots, p^n) = 0,$$

where F is of class C^∞ in an open subset U of $\mathbf{R}^{2n+1}$, and is such that the equation (18.17.2) defines a *submanifold* M_0 of U. We have seen (18.8.9) that the differential system associated with this equation corresponds to the differential ideal generated by the 0-form F, the 1-forms

$$dF = \sum_{i=1}^{n} \frac{\partial F}{\partial x^i} dx^i + \frac{\partial F}{\partial z} dz + \sum_{i=1}^{n} \frac{\partial F}{\partial p^i} dp^i,$$

$$\omega = dz - \sum_{i=1}^{n} p^i \, dx^i$$

and the 2-form

$$-d\omega = \sum_{i=1}^{n} dp^i \wedge dx^i.$$

By virtue of (18.10.8), at a point $y = (x^1, \ldots, x^n, z, p^1, \ldots, p^n) \in M_0$, a vector subspace $G_y \subset \{y\} \times \mathbf{R}^{2n+1}$ is an *integral element* of this differential system if all the vectors $(y, \mathbf{u}) \in G_y$, where

$$\mathbf{u} = (v^1, \ldots, v^n, s, q^1, \ldots. g^n),$$

satisfy the two relations

(18.17.3)
$$\sum_{i=1}^{n} \frac{\partial F}{\partial x^i} v^i + \frac{\partial F}{\partial z} s + \sum_{i=1}^{n} \frac{\partial F}{\partial p^i} q^i = 0$$

and

(18.17.4)
$$s = \sum_{i=1}^{n} p^i v^i,$$

and if *any* two vectors $\mathbf{u}', \mathbf{u}'' \in G'_y$ satisfy the equation

(18.17.5)
$$\sum_{i=1}^{n} (v'^i q''^i - v''^i q'^i) = 0.$$

It follows from (18.17.4) that if $\mathbf{w} = (v^1, \ldots, v^n, q^1, \ldots, q^n)$ is the projection of $\mathbf{u}$ on the subspace $\mathbf{R}^{2n}$ spanned by the first n and the last n vectors of the canonical basis of $\mathbf{R}^{2n+1}$, then G_y is uniquely determined by its projection (of the same dimension) G'_y in $\mathbf{R}^{2n}$, and the vectors $\mathbf{w} \in G'_y$ must satisfy the relation

(18.17.6)
$$\sum_{i=1}^{n} \left(\frac{\partial F}{\partial x^i} + p^i \frac{\partial F}{\partial z}\right) v^i + \sum_{i=1}^{n} \frac{\partial F}{\partial p^i} q^i = 0;$$

also any two vectors $\mathbf{w}', \mathbf{w}'' \in G'_y$ must satisfy $B(\mathbf{w}', \mathbf{w}'') = 0$, where B is the *alternating bilinear form* of rank $2n$ on $\mathbf{R}^{2n}$ which appears as the left-hand side of (18.17.5).

We shall consider only the points of M_0 at which the $2n$ coefficients in the linear form on the left-hand side of (18.17.6) are not all zero. Then the equation (18.17.6) defines a *hyperplane* H_y in $\mathbf{R}^{2n}$, and the preceding remarks show that the subspaces G'_y are the *totally isotropic subspaces* (relative to the form B) *contained in* H_y (A.16.3). The line D_y orthogonal to H_y (relative to B) evidently contains the vector

(18.17.7)
$$\mathbf{w}_0 = \left(\frac{\partial F}{\partial p^1}, \ldots, \frac{\partial F}{\partial p^n}, -\left(\frac{\partial F}{\partial x^1} + p^1 \frac{\partial F}{\partial z}\right), \ldots, -\left(\frac{\partial F}{\partial x^n} + p^n \frac{\partial F}{\partial z}\right)\right) \in H_y.$$

Hence (A.16.3) the totally isotropic subspaces contained in H_y are *of dimension* $\leqq n$; those which are of dimension n contain D_y, and if V_y is a supplement (of dimension $2n - 2$) of D_y in H_y, then their intersection with V_y is a totally isotropic subspace of V_y of *maximum* dimension $n - 1$.

(18.17.8) Since we are interested only in the integral manifolds of the above differential system which, when projected on $\mathbf{R}^{n+1}$, give the graph of a solution of (18.17.2), the only n-dimensional integral elements which need be considered are those whose projection G_y'' on $\mathbf{R}^n$ (identified with the subspace of $\mathbf{R}^{2n}$ spanned by the first n vectors of the canonical basis) has dimension n. In view of the expression (18.17.7), there exist no such subspaces G_y' unless at least one of the partial derivatives $\partial F/\partial p^i$ is *nonzero* at the point y under consideration.

The theory of alternating bilinear forms shows also that if linearly independent vectors $\mathbf{w}_1, \ldots, \mathbf{w}_{n-1}$ in $G_y' \cap V_y$ are determined successively as in (18.10.3), then we have $s_0 = 2$ and $s_1 = \cdots = s_{n-1} = 1$, and therefore every integral element contained in V_y is *regular*.

Moreover, it follows from (18.16) that in each integral element G_y of dimension n there exist *Cauchy characteristic vectors*, all of which are scalar multiples of the vector $(y, \mathbf{u}_0)$, where

(18.17.8.1)

$$\mathbf{u}_0 = \left(\frac{\partial F}{\partial p^1}, \ldots, \frac{\partial F}{\partial p^n}, \sum_{i=1}^{n} p^i \frac{\partial F}{\partial p^i}, -\left(\frac{\partial F}{\partial x^1} + p^1 \frac{\partial F}{\partial z}\right), \ldots, -\left(\frac{\partial F}{\partial x^n} + p^n \frac{\partial F}{\partial z}\right)\right).$$

As y runs through M_0, these vectors therefore form a C^∞ vector field, called the *characteristic field*.

(18.17.9) This will enable us to show, by using (18.16.8), that for a first-order equation (18.17.2) *Cauchy's problem can be solved without assuming that the data are analytic*. We shall pose this problem in a more general form than in (18.12.1): we consider in $\mathbf{R}^{n+1}$ a *submanifold* T *of dimension* $n-1$, and we wish to determine whether there exists a submanifold S of dimension n *containing* T, defined by an equation $z = f(x^1, \ldots, x^n)$, where f is a *solution of* (18.17.2). Clearly it is necessary to assume that the restriction to T of the projection on $\mathbf{R}^n$ is injective. In fact we shall assume that it is an *injective immersion*.

We shall see that in general it is possible, by restricting T to a neighborhood of one of its points, to find in $\mathbf{R}^{2n+1}$ an *integral manifold* T′ *of dimension* $n-1$ of the differential system associated with the equation (18.17.2) which has T as its *projection* on $\mathbf{R}^{n+1}$. For this purpose consider a chart of T, and let $\mathbf{t} = (t^1, \ldots, t^{n-1})$ be the corresponding system of local coordinates of the point $x = (x^1, \ldots, x^n)$. Then we have say

$$x^i = g^i(t^1, \ldots, t^{n-1}) \qquad (1 \leqq i \leqq n),$$
$$z = h(t^1, \ldots, t^{n-1}),$$

and by hypothesis the Jacobian matrix of the n functions g^i is of maximum rank $n-1$. The problem is then to determine the p^i $(1 \leqq i \leqq n)$ as functions of $(t^1, \dots, t^{n-1})$ such that we have identically

$$\begin{cases} \sum_{i=1}^{n} p^i \dfrac{\partial g^i}{\partial t^k} = \dfrac{\partial h}{\partial t^k} \qquad (1 \leqq k \leqq n-1), \\ \mathrm{F}(g^1(\mathbf{t}), \dots, g^n(\mathbf{t}), h(\mathbf{t}), p^1, \dots, p^n) = 0. \end{cases} \tag{18.17.10}$$

We shall *assume* that there exists at least one solution $(p_0^1, \dots, p_0^n)$ of this system for a value $\mathbf{t}_0 = (t_0^1, \dots, t_0^{n-1})$, such that at the point

$$(t_0^1, \dots, t_0^{n-1}, p_0^1, \dots, p_0^n)$$

the determinant

$$\begin{vmatrix} \partial g^1/\partial t^1 & \cdots & \partial g^1/\partial t^{n-1} & \partial \mathrm{F}/\partial p^1 \\ \vdots & \cdots & \vdots & \vdots \\ \partial g^n/\partial t^1 & \cdots & \partial g^n/\partial t^{n-1} & \partial \mathrm{F}/\partial p^n \end{vmatrix} \tag{18.17.11}$$

does not vanish; the implicit function theorem (10.2.3) then establishes the existence of functions $p^i = q^i(t^1, \dots, t^{n-1})$ of class C^∞ with the desired properties, and such that the determinant (18.17.11) is nonzero on a nonempty open set in $\mathbf{R}^{n-1}$. This latter condition means that the n columns of the determinant are linearly independent, that is to say, that the vector

$$(\partial \mathrm{F}/\partial p^1, \dots, \partial \mathrm{F}/\partial p^n)$$

and the projection on $\mathbf{R}^n$ of the tangent space to T span $\mathbf{R}^n$; and this implies *a fortiori* that the characteristic vector $\mathbf{u}_0$ is not contained in the tangent space to T′, and hence (18.16.8) may be applied to T′. This gives us an n-dimensional integral manifold S′ of the differential system associated to (18.17.2), and by reason of the remark above, the projection S of S′ on $\mathbf{R}^{n+1}$ is of dimension n, and the projection of S on $\mathbf{R}^n$ is a local diffeomorphism. Consequently S is locally of the form $z = f(x^1, \dots, x^n)$, and Cauchy's problem is therefore solved.

Remarks

(18.17.12) When there exists *only one* solution $(p_0^1, \dots, p_0^n)$ of the system (18.17.10) for a given $\mathbf{t}_0$, such that this solution does not make the determinant (18.17.11) vanish, then the solution of Cauchy's problem considered above is *unique* (in a sufficiently small neighborhood of the point of T corresponding to $\mathbf{t}_0$). This follows from the proof of (18.16.8), which shows that every integral manifold of *maximum* dimension of (18.16.7.1) is necessarily of the form $\pi^{-1}(\mathrm{L})$, where L is an integral manifold of maximum dimension of the Pfaffian system (18.16.7.1) considered as a Pfaffian system *in* P.

On the other hand, when the determinant (18.17.11) is *zero*, it can happen that Cauchy's problem has *no* solution or that it has *infinitely many* solutions. Take, for example, $n = 2$ and consider the first-order equation

$$p^1 x^1 + p^2 x^2 = 0.$$

One finds immediately that the projections on $\mathbf{R}^3$ of the Cauchy characteristics are the lines

(18.17.12.1) $$ax^1 + bx^2 = 0, \qquad z = c,$$

where a, b, c are three constants (and a, b are not both zero). Here T must be taken to be a curve in $\mathbf{R}^3$, and the surface S which is to be the solution of Cauchy's problem must be obtained as a union of lines (18.17.12.1) which *meet* T. But if, for example, T is the curve $x^2 = (x^1)^2$, $z = (x^1)^3$, then the set S obtained by this procedure fails to be a submanifold of $\mathbf{R}^3$ at the point $(0, 0, 0)$ (i.e., this point is "singular"); hence Cauchy's problem has no solution in a neighborhood of this point. If on the other hand T is one of the lines (18.17.12.1), then Cauchy's problem has infinitely many solutions.

PROBLEMS

1. Consider a system of m first-order partial differential equations in one unknown function z of n real variables $x^1, \ldots, x^n$:

(1) $$\mathrm{F}_j(x^1, \ldots, x^n, z, p^1, \ldots, p^n) = 0 \qquad (1 \leq j \leq m),$$

where the F_j are of class C^∞ in an open set $\mathrm{U} \subset \mathbf{R}^{2n+1}$ and are such that the equations (1) define a submanifold M_0 of U. For each point $y \in \mathrm{M}_0$ consider the m vectors in $\mathbf{R}^{2n+1}$:

$$\mathbf{w}_0^{(j)} = \left(\frac{\partial \mathrm{F}_j}{\partial p^1}, \ldots, \frac{\partial \mathrm{F}_j}{\partial p^n}, -\left(\frac{\partial \mathrm{F}_j}{\partial x^1} + p^1 \frac{\partial \mathrm{F}_j}{\partial z}\right), \ldots, -\left(\frac{\partial \mathrm{F}_j}{\partial x^n} + p^n \frac{\partial \mathrm{F}_j}{\partial z}\right)\right) \qquad (1 \leq j \leq m),$$

which we assume are all $\neq 0$. In order that there should exist an n-dimensional integral element of the system (1) passing through y, it is necessary that the vectors $\mathbf{w}_0^{(j)}$ should be *pairwise orthogonal* relative to the alternating bilinear form defined in (18.17.5). In other words, there can exist no C^∞-function z satisfying (1) in an open set $\mathrm{V} \subset \mathbf{R}^n$ (with $p^i = \partial z/\partial x^i$), unless z is also a solution of the partial differential equations obtained by setting equal to zero the *nonhomogeneous Poisson brackets* (Section 17.15, Problem 9)

$$\{\mathrm{F}_j, \mathrm{F}_k\} = -\sum_{i=1}^{n}\left(\frac{\partial \mathrm{F}_j}{\partial p^i}\left(\frac{\partial \mathrm{F}_k}{\partial x^i} + p^i \frac{\partial \mathrm{F}_k}{\partial z}\right) - \frac{\partial \mathrm{F}_k}{\partial p^i}\left(\frac{\partial \mathrm{F}_j}{\partial x^i} + p^i \frac{\partial \mathrm{F}_j}{\partial z}\right)\right) \qquad (1 \leq j < k \leq m).$$

Suppose that the brackets $\{F_i, F_j\}$ are identically zero in M_0 and that the m vectors $\mathbf{w}_0^{(j)}$ are such that their projections on $\mathbf{R}^n$,

$$\mathbf{v}_0^{(j)} = (\partial F_j/\partial p^1, \ldots, \partial F_j/\partial p^n),$$

are linearly independent. Let N_x denote the space of Cauchy characteristic vectors at $x \in M_0$. Show that N_x is of dimension m, and that for each submanifold T of dimension $n-m$ in $\mathbf{R}^{n+1}$ whose projection on $\mathbf{R}^n$ is an immersion and which is such that the subspace of dimension m spanned by the $\mathbf{v}_0^{(j)}$ is supplementary to the tangent space to the projection T′ of T, there exists an n-dimensional integral manifold S, containing the intersection of T with a sufficiently small open set, and defined by an equation of the form $z = f(x^1, \ldots, x^n)$.

Consider the case where z is absent from the equations (1). (The Poisson brackets are then homogeneous.)

Consider also the particular case of systems of the form $\theta_{X_j} \cdot f = 0$, giving simultaneous *first integrals* of a system of vector fields X_j $(1 \leq j \leq m)$ on M, linearly independent at each point of M (Section 18.2, Problem 12). In order that there should exist a first integral taking an assigned value at any point of M, it is necessary and sufficient that the field of m-directions $x \mapsto L_x$ should be *completely integrable*, where L_x is the subspace of $T_x(M)$ spanned by the vectors $X_j(x)$. (The system of partial differential equations $\theta_{X_j} \cdot f = 0$ $(1 \leq j \leq m)$ is said to be a *complete system* when this condition is satisfied.)

2. Given a first-order partial differential equation (18.17.2), a *complete integral* of the equation in an open set $V \subset \mathbf{R}^n$ is by definition a C^∞-mapping

$$(x^1, \ldots, x^n, X^1, \ldots, X^n) \mapsto \Phi(x^1, \ldots, x^n, X^1, \ldots, X^n)$$

of $V \times W$ into $\mathbf{R}$ (where W is open in $\mathbf{R}^n$) such that:

(1) for each point $(X^1, \ldots, X^n) \in W$, the mapping

$$(x^1, \ldots, x^n) \mapsto \Phi(x^1, \ldots, x^n, X^1, \ldots, X^n)$$

is a solution of (18.17.2) in V;

(2) the determinant of the matrix $(\partial^2\Phi/\partial x^i\, \partial X^j)$ does not vanish in $V \times W$.

(a) Show that in $\mathbf{R}^{4n+1}$ the equations

$$\begin{aligned} z - \Phi(x^1, \ldots, x^n, X^1, \ldots, X^n) &= 0 \\ p^i - \partial\Phi/\partial x^i &= 0 \qquad (1 \leq i \leq n), \\ Y^i - \partial\Phi/\partial X^i &= 0 \qquad (1 \leq i \leq n) \end{aligned}$$

define a submanifold G of dimension $2n$, whose projection on $\mathbf{R}^{2n+1}$ is the submanifold $M_0 \subset U$ defined by (18.17.2); furthermore, in a neighborhood of each of its points, G is the graph of a local diffeomorphism π of an open subset of M_0 into $\mathbf{R}^{2n}$, and we have

$${}^t\pi\left(\sum_{i=1}^{n} Y^i\, dX^i\right) = {}^t j(\omega),$$

where $j : M_0 \to \mathbf{R}^{2n+1}$ is the canonical injection.

(b) Show that the knowledge of a complete integral permits the solution of Cauchy's problem without integration of differential equations. With the notation of (**18.17.9**) and (**18.17.10**), the equations

$$\begin{aligned} x^i &= g^i(t^1, \ldots, t^{n-1}) \qquad (1 \leqq i \leqq n), \\ z &= h(t^1, \ldots, t^{n-1}), \\ p^i &= q^i(t^1, \ldots, t^{n-1}) \qquad (1 \leqq i \leqq n), \end{aligned}$$

define a submanifold N of M_0 of dimension $n-1$; hence (restricting M_0 if necessary) $N' = \pi(N)$ in $\mathbf{R}^{2n}$ is a submanifold, is defined by equations of the form

$$X^i = G^i(t^1, \ldots, t^{n-1}), \qquad Y^i = Q^i(t^1, \ldots, t^{n-1}) \qquad (1 \leqq i \leqq n);$$

the n equations

$$\begin{aligned} &\sum_{i=1}^{n} Q_i \frac{\partial G^i}{\partial t^j} = 0 \qquad (1 \leqq j \leqq n-1), \\ &z - \Phi(x^1, \ldots, x^n, G^1, \ldots, G^n) = 0 \end{aligned}$$

then define a submanifold of the space $\mathbf{R}^{n-1} \times \mathbf{R}^{n+1}$ whose projection on $\mathbf{R}^{n+1}$ is the required integral manifold.

(c) Suppose that n first integrals (Section **18.2**, Problem 12) $f_1, \ldots, f_n$ of the field of characteristic vectors (**18.17.8.1**) are such that the functional determinant

$$\partial(f_1, \ldots, f_n, F)/\partial(p^1, \ldots, p^n, z)$$

does not vanish in M. Show that in a sufficiently small open subset of $\mathbf{R}^{3n+1}$, the differential manifold defined by the $n+1$ equations

$$\begin{aligned} f_j(x^1, \ldots, x^n, z, p^1, \ldots, p^n) - X^j &= 0 \qquad (1 \leqq j \leqq n), \\ F(x^1, \ldots, x^n, z, p^1, \ldots, p^n) &= 0 \end{aligned}$$

is the graph of a complete integral of (**18.17.2**).

3. (a) With the notation of Section **16.20**, Problem 3, show that for two functions F, G defined on V′ we have

$$\{F \circ f, G \circ f\} = \rho^{-1}(\{F, G\} \circ f)$$

for the nonhomogeneous Poisson brackets (Section **17.15**, Problem 9). In particular, if we put $X^i = x'^i \circ f$, $Y_i = y'_i \circ f$, $Z = z' \circ f$ (the scalar components of f), we have

$$\begin{aligned} &\{Z, X^i\} = \{X^i, X^k\} = \{X^i, Y_k\} = \{Y_i, Y_k\} = 0 \qquad \text{for } 1 \leqq i < k \leqq n, \\ &\{Z, Y_i\} = \rho^{-1} Y_i, \qquad \{Y_i, X^i\} = -\rho^{-1} \qquad \text{for } 1 \leqq i \leqq n. \end{aligned}$$

(b) Conversely, let X^i $(1 \leqq j \leqq n)$ and Z be $n+1$ scalar functions defined on V, such that $\{Z, X^i\} = \{X^i, X^k\}$ for $1 \leqq i, k \leqq n$. Show that if the differentials dZ, dX^i $(1 \leqq i \leqq n)$ are linearly independent at each point of V, then in each sufficiently small neighborhood of a point of V there exist n functions Y_i such that $f = (Z, (X^i), (Y_i))$ is a nonhomogeneous contact transformation defined in this neighborhood. (Use Problem 1.)

(c) Consider likewise the case of homogeneous contact transformations (**16.20.12**).

18. EXAMPLES: II. SECOND-ORDER PARTIAL DIFFERENTIAL EQUATIONS

Consider now a second-order partial differential equation

$$\text{(18.18.1)} \qquad F(x^1, \ldots, x^n, z, p^1, \ldots, p^n, (p^{ij})_{1 \leqq i \leqq j \leqq n}) = 0,$$

where F is a real-valued C^∞-function on an open set $U \subset \mathbf{R}^N$, and $N = 2n + 1 + \frac{1}{2}n(n+1)$. A *solution* of this equation is a C^2-function $v(x^1, \ldots, x^n)$, defined on an open set $V \subset \mathbf{R}^n$, such that the point of $\mathbf{R}^N$ with coordinates

$$x^1, \ldots, x^n, v(x^1, \ldots, x^n), \frac{\partial v}{\partial x^1}, \ldots, \frac{\partial v}{\partial x^n}, \left(\frac{\partial^2 v}{\partial x^i\, \partial x^j}\right)_{1 \leqq i \leqq j \leqq n}$$

belongs to U and satisfies (18.18.1) for all $(x^1, \ldots, x^n) \in V$. We shall assume that (18.18.1) defines a *submanifold* M_0 of U. The general method (18.8.10) consists in associating with the equation (18.18.1) the differential system corresponding to the differential ideal generated by the 0-form F, the 1-forms

$$dF = \sum_{i=1}^{n} \frac{\partial F}{\partial x^i} dx^i + \frac{\partial F}{\partial z} dz + \sum_{i=1}^{n} \frac{\partial F}{\partial p^i} dp^i + \sum_{i \leqq j} \frac{\partial F}{\partial p^{ij}} dp^{ij},$$

$$\omega = dz - \sum_{i=1}^{n} p^i\, dx^i,$$

$$\omega_i = dp^i - \sum_{j=1}^{n} p^{ij}\, dx^i \qquad (1 \leqq i \leqq n)$$

(where we have put $p^{ij} = p^{ji}$ for $i > j$), and the exterior differentials of the forms ω and ω_i. Since

$$p^{ij}\, dx^j \wedge dx^i + p^{ji}\, dx^i \wedge dx^j = 0,$$

we have

$$-d\omega = \sum_{i=1}^{n} dp^i \wedge dx^i = \sum_{i=1}^{n} \omega_i \wedge dx^i,$$

and therefore $d\omega$ already belongs to the differential ideal generated by the ω_i $(1 \leqq i \leqq n)$.

The same reasoning as in (18.17), using the 2-form $d\omega$, shows first of all that the integral elements of this differential system, being totally isotropic subspaces relative to an alternating bilinear form of rank $2n$, are *of dimension at most n*.

We shall show that in general there exist integral elements of dimension n whose projection on $\mathbf{R}^n$ (identified with the subspace of $\mathbf{R}^N$ spanned by the first n vectors of the canonical basis) is surjective. At the same time we shall show that in general there exist integral manifolds of the differential system associated with (18.18.1) *of dimension* $n-1$, and that for such a manifold a *Cauchy problem* can be posed (but in general this Cauchy problem *does not necessarily have a solution*: cf. Chapter XXIII).

Consider, then, as in (18.17), a submanifold T of dimension $n-1$ in $\mathbf{R}^{n+1}$. We shall assume that its projection on $\mathbf{R}^n$ is an injective immersion, and we seek to construct an integral manifold T′ of dimension $n-1$ in $\mathbf{R}^N$ which projects onto T. By restricting U if necessary, we may suppose that T is defined on the projection of U on $\mathbf{R}^{n+1}$ by

(18.18.2)
$$\begin{cases} x^i = g^i(t^1, \ldots, t^{n-1}) \qquad (1 \leqq i \leqq n), \\ z = h(t^1, \ldots, t^{n-1}), \end{cases}$$

where $\mathbf{t} = (t^1, \ldots, t^{n-1})$ is a system of local coordinates and the Jacobian matrix of the g^i is of maximum rank $n-1$. Then we have to determine the p^i $(1 \leqq i \leqq n)$ and the p^{ij} $(1 \leqq i \leqq j \leqq n)$ as functions of $\mathbf{t} = (t^1, \ldots, t^{n-1})$ such that we have identically

(18.18.3)
$$\begin{cases} \sum_{i=1}^{n} p^i \dfrac{\partial g^i}{\partial t^k} = \dfrac{\partial h}{\partial t^k} \qquad (1 \leqq k \leqq n-1), \\ \sum_{i=1}^{n} p^{ij} \dfrac{\partial g^j}{\partial t^k} = \dfrac{\partial p^i}{\partial t^k} \qquad (1 \leqq k \leqq n-1, \quad 1 \leqq i \leqq n), \\ \mathrm{F}(g^1(\mathbf{t}), \ldots, g^n(\mathbf{t}), h(\mathbf{t}), p^1, \ldots, p^n, (p^{ij})) = 0 \end{cases}$$

(where $p^{ij} = p^{ji}$ when $i > j$).

Let us show that this is possible in general. Without loss of generality, we may limit ourselves to the case where $x^i = t^i$ for $1 \leqq i \leqq n-1$; then the equations (18.18.3), apart from the last one, become

(18.18.4)
$$p^k + p^n \frac{\partial g^n}{\partial x^k} = \frac{\partial h}{\partial x^k} \qquad (1 \leqq k \leqq n-1),$$

(18.18.5)
$$p^{ik} + p^{in} \frac{\partial g^n}{\partial x^k} = \frac{\partial p^i}{\partial x^k} \qquad (1 \leqq k \leqq n-1, \quad 1 \leqq i \leqq n).$$

The equations (18.18.4) determine the p^i for $i \leqq n-1$, once p^n is known. Equations (18.18.5), when $i = n$, give

(18.18.6)
$$p^{nk} + p^{nn} \frac{\partial g^n}{\partial x^k} = \frac{\partial p^n}{\partial x^k} \qquad (1 \leqq k \leqq n-1)$$

and then, for $i < n$, by substituting the value of $p^{in} = p^{ni}$ obtained from (18.18.6) into (18.18.5),

$$p^{ik} = \frac{\partial p^i}{\partial x^k} - \frac{\partial p^n}{\partial x^i}\frac{\partial g^n}{\partial x^k} + p^{nn}\frac{\partial g^n}{\partial x^i}\frac{\partial g^n}{\partial x^k} \qquad (1 \leqq i \leqq k \leqq n-1). \tag{18.18.7}$$

Hence all the p^{ik} are known once p^{nn} is known. This being so, consider a point of the manifold M_0,

$$y_0 = (x_0^1, \ldots, x_0^n, z_0, p_0^1, \ldots, p_0^n, (p_0^{ij})_{1 \leqq i \leqq j \leqq n}).$$

We can determine g^n so that $g^n(x_0^1, \ldots, x_0^{n-1}) = x_0^n$, the values ζ_k of the derivatives $\partial g^n/\partial x^k$ at the point $w_0 = (x_0^1, \ldots, x_0^{n-1})$ being for the moment arbitrary. The equations (18.18.4) then give the values of the $\partial h/\partial x^k$ at this point, and also we have $h(x_0^1, \ldots, x_0^{n-1}) = z_0$. Clearly we can construct (in infinitely many ways) functions g^n, h, φ^k $(1 \leqq k \leqq n)$ of class C^∞ whose values, and the values of their first derivatives, at the point w_0, are the numbers above; and it now remains to show that there exists a function p_0^{nn} defined in a neighborhood W of w_0 which takes the value p_0^{nn} at this point and is such that

$$\Phi(x^1, \ldots, x^{n-1}, p^{nn}) = 0$$

identically in W, where Φ is the function obtained from F by replacing x^n, z, p^k $(1 \leqq k \leqq n)$ by the functions g^n, h, φ^k, and the p^{ik} other than p^{nn} by the functions determined by (18.18.6) and (18.18.7). The implicit function theorem then guarantees the existence and uniqueness of p^{nn} under the hypothesis

$$\partial \Phi/\partial p^{nn} \neq 0$$

at the point w_0; in view of (18.18.6) and (18.18.7), this is equivalent to

$$\left(\frac{\partial F}{\partial p^{nn}}\right)_0 - \sum_{k=1}^{n-1}\left(\frac{\partial F}{\partial p^{kn}}\right)_0 \zeta_k + \sum_{1 \leqq i \leqq k \leqq n-1}\left(\frac{\partial F}{\partial p^{ik}}\right)_0 \zeta_i \zeta_k \neq 0, \tag{18.18.8}$$

the derivatives of F being evaluated at the point y_0. It will always be possible to satisfy this inequality provided that these derivatives are not all zero, and this establishes our assertion.

We shall now show that, under the assumptions we have made, there is a *unique* integral element of dimension n containing the tangent space $T_{y_0}(T')$ at the point y_0. Let $(\mathbf{e}_j)_{1 \leqq j \leqq n}$ be the canonical basis of $\mathbf{R}^n$, and for $1 \leqq j \leqq n-1$ let $\mathbf{u}_j$ be the vector in $T_{y_0}(T')$ whose projection on $\mathbf{R}^n$ is $\mathbf{e}_j + \zeta_j \mathbf{e}_n$. These $n-1$ vectors form a basis of $T_{y_0}(T')$, and it is sufficient to show that there is only *one* vector $\mathbf{u}$ in the integral element in question whose projection on $\mathbf{R}^n$ is $\mathbf{e}_n$. Let

$$\mathbf{u} = (0, \ldots, 0, 1, v, q^1, \ldots, q^n, (q^{ij})_{1 \leqq i \leqq j \leqq n}).$$

The equation $\langle \mathbf{u}, \omega(y_0) \rangle = 0$ gives $v = p_0^n$, and the equations

$$\langle \mathbf{u}, \omega_i(y_0) \rangle = 0 \qquad (1 \leqq i \leqq n)$$

give $q^i = p_0^{in}$. Next consider the $n(n-1)$ relations

$$\langle \mathbf{u}_j \wedge \mathbf{u}, d\omega_i(y_0) \rangle = 0 \qquad (1 \leqq j \leqq n-1, \quad 1 \leqq i \leqq n);$$

for $1 \leqq j \leqq n - 1$, we have

$$\begin{aligned} &\langle dx^k, \mathbf{u}_j \rangle = 0 \qquad \text{for} \quad k \leqq n-1 \quad \text{and} \quad k \neq j, \\ &\langle dx^j, \mathbf{u}_j \rangle = 1, \\ &\langle dx^n, \mathbf{u}_j \rangle = \zeta_j; \end{aligned}$$

on the other hand, $\langle dx^k, \mathbf{u} \rangle = 0$ for $k \leqq n-1$ and $\langle dx^n, \mathbf{u} \rangle = 1$. Hence

$$\langle dp^{ij}, \mathbf{u} \rangle + \langle dp^{in}, \mathbf{u} \rangle \zeta_j - \langle dp^{in}, \mathbf{u}_j \rangle = 0$$

which therefore determines $q^{ij} + q^{in}\zeta_j = \langle dp^{in}, \mathbf{u}_j \rangle$ for $i \leqq n$ and $j \leqq n-1$. Just as in (18.18.7), we deduce the values of the q^{ij} as functions of the single number q^{nn}; but in the equation of the first degree so obtained for q^{nn}, the coefficient of q^{nn} is the expression (18.18.8). This establishes the uniqueness of the integral element of dimension n contining $T_{y_0}(T')$. Moreover, this element is *regular*, since the condition (18.18.8) remains satisfied for points of M_0 close to y_0 and values of ζ_i close to the original values.

By contrast with the case of first-order equations (18.17), there does *not* in general exist an integral manifold of the equation which contains T′, except when the data are analytic (in which case the Cartan–Kähler theorem (18.13.8) applies).

The same calculation shows that if E is a *regular* integral element of dimension n at the point y_0 for the differential system associated with (18.18.1), then the *singular* integral elements F of dimension $n - 1$ are the intersections of E with hyperplanes $\{y_0\} \times (H \times \mathbf{R}^{N-n})$, where $H \subset \mathbf{R}^n$ is a hyperplane given by the equation $\xi_1 x^1 + \cdots + \xi_n x^n = 0$, satisfying the condition

(18.18.9)
$$\sum_{1 \leqq i \leqq j \leqq n} \left(\frac{\partial F}{\partial p^{ij}} \right)_0 \xi_i \xi_j = 0;$$

for we may, for example, suppose that $\xi_n = 1$, and then H is spanned by the $n - 1$ vectors $\mathbf{e}_j - \xi_j \mathbf{e}_n$ $(1 \leqq j \leqq n-1)$, and the left-hand side of (18.18.9) is then identical with the left-hand side of (18.18.8) if we replace ζ_j by $-\xi_j$ for $1 \leqq j \leqq n-1$.

Since there is at least the integral element E of dimension n containing F, it follows that there are *infinitely many* such elements, and therefore F is indeed singular.

Suppose that N is a *regular* integral manifold of the differential system associated with (18.18.1), and that the projection of N on $\mathbf{R}^{n+1}$ is the graph of a solution of (18.18.1). Then the *characteristic manifolds* of dimension $n-1$ contained in N are obtained by intersecting N with the manifolds $Q \times \mathbf{R}^{N-n}$, where Q is an $(n-1)$-dimensional integral manifold of the *first-order partial differential equation*

$$\sum_{1 \leqq i \leqq j \leqq n} A_{ij}(x^1, \ldots, x^n) p^i p^j = 0 \tag{18.18.10}$$

in which A_{ij} is the function obtained from $\partial F/\partial p^{ij}$ by replacing z, p^i and p^{ij} by their expressions derived from the equations of N as functions of $x^1, \ldots, x^n$. It should be observed carefully that this equation will *in general depend on the integral manifold* N under consideration.

These results are easily generalized to partial differential equations of arbitrary order $m \geqq 2$,

$$F(x^1, \ldots, x^n, z, (p^\alpha)_{|\alpha| \leqq m}) = 0, \tag{18.18.11}$$

in one unknown function; a solution is a function $v(x^1, \ldots, x^n)$ such that when z is replaced by v, and p^α by $D^\alpha v$ in (18.18.11), the result is identically zero in an open subset of $\mathbf{R}^n$. The condition (18.18.9) is replaced by

$$\sum_{|\alpha| = m} \left(\frac{\partial F}{\partial p^\alpha}\right)_0 \xi^\alpha = 0, \tag{18.18.12}$$

where $\xi^\alpha = \xi_1^{\alpha_1} \cdots \xi_n^{\alpha_n}$, if α is the multi-index $(\alpha_1, \ldots, \alpha_n)$.

CHAPTER XIX

LIE GROUPS AND LIE ALGEBRAS

Our first application of the techniques of "analysis on differential manifolds" developed throughout the last three chapters will be to the study of the structure of Lie groups. The fact that makes this possible is that the translations (on the left, say) in a Lie group G are diffeomorphisms, and therefore act via transport of structure on every object intrinsically attached to the manifold structure of the group G. In particular, we obtain in this way a generalization of the operators of differentiation (of arbitrary order) in $\mathbf{R}^n$ with respect to the coordinates (or, more accurately, linear combinations of these operators, with constant coefficients): they are the differential operators which are *invariant* under left translations (a notion which obviously has no meaning on an arbitrary differential manifold). These operators form an (associative) subalgebra $\mathfrak{G}$ of the algebra Diff(G) of all differential operators on G; but whereas in the case of $\mathbf{R}^n$ this algebra is commutative (and, in fact, isomorphic to a polynomial algebra), in the case of a general Lie group G its structure reflects very faithfully what may be called the "localized" group structure in a neighborhood of the neutral element (**19.5.9**). This fact naturally suggests replacing the study of G by the purely algebraic investigation of the algebra $\mathfrak{G}$.

At this point, a remarkable phenomenon intervenes. The *first-order* differential operators belonging to $\mathfrak{G}$ no longer form an associative algebra, but they do form a Lie algebra for the bracket $[X, Y] = XY - YX$; this Lie algebra determines completely the associative algebra $\mathfrak{G}$, and consequently the local structure of the group G. More precisely, there is a canonical one-to-one correspondence between simply connected (real) *Lie groups* and finite-dimensional real *Lie algebras* (up to isomorphism). This result, which requires a rather deep analysis of the structure of Lie algebras, will not be completely proved until Chapter XXI, but already in this chapter we shall be able to compile a "dictionary" which will associate with each of the classical notions of group-theory (homomorphism, kernel, image, normalizer, centralizer, commutator subgroup, semidirect product, etc.) a corresponding notion for Lie algebras. The point of this is that it enables us to reduce to problems of *linear algebra* the study of these notions in Lie groups, because we can "climb back" from the Lie algebra to the Lie group from which it comes, with the help of the theory of completely integrable systems (**19.7.4**).

1. EQUIVARIANT ACTIONS OF LIE GROUPS ON FIBER BUNDLES

Let G be a Lie group and let E be a fiber bundle with base B and projection π. We recall that G is said to act *equivariantly* (relative to π) on E and B if G acts differentiably on E and on B and if, for each $x \in \mathrm{E}$ and each $s \in \mathrm{G}$, we have

(**19.1.1**) $$\pi(s \cdot x) = s \cdot \pi(x).$$

It follows that for each $b \in \mathrm{B}$ we have $s \cdot \mathrm{E}_b \subset \mathrm{E}_{s \cdot b}$ and $s^{-1} \cdot \mathrm{E}_{s \cdot b} \subset \mathrm{E}_b$, so that $s \cdot \mathrm{E}_b = \mathrm{E}_{s \cdot b}$, and $x \mapsto s \cdot x$ is a diffeomorphism of E_b onto $\mathrm{E}_{s \cdot b}$.

For example, if G acts on two manifolds X, Y, it acts (differentiably) on $\mathrm{X} \times \mathrm{Y}$ by $(s, (x, y)) \mapsto (s \cdot x, s \cdot y)$; considering $\mathrm{X} \times \mathrm{Y}$ as a trivial bundle over X, the group G acts equivariantly on $\mathrm{X} \times \mathrm{Y}$ and X.

Again, if P is a principal bundle over B, with group G (**16.14**), and if we make G act trivially on B, then G acts equivariantly on P and B.

Let E, F be two fiber bundles over the same base B, and suppose that G acts equivariantly on E and B, and also on F and B (the two actions of G on B being the same). Then G acts by transport of structure (**16.16.6**) on the set Mor(E, F) of B-morphisms of E into F: if $f \in \mathrm{Mor(E, F)}$ we have

(**19.1.2**) $$(s \cdot f)(x) = s \cdot f(s^{-1} \cdot x)$$

for all $s \in \mathrm{G}$ and all $x \in \mathrm{E}$.

The B-morphism f is said to be G-*invariant* if $s \cdot f = f$ for all $s \in \mathrm{G}$, in other words if

(**19.1.2.1**) $$f(s \cdot x) = s \cdot f(x)$$

for all $x \in \mathrm{E}$ and all $s \in \mathrm{G}$.

In particular, G acts on the set $\Gamma(\mathrm{B, E})$ of C^∞ global *sections* of E. In this case we write $\gamma(s)f$ in place of $s \cdot f$. More particularly, if X and Y are two manifolds on which G acts, then G also acts on the set of C^∞ *mappings* of

X into Y, which may be considered as the sections of the trivial bundle X × Y over X; if G acts *trivially* on Y, the formula (19.1.2) becomes

(19.1.2.2) $$(s \cdot f)(x) = f(s^{-1} \cdot x),$$

and to say that f is G-invariant means that $f(s \cdot x) = f(x)$ for all $s \in G$ and all $x \in X$.

If H is another bundle over B and if G acts equivariantly on H and B (the action on B being the same as before), and if g is a B-morphism of F into H, then we have

$$s \cdot (g \circ f) = (s \cdot g) \circ (s \cdot f).$$

(19.1.3) Suppose now that E is a *vector* bundle over B. Then G is said to act *equilinearly* on E and B if it acts equivariantly on E and B, and if moreover, for each $b \in B$, the (bijective) mapping $\mathbf{u}_b\, s \mapsto \cdot\, \mathbf{u}_b$ of E_b onto $E_{s \cdot b}$ is *linear*.

Example

(19.1.4) Let M be a differential manifold on which G acts (differentiably); then G acts *equilinearly* on T(M) and M by the action $(s, \mathbf{k}_x) \mapsto s \cdot \mathbf{k}_x$ defined in (16.10).

For, by virtue of (16.10.1), G acts on T(M), and it is enough to show that the action is *differentiable*. If $s_0 \in G$ and $a \in M$, and if $(\varphi^i)_{1 \leq i \leq n}$ (resp. $(\psi^i)_{1 \leq i \leq n}$) is a local coordinate system at the point a (resp. the point $s_0 \cdot a$), then the functions $\psi^i(s \cdot x) = g_i(s, \xi^1, \ldots, \xi^n)$, where $\xi^j = \varphi^j(x)$, are of class C^∞ in a neighborhood of $(s_0, \varphi(a))$, and the mapping $s \mapsto s \cdot \mathbf{k}_x$ has as local expression (16.3.1)

$$s \mapsto \left(\sum_{j=1}^{n} \frac{\partial g^i}{\partial \xi^j}(s, \xi^1, \ldots, \xi^n) k^j \right)_{1 \leq i \leq n},$$

which proves the assertion.

(19.1.5) If G acts equilinearly on E and B, and also on F and B, where F is another vector bundle over B (the action of G on B being the same in each case), then G also acts equilinearly on $E \oplus F$, $E \otimes F$, $\bigwedge^p E$ and Hom(E, F), by virtue of (16.16.6). To verify, for example, that we have a *differentiable* action of G on $E \otimes F$, we may assume that E and F are trivial, say $E = B \times \mathbf{R}^n$ and $F = B \times \mathbf{R}^m$; and then, if the actions of G on E and F are defined by

$$(s, (b, \mathbf{u})) \mapsto (s \cdot b, A(s, b) \cdot \mathbf{u}), \qquad (s, (b, \mathbf{v})) \mapsto (s \cdot b, B(s, b) \cdot \mathbf{v}),$$

where A and B are $n \times n$ and $m \times m$ matrices, respectively, whose elements are C^∞ functions on G × B, then the action of G on $B \times \mathbf{R}^{mn}$ is given by

$$(s, (b, \mathbf{w})) \mapsto (s \cdot b, (A(s, b) \otimes B(s, b)) \cdot \mathbf{w}),$$

and the result follows. The proof is analogous for $E \oplus F$ and $\bigwedge^p E$. In the case of Hom(E, F), we may again assume that E and F are trivial in order to verify that the action is differentiable; then Hom(E, F) may be identified with $B \times \mathrm{Hom}(\mathbf{R}^n, \mathbf{R}^m)$, and with the notation introduced above the action in question is given by

$$(s, (b, U)) \mapsto (s \cdot b, B(s, b)UA(s^{-1}, s \cdot b)),$$

from which the result again follows.

It is easily verified that the canonical morphisms defined in (16.18) are *invariant* under the action of G.

(19.1.6) We have seen (19.1.2) that if G acts equilinearly on E and B, then it acts on the set $\Gamma(B, E)$; likewise it acts on $\Gamma^{(r)}(B, E)$ for each integer $r > 0$, and moreover for each $s \in G$ the mapping $f \mapsto s \cdot f$ of $\Gamma(B, E)$ (resp. $\Gamma^{(r)}(B, E)$) into itself is *linear* and *continuous* for the topology defined in (17.2). For, by virtue of (3.13.14) and (17.2), we may assume that $E = B \times \mathbf{R}$, and then it is enough to show that if a sequence (f_k) of functions in $\mathscr{E}(B)$ (resp. $\mathscr{E}^{(r)}(B)$) tends uniformly to 0 together with all its derivatives (resp. all its derivatives of order $\leq r$) in a compact subset K of B, then the sequence $(s \cdot f_k)$ has the same property relative to the compact set $s \cdot K$; but this is immediate, by virtue of the formula (19.1.2) and Leibniz's formula.

(19.1.7) Consider again the situation of (19.1.5). Then G acts by transport of structure on the set of C^∞ *differential operators* from E to F (17.13.1). For such an operator P and all for $s \in G$, $\gamma(s)P$ is defined by

(19.1.7.1) $$(\gamma(s)P) \cdot f = \gamma(s)(P \cdot \gamma(s^{-1})f)$$

for all $f \in \Gamma(B, E)$. The operator P is said to be G-*invariant* if $\gamma(s)P = P$ for all $s \in G$.

If Q is a C^∞ differential operator of F into a third vector bundle H over B on which G acts equilinearly (with the same action of G on B as before), then we have

(19.1.8) $$\gamma(s)(Q \circ P) = (\gamma(s)Q) \circ (\gamma(s)P).$$

Examples

(19.1.9) Consider again the example (19.1.4), with the notation introduced there. For each covector $\mathbf{z}_x^* \in T_x(M)^*$ and each $s \in G$, the covector

$$s \cdot \mathbf{z}_x^* \in T_{s \cdot x}(M)^*$$

is defined by the relation

(19.1.9.1) $$\langle s \cdot \mathbf{z}_x^*, \mathbf{h}_{s \cdot x} \rangle = \langle \mathbf{z}_x^*, s^{-1} \cdot \mathbf{h}_{s \cdot x} \rangle$$

for all $\mathbf{h}_{s \cdot x} \in T_{s \cdot x}(M)$ by virtue of (19.1.5). For each (scalar) differential 1-form ω on M, it follows that $\gamma(s)\omega$ (or $s \cdot \omega$) is defined by the relation

(19.1.9.2) $$\langle (s \cdot \omega)(x), \mathbf{h}_x \rangle = \langle \omega(s^{-1} \cdot x), s^{-1} \cdot \mathbf{h}_x \rangle$$

by virtue of (19.1.9.1) and (19.1.5). By transport of structure (17.15.3.2) we have for a p-form α of class C^1

(19.1.9.3) $$d(\gamma(s)\alpha) = \gamma(s)(d\alpha).$$

Suppose now that $\boldsymbol{\alpha}$ is a *vector-valued* differential p-form with values in a finite-dimensional vector space V on which G acts by a *linear representation* $\rho : G \to \mathbf{GL}(V)$ of class C^∞. If $(\mathbf{e}_j)$ is a basis of V, we may write

$$\boldsymbol{\alpha} = \sum_j \alpha_j \mathbf{e}_j,$$

where the α_j are scalar differential p-forms. By definition (19.1.2), we have

$$(\gamma(s)\boldsymbol{\alpha})(x) = (s \cdot \boldsymbol{\alpha})(x) = \rho(s) \cdot \boldsymbol{\alpha}(s^{-1} \cdot x) = \rho(s)\Big(\sum_j \alpha_j(s^{-1} \cdot x)\mathbf{e}_j\Big)$$

or

(19.1.9.4) $$\gamma(s)\boldsymbol{\alpha} = \sum_j (\gamma(s)\alpha_j)(\rho(s) \cdot \mathbf{e}_j).$$

Also, again by transport of structure, we have

(19.1.9.5) $$d(\gamma(s)\boldsymbol{\alpha}) = \gamma(s)(d\boldsymbol{\alpha})$$

if $\boldsymbol{\alpha}$ is of class C^1.

Let X be a C^∞ *vector field* on M. For each $s \in G$, the vector field $\gamma(s)X$ is, by definition (19.1.2), given by

(19.1.9.6) $$(\gamma(s)X)(x) = s \cdot X(s^{-1} \cdot x).$$

Consider the corresponding differential operator θ_X; by transport of structure (17.14.8), we have

(19.1.9.7) $$\gamma(s)\theta_X = \theta_{\gamma(s)X},$$

and, for two C^∞ vector fields X, Y on M,

(19.1.9.8) $$\gamma(s)[X, Y] = [\gamma(s)X, \gamma(s)Y].$$

Remarks

(19.1.10) (i) Let E be a vector bundle over B, and suppose that G acts equilinearly on E and B. Then G acts equilinearly on T(E) and E (19.1.4), and it follows immediately that if T(E) is regarded as a fiber bundle over B, then G acts *equivariantly* on T(E) and B. Since however there is no canonical vector bundle structure on T(E) as a bundle over B, we cannot say that G acts *equilinearly* on T(E) and B.

(ii) The definitions and results are analogous when G acts *on the right.*

PROBLEMS

1. Let G be a Lie group, H a Lie subgroup of G, and let $X = G/H$ be the homogeneous space. Let (E, X, π) be a fibration over X, and let x_0 denote the point $eH = H \in G/H = X$, and $E_0 = \pi^{-1}(x_0)$ the fiber of E at x_0.

(a) Suppose that G acts equivariantly on E and X (the action of G on X being the canonical action, by left translations). Then the mapping $(t, y) \mapsto t \cdot y$ defines a left action of H on E_0. Show that if G is regarded as a principal bundle with base X and group H (16.14.2), there exists an X-isomorphism f of $G \times^H E_0$ onto E, such that $f(s \cdot z) = s \cdot f(z)$ for all $s \in G$ and all $z \in G \times^H E_0$ (the left action of G on $G \times^H E_0$ being defined by $s \cdot (t \cdot y) = (st) \cdot y$). (Consider the mapping $(s, y) \mapsto s \cdot y$ of $G \times E_0$ into E.)

Consider the converse, and the case where E is a vector bundle over X, and G acts equilinearly.

(b) Let (E', X, π') be another fibration over X on which G acts equivariantly, and let $E'_0 = \pi'^{-1}(x_0)$ be the fiber of E' at x_0. Let $u : E_0 \to E'_0$ be a C^∞ mapping such that $u(t \cdot y) = t \cdot u(y)$ for all $t \in H$. Show that there exists a unique X-morphism $\bar{u} : E \to E'$ extending u such that $\bar{u}(s \cdot z) = s \cdot \bar{u}(z)$ for all $s \in G$ and all $z \in E$.

(c) Let I be the set of points $y \in E_0$ that are invariant under H. For each $y \in I$, let $\sigma_y : X \to E$ be the mapping defined by $\sigma_y(s \cdot x_0) = s \cdot y$ for all $s \in G$. Show that every G-invariant section of E over X is of class C^∞, and that the mapping $y \mapsto \sigma_y$ is a bijection of I onto the set of these sections.

2. Let G, G′ be two Lie groups, $\rho : G \to G'$ a Lie group homomorphism, X (resp. X′) a differential manifold on which G (resp. G′) acts differentiably, and $f : X \to X'$ a C^∞ mapping such that $f(s \cdot x) = \rho(s) \cdot f(x)$ for all $x \in X$ and all $s \in G$. Let E′ be a fiber bundle over X′, and suppose that G′ acts equivariantly on E′ and X′. Show that there exists a unique differentiable action of G on the inverse image $E = X \times_{X'} E'$ of E′ under f such that if (f, g) is the canonical morphism of E onto E′ (**16.12.8**), then $g(s \cdot z) = \rho(s) \cdot g(z)$ for all $z \in E$ and all $s \in G$.

3. (a) Let G be a *compact* Lie group, X a *compact* differential manifold on which G acts differentiably, and E a vector bundle over X such that G acts equilinearly on E and X. The group G then acts continuously and linearly on the Fréchet space $\Gamma(X, E)$ (resp. the Banach space $\Gamma^{(r)}(X, E)$, for each integer $r \geqq 0$; cf. (**17.2.2**)). For each section $\sigma \in \Gamma^{(r)}(X, E)$ (r an integer $\geqq 0$ or $+\infty$), the set of sections $s \cdot \sigma$ (**19.1.2**), where $s \in G$, has a compact closed convex hull in $\Gamma^{(r)}(X, E)$ (Section **12.14**, Problem 13). Deduce that the integral

$$\int_G (s \cdot \sigma)\, d\beta(s),$$

where β is a Haar measure on G, has a meaning (Section **13.10**, Problem 2) and is a G-*invariant* section of class C^r.
(b) Let A be a submanifold of X which is stable under the action of G. Show that each C^r section (r an integer $\geqq 0$ or $+\infty$) of E over A that is invariant under G can be extended to a G-invariant global C^r section of E. (Use (**16.12.11**) and part (a).)
(c) Deduce from (b) that if F is another vector bundle over X on which G acts equilinearly, and if there exists a G-*isomorphism* f of $E|A$ onto $F|A$ (that is, an A-isomorphism such that $f(s \cdot z) = s \cdot f(z)$ for all $z \in E$ and $s \in G$), then there exists an open neighborhood U of A in X which is stable under G, and a G-isomorphism of $E|U$ onto $F|U$ which extends f. (Apply (b) to the vector bundles Hom (E, F) and Hom(F, E), and then use (**8.3.2.1**) applied to the Banach spaces

$$\Gamma^{(0)}(X, \mathrm{End}(E)) \qquad \text{and} \qquad \Gamma^{(0)}(X, \mathrm{End}(F)).)$$

(d) Let Y be a differential manifold on which G acts differentiably, and let

$$\varphi : Y \times J \to X$$

be a C^∞ mapping (where J is an open interval in **R**, containing [0, 1]) such that

$$\varphi(s \cdot y, t) = s \cdot \varphi(y, t)$$

for $s \in G$, $y \in Y$, and $t \in J$. If E_0, E_1 are the inverse images of E under $f_0 = \varphi(., 0)$ and $f_1 = \varphi(., 1)$, show that E_0 and E_1 are G-isomorphic. (Use (c).)

4. Let G be a Lie group and X a principal bundle with base G and group G.

(a) Let G act on the base of X by left translations. Show that if there exists a differentiable action of G on X such that G acts equivariantly on the bundle X and the base G, then X is trivializable. (Observe that a G-orbit in X is a section of X.)
(b) Give an example of a principal bundle with base G and group G which is not trivializable. (Consider the Klein bottle (**16.14.10**).)

5. Let G be a *compact* Lie group, M a compact differential manifold, J an open interval in **R** containing 0. Suppose that G acts differentiably on M × J in such a way that

$$s \cdot (x, \xi) = (m_\xi(s, x), \xi),$$

where for each $\xi \in$ J, m_ξ is a differentiable action of G on M.

(a) Show that the vector field

$$(X, E) = \int_G \gamma(s)(0, E)\, d\beta(s)$$

on M × J, where β is a Haar measure of total mass 1 on G and E is the unit vector field on **R**, is invariant under the action of G on M × J.

(b) Deduce from (a) that there exists a diffeomorphism $(x, \xi) \mapsto (h_\xi(x), \xi)$ of M × J onto itself (so that h_ξ is a diffeomorphism of M onto itself, for each $\xi \in$ J) for which

$$m_\xi(s, x) = h_\xi(m_0(s, h_\xi^{-1}(x))).$$

In other words, the actions m_ξ are *isotopic*. (Consider the flow of the vector field (X, E).)

6. Let G be a *compact* Lie group, acting differentiably on a differential manifold M, and let $x_0 \in$ M be a point *fixed* by G. Then G acts linearly on the tangent space $T_{x_0}(M)$ by $(s, \mathbf{h}_{x_0}) \mapsto s \cdot \mathbf{h}_{x_0}$ (16.10.1). The point x_0 has arbitrarily small G-stable open neighborhoods (12.10.5). If V is one, let $\mathbf{f}_0 : V \to T_{x_0}(M)$ be a C^∞ mapping such that $\mathbf{f}_0(x_0) = \mathbf{0}_{x_0}$ and such that $T_{x_0}(\mathbf{f}_0)$ is the identity. If β is a Haar measure on G, show that the mapping $\mathbf{f} : V \to T_{x_0}(M)$ defined by

$$\mathbf{f}(x) = \int_G t \cdot \mathbf{f}_0(t^{-1} \cdot x)\, d\beta(t)$$

is of class C^∞, that $\mathbf{f}(x_0) = \mathbf{0}_{x_0}$, and that $T_{x_0}(\mathbf{f})$ is the identity; also show that $\mathbf{f}(s \cdot x) = s \cdot \mathbf{f}(x)$ for all $s \in$ G. Deduce that there exists a G-stable open neighborhood W ⊂ V of x_0 and a chart of W such that the local expressions of the diffeomorphisms $x \mapsto s \cdot x$ of W onto itself for all $s \in$ G, relative to this chart, are *linear* transformations (*Bochner's theorem*).

2. ACTIONS OF A LIE GROUP G ON BUNDLES OVER G

If G is a Lie group, then G acts on itself by left translations $x \mapsto sx$. The definitions of (19.1) may therefore be applied to fiber bundles with *base* G, and we shall say that G acts *left-equivariantly* on a bundle E over G if G acts equivariantly on E and G (the action of G on G being left-translation). Likewise we define a *right-equivariant* action of G on E: the formula which replaces (19.1.2) in this case is

$$(f \cdot s)(x) = f(x \cdot s^{-1}) \cdot s,$$

and we shall write $\delta(s^{-1})f$ in place of $f \cdot s$. Similarly, for differential operators, we shall put $(\delta(s)P) \cdot f = \delta(s)(P \cdot \delta(s^{-1})f)$.

(19.2.1) For example, G acts left-equivariantly on the tangent bundle T(G) by $(s, \mathbf{h}_x) \mapsto s \cdot \mathbf{h}_x$, and right-equivariantly by $(s, \mathbf{h}_x) \mapsto \mathbf{h}_x \cdot s$. Moreover, G acts *equilinearly* in both cases, and we have $s \cdot (\mathbf{h}_x \cdot t) = (s \cdot \mathbf{h}_x) \cdot t$ (16.9.8). By virtue of (19.1.5) there are analogous statements for the tensor bundles $\mathbf{T}_s^r(G)$ and the exterior powers of T(G) and T(G)*.

For such bundles it is clear that

$$\gamma(s)\delta(t) = \delta(t)\gamma(s)$$

for all s, t in G, for the actions of G on the sections or on the differential operators.

In particular, for each $s \in G$, the mapping

$$\mathbf{h}_e \mapsto s \cdot \mathbf{h}_e \cdot s^{-1}$$

is an *automorphism* of the tangent space $T_e(G)$ at the neutral element e of G. It is denoted by Ad(s), and it is clear that

(19.2.1.1) $$\mathrm{Ad}(s) = T_e(\mathrm{Int}(s)),$$

where Int(s) is the inner automorphism $x \mapsto sxs^{-1}$ of G. By (19.1.4) the mapping $s \mapsto \mathrm{Ad}(s)$ is a *Lie group homomorphism* of G into $\mathbf{GL}(T_e(G))$, called the *adjoint representation* of G.

(19.2.2) Let E be a bundle over G, on which G acts left-equivariantly. Then every *invariant section* f of E over G is uniquely determined by its value $f(e) \in E_e$ at the neutral element of G, because we must have $f(s) = s \cdot f(e)$ for all $s \in G$. Conversely, it is clear that, for each element $\mathbf{u}_e \in E_e$, the mapping $s \mapsto s \cdot \mathbf{u}_e$ is a G-invariant section of E, of class C^∞.

In particular, if E is a vector bundle of rank n over G, and if G acts equilinearly on E, then the mapping which assigns to each $\mathbf{u}_e \in E_e$ the invariant section $s \mapsto s \cdot \mathbf{u}_e$ is an *isomorphism* of the vector space E_e onto the vector subspace I of $\Gamma(G, E)$ consisting of the G-invariant sections (and therefore I has dimension n). Furthermore, if $(\mathbf{f}_i)_{1 \leq i \leq n}$ is a basis over **R** of this vector space, the sections $\mathbf{f}_i$ form a *frame* of E over G. Hence every vector bundle over G on which G acts *equilinearly* is *trivializable* (16.15.3).

(19.2.3) Let E, F be two bundles over G on which G acts left-equivariantly. Each *invariant morphism* g of E into F is uniquely determined by its restriction $g_e : E_e \to F_e$ to the fiber E_e at the neutral element; for it follows from (19.1.2) that $g_s(\mathbf{u}_s) = s \cdot g_e(s^{-1} \cdot \mathbf{u}_s)$ for all $s \in G$. Conversely, given any C^∞ mapping $g_e : E_e \to F_e$, if we define g_s for each $s \in G$ by this formula, it is clear that the mapping $g : E \to F$ which is equal to g_s on each fiber E_s is an invariant morphism.

(19.2.4) Finally, if E and F are two vector bundles over G, on which G acts left-equivariantly and equilinearly, then every G-*invariant* differential operator P from E to F is uniquely determined by the continuous linear mapping $f \mapsto (P \cdot f)(e)$ of $\Gamma(G, E)$ into F_e. For by (19.1.8) and (19.1.2), we have $(P \cdot f)(s) = s \cdot ((P \cdot \gamma(s^{-1})f)(e))$ for all $s \in G$. We shall consider more particularly the case where $E = F = G \times \mathbf{R}$, in other words the *fields of real point-distributions* on G (17.13.6) which are *left-invariant* (i.e., invariant under left translations of G).

3. THE INFINITESIMAL ALGEBRA AND THE LIE ALGEBRA OF A LIE GROUP

(19.3.1) Let G be a Lie group. The set of real differential operators $P \in \text{Diff}(G)$ (17.13.6) which are *left-invariant* forms is, by virtue of (19.1.7), a sub-**R**-algebra $\mathfrak{G}$ of the (associative) algebra Diff(G), called the *infinitesimal algebra* of G. Its identity element is the identity mapping of $\mathscr{E}(G)$.

We have seen (19.2.4) that the mapping $P \mapsto P(e)$ is a *bijection* of $\mathfrak{G}$ onto the set of real distributions with support contained in $\{e\}$; also we have the relation

$$(P \circ Q)(e) = P(e) * Q(e), \tag{19.3.2}$$

which shows that $P \mapsto P(e)$ is an *isomorphism* of the *algebra* $\mathfrak{G}$ onto the *algebra* of real distributions with support contained in $\{e\}$ (the multiplication in this algebra being convolution), which we shall denote by $\mathfrak{G}_e$.

To prove (19.3.2), consider an arbitrary function $f \in \mathscr{E}(G)$. We have

$$((P \circ Q)(e)) \cdot f = ((P \circ Q) \cdot f)(e) = (P \cdot (Q \cdot f))(e);$$

and by virtue of the invariance of Q, we have (19.1.8) for each $x \in G$,

$$(Q \cdot f)(x) = (Q \cdot \gamma(x^{-1})f)(e) = \langle Q(e), \gamma(x^{-1})f \rangle;$$

putting $Q(e) = S$, this last expression may also be written in the form $\int f(xy)\, dS(y)$ (17.3.8.1), and therefore

$$(P \cdot (Q \cdot f))(e) = P(e) \cdot (Q \cdot f) = \int dR(x) \int f(xy)\, dS(y),$$

where $R = P(e)$. By virtue of (17.11.1) and (17.10.3), this proves (19.3.2).

(19.3.3) The set $\mathfrak{g} \subset \mathfrak{G}$ of left-invariant differential operators which are *of order* ≤ 1 and annihilate the constants may be identified (via the mapping $X \mapsto \theta_X$) with the set of *left-invariant* C^∞ *vector fields* on G (in other words, the invariant sections of the tangent bundle T(G)) (17.14.2). It follows from (19.1.8) that $\mathfrak{g}$ is a Lie subalgebra of the Lie algebra of all C^∞ vector fields on G (17.14.3), in which the bracket is defined by $[X, Y] = X \circ Y - Y \circ X$.

The mapping $X \mapsto X(e)$ is an isomorphism of the *vector space* $\mathfrak{g}$ onto the tangent space $T_e(G)$ (19.2.2); the inverse isomorphism assigns to a tangent vector $\mathbf{u} \in T_e(G)$ the invariant vector field

(19.3.3.1) $$X_{\mathbf{u}} : s \mapsto s \cdot \mathbf{u}.$$

The distribution $X_{\mathbf{u}}(e)$ is just the *differentiation* $\theta_{\mathbf{u}}$ *in the direction of the tangent vector* $\mathbf{u}$ *at the point e* (17.14.1). By reason of the above isomorphism, given any two tangent vectors $\mathbf{u}, \mathbf{v} \in T_e(G)$, there exists a unique tangent vector, denoted by $[\mathbf{u}, \mathbf{v}]$, with the property that

(19.3.3.2) $$\theta_{[\mathbf{u}, \mathbf{v}]} = \theta_{\mathbf{u}} * \theta_{\mathbf{v}} - \theta_{\mathbf{v}} * \theta_{\mathbf{u}},$$

so that we have

(19.3.3.3) $$[X_{\mathbf{u}}, X_{\mathbf{v}}] = X_{[\mathbf{u}, \mathbf{v}]}.$$

The tangent vector $[\mathbf{u}, \mathbf{v}]$ is called the *Lie bracket* of the two tangent vectors $\mathbf{u}, \mathbf{v}$. The tangent space $T_e(G)$, endowed with this law of composition, is clearly a *Lie algebra* $\mathfrak{g}_e$, and the mapping $X \mapsto X(e)$ is a *Lie algebra isomorphism* of $\mathfrak{g}$ onto $\mathfrak{g}_e$. The algebra $\mathfrak{g}_e$ (or $\mathfrak{g}$) is called the *Lie algebra of the Lie group* G, and will sometimes be denoted by Lie(G). A little further on we shall see that $\mathfrak{g}_e$ and the identity element *generate* the (associative) algebra $\mathfrak{G}_e$ (19.6.2).

(19.3.4) Let G, G′ be two Lie groups, e, e' their respective neutral elements, $f: G \to G'$ a Lie group *homomorphism* (16.9.7), and $\mathfrak{G}, \mathfrak{G}'$ the infinitesimal algebras of G, G′, respectively. For each field of point distributions $P \in \mathfrak{G}$, the *image* $f(P(e))$ is a point distribution belonging to $\mathfrak{G}'_{e'}$ (17.7.1); hence there exists a unique invariant operator $P' \in \mathfrak{G}'$ such that

(19.3.4.1) $$P'(e') = f(P(e))$$

The operator P' is called the *image* of P under f, and will sometimes be denoted by $f_*(P)$. By definition, we have for each $s \in G$

$$f_*(P)(f(s)) = \gamma(f(s))f(P(e))$$

and hence, for each function $u \in \mathscr{E}(G')$,

$$\langle u, f_*(P)(f(s))\rangle = \langle u, \gamma(f(s))f(P(e))\rangle = \langle \gamma(f(s)^{-1})u, f(P(e))\rangle.$$

By definition, $v = \gamma(f(s)^{-1})u$ is the function $t' \mapsto u(f(s)t')$, and therefore the composition $v \circ f$ is the function

$$t \mapsto u(f(s)f(t)) = u(f(st)),$$

so that $v \circ f = \gamma(s^{-1})(u \circ f)$. Hence we have

$$\langle \gamma(f(s)^{-1})u, f(P(e))\rangle = \langle \gamma(s^{-1})(u \circ f), P(e)\rangle = \langle u \circ f, P(s)\rangle = \langle u, f(P(s))\rangle,$$

and therefore, finally,

$$f_*(P)(f(s)) = f(P(s)), \tag{19.3.4.2}$$

which justifies the name of *image*, by showing that $f(P(s))$ depends only on $f(s)$ and not on s itself (but it should be observed that the distributions $f_*(P)(s')$ are defined also at points $s' \in G'$ not in $f(G)$.) Furthermore, we have

$$f_*(P \circ Q) = f_*(P) \circ f_*(Q). \tag{19.3.4.3}$$

For

$$f_*(P \circ Q)(e') = f((P \circ Q)(e)) = f(P(e) * Q(e)) = f(P(e)) * f(Q(e))$$

by (17.11.10), and by definition this last distribution is equal to

$$(f_*(P)(e')) * (f_*(Q)(e')),$$

whence (19.3.4.3) follows. In other words, f_* is a *homomorphism of the algebra* $\mathfrak{G}$ *into the algebra* $\mathfrak{G}'$, called the *derived homomorphism* of f. If $g : G' \to G''$ is another Lie group homomorphism, then it is immediately verified that

$$(g \circ f)_* = g_* \circ f_*. \tag{19.3.4.4}$$

(19.3.5) Since the image under f of a distribution of order $\leqq 1$ is a distribution of order $\leqq 1$, it is clear that if $\mathfrak{g}$, $\mathfrak{g}'$ are the Lie algebras of left-invariant C^∞ vector fields on G, G', respectively, then we have $f_*(\mathfrak{g}) \subset \mathfrak{g}'$, and the restriction of f_* to $\mathfrak{g}$ is a *Lie algebra homomorphism* of $\mathfrak{g}$ into $\mathfrak{g}'$, by the definition of the Lie bracket. Under the canonical identification of $\mathfrak{g}$ with $\mathfrak{g}_e$, and $\mathfrak{g}'$ with $\mathfrak{g}'_{e'}$, the homomorphism f_* is identified with the tangent linear mapping $T_e(f)$, which we shall denote also by f_* or Lie(f). Hence, if $\mathbf{u}$, $\mathbf{v}$ are any two vectors in $\mathfrak{g}_e$, we have

(19.3.5.1) $$f_*([\mathbf{u}, \mathbf{v}]) = [f_*(\mathbf{u}), f_*(\mathbf{v})].$$

In particular, for each $s \in G$, we have (19.2.1.1)

(19.3.5.2) $$\mathrm{Ad}(s) \cdot [\mathbf{u}, \mathbf{v}] = [\mathrm{Ad}(s) \cdot \mathbf{u}, \mathrm{Ad}(s) \cdot \mathbf{v}].$$

(19.3.6) If we suppose merely that f is a *local homomorphism* (16.9.9.4), the mapping $S \mapsto f(S)$ is still a homomorphism of the algebra $\mathfrak{G}_e$ into the algebra $\mathfrak{G}'_{e'}$ (17.11.10.2), which is again denoted by f_*; its restriction to $\mathfrak{g}_e$ (which may be identified with $T_e(f)$) is therefore again a Lie algebra homomorphism of $\mathfrak{g}_e$ into $\mathfrak{g}'_{e'}$.

Remark

(19.3.7) Let G be a Lie group acting differentiably *on the right* on a differential manifold M. We shall show that to each (*left*-invariant) differential operator $P \in \mathfrak{G}$ on G, there is canonically associated a differential operator P_M on M, whose order is at most equal to that of P. For each $x \in M$, let σ_x denote the mapping $s \mapsto x \cdot s$ of G into M, which is of class C^∞, and put

(19.3.7.1) $$\langle P_M(x), f\rangle = \langle P(e), f \circ \sigma_x\rangle$$

for all $f \in \mathscr{E}(M)$, or equivalently $P_M(x) = (\sigma_x)_*(P(e))$ (17.7.1). From the local expression (17.13.3) of $P(e)$ it is immediately verified that P_M is a differential operator of order at most equal to the order of P. Moreover, for each $s \in G$, we have

(19.3.7.2) $$\langle P_M(x \cdot s), f\rangle = \langle P(s), f \circ \sigma_x\rangle,$$

because the right-hand side is equal to $\langle P(e), \gamma(s^{-1})(f \circ \sigma_x)\rangle$, and $\gamma(s^{-1})(f \circ \sigma_x)$ is equal to $f \circ \sigma_{x \cdot s}$. From this we deduce that $P \mapsto P_M$ is a *homomorphism* of the algebra $\mathfrak{G}$ into the algebra Diff(M) of differential operators on M (17.13.6); that is to say,

(19.3.7.3) $$(P \circ Q)_M = P_M \circ Q_M$$

for P, $Q \in \mathfrak{G}$. For if we put $\mathrm{R} = P(e)$ and $\mathrm{S} = Q(e)$, then for all $f \in \mathscr{E}(\mathrm{M})$ (with the notation of (17.10.3.2)), we have

$$\begin{aligned}((P_{\mathrm{M}} \circ Q_{\mathrm{M}}) \cdot f)(x) &= \int (Q_{\mathrm{M}} \cdot f)(x \cdot s)\, d\mathrm{R}(s) \\ &= \int d\mathrm{R}(s) \int f((x \cdot s) \cdot t)\, d\mathrm{S}(t) \\ &= \iint f(x \cdot (st))\, d\mathrm{R}(s)\, d\mathrm{S}(t) \\ &= \int f(x \cdot v)\, d(\mathrm{R} * \mathrm{S})(v) \\ &= ((P \circ Q)_{\mathrm{M}} \cdot f)(x)\end{aligned}$$

since $(P \circ Q)(e) = \mathrm{R} * \mathrm{S}$.

In particular, for $\mathbf{u} \in \mathfrak{g}_e$, the operator $(\theta_{X_{\mathbf{u}}})_{\mathrm{M}}$ is equal to $\theta_{Z_{\mathbf{u}, \mathrm{M}}}$, where $Z_{\mathbf{u}, \mathrm{M}}$ (also denoted by $Z_{\mathbf{u}}$) is the vector field on M

(19.3.7.4) $$x \mapsto x \cdot \mathbf{u}$$

(cf. (16.10.1)). For by definition, we have

$$\begin{aligned}((\theta_{X_{\mathbf{u}}})_{\mathrm{M}} \cdot f)(x) = (\theta_{X_{\mathbf{u}}} \cdot (f \circ \sigma_x))(e) &= \langle d_e(f \circ \sigma_x), \mathbf{u} \rangle \\ = \langle d_x f, \mathrm{T}_e(\sigma_x) \cdot \mathbf{u} \rangle &= \langle d_x f, x \cdot \mathbf{u} \rangle,\end{aligned}$$

from which the assertion follows.

The operator P_{M} (resp. the vector field $Z_{\mathbf{u}, \mathrm{M}}$) is called the *transport* of P (resp. $X_{\mathbf{u}}$) by the action of G on M. By virtue of (19.3.7.3) we have

(19.3.7.5) $$Z_{[\mathbf{u}, \mathbf{v}]} = [Z_{\mathbf{u}}, Z_{\mathbf{v}}]$$

so that the mapping $\mathbf{u} \mapsto Z_{\mathbf{u}}$ is a *Lie algebra homomorphism* of $\mathfrak{g}_e$ into the Lie algebra $\mathscr{T}_0^1(\mathrm{M})$ of C^∞ vector fields on M. The field $Z_{\mathbf{u}}$ is also called the *Killing field* on M corresponding to $\mathbf{u}$. For each $s \in \mathrm{G}$ we have

(19.3.7.6) $$Z_{\mathbf{u}}(x \cdot s) = Z_{\mathrm{Ad}(s) \cdot \mathbf{u}}(x) \cdot s.$$

For the left-hand side is $(x \cdot s) \cdot \mathbf{u} = (x \cdot (s \cdot \mathbf{u} \cdot s^{-1})) \cdot s$ (16.10.1).

PROBLEMS

1. Let G be a Lie group and X a differential manifold. A *partial right action* of G on X is by definition a C^∞ mapping $\psi : \Omega \to X$, where Ω is an open subset of $G \times X$ containing $\{e\} \times X$, such that:

(1) $\psi(e, x) = x$ for all $x \in X$;
(2) there exists a neighborhood Ω_1 of $\{e\} \times \{e\} \times X$ in $G \times G \times X$ such that, for all $(s, t, x) \in \Omega_1$, we have $(t, x) \in \Omega$, $(ts, x) \in \Omega$, $(s, \psi(t, x)) \in \Omega$, and

$$\psi(s, \psi(t, x)) = \psi(ts, x).$$

We write $\psi(s, x) = x \cdot s$ for $(s, x) \in \Omega$.

Let (X_n) be a locally finite denumerable open covering of X. For each n, let ψ_n be a partial right action of G on X_n, and suppose that for each pair of integers p, q and each $x \in X_p \cap X_q$, the actions ψ_p and ψ_q coincide in some neighborhood of (e, x). Show that there exists a partial right action ψ of G on X such that, for each n and each $x \in X_n$, the partial actions ψ and ψ_n coincide on some neighborhood of (e, x). (Use Problem 5 of Section **18.14**.)

2. Let G be a Lie group, M a differential manifold, and let $(s, x) \mapsto x \cdot s$ be a partial right action of G on M (Problem 1). Show that, for each differential operator $P \in \mathfrak{G}$, the definition (**19.3.7.1**) remains valid; likewise the formula (**19.3.7.3**), and the formula (**19.3.7.2**) for all pairs $(s, x) \in \Omega$. In particular, the mapping $\mathbf{u} \mapsto Z_{\mathbf{u}}$ is a Lie algebra homomorphism of $\mathfrak{g}_e$ into the Lie algebra $\mathscr{T}_0^1(M)$ of C^∞ vector fields on M. Such a homomorphism is called an *infinitesimal action* of $\mathfrak{g}_e$ on M.

Conversely, suppose we are given an infinitesimal action $\mathbf{u} \mapsto Y_{\mathbf{u}}$ of $\mathfrak{g}_e$ on M. Show that there exists a partial right action of G on M such that $Z_{\mathbf{u}} = Y_{\mathbf{u}}$ for all $\mathbf{u} \in \mathfrak{g}_e$. (For each point $(s, x) \in G \times M$, consider the n-direction $L_{(s, x)}$ in $T_{(s, x)}(G \times M) = T_s(G) \times T_x(M)$ (where $n = \dim(G)$) generated by the tangent vectors $(X_{\mathbf{u}}(s), Y_{\mathbf{u}}(x))$ as $\mathbf{u}$ runs through $\mathfrak{g}_e$. Show that this field of n-directions defines a completely integrable Pfaffian system, and then use Problem 1.)

3. Give an example of a partial right action of G on a differential manifold M which cannot be extended to a differentiable action of G on M (**16.10**). (Observe that if U is an open subset of M, and $\mu : (s, x) \mapsto x \cdot s$ an action of G on M, then there exists a neighborhood Ω of $\{e\} \times U$ in $G \times U$ such that $\mu(\Omega) \subset U$.) (see Section **19.8**, Problem 3).

4. Let G be a separable metrizable connected topological group, and suppose that we are given, on a symmetric open neighborhood U of e, a structure of a differential (resp. analytic) manifold with the following property: if V is a symmetric open neighborhood of e such that $V^2 \subset U$, then the mapping $(s, t) \mapsto st^{-1}$ of $V \times V$ into U is of class C^∞ (resp. analytic). Show that there exists a unique structure of differential (resp. analytic) manifold on G, compatible with the group structure of G, which induces on each symmetric open neighborhood W of e such that $W^2 \subset U$ the manifold structure induced by that given on U. (Use (**16.2.5**) to define the differential (resp. analytic) manifold structure on G. Then show that for each $s \in G$ there exists a neighborhood

Z_s of e, contained in W, such that $sZ_s s^{-1} \subset V$ and such that the mapping $x \mapsto sxs^{-1}$ of Z_s into V is of class C^∞ (resp. analytic). For this purpose, use the fact that s can be written as a product $s_1 s_2 \cdots s_r$ of elements of W.)

5. Let G be a connected Lie group and M a connected differential manifold on which G acts differentiably on the right.

(a) For G to act transitively on M it is necessary and sufficient that, for each $x \in M$, the mapping $\mathbf{u} \mapsto x \cdot \mathbf{u} = Z_{\mathbf{u}}(x)$ of $\mathfrak{g}_e$ into $T_x(M)$ should be surjective. (To show that the condition is sufficient, observe that if it is satisfied every G-orbit is open in M.)
(b) Suppose that G acts transitively on M, so that M may be identified with a homogeneous space H\G, where H is the stabilizer of some point $x_0 \in M$. The kernel of the homomorphism $\mathbf{u} \mapsto Z_{\mathbf{u}}$ of $\mathfrak{g}_e$ into the Lie algebra $\mathscr{T}_0^1(M)$ is then the Lie algebra of the largest normal subgroup K of G contained in H, namely, the intersection of the conjugates sHs^{-1} of H (cf. **19.8.11**); M may be identified with the homogeneous space (K\H)\(K\G).
(c) With the hypotheses of (b), the action of G on M is said to be *imprimitive* if there exists a closed submanifold V of M such that $0 < \dim(V) < \dim(M)$ and such that each transform $V \cdot s$ of V by an element of G is either equal to or disjoint from V. For this to be the case, it is necessary and sufficient that there should exist a closed subgroup L of G such that $H \subset L \subset G$ and $\dim(H) < \dim(L) < \dim(G)$. If it is not the case, the action of G on M is said to be *primitive*; this will occur whenever there exists no Lie subalgebra of $\mathfrak{g}_e$ containing the Lie algebra $\mathfrak{h}_e$ of H, other than $\mathfrak{g}_e$ and $\mathfrak{h}_e$.
(d) With the hypotheses of (b), let $\mathfrak{L}$ be the image of $\mathfrak{g}_e$ under the homomorphism $\mathbf{u} \mapsto Z_{\mathbf{u}}$. Let $\mathscr{E}$ be the algebra of C^∞-functions on M, and let $\mathfrak{m}$ be the maximal ideal of $\mathscr{E}$ consisting of the functions vanishing at x_0. For $p = -1, 0, 1, 2, \ldots$, let $\mathfrak{L}_p$ denote the set of vector fields $X = Z_{\mathbf{u}} \in \mathfrak{L}$ such that $\theta_X \cdot f \in \mathfrak{m}^{p+1}$ for all $f \in \mathscr{E}$, so that $\mathfrak{L}_{-1} = \mathfrak{L}$, and $\mathfrak{L}_0$ is the image of the Lie algebra $\mathfrak{h}_e$ of H. If $c = (U, \varphi, n)$ is a chart of M at the point x_0 such that $\varphi(x_0) = 0$, and if $(X_i)_{1 \leq i \leq n}$ are the vector fields associated with this chart (**16.15.4.2**), then the elements of $\mathfrak{L}_p$ are the vector fields $\sum_{i=1}^n a_i X_i$, where $a_i \in \mathfrak{m}^{p+1}$ $(1 \leq i \leq n)$. Show that $[\mathfrak{L}_p, \mathfrak{L}_q] \subset \mathfrak{L}_{p+q}$ (with the convention that $\mathfrak{L}_{-2} = \mathfrak{L}$), Furthermore, if there exists a vector field $Y \in \mathfrak{L}_p$ (for $p \geq 0$) such that $Y \notin \mathfrak{L}_{p+1}$. then there exists $X \in \mathfrak{L}$ such that $[Y, X] \notin \mathfrak{L}_p$ (but we have $[Y, X] \in \mathfrak{L}_{p-1}$). (Observe that, by (a), for each index i there exists $X \in \mathfrak{L}$ such that

$$(\theta_X \cdot \varphi^i)(x_0) \neq 0 \quad \text{and} \quad (\theta_X \cdot \varphi^j)(x_0) = 0 \quad \text{for } j \neq i.)$$

The $\mathfrak{L}_p$ are Lie subalgebras of $\mathfrak{L}$ and are stable under the mappings $X \mapsto \boldsymbol{\delta}(t)X$ forall $t \in H$. For each $p \geq 0$, $\mathfrak{L}_p$ is an ideal in $\mathfrak{L}_0$. If, for each $t \in H$, $\rho(t)$ denotes the endomorphism $\mathbf{h} \mapsto \mathbf{h} \cdot t$ of $T_{x_0}(M)$, then ρ is a linear representation of H on $T_{x_0}(M)$, and if $\tilde{H}$ is the image $\rho(H)$ of H in $\mathbf{GL}(T_{x_0}(M))$, the Lie algebra of $\tilde{H}$ is isomorphic to $\mathfrak{L}_0/\mathfrak{L}_1$.
(e) Suppose further that $\bigcap_p \mathfrak{L}_p = \{0\}$. Then there exists a largest index r such that $\mathfrak{L}_r \neq \{0\}$ and all the $\mathfrak{L}_j$ with $j \leq r$ are distinct. For each $p \geq 0$, we have

$$\dim(\mathfrak{L}_p/\mathfrak{L}_{p+1}) \leq \binom{n+p}{p+1}.$$

(f) If G is a real (resp. complex) analytic group acting analytically on a real (resp. complex) analytic manifold M, the condition $\bigcap_p \mathfrak{L}_p = \{0\}$ is always satisfied.

6. Let $\mathbf{u} \mapsto Y_{\mathbf{u}}$ be an infinitesimal action (Problem 2) of the Lie algebra $\mathfrak{g}_e$ of a Lie group G on a differential manifold M. Given a point $x_0 \in \mathrm{M}$ we may then define the $\mathfrak{L}_p$ $(p \geqq -1)$ as in Problem 5(d). The action of $\mathfrak{g}_e$ is said to be *transitive at* x_0 (or the *Lie algebra* $\mathfrak{L}$ is *transitive at* x_0) if $\dim(\mathfrak{L}_{-1}/\mathfrak{L}_0) = \dim_{x_0}(\mathrm{M})$. An infinitesimal action which is transitive at x_0 is said to be *primitive at* x_0 if there exists no Lie subalgebra of $\mathfrak{L}$ containing $\mathfrak{L}_0$, other than $\mathfrak{L}$ and $\mathfrak{L}_0$. If the infinitesimal action $\mathbf{u} \mapsto Y_{\mathbf{u}}$ is transitive (resp. primitive) at the point x_0, what can be said about the corresponding partial action (Problem 2)?

7. (a) Let M be a differential manifold of *dimension* 1. If the Lie subalgebra $\mathfrak{L}$ of $\mathscr{T}_0^1(\mathrm{M})$ is transitive at a point x_0 (Problem 6), then the condition $\bigcap_p \mathfrak{L}_p = \{0\}$ is satisfied. (Identifying a neighborhood of x_0 in M with an interval in $\mathbf{R}$ containing $x_0 = 0$, the restrictions to this neighborhood of the vector fields belonging to $\mathfrak{L}$ are of the form $x \mapsto f(x)E$ (in the notation of (18.1.1)) and at least one of the functions f does not vanish at 0. By change of variable, we may assume that $E \in \mathfrak{L}$. Deduce that, if $\dim(\mathfrak{L}) = m$, each of these functions f satisfies a homogeneous linear differential equation of order $\leqq m$, with constant coefficients, and hence is *analytic*.)
(b) With the hypotheses of (a), show that $\dim(\mathfrak{L}) \leqq 3$ and that the partial action corresponding to $\mathfrak{L}$ is necessarily of one of the following types:

(1) $\dim(\mathfrak{L}) = 1$, $\mathrm{G} = \mathbf{R}$, and the partial action is

$$(t, x) \mapsto x + t \qquad (t \in \mathrm{G}, \quad x \in \mathrm{M}).$$

(2) $\dim(\mathfrak{L}) = 2$, G is the group defined in Example (19.5.11), and the partial action is

$$((t_1, t_2), x) \mapsto t_1 x + t_2 .$$

(3) $\dim(\mathfrak{L}) = 3$, $\mathrm{G} = \mathbf{PGL}(2, \mathbf{R})$, and the partial action is

$$\left(\begin{pmatrix} t_1 & t_2 \\ t_3 & 1 \end{pmatrix}, x\right) \mapsto \frac{t_1 x + t_2}{t_3 x + 1}$$

(t_1 close to 1, and t_2, t_3, x close to 0). (Observe that $[\mathfrak{L}_{r-1}, \mathfrak{L}_r] \neq \{0\}$ and deduce that $r \leqq 1$.)

If M is connected (and therefore diffeomorphic to either $\mathbf{R}$ or $\mathbf{S}_1$), which of the partial actions defined above can be extended to an action of G (or a connected group locally isomorphic to G) on M?
(c) Give examples in which $\mathfrak{L}$ is not transitive at a point and has arbitrarily large dimension. (Observe that there exist C^∞-functions f, g on $\mathbf{R}$, with nonempty compact support, such that $fg' - gf' = f$.)

8. For all integers $r \geqq 0$, the vector fields (in the notation of (18.7.1))

$$E_1, E_2, x^1 E_2, (x^1)^2 E_2, \ldots, (x^1)^r E_2$$

form a basis of a Lie subalgebra of $\mathscr{T}_0^1(\mathbf{R}^2)$ defining an infinitesimal action which is transitive and imprimitive at every point.

9. (a) With the notation of Problem 6, for each $p \geqq 0$ such that $\mathfrak{L}_p \neq \{0\}$, the vectors $X(x)$ with $X \in \mathfrak{L}_p$ generate a vector subspace $\mathrm{E}_p(x)$ of $\mathrm{T}_x(\mathrm{M})$ at each point $x \neq x_0$ of a neighborhood V of x_0. There exists a nonempty open subset $\mathrm{U} \subset \mathrm{V} - \{x_0\}$

such that $x_0 \in \bar{U}$ and such that $\dim(E_p(x))$ is constant and nonzero on U; the field of directions $x \mapsto E_p(x)$ in U is completely integrable. If this field of directions is invariant under the partial action on U corresponding to the given infinitesimal action, and if the latter is transitive at the point x_0 (and therefore in the neighborhood V, for sufficiently small V), show that it cannot be primitive at the points of U unless $\dim(E_p(x)) = \dim_x(M)$ at these points.

(b) If the infinitesimal action defined by $\mathfrak{L}$ is transitive at the point x_0 and if the Lie algebra $\mathfrak{L}$ is commutative, then $\mathfrak{L}_0 = \{0\}$ (replacing M if necessary by a neighborhood of x_0).

(c) Suppose that $\mathfrak{L}_r \neq \{0\}$ and $\mathfrak{L}_{r+1} = \{0\}$ for some integer $r \geqq 0$, and that the infinitesimal action defined by $\mathfrak{L}$ is transitive at the point x_0 and primitive at all points of a neighborhood of x_0. Let $n = \dim_{x_0}(M)$. Show that, if $r > n$, there exists a nonempty open set $U \subset V - \{x_0\}$ such that $x_0 \in \bar{U}$ and $\dim(E_{r-n+1}(x)) = n$ for all $x \in U$. (If for some $p \leqq r$ we have $\dim(E_p(x)) < n$ at all points of an open set whose closure contains x_0, show that on another such open set we have

$$\dim(E_{p-1}(x)) \geqq \dim(E_p(x)) + 1,$$

using (a) above and arguing by contradiction.)

(d) Deduce from (c) that under the same hypotheses we have $r \leqq 2n + 1$, and hence that

$$\dim(\mathfrak{L}) \leqq n + \binom{3n+3}{2n+3}.$$

(Observe that $\mathfrak{L}_{r-n}$ is commutative if $r > 2n + 1$ and that by (c), there exists an open set, containing x_0 in its closure, on which the infinitesimal action defined by $\mathfrak{L}_{r-n}$ is transitive; then obtain a contradiction by using (b) and a lower bound for $\dim(\mathfrak{L}_{r-n})$.)

10. With the notation of Problem 6, suppose that $\mathfrak{L}$ is transitive at the point x_0, and that $\mathfrak{L}_r \neq \{0\}$, $\mathfrak{L}_{r+1} = \{0\}$ for some integer $r \geqq 0$; then $\mathfrak{L}_r$ is a commutative Lie algebra. Show that if the infinitesimal action defined by $\mathfrak{L}_r$ is transitive on some open subset of $M - \{x_0\}$ whose closure contains x_0, then we have $r = 1$ and $\dim(\mathfrak{L}_r) = n$. (Observe that if $r > 1$, the algebra $\mathfrak{L}_{r-1}$ would also be commutative, and show that this would contradict Problem 9(b).) Deduce that $\dim(\mathfrak{L}) \leqq n(n+2)$.

11. Let G be a connected Lie group acting differentiably and transitively on a differential manifold M. For each integer $k \geqq 2$, the action of G on M is said to be *k-ply transitive* if there exists an open orbit for the action $(s, (x_1, \ldots, x_k)) \mapsto (s \cdot x_1, \ldots, s \cdot x_k)$ of G on M^k. If so, we have $\dim(G) \geqq kn$. If H is the stabilizer of a point of M, then for the action of G on M to be k-ply transitive it is necessary and sufficient that there should exist k conjugates $H_1, \ldots, H_k$ of H such that

$$\dim(G) - \dim(H_1 \cap \cdots \cap H_k) = kn.$$

When this condition is satisfied, there exists an orbit of the action of H on M, on which the action of H is $(k-1)$-ply transitive. Moreover, the action of G on M is primitive (Problem 5).

12. (a) State the definitions and results corresponding to those of Problem 11 for partial actions and infinitesimal actions.

(b) With the notation of Problem 6, suppose that the infinitesimal action defined by $\mathfrak{L}$ is k-ply transitive in a neighborhood of x_0. Suppose also that there exists an integer $r \geqq 0$ such that $\mathfrak{L}_r \neq \{0\}$ and $\mathfrak{L}_{r+1} = \{0\}$. Show that $k \leqq n+2$, where $n = \dim_{x_0}(\mathrm{M})$. (We may assume that $k \geqq 3$. The infinitesimal action defined by $\mathfrak{L}_0$ in an open subset of $\mathrm{M} - \{x_0\}$ whose closure contains x_0 cannot be $(k-1)$-ply transitive unless, in the notation of Problem 9(a), we have $\dim(\mathrm{E}_r(x)) = n$ in this open set. Deduce from Problem 10 that $\dim(\mathfrak{L}) \leqq n(n+2)$.)

4. EXAMPLES

(19.4.1) We shall begin by determining the infinitesimal algebra $\mathfrak{G}$ of the commutative Lie group $\mathrm{G} = \mathbf{R}^n$. From (17.7.3), the space $\mathfrak{G}_e$ of distributions with support contained in $\{0\}$ is the set of all $p(\mathrm{D})\varepsilon_e$, where $p(\mathrm{D}) = \sum_\lambda c_\lambda \mathrm{D}^\lambda$ is a polynomial in the partial differentiation operators $\mathrm{D}_i = \partial/\partial x^i$ $(1 \leqq i \leqq n)$. Also, for any two polynomials p, q, we have (17.11.11.2)

$$p(\mathrm{D})\varepsilon_e * q(\mathrm{D})\varepsilon_e = (p(\mathrm{D})q(\mathrm{D}))\varepsilon_e,$$

which proves that the algebra $\mathfrak{G}_e$ is isomorphic to the *algebra* $\mathbf{R}[\mathrm{X}_1, \dots, \mathrm{X}_n]$ *of polynomials in n indeterminates over* $\mathbf{R}$. The invariant field of distributions P corresponding to $p(\mathrm{D})\varepsilon_e$ is such that, for all $f \in \mathscr{E}(\mathbf{R}^n)$ and all $x \in \mathbf{R}^n$, we have

$$(P \cdot f)(x) = \langle p(\mathrm{D})\varepsilon_e, \gamma(-x)f\rangle = (p(\mathrm{D})f)(x),$$

so that P is just the *differential operator* $p(\mathrm{D})$.

The Lie algebra $\mathfrak{g}_e$ may be canonically identified with $\mathbf{R}^n$; it is obviously *commutative*, i.e., $[\mathbf{u}, \mathbf{v}] = 0$ for all vectors $\mathbf{u}$, $\mathbf{v}$.

(19.4.2) Let A be a finite-dimensional (associative) $\mathbf{R}$-algebra with an identity element e, and let A^* be the Lie group of invertible elements of A (16.9.3). Since A^* is an open subset of the vector space A (15.2.4), the tangent space $\mathrm{T}_e(\mathrm{A}^*)$ may be canonically identified with the tangent space $\mathrm{T}_e(\mathrm{A})$ (16.8.6), and we have a canonical linear bijection (16.5.2) $\tau_e : \mathrm{T}_e(\mathrm{A}) \to \mathrm{A}$. We shall show that

(19.4.2.1) $$\tau_e([\mathbf{u}, \mathbf{v}]) = [\tau_e(\mathbf{u}), \tau_e(\mathbf{v})] = \tau_e(\mathbf{u})\tau_e(\mathbf{v}) - \tau_e(\mathbf{v})\tau_e(\mathbf{u}),$$

which will allow us to identify the Lie algebra $\mathrm{Lie}(\mathrm{A}^*)$ with the vector-space A endowed with the bracket operation $[x, y] = xy - yx$. It is enough to prove that the values taken by an arbitrary *linear form f* on A at the vectors $\tau_e([\mathbf{u}, \mathbf{v}])$ and $\tau_e(\mathbf{u})\tau_e(\mathbf{v}) - \tau_e(\mathbf{v})\tau_e(\mathbf{u})$ are the same. Since $\mathrm{D}f = f$ (8.1.3), we have

$$f(\tau_e([\mathbf{u}, \mathbf{v}])) = \mathrm{D}f \cdot \tau_e([\mathbf{u}, \mathbf{v}]) = \theta_{[\mathbf{u}, \mathbf{v}]} \cdot f$$

by (17.4.1). By virtue of (19.3.3.2), we have to calculate

$$(\theta_{\mathbf{u}} * \theta_{\mathbf{v}}) \cdot f = \int d\theta_{\mathbf{u}}(x) \int f(xy)\, d\theta_{\mathbf{v}}(y);$$

but $\int f(xy)\, d\theta_{\mathbf{v}}(y)$ is by definition the derivative of the linear form $y \mapsto f(xy)$ at the point e in the direction of the vector $\mathbf{v}$, hence (8.1.3) is equal to $f(x\tau_e(\mathbf{v}))$. The same remark applies to $\int f(x\tau_e(\mathbf{v}))\, d\theta_{\mathbf{u}}(x)$, and so we obtain $f(\tau_e(\mathbf{u})\tau_e(\mathbf{v}))$ and hence the formula (19.4.2.1).

In particular, if E is a real vector space of dimension n and if A = End(E), we have $A^* = \mathbf{GL}(E)$ (16.9.3). The Lie algebra of this group is denoted by $\mathfrak{gl}(E)$, and may be identified with End(E) endowed with the bracket operation. More particularly, if $E = \mathbf{R}^n$, we denote by $\mathfrak{gl}(n, \mathbf{R})$ the Lie algebra of the group $\mathbf{GL}(n, \mathbf{R})$; it has as a basis over $\mathbf{R}$ the *canonical basis* (E_{ij}) of the matrix algebra $\mathbf{M}_n(\mathbf{R})$ (where E_{ij} is the matrix of the endomorphism u_{ij} of $\mathbf{R}^n$ defined by $u_{ij}(\mathbf{e}_j) = \mathbf{e}_i$, $u_{ij}(\mathbf{e}_k) = 0$ for $k \neq j$) with the following multiplication table:

(19.4.2.2)
$$\begin{cases} [E_{ij}, E_{hk}] = 0 & \text{if } j \neq h \text{ and } k \neq i, \\ [E_{ij}, E_{jk}] = E_{ik} & \text{if } k \neq i, \\ [E_{ij}, E_{hi}] = -E_{hj} & \text{if } h \neq j, \\ [E_{ij}, E_{ji}] = E_{ii} - E_{jj}. \end{cases}$$

Likewise we denote by $\mathfrak{gl}(n, \mathbf{C})$ and $\mathfrak{gl}(n, \mathbf{H})$ the Lie algebras of the groups $\mathbf{GL}(n, \mathbf{C})$ and $\mathbf{GL}(n, \mathbf{H})$ (16.9.3). The matrices E_{ij} again form a basis of $\mathfrak{gl}(n, \mathbf{C})$ as a vector-space over $\mathbf{C}$ (resp. of $\mathfrak{gl}(n, \mathbf{H})$ as a left-vector-space over $\mathbf{H}$).

(19.4.3) Let G be a Lie group, H a Lie subgroup of G. Since the tangent space $T_e(H)$ may be identified with a subspace of $T_e(G)$, it follows that the Lie algebra $\mathfrak{h}_e$ of H is thereby identified with a Lie subalgebra of $\mathfrak{g}_e$.

Consider in particular a finite-dimensional **R**-algebra A with identity element, and suppose that we are given an *involution* $x \mapsto x^*$ on A. (The definition is the same as in (15.4), except that here $(\lambda x)^* = \lambda x^*$ for $\lambda \in \mathbf{R}$, i.e., the mapping $x \mapsto x^*$ is *linear*; an involution in a **C**-algebra is also an involution for the underlying **R**-algebra.) Let a be an invertible element; we shall show that the set H of elements $x \in A^*$ such that

(19.4.3.1)
$$x^*ax = a$$

is a *Lie subgroup* of $G = A^*$. Clearly H is a subgroup of G, so that it has to be shown that H is a submanifold of G. For this it is enough to show that the mapping $x \mapsto x^*ax$ is a *subimmersion* of A into itself (16.8.8). Since the mapping $x \mapsto x^*$ is linear, and the mapping $(x, y) \mapsto yax$ bilinear, it follows from (8.1.3), (8.1.4), and (8.2.1) that the derivative of the mapping $x \mapsto x^*ax$ at $x_0 \in A^*$ is

$$\mathbf{h} \mapsto \mathbf{h}^*ax_0 + x_0^*a\mathbf{h} = x_0^*((\mathbf{h}x_0^{-1})^*a + a(\mathbf{h}x_0^{-1}))x_0.$$

The assertion now follows, because $\mathbf{h} \mapsto \mathbf{h}x_0^{-1}$ is a bijective linear mapping and therefore the rank of the derivative at the point x_0 is equal to that of the linear mapping $\mathbf{h} \mapsto \mathbf{h}^*a + a\mathbf{h}$, which is independent of x_0. The same calculation shows (16.8.8) that the Lie algebra of H may be identified with the vector subspace of A defined by the equation

(19.4.3.2) $$x^*a + ax = 0.$$

This applies in particular to $A = \mathbf{M}_n(\mathbf{R})$, $\mathbf{M}_n(\mathbf{C})$, or $\mathbf{M}_n(\mathbf{H})$, the involution being $X \mapsto {}^t\overline{X}$; in this way we obtain the Lie algebras of the groups $\mathbf{O}(n)$, $\mathbf{U}(n, \mathbf{C})$ and $\mathbf{U}(n, \mathbf{H})$ (16.11.2 and 16.11.3).

More generally, consider a nondegenerate symmetric or alternating bilinear form Φ on $\mathbf{R}^n$,and let S be its matrix relative to the canonical basis. The group of endomorphisms of $\mathbf{R}^n$ which leave Φ invariant may be identified with the group of matrices $X \in \mathbf{M}_n(\mathbf{R})$ such that ${}^tXSX = S$; the Lie algebra of this group is then identified with the Lie algebra of matrices $X \in \mathbf{M}_n(\mathbf{R})$ such that

(19.4.3.3) $${}^tX \cdot S + S \cdot X = 0$$

(the bracket operation being $[X, Y] = XY - YX$).

Remark

(19.4.4) Consider $\mathbf{GL}(E)$ as acting on E by the canonical action $(S, \mathbf{v}) \mapsto S \cdot \mathbf{v}$, the product of the automorphism S and the vector $\mathbf{v}$. Fix a vector $\mathbf{v}_0 \in E$ and consider the mapping $g : S \mapsto S \cdot \mathbf{v}_0$ of $\mathbf{GL}(E)$ into E; it follows immediately from (19.4.2) that the differential of g at the neutral element I of $\mathbf{GL}(E)$ is given by

(19.4.4.1) $$d_I g \cdot U = U \cdot \mathbf{v}_0$$

for $U \in \mathrm{End}(E)$ (identified with the Lie algebra of $\mathbf{GL}(E)$).

Consider now a Lie group G, a C^∞ linear representation $\rho : G \to \mathbf{GL}(E)$ and a differential manifold M on which G acts differentiably on the right. Consider G as acting on the right on E:

$$(s, \mathbf{v}) \mapsto \rho(s^{-1}) \cdot \mathbf{v}.$$

Let $\mathbf{f}: \mathrm{M} \to \mathrm{E}$ be a G-*invariant* C^∞ mapping, i.e. (19.1.2.1) such that

(19.4.4.2) $$\mathbf{f}(x \cdot s) = \rho(s^{-1}) \cdot \mathbf{f}(x).$$

Then for each vector $\mathbf{u} \in \mathfrak{g}_e$, in the notation of (19.3.7), we have

(19.4.4.3) $$(\theta_{Z_\mathbf{u}} \cdot \mathbf{f})(x) = -\rho_*(\mathbf{u}) \cdot \mathbf{f}(x),$$

$\rho_*(\mathbf{u})$ being the image in $\mathfrak{gl}(\mathrm{E}) = \mathrm{End}(\mathrm{E})$ of the vector $\mathbf{u}$ under the derived homomorphism of ρ. For by the definition of a Killing field and by (17.14.9) we have

$$(\theta_{Z_\mathbf{u}} \cdot \mathbf{f})(x) = d_x \mathbf{f} \cdot (x \cdot \mathbf{u}).$$

On the other hand, if we evaluate at $\mathbf{u}$ the differentials at the point e of the two sides of (19.4.4.2), considered as mappings of G into E (for fixed x), we obtain on the left-hand side $d_x \mathbf{f} \cdot (x \cdot \mathbf{u})$ (16.5.8.5); and since $\mathrm{T}_e(\rho) \cdot \mathbf{h} = \rho_*(\mathbf{h})$ by definition, it follows from (19.4.4.1), (16.5.8.5), and (16.9.9(i)) that on the right-hand side we obtain $-\rho_*(\mathbf{u}) \cdot \mathbf{f}(x)$.

5. TAYLOR'S FORMULA IN A LIE GROUP

We shall first give some supplementary results on Taylor expansions (8.14.3).

(19.5.1) Let U be an open neighborhood of 0 in $\mathbf{R}^n$, and let $\|\mathbf{x}\|$ be a norm on $\mathbf{R}^n$ compatible with the topology (e.g., the Euclidean norm). In the ring $\mathscr{E}_\mathbf{R}(\mathrm{U})$ of real-valued C^∞ functions on U, we denote by $o_m(\mathrm{U})$ (or simply o_m), for each integer $m \geqq 0$, the set of all $f \in \mathscr{E}_\mathbf{R}(\mathrm{U})$ such that $f(\mathbf{x})/\|\mathbf{x}\|^{m+1}$ remains bounded as $\mathbf{x} \to 0$ (and $\mathbf{x} \neq 0$). It is clear that $o_m(\mathrm{U})$ is an ideal in $\mathscr{E}_\mathbf{R}(\mathrm{U})$, and that o_0 is the set of C^∞-functions which vanish at the origin. We shall use the same notation $o_m(\mathrm{U})$ to denote the set of C^∞-functions $\mathbf{f}$ on U with values in a finite-dimensional real vector space F, such that $\|\mathbf{f}(\mathbf{x})\|/\|\mathbf{x}\|^{m+1}$ remains bounded as $\mathbf{x} \to 0$ (and $\mathbf{x} \neq 0$); or, equivalently, such that the components of $\mathbf{f}$ relative to a basis of F are functions in $\mathscr{E}_\mathbf{R}(\mathrm{U})$ belonging to $o_m(\mathrm{U})$.

(19.5.2) For each function $f \in \mathscr{E}_\mathbf{R}(\mathrm{U})$ and each integer $m \geqq 0$, the sum of the first $m + 1$ terms of Taylor's formula (8.14.3) for f:

$$\mathrm{P}_m(\mathbf{x}) = f(0) + \frac{1}{1!} f'(0) \cdot \mathbf{x} + \cdots + \frac{1}{m!} f^{(m)}(0) \cdot \mathbf{x}^{(m)}$$

is called the *Taylor polynomial of degree* $\leqq m$ of the function f. Since

$$f^{(p)}(0) \cdot \mathbf{x}^{(p)} = (x^1 D_1 + \cdots + x^n D_n)^p f(0)$$

by virtue of (8.13), we may also write

(19.5.2.1) $$P_m(\mathbf{x}) = \sum_{|\alpha| \leqq m} \frac{D^\alpha f(0)}{\alpha!} \mathbf{x}^\alpha$$

with the notation of (17.1).

(19.5.3) *The Taylor polynomial* P_m *is the unique polynomial* P *of degree* $\leqq m$ *such that*

(19.5.3.1) $$f - P \in o_m(U).$$

It follows from Taylor's formula (8.14.3) that $f - P_m \in o_m(U)$. Suppose that there exists a polynomial $P \neq P_m$ of degree $\leqq m$ satisfying (19.5.3.1). Then the polynomial $Q = P - P_m$ belongs to $o_m(U)$. Writing $Q = Q_0 + Q_1 + \cdots + Q_m$, where Q_k is homogeneous of degree k $(0 \leqq k \leqq m)$, suppose that p is the smallest integer such that $Q_p \neq 0$. Then, for $\mathbf{x} \neq 0$, putting $\mathbf{x} = \|\mathbf{x}\|\mathbf{z}$, we shall have

$$Q(\mathbf{x})/\|\mathbf{x}\|^m = \|\mathbf{x}\|^{p-m}(Q_p(\mathbf{z}) + \|\mathbf{x}\|Q_{p+1}(\mathbf{z}) + \cdots + \|\mathbf{x}\|^{m-p}Q_m(\mathbf{z})),$$

and since $Q_p \neq 0$, there exists $\mathbf{z} \in S_{n-1}$ such that $Q_p(\mathbf{z}) \neq 0$; but then, in the formula above, when $\mathbf{z}$ is fixed and $\mathbf{x} \to 0$, the expression in brackets tends to $Q_p(\mathbf{z}) \neq 0$, and the absolute value of the left-hand side would tend to $+\infty$, which is absurd.

(19.5.4) For each function $f \in \mathscr{E}_{\mathbf{R}}(U)$, put

$$\tilde{f}(X_1, \ldots, X_n) = \sum_\alpha \frac{1}{\alpha!} D^\alpha f(0) X^\alpha,$$

a *formal power series* belonging to $\mathbf{R}[[X_1, \ldots, X_n]]$, where $X^\alpha = X_1^{\alpha_1} \cdots X_n^{\alpha_n}$ if $\alpha = (\alpha_1, \ldots, \alpha_n)$ (A.21.2).

(19.5.5) *Let* V *be an open neighborhood of* 0 *in* $\mathbf{R}^p$, *and let* $\mathbf{g} = (g_1, \ldots, g_n)$ *be a* C^∞ *mapping of* V *into* U *such that* $\mathbf{g}(0) = 0$. *Then for each function* $f \in \mathscr{E}_{\mathbf{R}}(U)$, *if* $h = f \circ \mathbf{g}$, *we have*

(19.5.5.1) $$\tilde{h}(Y_1, \ldots, Y_p) = \tilde{f}(\tilde{g}_1(Y_1, \ldots, Y_p), \ldots, \tilde{g}_n(Y_1, \ldots, Y_p)).$$

By virtue of (19.5.3) it has to be shown that, for each integer $m \geqq 0$, if S_m is the sum of the terms of degree $\leqq m$ in the formal power series on the right-hand side of (19.5.5.1), then $h - S_m \in o_m(V)$. Let $P_m, Q_{1m}, \ldots, Q_{nm}$ be the Taylor polynomials of degree $\leqq m$ of $f, g_1, \ldots, g_n$, respectively. Since the formal power series $\tilde{g}_1, \ldots, \tilde{g}_n$ have constant terms equal to 0, it follows from the properties of formal power series (A.21.3) that S_m is the sum of the terms of degree $\leqq m$ in the polynomial

$$R_m(Y_1, \ldots, Y_p) = P_m(Q_{1m}(Y_1, \ldots, Y_p), \ldots, Q_{nm}(Y_1, \ldots, Y_p)).$$

Since the function $\mathbf{y} \mapsto R_m(\mathbf{y}) - S_m(\mathbf{y})$ belongs to $o_m(V)$, it is enough to show that $h - R_m \in o_m(V)$. We may write

$$g_j(\mathbf{y}) = Q_{jm}(\mathbf{y}) + r_j(\mathbf{y}),$$

where $r_j \in o_m(V)$, and therefore the function

$$\begin{aligned} \mathbf{y} \mapsto P_m(g_1(\mathbf{y}), \ldots, g_n(\mathbf{y})) &- P_m(Q_{1m}(\mathbf{y}), \ldots, Q_{nm}(\mathbf{y})) \\ = P_m(g_1(\mathbf{y}), \ldots, g_n(\mathbf{y})) &- R_m(\mathbf{y}) \end{aligned}$$

belongs to $o_m(V)$, since the latter is an ideal in $\mathscr{E}_{\mathbf{R}}(V)$. Hence it is enough to show that the function

$$\mathbf{y} \mapsto f(g_1(\mathbf{y}), \ldots, g_n(\mathbf{y})) - P_m(g_1(\mathbf{y}), \ldots, g_n(\mathbf{y}))$$

belongs to $o_m(V)$. Now, the hypothesis $\mathbf{g}(0) = 0$ implies that there exists a neighborhood $W_0 \subset V$ of 0 and a number $k \geqq 0$ such that $\|\mathbf{g}(\mathbf{y})\| \leqq k\|\mathbf{y}\|$ for all $\mathbf{y} \in W_0$. On the other hand, by the definition of P_m, there exists a neighborhood $W \subset W_0$ of 0 and a constant $A > 0$ such that, for all $\mathbf{y} \in W$, we have

$$|f(g_1(\mathbf{y}), \ldots, g_n(\mathbf{y})) - P_m(g_1(\mathbf{y}), \ldots, g_n(\mathbf{y}))| \leqq A\|\mathbf{g}(\mathbf{y})\|^{m+1} \leqq Ak^{m+1}\|\mathbf{y}\|^{m+1}.$$

This completes the proof.

(19.5.6) We shall use the following notation to express that the formal power series $\sum_\alpha a_\alpha X^\alpha$ is equal to the formal power series $\tilde{f}(X_1, \ldots, X_n)$:

(19.5.6.1) $$f(\mathbf{x}) \sim \sum_\alpha a_\alpha \mathbf{x}^\alpha.$$

(The use of this notation does *not* imply that the series on the right-hand side converges for any $\mathbf{x} \neq 0$, nor that if it does converge its sum is equal to $f(\mathbf{x})$.) The right-hand side of (19.5.6.1) is called the (infinite) *Taylor expansion* of f at the point 0.

(19.5.7) After these preliminaries, let G be a Lie group and let $c = (U, \varphi, n)$ be a chart of G at the neutral element e, such that $\varphi(e) = 0$. Let V be a symmetric open neighborhood of e in G such that $V^3 \subset U$. Then the function

(19.5.7.1) $$(\mathbf{x}, \mathbf{y}) \mapsto \mathbf{m}(\mathbf{x}, \mathbf{y}) = \varphi(\varphi^{-1}(\mathbf{x})\varphi^{-1}(\mathbf{y})) = (m_i(\mathbf{x}, \mathbf{y}))_{1 \leqq i \leqq n} \in \mathbf{R}^n$$

is defined and of class C^∞ on the open set $\varphi(V) \times \varphi(V) \subset \mathbf{R}^{2n}$; it is called the *local expression* of the multiplication law in G, relative to the chart c. As functions of $(\mathbf{x}, \mathbf{y})$, for $1 \leqq i \leqq n$, we may write

(19.5.7.2) $$m_i(\mathbf{x}, \mathbf{y}) \sim \sum_{\alpha, \beta} b_{\alpha\beta}^{(i)} \mathbf{x}^\alpha \mathbf{y}^\beta$$

in order to express in a concise form the Taylor expansions of the functions m_i. The fact that e is the neutral element of G is expressed by the conditions

$$\mathbf{m}(0, \mathbf{y}) = \mathbf{y}, \qquad \mathbf{m}(\mathbf{x}, 0) = \mathbf{x}$$

for $\mathbf{x}, \mathbf{y} \in V$. Consequently, the formal power series on the right-hand side of (19.5.7.2) are of the form

(19.5.7.3) $$X_i + Y_i + \sum_{|\alpha| \geqq 1, |\beta| \geqq 1} b_{\alpha\beta}^{(i)} X^\alpha Y^\beta.$$

For each multi-index $\gamma = (\gamma_1, \ldots, \gamma_n)$, let

(19.5.7.4) $$(\mathbf{m}(\mathbf{x}, \mathbf{y}))^\gamma = (m_1(\mathbf{x}, \mathbf{y}))^{\gamma_1} \cdots (m_n(\mathbf{x}, \mathbf{y}))^{\gamma_n}$$

and write its Taylor expansion in the form

(19.5.7.5) $$(\mathbf{m}(\mathbf{x}, \mathbf{y}))^\gamma \sim \sum_{\alpha, \beta} c_{\alpha\beta\gamma} \mathbf{x}^\alpha \mathbf{y}^\beta,$$

so that, by virtue of (19.5.7.3) and (19.5.5), we have

$$\sum_{\alpha, \beta} c_{\alpha\beta\gamma} X^\alpha Y^\beta = \Big(X_1 + Y_1 + \sum_{\alpha, \beta} b_{\alpha\beta}^{(1)} X^\alpha Y^\beta\Big)^{\gamma_1} \cdots \Big(X_n + Y_n + \sum_{\alpha, \beta} b_{\alpha\beta}^{(n)} X^\alpha Y^\beta\Big)^{\gamma_n}.$$

This shows immediately that

(19.5.7.6) $$c_{\alpha\beta\gamma} = 0 \quad \text{if} \quad |\alpha| + |\beta| < |\gamma|$$

and that the coefficients $c_{\alpha\beta\gamma}$ for which $|\alpha| + |\beta| = |\gamma|$ are those of the polynomial

$$(X_1 + Y_1)^{\gamma_1} \cdots (X_n + Y_n)^{\gamma_n};$$

in other words, the only nonzero coefficients $c_{\alpha\beta\gamma}$ with $|\alpha| + |\beta| = |\gamma|$ are those for which $\alpha + \beta = \gamma$, and

(19.5.7.7) $$c_{\alpha, \beta, \alpha+\beta} = (\alpha + \beta)!/\alpha!\beta!.$$

(19.5.8) Consider now, for each multi-index $\alpha = (\alpha_1, \ldots, \alpha_n)$, the real distribution Δ_α with support $\{e\}$ on G defined by

(19.5.8.1) $$\Delta_\alpha \cdot f = \frac{1}{\alpha!} D^\alpha(f \circ \varphi^{-1})(0)$$

for all $f \in \mathscr{E}_{\mathbf{R}}(U)$. In other words, $\Delta_\alpha = \varphi^{-1}\left(\frac{1}{\alpha!} D^\alpha \varepsilon_0\right)$. Since the distributions $D^\alpha \varepsilon_0$ form a *basis* over **R** of the space of real distributions on $\mathbf{R}^n$ with support contained in $\{0\}$ (17.7.3), it follows that the Δ_α form a *basis of the algebra* $\mathfrak{G}_e$. If $Z_\alpha \in \text{Diff}(G)$ is the left-invariant differential operator on G which reduces to Δ_α at the point e (19.2.4), the Z_α form a *basis of the infinitesimal algebra* $\mathfrak{G}$, and we have

(19.5.8.2) $$(Z_\alpha \cdot f)(s) = \Delta_\alpha \cdot \gamma(s^{-1})f.$$

We have now the following result:

(19.5.9) *For each $s \in V$, the function $\mathbf{y} \mapsto f(s\varphi^{-1}(\mathbf{y}))$, defined on $\varphi(V)$, has the Taylor expanison*

(19.5.9.1) $$f(s\varphi^{-1}(\mathbf{y})) \sim \sum_\alpha (Z_\alpha \cdot f)(s)\mathbf{y}^\alpha$$

and the multiplication table for the basis (Z_α) of $\mathfrak{G}$ is given by

(19.5.9.2) $$Z_\alpha \circ Z_\beta = \sum_\gamma c_{\alpha\beta\gamma} Z_\gamma,$$

where the coefficients $c_{\alpha\beta\gamma}$ are defined by (19.5.7.5).

Let us write the Taylor expansion of the function $\mathbf{y} \mapsto f(s\varphi^{-1}(\mathbf{y}))$ in the form

(19.5.9.3) $$f(s\varphi^{-1}(\mathbf{y})) \sim \sum_\alpha (P_\alpha \cdot f)(s)\mathbf{y}^\alpha,$$

where the P_α are differential operators; we have

$$(P_\alpha \cdot f)(e) = \frac{1}{\alpha!} \mathrm{D}^\alpha(f \circ \varphi^{-1})(0) = \Delta_\alpha \cdot f.$$

Next, we may write $f(s\varphi^{-1}(\mathbf{y})) = (\gamma(s^{-1})f)(e\varphi^{-1}(\mathbf{y}))$, and so by replacing f by $\gamma(s^{-1})f$ in the formula above we obtain another Taylor expansion:

(19.5.9.4) $$f(s\varphi^{-1}(\mathbf{y})) \sim \sum_\alpha (P_\alpha \cdot \gamma(s^{-1})f)(e)\mathbf{y}^\alpha.$$

Comparing the two expansions and bearing in mind (19.5.8.2), we obtain

$$(P_\alpha \cdot f)(s) = (Z_\alpha \cdot f)(s)$$

for all $s \in \mathrm{V}$.

If we replace f by $Z_\beta \cdot f$ in (19.5.9.1), we get

(19.5.9.5) $$(Z_\beta \cdot f)(s\varphi^{-1}(\mathbf{y})) \sim \sum_\alpha ((Z_\alpha \circ Z_\beta) \cdot f)(s)\mathbf{y}^\alpha.$$

The Taylor expansion of the function

$$(\mathbf{x}, \mathbf{y}) \mapsto f(s\varphi^{-1}(\mathbf{x})\varphi^{-1}(\mathbf{y}))$$

on $\varphi(\mathrm{V}) \times \varphi(\mathrm{V})$, where $s \in \mathrm{V}$, is obtained by replacing s by $s\varphi^{-1}(\mathbf{x})$ in the right-hand side of (19.5.9.1) and is therefore, by (19.5.9.5),

(19.5.9.6) $$f(s\varphi^{-1}(\mathbf{x})\varphi^{-1}(\mathbf{y})) \sim \sum_{\alpha, \beta} ((Z_\alpha \circ Z_\beta) \cdot f)(s)\mathbf{x}^\alpha \mathbf{y}^\beta.$$

On the other hand, $\varphi^{-1}(\mathbf{x})\varphi^{-1}(\mathbf{y}) = \varphi^{-1}(\mathbf{m}(\mathbf{x}, \mathbf{y}))$ and therefore, by virtue of (19.5.5), we obtain the Taylor expansion of the function

$$(\mathbf{x}, \mathbf{y}) \mapsto f(s\varphi^{-1}(\mathbf{m}(\mathbf{x}, \mathbf{y})))$$

by substituting, for each γ, the formal power series

$$\sum_{\alpha, \beta} c_{\alpha\beta\gamma} \mathrm{X}^\alpha \mathrm{Y}^\beta$$

(the Taylor expansion of $(\mathbf{m}(\mathbf{x}, \mathbf{y}))^\gamma$) for Y^γ in the formal power series $\sum_\gamma (Z_\gamma \cdot f)(s)\mathrm{Y}^\gamma$. This gives rise to the series

(19.5.9.7) $$\sum_{\alpha, \beta} \left(\sum_\gamma c_{\alpha\beta\gamma}(Z_\gamma \cdot f)(s) \right) \mathrm{X}^\alpha \mathrm{Y}^\beta$$

and now a comparison of this series with (19.5.9.6) gives the relation (19.5.9.2) for each pair of multi-indices (α, β).

The formula (19.5.9.1) is called *Taylor's formula* at the point e in G, relative to the chart $c = (\mathrm{U}, \varphi, n)$.

(19.5.10) Let ε_i denote the multi-index $(\delta_{ij})_{1 \leqq j \leqq n}$, where δ_{ij} is the Kronecker delta. Then the invariant vector fields $Z_{\varepsilon_i} = X_i$ $(1 \leqq i \leqq n)$ form a *basis* of the Lie algebra $\mathfrak{g}$ of invariant vector fields. Moreover, by (19.5.7.6) and (19.5.7.7), we have

$$\begin{aligned} c_{\varepsilon_i \varepsilon_j \gamma} &= 0 \qquad \text{for} \quad |\gamma| > 2, \\ c_{\varepsilon_i, \varepsilon_j, \varepsilon_i + \varepsilon_j} &= 1 \qquad \text{for} \quad i \neq j, \\ c_{\varepsilon_i, \varepsilon_j, \varepsilon_k} &= b^{(k)}_{\varepsilon_i \varepsilon_j} \end{aligned}$$

(which we shall write as $b_{ij}^{(k)}$ for simplicity). Hence it follows from (19.5.9.2) that, for $i \neq j$, we have

(19.5.10.1) $$X_i \circ X_j = Z_{\varepsilon_i + \varepsilon_j} + \sum_{k=1}^{n} b_{ij}^{(k)} X_k,$$

and therefore the multiplication table for the basis $(X_i)_{1 \leqq i \leqq n}$ of the Lie algebra $\mathfrak{g}$ is given by

(19.5.10.2) $$[X_i, X_j] = \sum_{k=1}^{n} (b_{ij}^{(k)} - b_{ji}^{(k)}) X_k \qquad (1 \leqq i, j \leqq n).$$

Since $X_i(e) = \mathbf{u}_i = (d_e \varphi)^{-1} \cdot \mathbf{e}_i$ (16.5.7), the basis $(\mathbf{u}_i)$ of the Lie algebra $\mathfrak{g}_e$ of the group G has the same multiplication table:

(19.5.10.3) $$[\mathbf{u}_i, \mathbf{u}_j] = \sum_{k=1}^{n} (b_{ij}^{(k)} - b_{ji}^{(k)}) \mathbf{u}_k \qquad (1 \leqq i, j \leqq n).$$

Example

(19.5.11) Consider the Lie group G whose underlying manifold is $\mathbf{R}^* \times \mathbf{R}$, and multiplication given by

(19.5.11.1) $$(s,^1 s^2)(t^1, t^2) = (s^1 t^1, s^1 t^2 + s^2),$$

so that the neutral element is $e = (1, 0)$. As chart we take $(\mathrm{G}, \varphi, 2)$, where φ is the translation in $\mathbf{R}^2$ which takes $(1, 0)$ to $(0, 0)$. Then, with the notation of (19.5.7), we have in this case

(19.5.11.2) $$\begin{cases} m_1(\mathbf{x}, \mathbf{y}) = x^1 + y^1 + x^1 y^1, \\ m_2(\mathbf{x}, \mathbf{y}) = x^2 + y^2 + x^1 y^2, \end{cases}$$

and by virtue of (19.5.10.3) the Lie algebra of G has the multiplication table

(19.5.11.3) $$[\mathbf{u}_1, \mathbf{u}_2] = \mathbf{u}_2 .$$

PROBLEMS

1. Let G be the Lie subgroup of **GL**(3, **R**) consisting of the matrices

$$\begin{pmatrix} 1 & x & z \\ 0 & 1 & y \\ 0 & 0 & 1 \end{pmatrix}.$$

Show that the Lie algebra $\mathfrak{g}_e$ has a basis (**u**, **v**, **w**) for which the multiplication table is

$$[\mathbf{u}, \mathbf{v}] = \mathbf{w}, \qquad [\mathbf{u}, \mathbf{w}] = 0, \qquad [\mathbf{v}, \mathbf{w}] = 0.$$

2. Show that the Lie algebra of the group **SL**(2, **R**) has a basis $(\mathbf{u}_1, \mathbf{u}_2, \mathbf{u}_3)$ for which the multiplication table is

$$[\mathbf{u}_1, \mathbf{u}_2] = 2\mathbf{u}_2, \qquad [\mathbf{u}_1, \mathbf{u}_3] = -2\mathbf{u}_3, \qquad [\mathbf{u}_2, \mathbf{u}_3] = \mathbf{u}_1.$$

6. THE ENVELOPING ALGEBRA OF THE LIE ALGEBRA OF A LIE GROUP

We retain the hypotheses and notation introduced in (19.5.7), (19.5.8), and (19.5.10), and we denote the composition $X \circ Y$ in $\mathfrak{G}$ by XY. For each multi-index $\alpha = (\alpha_1, \ldots, \alpha_n)$ we shall write

(19.6.1) $$X_\alpha = X_1^{\alpha_1} X_2^{\alpha_2} \cdots X_n^{\alpha_n},$$

with the convention that $X_0 = 1$. It should be remarked that it is *not* legitimate to permute the X_i in this product, because in general the algebra $\mathfrak{G}$ is not commutative.

(19.6.2) *The operators X_α form a basis of the infinitesimal algebra $\mathfrak{G}$ of the group* G.

For each integer $m > 0$, let $\mathfrak{G}_m$ denote the vector subspace of $\mathfrak{G}$ formed by the invariant operators of order $\leqq m$. By virtue of (19.5.8.1) and (19.5.9.2), the space $\mathfrak{G}_m$ has as a *basis* the set of the Z_α such that $|\alpha| \leqq m$. Since the X_α and the Z_α have the same set of indices, it is enough to show that the X_α

with $|\alpha| \leqq m$ *span* $\mathfrak{G}_m$ (A.4.8). For $m = 1$, this is clear from the definition of the X_i, and therefore it will be enough to prove, by induction on m, that for $|\alpha| = m$, we have

(19.6.2.1)
$$X_\alpha = \alpha! Z_\alpha + \sum_{|\lambda| < m} q_{\alpha\lambda} Z_\lambda$$

with coefficients $q_{\alpha\lambda} \in \mathbf{R}$. Now, if $\alpha = (\alpha_1, \ldots, \alpha_n)$, let i be the first index such that $\alpha_i > 0$; then by definition we have $X_\alpha = X_i X_{\alpha - \varepsilon_i}$, and the inductive hypothesis gives

(19.6.2.2)
$$X_{\alpha - \varepsilon_i} = (\alpha - \varepsilon_i)!\, Z_{\alpha - \varepsilon_i} + \sum_{|\mu| < m-1} q_{\alpha - \varepsilon_i, \mu} Z_\mu .$$

From (19.5.9.2), (19.5.7.6), and (19.5.7.7), we obtain

(19.6.2.3)
$$X_i Z_{\alpha - \varepsilon_i} = \alpha_i Z_\alpha + \sum_{|\lambda| < m} c_{\varepsilon_i, \alpha - \varepsilon_i, \lambda} Z_\lambda ,$$

and likewise, for $|\mu| < m - 1$,

(19.6.2.4)
$$X_i Z_\mu = \sum_{|\lambda| < m} c_{\varepsilon_i, \mu, \lambda} Z_\lambda .$$

Now substitute these expressions into the right-hand side of (19.6.2.2) multiplied on the left by X_i, and we obtain (19.6.2.1).

(19.6.3) We shall show that the associative algebra $\mathfrak{G}$ is the "enveloping" algebra of the Lie algebra $\mathfrak{g}$, in the following sense:

(19.6.4) *For each (associative)* **R**-*algebra* B *with identity element, and each linear mapping* $f: \mathfrak{g} \to \mathrm{B}$ *satisfying the relation*

(19.6.4.1)
$$f([X, Y]) = f(X)f(Y) - f(Y)f(X)$$

for all $X, Y \in \mathfrak{g}$, *there exists a unique homomorphism* h *of the algebra* $\mathfrak{G}$ *into the algebra* B *which extends* f *and is such that* $h(1) = 1$.

The uniqueness of h is clear, because $\mathfrak{g}$ and the identity element generate $\mathfrak{G}$, by (19.6.2). As to the existence of h, we remark that for each multi-index α we must have

(19.6.4.2)
$$h(X_\alpha) = (f(X_1))^{\alpha_1} (f(X_2))^{\alpha_2} \cdots (f(X_n))^{\alpha_n}$$

with the notation introduced above. There exists a unique *linear* mapping $h : \mathfrak{G} \to \mathrm{B}$ satisfying (19.6.4.2) for each α, and to prove that h is an algebra homomorphism it is enough to verify that for all α and β, we have

(19.6.4.3) $$h(X_\alpha X_\beta) = h(X_\alpha)h(X_\beta).$$

Consider first the case where $\alpha = \varepsilon_i$ and $\beta = \varepsilon_j$. If $i \leqq j$, we have $X_i X_j = X_{\varepsilon_i + \varepsilon_j}$, and the relation (19.6.4.3) follows from the definition (19.6.4.2); whereas if $i > j$, we write

$$X_i X_j = [X_i, X_j] + X_j X_i,$$

and then the relation (19.6.4.3) follows from the previous case and the hypothesis (19.6.4.1).

Consider next the case where $\alpha = \varepsilon_i$ and β is arbitrary. We shall proceed by induction on $|\beta| = m$, and by induction on i. There exists an index j such that $X_\beta = X_j X_\gamma$ with $\gamma = \beta - \varepsilon_j$. If $i \leqq j$ (which in particular will be the case when $i = 1$), we have $X_i X_\beta = X_{\beta + \varepsilon_i}$, and the relation (19.6.4.3) again follows from the definition (19.6.4.2); whereas if $i > j$, we have

$$h(X_i X_j X_\gamma) = h([X_i, X_j]X_\gamma) + h(X_j X_i X_\gamma),$$

and since $[X_i, X_j]$ is a linear combination of the X_k, and $|\gamma| = m - 1$, it follows from the inductive hypothesis that

$$h([X_i, X_j]X_\gamma) = h([X_i, X_j])h(X_\gamma);$$

also

$$h([X_i, X_j]) = h(X_i)h(X_j) - h(X_j)h(X_i),$$
$$h(X_j)h(X_\gamma) = h(X_\beta),$$

and therefore

$$h([X_i, X_j]X_\gamma) = h(X_i)h(X_\beta) - h(X_j)h(X_i)h(X_\gamma).$$

On the other hand, we may write $X_i X_\gamma = \sum_\lambda r_\lambda X_\lambda$, where the multi-indices λ satisfy $|\lambda| \leqq m$, and therefore

$$h(X_j X_i X_\gamma) = \sum_\lambda r_\lambda h(X_j X_\lambda) = \sum_\lambda r_\lambda h(X_j)h(X_\lambda)$$
$$= h(X_j)h(X_i X_\gamma)$$

since $j < i$. Since $|\gamma| = m - 1$, we have $h(X_i X_\gamma) = h(X_i)h(X_\gamma)$ by the inductive hypothesis. Hence we obtain $h(X_i X_\beta) = h(X_i)h(X_\beta)$ for all i and all β, as desired.

Consider finally the general case, by induction on $|\alpha| = m$. We may again write $X_\alpha = X_i X_\gamma$, where $\gamma = \alpha - \varepsilon_i$, for some index i. We have then

$$h(X_\alpha X_\beta) = h(X_i X_\gamma X_\beta),$$

and since $X_\gamma X_\beta = \sum_\lambda r_\lambda X_\lambda$, it follows from above that

$$h(X_i X_\gamma X_\beta) = \sum_\lambda r_\lambda h(X_i X_\lambda) = \sum_\lambda r_\lambda h(X_i)h(X_\lambda)$$
$$= h(X_i)h(X_\gamma X_\beta).$$

Finally, since $|\gamma| = m - 1$, the inductive hypothesis implies that

$$h(X_\gamma X_\beta) = h(X_\gamma)h(X_\beta).$$

This completes the proof, because $h(X_i)h(X_\gamma) = h(X_\alpha)$.

The associative algebra $\mathfrak{G}$ is also denoted by $\mathrm{U}(\mathfrak{g})$.

PROBLEM

Let $\mathbf{T}^n(\mathfrak{g})$ denote the real vector space $\otimes^n\mathfrak{g}$, the nth tensor power of the Lie algebra $\mathfrak{g}$ of a Lie group G, and let $\mathbf{T}(\mathfrak{g})$ be the tensor algebra of $\mathfrak{g}$, i.e., the direct sum of the $\mathbf{T}^n(\mathfrak{g})$ for all $n \geqq 0$.

(a) Show that the enveloping algebra $\mathrm{U}(\mathfrak{g})$ is isomorphic to the quotient of $\mathbf{T}(\mathfrak{g})$ by the two-sided ideal $\mathfrak{J}$ generated by elements of the form $X \otimes Y - Y \otimes X - [X, Y]$ in $\mathbf{T}(\mathfrak{g})$, for all $X, Y \in \mathfrak{g}$.
(b) Let T_n be the direct sum of the $\mathbf{T}^k(\mathfrak{g})$ for $0 \leqq k \leqq n$. The canonical image U_n of T_n in $\mathrm{U}(\mathfrak{g})$ is the vector subspace spanned by products of n elements which are either scalars or elements of $\mathfrak{g}$. Let $\mathrm{P}_n = \mathrm{U}_n/\mathrm{U}_{n-1}$ for $n \geqq 0$ (with the convention that $\mathrm{U}_{-1} = \{0\}$), and let $\mathrm{P}(\mathfrak{g})$ be the vector space direct sum of the P_n. The multiplication in $\mathrm{U}(\mathfrak{g})$ induces bilinear mappings $\mathrm{P}_m \times \mathrm{P}_n \to \mathrm{P}_{m+n}$ for all pairs of integers $m, n \geqq 0$, and in this way $\mathrm{P}(\mathfrak{g})$ becomes a graded associative $\mathbf{R}$-algebra. If $\dim(\mathfrak{g}) = m$, show that $\mathrm{P}(\mathfrak{g})$ is isomorphic to the polynomial algebra $\mathbf{R}[\mathrm{T}_1, \dots, \mathrm{T}_m]$ in m independent indeterminates.
(c) Deduce from (b) that the algebra $\mathrm{U}(\mathfrak{g})$ is left- and right-Noetherian. (If $\mathfrak{a}$ is a left ideal of $\mathrm{U}(\mathfrak{g})$, consider the graded ideal of $\mathrm{P}(\mathfrak{g})$ generated by the canonical images of $\mathfrak{a} \cap \mathrm{U}_n$ in P_n, and lift a finite system of generators of this ideal.)
(d) Let $\mathfrak{a}_1, \dots, \mathfrak{a}_p$ be left ideals of finite codimension in $\mathrm{U}(\mathfrak{g})$. Show that the left ideal $\mathfrak{a}_1 \mathfrak{a}_2 \cdots \mathfrak{a}_p$ is of finite codimension. (Induction on p, using (c).)
(e) Show that the algebra $\mathrm{U}(\mathfrak{g})$ has no zero-divisors (use a method analogous to (c)), and that the only invertible elements in $\mathrm{U}(\mathfrak{g})$ are the scalars.
(f) Let A, B be two nonzero elements of U_p. Show that there exist two nonzero elements C, D of $\mathrm{U}(\mathfrak{g})$ such that $CA = DB$. (Compare the dimensions of $\mathrm{U}_n A$, $\mathrm{U}_n B$, and U_{n+p}.) Deduce that $\mathrm{U}(\mathfrak{g})$ is isomorphic to a subalgebra U' of a division ring K (which is an $\mathbf{R}$-algebra), such that every element of K is of the form $\xi^{-1}\eta$ and of the

form $\eta'\xi'^{-1}$, where ξ, ξ', η, η' are elements of U′. (Imitate the construction of the field of fractions of a commutative integral domain, by considering in the product

$$\mathrm{U}(\mathfrak{g}) \times (\mathrm{U}(\mathfrak{g}) - \{0\})$$

the following relation between (A, A') and (B, B'): for each pair (C, D) of nonzero elements of $\mathrm{U}(\mathfrak{g})$ such that $CA' = DB'$, we have $CA = DB$.)

7. IMMERSED LIE GROUPS AND LIE SUBALGEBRAS

In the remainder of this chapter we shall see how the theory of Lie groups may in large measure be reduced to the theory of (finite-dimensional) *Lie algebras* over **R**.

(19.7.1) *Let* G, G′ *be two Lie groups with neutral elements* e, e', *respectively, and Lie algebras* $\mathfrak{g}_e = \mathrm{Lie}(\mathrm{G})$, $\mathfrak{g}'_{e'} = \mathrm{Lie}(\mathrm{G}')$. *Let* $f : \mathrm{G} \to \mathrm{G}'$ *be a Lie group homomorphism and let* $f_* = \mathrm{Lie}(f) = \mathrm{T}_e(f) : \mathfrak{g}_e \to \mathfrak{g}'_{e'}$, *which is a Lie algebra homomorphism, whose image* $f_*(\mathfrak{g}_e)$ *is a Lie subalgebra of* $\mathfrak{g}'_{e'}$, *and whose kernel* $\mathfrak{n}_e$ *is an ideal of the Lie algebra* $\mathfrak{g}_e$, *the quotient* $\mathfrak{g}_e/\mathfrak{n}_e$ *being isomorphic to* $f_*(\mathfrak{g}_e)$.

(i) *If* $f(\mathrm{G})$ *is a Lie subgroup of* G′, *then its Lie algebra may be identified with* $f_*(\mathfrak{g}_e)$ (cf. (19.7.5)).

(ii) *The kernel* N *of* f *is a Lie subgroup of* G *whose Lie algebra may be identified with* $\mathfrak{n}_e$.

(iii) *In order that* f *should be an immersion* (resp. *a submersion*, resp. *a local diffeomorphism*), *it is necessary and sufficient that* f_* *should be injective* (resp. *surjective*, resp. *bijective*). *If* f *is a submersion, then* $f(\mathrm{G})$ *is an open subgroup of* G′, *and* G/N *is locally isomorphic to* G′.

The assertions (ii) and (iii) are immediate consequences of (16.9.9(iii)) and (16.7.5), and (i) follows from (iii) by considering f as a submersion of G onto $f(\mathrm{G})$.

Example

(19.7.1.1) Consider the Lie group homomorphism $f : X \mapsto \det(X)$ of $\mathbf{GL}(n, \mathbf{R})$ into $\mathbf{R}^*$. By virtue of the identifications of (19.4.2), the derived homomorphism f_* is identified with the *derivative* $\mathrm{D}f$ of f at the point I. The expansion of a determinant shows immediately that

$$\det(I + Z) = 1 + \mathrm{Tr}(Z) + r_2(Z)$$

with $r_2(Z) \in o_2$, in the notation of (19.5.1). It follows that f_* is the trace mapping $Z \mapsto \operatorname{Tr}(Z)$ of $\mathbf{M}_n(\mathbf{R}) = \mathfrak{gl}(n, \mathbf{R})$ into $\mathbf{R} = \operatorname{Lie}(\mathbf{R}^*)$. Since the kernel of f is the *special linear group* $\mathbf{SL}(n, \mathbf{R})$, it follows from (19.7.1) that the Lie algebra $\mathfrak{sl}(n, \mathbf{R})$ of $\mathbf{SL}(n, \mathbf{R})$ is the Lie subalgebra of $\mathfrak{gl}(n, \mathbf{R})$ consisting of the *matrices of trace* 0.

(19.7.2) *Let* G_1, G_2 *be two Lie groups and* $G = G_1 \times G_2$ *their product* (16.9.4). *Then the mapping* $\mathbf{u} \mapsto (T_e(\mathrm{pr}_1) \cdot \mathbf{u}, T_e(\mathrm{pr}_2) \cdot \mathbf{u})$ *is an isomorphism of* $\operatorname{Lie}(G)$ *onto the product Lie algebra* $\operatorname{Lie}(G_1) \times \operatorname{Lie}(G_2)$.

For by (16.6.2) this mapping is bijective, and evidently it is a Lie algebra homomorphism.

We shall usually identify the invariant vector fields on $G_1 \times G_2$ with pairs (X', X''), where X' (resp. X'') is an invariant vector field on G_1 (resp. G_2), the vector field X' being identified with $(X', 0)$ and X'' with $(0, X'')$, so that (X', X'') may also be identified with $X' + X''$. Let $(X'_i)_{1 \leq i \leq m}$ (resp. $(X''_j)_{1 \leq j \leq n}$) be a basis of the Lie algebra $\mathfrak{g}_1$ (resp. $\mathfrak{g}_2$) of invariant vector fields on G_1 (resp. G_2).

Since X'_i and X''_j commute, it follows that the infinitesimal algebra of $G_1 \times G_2$ has a basis consisting of all products $X'_\alpha X''_\beta$ with $\alpha \in \mathbf{N}^m$ and $\beta \in \mathbf{N}^n$ (in the notation of (19.6.1)), and that we have

$$(X'_\alpha X''_\beta)(X'_\lambda X''_\mu) = (X'_\alpha X'_\lambda)(X''_\beta X''_\mu);$$

in other words, the infinitesimal algebra of $G_1 \times G_2$ may be identified with the *tensor product* $\mathfrak{G}_1 \otimes \mathfrak{G}_2$ of the infinitesimal algebras of G_1 and G_2 (A.20.4).

It should be remarked that if two connected Lie subgroups H_1, H_2 of a connected Lie group G are such that the Lie algebra $\mathfrak{g}_e$ of G is the *direct sum* of the Lie algebras of H_1 and H_2, it does not necessarily follow that G is isomorphic to the product $H_1 \times H_2$; all that can be asserted is that G is *locally isomorphic* to $H_1 \times H_2$ (cf. (19.7.6) and Problem 1).

(19.7.3) We have seen (19.4.3) that the Lie algebra of a Lie subgroup of a Lie group G may be identified with a Lie subalgebra of $\operatorname{Lie}(G) = \mathfrak{g}_e$. But, conversely, an arbitrary Lie subalgebra $\mathfrak{h}_e$ of $\mathfrak{g}_e$ is *not* necessarily the Lie algebra of a Lie subgroup of G (which is necessarily *closed*). However, there is the following proposition:

(19.7.4) *Let* G *be a Lie group,* $\mathfrak{g}_e$ *its Lie algebra. For each Lie subalgebra* $\mathfrak{h}_e$ *of* $\mathfrak{g}_e$, *there exists a connected Lie group* H *and an injective Lie group homo-*

morphism $j : \mathrm{H} \to \mathrm{G}$ *such that* j_* *is an isomorphism of* Lie(H) *onto* $\mathfrak{h}_e$. *Moreover,* H *and* j *are determined up to isomorphism by these conditions: if* $j' : \mathrm{H}' \to \mathrm{G}$ *is an injective homomorphism of a connected Lie group* H′ *into* G, *such that* j'_* *is an isomorphism of* Lie(H′) *onto* $\mathfrak{h}_e$, *then there exists a unique isomorphism* u *of* H′ *onto* H *such that* $j' = j \circ u$.

Let $(\mathbf{u}_i)_{1 \leqq i \leqq m}$ be a basis of $\mathfrak{h}_e$, and for each index i let X_i be the left-invariant vector field on G such that $X_i(e) = \mathbf{u}_i$. Since $\mathfrak{h}_e$ is a Lie subalgebra of $\mathfrak{g}_e$, the brackets $[\mathbf{u}_i, \mathbf{u}_j]$ are linear combinations of the $\mathbf{u}_k$ with real coefficients, hence (19.3.3) the brackets $[X_i, X_j]$ are linear combinations of the X_k with *constant* coefficients. *A fortiori* (18.14.5), if L_x denotes the subspace of $\mathrm{T}_x(\mathrm{G})$ spanned by the vectors $X_j(x)$ (so that $\mathrm{L}_x = x \cdot \mathfrak{h}_e$), the field of m-directions $x \mapsto \mathrm{L}_x$ is *completely integrable.* Consider then the set $\mathfrak{M}$ of *maximal integral manifolds* of this field (18.14.6) (we recall that they are *not* in general *submanifolds* of G). Since the field of directions $x \mapsto x\mathfrak{h}_e$ is invariant under left translations, it follows that the left translations by elements of G are homeomorphisms of G onto itself for the topology $\mathscr{T}$ defined in (18.14.6) relative to this field of directions. Hence, for each maximal integral manifold $\mathrm{C} \in \mathfrak{M}$ and each $s \in \mathrm{G}$, the translate $s\mathrm{C}$ is a connected integral manifold for the topology $\mathscr{T}$, hence contained in some $\mathrm{C}' \in \mathfrak{M}$; and conversely $s^{-1}\mathrm{C}'$ is a connected integral manifold which intersects C, hence is contained in C; in other words, $s\mathrm{C}$ is a maximal integral manifold.

Now let H be the maximal integral manifold containing the neutral element e. We shall first show that H is a subgroup of G. For if s, t are any two points of H, then $st^{-1}\mathrm{H}$ is a maximal integral manifold containing the point s, hence $st^{-1}\mathrm{H} = \mathrm{H}$; since $e \in \mathrm{H}$, this shows that $st^{-1} \in \mathrm{H}$ and therefore H is a subgroup of G. We now give H the topology induced by $\mathscr{T}$ and the structure of a differential manifold defined in (18.4.6), and we shall show that H endowed with these structures (and its group structure) is a *Lie group.*

In order to prove that $(s, t) \mapsto st^{-1}$ is a C^∞ mapping of $\mathrm{H} \times \mathrm{H}$ into H, we may assume that s (resp. t) is in an arbitrarily small neighborhood of a point s_0 (resp. t_0). Now there exist such open neighborhoods U, V on which the manifold structure induced by that of H is the structure of a submanifold of G, and such that the image of $\mathrm{U} \times \mathrm{V}$ under $(s, t) \mapsto st^{-1}$ is contained in a neighborhood W of $s_0 t_0^{-1}$ in H having the same properties. The assertion is now obvious, and it is also clear that j is an immersion and that $j_*(\mathrm{Lie}(\mathrm{H})) = \mathfrak{h}_e$.

Finally, the uniqueness: by hypothesis, H′ has the same dimension as H, and therefore (in view of (16.9.9)) the mapping j' factorizes as $j \circ u$, where u is a diffeomorphism of H′ onto an open subset of H (for the topology $\mathscr{T}$) (18.14.6). Since j' is given to be a group homomorphism, the same is true of u, and therefore $u(\mathrm{H}')$ is an open subgroup of H (for the topology $\mathscr{T}$). But H is connected for the topology $\mathscr{T}$, hence $u(\mathrm{H}') = \mathrm{H}$. Q.E.D.

The Lie group H defined in the proof of (19.7.4) is called the connected Lie group *immersed in* G corresponding to $\mathfrak{h}_e$, when H is identified with the subgroup $j(\mathrm{H})$ of G. In general it is not closed in G, even if G is simply-connected (cf. Problem 2). Whenever we consider H as a topological group (and *a fortiori* as a Lie group) it is always the topology $\mathscr{T}$ defined here (called the *proper* topology of H) that is to be understood, and *not* the topology induced by that of G, unless it should happen that the two topologies coincide, in which case, H is a *submanifold* of G (16.8.4), and hence a Lie subgroup of G and *closed* in G (16.9.6).

The uniqueness assertion of (19.7.4) shows immediately that if $\mathfrak{h}'_e$ is a Lie subalgebra of $\mathfrak{h}_e$, then the connected Lie group immersed *in* H, corresponding to $\mathfrak{h}'_e$, is the same as the connected Lie group immersed *in* G, corresponding to $\mathfrak{h}'_e$.

(19.7.5) *Let* G, G′ *be two Lie groups,* $f : \mathrm{G} \to \mathrm{G}'$ *a Lie group homomorphism. If* G *is connected, then* $f(\mathrm{G})$ *is a Lie group immersed in* G′, *corresponding to the Lie subalgebra* $f_*(\mathfrak{g}_e)$ *of* $\mathfrak{g}'_{e'}$.

If $\mathrm{N} = f^{-1}(e')$ is the kernel of f, then f factorizes into

$$\mathrm{G} \xrightarrow{p} \mathrm{G/N} \xrightarrow{u} \mathrm{G}',$$

where p is the canonical homomorphism and u is an injective Lie group homomorphism (16.10.4). Hence we may restrict our attention to the case where f is injective, and therefore an immersion (16.9.9). Because f_* is a Lie algebra homomorphism, $f_*(\mathfrak{g}_e)$ is a Lie subalgebra of $\mathfrak{g}'_{e'}$, to which there corresponds a connected Lie group H immersed in G′. Since, moreover, f_* is injective, it is an isomorphism of $\mathfrak{g}_e$ onto $f_*(\mathfrak{g}_e)$; now apply the uniqueness assertion of (19.7.4), and the proof is complete.

(19.7.6) *Let* G, G′ *be two Lie groups, and* $\mathfrak{g}_e$, $\mathfrak{g}'_{e'}$ *their respective Lie algebras. For each Lie algebra homomorphism* $u : \mathfrak{g}_e \to \mathfrak{g}'_{e'}$, *there exists a* C^∞ *local homomorphism* h *from* G *to* G′ (16.9.9.4), *such that* $\mathrm{T}_e(h) = u$. *Moreover, any* C^∞ *local homomorphism* h_1 *from* G *to* G′ *such that* $\mathrm{T}_e(h_1) = u$ *coincides with* h *on some neighborhood of the neutral element* e *of* G. *In particular, if* G *is connected, the mapping* $h \mapsto h_*$ *of the set of homomorphisms of* G *into* G′, *into the set of homomorphisms of* $\mathfrak{g}_e$ *into* $\mathfrak{g}'_{e'}$, *is injective; and if moreover* G *is simply-connected, this mapping is bijective.*

Consider the Lie group $\mathrm{G} \times \mathrm{G}'$, whose Lie algebra may be identified with $\mathfrak{g}_e \times \mathfrak{g}'_{e'}$ (19.7.2). It is immediately verified that the graph Γ_u of u is a Lie subalgebra of $\mathfrak{g}_e \times \mathfrak{g}'_{e'}$. Let H be the connected Lie group immersed in $\mathrm{G} \times \mathrm{G}'$, corresponding to Γ_u, and let $j : \mathrm{H} \to \mathrm{G} \times \mathrm{G}'$ be the canonical injection. We have $(\mathrm{pr}_1 \circ j)_* = (\mathrm{pr}_1)_* \circ j_*$; but $(\mathrm{pr}_1)_*$ is the first projection $\mathfrak{g}_e \times \mathfrak{g}'_{e'} \to \mathfrak{g}_e$, and $j_*(\mathrm{Lie(H)}) = \Gamma_u$, so that $(\mathrm{pr}_1 \circ j)_*$ is an *isomorphism* of Lie(H) onto $\mathrm{Lie(G)} = \mathfrak{g}_e$. It follows (19.7.1) that the restriction v of $\mathrm{pr}_1 \circ j$ to a sufficiently small open neighborhood U of (e, e') in H is a *local isomorphism* of H with G (16.9.9.4). Consequently (*loc. cit.*) there exists an open neighborhood $\mathrm{V} \subset v(\mathrm{U})$ of e in G such that $v^{-1} \mid \mathrm{V}$ is a local isomorphism of G with H, and $j_* \circ (v^{-1})_*$ is the mapping $x \mapsto (x, u(x))$ of $\mathfrak{g}_e$ onto Γ_u. It is clear that $h = \mathrm{pr}_2 \circ j \circ (v^{-1} \mid \mathrm{V})$ is a local homomorphism from G to G′ such that $h_* = u$.

Conversely, if h_1 is a local homomorphism from G to G′ such that $h_{1*} = u$, then $g : x \mapsto (x, h_1(x))$ is a local homomorphism from G to $\mathrm{G} \times \mathrm{G}'$ such that $g_*(\mathfrak{g}_e)$ is the Lie subalgebra Γ_u. We deduce that, in a neighborhood of (e, e') in $\mathrm{G} \times \mathrm{G}'$, the graph of h_1 is an integral manifold of the field of directions formed by the translates of Γ_u; since H is a maximal integral manifold of this field and contains (e, e'), it follows that the graph of h_1 is contained in H, and therefore h_1 and h coincide in a neighborhood of e.

If G is connected, any two homomorphisms of G into G′ which coincide in a neighborhood of e are equal, because each neighborhood of e generates G (12.8.8). Finally, if G is connected and simply-connected, then every local homomorphism from G to G′ extends to a Lie group homomorphism of G into G′ (16.30.7).

(19.7.7) *In order that two Lie groups* G, G′ *should be locally isomorphic, it is necessary and sufficient that their Lie algebras should be isomorphic. If* G, G′ *are connected, simply-connected Lie groups whose Lie algebras are isomorphic, then for each isomorphism* $u : \mathfrak{g}_e \to \mathfrak{g}'_{e'}$ *of the Lie algebras there exists a unique Lie group isomorphism* $f : \mathrm{G} \to \mathrm{G}'$ *such that* $f_* = u$.

This is an immediate consequence of (19.7.6).

(19.7.8) We shall see later (Chapter XXI) that *every* finite-dimensional Lie algebra over **R** is the Lie algebra of some Lie group. There is, therefore, by virtue of (19.7.7), a one-to-one correspondence between isomorphism classes of finite-dimensional real Lie algebras and isomorphism classes of *connected and simply-connected* Lie groups. Furthermore, we obtain *all* the *connected* Lie groups with a given Lie algebra $\mathfrak{g}$ by taking the connected and simply connected group G (which is determined up to isomorphism) whose Lie algebra is $\mathfrak{g}$, and forming the quotients G/D of G by the discrete subgroups D contained in the center of G (16.30.2).

Example: Connected Commutative Lie Groups

(19.7.9) If a Lie group G is commutative, then so is its infinitesimal algebra

$\mathfrak{G}$ (17.11.8) and therefore, if $\dim(G) = n$, the Lie algebra of G is isomorphic to $\mathbf{R}^n$ (19.4.1). Every connected commutative Lie group of dimension n is therefore isomorphic to a quotient group $\mathbf{R}^n/D$, where D is a discrete subgroup (hence closed (12.8.7)) of $\mathbf{R}^n$ (16.30.2). We shall determine all these subgroups D, up to isomorphism.

(19.7.9.1) *For each closed subgroup* F *of* $\mathbf{R}^n$, *there exists an automorphism* u *of the vector space* $\mathbf{R}^n$ *such that* $u(F) = \mathbf{Z}^p \times \mathbf{R}^r$, *where* $p \geqq 0$, $r \geqq 0$, *and* $p + r \leqq n$, $\mathbf{Z}^p$ *is the discrete* **Z**-*module with basis* $\mathbf{e}_1, \ldots, \mathbf{e}_p$, *and* $\mathbf{R}^r$ *is the vector subspace spanned by* $\mathbf{e}_{p+1}, \ldots, \mathbf{e}_{p+r}$ ($(\mathbf{e}_i)_{1 \leqq i \leqq n}$ *being the canonical basis of* $\mathbf{R}^n$).

We shall begin by showing that if F is not discrete, it contains at least one line $\mathbf{Rb}$ (with $\mathbf{b} \neq 0$). Let $\|\mathbf{x}\|$ be the Euclidean norm on $\mathbf{R}^n$; by hypothesis, there exists a sequence $(\mathbf{a}_m)$ of points of F such that $0 < \|\mathbf{a}_m\| < 1/m$ for each $m > 0$. Let $\mathbf{b}_m = \mathbf{a}_m/\|\mathbf{a}_m\|$, so that $\|\mathbf{b}_m\| = 1$. Since the sphere S_{n-1} is compact, the sequence $(\mathbf{b}_m)$ has a subsequence $(\mathbf{b}_{m_k})$ which converges to a limit $\mathbf{b} \in S_{n-1}$. We shall see that $\mathbf{Rb} \subset F$. For this purpose let t be any real number, and for each m_k let t_k be the unique integer such that

$$t_k\|\mathbf{a}_{m_k}\| \leqq t < (t_k + 1)\|\mathbf{a}_{m_k}\|.$$

The point $t_k \mathbf{a}_{m_k}$ belongs to F, and we have

$$\begin{aligned}
\|t_k \mathbf{a}_{m_k} - t\mathbf{b}\| &= \|(t_k\|\mathbf{a}_{m_k}\|)\mathbf{b}_{m_k} - t\mathbf{b}\| \\
&\leqq |t_k| \cdot \|\mathbf{a}_{m_k}\| \cdot \|\mathbf{b}_{m_k} - \mathbf{b}\| + |t_k\|\mathbf{a}_{m_k}\| - t| \cdot \|\mathbf{b}\| \\
&\leqq \left(|t| + \frac{1}{m_k}\right) \cdot \|\mathbf{b}_{m_k} - \mathbf{b}\| + \frac{1}{m_k}.
\end{aligned}$$

Since F is closed, it follows that $t\mathbf{b} \in F$.

It is clear that the sum of all the vector subspaces contained in F is the *largest* vector subspace M contained in F. If N is a supplement of M in $\mathbf{R}^n$, we have $F = M \oplus (F \cap N)$, and $F \cap N$ is discrete, because it is closed and contains no vector subspace other than $\{0\}$. Among the vector subspaces P of N such that $F \cap P$ is a free **Z**-module generated by a basis of P, choose, say, P_0 of *maximum* dimension p, and let $(\mathbf{c}_1, \ldots, \mathbf{c}_p)$ be a basis of P_0 which generates $F \cap P_0$. We shall show that $F \cap N \subset P_0$ and hence $F \cap N = F \cap P_0$.

Suppose that $F \cap N \not\subset P_0$, and let $\mathbf{a} \in F \cap N$ be such that $\mathbf{a} \notin P_0$. Let K be the set of points of N of the form

$$t\mathbf{a} + \sum_{k=1}^{p} t_k \mathbf{c}_k,$$

where t and the t_k vary in the interval $[0, 1] \subset \mathbf{R}$. The set K is compact, hence $K \cap (F \cap N)$ is compact and discrete, i.e., *finite* (3.16.3). Since it is not contained in P_0, there exists among the points of $K \cap (F \cap N) \cap \complement P_0$ a point

$$\mathbf{c} = t_0 \mathbf{a} + \sum_{k=1}^{p} t_k \mathbf{c}_k,$$

for which $t_0 > 0$ has the least possible (strictly positive) value. We shall show that

$$F \cap (P_0 + \mathbf{R}\mathbf{c}) = (F \cap P_0) + \mathbf{Z}\mathbf{c}.$$

For, by subtracting from a point $\mathbf{x} \in F \cap (P_0 + \mathbf{R}\mathbf{c})$ a point of the form $m\mathbf{a} + \sum_{k=1}^{p} m_k \mathbf{c}_k$, where m and the m_k are suitably chosen rational integers, we may assume that $\mathbf{x} \in K \cap (F \cap N)$ and hence

$$\mathbf{x} = r\mathbf{a} + \sum_{k=1}^{p} r_k \mathbf{c}_k,$$

with $0 \leqq r \leqq 1$ and $0 \leqq r_k \leqq 1$ $(1 \leqq k \leqq p)$. If r is not an integer multiple of t_0, there will exist an integer q such that $qt_0 < r < (q+1)t_0$, and then by subtracting from $\mathbf{x} - q\mathbf{c}$ a suitably chosen vector of $F \cap P_0$, we shall obtain a point of $K \cap (F \cap N)$ of the form

$$(r - qt_0)\mathbf{a} + \sum_{k=1}^{p} t_k \mathbf{c}_k,$$

which is absurd. But then $P_0 + \mathbf{R}\mathbf{c}$ is a subspace of N of dimension $p + 1$ such that $F \cap (P_0 + \mathbf{R}\mathbf{c})$ is the $\mathbf{Z}$-module generated by the basis of $P_0 + \mathbf{R}\mathbf{c}$ consisting of $\mathbf{c}$ and the $\mathbf{c}_k$, which contradicts the definition of P_0.

To complete the proof of (19.7.9.1), we have only to choose an automorphism u of $\mathbf{R}^n$ which takes $\mathbf{c}_i$ to $\mathbf{e}_i$ for $1 \leqq i \leqq p$, and a basis of M to $\mathbf{e}_{p+1}, \ldots, \mathbf{e}_{p+r}$, and this is clearly possible.

As a consequence, we obtain the desired classification of all commutative Lie groups $\mathbf{R}^n/D$:

(19.7.9.2) *Every connected commutative Lie group of dimension n is isomorphic to one of the products $\mathbf{T}^p \times \mathbf{R}^{n-p}$, where $0 \leqq p \leqq n$.*

For, by virtue of (19.7.9.1), we may assume that $D = \mathbf{Z}^p$, so that $\mathbf{R}^n/D = (\mathbf{R}^p/\mathbf{Z}^p) \times \mathbf{R}^{n-p}$ up to isomorphism (16.10.5); for the same reason, $\mathbf{R}^p/\mathbf{Z}^p$ is isomorphic to $(\mathbf{R}/\mathbf{Z})^p = \mathbf{T}^p$.

We remark that $\mathbf{T}^p \times \{0\}$ is the *largest* compact subgroup of $\mathbf{T}^p \times \mathbf{R}^{n-p}$, because $\{0\}$ is the only compact subgroup of $\mathbf{R}^{n-p}$.

PROBLEMS

1. Let Z be the center of the unitary group $\mathbf{U}(2, \mathbf{C})$. Show that in the Lie algebra of $\mathbf{U}(2, \mathbf{C})$, the Lie algebras of the connected subgroups Z and $\mathbf{SU}(2, \mathbf{C})$ are supplementary ideals, but that $\mathbf{U}(2, \mathbf{C})$ is not isomorphic to the product $\mathrm{Z} \times \mathbf{SU}(2, \mathbf{C})$.

2. In the simply-connected group $\mathbf{SU}(3, \mathbf{C})$, give examples of one-parameter subgroups which are not closed.

3. In the Lie group, $\mathrm{G} = \mathbf{GL}(2, \mathbf{R})$, let A be the Lie subgroup of all matrices $\begin{pmatrix} x & y \\ 0 & z \end{pmatrix}$, B the Lie subgroup of all matrices $\begin{pmatrix} 1 & 0 \\ t & 1 \end{pmatrix}$. Show that $\mathrm{Lie(G)} = \mathrm{Lie(A)} \oplus \mathrm{Lie(B)}$, but $\mathrm{AB} \neq \mathrm{BA}$ and AB is not the whole of G.

4. Let F, G be two closed subgroups of $\mathbf{R}^n$ such that $\mathrm{G} \subset \mathrm{F}$; let U, V be the identity components of F, G, respectively, so that U, V are vector subspaces of $\mathbf{R}^n$ and $\mathrm{V} \subset \mathrm{U}$. Let W be a vector space supplement of U in $\mathbf{R}^n$, so that F is isomorphic to $\mathrm{U} \times (\mathrm{W} \cap \mathrm{F})$, and $\mathrm{W} \cap \mathrm{F}$ is discrete. If S is a supplement of V in U, show that there exists a supplement W_0 of U in $\mathbf{R}^n$ such that

$$\mathrm{G} = \mathrm{V} \oplus (\mathrm{S} \cap \mathrm{G}) \oplus (\mathrm{W}_0 \cap \mathrm{G}),$$

where $\mathrm{S} \cap \mathrm{G}$ and $\mathrm{W}_0 \cap \mathrm{G} \subset \mathrm{W}_0 \cap \mathrm{F}$ are discrete. (Project G on $\mathrm{W} \cap \mathrm{F}$ parallel to U, and apply the theory of invariant factors to this subgroup of $\mathrm{W} \cap \mathrm{F}$; then lift the basis so obtained up to G.)

5. A commutative Lie group A is said to be *elementary* if it is isomorphic to a product

$$\mathbf{R}^m \times \mathbf{T}^n \times \mathbf{Z}^p \times \mathrm{F},$$

where F is a finite commutative group. Show that every closed subgroup and every quotient by a closed subgroup of an elementary group is elementary. (Use Problem 4.)

6. (a) Every continuous homomorphism from $\mathbf{R}^m$ to $\mathbf{T}^n$ is of the form $\pi_n \circ g$, where

$$g : \mathbf{R}^m \to \mathbf{R}^n$$

is $\mathbf{R}$-*linear*, and $\pi_n : \mathbf{R}^n \to \mathbf{T}^n$ is the canonical homomorphism ((16.30.4) and (4.1.3)).
(b) If $f : \mathbf{T}^m \to \mathbf{T}^n$ is a continuous homomorphism, there exists an $\mathbf{R}$-linear mapping $g : \mathbf{R}^m \to \mathbf{R}^n$ such that $g(\mathbf{Z}^m) \subset \mathbf{Z}^n$ and $\pi_n \circ g = f \circ \pi_m$. (Use (a).)
(c) Let $r_1, \ldots, r_n$ be real numbers such that the $n + 1$ numbers $1, r_1, \ldots, r_n$ are linearly independent over $\mathbf{Q}$. Let $z \in \mathbf{T}^n$ be the canonical image of $(r_1, \ldots, r_n) \in \mathbf{R}^n$. Show that the subgroup G of $\mathbf{T}^n$ generated by z is *dense* in $\mathbf{T}^n$. (If not, then $\mathbf{T}^n/\bar{\mathrm{G}}$ would be isomorphic to a torus $\mathbf{T}^p$ with $1 \leq p \leq n$, by virtue of Problem 5; hence there would exist a continuous homomorphism g of $\mathbf{T}^n$ onto $\mathbf{T}$ such that $g \mid \mathrm{G} = 0$. Now use (b) to obtain a contradiction.)

7. Let G be a Lie group, H a Lie subgroup of G, and $\pi : \mathrm{G} \to \mathrm{G}/\mathrm{H}$ the canonical submersion. For each tangent vector $\mathbf{u}_e \in \mathfrak{g}_e$ and each $t \in \mathrm{H}$, we have $\mathrm{T}(\pi) \cdot (\mathbf{u}_e \cdot t) =$

$T(\pi) \cdot \mathbf{u}_e$. Let $\mathfrak{m}$ be a vector subspace of $\mathfrak{g}_e = \text{Lie}(G)$ such that $\mathfrak{m} \cap \text{Lie}(H) = \{0\}$ and $\text{Ad}(t) \cdot \mathfrak{m} = \mathfrak{m}$ for all $t \in H$. Show that in $T(G/H)$, the union of the vector subspaces

$$s \cdot (T(\pi) \cdot \mathfrak{m}) \subset T_{s \cdot x_0}(G/H)$$

for $s \in G$, where $x_0 = \pi(e)$, is a vector subbundle of $T(G/H)$ of rank equal to the dimension of $\mathfrak{m}$, and on which G acts equilinearly.

8. The Stiefel manifold $\mathbf{S}_{n,p}$ may be canonically identified with a submanifold of $(\mathbf{S}_{n-1})^p$, which in turn is a submanifold of $\mathbf{R}^{np}$. If $j : \mathbf{S}_{n,p} \to (\mathbf{S}_{n-1})^p$ is the canonical injection, it follows that $T(\mathbf{S}_{n,p})$ is a subbundle of $j^*(T((\mathbf{S}_{n-1})^p))$, which in turn is a subbundle of the trivial bundle $\mathbf{S}_{n,p} \times \mathbf{R}^{np}$. The fiber $T_{\mathbf{z}}((\mathbf{S}_{n-1})^p)$ at a point $\mathbf{z} = (\mathbf{x}_1, \dots, \mathbf{x}_p) \in (\mathbf{S}_{n-1})^p$ consists of the vectors $(\mathbf{z}, \mathbf{u}_1, \dots, \mathbf{u}_p) \in \{\mathbf{z}\} \times \mathbf{R}^{np}$ such that $(\mathbf{x}_j | \mathbf{u}_j) = 0$ for $1 \leqq j \leqq p$.

(a) Let $\mathbf{z}_0 = (\mathbf{e}_1, \dots, \mathbf{e}_p)$ be the image in $\mathbf{S}_{n,p}$ of the identity element of $\mathbf{SO}(n, \mathbf{R})$. Show that the fiber $T_{\mathbf{z}_0}((\mathbf{S}_{n-1})^p)$, identified with a vector subspace of $\{\mathbf{z}_0\} \times \mathbf{R}^{np}$, is the direct sum of the following mutually orthogonal subspaces, each of which is stable under $\text{Ad}(t)$ for all $t \in \mathbf{SO}(n-p, \mathbf{R})$:

(1) for each pair of integers r, s such that $1 \leqq r < s \leqq p$, the 1-dimensional space spanned by the vector $(\mathbf{z}_0, \mathbf{u}_1, \dots, \mathbf{u}_p)$, where $\mathbf{u}_j = 0$ for $j \neq r, s$, and $\mathbf{u}_r = \mathbf{e}_s$, $\mathbf{u}_s = \mathbf{e}_r$ (altogether $\frac{1}{2}p(p-1)$ subspaces);

(2) for each pair of integers r, s such that $1 \leqq r < s \leqq p$, the 1-dimensional space spanned by the vector $(\mathbf{z}_0, \mathbf{u}_1, \dots, \mathbf{u}_p)$, where $\mathbf{u}_j = 0$ for $j \neq r, s$, and $\mathbf{u}_r = \mathbf{e}_s$, $\mathbf{u}_s = -\mathbf{e}_r$ (altogether $\frac{1}{2}p(p-1)$ subspaces);

(3) for each integer r such that $1 \leqq r \leqq p$, the $(n-p)$-dimensional space spanned by the vectors $(\mathbf{z}_0, \mathbf{u}_1, \dots, \mathbf{u}_p)$, where $\mathbf{u}_j = 0$ if $j \neq r$, and $\mathbf{u}_r$ is a linear combination of $\mathbf{e}_{p+1}, \dots, \mathbf{e}_n$ (altogether p subspaces).

Let $A_{r,s}$, $B_{r,s}$, and C_r denote the vector subbundles of $j^*(T((\mathbf{S}_{n-1})^p))$ which are the unions of the images of these three types of subspace under the action of $\mathbf{SO}(n, \mathbf{R})$ on $(\mathbf{S}_{n-1})^p$. The vector bundles $A_{r,s}$ and $B_{r,s}$ are isomorphic to the trivial bundle

$$I = \mathbf{S}_{n,p} \times \mathbf{R},$$

and all the bundles C_r are isomorphic. Furthermore, the *normal* bundle **(16.19.2)** of $\mathbf{S}_{n,p}$ in $(\mathbf{S}_{n-1})^p$ is the direct sum of the $A_{r,s}$, hence is *trivializable*, and the tangent bundle $T(\mathbf{S}_{n,p})$ is the direct sum of the $B_{r,s}$ and the C_r (Problem 7). (Use the description **(19.4.3.3)** of the Lie algebra of $\mathbf{SO}(n, \mathbf{R})$.)

(b) Consider the trivial vector bundle $nI = \mathbf{S}_{n,p} \times \mathbf{R}^n$. Show that the fiber $(nI)_{\mathbf{z}_0}$ is the direct sum of the following two orthogonal vector subspaces, each of which is stable under $\text{Ad}(t)$ for all $t \in \mathbf{SO}(n-p, \mathbf{R})$:

(1) the p-dimensional subspace spanned by the vectors $(\mathbf{z}_0, \mathbf{e}_k)$ for $1 \leqq k \leqq p$;

(2) the $(n-p)$-dimensional subspace spanned by the vectors $(\mathbf{z}_0, \mathbf{e}_k)$ for $p+1 \leqq k \leqq n$.

Let L, C denote the vector subbundles of nI that are the unions of the images of these two subspaces under the action of $\mathbf{SO}(n, \mathbf{R})$. Then L is trivializable (and isomorphic to pI), and C is isomorphic to the bundles C_r in (a). Hence we have $pI \oplus C = nI$, up to isomorphism.

(c) Deduce from (a) and (b) that $qC + mI$ is a trivializable bundle for each integer $q \geqq 1$ and each integer $m \geqq p$. Hence show that for $p \geqq 3$ the tangent bundle $T(\mathbf{S}_{n,p})$ is *trivializable*, i.e., that the Stiefel manifold $\mathbf{S}_{n,p}$ is *parallelizable*.

9. Regarding the Stiefel manifold $\mathbf{S}_{n,p}$ as embedded in $\mathbf{R}^{np}$ as in Problem 8, the group $\mathbf{O}(p, \mathbf{R})$ acts freely on $\mathbf{S}_{n,p}$ on the right, so that $\mathbf{S}_{n,p}$ is a principal bundle with group $\mathbf{O}(p, \mathbf{R})$ and base $\mathbf{G}_{n,p}$ (**16.14.10**).

(a) Let $\mathrm{N} = \mathrm{N}(\mathbf{S}_{n,p})$ be the subbundle of $i^*(\mathrm{T}(\mathbf{R}^{np}))$ (where $i: \mathbf{S}_{n,p} \to \mathbf{R}^{np}$ is the canonical injection) which may be identified with the normal bundle of $\mathbf{S}_{n,p}$ in $\mathbf{R}^{np}$ (Section **16.19**, Problem 5(a)). Let $\mathrm{V} = \mathrm{V}(\mathbf{S}_{n,p})$ be the subbundle of vertical tangent vectors ($\mathbf{S}_{n,p}$ being regarded as fibered over $\mathbf{G}_{n,p}$). If we identify $\mathbf{R}^{np}$ with the space of all $n \times p$ matrices, and $\mathrm{T}(\mathbf{R}^{np})$ with $\mathbf{R}^{np} \times \mathbf{R}^{np}$, show that the vector bundle $\mathrm{V} \oplus \mathrm{N}$ over $\mathbf{S}_{n,p}$ is identified with the space of all pairs of matrices

$$\left(S \cdot E, S \cdot \begin{pmatrix} X \\ 0 \end{pmatrix}\right),$$

where $E = \begin{pmatrix} I_p \\ 0 \end{pmatrix}$ is the $n \times p$ matrix whose p columns are the vectors $\mathbf{e}_1, \ldots, \mathbf{e}_p$ of the canonical basis of $\mathbf{R}^n$, the matrix S runs through $\mathbf{O}(n, \mathbf{R})$, and X runs through the set $\mathbf{M}_p(\mathbf{R})$ of all $p \times p$ real matrices. (Use the results of Problem 8.) The group $\mathbf{O}(p, \mathbf{R})$ acts on $\mathrm{V} \oplus \mathrm{N}$ on the right, as follows:

$$\left(\left(S \cdot E, S \cdot \begin{pmatrix} X \\ 0 \end{pmatrix}\right), T\right) \mapsto \left(S \cdot E \cdot T, S \cdot \begin{pmatrix} XT \\ 0 \end{pmatrix}\right).$$

(b) Deduce from (a) that the orbit space $\mathbf{O}(p, \mathbf{R}) \backslash (\mathrm{V} \oplus \mathrm{N})$ is a vector bundle over $\mathbf{G}_{n,p}$ isomorphic to $\mathrm{End}(\mathrm{U}_{n,p})$, where $\mathrm{U}_{n,p}$ is the canonical vector bundle over $\mathbf{G}_{n,p}$ (Section **16.16**, Problem 1). (Observe that we may write

$$S \cdot \begin{pmatrix} XT \\ 0 \end{pmatrix} = (S \cdot E \cdot T) \cdot \begin{pmatrix} T^{-1}XT \\ 0 \end{pmatrix}$$

and hence that $\mathbf{O}(p, \mathbf{R}) \backslash (\mathrm{V} \oplus \mathrm{N})$ may be identified with the associated bundle

$$\mathbf{S}_{n,p} \times^{\mathbf{O}(p, \mathbf{R})} \mathrm{End}(\mathbf{R}^p),$$

where $\mathbf{O}(p, \mathbf{R})$ acts on the left on $\mathrm{End}(\mathbf{R}^p)$ by inner automorphisms.)

(c) Show that the associated bundle $\mathbf{S}_{n,p} \times^{\mathbf{O}(p,\mathbf{R})} \mathbf{R}^{np}$ (where $\mathbf{O}(p, \mathbf{R})$ acts on the left on $\mathbf{R}^{np}$ by $(Z, T) \mapsto Z \cdot T^{-1}$) is isomorphic to $n\mathrm{U}_{n,p}$.

(d) Deduce from (b) and (c) and Section **16.19**, Problem 6, that

$$\mathrm{T}(\mathbf{G}_{n,p}) \oplus \mathrm{End}(\mathrm{U}_{n,p}) = n\mathrm{U}_{n,p}$$

as bundles over $\mathbf{G}_{n,p}$, up to isomorphism.

10. Since the diagonal mapping $\Delta: \mathrm{G} \to \mathrm{G} \times \mathrm{G}$ of a Lie group G into the Lie group $\mathrm{G} \times \mathrm{G}$ is a Lie group homomorphism, the derived homomorphism $\Delta_*: \mathfrak{G} \to \mathfrak{G} \otimes_{\mathbf{R}} \mathfrak{G}$ is a homomorphism of associative algebras having the following properties

(1) $(\Delta_* \otimes 1_{\mathfrak{G}}) \circ \Delta_* = (1_{\mathfrak{G}} \otimes \Delta_*) \circ \Delta_*$;

(2) If $\sigma: \mathfrak{G} \otimes \mathfrak{G} \to \mathfrak{G} \otimes \mathfrak{G}$ is the canonical isomorphism which takes $Z \otimes Z'$ to $Z' \otimes Z$, then $\sigma \circ \Delta_* = \Delta_*$.

With the notation of (**19.5.8.2**), we have

$$\Delta_*(Z_\alpha) = \sum_{\beta + \gamma = \alpha} Z_\beta \otimes Z_\gamma.$$

The elements of the Lie algebra $\mathfrak{g} \subset \mathfrak{G}$ are characterized by the relation

$$\Delta_*(Z) = 1 \otimes Z + Z \otimes 1.$$

If $u : G \to G'$ is a Lie group homomorphism, we have

$$(u_* \otimes u_*) \circ \Delta_* = \Delta'_* \circ u_*,$$

where $\Delta'_* : \mathfrak{G}' \to \mathfrak{G}' \otimes \mathfrak{G}'$ is the derived homomorphism of the diagonal homomorphism $\Delta' : G' \to G' \times G'$.

In order that a subalgebra $\mathfrak{H}$ of $\mathfrak{G}$ should be the infinitesimal algebra of a Lie group H immersed in G, it is necessary and sufficient that $\Delta_*(\mathfrak{H}) \subset \mathfrak{H} \otimes \mathfrak{H}$. When this condition is satisfied, the basis (X_α) of $\mathfrak{G}$ (19.6.2.1) may be chosen so that the X_α for which $\alpha_j = 0$ for $j > p$ (where $p = \dim(H)$) form a basis of $\mathfrak{H}$.

8. INVARIANT CONNECTIONS, ONE-PARAMETER SUBGROUPS, AND THE EXPONENTIAL MAPPING

Let G be a Lie group and let $\mathbf{u}$ be a nonzero vector in the Lie algebra $\mathfrak{g}_e$ of G. Since the additive group **R** has **R** as (commutative) Lie algebra, the mapping $w : t \mapsto t\mathbf{u}$ is a *Lie algebra homomorphism* of Lie(**R**) into Lie(G). Since the group **R** is simply-connected (16.27.7), there exists a unique Lie group homomorphism $v : \mathbf{R} \to G$ such that $v_* = w$ (19.7.6). Let $X_\mathbf{u}$ denote the left-invariant vector field on G such that $X_\mathbf{u}(e) = \mathbf{u}$ (19.3.3). We shall show that v is the *integral curve of the field* $X_\mathbf{u}$ which passes through the point e; in other words (18.1.2.4), that

(19.8.1) $$v'(t) = X_\mathbf{u}(v(t)) = v(t) \cdot \mathbf{u}$$

for *all* $t \in \mathbf{R}$. For, by (19.7.6), the graph Γ_v is the image of the integral curve through $(0, e)$ of the left-invariant vector-field on $\mathbf{R} \times G$ which takes the value $(1, \mathbf{u})$ at the point $(0, e)$, i.e., the vector field $(t, x) \mapsto (E(t), X_\mathbf{u}(x))$. This establishes our assertion.

The image of the integral curve $t \mapsto v(t)$ is called the *one-parameter subgroup* of G corresponding to the vector $\mathbf{u} \in \mathfrak{g}_e$.

(19.8.2) We shall show that the one-parameter subgroups and their left-translates (i.e., the left cosets of the one-parameter subgroups) are the geodesic trajectories of certain linear connections on G (18.6.1). Since a connection **C** on G is a morphism of $T(G) \oplus T(G)$ into $T(T(G))$, we shall say that **C** is *invariant* if this morphism is invariant (19.1.2). In other words, for all elements s, x in G, and all vectors $\mathbf{h}_x$ and $\mathbf{k}_x$ in $T_x(G)$, we must have

$$\mathbf{C}_{s \cdot x}(s \cdot \mathbf{h}_x, s \cdot \mathbf{k}_x) = s \cdot \mathbf{C}_x(\mathbf{h}_x, \mathbf{k}_x),$$

which determines $\mathbf{C}$ completely once the mapping $(\mathbf{u}, \mathbf{v}) \mapsto \mathbf{C}_e(\mathbf{u}, \mathbf{v})$ of $\mathfrak{g}_e \times \mathfrak{g}_e$ into $(T(T(G)))_e$ is known: it must of course satisfy the conditions (17.16.3.2)–(17.16.3.4) for $x = e$. Conversely, given such a mapping (which is equivalent, once a chart at the point e of G has been chosen, to being given n arbitrary bilinear forms Γ^j on $\mathfrak{g}_e \times \mathfrak{g}_e$, by virtue of (17.16.4)), we may define an invariant linear connection by the formula

(19.8.2.1) $$\mathbf{C}_s(s \cdot \mathbf{h}_e, s \cdot \mathbf{k}_e) = s \cdot \mathbf{C}_e(\mathbf{h}_e, \mathbf{k}_e),$$

by virtue of the fact that G acts equilinearly on T(G) and G, and also on T(T(G)) and T(G).

(19.8.3) Relative to an invariant linear connection $\mathbf{C}$ on G, if X, Y are two invariant vector fields, then the field $\nabla_X \cdot Y$ is also *invariant*. This follows from the formula (17.17.2.1) which defines $\nabla_X \cdot Y$, and the relation

$$T_{sx}(s \cdot Y)(s \cdot \mathbf{h}_x) = s \cdot (T_x(Y) \cdot \mathbf{h}_x)$$

for $s \in G$. Furthermore, it is clear that the mapping $(X, Y) \mapsto \nabla_X \cdot Y$ of $\mathfrak{g} \times \mathfrak{g}$ into $\mathfrak{g}$ is **R**-*bilinear* (17.18.1). Conversely, if we are given an *arbitrary* **R**-bilinear mapping $\beta : \mathfrak{g} \times \mathfrak{g} \to \mathfrak{g}$, there exists a unique invariant connection $\mathbf{C}$ for which $\nabla_X \cdot Y = \beta(X, Y)$. This follows from the formula (17.17.2.1) and the fact that, for all $x \in G$, there exists an open neighborhood U of x such that the restrictions to U of a basis of $\mathfrak{g}$ over **R** form a basis of $\Gamma(U, T(G))$ over $\mathscr{E}(U)$ (19.2.2).

(19.8.4) *In order that an invariant linear connection* $\mathbf{C}$ *on* G *should be such that, for each* $\mathbf{u} \in \mathfrak{g}_e$, *the integral curves of the vector field* $X_{\mathbf{u}}$ *are geodesics of* $\mathbf{C}$, *it is necessary and sufficient that* $\nabla_X \cdot X = 0$ *for all invariant vector fields* X *on* G (in other words, the **R**-bilinear mapping $(X, Y) \mapsto \nabla_X \cdot Y$ of $\mathfrak{g} \times \mathfrak{g}$ into $\mathfrak{g}$ must be *alternating*).

If moreover the torsion of $\mathbf{C}$ *is* zero, *then for any two* invariant *vector fields* X, Y, *we have*

(19.8.4.1) $$\nabla_X \cdot Y = \tfrac{1}{2}[X, Y]$$

and for any three invariant *vector fields* X, Y, Z, *the curvature morphism of* $\mathbf{C}$ *is given by*

(19.8.4.2) $$(\boldsymbol{r} \cdot (X \wedge Y)) \cdot Z = -\tfrac{1}{4}[[X, Y], Z].$$

To say that a mapping $t \mapsto v(t)$ of **R** into G is a geodesic signifies that $\nabla_{v'(t)} \cdot v'(t) = 0$ for all $t \in \mathbf{R}$ (18.16.1.2). By virtue of (19.8.1), to say that every integral curve of $X_{\mathbf{u}}$ is a geodesic therefore signifies that the field $\nabla_{X_{\mathbf{u}}} \cdot X_{\mathbf{u}}$ is zero at all points of the one-parameter subgroup corresponding to $\mathbf{u}$. But since this vector field is invariant, it is enough that it should be zero at one point to be zero everywhere on G. This establishes the first assertion.

Since we now have $\nabla_Y \cdot X = -\nabla_X \cdot Y$, the formula (19.8.4.1) (when the torsion of $\mathbf{C}$ is zero) follows from the definition of the torsion (17.20.6.1). Then the definition of the curvature (17.20.4.1) shows that, by virtue of (19.8.4.1),

$$(\mathbf{r} \cdot (X \wedge Y)) \cdot Z = \tfrac{1}{4}[X, [Y, Z]] - \tfrac{1}{4}[Y, [X, Z]] - \tfrac{1}{2}[[X, Y], Z],$$

which is equal to $-\frac{1}{4}[[X, Y], Z]$ by Jacobi's identity.

(19.8.5) The solution v of (19.8.1) which takes the value e at $t = 0$ may therefore be written

(19.8.5.1) $$v(t) = \exp(t\mathbf{u})$$

by virtue of (18.4.4) since $X_{\mathbf{u}}(e) = \mathbf{u}$. The mapping $\mathbf{u} \mapsto \exp(\mathbf{u})$ (also denoted by $\exp_G$) is called the *exponential mapping* of the Lie algebra $\mathfrak{g}_e$ into the group G. We have

(19.8.5.2) $$\exp(s(t\mathbf{u})) = \exp((st)\mathbf{u}).$$

(19.8.5.3) $$\exp((s + t)\mathbf{u}) = \exp(s\mathbf{u}) \exp(t\mathbf{u}),$$

for all $s, t \in \mathbf{R}$, which justifies the notation; but in general

$$\exp(\mathbf{u} + \mathbf{v}) \neq \exp(\mathbf{u}) \exp(\mathbf{v})$$

for $\mathbf{u}, \mathbf{v} \in \mathfrak{g}_e$.

(19.8.6) *There exists an open neighborhood* U *of* 0 *in* $\mathfrak{g}_e$ *such that the exponential mapping is a diffeomorphism of* U *onto an open neighborhood of* e *in* G.

In view of the remarks above, this is a particular case of (18.4.6).

The inverse diffeomorphism, of exp(U) onto U, is denoted by $x \mapsto \log x$ (or $\log_G x$). Given a *basis* of $\mathfrak{g}_e$, the composition of $\log_G$ and the bijection of $\mathfrak{g}_e$ onto $\mathbf{R}^n$ determined by the basis defines a chart of G at the point e, called the *canonical chart* relative to U and the chosen basis of $\mathfrak{g}_e$. The local coordinates corresponding to this chart are called *canonical coordinates* (or *canonical coordinates of the first kind*) in U, relative to the chosen basis of $\mathfrak{g}_e$.

It follows from (18.4.5) that at the point $\mathbf{0}_e \in \mathfrak{g}_e$ we have

(19.8.6.1) $$T_{\mathbf{0}_e}(\exp) = 1_{\mathfrak{g}_e}.$$

Examples

(19.8.7.1) If $G = \mathbf{R}^n$, it is immediately seen that if we take as chart the identity mapping, the local expression of the equation (19.8.1) is $D\mathbf{x} = \mathbf{u}$. Hence $\exp(t\mathbf{u}) = t\mathbf{u}$, i.e., the exponential mapping is in this case the *identity mapping* 1_G.

(19.8.7.2) Consider next the multiplicative group A^* of a finite-dimensional $\mathbf{R}$-algebra A with identity element. For each $s \in A^*$, the mapping $x \mapsto sx$ of A into A is linear, hence equal to its derivative (8.1.3), and therefore, for each tangent vector $\mathbf{u} \in \mathfrak{g}_e$ the local expression of the invariant vector field $X_\mathbf{u}$ is $x \mapsto (x, xu)$, where $u = \tau_e(\mathbf{u}) \in A$ (19.4.2). Consequently the local expression of the differential equation (19.8.1) is

$$Dx = xu.$$

The solution of this equation, which takes the value e at $t = 0$, is the exponential series

$$t \mapsto e + \frac{t}{1!}u + \frac{t^2}{2!}u^2 + \cdots + \frac{t^n}{n!}u^n + \cdots,$$

which is normally convergent in every bounded interval of $\mathbf{R}$, relative to any norm for which A is a Banach algebra (15.1.3). The notation (19.8.5.1) for this series is therefore consistent with the usual notation.

Remark

(19.8.8) Take for example $A = \mathbf{M}_2(\mathbf{R})$, so that $A^* = \mathbf{GL}(2, \mathbf{R})$. Regarding A as a subalgebra of the complex matrix algebra $\mathbf{M}_2(\mathbf{C})$, if $Y \in A$ is any real matrix there exists an invertible complex matrix P such that $X = PYP^{-1}$ is either a real triangular matrix

$$\begin{pmatrix} \lambda & 0 \\ \nu & \mu \end{pmatrix} \quad \text{or a diagonal matrix} \quad \begin{pmatrix} \alpha + i\beta & 0 \\ 0 & \alpha - i\beta \end{pmatrix}$$

with complex numbers on the diagonal (i.e., $\beta \neq 0$). In the first case, $\exp(tX)$ is a triangular matrix with diagonal entries $e^{\lambda t}$, $e^{\mu t}$, and in the second case $\exp(tX)$ is the diagonal matrix

$$\begin{pmatrix} e^{(\alpha+i\beta)t} & 0 \\ 0 & e^{(\alpha-i\beta)t} \end{pmatrix}.$$

In either case, the matrix

$$P^{-1}\exp(tX)P = \exp(tP^{-1}XP) = \exp tY$$

(where t is real) cannot be equal to the matrix

$$\begin{pmatrix} -\lambda & 0 \\ 0 & -\lambda^{-1} \end{pmatrix}$$

for $\lambda > 0$ and $\lambda \neq 1$, which belongs to the identity component $\mathbf{GL}^+(2, \mathbf{R})$ of $\mathbf{GL}(2, \mathbf{R})$. This example therefore shows that, for a connected Lie group, the exponential mapping *is not necessarily surjective*. We have already seen (18.7.13) that it is not necessarily injective (cf. Section 19.14, Problem 4).

It can be shown (Problem 2) that in this case the matrices in $\mathbf{GL}^+(2, \mathbf{R})$ which are of the form $\exp(tY)$ are the matrices

$$\begin{pmatrix} a & b \\ c & d \end{pmatrix}$$

with $\Delta = ad - bc > 0$ and $a + d > -2\sqrt{\Delta}$, and the matrices

$$\begin{pmatrix} -\lambda & 0 \\ 0 & -\lambda \end{pmatrix}$$

with $\lambda > 0$. This set of matrices is therefore *neither open nor closed in*

$$\mathbf{GL}^+(2, \mathbf{R}).$$

(19.8.9) *Let* G, G′ *be two Lie groups, and let* $f : \mathrm{G} \to \mathrm{G}'$ *be a Lie group homomorphism. Then for each vector* $\mathbf{u} \in \mathfrak{g}_e$ *and each* $t \in \mathbf{R}$ *we have*

(19.8.9.1) $$f(\exp(t\mathbf{u})) = \exp(tf_*(\mathbf{u})).$$

The mappings $v : t \mapsto f(\exp(t\mathbf{u}))$ and $w : t \mapsto \exp(tf_*(\mathbf{u}))$ are homomorphisms of $\mathbf{R}$ into G′ such that $v'(0) = w'(0) = f_*(\mathbf{u})$, by virtue of (16.5.4). The result therefore follows from the uniqueness assertion of (19.7.6).

(19.8.10) *Let* G *be a Lie group,* H *a connected Lie group immersed in* G, *and* $\mathfrak{h}_e$ *the Lie algebra of* H, *identified with a Lie subalgebra of* $\mathfrak{g}_e$. *In order that a vector* $\mathbf{u} \in \mathfrak{g}_e$ *should belong to* $\mathfrak{h}_e$, *it is necessary and sufficient that* $\exp(t\mathbf{u}) \in \mathrm{H}$ *for all* $t \in \mathbf{R}$.

The necessity of the condition follows from (19.8.9) applied to the canonical injection $j : \mathrm{H} \to \mathrm{G}$. Conversely, if the image of $v : t \mapsto \exp(t\mathbf{u})$ is contained in H, then since this mapping is of class C^∞ as a mapping of $\mathbf{R}$ into G, it is also of class C^∞ when regarded as a mapping of $\mathbf{R}$ into H, for the manifold structure of H (18.14.7). Hence we have $\mathbf{u} = v'(0) \in \mathrm{T}_e(\mathrm{H}) = \mathfrak{h}_e$.

Remark

(19.8.11) Let M be a differential manifold on which G acts differentiably on the *right*, and consider for each vector $\mathbf{u} \in \mathfrak{g}_e$ the Killing vector field $Z_{\mathbf{u}}$ on M, which is defined by $Z_{\mathbf{u}}(x) = x \cdot \mathbf{u}$ (19.3.7.4). Then the maximal integral curve of this vector field with origin $x_0 \in \mathrm{M}$ (18.2.2) is defined on the whole of $\mathbf{R}$ and is given by

(19.8.11.1) $$t \mapsto x_0 \cdot \exp(t\mathbf{u}).$$

For if we put $v(t) = \exp(t\mathbf{u})$ and $w(t) = x_0 \cdot v(t)$, then by (16.10.1) we have

(19.8.11.2) $$\mathrm{T}(w) \cdot E(t) = x_0 \cdot (\mathrm{T}(v) \cdot E(t)) = x_0 \cdot (v(t) \cdot \mathbf{u}) = w(t) \cdot \mathbf{u}$$

by virtue of the definition of $\exp(t\mathbf{u})$ (19.8.1).

From this we shall deduce that, in order that a tensor field $\mathbf{S} \in \mathscr{T}_s^r(\mathrm{M})$ should be G-*invariant*, it is necessary that

(19.8.11.3) $$\theta_{Z_{\mathbf{u}}} \cdot \mathbf{S} = \mathbf{0}$$

for all $\mathbf{u} \in \mathfrak{g}_e$, and moreover this condition is sufficient if G is *connected*.

We have seen (18.2.14.8) that the value of $\theta_{Z_{\mathbf{u}}} \cdot \mathbf{S}$ at a point $x \in \mathrm{M}$ is the limit as $t \to 0$ of the tensor

$$\frac{1}{t}(\mathbf{S}(x \cdot \exp(t\mathbf{u})) \cdot \exp(-t\mathbf{u}) - \mathbf{S}(x)) \in (\mathbf{T}_s^r(\mathrm{M}))_x,$$

which may also be written in the form

$$\frac{1}{t}(\mathbf{S}(x \cdot \exp(t\mathbf{u})) - \mathbf{S}(x) \cdot \exp(t\mathbf{u})) \cdot \exp(-t\mathbf{u}).$$

Since the question is local, we may identify T(M) with $\mathrm{M} \times \mathbf{R}^n$ in a neighborhood of x; by considering the vector part (16.15.1.3) of the above tensor, it follows that if we put

(19.8.11.4) $$\mathbf{F}(t) = \mathbf{S}(x \cdot \exp(t\mathbf{u})) - \mathbf{S}(x) \cdot \exp(t\mathbf{u})$$

(which is an element of $(\mathbf{T}_s^r(\mathrm{M}))_{x \cdot \exp(t\mathbf{u})}$), the derivative $\mathbf{F}'(0) = \mathrm{T}(\mathbf{F}) \cdot E(0)$ is given by the expression

(19.8.11.5) $$\mathbf{F}'(0) = \tau_{\mathbf{S}(x)}^{-1}((\theta_{Z_{\mathbf{u}}} \cdot \mathbf{S})(x)).$$

If $\boldsymbol{S}$ is G-invariant, that is to say if $\boldsymbol{S}(x \cdot s) = \boldsymbol{S}(x) \cdot s$ for all $s \in \mathrm{G}$ and all $x \in \mathrm{M}$, then we have $\boldsymbol{F}(t) = \boldsymbol{0}$ for all $t \in \mathbf{R}$ and hence the relation (19.8.11.3). Conversely, if this relation is satisfied, let $x_0 \in \mathrm{M}$, and put

(19.8.11.6) $$\mathbf{F}_0(t) = \boldsymbol{S}(x_0 \cdot \exp(t\mathbf{u})) - \boldsymbol{S}(x_0) \cdot \exp(t\mathbf{u}),$$

so that $t \mapsto \mathbf{F}_0(t)$ is a lifting to $\mathbf{T}_s^r(\mathrm{M})$ of the mapping (19.8.11.1). Writing $x = x_0 \cdot \exp(t\mathbf{u})$, we have

$$\mathbf{F}_0(t + t') = \mathbf{F}_0(t) \cdot \exp(t'\mathbf{u}) + (\boldsymbol{S}(x \cdot \exp(t'\mathbf{u})) - \boldsymbol{S}(x) \cdot \exp(t'\mathbf{u})),$$

from which it follows, by use of (19.8.11.2) and (19.8.11.5) and the hypothesis (19.8.11.3), that

(19.8.11.7) $$\mathbf{F}_0'(t) = \mathbf{F}_0(t) \cdot \mathbf{u},$$

the right-hand side of which corresponds (16.10.1) to the action of G on $\mathbf{T}_s^r(\mathrm{M})$ induced from the action of G on M ((19.1.3) and (19.1.5)). Since $\mathbf{F}_0(0) = \mathbf{0}_{x_0}$, the only solution of (19.8.11.7) is evidently $\boldsymbol{0}$ (18.2.2). By virtue of (19.8.6), we have, therefore, $\boldsymbol{S}(x_0 \cdot s) = \boldsymbol{S}(x_0) \cdot s$ for all s in some neighborhood of e in G. The hypothesis that G is connected now implies that this relation is valid for *all* $s \in \mathrm{G}$, by virtue of (12.8.8).

PROBLEMS

1. With the hypotheses and notation of (19.8.7.2), give the series expansion of $\log(e + x)$ (for x in a sufficiently small neighborhood of 0 in A) in powers of x.

2. Determine the image of the exponential mapping for the groups $\mathbf{GL}(2, \mathbf{R})$, $\mathbf{SL}(2, \mathbf{R})$, $\mathbf{GL}(2, \mathbf{C})$, and $\mathbf{SL}(2, \mathbf{C})$ (use the reduction of a matrix to Jordan form). Deduce that the exponential mapping for $\mathbf{GL}(n, \mathbf{C})$ is surjective, and hence also for $\mathbf{PGL}(n, \mathbf{C})$, the quotient of $\mathbf{GL}(n, \mathbf{C})$ by its center. On the other hand, the exponential mapping for $\mathbf{SL}(n, \mathbf{C})$ is *not* surjective, although the quotient $\mathbf{PSL}(n, \mathbf{C})$ of $\mathbf{SL}(n, \mathbf{C})$ by its center is isomorphic to $\mathbf{PGL}(n, \mathbf{C})$.

3. Let G be a simply-connected Lie group, and M a compact differential manifold. Show that every infinitesimal action of $\mathfrak{g}_e$ on M arises, by the formula (19.3.7.4), from a differentiable action of G on M. (With the notation of Section 19.3, Problem 2, consider a maximal integral manifold of the field of n-directions $(s, x) \mapsto \mathrm{L}_{(s, x)}$ on $\mathrm{G} \times \mathrm{M}$; then use Problem 5 of Section 16.29, together with (18.2.11).) Does the above result remain true if G is connected but not simply-connected? (Consider the case where M is a compact Lie group G′, and $\mathrm{G} = \mathrm{G}'/\mathrm{D}$ is the quotient of G′ by a finite subgroup $\mathrm{D} \neq \{e\}$ of the center of G′.) (Cf. Section 19.9, Problem 9.)

4. Show by use of the exponential mapping that a Lie group has no arbitrarily small subgroups (Section **12.9**, Problem 6).

5. Let G be a separable, metrizable, locally compact group with no arbitrarily small subgroups. We shall use the notation and terminology of Section **14.11**, Problems 9–14. Let L_G be the set of one-parameter subgroups of G (Section **14.11**, Problem 12). Let V be a symmetric compact neighborhood of e in G which contains no subgroup $\neq \{e\}$ of G, and such that for all $x, y \in V$ the relation $x^2 = y^2$ implies $x = y$. Let K denote the set of all $X \in L_G$ such that $X(r) \in V$ for $|r| \leq 1$. Denote by exp the mapping $X \mapsto X(1)$ of L_G into G, and put $K_1 = \exp(K)$. The mapping exp is injective on K.

(a) Show that K_1 is closed (hence compact) in G. (If a sequence (X_i) in K is such that the sequence $(X_i(1))$ tends to a limit a, remark that the sequence $(X_i(1/i)^i)$ has limit a, and use Section **12.9**, Problem 7.)
(b) Let g be the function defined in Section **14.11**, Problem 14(b). Let $X_1, \ldots, X_m$ be one-parameter subgroups of G such that

$$\sum_{k=1}^{m} D_{X_k} g = 0.$$

Show that for each function f that is X_k-differentiable (Section **14.11**, Problem 12) for each k, we have

$$\sum_{k=1}^{m} D_{X_k} f = 0.$$

(Consider the sequence of elements $b_i = X_1(1/i) \cdots X_m(1/i) \in G$, and show that the functions $i(\gamma(b_i)g - g)$ tend uniformly to 0 as $i \to +\infty$, by using the formula

$$(*) \qquad \gamma(st)g - g = \gamma(s)(\gamma(t)g - g) + (\gamma(s)g - g).$$

Use Problem 13(a) of Section **14.11** to show that a subsequence of the sequence $(b_i^{[ri]})$ (independent of $r \in \mathbf{R}$) converges to $X(r)$, where X is a one-parameter subgroup, and show that $X = 0$ by using Problem 12 of Section **14.11**. Conclude that the sequence $(b_i^{[ri]})$ converges to e in G.)
(c) With the notation of Section **14.11**, Problem 14, let (a_j) be a sequence in G converging to e and let (m_j) be a sequence of integers tending to $+\infty$. Show that if the sequence of functions $m_j(\gamma(a_j)g_j - g_j)$ converges uniformly to a function h, then the sequence $(a_j^{[rm_j]})$ converges to $X(r)$, where X is a one-parameter subgroup, and that $D_X g = h$. (Use Problem 13(a) of Section **14.11** and Problem 7 of Section **12.9** to show that there exists a subsequence of the sequence $(a_j^{[rm_j]})$ which converges to a one-parameter subgroup $X(r)$; then use part (b) to show that this one-parameter subgroup is independent of the choice of subsequence.)
(d) Given two one-parameter subgroups X, Y, show that there exists a unique one-parameter subgroup such that each function f which is both X-differentiable and Y-differentiable is Z-differentiable, and

$$D_Z f = D_X f + D_Y f.$$

(With the same notation as before, consider the sequence of functions

$$i(\gamma(X(1/i))\gamma(Y(1/i))g - g)$$

and use the formula $(*)$. Then use part (c) to show that the limit of the sequence $((X(1/i)Y(1/i))^{[ri]})$ exists and is the one-parameter subgroup $Z(r)$ required.) We have therefore equipped L_G with a structure of a *real vector space*.

6. The hypotheses and notation are as in Problem 5. For each $X \in \mathrm{L_G}$, put $\|X\| = \|D_X g\|$.

(a) Show that $\|X\|$ is a *norm* on $\mathrm{L_G}$, and that the set K (Problem 5) is a neighborhood of 0 relative to this norm. (Argue as in Section **14.11**, Problem 13(a), and use Section **14.11**, Problem 11(c).)
(b) Let (X_i) be a sequence in K. Show that if the sequence $(X_i(1))$ converges to $X(1)$, then the sequence $(\mathrm{D}_{X_i} g)$ converges uniformly to $\mathrm{D}_X g$. (Remark that, by virtue of Section **14.11**, Problem 14(b), the set of functions $\mathrm{D}_Y g$, where Y runs through K, is uniformly equicontinuous and uniformly bounded. Hence there exists a sequence of integers h_i tending to $+\infty$ such that some subsequence of the sequence

$$h_i(\gamma(X_i(1/h_i))g - g)$$

converges uniformly. By using Problem 5(c), show that the limit of this subsequence is necessarily $\mathrm{D}_X g$.
(c) Prove that $\mathrm{L_G}$ is a *finite-dimensional* vector space. (Use (a) and (b), the compactness of K_1, and F. Riesz's theorem (**5.9.4**).)
(d) If G is a Lie group, show that $\mathrm{L_G}$ may be canonically identified with Lie(G), and exp with the exponential mapping defined in (**19.8.5**). (Use (**19.10.2**).)

7. (a) Let G, H be two separable, metrizable, locally compact groups with no arbitrarily small subgroups. For each continuous homomorphism f: $\mathrm{G} \to \mathrm{H}$, and each one-parameter subgroup $X \in \mathrm{L_G}$, the group $f_*(X) = X \circ f$ is a one-parameter subgroup of H. Deduce from Problem 5(d) that f_* is a linear mapping of $\mathrm{L_G}$ into $\mathrm{L_H}$.
(b) In particular, an inner automorphism Int(w) of G induces in this way an automorphism $\mathrm{Ad}(w) = \mathrm{Int}(w)_*$ of the vector space $\mathrm{L_G}$, for each element w of G. Show that $w \mapsto \mathrm{Ad}(w)$ is a continuous homomorphism of G into $\mathbf{GL}(\mathrm{L_G})$. (It is enough to prove that $w \mapsto \mathrm{Ad}(w) \cdot X$ is a continuous mapping of G into $\mathrm{L_G}$, for each $X \in \mathrm{L_G}$; bearing in mind the definition of the norm in $\mathrm{L_G}$ (Problem 6), use Section **14.11**, Problem 11(c).)
(c) Put $\mathrm{ad} = \mathrm{Ad}_*$, which is a linear mapping of $\mathrm{L_G}$ into $\mathfrak{gl}(\mathrm{L_G}) = \mathrm{End}(\mathrm{L_G})$. Show that for each $X \in \mathrm{L_G}$ the linear mapping $\mathrm{ad}(X)$ of $\mathrm{L_G}$ into $\mathrm{L_G}$ is the limit in $\mathrm{End}(\mathrm{L_G})$ of the sequence $i(\mathrm{Ad}(X(1/i)) - I)$ as $i \to +\infty$. The convergence is uniform on compact subsets of $\mathrm{L_G}$.
(d) Show that for all X, $Y \in \mathrm{L_G}$ and all $r \in \mathbf{R}$, we have

$$(\mathrm{ad}(Y) \cdot X)(r) = \lim_{j \to \infty} \left(\lim_{i \to \infty} (Y(1/j)X(1/i)Y(-1/j)X(-1/i))^{[rij]} \right).$$

(Use part (c) and Problem 5(d).) Deduce that $\mathrm{ad}(X) \cdot X = 0$, and hence that

$$(X, Y) \mapsto \mathrm{ad}(X) \cdot Y$$

is an alternating bilinear mapping of $\mathrm{L_G} \times \mathrm{L_G}$ into $\mathrm{L_G}$.
(e) With the hypothesis of (a), show that

$$f_*(\mathrm{Ad}(w) \cdot Y) = \mathrm{Ad}(f(w)) \cdot f_*(Y)$$

for all $Y \in \mathrm{L_G}$ and $w \in \mathrm{G}$, and deduce that

$$f_*(\mathrm{ad}(X) \cdot Y) = \mathrm{ad}(f_*(X)) \cdot f_*(Y)$$

for all $X, Y \in \mathrm{L_G}$. Hence show that if we define

$$[X, Y] = \mathrm{ad}(X) \cdot Y,$$

the vector space $\mathrm{L_G}$ together with this bracket operation is a real *Lie algebra*. (Take $\mathrm{H} = \mathbf{GL}(\mathrm{L_G})$ and $f_* = \mathrm{ad}(Z)$ in the formula above.)

(f) Show that, for all $X \in \mathrm{L_G}$ and $t \in \mathbf{R}$, we have

$$\mathrm{Ad}(X(t)) = \exp(\mathrm{ad}(tX))$$

in $\mathbf{GL}(\mathrm{L_G})$, and

$$(\mathrm{Ad}(X(t)) \cdot Y)(r) = X(t)Y(r)X(-t)$$

in G for all $r \in \mathbf{R}$.

8. The hypotheses and notation remain the same.

(a) Let (Y_i) be a sequence of elements of $\mathrm{L_G}$ tending to Y, and let $X \in \mathrm{L_G}$. For each integer m, let

$$b_{mi} = Y_i(-1)(Y_i(1/m)X(1/mi))^m.$$

Show that for each $r \in \mathbf{R}$, we have

$$\lim_{i\to\infty}\left(\lim_{m\to\infty} b^{[ri]}\right) = \lim_{m\to\infty}\left(\lim_{i\to\infty} b^{[ri]}\right)$$

(all the limits exist in G). (Use Problem 5(c) and Section **14.11**, Problem 14(b), together with (7.5.6), in order to reduce the problem to the existence of the repeated limit

$$\lim_{m\to\infty}\left(\lim_{i\to\infty} i(\gamma(b_{mi})g - g)\right).$$

Observe that if we put

$$X_{i,t}(r) = Y_i(-t/m)X(r)Y_i(t/m),$$

we may write

$$b_{mi} = X_{i,\,m-1}(1/mi) \cdots X_{i,\,0}(1/mi).$$

Use the relation (*) of Problem 5(b) to show that

$$\lim_{i\to\infty} i(\gamma(b_{mi})g - g) = \frac{1}{m}\sum_{\nu=0}^{m-1} \mathrm{D}_{X_{\nu/m}}g,$$

where $X_t(r) = Y(-t)X(r)Y(t)$.)

(b) Deduce from (a) that if $Y = 0$ we have

$$\lim_{i\to\infty}\left(\exp(-Y_i)\exp\left(Y_i + \frac{1}{i}X\right)\right)^{[ri]} = X(r).$$

(Use Problem 5(d), which defines $Y_i + (1/i)X$.)

(c) With the notation of Section **14.11**, Problem 11, let (m_j) be a sequence of integers tending to $+\infty$, and for each j let a_j, b_j be two elements of U_{m_j}. Suppose that the sequence $(a_j^{[rm_j]})$ converges to $X(r)$, and the sequence $(b_j^{[rm_j]})$ to $Y(r)$, where X and Y

belong to L_G. Show that, for all sufficiently small $r \in \mathbf{R}$, the sequence $((a_j b_j)^{[rm_j]})$ converges to $(X+Y)(r)$. (Use Section 14.11, Problems 14(a) and 14(b).)

(d) Show that $K_1 = \exp(K)$ (Problem 5) is a *neighborhood of e in* G. (For each $x \in G$, let $\nu(x)$ be the least integer $n \geq 0$ such that $x^{n+1} \notin V$. Suppose that there exists a sequence (a_j) in G, converging to e, such that $a_j \notin K_1$ for all j. Let $Y_i \in K$ be such that $\nu(Y_i(-1)a_i)$ takes the greatest value of all the $\nu(x_i^{-1}a_i)$ as x_i runs through K_1, and let ν_i be this greatest value; we have $\nu_i \to \infty$ as $i \to \infty$. We may assume, by passing to a subsequence if necessary, that the sequence $((Y_i(-1)a_i)^{[r\nu_i]})$ converges to $X(r)$, where $X \neq 0$ in L_G (Section 12.9, Problem 7). The sequence (Y_i) then converges to 0 in L_G, and therefore we have

$$Y_i + \frac{1}{\nu_i} X \in K$$

for all sufficiently large i. Put

$$b_i = \exp\left(-Y_i - \frac{1}{\nu_i} X\right) a_i = c_i^{-1}(Y_i(-1)a_i).$$

Deduce from (b) that $c_i^{[r\nu_i]}$ converges to $X(r)$, and from (c) that the sequence $(b_i^{[r\nu_i]})$ converges to e. This contradicts the definition of $\nu(b_i)$ and ν_i.)

9. The hypotheses and notation remain the same as in Problems 5–8.

(a) Let $Y \in L_G$, so that (Problem 7(f)) we have

$$\mathrm{Ad}(Y(t)) = \sum_{k=0}^{\infty} \frac{1}{k!} (\mathrm{ad}(Y))^k t^k$$

in $\mathbf{GL}(L_G)$. Put

$$S_Y = \int_0^1 \mathrm{Ad}(Y(t))\, dt = \sum_{k=0}^{\infty} \frac{1}{(k+1)!} (\mathrm{ad}(Y))^k$$

in $\mathrm{End}(L_G)$; S_Y is invertible whenever Y is sufficiently close to 0. Show that the sequence

$$\left(\exp(-Y)\exp\left(Y + \frac{1}{i} X\right)\right)^{[ri]}$$

in G tends to $(S_Y \cdot X)(r)$, uniformly in X for X near 0. (Use Problem 8(b).)

(b) Use Problem 8(d) to show that in a sufficiently small neighborhood K' of 0 in L_G, a law of composition $(X, Y) \mapsto X \cdot Y$ may be defined by

$$\exp(X \cdot Y) = \exp(X)\exp(Y).$$

We have $X \cdot 0 = 0 \cdot X = X$, and $X \cdot (-X) = (-X) \cdot X = 0$. Moreover, there exists a neighborhood $K'' \subset K'$ of 0 in L_G such that the products $X \cdot (Y \cdot Z)$ and $(X \cdot Y) \cdot Z$ are defined and equal for all X, Y, Z in K'', and we have

$$X \cdot ((t + t')Y) = (X \cdot (tY)) \cdot (t'Y)$$

for all sufficiently small $t, t' \in \mathbf{R}$.

(c) With the above notation, show that in L_G we have

$$\lim_{t \to 0} \frac{1}{t} ((-Y) \cdot (Y + t(S_Y^{-1} \cdot Z))) = Z$$

for Y fixed and sufficiently small, uniformly in Z for Z sufficiently close to 0. Deduce that

$$\lim_{t \to 0} \frac{1}{t}((Y \cdot tZ) - Y) = S_Y^{-1} \cdot Z.$$

Hence show that the function $t \mapsto W(t) = Y \cdot tZ$ satisfies the differential equation

$$\frac{dW}{dt} = S_{W(t)}^{-1} \cdot Z.$$

(d) Deduce from (c) and from the expression for S_Y given in (a) that there exists a *Lie group structure* on the neutral component of G, for which the underlying topological group structure is the given one (*Gleason–Yamabe theorem*).† (Use Section 19.3, Problem 4.)

10. Let G be a metrizable topological group and N a closed normal subgroup of G.

(a) If N and G/N are locally compact, then G is locally compact. (Let V_0 be a symmetric closed neighborhood of e in G such that $V_0 \cap N$ is compact. If V_1 is a symmetric closed neighborhood of e in G such that $V_1^2 \subset V_0$, then $V_1 \cap xN$ is compact for each $x \in V_1$. Let $\pi : G \to G/N$ be the canonical homomorphism, and let C be a compact neighborhood of $\pi(e)$ in G/N, contained in $\pi(V_1)$. If V_2 is a symmetric closed neighborhood of e in G such that $V_2^2 \subset V_1$, show that $W = V_2 \cap \pi^{-1}(C)$ is compact in G. For this purpose, if $\mathfrak{R}$ is a covering of W by open subsets of G, then for each $y \in C$ there is a finite number of sets of $\mathfrak{R}$ which cover $V_1 \cap \pi^{-1}(y)$; if T_y is their union, show that there exists a neighborhood S_y of y in G/N such that

$$V_2 \cap \pi^{-1}(C \cap S_y) \subset T_y,$$

and finally use the compactness of C.)

(b) Deduce from (a) and Problem 9 that if N and G/N are the underlying topological groups of Lie groups, then the same is true of G.

9. PROPERTIES OF THE EXPONENTIAL MAPPING

(19.9.1) *Let G be a Lie group and let* $\mathbf{u} \in \mathfrak{g}_e$. *Let* $f \in \mathscr{E}(G)$, *and put* $g(t) = f(\exp(t\mathbf{u}))$ *for* $t \in \mathbf{R}$. *Then (writing* $X \cdot f$ *for* $\theta_X \cdot f$*) we have*

$$g^{(m)}(t) = (X_{\mathbf{u}}^m \cdot f)(\exp(t\mathbf{u})).$$

Clearly it is enough to prove the formula when $m = 1$. Putting $s = \exp(t\mathbf{u})$, we have by virtue of (16.5.4) and the definition of a differential (16.5.7),

$$g'(t) = \langle d_s f, v'(t)\rangle = \langle d_s f, X_{\mathbf{u}}(s)\rangle = (X_{\mathbf{u}} \cdot f)(s)$$

by (17.14.1.1) and (19.8.1).

† The method of proof is taken from unpublished lecture notes of the late Yamabe.

Hence we obtain the Taylor expansion

(19.9.2) $$f(\exp(t\mathbf{u})) \sim \sum_{m=0}^{\infty} \frac{t^m}{m!}(X_{\mathbf{u}}^m \cdot f)(e).$$

Choose a basis $(\mathbf{u}_i)_{1 \leqq i \leqq n}$ of $\mathfrak{g}_e$. Then, in an open neighborhood exp(U) of e in G on which the mapping $\log_G$ is defined, the canonical coordinates of $\exp\left(\sum_{i=1}^{n} t_i \mathbf{u}_i\right)$, corresponding to the given choice of basis of $\mathfrak{g}_e$, are $t_1, \ldots, t_n$. If we put $X_i = X_{\mathbf{u}_i}$ $(1 \leqq i \leqq n)$, the formula (19.9.2) gives

(19.9.3) $$f\left(\exp\left(\sum_{i=1}^{n} t_i \mathbf{u}_i\right)\right) \sim \sum_{m=0}^{\infty}\left(\left(\sum_{i=1}^{n} t_i X_i\right)^m \cdot f\right)(e),$$

on the right-hand side of which the coefficient of t^α (for an arbitrary multi-index α) comes only from the power $\left(\sum_{i=1}^{n} t_i X_i\right)^{|\alpha|}$. Now the coefficient of t^α in this operator is

(19.9.4) $$S_\alpha = \sum X_{i_1} X_{i_2} \cdots X_{i_{|\alpha|}},$$

the sum being over all sequences $(i_1, \ldots, i_{|\alpha|})$ of $|\alpha|$ integers between 1 and n, in which the number of terms equal to k is α_k, for $1 \leqq k \leqq n$. Hence, with the notation of (19.5.8), we have the following explicit expression *relative to canonical coordinates* (of the first kind) *at the point e*:

(19.9.5) $$Z_\alpha = \frac{\alpha!}{|\alpha|!} S_\alpha .$$

For example, in the notation of (19.5.10), we have

(19.9.6) $$Z_{\varepsilon_i + \varepsilon_j} = \tfrac{1}{2}(X_i X_j + X_j X_i)$$

for all i, j.

By comparing this formula with (19.5.10.1) we see that, *relative to canonical coordinates* (of the first kind) at the point e, we have

(19.9.7) $$b_{ij}^{(k)} = -b_{ji}^{(k)}$$

for $i \neq j$. Hence the multiplication table for the basis $(\mathbf{u}_i)$ of the Lie algebra can be written in the form

(19.9.8) $$[\mathbf{u}_i, \mathbf{u}_j] = 2\sum_{k=1}^{n} b_{ij}^{(k)} \mathbf{u}_k \qquad (i \neq j).$$

Relative to a system of canonical coordinates (of the first kind), the function $\mathbf{m}(\mathbf{x}, \mathbf{y})$ of (19.5.7.1) is $\log(\exp(\mathbf{x})\exp(\mathbf{y}))$. Comparison of the formulas (19.9.8) and (19.5.7.2) therefore shows that

(19.9.9)
$$\log(\exp(\mathbf{x})\exp(\mathbf{y})) = \mathbf{x} + \mathbf{y} + \tfrac{1}{2}[\mathbf{x}, \mathbf{y}] + \mathbf{r}_2(\mathbf{x}, \mathbf{y}),$$

where $\mathbf{r} \in o_2(\mathrm{U} \times \mathrm{U})$ (19.5.1). Equivalently,

(19.9.10)
$$\exp(\mathbf{x})\exp(\mathbf{y}) = \exp(\mathbf{x} + \mathbf{y} + \tfrac{1}{2}[\mathbf{x}, \mathbf{y}] + \mathbf{r}_2(\mathbf{x}, \mathbf{y})).$$

It follows by induction that, for each integer $k > 2$,

(19.9.11)
$$\exp(\mathbf{x}_1)\exp(\mathbf{x}_2)\cdots\exp(\mathbf{x}_k) = \exp\Big(\mathbf{x}_1 + \cdots + \mathbf{x}_k + \tfrac{1}{2}\sum_{h<k}[\mathbf{x}_h, \mathbf{x}_k] + \mathbf{r}_2(\mathbf{x}_1, \ldots, \mathbf{x}_k)\Big),$$

where $\mathbf{r}_2 \in o_2(\mathrm{U}^k)$. In particular, since $\exp(-\mathbf{x}) = \exp(\mathbf{x})^{-1}$,

(19.9.12)
$$\exp(\mathbf{x})\exp(\mathbf{y})\exp(\mathbf{x})^{-1} = \exp(\mathbf{y} + [\mathbf{x}, \mathbf{y}] + \mathbf{r}_2'(\mathbf{x}, \mathbf{y})),$$

(19.9.13)
$$\exp(\mathbf{x})\exp(\mathbf{y})\exp(\mathbf{x})^{-1}\exp(\mathbf{y})^{-1} = \exp([\mathbf{x}, \mathbf{y}] + \mathbf{r}_2''(\mathbf{x}, \mathbf{y})),$$

with $\mathbf{r}_2', \mathbf{r}_2'' \in o_2(\mathrm{U} \times \mathrm{U})$.

(19.9.14) *Suppose that $\mathfrak{g}_e$ is the direct sum of k vector subspaces $\mathfrak{v}_1, \ldots, \mathfrak{v}_k$. Then the C^∞ mapping*

$$(\mathbf{x}_1, \mathbf{x}_2, \ldots, \mathbf{x}_k) \mapsto \exp(\mathbf{x}_1)\exp(\mathbf{x}_2)\cdots\exp(\mathbf{x}_k)$$

of $\mathfrak{g}_e = \mathfrak{v}_1 \times \cdots \times \mathfrak{v}_k$ into G is a diffeomorphism of a neighborhood in 0 in $\mathfrak{g}_e$ onto a neighborhood of e in G.

It follows from (19.9.11) that the tangent linear mapping at the point 0 is the identity mapping of $\mathfrak{g}_e$ (identified with $\mathfrak{v}_1 \times \cdots \times \mathfrak{v}_k$), and the result therefore follows from (16.5.6).

(19.9.15) Consider in particular a basis $(\mathbf{u}_i)_{1 \leq i \leq n}$ of $\mathfrak{g}_e$, and take $\mathfrak{v}_j = \mathbf{R}\mathbf{u}_j$ in (19.9.14) for $1 \leq j \leq n$. Then the mapping

$$(t_1, \ldots, t_n) \mapsto \exp(t_1\mathbf{u}_1)\exp(t_2\mathbf{u}_2)\cdots\exp(t_n\mathbf{u}_n)$$

is a diffeomorphism of an open neighborhood of 0 in $\mathbf{R}^n$ onto an open neighborhood U of e in G. The inverse mapping φ is therefore a *chart* of G at the point e, and the coordinates $t_1, \ldots, t_n$ of $\varphi(s)$ for a point $s \in$ U are called the *canonical coordinates of the second kind* of the point s, relative to the basis $(\mathbf{u}_i)$ of $\mathfrak{g}_e$.

PROBLEMS

1. Let G be a Lie group, $\mathbf{u}$ a vector in $\mathfrak{g}_e$, and x a point of G. Let $f \in \mathscr{E}(G)$ and put $g(t) = f(x \cdot \exp(t\mathbf{u}))$ for $t \in \mathbf{R}$. Then we have

$$g^{(m)}(t) = (X_{\mathbf{u}}^m \cdot f)(x \cdot \exp(t\mathbf{u})).$$

Deduce that, relative to canonical coordinates of the second kind **(19.9.15)**, we have

$$Z = \frac{1}{\alpha!} X_\alpha$$

in the notation of **(19.6.1)**.

2. Let G be a Lie group. Show that there exists a neighbourhood U of e in G having the following property: for each sequence (x_n) of elements of U, if we define inductively $y_1 = x_1$, $y_n = (x_n, y_{n-1})$, the sequence of commutators (y_n) converges to e (cf. Problem 6).

3. (a) Show that in the unitary group $\mathbf{U}(n)$ there exists a compact neighborhood V of e which has the property of Problem 2, is stable under all inner automorphisms, and is such that if $x, y \in$ V do not commute, then x and the commutator (x, y) do not commute. (For the third property, see Section **16.11**, Problem 1. For the second property, remark that for each neighborhood W of e in a compact group G, and each $s \in$ G, there exists a neighborhood $V_s \subset W$ of e and a neighborhood T_s of s such that the relation $t \in T_s$ implies $tV_st^{-1} \subset W$.)
(b) Let β by the normalized Haar measure **(14.3)** on $\mathbf{U}(n)$, and let V_1 be a compact symmetric neighborhood of e in $\mathbf{U}(n)$ such that $V_1^2 \subset V$. Let $f(n)$ be the smallest integer such that $\beta(V_1) > 1/f(n)$. Show that, for every *finite* subgroup F of $\mathbf{U}(n)$, there exists a *commutative normal* subgroup A(F) of F such that

$$(F : A(F)) \leq f(n) \qquad \textit{(Jordan's theorem)}.$$

(Consider the subgroup A(F) generated by $F \cap V$, and use (a).)

4. Let G be a connected Lie group and let $\mathbf{u}, \mathbf{v} \in \mathfrak{g}_e$. If $[\mathbf{u}, \mathbf{v}] = 0$, show that $\exp(t'\mathbf{u})$ and $\exp(t''\mathbf{v})$ commute for all $t', t'' \in \mathbf{R}$, and that

$$\exp(t'\mathbf{u}) \exp(t''\mathbf{v}) = \exp(t'\mathbf{u} + t''\mathbf{v}).$$

5. Let G be a Lie group and let U be a neighborhood of e in G of the form exp(V), where V is an open ball in $\mathfrak{g}_e$ with center 0 (relative to some norm on $\mathfrak{g}_e$), such that the exponential mapping is a diffeomorphism of V onto exp(V). Let $W \subset V$ be an open ball with center 0 such that $(\exp(W))^3 \subset U$. Show that if $\mathbf{u}, \mathbf{v} \in W$ are such that $\exp(\mathbf{u})$ and $\exp(\mathbf{v})$ commute, then $[\mathbf{u}, \mathbf{v}] = 0$. (Consider the image of the one-parameter subgroup corresponding to $\mathbf{u}$ under the inner automorphism $x \mapsto \exp(\mathbf{v}) \cdot x \cdot \exp(-\mathbf{v})$.)

6. Let G be a Lie group. Show that for each number $\rho \in]0, 1[$ there exists a neighborhood U of e on which the log function is defined and such that, for a given norm on $\mathfrak{g}_e$, we have for all $x, y \in U$

$$xyx^{-1}y^{-1} \in U \quad \text{and} \quad \|\log(xyx^{-1}y^{-1})\| \leq \rho \cdot \inf(\|\log x\|, \|\log y\|).$$

7. Let G be a connected Lie group and Z a connected Lie subgroup of G contained in the center of G. Suppose that the exponential mappings $\exp_Z$ and $\exp_{G/Z}$ are diffeomorphisms. Show that $\exp_G$ is a diffeomorphism. (Let $\pi : G \to G/Z$ be the canonical homomorphism, $\pi_* : \mathfrak{g}_e \to \mathfrak{g}_e/\mathfrak{z}_e$ the derived homomorphism, and let σ be a linear mapping of $\mathfrak{g}_e/\mathfrak{z}_e$ into $\mathfrak{g}_e$ such that $\pi_* \circ \sigma = 1_{\mathfrak{g}_e/\mathfrak{z}_e}$. Consider the mapping $\alpha : x \mapsto \sigma(\log_{G/Z}(\pi(x)))$ of G into $\mathfrak{g}_e$: show that $\exp_G(-\alpha(x))x \in Z$, by showing that its image under π is the neutral element of G/Z. Then put $\beta(x) = \alpha(x) + \log_Z(\exp_G(-\alpha(x))x)$, and show that $\exp_G(\beta(x)) = x$ and $\beta(\exp_G(\mathbf{u})) = \mathbf{u}$ for all $x \in G$ and $\mathbf{u} \in \mathfrak{g}_e$.

8. Let G be a Lie group and $\mathfrak{g}_e$ its Lie algebra. Let μ denote the C^∞ mapping $(s, \mathbf{u}) \mapsto s \cdot \mathbf{u}$ of $G \times \mathfrak{g}_e$ into T(G).

(a) Show that if T(G) is trivialized over an open set U on which $\log_G$ is defined, by means of the corresponding canonical chart (so that U is identified with an open subset of $\mathfrak{g}_e$, and T(U) with $U \times \mathfrak{g}_e$), then for all $\mathbf{u}, \mathbf{v}, \mathbf{w} \in \mathfrak{g}_e$ we have

$$T_{(e,\mathbf{u})}(\mu) \cdot ((e, \mathbf{u}), (\mathbf{v}, \mathbf{w})) = ((e, \mathbf{u}), (\mathbf{v}, \mathbf{w} + \tfrac{1}{2}[\mathbf{v}, \mathbf{u}])).$$

(Use (**19.9.10**).)
(b) Deduce that if M is a differential manifold and if $f : M \to G$, $\mathbf{u} : M \to \mathfrak{g}_e$ are C^1 mappings, then we have

$$\nabla_{\mathbf{h}_z} \cdot (f \cdot \mathbf{u}) = f(z) \cdot (T_z(\mathbf{u}) \cdot \mathbf{h}_z + \tfrac{1}{2}[f(z)^{-1} \cdot (T_z(f) \cdot \mathbf{h}_z), \mathbf{u}(z)])$$

relative to an invariant linear connection **C** on G satisfying (**19.8.4.1**).

9. Let G be a simply-connected Lie group of dimension n, let $\mathfrak{g}_e$ be its Lie algebra, and let $\mathbf{u} \mapsto Y_\mathbf{u}$ be an *isomorphism of* $\mathfrak{g}_e$ onto a Lie subalgebra of the Lie algebra $\mathcal{T}_0^1(M)$ of C^∞ vector fields on a differential manifold M. For each point $(s, x) \in G \times M$, let

$$L_{(s,x)} \subset T_{(s,x)}(G \times M)$$

be the n-direction spanned by the vectors $(X_\mathbf{u}(s), Y_\mathbf{u}(x))$ for all $\mathbf{u} \in \mathfrak{g}_e$ (Section **19.3**, Problem 2), and for each point $x \in M$ let N_x be the maximal integral manifold of this completely integrable field which contains the point (e, x) (**18.14.6**). If G acts on $G \times M$ by left-translation on G, the image $s \cdot N$ of any maximal integral manifold N is another such.

(a) The field $Y_{\mathbf{u}}$ is said to be *complete* if every maximal integral curve of the field (18.2.2) is defined on the whole of $\mathbf{R}$. When this is so, we put $f_{t,\mathbf{u}}(x) = \mathrm{F}_{Y_{\mathbf{u}}}(x, t)$ for $x \in \mathrm{M}$ and $t \in \mathbf{R}$, in the notation of (18.2.3). We have $f_{t,\mathbf{u}} \circ f_{t',\mathbf{u}} = f_{t+t',\mathbf{u}}$. Show that if the fields $Y_{\mathbf{u}_1}, \dots, Y_{\mathbf{u}_m}$ are complete, the mapping

$$(t_1, \dots, t_m) \mapsto (\exp(t_1\mathbf{u}_1)\cdots\exp(t_m\mathbf{u}_m), (f_{t_m,\mathbf{u}_m} \circ \cdots \circ f_{t_1,\mathbf{u}_1})(x))$$

of $\mathbf{R}^m$ into $\mathrm{G} \times \mathrm{M}$ is of class C^∞, and its image is contained in N_x (cf. (18.14.7)). Show that if $\mathbf{u}$, $\mathbf{v}$ are such that $Y_{\mathbf{u}}$ and $Y_{\mathbf{v}}$ are complete, then so also is $Y_{[\mathbf{u},\mathbf{v}]}$. (Consider the mapping

$$t \mapsto (\exp(\mathbf{u})\exp(t\mathbf{v})\exp(-\mathbf{u}), (f_{-1,\mathbf{u}} \circ f_{t,\mathbf{v}} \circ f_{1,\mathbf{u}})(x))$$

and use (19.11.2.2) and (19.11.2.3).)

(b) Suppose that there exists a system of generators of the Lie algebra $\mathfrak{g}_e$ such that the fields $Y_{\mathbf{u}}$ corresponding to the generators are complete. Then it follows from (a) that there exists a *basis* $(\mathbf{u}_j)_{1 \leq j \leq n}$ of $\mathfrak{g}_e$ having this property. Let W be a connected neighborhood of 0 in $\mathbf{R}^n$ such that the mapping

$$w : (t_1, \dots, t_n) \mapsto \exp(t_1\mathbf{u}_1)\exp(t_2\mathbf{u}_2)\cdots\exp(t_n\mathbf{u}_n)$$

is a diffeomorphism of W onto a symmetric neighborhood V of e in G. For each $s \in \mathrm{V}$, put

$$h_s = f_{t_n,\mathbf{u}_n} \circ \cdots \circ f_{t_1,\mathbf{u}_1},$$

where $(t_1, \dots, t_n) = w^{-1}(s)$. Show that the mapping $\Phi : (s, x) \mapsto (s, h_s(x))$ is a diffeomorphism of $\mathrm{V} \times \mathrm{M}$ onto itself and that, for each $x \in \mathrm{M}$, the image of $\mathrm{V} \times \{x\}$ under Φ is the connected component of (e, x) in $\mathrm{N}_x \cap (\mathrm{V} \times \mathrm{M})$. (Show that this set is both open and closed in $\mathrm{N}_x \cap (\mathrm{V} \times \mathrm{M})$ with respect to the proper topology of N_x.)

(c) Deduce from (b) that the restriction of the projection pr_1 to each N_x is a diffeomorphism of N_x onto G. (Prove that this mapping makes N_x a covering of G, when N_x is endowed with its proper topology.) Hence show that Φ is a differentiable action of G on M, such that for this action we have $Z_{\mathbf{u}} = Y_{\mathbf{u}}$ for all $\mathbf{u} \in \mathfrak{g}_e$.

10. Let G be a Lie group, $\mathfrak{g}_e$ its Lie algebra, and let u be a mapping of $\mathfrak{g}_e$ into G such that $u(n\mathbf{x}) = (u(\mathbf{x}))^n$ for all $n \in \mathbf{Z}$ and all $\mathbf{x} \in \mathfrak{g}_e$. Suppose also that u is continuous at the point 0 and that if U is a neighborhood of 0 in $\mathfrak{g}_e$ such that $u(\mathrm{U})$ is contained in a neighborhood of e on which $\log_{\mathrm{G}}$ is defined, then the function $\mathbf{x} \mapsto \log(u(\mathbf{x})) = v(\mathbf{x})$ is differentiable at the point 0 and has as derivative the identity mapping. Show that $u = \exp_{\mathrm{G}}$. (Observe that $v(\mathbf{x}/n) = v(\mathbf{x})/n$ for all integers $n > 0$, and deduce that $v(\mathbf{x}) = \mathbf{x}$.)

10. CLOSED SUBGROUPS OF REAL LIE GROUPS

(19.10.1) (E. Cartan's theorem) *Every closed subgroup* H *of a real Lie group* G *is a Lie subgroup of* G (in other words, the subspace H of G is the underlying space of a *submanifold* of G).

Let $\mathfrak{h}_e$ denote the subset of the Lie algebra $\mathfrak{g}_e$ of G consisting of all vectors $\mathbf{u} \in \mathfrak{g}_e$ such that $\exp(t\mathbf{u}) \in \mathrm{H}$ for all $t \in \mathbf{R}$. We shall begin by proving that

(19.10.1.1) *$\mathfrak{h}_e$ is a Lie subalgebra of $\mathfrak{g}_e$.*

It is clear that if $\mathbf{u} \in \mathfrak{h}_e$, then $t\mathbf{u} \in \mathfrak{h}_e$ for all $t \in \mathbf{R}$. Let us show that if $\mathbf{u}$, $\mathbf{v}$ are two vectors in $\mathfrak{h}_e$, then $\mathbf{u} + \mathbf{v} \in \mathfrak{h}_e$; this will prove that $\mathfrak{h}_e$ is a *vector subspace* of $\mathfrak{g}_e$. Now, if $t \in \mathbf{R}$ and if n is an integer >0, then by virtue of (19.9.10) we may write

$$\exp\left(\frac{t}{n}\mathbf{u}\right)\exp\left(\frac{t}{n}\mathbf{v}\right) = \exp\left(\frac{t}{n}(\mathbf{u}+\mathbf{v}) + \frac{t^2}{2n^2}[\mathbf{u}, \mathbf{v}] + \frac{1}{n^3}\mathbf{w}_n(t, \mathbf{u}, \mathbf{v})\right),$$

where, for fixed t, $\mathbf{u}$, $\mathbf{v}$, the $\|\mathbf{w}_n(t, \mathbf{u}, \mathbf{v})\|$ form a bounded set. It follows that

$$\text{(19.10.1.2)} \qquad \left(\exp\left(\frac{t}{n}\mathbf{u}\right)\exp\left(\frac{t}{n}\mathbf{v}\right)\right)^n = \exp\left(t(\mathbf{u}+\mathbf{v}) + \frac{1}{n}\mathbf{a}_n\right),$$

where the sequence of numbers $\|\mathbf{a}_n\|$ is bounded (for fixed t, $\mathbf{u}$, $\mathbf{v}$). By definition, the left-hand side of (19.10.1.2) belongs to H, and as $n \to \infty$ it tends to the limit $\exp(t(\mathbf{u}+\mathbf{v}))$; since H is *closed*, it follows that $\mathbf{u} + \mathbf{v} \in \mathfrak{h}_e$.

Next, let us show that $[\mathbf{u}, \mathbf{v}] \in \mathfrak{h}_e$. The formula (19.9.13) gives

$$\exp\left(\frac{t}{n}\mathbf{u}\right)\exp\left(\frac{t}{n}\mathbf{v}\right)\exp\left(-\frac{t}{n}\mathbf{u}\right)\exp\left(-\frac{t}{n}\mathbf{v}\right) = \exp\left(\frac{t^2}{n^2}[\mathbf{u}, \mathbf{v}] + \frac{1}{n^3}\mathbf{b}_n\right),$$

where the sequence of numbers $\|\mathbf{b}_n\|$ is bounded (for fixed t, $\mathbf{u}$, $\mathbf{v}$). We deduce that

(19.10.1.3)

$$\left(\exp\left(\frac{t}{n}\mathbf{u}\right)\exp\left(\frac{t}{n}\mathbf{v}\right)\exp\left(-\frac{t}{n}\mathbf{u}\right)\exp\left(-\frac{t}{n}\mathbf{v}\right)\right)^{n^2} = \exp\left(t^2[\mathbf{u}, \mathbf{v}] + \frac{1}{n}\mathbf{b}_n\right)$$

and the same argument as before shows that $[\mathbf{u}, \mathbf{v}] \in \mathfrak{h}_e$.

We may therefore consider the *connected Lie group* K *immersed in* G (19.7.4) which corresponds to the Lie subalgebra $\mathfrak{h}_e$ of $\mathfrak{g}_e$. Since $\exp(\mathfrak{h}_e)$ is a neighborhood of e in K (for the proper topology of K) (19.8.6), and since K is generated by any neighborhood of e (12.8.8), it follows that $\mathrm{K} \subset \mathrm{H}$ by the definition of $\mathfrak{h}_e$. It will therefore be enough to prove that

(19.10.1.4) *The subgroup* K *is open in* H (for the topology induced by that of G), *and the topology induced on* K *by that of* G *is the proper topology of* K.

For it will then follow from (16.8.4) that K is a submanifold of G; moreover, K is the identity component of H, and therefore by translation H is a submanifold of G, and hence is a Lie subgroup of G.

To prove (19.10.1.4), it is enough to show that every neighborhood N of e in K (for the proper topology of K) is a neighborhood of e in H (for the topology induced by that of G): for by translation, the same will then be true for every neighborhood of any point of K.

Suppose therefore that there exists a neighborhood N of e in K that is not a neighborhood of e in H. Then there exists a sequence (a_n) of points of $H - N$ which tends to e in G. Let us decompose the Lie algebra $\mathfrak{g}_e$ as a direct sum $\mathfrak{h}_e \oplus \mathfrak{m}$, where $\mathfrak{m}$ is a vector subspace of $\mathfrak{g}_e$. There exists a bounded neighborhood V of 0 in $\mathfrak{h}_e$ and a bounded neighborhood W of 0 in $\mathfrak{m}$ such that the mapping $(\mathbf{x}, \mathbf{y}) \mapsto \exp(\mathbf{x}) \exp(\mathbf{y})$ is a homeomorphism of $V \times W$ onto a neighborhood U of e in G (19.9.14). We may assume that $a_n \in U$ for all n, and hence for each n there exist well-defined vectors $\mathbf{x}_n \in V$ and $\mathbf{y}_n \in W$ such that $a_n = \exp(\mathbf{x}_n) \exp(\mathbf{y}_n)$. Moreover, by replacing V by a smaller neighborhood, we may assume that $\exp(\mathbf{x}_n) \in N$, since N is a neighborhood of e in K (for the proper topology of K). Since $a_n \in H - N$, we must have $\mathbf{y}_n \neq 0$ and $\lim_{n\to\infty} \mathbf{y}_n = 0$. As W is bounded and $\mathbf{y}_n \neq 0$, there exists an integer $r_n > 0$ such that $r_n \mathbf{y}_n \in W$ and $(r_n + 1)\mathbf{y}_n \notin W$. Furthermore, since W is relatively compact in $\mathfrak{m}$, we may, by passing to a subsequence of (a_n), assume that the sequence $(r_n \mathbf{y}_n)$ has a limit $\mathbf{y} \in \overline{W}$. Since $\lim_{n\to\infty} \mathbf{y}_n = 0$, the sequence $((r_n + 1)\mathbf{y}_n)$ also tends to $\mathbf{y}$; but because $(r_n + 1)\mathbf{y}_n \notin W$, this shows that $\mathbf{y}$ belongs to the *frontier* Fr(W) of W in $\mathfrak{m}$, and therefore $\mathbf{y} \neq 0$, so that $\mathbf{y} \notin \mathfrak{h}_e$.

We shall now show that $\exp(t\mathbf{y}) \in H$ for all $t \in \mathbf{R}$; by the definition of $\mathfrak{h}_e$, this will imply that $\mathbf{y} \in \mathfrak{h}_e$ and will give the desired contradiction. Since H is *closed*, it is enough to show that $\exp\left(\frac{p}{q}\mathbf{y}\right) \in H$ for all rational integers p, q (where $q > 0$). Now, we can write $pr_n = qs_n + u_n$, where s_n and u_n are integers and $0 \leq u_n < q$. This implies that

$$\lim_{n\to\infty} \frac{u_n}{q} \mathbf{y}_n = 0$$

and hence, in G,

$$\exp\left(\frac{p}{q}\mathbf{y}\right) = \lim_{n\to\infty} \left(\exp\left(\frac{pr_n}{q}\mathbf{y}_n\right)\right) = \lim_{n\to\infty} (\exp(\mathbf{y}_n))^{s_n}.$$

But since $\exp(\mathbf{x}_n) \in N \subset H$ and $a_n \in H$, we have $\exp(\mathbf{y}_n) \in H$. Consequently, as H is *closed* in G, we have

$$\exp\left(\frac{p}{q}\mathbf{y}\right) \in H,$$

and the proof of (19.10.1) is complete.

(19.10.2) *Let* G, G′ *be two Lie groups. Every* continuous *homomorphism* $f: G \to G'$ *is a homomorphism of Lie groups* (i.e., is of class C^∞).

For the graph Γ_f of f is a *closed* subgroup of the Lie group $G \times G'$ (12.3.5), hence is a *submanifold* of $G \times G'$ by (19.10.1). The result now follows from (16.9.10).

In particular:

(19.10.3) *Two structures of differential manifold which are compatible* (16.9.1) *with the same structure of topological group are identical.*

This follows from (19.10.2), applied to the identity mapping of the group in question.

Remark

(19.10.4) Let G be a Lie group and let $(H_\lambda)_{\lambda \in L}$ be any family of Lie subgroups of G. Since $H = \bigcap_{\lambda \in L} H_\lambda$ is closed in G, it is a Lie subgroup of G, and it follows immediately from (19.8.10) that the Lie algebra of H is the *intersection* of the Lie algebras of the H_λ.

PROBLEMS

1. Show that if a one-parameter subgroup of a Lie group G is not closed, then its closure in G is *compact* (hence a torus). (Use (19.10.1) to reduce to the case where G is commutative and connected, and then use (19.7.9.2).)

2. (a) Let H be a closed subgroup of a Lie group G, and let L be a connected Lie group immersed in G. If the intersection of the Lie algebras of H and L is zero, show that $H \cap L$ is discrete in L (for the proper topology of L). Give an example in which $H \cap L$ is dense in H.
(b) Give an example of two connected Lie groups L, L′ immersed in G, such that the Lie algebras of L, L′ have zero intersection and $L \cap L'$ is dense in G.

3. (a) Let G be a connected Lie group, $\mathfrak{g}_e$ its Lie algebra. If $\mathbf{u} \in \mathfrak{g}_e$ is such that the one-parameter subgroup $\exp(\mathbf{R}\mathbf{u})$ of G is not closed in G, show that there exists a vector $\mathbf{v} \in \mathfrak{g}_e$ arbitrarily close to $\mathbf{u}$ (for the canonical topology of the vector space $\mathfrak{g}_e$) such that $\exp(\mathbf{R}\mathbf{v})$ is closed in G. (Use Problem 1 to reduce to the case where G is a torus.)
(b) Deduce from (a) that there exists a basis $(\mathbf{u}_j)_{1 \leq j \leq n}$ of $\mathfrak{g}_e$ such that each of the one-parameter subgroups $\exp(\mathbf{R}\mathbf{u}_j)$ is closed in G.

4. Let G, G′ be two Lie groups and $u : G \to G'$ a homomorphism (of abstract groups). Suppose that, for each continuous homomorphism $v : \mathbf{R} \to G$, the composite homomorphism $u \circ v : \mathbf{R} \to G'$ is continuous. Prove that u is a homomorphism of Lie groups. (Use (19.9.15).)

5. In this problem, assume the theorem that every finite-dimensional Lie algebra over **R** is the Lie algebra of some Lie group (cf. Chapter XXI). Let M be a connected differential manifold and Γ a group of diffeomorphisms of M. Let S be the set of all vector fields $Y \in \mathscr{T}_0^1(M)$ that are *complete* (Section 19.9, Problem 9) and such that for each $t \in \mathbf{R}$ the diffeomorphism $x \mapsto F_Y(x, t)$ of M onto itself (18.2.8) belongs to Γ. Assume that the Lie subalgebra $\mathfrak{g}$ of $\mathscr{T}_0^1(M)$ generated by S is finite-dimensional.

(a) Let G be a simply connected Lie group whose Lie algebra is isomorphic to $\mathfrak{g}$. Show that there exists a homomorphism (of abstract groups) $h : G \to \Gamma$ such that, for each vector $\mathbf{u}$ in the Lie algebra $\mathfrak{g}_e$ of G, if $Y_{\mathbf{u}} \in \mathfrak{g}$ is the vector field corresponding to $\mathbf{u}$, we have $h(\exp(t\mathbf{u})) \cdot x = F_{Y_{\mathbf{u}}}(x, t)$ for all $t \in \mathbf{R}$ and all $x \in M$. Furthermore, there exists a neighborhood V of the identity element of G such that the restriction of h to V is injective. (Use Section 19.9, Problem 9.)
(b) Show that there exists on $G_0 = h(G) \subset \Gamma$ a unique structure of Lie group such that h is a surjective Lie group homomorphism of G onto G_0. The group G_0 is normal in Γ, and for each $w \in \Gamma$, the mapping $u \mapsto w \circ u \circ w^{-1}$ is an automorphism of the Lie group G_0. (Use Problem 4.) Deduce that there exists on Γ a unique topology $\mathscr{T}$ which is compatible with the group structure of Γ, induces on G_0 the topology defined above, and for which G_0 is open in Γ (and therefore the identity component of Γ).
(c) Show that the topology $\mathscr{T}$ has a basis of sets W(U, K), where U (resp. K) runs through the open (resp. compact) subsets of M, and W(U, K) is the set of all $w \in \Gamma$ such that $w(K) \subset U$. (Reduce to proving that the topology induced by $\mathscr{T}$ on $h(V)$ can be defined in this way.) Deduce that Γ, endowed with the topology $\mathscr{T}$, is metrizable and separable, and hence that Γ is a *Lie group* acting differentiably on M (*Palais' theorem*).

6. Let M be a *parallelizable* connected differential manifold (Section 16.15, Problem 1) of dimension n, so that there exist n vector fields $X_j \in \mathscr{T}_0^1(M)$ $(1 \leq j \leq n)$ such that at each point the n vectors $X_j(x)$ form a basis of $T_x(M)$. Let Γ denote the group of diffeomorphisms of M which leave *invariant* each of the fields X_j.

(a) With the notation of Problem 5, show that S is contained in the set $\mathfrak{a}$ of vector fields $Y \in \mathscr{T}_0^1(M)$ such that $[Y, X_j] = 0$ for $1 \leq j \leq n$.
(b) For each point $\mathbf{u} = (u_1, \ldots, u_n) \in \mathbf{R}^n$, put $X(\mathbf{u}) = \sum_j u_j X_j$. For each $x \in M$, there exists a neighborhood V of 0 in $\mathbf{R}^n$ such that $F_{X(\mathbf{u})}(x, 1)$ is defined for all $\mathbf{u} \in V$ and such that $\mathbf{u} \mapsto F_{X(\mathbf{u})}(x, 1)$ is a diffeomorphism of V onto a neighborhood of x in M. Deduce that if $Y \in \mathfrak{a}$, the set of points $x \in M$ such that $Y(x) = 0$ is both open and closed in M, and hence that for each $x \in M$ the mapping $Y \mapsto Y(x)$ of $\mathfrak{a}$ into $T_x(M)$ is injective. (Observe that $[Y, X(\mathbf{u})] = 0$ for all $\mathbf{u} \in V$.)
(c) Deduce that, for the topology $\mathscr{T}$ described in Problem 5(c), Γ can be endowed with a manifold structure which makes it a Lie group acting differentiably on M, with $\mathscr{T}$ as underlying topology. (Use Problem 5.)

11. THE ADJOINT REPRESENTATION. NORMALIZERS AND CENTRALIZERS

(19.11.1) Recall (16.9) that a *linear representation* of a Lie group G is a Lie group homomorphism $f: \mathrm{G} \to \mathbf{GL}(\mathrm{E})$, where E is a finite-dimensional real vector space. The tangent linear mapping to f at the point e is therefore a Lie algebra homomorphism

$$f_* : \mathfrak{g}_e \to \mathfrak{gl}(\mathrm{E})$$

(19.3.5); it follows in particular, by virtue of (19.8.9) and (19.9.1) that for all $\mathbf{x} \in \mathfrak{g}_e$, we have

(19.11.1.1) $$f(\exp(\mathbf{x})) = \exp(f_*(\mathbf{x})) = \sum_{m=0}^{\infty} \frac{1}{m!} (f_*(\mathbf{x}))^m$$

(recall that $f_*(\mathbf{x}) \in \mathfrak{gl}(\mathrm{E})$ may be canonically identified with an element of the ring End(E) (19.4.2)).

We shall consider in particular the *adjoint representation* $s \mapsto \mathrm{Ad}(s)$ of a Lie group G in its Lie algebra $\mathfrak{g}_e$ (19.2.1).

(19.11.2) *The tangent linear mapping at the point e to the adjoint representation* $s \mapsto \mathrm{Ad}(s)$ *is the homomorphism* $\mathbf{x} \mapsto \mathrm{ad}(\mathbf{x})$ *of* $\mathfrak{g}_e$ *into* $\mathfrak{gl}(\mathfrak{g}_e)$ (we recall that $\mathrm{ad}(\mathbf{x}) \cdot \mathbf{y} = [\mathbf{x}, \mathbf{y}]$; the homomorphism $\mathbf{x} \mapsto \mathrm{ad}(\mathbf{x})$ is called the *adjoint representation* of $\mathfrak{g}_e$).

For each $\mathbf{x} \in \mathfrak{g}_e$, the mapping $\mathbf{y} \mapsto \mathrm{Ad}(\exp(\mathbf{x})) \cdot \mathbf{y} = \exp(\mathbf{x}) \cdot \mathbf{y} \cdot \exp(\mathbf{x})^{-1}$ of $\mathfrak{g}_e$ into itself is linear. For fixed $\mathbf{y}$, the derivative of the mapping

$$\mathbf{x} \mapsto \mathrm{Ad}(\exp(\mathbf{x})) \cdot \mathbf{y}$$

at the point $\mathbf{0}_e$ is therefore ((8.1.3) and (8.2.1))

$$\mathbf{h} \mapsto ((\mathrm{T}_e(\mathrm{Ad}) \circ \mathrm{T}_{\mathbf{0}_e}(\exp)) \cdot \mathbf{h}) \cdot \mathbf{y}.$$

Since $\mathrm{T}_{\mathbf{0}_e}(\exp)$ is the identity mapping of $\mathfrak{g}_e$, we obtain

$$\mathbf{h} \mapsto (\mathrm{T}_e(\mathrm{Ad}) \cdot \mathbf{h}) \cdot \mathbf{y}.$$

But $\mathbf{y} \mapsto \mathrm{Ad}(\exp(\mathbf{x})) \cdot \mathbf{y}$ is the derivative at the point $\mathbf{0}_e$ of the mapping

(19.11.2.1) $$\mathbf{y} \mapsto \log(\exp(\mathbf{x}) \exp(\mathbf{y}) \exp(\mathbf{x})^{-1})$$

of $\mathfrak{g}_e$ into itself, by virtue of the definition of Ad(s) (19.2.1.1) and of the fact that $\mathrm{T}_e(\log)$ is the identity of $\mathfrak{g}_e$. Now it follows from Taylor's formula (19.9.12) that the derivative of (19.11.2.1) at the point $\mathbf{0}_e$ is the linear mapping

$$\mathbf{y} \mapsto \mathbf{y} + [\mathbf{x}, \mathbf{y}] + r_1(\mathbf{x}) \cdot \mathbf{y},$$

where $r_1(\mathbf{x}) \in \mathrm{End}(\mathfrak{g}_e)$ is such that $r_1(\mathbf{x})/\|\mathbf{x}\|$ tends to 0 as $\|\mathbf{x}\| \to 0$ (16.8.9.1). For each fixed $\mathbf{y}$, the derivative at $\mathbf{0}_e$ of the mapping (of $\mathfrak{g}_e$ into itself)

$$\mathbf{x} \mapsto \mathbf{y} + [\mathbf{x}, \mathbf{y}] + r_1(\mathbf{x}) \cdot \mathbf{y}$$

is therefore $\mathbf{h} \mapsto [\mathbf{h}, \mathbf{y}]$, which proves (19.11.2).

Hence, by virtue of (19.11.1.1), we have for each vector $\mathbf{x} \in \mathfrak{g}_e$

(19.11.2.2) $$\mathrm{Ad}(\exp(\mathbf{x})) = \exp(\mathrm{ad}(\mathbf{x})) = \sum_{m=0}^{\infty} \frac{1}{m!} (\mathrm{ad}(\mathbf{x}))^m$$

in the algebra $\mathrm{End}(\mathfrak{g}_e)$.

Likewise, by applying the formula (19.8.9.1) to the case where f is the inner automorphism $\mathrm{Int}(s) : t \mapsto sts^{-1}$, we obtain, for $\mathbf{x} \in \mathfrak{g}_e$:

(19.11.2.3) $$\mathrm{Int}(s)(\exp(\mathbf{x})) = s(\exp(\mathbf{x}))s^{-1} = \exp(\mathrm{Ad}(s) \cdot \mathbf{x}).$$

Remark

(19.11.2.4) If A is a finite-dimensional **R**-algebra with identity element, it follows immediately from (8.1.3) that in the Lie group A* the adjoint representation is given by

(19.11.2.5) $$\mathrm{Ad}(s) : \mathbf{u} \mapsto s\mathbf{u}s^{-1}$$

(where A, endowed with the bracket operation, is identified with the Lie algebra of A* (19.4.2)).

(19.11.3) *Let* G *be a Lie group and let* $\mathfrak{m}$ *be a vector subspace of the Lie algebra* $\mathfrak{g}_e$ *of* G.

(i) *The set of all* $s \in$ G *such that* $\mathrm{Ad}(s) \cdot \mathfrak{m} \subset \mathfrak{m}$ (or equivalently such that $\mathrm{Ad}(s) \cdot \mathfrak{m} = \mathfrak{m}$, since Ad$(s)$ is an automorphism of $\mathfrak{g}_e$) *is a closed subgroup* H *of* G, *whose Lie algebra* $\mathfrak{h}$ *is the set of all* $\mathbf{u} \in \mathfrak{g}_e$ *such that* $\mathrm{ad}(\mathbf{u}) \cdot \mathfrak{m} \subset \mathfrak{m}$.

(ii) *The set of all* $s \in G$ *such that the restriction of* $\mathrm{Ad}(s)$ *to* $\mathfrak{m}$ *is the identity mapping* $1_{\mathfrak{m}}$ *is a closed normal subgroup* K *of* H, *whose Lie algebra* $\mathfrak{k}$ *is the set of all* $\mathbf{u} \in \mathfrak{g}_e$ *such that the restriction of* $\mathrm{ad}(\mathbf{u})$ *to* $\mathfrak{m}$ *is zero.*

(i) Choose a basis of $\mathfrak{g}_e$ containing a basis of $\mathfrak{m}$. Then to say that $\mathrm{Ad}(s) \cdot \mathfrak{m} \subset \mathfrak{m}$ is to say that certain of the entries in the matrix of $\mathrm{Ad}(s)$ relative to this basis are zero. From this it is clear that H is closed in G; and since $\mathrm{Ad}(s) \cdot \mathfrak{m} = \mathfrak{m}$ is equivalent to $\mathfrak{m} = (\mathrm{Ad}(s))^{-1} \cdot \mathfrak{m} = \mathrm{Ad}(s^{-1}) \cdot \mathfrak{m}$ it follows that H is a subgroup of G. Now let $\mathfrak{h} = \mathrm{Lie}(\mathrm{H})$. For each $\mathbf{y} \in \mathfrak{m}$, the derivative at the point $\mathbf{0}_e$ of the mapping $\mathbf{x} \mapsto \mathrm{Ad}(\exp(\mathbf{x})) \cdot \mathbf{y}$ of $\mathfrak{h}$ into $\mathfrak{m}$ must be an element of $\mathrm{Hom}(\mathfrak{h}, \mathfrak{m})$; but this derivative is the restriction to $\mathfrak{h}$ of the mapping $\mathbf{u} \mapsto [\mathbf{u}, \mathbf{y}]$ (19.11.2), hence we must have $[\mathbf{u}, \mathbf{y}] \in \mathfrak{m}$ for all $\mathbf{y} \in \mathfrak{m}$ and all $\mathbf{u} \in \mathfrak{h}$. Conversely, let $\mathfrak{h}'$ be the Lie subalgebra of $\mathfrak{g}_e$ consisting of all $\mathbf{u} \in \mathfrak{g}_e$ such that $\mathrm{ad}(\mathbf{u}) \cdot \mathfrak{m} \subset \mathfrak{m}$. For each $\mathbf{x} \in \mathfrak{h}'$, the restriction of $\mathrm{ad}(\mathbf{x})$ to $\mathfrak{m}$ is an endomorphism of this vector space, hence the same is true of $\exp(\mathrm{ad}(\mathbf{x}))$ (the exponential being taken in $\mathbf{GL}(\mathfrak{m})$). By virtue of (19.11.2.2), we have therefore $\mathrm{Ad}(\exp(\mathbf{x})) \cdot \mathbf{y} \in \mathfrak{m}$ for all $\mathbf{y} \in \mathfrak{m}$; in other words, $\exp(\mathfrak{h}') \subset \mathrm{H}$. Since $\mathfrak{h}' \supset \mathfrak{h}$, it follows that $\exp(\mathfrak{h}')$ is a neighborhood of e in H, which implies that $\mathfrak{h}' \subset \mathfrak{h}$ (19.8.10). Hence $\mathfrak{h}' = \mathfrak{h}$.

(ii) It is immediate that K is a closed normal subgroup of H. Let $\mathfrak{k} = \mathrm{Lie}(\mathrm{K})$. For each $\mathbf{y} \in \mathfrak{m}$, the mapping $\mathbf{x} \mapsto \mathrm{Ad}(\exp(\mathbf{x})) \cdot \mathbf{y}$ of $\mathfrak{k}$ into $\mathfrak{m}$ is constant (equal to $\mathbf{y}$), hence its derivative is zero, which as above gives $[\mathbf{x}, \mathbf{y}] = \mathbf{0}_e$. Conversely, let $\mathfrak{k}'$ be the Lie subalgebra of $\mathfrak{g}_e$ consisting of the vectors $\mathbf{u}$ such that the restriction of $\mathrm{ad}(\mathbf{u})$ to $\mathfrak{m}$ is zero. For each $\mathbf{x} \in \mathfrak{k}'$, the restriction of $\mathrm{ad}(\mathbf{x})$ to $\mathfrak{m}$ is zero, hence the restriction of $\mathrm{Ad}(\exp(\mathbf{x})) = \exp(\mathrm{ad}(\mathbf{x}))$ to $\mathfrak{m}$ is the identity mapping (19.11.2.2). Consequently, we have $\exp(\mathfrak{k}') \subset \mathrm{K}$, and since $\mathfrak{k}' \supset \mathfrak{k}$ it follows as before that $\mathfrak{k}' = \mathfrak{k}$.

The group H is called the *normalizer in* G *of the vector space* $\mathfrak{m} \subset \mathfrak{g}_e$, and is denoted by $\mathscr{N}(\mathfrak{m})$. Its Lie algebra $\mathfrak{h}$ is called the *normalizer of* $\mathfrak{m}$ in $\mathfrak{g}_e$, and is denoted by $\mathfrak{N}(\mathfrak{m})$. The group K is called the *centralizer in* G *of the vector space* $\mathfrak{m}$, and is denoted by $\mathscr{Z}(\mathfrak{m})$. Its Lie algebra $\mathfrak{k}$ is called the *centralizer of* $\mathfrak{m}$ *in* $\mathfrak{g}_e$, and is denoted by $\mathfrak{Z}(\mathfrak{m})$.

The connections between these notions and those of the normalizer and centralizer in G of a *subset of* G (12.8.6) are brought out in the following results.

(19.11.4) *Let* G *be a Lie group,* H *a connected Lie group immersed in* G (19.7.4), *corresponding to a Lie subalgebra* $\mathfrak{h}_e$ *of* $\mathfrak{g}_e$. *Then the normalizer* $\mathscr{N}(\mathrm{H})$ *of* H *in* G *is the closed subgroup* $\mathscr{N}(\mathfrak{h}_e)$, *whose Lie algebra is* $\mathfrak{N}(\mathfrak{h}_e)$.

It is clear that, for each $s \in G$, the group sHs^{-1} is a connected Lie group immersed in G, with Lie algebra $\mathrm{Ad}(s) \cdot \mathfrak{h}_e$ (19.2.1). Hence (19.7.4) we have $s \in \mathscr{N}(H)$ if and only if $s \in \mathscr{N}(\mathfrak{h}_e)$. The proposition is therefore a consequence of (19.11.3(i)).

(19.11.5) *Let* H *be a connected Lie group immersed in a connected Lie group* G. *Then* H *is normal in* G *if and only if its Lie algebra* $\mathfrak{h}_e$ *is an ideal in* $\mathfrak{g}_e$.

To say that H is normal signifies that $\mathscr{N}(H) = G$. This is equivalent to $\mathscr{N}(\mathfrak{h}_e) = G$, and implies that $\mathfrak{N}(\mathfrak{h}_e) = \mathfrak{g}_e$, i.e., that $\mathfrak{h}_e$ is an ideal in $\mathfrak{g}_e$; but since G is connected, the relation $\mathfrak{g}_e = \mathfrak{N}(\mathfrak{h}_e)$ also implies that $\mathscr{N}(\mathfrak{h}_e) = G$.

(19.11.6) *Let* G *be a Lie group,* H *a connected Lie group immersed in* G, *corresponding to a Lie subalgebra* $\mathfrak{h}_e$ *of* $\mathfrak{g}_e$. *Then the centralizer* $\mathscr{Z}(H)$ *of* H *in* G *is the closed subroup* $\mathscr{Z}(\mathfrak{h}_e)$, *whose Lie algebra is* $\mathfrak{Z}(\mathfrak{h}_e)$.

To say that $s \in \mathscr{Z}(H)$ signifies that the restriction of $\mathrm{Int}(s): t \mapsto sts^{-1}$ to H is the identity automorphism of H. Since H is connected, this is equivalent to saying that the restriction of $\mathrm{Ad}(s)$ to $\mathfrak{h}_e$ is the identity automorphism of $\mathfrak{h}_e$ (19.7.6). The proposition is therefore a consequence of (19.11.3(ii)).

In particular:

(19.11.7) *The Lie algebra of the center* $C = \mathscr{Z}(G)$ *of a connected Lie group* G *is the center* $\mathfrak{c}_e$ *of the Lie algebra* $\mathfrak{g}_e$ *of* G (i.e., is the set of all $\mathbf{x} \in \mathfrak{g}_e$ such that $[\mathbf{x}, \mathbf{y}] = \mathbf{0}_e$ for all $\mathbf{y} \in \mathfrak{g}_e$).

(19.11.8) We shall prove in Chapter XXI that if $\mathfrak{g}$ is any finite-dimensional Lie algebra over **R**, there exists a Lie group G such that Lie(G) is isomorphic to $\mathfrak{g}$. Here we shall prove a particular case of this theorem:

(19.11.9) *If* $\mathfrak{g}$ *is any finite-dimensional Lie algebra over* **R** *and if* $\mathfrak{c}$ *is the center of* $\mathfrak{g}$, *then there exists a Lie group* H *such that* Lie(H) *is isomorphic to the Lie algebra* $\mathfrak{g}/\mathfrak{c}$.

Consider the adjoint representation $\mathbf{x} \mapsto \mathrm{ad}(\mathbf{x})$ of $\mathfrak{g}$: its image is a Lie subalgebra $\mathfrak{h}$ of $\mathfrak{gl}(\mathfrak{g})$, isomorphic to $\mathfrak{g}/\mathfrak{c}$. If $n = \dim(\mathfrak{g})$, then $\mathfrak{gl}(\mathfrak{g})$ is isomorphic to the Lie algebra of the Lie group $\mathbf{GL}(n, \mathbf{R})$. Hence (19.7.4) there exists a connected Lie group H immersed in $\mathbf{GL}(n, \mathbf{R})$, whose Lie algebra is isomorphic to $\mathfrak{h}$, hence to $\mathfrak{g}/\mathfrak{c}$.

PROBLEMS

1. Let G be a connected Lie group, $\mathfrak{G}$ the infinitesimal algebra of G, formed by the left-invariant differential operators (**19.3.1**).

(a) For each left-invariant vector field $X \in \mathfrak{g}$, consider the derivation

$$\mathrm{ad}(X) : Z \mapsto X \circ Z - Z \circ X$$

of $\mathfrak{G}$. With the notation of (**19.6.2**), show that $\mathrm{ad}(X)$ is an endomorphism of each of the vector subspaces $\mathfrak{G}_m$ of $\mathfrak{G}$, and hence that $\exp(\mathrm{ad}(X))$ is an automorphism of each vector space $\mathfrak{G}_m$, and therefore an automorphism of the vector space $\mathfrak{G}$. Using the fact that $\mathrm{ad}(X)$ is a derivation, together with Leibniz's formula, show that $\exp(\mathrm{ad}(X))$ is an automorphism of the *algebra* $\mathfrak{G}$.

(b) For each differential operator $P \in \mathrm{Diff}(\mathrm{G})$ and each $s \in \mathrm{G}$, define

$$\mathrm{Ad}(s) \cdot P = \gamma(s)\delta(s^{-1})P.$$

If $P \in \mathfrak{G}$, we have $\mathrm{Ad}(s) \cdot P = \delta(s^{-1})P$, and $\mathrm{Ad}(s) \cdot P$ is the differential operator whose value at the point e is the image under $\mathrm{Int}(s)$ of the point-distribution $P(e)$. Show that for each $\mathbf{u} \in \mathfrak{g}_e$ we have

$$\mathrm{Ad}(\exp(\mathbf{u})) \cdot Z = \exp(\mathrm{ad}(X_{\mathbf{u}})) \cdot Z$$

for all $Z \in \mathfrak{G}$. (Observe that the two sides agree when $Z \in \mathfrak{g}$, and that $\mathfrak{g}$ generates the associative algebra $\mathfrak{G}$.)

(c) Deduce from (b) that the *center* of the algebra $\mathfrak{G}$ consists of the operators

$$P \in \mathrm{Diff}(\mathrm{G})$$

which are *both left- and right-invariant*. (Observe that in each space $\mathfrak{G}_m$, we have

$$\lim_{t \to 0} \frac{1}{t} (\exp(\mathrm{ad}(X)) \cdot Z - Z) = \mathrm{ad}(X) \cdot Z$$

for $X \in \mathfrak{g}$.) If $\mathfrak{c}$ is the center of the Lie algebra $\mathfrak{g}$, it is therefore contained in the center of $\mathfrak{G}$.

2. (a) For the group G considered in (**19.5.11**), show that the center of the infinitesimal algebra $\mathfrak{G}$ consists only of scalars.

(b) Let G be the Lie group $\mathbf{SL}(2, \mathbf{R})$, and take as basis for the Lie algebra $\mathfrak{sl}(2, \mathbf{R})$ the basis $(\mathbf{u}_1, \mathbf{u}_2, \mathbf{u}_3)$ of Section 19.5, Problem 2. Show that the center of this Lie algebra is zero, but that the element $\mathbf{u}_1^2 + 2(\mathbf{u}_2 \mathbf{u}_3 + \mathbf{u}_3 \mathbf{u}_2)$ belongs to the center of the infinitesimal algebra $\mathfrak{G}_e$.

3. Let G be a connected Lie group, H a connected Lie group immersed in G; let $\bar{\mathrm{H}} = \mathrm{H}'$ be the closure of H in G and let $\mathfrak{g}_e$, $\mathfrak{h}_e$, $\mathfrak{h}'_e$ be the Lie algebras of G, H, H′, respectively.

(a) Show that if a connected Lie group K immersed in H is normal in H, then it is also normal in H′. (If $\mathfrak{k}_e$ is the Lie algebra of K, observe that $\mathrm{Ad}(s) \cdot \mathfrak{k}_e \subset \mathfrak{k}_e$ for all $s \in \mathrm{H}$, and use (**19.11.4**).)

(b) Show that if a connected Lie group L is immersed in G and if $H \subset L \subset H' = \bar{H}$, then L is normal in H'.
(c) Suppose that $H' = G$ (i.e., that H is dense in G). Show that the Lie algebra $\mathfrak{g}_e/\mathfrak{h}_e$ is commutative. (Observe that every Lie subalgebra $\mathfrak{l}_e$ such that $\mathfrak{h}_e \subset \mathfrak{l}_e \subset \mathfrak{g}_e$ is an ideal of $\mathfrak{g}_e$.) (Cf. Section **19.16**, Problem 11.)
(d) Under the hypotheses of (c), show that $\mathfrak{g}_e$ is the direct sum of an ideal $\mathfrak{f}_e$ contained in the center $\mathfrak{c}_e$ of $\mathfrak{g}_e$, and an ideal $\mathfrak{l}_e \supset \mathfrak{h}_e$ such that $\mathfrak{l}_e \cap \mathfrak{c}_e \subset \mathfrak{h}_e$. (Decompose $\mathfrak{g}_e/\mathfrak{h}_e$ as the direct sum of $(\mathfrak{c}_e + \mathfrak{h}_e)/\mathfrak{h}_e$ and $\mathfrak{l}_e/\mathfrak{h}_e$, and $\mathfrak{c}_e$ as the direct sum of $\mathfrak{c}_e \cap \mathfrak{h}_e$ and $\mathfrak{l}_e$.) Furthermore, every element $\mathbf{u} \in \mathfrak{l}_e$ such that $[\mathbf{u}, \mathbf{v}] = 0$ for all $\mathbf{v} \in \mathfrak{h}_e$ belongs to $\mathfrak{c}_e \cap \mathfrak{h}_e$. (Remark that the hypothesis implies that $\mathrm{Ad}(s) \cdot \mathbf{u} = \mathbf{u}$ for all $s \in H$, and deduce that the same is true for all $s \in G$.)

4. Let G be a connected Lie group, H a connected Lie group immersed in G and *dense* in G. For each $s \in G$, the image of $\mathfrak{h}_e$ under $\mathrm{Ad}(s)$ is contained in $\mathfrak{h}_e$ (Problem 3(a)). Let $\mathrm{Ad}_H(s)$ denote the restriction of $\mathrm{Ad}(s)$ to $\mathfrak{h}_e$, which is an automorphism of the vector space $\mathfrak{h}_e$.

(a) Suppose that the image of H in $\mathbf{GL}(\mathfrak{h}_e)$ under the adjoint representation is a *closed* subgroup Ad(H) of $\mathbf{GL}(\mathfrak{h}_e)$. Show that Ad(H) is equal to the image $\mathrm{Ad}_H(G)$ of G under $s \mapsto \mathrm{Ad}_H(s)$. Deduce (in the notation of Problem 3(d)) that $\mathfrak{l}_e = \mathfrak{h}_e$ and hence that $\mathfrak{g}_e$ is the *direct sum* of $\mathfrak{h}_e$ and an ideal $\mathfrak{f}_e \subset \mathfrak{c}_e$.
(b) Deduce from (a) that the center $\mathscr{Z}(G)$ of G is the closure in G of the center $\mathscr{Z}(H)$ of H. (Let $s \in \mathscr{Z}(G)$ be the limit of a sequence (s_n) of elements of H. Show that there exist elements $t_n \in \mathscr{Z}(H)$ such that $x_n = s_n t_n^{-1}$ tends to e in H; for this purpose, observe that Ad(H) is isomorphic to $H/\mathscr{Z}(H)$, where H is endowed with its proper topology.) Show that $G = H\mathscr{Z}(G)$ and that $\mathrm{Ad}(G) = \mathrm{Ad}(H)$ is closed in $\mathbf{GL}(\mathfrak{g}_e)$.
(c) Under the hypotheses of (a), deduce from (b) that for H to be equal to G, it is necessary and sufficient that $\mathscr{Z}(H)$ be closed in G.

5. Let G be a connected Lie group, $u : G \to G'$ a homomorphism of Lie groups. Assume that Ad(G) is a closed subgroup of $\mathbf{GL}(\mathfrak{g}_e)$. Then $u(G)$ is closed in G' if and only if $u(\mathscr{Z}(G))$ is closed in G'. (Use Problem 4.)

6. Let G be the closed subgroup of $\mathbf{GL}(3, \mathbf{C})$ consisting of the matrices

$$\begin{pmatrix} e^{is} & 0 & x \\ 0 & e^{it} & y \\ 0 & 0 & 1 \end{pmatrix},$$

where s, t take all values in $\mathbf{R}$ and x, y all values in $\mathbf{C}$. Show that $\mathscr{Z}(G)$ consists of the identity element. Let α be a given irrational number and let H be the subgroup of all elements of G such that $t = \alpha s$. Show that Ad(H) is not closed in $\mathbf{GL}(\mathfrak{h}_e)$. (Use Problem 4.)

7. Let G be a Lie group, H *any* subgroup of G. For each $x \in H$, let $\mathfrak{h}_x$ denote the set of vectors $\mathbf{u}_x \in T_x(G)$ for which there exists an open interval $I \subset \mathbf{R}$ containing 0 and a C^∞ mapping $f : I \to G$ such that $f(0) = x$, $f(I) \subset H$, and $T_0(f) \cdot E(0) = \mathbf{u}_x$.

(a) Show that for each $x \in H$ we have $\mathfrak{h}_x = x \cdot \mathfrak{h}_e = \mathfrak{h}_e \cdot x$, and that $\mathrm{Ad}(x) \cdot \mathfrak{h}_e \subset \mathfrak{h}_e$.
(b) Show that $\mathfrak{h}_e$ is a Lie subalgebra of $\mathfrak{g}_e = \mathrm{Lie}(G)$. (Use (**16.9.9**(ii)) to prove that $\mathfrak{h}_e$ is a vector subspace of $\mathfrak{g}_e$. Then observe that if $\mathbf{u}, \mathbf{v}$ are two vectors in $\mathfrak{h}_e$ and if

$f: \mathrm{I} \to \mathrm{G}$ is a C^∞ mapping of a neighborhood I of 0 in **R** such that $f(0) = e$, $f(\mathrm{I}) \subset \mathrm{H}$, and $\mathrm{T}_0(f) \cdot E(0) = \mathbf{u}$, then we have $\mathrm{Ad}(f(t)) \cdot \mathbf{v} \in \mathfrak{h}_e$ for all $t \in \mathrm{I}$. Finally use **(19.11.2)**.)
(c) Let H_0 be the connected Lie group immersed in G, such that $\mathrm{Lie}(\mathrm{H}_0) = \mathfrak{h}_e$. Let M be a differential manifold, $f: \mathrm{M} \to \mathrm{G}$ a C^∞ mapping such that $f(z_0) = e$ for some $z_0 \in \mathrm{M}$, and $f(\mathrm{M}) \subset \mathrm{H}$. Show that there exists an open neighborhood U of z_0 in M such that $f(\mathrm{U}) \subset \mathrm{H}_0$. (Use **(18.14.7)**.) In particular, H_0 is a normal subgroup of H. (Use the definition of $\mathfrak{h}_e$ and **(19.9.15)**.) Deduce that there exists a unique topology on H which is compatible with its group structure, which induces on H_0 its proper topology **(19.7.4)**, and for which H_0 is the identity component of H. (Cf. Section **12.8**, Problem 1.)

8. Let M be a differential manifold. A subset A of M is said to be C^∞-*connected* if, for all $x, y \in \mathrm{A}$, there exists a sequence $(z_j)_{0 \leq j \leq n}$ of points of A such that $z_0 = x$, $z_n = y$, and for $1 \leq j \leq n$, a C^∞ mapping f_j of an open interval $\mathrm{I} \subset \mathbf{R}$ into M such that $f_j(\mathrm{I})$ contains z_{j-1} and z_j and is contained in A.

(a) Show that if a subgroup H of a Lie group G is C^∞-connected, then H is a connected Lie group immersed in G. (With the notation of Problem 7, show that $\mathrm{H}_0 = \mathrm{H}$.)
(b) Deduce from (a) that if A, B are two connected Lie groups immersed in a Lie group G, then the subgroup H of G generated by $\mathrm{A} \cup \mathrm{B}$ is a connected Lie group immersed in G, whose Lie algebra is generated by the union of the Lie algebras of A and B. (To establish the latter point, show that if $\mathbf{u} \in \mathrm{Lie}(\mathrm{A})$ and $\mathbf{v} \in \mathrm{Lie}(\mathrm{B})$, then $[\mathbf{u}, \mathbf{v}] \in \mathrm{Lie}(\mathrm{H})$, by using Problem 7 and **(19.9.13)**.

9. Let A and B be two connected Lie groups immersed in a Lie group G, and let $\mathfrak{a}_e, \mathfrak{b}_e$ be their Lie algebras.

(a) Show that the mapping $(x, y) \mapsto xy$ of $\mathrm{A} \times \mathrm{B}$ into G (where A, B are endowed with their proper topologies and their Lie group structures) is a subimmersion of constant rank $\dim(\mathfrak{a}_e + \mathfrak{b}_e) = \dim \mathfrak{a}_e + \dim \mathfrak{b}_e - \dim (\mathfrak{a}_e \cap \mathfrak{b}_e)$. Moreover, if AB is a locally closed subspace of G, then AB is a submanifold of G. (Consider the left action of $\mathrm{A} \times \mathrm{B}$ on G defined by $((x, y), s) = xsy^{-1}$ and use **(16.10.2)** and **(16.10.7)**.) Give an example where AB is open and dense in G but is not closed. Section **19.7**, Problem 3.
(b) Suppose that $\mathrm{A} \cap \mathrm{B}$ is closed in G and normal in B, and that AB is a submanifold of G. The group $\mathrm{A} \cap \mathrm{B}$ acts on the manifold AB by right translations, and the orbit manifold $\mathrm{AB}/(\mathrm{A} \cap \mathrm{B})$ exists. Show that it is canonically diffeomorphic to

$$(\mathrm{A}/(\mathrm{A} \cap \mathrm{B})) \times (\mathrm{B}/(\mathrm{A} \cap \mathrm{B})).$$

(Use **(16.10.4)**.) Deduce that A and B are Lie subgroups of G.
(c) Show that if $\mathrm{AB} = \mathrm{BA}$, $\mathfrak{a}_e + \mathfrak{b}_e$ is a Lie subalgebra of $\mathfrak{g}_e = \mathrm{Lie}(\mathrm{G})$. Use (a) and Problem 8(b); show that the immersed Lie group AB cannot have dimension $> \dim(\mathfrak{a}_e + \mathfrak{b}_e)$ by using **(18.14.7)** and Baire's theorem **(12.16.1)**.)
(d) Suppose that B is a normal Lie subgroup of G. Show that when A and AB are endowed with their proper topologies, the quotient groups $\mathrm{A}/(\mathrm{A} \cap \mathrm{B})$ and AB/B are canonically isomorphic.

10. The automorphism group L of **T** is a discrete group of two elements, the identity and the automorphism $x \mapsto -x$ (Section **19.7**, Problem 6). Consider the nonconnected

Lie group $G = L \times_\sigma T$, the semidirect product of L and T relative to the identity automorphism $\sigma : L \to \text{Aut}(T)$ (19.4.5). Show that the center of G is discrete. (Compare with (19.11.7).)

12. THE LIE ALGEBRA OF THE COMMUTATOR GROUP

Let G be a group and let H, K be two normal subgroups of G. The subgroup of G generated by the commutators $hkh^{-1}k^{-1}$, where $h \in H$ and $k \in K$ is called the *commutator group of* H *and* K and is denoted by (H, K). It is clear that (H, K) is a *normal* subgroup of G and is contained in $H \cap K$. The *commutator group of* G is the group (G, G), which is also denoted by $\mathscr{D}(G)$ and is also called the *derived group* of G. It is the smallest normal subgroup N of G such that G/N is commutative.

We recall also that if $\mathfrak{h}$, $\mathfrak{k}$ are two ideals in a Lie algebra $\mathfrak{g}$, the vector subspace generated by the elements $[\mathbf{x}, \mathbf{y}]$, where $\mathbf{x} \in \mathfrak{h}$ and $\mathbf{y} \in \mathfrak{k}$, is denoted by $[\mathfrak{h}, \mathfrak{k}]$. It follows immediately from the Jacobi identity that $[\mathfrak{h}, \mathfrak{k}]$ is an *ideal* of $\mathfrak{g}$ contained in $\mathfrak{h} \cap \mathfrak{k}$. In particular, the ideal $[\mathfrak{g}, \mathfrak{g}]$ is denoted by $\mathfrak{D}(\mathfrak{g})$ and is called the *derived ideal* of $\mathfrak{g}$.

For Lie groups, these notions are connected by the following theorem:

(19.12.1) *Let* G *be a connected Lie group,* $\mathfrak{g}_e$ *its Lie algebra. Then the derived group* $\mathscr{D}(G)$ *is the underlying group of the connected Lie group immersed in* G *which corresponds to the Lie subalgebra* $\mathfrak{D}(\mathfrak{g}_e)$.

Let G′ denote the connected Lie group immersed in G which corresponds to $\mathfrak{D}(\mathfrak{g}_e)$; then G′ is *normal* in G (19.11.3). We shall first show that $\mathscr{D}(G) \subset G'$.

The quotient Lie algebra $\mathfrak{h} = \mathfrak{g}_e/\mathfrak{D}(\mathfrak{g}_e)$ is commutative, hence is the Lie algebra of a group H isomorphic to $\mathbf{R}^m$ for some $m \geqq 0$. If $\omega : \mathfrak{g}_e \to \mathfrak{h} = \mathfrak{g}_e/\mathfrak{D}(\mathfrak{g}_e)$ is the canonical homomorphism, there exists a local homomorphism w from G to H of class C^∞ such that $T_e(w) = \omega$ (19.7.6). Let U be a symmetric neighborhood of e in G on which the function $\log_G$ and the local homomorphism w are defined, and let V be a symmetric neighborhood of e such that $V^4 \subset U$. Since H is commutative, the relations $s \in V$, $t \in V$ imply $w(st) = w(ts)$, hence $w(sts^{-1}t^{-1}) = 0$. But since the function $\log_G$ is defined on U, we have $sts^{-1}t^{-1} = \exp(\log(sts^{-1}t^{-1}))$, and therefore (19.8.9) $w(sts^{-1}t^{-1}) = \exp(\omega(\log(sts^{-1}t^{-1}))) = 0$. By definition, this signifies that $\log(sts^{-1}t^{-1}) \in \mathfrak{D}(\mathfrak{g}_e)$ and hence implies that $sts^{-1}t^{-1} \in G'$. Since G is connected, the neighborhood V generates G (12.8.8); hence the formula

$$(st)u(st)^{-1}u^{-1} = s(tut^{-1}u^{-1})s^{-1}(sus^{-1}u^{-1})$$

and the normality of G′ (19.11.3) show that every commutator $sts^{-1}t^{-1}$ belongs to G′, i.e., $\mathscr{D}(G) \subset G'$.

Conversely, we shall show that $G' \subset \mathscr{D}(G)$. Since G′ is connected, it will be enough to show that there exists a neighborhood of e in G′ (for the Lie group topology of G′) contained in $\mathscr{D}(G)$ (12.8.8). Let $n = \dim \mathfrak{g}_e$, $r = \dim \mathfrak{D}(\mathfrak{g}_e) \leq n$. Then there exist r pairs $(\mathbf{a}_j, \mathbf{b}_j)$ of elements of $\mathfrak{g}_e$ such that the vectors $\mathbf{c}_j = [\mathbf{a}_j, \mathbf{b}_j]$ $(1 \leq j \leq r)$ form a basis of $\mathfrak{D}(\mathfrak{g}_e)$. Complete this basis, by adjoining $n - r$ elements $\mathbf{c}_{r+1}, \ldots, \mathbf{c}_n$, to a basis of $\mathfrak{g}_e$. Let V be a symmetric neighborhood of e in G, which is such that the function $\log_G$ is defined on V^4. Let $(u_j)_{1 \leq j \leq n}$ be the system of canonical local coordinates of the first kind at e corresponding to the basis $(\mathbf{c}_j)$, so that $u_j(x)$ is the jth coordinate of $\log(s)$. Let I be an open neighborhood of 0 in $\mathbf{R}$ such that, for $\lambda, \mu \in I$, the points $s_j(\lambda) = \exp(\lambda \mathbf{a}_j)$ and $t_j(\mu) = \exp(\mu \mathbf{b}_j)$ belong to V for $1 \leq j \leq r$. Then the point

$$g_j(\lambda, \mu) = s_j(\lambda) t_j(\mu) s_j(-\lambda) t_j(-\mu)$$

belongs to V^4, and it follows from (19.9.13) that we may write

(19.12.1.1) $$g_j(\lambda, \mu) = \exp(\lambda\mu[\mathbf{a}_j, \mathbf{b}_j] + \lambda^2 \mu \mathbf{A}_j(\lambda, \mu) + \lambda \mu^2 \mathbf{B}_j(\lambda, \mu)),$$

where $\mathbf{A}_j$ and $\mathbf{B}_j$ are bounded in $I \times I$. Taylor's formula, applied to the functions of one variable $\mu \mapsto u_i(g_j(\lambda, \mu))$, now gives

(19.12.1.2) $$u_i(g_j(\lambda, \mu)) = \mu v_{ij}(\lambda) + \mu^2 w_{ij}(\lambda, \mu),$$

where v_{ij}, w_{ij} are C^∞ functions on $I \times I$. By virtue of (19.12.1.1) and the definition of the $\mathbf{c}_j$ $(1 \leq j \leq r)$, we have

(19.12.1.3) $$v_{ij}(\lambda) = \delta_{ij} \lambda + \lambda^2 C_{ij}(\lambda)$$

for $1 \leq i, j \leq r$ (δ_{ij} being the Kronecker delta), where the C_{ij} are bounded in I. Hence there exists $\lambda_0 \neq 0$ in I such that $\det(v_{ij}(\lambda_0)) \neq 0$ $(1 \leq i, j \leq r)$. Now define, for $\mu_1, \mu_2, \ldots, \mu_r$ in I,

(19.12.1.4) $$g(\mu_1, \mu_2, \ldots, \mu_r) = g_1(\lambda_0, \mu_1) g_2(\lambda_0, \mu_2) \cdots g_r(\lambda_0, \mu_r).$$

It follows from (19.9.11) and (19.12.1.2) that the mapping

$$(\mu_1, \ldots, \mu_r) \mapsto \log(g(\mu_1, \ldots, \mu_r))$$

of I^r into $\mathbf{R}^r$ has Jacobian equal to $\det(v_{ij}(\lambda_0))$ at the origin; hence g is a diffeomorphism of a neighborhood W of 0 in $\mathbf{R}^r$ onto a *neighborhood of* e in G′ (for the proper topology of G′); but, by definition, $g(\mu_1, \mu_2, \ldots, \mu_r) \in \mathscr{D}(G)$.

Q.E.D.

(19.12.2) In particular, we obtain the fact that the commutator group of a connected Lie group G is connected (cf. Section 12.8, Problem 4); on the other hand, it is not necessarily closed in G (Section 12.8, Problem 7).

(19.12.3) For any group G and any integer $n \geqq 1$, the nth *derived group* $\mathscr{D}^n(G)$ of G is defined inductively by the conditions

$$\mathscr{D}^1(G) = \mathscr{D}(G) = (G, G)$$

and, for $n \geqq 2$,

$$\mathscr{D}^n(G) = \mathscr{D}(\mathscr{D}^{n-1}(G)) = (\mathscr{D}^{n-1}(G), \mathscr{D}^{n-1}(G)).$$

It is clear that the sequence $(\mathscr{D}^n(G))$ is a decreasing sequence of *normal* subgroups of G. The group G is said to be *solvable* if there exists an integer $n \geqq 1$ such that $\mathscr{D}^n(G) = \{e\}$. If G is a connected Lie group, it follows from (19.12.1) that $\mathscr{D}^n(G)$ is the underlying group of a connected Lie group immersed in G, corresponding to the Lie subalgebra $\mathfrak{D}^n(\mathfrak{g}_e)$ of $\mathfrak{g}_e$. For G to be solvable, it is necessary and sufficient that $\mathfrak{D}^n(\mathfrak{g}_e) = \{0\}$ for some integer $n \geqq 1$, i.e., that the Lie algebra $\mathfrak{g}_e$ should be *solvable*.

PROBLEMS

1. Let $\mathfrak{g}$ be a Lie algebra of finite dimension over **R** or **C**.
   (a) If $\mathfrak{g}$ is solvable, then so is every Lie subalgebra of $\mathfrak{g}$.
   (b) Let $\mathfrak{a}$ be an ideal in $\mathfrak{g}$. Then $\mathfrak{g}$ is solvable if and only if the Lie algebras $\mathfrak{a}$ and $\mathfrak{g}/\mathfrak{a}$ are solvable.

2. Let $\mathfrak{a}$ be a Lie algebra of finite dimension over **R** or **C**. Show that the following properties are equivalent:

   (i) $\mathfrak{g}$ is solvable.
   (ii) There exists a decreasing sequence $\mathfrak{g} = \mathfrak{a}_0 \supset \mathfrak{a}_1 \supset \cdots \supset \mathfrak{a}_n = \{0\}$ of ideals in $\mathfrak{g}$ such that the algebras $\mathfrak{a}_{i-1}/\mathfrak{a}_i$ are commutative $(1 \leqq i \leqq n)$.
   (iii) There exists a decreasing sequence $\mathfrak{g} = \mathfrak{h}_0 \supset \mathfrak{h}_1 \supset \cdots \supset \mathfrak{h}_p = \{0\}$ of subalgebras of $\mathfrak{g}$ such that $\mathfrak{h}_i$ is an ideal in $\mathfrak{h}_{i-1}$ and the algebra $\mathfrak{h}_{i-1}/\mathfrak{h}_i$ is commutative for $1 \leqq i \leqq p$.
   (iv) There exists a decreasing sequence $\mathfrak{g} = \mathfrak{m}_0 \supset \mathfrak{m}_1 \supset \cdots \supset \mathfrak{m}_q = \{0\}$ of Lie subalgebras of $\mathfrak{g}$ such that $\mathfrak{m}_i$ is an ideal in $\mathfrak{m}_{i-1}$ and $\mathfrak{m}_{i-1}/\mathfrak{m}_i$ is of dimension 1 for $1 \leqq i \leqq q$. (To prove that (iv) implies (i), use Problem 1.)

3. For any Lie algebra $\mathfrak{g}$, the *descending central series* is the decreasing sequence of ideals $(\mathfrak{C}^p(\mathfrak{g}))_{p \geqq 1}$ defined by

$$\mathfrak{C}^1(\mathfrak{g}) = \mathfrak{g}, \qquad \mathfrak{C}^{p+1}(\mathfrak{g}) = [\mathfrak{g}, \mathfrak{C}^p(\mathfrak{g})].$$

   The *ascending central series* is the increasing sequence of ideals $(\mathfrak{C}_p(\mathfrak{g}))_{p \geqq 0}$, defined by

$$\mathfrak{C}_0(\mathfrak{g}) = \{0\}, \qquad \mathfrak{C}_{p+1}(\mathfrak{g})/\mathfrak{C}_p(\mathfrak{g}) = \text{center of } \mathfrak{g}/\mathfrak{C}_p(\mathfrak{g}).$$

Show that, for a finite-dimensional Lie algebra over **R** or **C**, the following conditions are equivalent:

(i) There exists a finite decreasing sequence $\mathfrak{g} = \mathfrak{a}_0 \supset \mathfrak{a}_1 \supset \cdots \supset \mathfrak{a}_p = \{0\}$ of ideals of $\mathfrak{g}$, such that $[\mathfrak{g}, \mathfrak{a}_{i-1}] \subset \mathfrak{a}_i$ for $1 \leqq i \leqq p$.
(ii) $\mathfrak{C}^k(\mathfrak{g}) = \{0\}$ for sufficiently large k.
(iii) $\mathfrak{C}_k(\mathfrak{g}) = \mathfrak{g}$ for sufficiently large k.
(iv) There exists an integer n such that, for all sequences $(\mathbf{x}_j)_{1 \leqq j \leqq n}$ of n elements of $\mathfrak{g}$, we have $\operatorname{ad}(\mathbf{x}_1) \circ \operatorname{ad}(\mathbf{x}_2) \circ \cdots \circ \operatorname{ad}(\mathbf{x}_n) = 0$ in $\operatorname{End}(\mathfrak{g})$.
(v) There exists a decreasing sequence of ideals $\mathfrak{g} = \mathfrak{a}_0 \supset \mathfrak{a}_1 \supset \cdots \supset \mathfrak{a}_q = \{0\}$ of $\mathfrak{g}$ such that $[\mathfrak{g}, \mathfrak{a}_{i-1}] \subset \mathfrak{a}_i$ and $\mathfrak{a}_{i-1}/\mathfrak{a}_i$ is of dimension 1 for $1 \leqq i \leqq q$.

(Observe that if (i) is satisfied, then $\mathfrak{a}_i \supset \mathfrak{C}^{i+1}(\mathfrak{g})$ and $\mathfrak{a}_{p-i} \subset \mathfrak{C}_i(\mathfrak{g})$.)

A Lie algebra $\mathfrak{g}$ satisfying these conditions is said to be *nilpotent*. If m (resp. n) is the smallest integer such that $\mathfrak{C}_m(\mathfrak{g}) = \mathfrak{g}$ (resp. $\mathfrak{C}^n(\mathfrak{g}) = \{0\}$), show that $n = m + 1$ and that $\mathfrak{C}_i(\mathfrak{g}) \supset \mathfrak{C}^{n-i}(\mathfrak{g})$ for $0 \leqq i \leqq n - 1$.

4. Let $\mathfrak{g}$ be a nilpotent Lie algebra of dimension n whose center $\mathfrak{C}_1(\mathfrak{g})$ has dimension 1. Show that there exists a basis of $\mathfrak{g}$ consisting of three elements **a**, **b**, **c** and a basis of a subspace $\mathfrak{w}$ of dimension $n - 3$, such that $[\mathbf{a}, \mathbf{b}] = \mathbf{c}$ and $[\mathbf{b}, \mathbf{u}] = 0$ for all $\mathbf{u} \in \mathfrak{w}$. (Take for **c** a basis of the center $\mathfrak{C}_1(\mathfrak{g})$, for **b** any element of $\mathfrak{C}_2(\mathfrak{g})$ not in $\mathfrak{C}_1(\mathfrak{g})$, and consider the subspace of $\mathfrak{g}$ which commutes with **b**.)

5. In a solvable but not nilpotent Lie algebra $\mathfrak{g}$ over **R**, there may exist a decreasing sequence of *ideals* $\mathfrak{g} = \mathfrak{a}_0 \supset \mathfrak{a}_1 \supset \cdots \supset \mathfrak{a}_n = \{0\}$ such that $\mathfrak{a}_{i-1}/\mathfrak{a}_i$ is of dimension 1 for $1 \leqq i \leqq n$. This is so, for example, for the Lie algebra of (**19.5.11**). Show, on the other hand, that there exists a solvable Lie algebra $\mathfrak{g}$ over **R** having a basis of three elements **u**, **v**, **w** such that $[\mathbf{u}, \mathbf{v}] = \mathbf{w}$, $[\mathbf{u}, \mathbf{w}] = -\mathbf{v}$, and $[\mathbf{v}, \mathbf{w}] = 0$, and that in this Lie algebra there exists no decreasing chain of ideals with the property stated above.

6. Let A, B be two connected Lie groups immersed in a Lie group G. Show that the group $C = (A, B)$ generated by the commutators $aba^{-1}b^{-1}$, where $a \in A$ and $b \in B$, is a connected Lie group immersed in G. (Use Problem 8(a) of Section **19.11**.) If $\mathfrak{a}_e$, $\mathfrak{b}_e$ are the Lie algebras of A, B respectively, show that the Lie algebra of C contains the vector subspace $[\mathfrak{a}_e, \mathfrak{b}_e]$ of $\mathfrak{g}_e = \operatorname{Lie}(G)$ generated by all $[\mathbf{u}, \mathbf{v}]$ with $\mathbf{u} \in \mathfrak{a}_e$ and $\mathbf{v} \in \mathfrak{b}_e$ (Section **19.11**, Problem 7). Given an example where A and B are closed and of dimension 1 in $G = \mathbf{GL}(2, \mathbf{R})$, and the Lie subalgebra generated by $[\mathfrak{a}_e, \mathfrak{b}_e]$ is distinct from Lie(C).

7. Let G be a Lie group, K a one-parameter *normal* subgroup of G. Show that the commutator group $\mathscr{D}(G)$ is contained in the centralizer of K.

13. AUTOMORPHISM GROUPS OF LIE GROUPS

(**19.13.1**) We recall that an automorphism of a Lie group is by definition an automorphism of the group structure of G which is also a diffeomorphism. These automorphisms form a group, denoted by Aut(G). If G_0 is the identity component of G, then it is clear that $u(G_0) = G_0$ for every $u \in \operatorname{Aut}(G)$, and the restriction u_0 of u to G_0 is an automorphism of the Lie group G_0. We have, therefore, a *homomorphism* $u \mapsto u_0$ of Aut(G) into Aut(G_0). This homomorphism is not necessarily injective (consider for example the case where G is discrete, so that G_0 and $\mathrm{Aut}(G_0)$ are reduced to the identity); nor is it necessarily surjective—in other words, an automorphism of G_0 cannot necessarily be extended to an automorphism of G (Problem 1).

(19.13.2) In what follows we shall restrict our attention to $\mathrm{Aut}(G_0)$, that is to say, we shall assume that the group G is *connected*. For each automorphism u of G, the tangent linear mapping u_* is then an *automorphism of the Lie algebra* $\mathfrak{g}_e$; for if v is the inverse of the automorphism u, we have $u_* \circ v_* = (u \circ v)_* = 1_{\mathfrak{g}_e}$, and likewise $v_* \circ u_* = 1_{\mathfrak{g}_e}$.

(19.13.3) *Let* G *be a connected Lie group.*

(i) *The mapping* $u \mapsto u_*$ *is an injective homomorphism of* Aut(G) *into the automorphism group* $\mathrm{Aut}(\mathfrak{g}_e)$ *of the Lie algebra of* G.

(ii) *If* G *is simply connected, then* $u \mapsto u_*$ *is an isomorphism of the group* Aut(G) *onto the group* $\mathrm{Aut}(\mathfrak{g}_e)$.

(iii) *In general, if* $G = \tilde{G}/D$, *where* $\tilde{G}$ *is the simply connected universal covering group of* G (16.30.1) *and* D *is a discrete subgroup of the center of* G, *then* Aut(G) *may be identified with the subgroup of* $\mathrm{Aut}(\tilde{G})$ *consisting of automorphisms* $\tilde{u}$ *such that* $\tilde{u}(D) = D$ (or, equivalently, $\tilde{u}(D) \subset D$).

The injectivity of $u \mapsto u_*$ in general, and the surjectivity when G is simply-connected, both follow from (19.7.6). Let $p : \tilde{G} \to G$ be the canonical homomorphism. For every (Lie group) homomorphism u of G into G, the mapping $u \circ p$ is a Lie group homomorphism of $\tilde{G}$ into G, and therefore (16.30.3) there exists a unique homomorphism $\tilde{u} : \tilde{G} \to \tilde{G}$ such that $p \circ \tilde{u} = u \circ p$. Moreover, if $v : G \to G$ is another homomorphism, we have $(v \circ u)^{\sim} = \tilde{v} \circ \tilde{u}$, because $p \circ (\tilde{v} \circ \tilde{u}) = (v \circ p) \circ \tilde{u} = v \circ (u \circ p)$. Consequently, if u is an automorphism of G, then $\tilde{u}$ is an automorphism of $\tilde{G}$, and $u \mapsto \tilde{u}$ is an injective homomorphism of Aut(G) into $\mathrm{Aut}(\tilde{G})$. Furthermore, the relation $p \circ \tilde{u} = u \circ p$ shows that we must have $\tilde{u}(D) = D$. Conversely, if $\tilde{u}$ has this property, there exists a homomorphism u of $\tilde{G}/D = G$ into itself that $p \circ \tilde{u} = u \circ p$, and this homomorphism is of class C^∞ (16.10.4). Likewise there is a Lie group homomorphism $v : G \to G$ such that $p \circ \tilde{u}^{-1} = v \circ p$. From this we deduce immediately that $v \circ u = u \circ v = 1_G$, and the proof is complete.

(19.13.4) In the notation of (19.13.3), the group $\mathrm{Aut}(\mathfrak{g}_e)$ is a *closed subgroup* of $\mathbf{GL}(\mathfrak{g}_e)$, hence a Lie subgroup of $\mathbf{GL}(\mathfrak{g}_e)$ (19.10.1). For if $(\mathbf{a}_j)_{1 \leqq j \leqq n}$ is a basis of $\mathfrak{g}_e$, an automorphism v of the vector space $\mathfrak{g}_e$ is also an automorphism of the *Lie algebra* $\mathfrak{g}_e$ if and only if v satisfies the conditions

$$v([\mathbf{a}_j, \mathbf{a}_k]) = [v(\mathbf{a}_j), v(\mathbf{a}_k)]$$

for all pairs of indices (j, k) such that $1 \leqq j < k \leqq n$. The coordinates (relative to the basis $(\mathbf{a}_j)$) of the two sides of this equation are polynomials in the elements of the matrix of v relative to the basis $(\mathbf{a}_j)$, and therefore (3.15.1) $\mathrm{Aut}(\mathfrak{g}_e)$ is closed in $\mathbf{GL}(\mathfrak{g}_e)$.

Now assume that G is simply-connected. Then (19.13.3) we have an isomorphism of groups $\varphi : u \mapsto u_*$ of Aut(G) onto $\mathrm{Aut}(\mathfrak{g}_e)$. By *transporting* via φ^{-1} the differential manifold structure of $\mathrm{Aut}(\mathfrak{g}_e)$ to Aut(G) (16.2.6), we obtain canonically a *Lie group* structure on Aut(G). In future, whenever we speak of Aut(G) as a Lie group, it is always this structure that is meant.

(19.13.5) *Let* G *be a connected Lie group,* $\tilde{G}$ *the simply-connected universal covering group of* G.

(i) *The group* Aut(G) *is closed in the Lie group* $\mathrm{Aut}(\tilde{G})$ (hence is a Lie group, by (19.10.1)).

(ii) *The mapping* $(u, x) \mapsto u(x)$ *of* Aut(G) × G *into* G *is of class* C^∞.

We shall begin by proving (ii) when G is simply connected, in which case $u \mapsto u_*$ is an isomorphism of Aut(G) onto $\mathrm{Aut}(\mathfrak{g}_e)$. Let $u_0 \in \mathrm{Aut(G)}$, and let U be a symmetric open neighborhood of e such that the function $\log_G$ is defined on an open set W containing $\bar{U}$ and $\overline{u_0(U)}$. Then there exists a neighborhood V_* of $(u_0)_*$ in $\mathrm{Aut}(\mathfrak{g}_e)$ such that, whenever $u_* \in V_*$ and $z \in U$, we have $\exp(u_*(\log z)) \in W$. If V is the inverse image of V_* in Aut(G), this shows (19.8.9) that $u(z) \in W$ whenever $u \in V$ and $z \in U$. Bearing in mind the definition of the Lie group structure of Aut(G), this proves that $(u, z) \mapsto u(z)$ is of class C^∞ on V × U. Now let x_0 be any point of G; then there exists a finite sequence $(a_j)_{1 \leq j \leq n}$ of points of U such that $x_0 = a_1 a_2 \cdots a_n$ (12.8.8). For each $x \in x_0 U$ we may write $u(x) = u(a_1)u(a_2) \cdots u(a_n)u(x_0^{-1}x)$. Now, each of the mappings $u \mapsto u(a_j)$ is of class C^∞ on V, and the mapping $(u, x) \mapsto u(x_0^{-1}x)$ is of class C^∞ on V × U. Hence $(u, x) \mapsto u(x)$ is of class C^∞ on Aut(G) × G.

If we now drop the assumption that G is simply-connected, so that $G = \tilde{G}/D$, then (19.13.3) Aut(G) may be identified with the subgroup of $\mathrm{Aut}(\tilde{G})$ consisting of the automorphisms u of $\tilde{G}$ such that $u(D) \subset D$. For each $z \in D$, the set F_z of automorphisms u of $\tilde{G}$ such that $u(z) \in D$ is closed, because D is closed and $u \mapsto u(z)$ is continuous (3.11.4). Since Aut(G) is the intersection of the sets F_z $(z \in D)$, this proves (i). The mapping $u \mapsto u_*$ is therefore an isomorphism of the Lie subgroup Aut(G) of $\mathrm{Aut}(\tilde{G})$ onto a Lie subgroup of $\mathrm{Aut}(\mathfrak{g}_e)$. The argument of the previous paragraph can now be used without any changes to prove (ii) in the general case.

Example

(19.13.6) Let $G = \mathbf{T}^n$, so that $\tilde{G} = \mathbf{R}^n$ and $D = \mathbf{Z}^n$. The Lie algebra $\mathfrak{g}_e$ is commutative, and therefore $\mathrm{Aut}(\tilde{G}) = \mathrm{Aut}(\mathfrak{g}_e)$ is the general linear group $\mathbf{GL}(n, \mathbf{R})$. Now an automorphism of the vector space $\mathbf{R}^n$ maps $\mathbf{Z}^n$ into itself if and only if its matrix, relative to the canonical basis, is a matrix of *integers*; it follows therefore that Aut(G) is the *discrete* subgroup $\mathbf{GL}(n, \mathbf{Z})$ of $\mathbf{GL}(n,\mathbf{R}) =$ Aut($\tilde{G}$).

(19.13.7) With the same notation, let us now determine the *Lie algebra* of the Lie group Aut($\mathfrak{g}_e$). Since this group is a closed subgroup of the linear group $\mathbf{GL}(\mathfrak{g}_e)$, its Lie algebra $\mathfrak{a}$ may be characterized as the set of endomorphisms U of the vector space $\mathfrak{g}_e$ such that $\exp(tU) \in \mathrm{Aut}(\mathfrak{g}_e)$ for all $t \in \mathbf{R}$ (19.8.10), i.e., such that

(19.13.7.1) $$\exp(tU) \cdot [\mathbf{x}, \mathbf{y}] = [\exp(tU) \cdot \mathbf{x}, \exp(tU) \cdot \mathbf{y}]$$

for all $t \in \mathbf{R}$ and all $\mathbf{x}, \mathbf{y} \in \mathfrak{g}_e$.

(19.13.8) *The Lie algebra* $\mathfrak{a}$ *of the Lie group* Aut($\mathfrak{g}_e$) *is the Lie algebra* Der($\mathfrak{g}_e$) *of derivations of* $\mathfrak{g}_e$.

Since the two sides of (19.13.7.1) are equal at $t = 0$, it is sufficient to express that their derivatives are equal. Since the derivative of $t \mapsto \exp(tU)$ is $U \exp(tU) = \exp(tU)\, U$, we obtain the equation

(19.13.8.1)

$$\exp(tU) \cdot (U \cdot [\mathbf{x}, \mathbf{y}]) = [U \cdot (\exp(tU) \cdot \mathbf{x}), \exp(tU) \cdot \mathbf{y}] + [\exp(tU) \cdot \mathbf{x}, U \cdot (\exp(tU) \cdot \mathbf{y})],$$

which for $t = 0$ reduces to

(19.13.8.2) $$U \cdot [\mathbf{x}, \mathbf{y}] = [U \cdot \mathbf{x}, \mathbf{y}] + [\mathbf{x}, U \cdot \mathbf{y}].$$

This shows that U must be a derivation of $\mathfrak{g}_e$. Conversely, if this is the case, then the derivative of the right-hand side of (19.13.7.1) is equal to

$$U \cdot [\exp(tU) \cdot \mathbf{x}, \exp(tU) \cdot \mathbf{y}].$$

If $\mathbf{v}(t)$ denotes the difference between the two sides of (19.13.7.1), we have therefore $\mathbf{v}'(t) = U \cdot \mathbf{v}(t)$; and since $\mathbf{v}(0) = \mathbf{0}_e$, it follows from (10.8.4) that $\mathbf{v}(t) = \mathbf{0}_e$ for all t.

(19.13.9) Let G be a *connected* Lie group, $\mathfrak{g}_e$ its Lie algebra. Recall that for each $s \in G$, the inner automorphism $t \mapsto sts^{-1}$ of G is denoted by Int(s). Clearly $s \mapsto \mathrm{Int}(s)$ is a homomorphism (of abstract groups) of G onto a subgroup of Aut(G). This subgroup is denoted by Int(G) and is (algebraically) isomorphic to G/C, where C is the center of G. If now we endow Aut(G) with its Lie group structure (19.13.5), then the homomorphism $s \mapsto \text{Int}(s)$ is a *Lie group* homomorphism of G into Aut(G). For by composing this homomorphism with the isomorphism $u \mapsto u_*$ of Aut(G) onto a Lie subgroup of $\text{Aut}(\mathfrak{g}_e)$, we obtain the homomorphism $s \mapsto \text{Ad}(s)$, which is of class C^∞ (19.2.1). Hence (19.7.5) Int(G) is a connected Lie group immersed in Aut(G), and normal in Aut(G) (because $u \circ \text{Int}(s) \circ u^{-1} = \text{Int}(u(s))$ for any $u \in \text{Aut}(G)$). Identifying Aut(G) with a Lie subgroup of $\text{Aut}(\mathfrak{g}_e)$ via $u \mapsto u_*$, the group Int(G) is identified with the connected Lie group Ad(G) immersed in $\text{Aut}(\mathfrak{g}_e)$. As a Lie group, it is isomorphic to G/C. The subalgebra Lie(Int(G)) of the Lie algebra $\text{Der}(\mathfrak{g}_e)$ of $\text{Aut}(\mathfrak{g}_e)$ is, by virtue of (19.11.2), the image $\text{ad}(\mathfrak{g}_e)$ of $\mathfrak{g}_e$ under its adjoint representation, and is isomorphic to the quotient $\mathfrak{g}_e/\mathfrak{c}_e$ of $\mathfrak{g}_e$ by its center.

PROBLEMS

1. Let A be a connected commutative Lie group (written additively) which contains an element $a \neq 0$ of order 2 (for example, a torus $\mathbf{T}^n$, where $n \geqq 1$) Show that the manifold $G = A \times \{0, 1\}$ (which has two connected components) becomes a solvable Lie group if the multiplication is defined by

$$\begin{aligned}(x, 0)(y, 0) &= (x + y, 0),\\ (x, 0)(y, 1) &= (x + y, 1),\\ (x, 1)(y, 0) &= (x - y, 1),\\ (x, 1)(y, 1) &= (x - y + a, 0).\end{aligned}$$

If there exists an automorphism σ of A such that $\sigma(a) \neq a$ (which will be the case when $A = \mathbf{T}^n$, $n \geqq 2$), show that σ cannot be extended to an automorphism of G (the group A being identified with the identity component G_0 of G).

2. Let G be a connected Lie group, T a normal subgroup of G. If T is isomorphic to a torus, show that T is contained in the center of G. (Observe that the group Aut(T) is discrete, and consider the homomorphism $s \mapsto \text{Int}(s) \mid T$ of G into Aut(T).)

3. Give an example of a connected Lie group G such that Int(G) is not closed in Aut(G). (See Section 19.11, Problem 6.)

4. Let G be a connected Lie group. For each compact subset K of G and each neighborhood V of e in G, let W(K, V) denote the set of all $u \in \text{Aut}(G)$ such that $u(x)x^{-1} \in V$ for all $x \in K$. Show that the W(K, V) form a fundamental system of neighborhoods of the identity automorphism 1_G in Aut(G). (Use (12.8.8).)

5. Let G be a Lie group, K a one-parameter subgroup of G which is closed and *normal* in G. Let a be an element of G such that $K \not\subset \mathscr{Z}(a)$

(a) If $K \cong \mathbf{R}$, then $\mathscr{Z}(a) \cap K = \{e\}$. If $K \cong \mathbf{T}$, then $\mathscr{Z}(a) \cap K$ has two elements. (Consider the restriction of $\operatorname{Int}(a)$ to K and use Section **19.7**, Problem 6.) The second possibility is excluded if G is connected (Problem 2).
(b) If in addition G/K is commutative, show that $G = \mathscr{Z}(a)K$ and that, for each closed subgroup A of $\mathscr{Z}(a)$, the group AK is closed in G. (Observe that the mapping $x \mapsto x^{-1}a^{-1}xa$ of K into K is surjective.)

6. Let G be a Lie group, K a one-parameter subgroup of G which is *normal* in G (but not necessarily closed in G). Let A be a closed subgroup of G. Show that, if AK is not closed in G, then K is contained in the identity component of the center of $\overline{AK}$. (Reduce to the case where $G = \overline{AK}$ and G is connected. Then $B = \mathscr{Z}(K) \cap A$ is a closed normal subgroup of G. Replacing A, K by their images in G/B, we reduce to the situation $B = \{e\}$. Using Section **19.12**, Problem 7, show that A is then commutative. If K were closed in G, then AK would also be closed in G, by virtue of Problem 5. Hence K is *not* closed in G; now use Problem 2, and Section **19.10**, Problem 1.)
Give an example of two one-parameter subgroups A, K of the commutative group $\mathbf{R} \times \mathbf{T}^2$ such that A is closed and the product AK is not closed.

14. SEMIDIRECT PRODUCTS OF LIE GROUPS

(19.14.1) Let G be a group (not necessarily topologized), N a normal subgroup of G. Then for any subgroup L of G, we have $LN = NL$ (recall that if A, B are subsets of G, the notation AB denotes the set of all products xy, where $x \in A$ and $y \in B$). G is said to be the *semidirect product* of N and L if every $z \in G$ is *uniquely* expressible as $z = xy$ with $x \in N$ and $y \in L$; or, equivalently, if $G = NL$ and $N \cap L = \{e\}$. (For if $N \cap L \neq \{e\}$, an element $z \neq e$ in $N \cap L$ can be written as $z = ez = ze$; hence in two ways as a product of an element of N and an element of L. Conversely, if $x'y' = xy$, where $x', x \in N$ and $y', y \in L$, then we have $x^{-1}x' = yy'^{-1} \in N \cap L$, and hence if $N \cap L = \{e\}$ we must have $x' = x$ and $y' = y$.)

If $\pi : G \to G/N$ is the canonical homomorphism, the restriction of π to L is an *isomorphism* of L onto G/N; for the relation $G = LN$ shows that $\pi | L$ is surjective, and the relation $N \cap L = \{e\}$ shows that it is injective. It should be remarked that, for a given normal subgroup N of a group G, there need not exist a subgroup L of G such that G is the semidirect product of N and L (Problem 1).

(19.14.2) Suppose that G is the semidirect product of N and L. For each $y \in L$, the mapping $\sigma_y : x \mapsto yxy^{-1}$ is an *automorphism* of the group N. Moreover, for any two elements u, v of L we have $\sigma_{uv} = \sigma_u \circ \sigma_v$, so that $y \mapsto \sigma_y$ is a *homomorphism* of L into the group $\mathscr{A}(N)$ of automorphisms of N. Furthermore, this homomorphism and the laws of composition in N and L determine the law of composition in G, because for x, x' in N and y, y' in L we have

(19.14.2.1) $$(xy)(x'y') = (xyx'y^{-1})(yy') = (x\sigma_y(x'))(yy').$$

(19.14.3) Conversely, let N and L be *any* two groups and $y \mapsto \sigma_y$ a homomorphism of L into $\mathscr{A}(\text{N})$. Then we may define a group structure on the set $\text{S} = \text{N} \times \text{L}$ by the rule

(19.14.3.1) $$(x, y)(x', y') = (x\sigma_y(x'), yy').$$

Associativity follows from the formulas

$$((x, y)(x', y'))(x'', y'') = (x\sigma_y(x')\sigma_{yy'}(x''), yy'y''),$$
$$(x, y)((x', y')(x'', y'')) = (x\sigma_y(x'\sigma_{y'}(x'')), yy'y''),$$

and the relation $\sigma_y \circ \sigma_{y'} = \sigma_{yy'}$. If e' (resp. e'') is the neutral element of N (resp. L), it is clear that (e', e'') is the neutral element of S. Finally, we have

$$(x, y)(\sigma_{y^{-1}}(x^{-1}), y^{-1}) = (\sigma_{y^{-1}}(x^{-1}), y^{-1})(x, y) = (e', e''),$$

which completes the proof of our assertion. If N′ (resp. L′) is the set of all (x, e'') for $x \in \text{N}$ (resp. (e', y) for $y \in \text{L}$), it is immediately seen that the group S is the semidirect product of N′ and L′, which are isomorphic respectively to N and L. We shall sometimes use the notation $\text{S} = \text{N} \times_\sigma \text{L}$ (the direct product of N and L corresponds to the trivial homomorphism $\sigma : y \mapsto 1_\text{N}$ of L into $\mathscr{A}(\text{N})$).

If now G is the semidirect product of subgroups N and L, and if $y \mapsto \sigma_y$ is the corresponding homomorphism of L into $\mathscr{A}(\text{N})$, then it is clear that the mapping $(x, y) \mapsto xy$ of $\text{S} = \text{N} \times_\sigma \text{L}$ into G is an *isomorphism* of S onto G, by virtue of (19.14.2.1) and (19.14.3.1).

(19.14.4) Suppose now that G is a *connected Lie group*, N a *connected closed normal subgroup* of G and L a *connected Lie group immersed* in G, and suppose that G is the semidirect product of N and L. The restriction to L of the canonical homomorphism $\pi : \text{G} \to \text{G/N}$ is continuous for the topology induced on L by G, and *a fortiori* for the proper topology of L (since the latter is finer than the induced topology). It is therefore a Lie group homomorphism (19.10.2), and since it is bijective, it is an *isomorphism of Lie groups* (16.9.9). For each $y \in \text{L}$, it is clear that $\sigma_y : x \mapsto yxy^{-1}$ is a Lie group automorphism of N, i.e., is an element of $\text{Aut(N)} \subset \mathscr{A}(\text{N})$. Furthermore, $y \mapsto \sigma_y$ is a *Lie group homomorphism* of L into Aut(N). For, by virtue of the definition of the Lie group structure of Aut(N) (19.13.5), if $\mathfrak{n}_e$ is the Lie algebra of N, the tangent linear mapping $\text{T}_e(\sigma_y)$ is the restriction to $\mathfrak{n}_e$ of the automorphism

$\mathrm{Ad}(y)$ of $\mathfrak{g}_e$; and since $y \mapsto \mathrm{Ad}(y)$ is a Lie group homomorphism of G into $\mathrm{Aut}(\mathfrak{g}_e)$, it follows that $y \mapsto \mathrm{Ad}(y) | \mathfrak{n}_e$ is a Lie group homomorphism of L into $\mathrm{Aut}(\mathfrak{n}_e)$, which proves the assertion.

(19.14.5) Conversely, let N and L be two connected Lie groups and let $y \mapsto \sigma_y$ be a Lie group homomorphism of L into Aut(N). It follows from (19.13.5) that the mapping $(x, y) \mapsto \sigma_y(x)$ is of class C^∞. The group structure defined on the product $N \times L$ by (19.14.3.1) is then compatible with the product manifold structure on $N \times L$, and we denote by $N \times_\sigma L$ the *Lie group* so defined.

(19.14.6) *Every connected Lie group* G *which is the semidirect product of a connected closed normal subgroup* N *and a connected Lie group* L *immersed in* G, *is isomorphic (as a Lie group) to a Lie group of the form* $N \times_\sigma L$, *where* $y \mapsto \sigma_y$ *is a Lie group homomorphism of* L *into* Aut(N). *In particular,* L *is necessarily closed in* G.

For if σ_y is defined as in (19.14.4), we may construct the Lie group $N \times_\sigma L$ defined in (19.14.5) (L being taken with its Lie group structure). It is then clear that the mapping $(x, y) \mapsto xy$ of $N \times_\sigma L$ into G is a bijective Lie group homomorphism, hence is an isomorphism (16.9.9).

(19.14.7) In the Lie algebra $\mathfrak{g}_e$ of the semidirect product $G = N \times_\sigma L$ defined in (19.14.5), the Lie algebra $\mathfrak{n}_e$ of N is an ideal, and the Lie algebra $\mathfrak{l}_e$ of L is a subalgebra which (as vector space) is a *supplement* of $\mathfrak{n}_e$. We have seen (19.14.4) that, for each $y \in L$, the mapping $T_e(\sigma_y) = (\sigma_y)_*$ is an automorphism of $\mathfrak{n}_e$, i.e., is an element of $\mathrm{Aut}(\mathfrak{n}_e)$, and the mapping $y \mapsto (\sigma_y)_*$ is the same as $y \mapsto \mathrm{Ad}(y) | \mathfrak{n}_e$. Its tangent linear mapping is therefore a Lie algebra *homomorphism* $\varphi : \mathfrak{l}_e \to \mathrm{Der}(\mathfrak{n}_e)$, which is the restriction to $\mathfrak{l}_e$ of the homomorphism $\mathbf{v} \mapsto \mathrm{ad}(\mathbf{v}) | \mathfrak{n}_e$. We remark that knowledge of the homomorphism φ and of the Lie algebra structures of $\mathfrak{n}_e$ and $\mathfrak{l}_e$ completely determines the Lie algebra structure of $\mathfrak{g}_e$; for each $\mathbf{w} \in \mathfrak{g}_e$ can be uniquely written as $\mathbf{u} + \mathbf{v}$ with $\mathbf{u} \in \mathfrak{n}_e$ and $\mathbf{v} \in \mathfrak{l}_e$, and for two such elements $\mathbf{w} = \mathbf{u} + \mathbf{v}$, $\mathbf{w}' = \mathbf{u}' + \mathbf{v}'$, we have

(19.14.7.1)
$$\begin{aligned}[\mathbf{u} + \mathbf{v}, \mathbf{u}' + \mathbf{v}'] &= [\mathbf{u}, \mathbf{u}'] + [\mathbf{v}, \mathbf{u}'] + [\mathbf{u}, \mathbf{v}'] + [\mathbf{v}, \mathbf{v}'] \\ &= [\mathbf{u}, \mathbf{u}'] + \varphi(\mathbf{v}) \cdot \mathbf{u}' - \varphi(\mathbf{v}') \cdot \mathbf{u} + [\mathbf{v}, \mathbf{v}'].\end{aligned}$$

The algebra $\mathfrak{g}_e$ is called the *semidirect product* of $\mathfrak{g}_e$ and $\mathfrak{l}_e$ corresponding to φ, and is denoted by $\mathfrak{n}_e \times_\varphi \mathfrak{l}_e$.

(19.14.8) The construction of semidirect products will allow us to prove in Chapter XXI that every finite-dimensional Lie algebra over **R** is the Lie algebra of a Lie group, by virtue of the following result:

(19.14.9) *Let* $\mathfrak{g}$ *be a finite-dimensional Lie algebra over* **R**, *let* $\mathfrak{n}$ *be an ideal in* $\mathfrak{g}$, *and let* $\mathfrak{l}$ *be a subalgebra of* $\mathfrak{g}$ *supplementary to* $\mathfrak{n}$. *Suppose that there exists a simply-connected Lie group* N (resp. L) *such that* Lie(N) (resp. Lie(L)) *is isomorphic to* $\mathfrak{n}$ (resp. $\mathfrak{l}$). *Then there exists a simply-connected Lie group* G *whose Lie algebra is isomorphic to* $\mathfrak{g}$ *and such that the manifold underlying* G *is diffeomorphic to* N × L.

For each $\mathbf{v} \in \mathfrak{l}$, the restriction of $\operatorname{ad}(\mathbf{v})$ to $\mathfrak{n}$ is a derivation $\varphi(\mathbf{v})$ of $\mathfrak{n}$, and the mapping $\mathbf{v} \mapsto \varphi(\mathbf{v})$ is a Lie algebra homomorphism of $\mathfrak{l}$ into $\operatorname{Der}(\mathfrak{n})$. Since L is simply-connected, there exists a unique Lie group homomorphism $\psi : \mathrm{L} \to \operatorname{Aut}(\mathfrak{n})$ such that $\psi_* = \varphi$ ((19.7.6) and (19.13.8)). Furthermore, because N is simply-connected, the mapping $f \mapsto f_*$ of Aut(N) into $\operatorname{Aut}(\mathfrak{n})$ is an isomorphism of Lie groups (19.13.3); hence there exists a Lie group homomorphism $y \mapsto \sigma_y$ of L into Aut(N) such that $(\sigma_y)_* = \psi(y)$ for all $y \in \mathrm{L}$. If we now consider the semidirect product $\mathrm{G} = \mathrm{N} \times_\sigma \mathrm{L}$, it follows from (19.14.7) and the definition of $y \mapsto \sigma_y$ that the Lie algebra $\mathfrak{g}_e$ of G is isomorphic to $\mathfrak{g}$. In view of (16.27.10), the proof is complete.

This result gives in particular a partial answer to the problem raised in (19.14.8):

(19.14.10) *Every solvable Lie algebra of dimension n over* **R** *is isomorphic to the Lie algebra of a simply-connected solvable Lie group, which is diffeomorphic to* $\mathbf{R}^n$.

The proof is by induction on n: the result is trivial for $n = 1$, by (19.4.1). If $\mathfrak{g}$ is a solvable Lie algebra of dimension $n > 1$, then by definition the derived algebra $\mathfrak{g}' = [\mathfrak{g}, \mathfrak{g}]$ is not equal to $\mathfrak{g}$, hence is of dimension $<n$. Since $\mathfrak{g}/\mathfrak{g}'$ is commutative, every vector subspace of $\mathfrak{g}/\mathfrak{g}'$ is an *ideal* of $\mathfrak{g}/\mathfrak{g}'$, hence every vector subspace $\mathfrak{h} \supset \mathfrak{g}'$ in $\mathfrak{g}$ is an *ideal* in $\mathfrak{g}$. Choose such an ideal $\mathfrak{n}$ of dimension $n - 1$, and let $\mathfrak{l}$ be a vector subspace of $\mathfrak{g}$ supplementary to $\mathfrak{n}$. Every 1-dimensional subspace of $\mathfrak{g}$ is automatically a Lie subalgebra of $\mathfrak{g}$; hence we are in the situation of (19.14.9): for $\mathfrak{n}$ and $\mathfrak{l}$ are Lie subalgebras of $\mathfrak{g}$, hence are solvable and of dimension $<n$. The result therefore follows from (19.14.9) and the inductive hypothesis.

PROBLEMS

1. Show that in Section 19.13, Problem 1, the group G is not a semidirect product of the normal subgroup A and a group of two elements.

2. Show that there exists a 4-dimensional nilpotent Lie algebra $\mathfrak{g}$ over $\mathbf{R}$ with basis $(\mathbf{u}_1, \mathbf{u}_2, \mathbf{u}_3, \mathbf{u}_4)$ and multiplication table

$$[\mathbf{u}_1, \mathbf{u}_2] = \mathbf{u}_3, \qquad [\mathbf{u}_1, \mathbf{u}_4] = \mathbf{u}_3, \qquad [\mathbf{u}_i, \mathbf{u}_j] = 0 \quad \text{otherwise} \qquad (i < j).$$

Show that $\mathbf{u}_3$ and $\mathbf{u}_4$ generate a commutative ideal $\mathfrak{a}$ in $\mathfrak{g}$, and that there exists no Lie subalgebra of $\mathfrak{g}$ supplementary to $\mathfrak{a}$.

3. (a) Let G be a Lie group, N a normal Lie subgroup of G, and suppose that the principal bundle G over $B = G/N$, with projection π, is *trivializable*. Let $s : B \to G$ be a C^∞ section of this bundle. Put

$$F(u, v) = s(uv)^{-1}s(u)s(v) \qquad (u, v \in B).$$

Then F is a C^∞ mapping of $B \times B$ into N. Show that

$$\text{(1)} \qquad F(u, vw)F(v, w) = F(uv, w)(s(w)^{-1}F(u, v)s(w)).$$

(b) Suppose that N is *commutative*. Then we may define a differentiable action of B on N by the rule

$$u \cdot t = s(u)ts(u)^{-1}$$

for $u \in B$ and $t \in N$: the right-hand side is *independent* of the choice of the section s. If the group law in N is written additively, the equation (1) takes the form

$$\text{(2)} \qquad F(u, vw) + F(v, w) = F(uv, w) + w^{-1} \cdot F(u, v).$$

Every C^∞ section of G over B is of the form $u \mapsto s(u)f(u)$, where f is any C^∞ mapping of B into N. To this section there corresponds the function

$$\text{(3)} \qquad F(u, v) + v^{-1} \cdot f(u) + f(v) - f(uv).$$

In order that G should be a semidirect product of N and a topological subgroup L of G such that the mapping $(x, y) \mapsto xy$ of $N \times L$ into G is a homeomorphism, it is necessary and sufficient that the expression (3) should be identically zero for a suitable choice of f.

(c) Suppose that N is isomorphic to $\mathbf{R}^n$ (which by (16.12.12) implies that the principal bundle G over B is trivializable), and that $B = G/N$ is *compact*. Show that G is then a semi-direct product of N and a compact subgroup L. (Integrate both sides of the equation (2) with respect to u.) (Compare with Problems 1 and 2.)

4. (a) Let G be a simply-connected Lie group of dimension n in which there exists a closed normal subgroup $A \cong \mathbf{R}^{n-1}$. Then G is a semidirect product $G = A \times_\sigma \mathbf{R}$, where $\sigma : t \mapsto \exp(tU)$ is a homomorphism of $\mathbf{R}$ into $\mathrm{Aut}(A) = \mathbf{GL}(n-1, \mathbf{R})$. Assume that G is not commutative, i.e., that $U \in \mathrm{End}(\mathbf{R}^{n-1})$ is not zero. In order that two such groups G, G_1 should be isomorphic, it is necessary and sufficient that the corresponding endomorphisms should be similar, up to a nonzero scalar factor. The automorphism group Aut(G) may be identified with the subgroup of $\mathbf{GL}(n, \mathbf{R})$ consisting of the automorphisms V of $\mathbf{R}^n$ which are such that $V \cdot \mathbf{e}_n - \mathbf{e}_n \in \mathbf{R}^{n-1}$, which stabilize $\mathbf{R}^{n-1}$, and whose restriction to $\mathbf{R}^{n-1}$ commutes with U.

(b) The subgroup A is the largest connected commutative normal subgroup of G. The connected Lie subgroups of G contained in A are the U-stable vector subspaces of A. Under what conditions do there exist connected normal subgroups of G of dimension $n - 1$, other than A? Under what conditions is the Lie algebra of G nilpotent?

(c) For an element $(x, y) \in \mathrm{A} \times_\sigma \mathbf{R}$ to lie in the center of G, it is necessary and sufficient that $\sigma_y = 1_{\mathrm{A}}$ and $U \cdot x = 0$. For such an element to exist with $y \neq 0$ it it necessary and sufficient that the matrix of U, relative to some basis of $\mathbf{R}^{n-1}$, should be conjugate in $\mathbf{GL}(n - 1, \mathbf{C})$ to a matrix of the form

$$\alpha \operatorname{diag}(in_1, -in_1, in_2, -in_2, \ldots, in_p, -in_p, 0, \ldots, 0),$$

where the n_j are integers $\neq 0$ and α is a nonzero real number. In this case, if H is the kernel of U, the center of G is isomorphic to $\mathrm{H} \times_\sigma \mathbf{Z}$.

(d) Show that the exponential mapping of G is the mapping

$$(\mathbf{u}, \xi) \mapsto \left(\frac{I - e^{-U}}{U} \cdot \mathbf{u}, \xi\right)$$

of $\mathbf{R}^{n-1} \times \mathbf{R}$ into itself, where by abuse of notation $(I - e^{-U})/U$ denotes the element of the Banach algebra $\mathrm{End}(\mathbf{R}^{n-1})$ obtained by substituting U for z in the series $(1 - e^{-z})/z$. Under what condition is this mapping bijective? Show that $\exp_{\mathrm{G}}$ is neither injective nor surjective if the center of G contains an element $(0, y)$ with $y \neq 0$.

(e) Let Γ be a *compact* group of automorphisms of G, and let Γ_* be the group of automorphisms of $\mathfrak{g}_e$ induced by the $\gamma \in \Gamma$. Show that there exists a Γ_*-invariant vector $\mathbf{u} \in \mathfrak{g}_e$ not contained in $\mathbf{R}^{n-1}$. (Observe that Γ_* leaves $\mathbf{R}^{n-1}$ globally invariant, hence also a supplementary line, by complete reducibility.)

5. Let G be a simply-connected solvable Lie group of dimension n.

(a) Deduce from the proof of (**19.14.10**) that there exists a basis $(\mathbf{u}_1, \mathbf{u}_2, \ldots, \mathbf{u}_n)$ of the Lie algebra $\mathfrak{g}_e$ such that the mapping

$$(1) \qquad (t_1, t_2, \ldots, t_n) \mapsto \exp(t_1 \mathbf{u}_1) \exp(t_2 \mathbf{u}_2) \cdots \exp(t_n \mathbf{u}_n)$$

is a *diffeomorphism* of $\mathbf{R}^n$ *onto* G.

(b) Let $\mathfrak{h}_e$ be a Lie subalgebra of $\mathfrak{g}_e$, of dimension m, and let H be the connected Lie group immersed in G corresponding to $\mathfrak{h}_e$. Show that there exists a basis $(\mathbf{u}_1, \ldots, \mathbf{u}_n)$ of $\mathfrak{g}_e$ with the property of part (a), and which contains a basis of $\mathfrak{h}_e$. (Use property (iv) of Section **19.12**, Problem 2, and define the $\mathbf{u}_j$ inductively by considering the intersections $\mathfrak{h}_e \cap \mathfrak{m}_{n-j}$.) Deduce that H is *simply-connected* and *closed* in G, and that the homogeneous space G/H is *simply-connected.* (Cf. Section **16.30**, Problem 11.)

(c) Show that the only compact subgroup of G is the identity subgroup. (Use induction on the length of the sequence of derived groups of G, together with part (b).)

6. (a) A *connected* Lie group is said to be *nilpotent* if its Lie algebra is nilpotent (Section **19.12**, Problem 3; cf. Section **19.16**, Problem 6). Show that if G is nilpotent and simply connected, the exponential mapping $\exp_{\mathrm{G}}$ is a *diffeomorphism* of $\mathfrak{g}_e$ onto G. (Argue by induction on the dimension n of G. Observe that the identity component C_0 of the center of G has dimension $\geqq 1$ (Section **19.12**, Problem 3); then use Problem 5(b) and Section **19.9**, Problem 7.)

It follows that for a connected nilpotent Lie group, the exponential mapping is *surjective*.

(b) Let G be a connected nilpotent Lie group. For each $\mathbf{u} \in \mathfrak{g}_e$, the mapping

$$\operatorname{Ad}(\exp(\mathbf{u})) = \exp(\operatorname{ad}(\mathbf{u}))$$

is a *polynomial* in $\operatorname{ad}(\mathbf{u})$ (Section **19.12**, Problem 3), and we have

$$\operatorname{ad}(\mathbf{u}) = \log(\exp(\operatorname{ad}(\mathbf{u}))),$$

where

$$\log(I + N) = \sum_{n=1}^{\infty} (-1)^{n-1} N^n/n$$

for any nilpotent endomorphism N of $\mathfrak{g}_e$.

(c) For the nilpotent group G of Section **19.5**, Problem 1, show that the mapping

$$(\xi, \eta, \zeta) \mapsto \exp(\xi\mathbf{u}) \exp(\eta\mathbf{v}) \exp(\zeta(\mathbf{u} + \mathbf{w}))$$

of $\mathbf{R}^3$ into G is neither injective nor surjective. (Compare with Problem 5(a).)

7. (a) Let G be a nilpotent connected Lie group and H a connected Lie group immersed in G. Show that the normalizer $\mathscr{N}(\mathrm{H})$ of H in G is connected. (By Problem 6(a), any element of $\mathscr{N}(\mathrm{H})$ can be expressed as $\exp(\mathbf{u})$ for some $\mathbf{u} \in \mathfrak{g}_e$. Use Problem 6(b) to show that $\mathbf{u} \in \mathfrak{N}(\mathfrak{h}_e)$.)

(b) Show that the center of a nilpotent connected Lie group is connected. (Same method as in (a).)

(c) Let G be a nilpotent connected Lie group. Show that for every proper subgroup L of G, the normalizer $\mathscr{N}(\mathrm{L})$ of L in G is not equal to L. (Induction on the dimension of G: distinguish two cases, according as L does or does not contain the center C of G.) In particular, if H is a connected Lie group immersed in G and of codimension 1 in G, then H is normal in G.

8. (a) Let G be a simply-connected nilpotent Lie group and H a connected Lie subgroup of G. Show that there exists a basis $(\mathbf{u}_j)_{1 \leq j \leq n}$ of the Lie algebra $\mathfrak{g}_e$ of G such that (i) if $\mathfrak{m}_j$ is the subspace of $\mathfrak{g}_e$ spanned by the $\mathbf{u}_k$ such that $k \geqq j$, then $\mathfrak{m}$ is a Lie subalgebra of $\mathfrak{g}_e$, and $\mathfrak{m}_{j+1}$ is an ideal in $\mathfrak{m}_j$; (ii) there exists an integer $p \in [1, n]$ such that $\mathfrak{m}_p$ is the Lie algebra $\mathfrak{h}_e$ of H. (Form the successive normalizers $\mathscr{N}(\mathrm{H})$, $\mathscr{N}(\mathscr{N}(\mathrm{H})), \dots,$ and use Problems 7(a) and 7(c).) The mapping

$$(t_1, \dots, t_{p-1}, y) \mapsto \exp(t_1\mathbf{u}_1) \cdots \exp(t_{p-1}\mathbf{u}_{p-1})y$$

is then a diffeomorphism of $\mathbf{R}^{p-1} \times \mathrm{H}$ onto G. Deduce that, for each closed subgroup L of H, the space G/L is diffeomorphic to $\mathbf{R}^{p-1} \times (\mathrm{H/L})$.

(b) With the hypotheses of (a), let D be a *discrete* (hence closed (**12.8.7**)) subgroup of G. Show that there exists a basis $(\mathbf{u}_j)_{1 \leq j \leq n}$ of $\mathfrak{g}_e$ satisfying condition (i) of (a) above, and also: there exists an integer $p \in [1, n]$ such that $a_k = \exp(\mathbf{u}_k) \in \mathrm{D}$ for $p \leqq k \leqq n$, and every element of D is uniquely of the form $a_p^{\nu_p} a_{p+1}^{\nu_{p+1}} \cdots a_n^{\nu_n}$, where $\nu_j \in \mathbf{Z} (p \leqq j \leqq n)$. (Induction on $n = \dim(\mathrm{G})$. Let A be the identity component of the normalizer $\mathscr{N}(\mathrm{D})$ of D in G; then $\dim(\mathrm{A}) \geqq 1$ (Problem 7(b)). Reduce to the situation where $\mathrm{A} \neq \mathrm{G}$ (cf. (**16.30.2.2**)). By considering the normalizer $\mathrm{H} = \mathscr{N}(\mathrm{A})$ of A in G and using Problem 7(a), and the fact that $\mathrm{D} \subset \mathrm{H}$, reduce (with the help of (a)) to the case $\mathrm{H} = \mathrm{G}$, i.e., to the case where A is normal in G. Then G/A is simply-connected and $DA/A \subset \mathcal{N}(D)/A$ is a discrete subgroup of G/A, and the inductive hypothesis applies to G/A and DA/A. Lift to $\mathfrak{g}_e$ the basis of $\mathfrak{g}_e/\mathfrak{a}_e$ so obtained, and extend to a basis of $\mathfrak{g}_e$ by adjoining a suitably chosen basis of $\mathfrak{a}_e$, bearing in mind that $D \cap A$ is a discrete subgroup of the center of A.) Show that no element $\neq e$ in D has finite order.

(c) Let M_j denote the connected Lie subgroup of G whose Lie algebra is $\mathfrak{m}_j$ (Problem 5(b)). Show that the homogeneous space M_p/D is compact. (Show by descending induction that $M_{p+j}/(D \cap M_{p+j})$ is compact. For this purpose, let L be the one-parameter subgroup of G corresponding to the vector $\mathbf{u}_p$, and P the subgroup of $L \cap D$ generated by a_p. Let $\pi : M_{p+1} \to M_{p+1}/(D \cap M_{p+1})$ be the canonical mapping. Using the fact that M_{p+1} is normal in M_p, show that P acts differentiably on the left on $M_{p+1}/(D \cap M_{p+1})$ by the rule

$$a_p^k \cdot \pi(y) = \pi(a_p^k y a_p^{-k}),$$

where $y \in M_{p+1}$. Considering L as a principal bundle with structure group P, show that the manifold M_p/D may be identified with the associated bundle

$$L \times^{P} (M_{p+1}/(D \cap M_{p+1}))$$

(**16.14.7**), and finally remark that L/P is compact.)

9. Let G be a connected nilpotent Lie group. Show that every homogeneous space G/H is diffeomorphic to a product $\mathbf{R}^q \times (M/L)$, where L, M are Lie subgroups of G such that $L \subset M$, and M is connected, L discrete and M/L *compact*. (By considering the universal covering $\tilde{G}$ of G, which acts transitively on G/H, reduce to the case where G is simply-connected. Let H_0 be the identity component of H and let $N \supset H_0$ be the normalizer of H_0 in G. By using Problem 8(a) and remarking that N/H is diffeomorphic to $(N/H_0)/(H/H_0)$ and that N/H_0 is simply-connected, reduce to the case where H is discrete and use 8(a) and 8(c).)

10. Let G be a connected Lie group and D a discrete subgroup of G. Let U be a neighborhood of e in G on which the function $\log_G$ is defined and satisfies the inequality

$$\| \log(xyx^{-1}y^{-1}) \| \leqq \tfrac{1}{2} \inf(\| \log x \|, \| \log y \|)$$

for the Euclidean norm on $\mathfrak{g}_e$ (Section **19.9**, Problem 6). Let V be a symmetric neighborhood of e such that $V^2 \subset U$, and suppose that $V \cap D$ contains n points x_j $(1 \leqq j \leqq n)$ such that the vectors $\log x_j$ form a basis of the vector space $\mathfrak{g}_e$.

Show that there exists a point $a \neq e$ in $V \cap D$ which commutes with all elements of $V \cap D$ (choose a point $\neq e$ of minimal norm); if $a = \exp(\mathbf{u})$, deduce that $\mathbf{u}$ belongs to the center of G (cf. Section **19.9**, Problem 5).

11. Let G be a connected Lie group. Suppose that there exists an increasing sequence (D_m) of discrete subgroups of G such that the union of the D_m is dense in G. Show that G is nilpotent. (Using Problem 10, show that the center of G has dimension $\geqq 1$, and then proceed by induction on the dimension of G.)

12. Let G be a simply-connected solvable Lie group of dimension n, and let D be a discrete subgroup contained in the center of G. Show that there exists a basis $(\mathbf{u}_j)_{1 \leqq j \leqq n}$ of the Lie algebra $\mathfrak{g}_e$ of G with the following properties: (i) the mapping

$$(t_1, \ldots, t_n) \mapsto \exp(t_1 \mathbf{u}_1) \cdots \exp(t_n \mathbf{u}_n)$$

is a diffeomorphism of $\mathbf{R}^n$ onto G; (ii) for some integer $p \in [1, n]$ we have $[\mathbf{u}_i, \mathbf{u}_j] = 0$ for $1 \leqq i < j \leqq p$, and the elements $a_k = \exp(\mathbf{u}_k)$ $(1 \leqq k \leqq p)$ form a basis of D as a (multiplicative) **Z**-module. (Proceed by induction on the dimension $n \geqq 1$. Let $\mathfrak{a}_e$ be a commutative ideal of $\mathfrak{g}_e$ of maximum dimension q ($\geqq 1$), A the corresponding commutative subgroup of G, which is closed and isomorphic to $\mathbf{R}^q$ (Problem 5). The group G/A is solvable and simply-connected, of dimension $n-q$. Show that DA/A is a discrete subgroup of the center of G/A (observe that $\overline{\text{DA}}$ is a commutative normal Lie subgroup of G which contains A, hence that A is its identity component). By the inductive hypothesis, there exists a basis $(\overline{\mathbf{u}}_j)_{1 \leqq j \leqq n-q}$ of $\mathfrak{g}_e/\mathfrak{a}_e$ satisfying conditions (i) and (ii) for G/A and DA/A (the number p being replaced by some integer $r \in [1, n-q]$). Next, lift the $\overline{\mathbf{u}}_j$ to $\mathfrak{g}_e$ inductively, as follows: suppose that we have already lifted $\overline{\mathbf{u}}_j$ to $\mathbf{u}_j$ for $1 \leqq j \leqq s \leqq r$ in such a way that $\exp(\mathbf{u}_j) \in$ D and $[\mathbf{u}_i, \mathbf{u}_j] = 0$ for $1 \leqq i < j \leqq s$. Then consider the subalgebra $\mathfrak{h}_e$ of $\mathfrak{g}_e$ generated by $\mathfrak{a}_e$ and the coset $\overline{\mathbf{u}}_{s+1}$ of $\mathfrak{a}_e$ in $\mathfrak{g}_e$. The corresponding connected Lie subgroup H of G is of the type considered in Problem 4. Let L be the closed commutative subgroup of G consisting of all products

$$\exp(t_1 \mathbf{u}_1) \cdots \exp(t_s \mathbf{u}_s),$$

and Γ the subgroup of Aut(H) consisting of the restrictions to H of the inner automorphisms Int(x) for $x \in$ L. Show that Γ is compact, and apply the result of Problem 4(e) to this group in order to choose an element $\mathbf{u}_{s+1} \in \overline{\mathbf{u}}_{s+1}$ with the desired properties. Once $\mathbf{u}_1, \ldots, \mathbf{u}_r$ have been chosen, choose $\mathbf{u}_j$ arbitrarily in $\overline{\mathbf{u}}_j$ for

$$r+1 \leqq j \leqq n-q.$$

Complete the basis of $\mathfrak{g}_e$ by adjoining a basis $(\mathbf{u}_j)_{n-q+1 \leqq j \leqq n}$ of $\mathfrak{a}_e$ such that the $p-r$ elements $\exp(\mathbf{u}_j)$, where $n-q+1 \leqq j \leqq n-q+p-r$, form a basis of the **Z**-module D $\cap$ A. By observing that the restriction of Int(G) to A leaves invariant the points of D $\cap$ A, show that the $\mathbf{u}_j$ for which $n-q+1 \leqq j \leqq n-q+p-r$ belong to the center of $\mathfrak{g}_e$. Deduce that these elements may be interchanged with the $\mathbf{u}_j$ such that

$$r+1 \leqq j \leqq n-q$$

without disturbing property (i) of the basis $(\mathbf{u}_j)$.)

13. Let G be a solvable connected Lie group. Deduce from Problem 12 that there exists in G a subgroup T isomorphic to a torus and a closed submanifold E diffeomorphic to $\mathbf{R}^m$, such that the mapping $(x, y) \mapsto xy$ of T $\times$ E into G is a diffeomorphism of T $\times$ E onto G.

14. Generalize the result of Problem 3(c) as follows: if G is a Lie group, N a normal Lie subgroup of G which is *solvable* and *simply-connected*, and if the group B = G/N is compact, then G is a semidirect product of N and a compact subgroup. (Proceed by induction on the dimension of N, by considering a simply-connected commutative normal subgroup A of N of dimension $\geqq 1$, and using Problems 5(b) and 3(c).)

15. Let G be a solvable connected Lie group, H a connected Lie group immersed in G. Suppose that the closure in G of every one-parameter subgroup of H is contained in H. Then H is *closed* in G (*Malcev's theorem*). (We may assume that H is dense in G. Proceed by induction on the dimension of G:

(a) Let A be a connected commutative Lie group of dimension $\geqq 1$ immersed in H, and let B be the identity component of $\bar{\text{A}} \cap$ H (for the Lie group topology of H). Show that B satisfies the same condition as H, and deduce from the inductive hypothesis that H contains a subgroup N which is a connected Lie subgroup of G, commutative and normal in G, and of dimension $\geqq 1$.

(b) Prove that the subgroup H/N of G/N satisfies the same condition as H, and hence deduce from the inductive hypothesis that H/N = G/N, i.e., that H = G. For this purpose, reduce to the cases where N is either a torus or a vector group $\mathbf{R}^n$, and observe that every one-parameter subgroup of H/N is the image of a one-parameter subgroup of H. If $N \cong \mathbf{R}^n$, use Problem 3(c) together with Section 12.9, Problem 10.)

16. Let $I(n)$ be the group of *isometries* of Euclidean space $\mathbf{R}^n$ (relative to the usual scalar product $(\mathbf{x}|\mathbf{y}) = \sum_{i=1}^{n} \xi^i \eta^i$); it is isomorphic to the Lie subgroup of $\mathbf{GL}(n+1, \mathbf{R})$ consisting of the matrices

$$S = \begin{pmatrix} U & \mathbf{x} \\ 0 & 1 \end{pmatrix},$$

where $U \in \mathbf{O}(n)$ and $\mathbf{x}$ is an arbitrary vector (i.e., column matrix) in $\mathbf{R}^n$. We may identify $I(n)$ with the semidirect product $\mathbf{O}(n) \times_\sigma \mathbf{R}^n$, where $\sigma : \mathbf{O}(n) \to \mathbf{GL}(n, \mathbf{R})$ is the canonical injection: the matrix S is denoted by $(U, \mathbf{x})$. Let $p : I(n) \to \mathbf{O}(n)$ be the canonical projection, with kernel $\mathbf{R}^n$.

(a) Let Γ be a *discrete* subgroup of $I(n)$. Show that the identity component of the closure of $p(\Gamma)$ in $\mathbf{O}(n)$ is commutative. (Let V be a neighborhood of the identity matrix I in $\mathbf{O}(n)$ having the properties enunciated in Section 19.9, Problem 3(a), and such that $\|U - I\| \leq \frac{1}{4}$ for $U \in V$ (the norm (5.7.1) on $\mathrm{End}(\mathbf{R}^n)$ being that induced by the Euclidean norm $\|\mathbf{x}\| = (\mathbf{x}|\mathbf{x})^{1/2}$ on $\mathbf{R}^n$). Suppose that there exist two elements $S_1 = (U_1, \mathbf{x}_1)$, $S_2 = (U_2, \mathbf{x}_2)$, such that U_1, U_2 belong to V and do not commute. Show that we have

$$(S_1, S_2) = ((U_1, U_2), \mathbf{y})$$

with $\|\mathbf{y}\| \leq \frac{1}{4}(\|\mathbf{x}_1\| + \|\mathbf{x}_2\|)$. Then define S_k inductively by the rule $S_k = (S_1, S_{k-1})$ for $k \geq 2$. Show that, by virtue of the choice of V, the sequence (S_k) has infinitely many distinct terms and is bounded in $I(n)$, which is absurd.)

(b) The group Γ is said to be *cocompact* or *crystallographic* if the homogeneous space $I(n)/\Gamma$ is compact. Show that if Γ is crystallographic, then for each $x \in \mathbf{R}^n$ the affine-linear subspace L generated by the orbit of x under Γ is the whole of $\mathbf{R}^n$. (If not, then the orbit of a point $y \notin L$ would have all its points at the same distance from L, and since this distance could be arbitrarily large, $I(n)/\Gamma$ would not be compact.)

(c) Show that if Γ is a commutative crystallographic group, then it is contained in $\mathbf{R}^n$ (i.e., it consists of translations). (Observe first that if $s = (U, \mathbf{x}) \in \Gamma$ is such that the vector subspace $V \subset \mathbf{R}^n$ of points fixed by U is not the whole space $\mathbf{R}^n$, and if $V^\perp$ is the orthogonal supplement of V, then for each $S' = (U', \mathbf{x}') \in \Gamma$, the linear transformation U' leaves V and $V^\perp$ stable. On the other hand, since the restriction of U to $V^\perp$ has no fixed point $\neq 0$, show that this allows us to assume (by change of origin) that $\mathbf{x} \in V$. Using (b), there exists $S' = (U', \mathbf{x}') \in \Gamma$ such that the orthogonal projection $\mathbf{y}'$ of $\mathbf{x}'$ on $V^\perp$ is nonzero. By calculating $S \cdot \mathbf{x}'$ by the formula $S = S'SS'^{-1}$, deduce that $U \cdot \mathbf{y}' = \mathbf{y}'$, which contradicts the definition of V.)

(d) Show that if Γ is a crystallographic group, then $\Gamma \cap \mathbf{R}^n$ is a free $\mathbf{Z}$-module of rank n and that $\Gamma/(\Gamma \cap \mathbf{R}^n)$ is a *finite* group (*Bieberbach's theorem*). (Prove first that $\Gamma \cap \mathbf{R}^n \neq \{0\}$, as follows: If $\Gamma \cap \mathbf{R}^n = \{0\}$, the compact Lie group $p(\Gamma)$ has only finitely many connected components and hence, by (a), Γ contains a *commutative* normal subgroup Γ_1 of finite index in Γ; the group Γ_1 is therefore also crystallographic, which leads to a contradiction with (c). Next show that if W is the vector subspace of $\mathbf{R}^n$ generated by $\Gamma \cap \mathbf{R}^n$, then the group $p(\Gamma) \subset \mathbf{O}(n)$ leaves W stable and its restriction to W is necessarily a finite group: otherwise there would exist a sequence (S_m) in Γ such that if $S_m = (U_m, \mathbf{x}_m)$, the sequence of linear transformations $U_m = p(S_m)$ would have infinitely many distinct terms and would tend to I. Consider now the commutators $(I, \mathbf{a}_j)S_m(I, \mathbf{a}_j)^{-1}S_m^{-1}$, where $(\mathbf{a}_j)$ is a basis of the $\mathbf{Z}$-module $\Gamma \cap \mathbf{R}^n$, and obtain a contradiction to the hypothesis that Γ is discrete. Finally, to show that $W = \mathbf{R}^n$, prove that if $W \neq \mathbf{R}^n$, the action of Γ on $\mathbf{R}^n/W$ would be that of a crystallographic group having no translation $\neq 0$.)

17. Let G be a Lie group, $m : G \times G \to G$ the multiplication on G. Show that the mapping $T(m) : T(G) \times T(G) \to T(G)$ defines a Lie group structure on T(G). Every element of T(G) is uniquely of the form $\mathbf{u} \cdot x$, where $x \in G$ and $\mathbf{u} \in \mathfrak{g}_e = T_e(G)$, and we have (16.10.1)

$$(\mathbf{u} \cdot x)(\mathbf{v} \cdot y) = (\mathbf{u} + \operatorname{Ad}(x) \cdot \mathbf{v}) \cdot (xy):$$

in other words, T(G) is a semidirect product of the normal subgroup $\mathfrak{g}_e$ and G, relative to the mapping $x \mapsto \operatorname{Ad}(x)$ of G into $\operatorname{Aut}(\mathfrak{g}_e)$.

Deduce that T(T(G)) may be identified with $\mathfrak{g}_e^3 \times G$ as a manifold, and that the multiplication is given by the formula

$$(\mathbf{p}, \mathbf{h}, \mathbf{u}, x)(\mathbf{q}, \mathbf{k}, \mathbf{v}, y) = (\mathbf{p} + \operatorname{Ad}(x) \cdot \mathbf{q}, \mathbf{h} + [\mathbf{u}, \operatorname{Ad}(x) \cdot \mathbf{k}], \mathbf{u} + \operatorname{Ad}(x) \cdot \mathbf{v}, xy).$$

15. DIFFERENTIAL OF A MAPPING INTO A LIE GROUP

(19.15.1) Let E be a finite-dimensional real vector space, M a differential manifold, $\mathbf{f} : M \to E$ a mapping of class C^1. We recall that for each $x \in M$ the *differential* $d_x \mathbf{f}$ of $\mathbf{f}$ at the point x is defined to be a linear mapping of $T_x(M)$ *into* E (not $T_{\mathbf{f}(x)}(E)$); the definition rests on the existence of a canonical linear mapping $\boldsymbol{\tau}_z : T_z(E) \to E$ for each $z \in E$ ((16.5.2) and (16.5.7)). It is immediately verified that the mapping $\boldsymbol{\tau}_0^{-1} \circ \boldsymbol{\tau}_z$ is the tangent linear mapping

$$T_z(\gamma(-z)) : T_z(E) \to T_0(E)$$

of the translation $t \mapsto t - z$. If we identify E with $T_0(E)$ by means of $\boldsymbol{\tau}_0$, we may regard $d_x \mathbf{f}$ as a linear mapping of $T_x(M)$ into the *Lie algebra* $T_0(E)$ of the Lie group E.

(19.15.2) It is now an easy matter to generalize the notion of a differential to the case of a C^1 mapping f of M into any *Lie group* G: for each $x \in M$, the differential $d_x f$ is the *linear mapping*

(19.15.2.1) $$\mathbf{h}_x \mapsto f(x)^{-1} \cdot (T_x(f) \cdot \mathbf{h}_x)$$

(cf. (16.9.8)) of $T_x(M)$ into the Lie algebra $T_e(G) = \mathfrak{g}_e$. The local expression of $T_x(f)$ (16.5) shows immediately that if f is of class C^r, then $x \mapsto d_x f$ is a *vector-valued differential* 1-*form* of class C^{r-1} on M (16.20.15) with values in $\mathfrak{g}_e$. We denote this 1-form by df (when there is no risk of confusion) and call it the *differential of* f.

Strictly speaking, what we have just defined is the "left differential"; there is also a "right differential," namely,

$$\mathbf{h}_x \mapsto (\mathrm{T}_x(f) \cdot \mathbf{h}_x) \cdot f(x)^{-1},$$

which can also be written in the form

(19.15.2.2) $$\mathbf{h}_x \mapsto \mathrm{Ad}(f(x)) \cdot (d_x f \cdot \mathbf{h}_x).$$

In what follows, we shall consider only left differentials.

Example

(19.15.3) Let A be a finite-dimensional **R**-algebra with identity element, and let $\mathrm{G} = \mathrm{A}^*$. In this situation it follows from (19.4.1) that the left differential in the sense of (19.15.2) is the mapping $x \mapsto f(x)^{-1} d_x f$ of M into A, where $d_x f$ denotes the differential *in the sense of* (16.5.7) (the Lie algebra of A^* being canonically identified with A). We shall therefore denote it by $f^{-1} df$ and refer to it as the (left) *logarithmic differential* of f, in order to avoid any ambiguity.

(19.15.4) *Let f, g be two C^1 mappings of a differential manifold* M *into a Lie group* G. *Then the differentials of the mappings*

$$fg : x \mapsto f(x)g(x) \qquad \textit{and} \qquad f^{-1} : x \mapsto f(x)^{-1}$$

of M *into* G *are given by*

(19.15.4.1) $$d_x(fg) = \mathrm{Ad}(g(x)^{-1}) \circ d_x f + d_x g,$$

(19.15.4.2) $$d_x(f^{-1}) = -\mathrm{Ad}(f(x)) \circ d_x f.$$

It follows from (16.9.9) and the composite function theorem (16.5.4) that for all $\mathbf{h}_x \in \mathrm{T}_x(\mathrm{M})$ we have

$$\mathrm{T}_x(fg) \cdot \mathbf{h}_x = (\mathrm{T}_x(f) \cdot \mathbf{h}_x) \cdot g(x) + f(x) \cdot (\mathrm{T}_x(g) \cdot \mathbf{h}_x),$$

and since $d_x(fg) \cdot \mathbf{h}_x = (f(x)g(x))^{-1} \cdot (\mathrm{T}_x(fg) \cdot \mathbf{h}_x)$, the formula (19.15.4.1) follows, by virtue of the definition of $\mathrm{Ad}(s)$ (19.2.1). We then derive (19.15.4.2) by replacing g by f^{-1} in (19.15.4.1) and using the fact that $s \mapsto \mathrm{Ad}(s)$ is a homomorphism.

In particular, if we take f or g to be the *constant* mapping $x \mapsto s$ of M into G, we obtain

(19.15.4.3) $$d_x(sg) = d_x g,$$

(19.15.4.4) $$d_x(gs) = \mathrm{Ad}(s^{-1}) \circ d_x g.$$

(19.15.5) *Let f, g be two mappings of class* C^1 *of* M *into a Lie group* G. *Then* fg^{-1} *is locally constant if and only if* $df = dg$.

For to say that fg^{-1} is locally constant is equivalent to saying that $T_x(fg^{-1}) = 0$ for all $x \in M$ (16.5.5), which in turn is equivalent to $d_x(fg^{-1}) = 0$. But, by virtue of (19.15.4), we have

$$d_x(fg^{-1}) = \mathrm{Ad}(g(x)) \circ (d_x f - d_x g),$$

whence the result follows.

(19.15.6) *Let f be a mapping of class* C^1 *of* M *into a Lie group* G, *and let* $u : G \to G'$ *be a homomorphism of Lie groups. Then for* $s \in G$ *and* $x \in M$ *we have*

(19.15.6.1) $$(d_s u) \cdot \mathbf{k}_s = u_*(s^{-1} \cdot \mathbf{k}_s),$$

(19.15.6.2) $$d_x(u \circ f) = u_* \circ d_x f.$$

The formula (19.15.6.1) follows directly from the definitions and from (16.9.9). As to (19.15.6.2), we have, again from (16.9.9),

$$\begin{aligned} d_x(u \circ f) \cdot \mathbf{h}_x &= (u(f(x)))^{-1} \cdot (T_{f(x)}(u) \cdot (T_x(f) \cdot \mathbf{h}_x)) \\ &= T_e(u) \cdot (f(x)^{-1} \cdot (T_x(f) \cdot \mathbf{h}_x)) \\ &= u_*(d_x f(\mathbf{h}_x)). \end{aligned}$$

16. INVARIANT DIFFERENTIAL FORMS AND HAAR MEASURE ON A LIE GROUP

(19.16.1) On a Lie group G, the *canonical differential form* ω or ω_G is by definition the (left) differential $d(1_G)$ of the identity mapping $1_G : G \to G$: it is therefore a *vector-valued* 1-*form* with values in the Lie algebra $\mathfrak{g}_e$. For each $x \in G$ we have, by (19.15.6.1),

(19.16.1.1) $$\omega(x) \cdot \mathbf{h}_x = x^{-1} \cdot \mathbf{h}_x.$$

Notice that the knowledge of this form determines the differential of *any* C^∞ mapping $f : M \to G$ of a differential manifold M into G. For the definition (19.15.2.1) can be rewritten in the form

$$d_x f = \omega(f(x)) \circ T_x(f),$$

or equivalently (16.20.15.3),

(19.16.1.2) $$df = {}^t f(\boldsymbol{\omega}).$$

(19.16.2) *For each $s \in G$, we have*

(19.16.2.1) $$\gamma(s)\boldsymbol{\omega} = \boldsymbol{\omega},$$

(19.16.2.2) $$\delta(s)\boldsymbol{\omega} = \mathrm{Ad}(s^{-1}) \circ \boldsymbol{\omega}.$$

For by definition ((19.1.2) and (19.2)) we have, for each $x \in G$,

$$(\gamma(s)\boldsymbol{\omega})(x) = s \cdot \boldsymbol{\omega}(s^{-1}x), \qquad (\delta(s^{-1})\boldsymbol{\omega})(x) = \boldsymbol{\omega}(xs^{-1}) \cdot s.$$

But also, from the same definitions.

$$\begin{aligned}(s \cdot \boldsymbol{\omega}(s^{-1}x)) \cdot \mathbf{h}_x &= \boldsymbol{\omega}(s^{-1}x) \cdot (s^{-1} \cdot \mathbf{h}_x) = (x^{-1}s) \cdot (s^{-1} \cdot \mathbf{h}_x)\\ &= x^{-1} \cdot \mathbf{h}_x = \boldsymbol{\omega}(x) \cdot \mathbf{h}_x\end{aligned}$$

because G acts here trivially on $\mathfrak{g}_e$; and likewise

$$\begin{aligned}(\boldsymbol{\omega}(xs^{-1}) \cdot s) \cdot \mathbf{h}_x &= \boldsymbol{\omega}(xs^{-1}) \cdot (\mathbf{h}_x \cdot s^{-1}) = (sx^{-1}) \cdot (\mathbf{h}_x \cdot s^{-1})\\ &= s \cdot (\boldsymbol{\omega}(x) \cdot \mathbf{h}_x) \cdot s^{-1} = \mathrm{Ad}(s) \cdot (\boldsymbol{\omega}(x) \cdot \mathbf{h}_x).\end{aligned}$$

(19.16.3) Let $(\mathbf{e}_i)_{1 \leqq i \leqq n}$ be a basis of the Lie algebra $\mathfrak{g}_e$. Then we can write

(19.16.3.1) $$\boldsymbol{\omega}(x) = \sum_{i=1}^{n} \omega_i(x)\mathbf{e}_i,$$

and since $\boldsymbol{\omega}(x)$ is a bijection of $T_x(G)$ onto $\mathfrak{g}_e = T_e(G)$, the (scalar-valued) differential 1-forms ω_i are *linearly independent*. Since they are left-invariant by (19.16.2), they form a *basis over* **R** of the vector space of *left-invariant differential* 1-*forms on* G, since $T_e(G)^*$ is of dimension n (19.2.2).

It follows (19.2.2) that, for each p such that $1 \leqq p \leqq n$, the differential p-forms on G

(19.16.3.2) $$\omega_{i_1} \wedge \omega_{i_2} \wedge \cdots \wedge \omega_{i_p} \qquad (i_1 < i_2 < \cdots < i_p)$$

form a *basis* over **R** for the vector space of left-invariant differential p-forms.

(19.16.4) In particular, the vector space of *left-invariant differential n-forms on* G, which is of dimension 1, is generated by the n-form

(19.16.4.1) $$\Omega = \omega_1 \wedge \omega_2 \wedge \cdots \wedge \omega_n.$$

To this corresponds (16.24.1) a *left-invariant positive Lebesgue measure* μ_Ω on G, that is to say, a *Haar measure* (14.1.2). Hence, on a Lie group, every

Haar measure is a Lebesgue measure. By virtue of (19.16.2.2) and (19.16.4.1), we have

(19.16.4.2) $$\Omega(xs^{-1}) = \det(\mathrm{Ad}(s))\Omega(x)$$

for all $s \in \mathrm{G}$. Since by (16.24.5.1) we have

$$\int f(xs)\,\Omega(x) = \int f(x)\,\Omega(xs^{-1})$$

for all $f \in \mathscr{K}(\mathrm{G})$, it follows (14.3.1.1) that the value of the *modulus function* on G is given by

(19.16.4.3) $$\Delta_{\mathrm{G}}(s) = |\det(\mathrm{Ad}(s^{-1}))|.$$

(19.16.5) *For each* $\mathbf{x} \in \mathfrak{g}_e$, *the differential at* $\mathbf{x}$ (19.15.2) *of the exponential mapping of* $\mathfrak{g}_e$ *into* G *is given by*

(19.16.5.1) $$d_{\mathbf{x}}(\exp) = \sum_{p=0}^{\infty} \frac{1}{(p+1)!}(\mathrm{ad}(-\mathbf{x}))^p.$$

For by definition, this differential is an endomorphism $M(\mathbf{x})$ of the vector space $\mathfrak{g}_e$; since $\mathrm{T}_{\mathbf{0}}(\exp)$ is the identity (19.8.6.1), we have

(19.16.5.2) $$M(\mathbf{0}) = I,$$

where I is the identity mapping of $\mathfrak{g}_e$. Let ξ, η be two real numbers, and consider the C^∞ mapping

$$\mathbf{x} \mapsto \exp((\xi + \eta)\mathbf{x}) = \exp(\xi\mathbf{x})\exp(\eta\mathbf{x})$$

of $\mathfrak{g}_e$ into G. Applying the formula (19.15.4.1), we obtain

$$(\xi + \eta)M((\xi + \eta)\mathbf{x}) = \mathrm{Ad}(\exp(-\eta\mathbf{x}))\xi M(\xi\mathbf{x}) + \eta M(\eta\mathbf{x}),$$

or, writing $N(\xi) = \xi M(\xi\mathbf{x}) \in \mathrm{End}(\mathfrak{g}_e)$,

(19.16.5.3) $$\begin{aligned} N(\xi + \eta) &= \mathrm{Ad}(\exp(-\eta\mathbf{x}))N(\xi) + N(\eta) \\ &= \exp(\mathrm{ad}(-\eta\mathbf{x}))N(\xi) + N(\eta) \end{aligned}$$

by virtue of (19.11.2.2). Fix ξ and take the derivatives at 0 of the two sides of (19.16.5.3), considered as mappings of $\mathbf{R}$ into $\mathrm{End}(\mathfrak{g}_e)$: since the derivative of $\eta \mapsto \exp(\mathrm{ad}(-\eta\mathbf{x}))$ at the point 0 is evidently $\mathrm{ad}(-\mathbf{x})$, we obtain by use of (19.16.5.2)

$$N'(\xi) = \mathrm{ad}(-\mathbf{x})N(\xi) + I$$

and it is clear that the entire function

$$\xi \mapsto \sum_{p=0}^{\infty} \frac{1}{(p+1)!} (\mathrm{ad}(-\mathbf{x}))^p \xi^{p+1}$$

is the unique solution of this linear differential equation which takes the value 0 at $\xi = 0$. To obtain the formula (19.16.5.1), we have merely to take $\xi = 1$.

(19.16.6) *In order that the tangent linear mapping* $\mathrm{T}_{\mathbf{x}}(\exp)$ *at a point* $\mathbf{x} \in \mathfrak{g}_e$ *should be bijective, it is necessary and sufficient that the characteristic polynomial*

$$\det(\mathrm{ad}(\mathbf{x}) - \lambda \cdot 1_{\mathfrak{g}_e})$$

of the endomorphism $\mathrm{ad}(\mathbf{x})$ *should have no nonzero root of the form*

$$2\pi i k \qquad (k \in \mathbf{Z}).$$

Since $d_{\mathbf{x}}(\exp) = (\exp \mathbf{x})^{-1} \cdot \mathrm{T}_{\mathbf{x}}(\exp)$, the mapping $\mathrm{T}_{\mathbf{x}}(\exp)$ is bijective if and only if the endomorphism $d_{\mathbf{x}}(\exp)$ of $\mathfrak{g}_e$ is bijective, i.e., if and only if 0 is not a root of the characteristic polynomial of this endomorphism. Let M be the matrix of $\mathrm{ad}(\mathbf{x})$ relative to some basis of $\mathfrak{g}_e$. Considering M as a matrix of complex numbers, there exists an invertible complex matrix P such that PMP^{-1} is lower triangular, with the eigenvalues $\lambda_1, \ldots, \lambda_n$ of $\mathrm{ad}(\mathbf{x})$ ranged down the diagonal, each eigenvalue appearing a number of times equal to its multiplicity as a root of the characteristic polynomial (A.6.10). It follows immediately that the matrix E of the right-hand side of (19.16.5.1) is such that PEP^{-1} is lower triangular, with diagonal elements

$$\sum_{p=0}^{\infty} \frac{1}{(p+1)!} (-\lambda_j)^p = \frac{1 - e^{-\lambda_j}}{\lambda_j} \qquad (1 \leq j \leq n)$$

(provided $(1 - e^0)/0$ is replaced by 1). The result now follows immediately.

PROBLEMS

1. Let G be a Lie group, $\mathfrak{g}_e$ its Lie algebra, and let $c = (\mathrm{U}, \varphi, n)$ be a chart of G at the identity element e such that $\varphi(e) = 0$; also let $(\mathbf{x}, \mathbf{y}) \mapsto \mathbf{m}(\mathbf{x}, \mathbf{y})$ be the local expression of the multiplication on G corresponding to the chart c (19.5.7).

   (a) Let (ω_i) be a basis of the vector space of left-invariant differential 1-forms on G, and let

$$^t\varphi^{-1}(\omega_i | \mathrm{U}) = \sum_{j=1}^{n} \mathrm{A}_{ij}(\xi^1, \ldots, \xi^n)\, d\xi^j$$

in $\varphi(U) \subset \mathbf{R}^n$. Let V be a symmetric neighborhood of e such that $V^2 \subset U$. Show that for each $\mathbf{x}_0 \in \varphi(V)$, if we put $\eta^i(\mathbf{x}) = m_i(\mathbf{x}_0, \mathbf{x})$ $(1 \leq i \leq n)$ for $\mathbf{x} \in \varphi(V)$, the graph of the mapping

$$(\xi^1, \ldots, \xi^n) \mapsto (\eta^1(\xi^1, \ldots, \xi^n), \ldots, \eta^n(\xi^1, \ldots, \xi^n))$$

of $\varphi(V)$ into $\varphi(U)$ is an integral manifold of the completely integrable Pfaffian system (in $\varphi(V) \times \varphi(U)$)

$$\sum_{j=1}^{n} A_{ij}(\eta^1, \ldots, \eta^n)\, d\eta^j = \sum_{j=1}^{n} A_{ij}(\xi^1, \ldots, \xi^n)\, d\xi^j \qquad (1 \leq i \leq n)$$

which passes through the point $(0, \mathbf{x}_0)$.

(b) Identify $\mathfrak{g}_e$ with $\mathbf{R}^n$ by choosing a basis, and suppose that the ω_i are the components of $\boldsymbol{\omega}$ relative to this basis. Take for chart $c = (U, \varphi, n)$ a canonical chart with $\varphi = \log_G$. Using the formula $d(\exp) = {}^t\exp(\boldsymbol{\omega})$, show that the matrix

$$(A_{ij}(\xi^1, \ldots, \xi^n))$$

is the matrix of the endomorphism

$$\sum_{p=0}^{\infty} \frac{1}{(p+1)!} (\operatorname{ad}(-\mathbf{x}))^p$$

of $\mathfrak{g}_e$, and hence that the A_{ij} are *analytic* functions of $\mathbf{x}$ in $\log_G(U)$. Deduce from this and from (10.9.5) that the local expression $\mathbf{m}(\mathbf{x}, \mathbf{y})$ is an *analytic* function of $(\mathbf{x}, \mathbf{y})$ in $\varphi(V) \times \varphi(V)$, and hence that there exists on G a structure of real-analytic manifold compatible with the group structure of G, and such that the underlying structure of differential manifold is the given one (cf. Section 19.3, Problem 4).

2. (a) Let $u : G \to G'$ be a homomorphism of Lie groups. Show that $u_*(\boldsymbol{\omega}_G(x)) = \boldsymbol{\omega}_{G'}(u(x))$ for all $x \in G$.
(b) With the notation of Problem 1(a), suppose that the ω_i are the components of $\boldsymbol{\omega}_G$ relative to a basis $(\mathbf{a}_i)_{1 \leq i \leq n}$ of $\mathfrak{g}_e$. Show that for each vector $\mathbf{u} = \sum_{j=1}^{n} \alpha_j \mathbf{a}_j$ in $\mathfrak{g}_e$, the mapping $t \mapsto \exp(t\mathbf{u})$ has as local expression in a neighborhood of 0 (relative to the chart c) the solution of the system of differential equations

$$\sum_{j=1}^{n} A_{ij}(\xi^1, \ldots, \xi^n) \frac{d\xi^j}{dt} = \alpha_i \qquad (1 \leq i \leq n)$$

which takes the value 0 at $t = 0$.

3. Let f be a C^1 mapping of an open set in $\mathbf{R}^2$ into a Lie group G. As in (18.7.1), we put

$$f'_\xi(\xi, \eta) = T(f) \cdot E_1(\xi, \eta),$$
$$f'_\eta(\xi, \eta) = T(f) \cdot E_2(\xi, \eta).$$

Prove that

$$\frac{\partial}{\partial \xi}(f^{-1} \cdot f'_\eta) - \frac{\partial}{\partial \eta}(f^{-1} \cdot f'_\xi) = [f^{-1} \cdot f'_\eta, f^{-1} \cdot f'_\xi].$$

(Use Problem 8(b) of Section 19.9 and the formula (18.7.4.2), which is valid because the connection **C** has no torsion.)

4. The algebra $\mathbf{M}_n(\mathbf{R})$ (or $\mathbf{M}_n(\mathbf{C})$) is assumed to be endowed with a norm that makes it a Banach algebra. Show that there exists a number $r > 0$ such that, whenever $\|S\| < r$ and $\|T\| < r$, the family of matrices

$$(*) \qquad \frac{(-1)^{n+1}}{n} \cdot \frac{S^{p_1}T^{q_1}S^{p_2}T^{q_2} \cdots S^{p_n}T^{q_n}}{p_1! \cdots p_n! q_1! \cdots q_n!},$$

where n takes all integer values $\geqq 1$, and for each n the pairs (p_i, q_i) $(1 \leqq i \leqq n)$ take all values in $\mathbf{N} \times \mathbf{N}$ except $(0, 0)$, is *absolutely summable* (5.3.3). Likewise the series

$$-\sum_{n=1}^{\infty} \frac{1}{n} (I - \exp(S)\exp(T))^n$$

is absolutely convergent, and its sum is equal to that of the family $(*)$, and to the matrix $\log((\exp S)(\exp T))$. (Reduce to (9.2.1) by majorizing the norms of the elements of the family $(*)$.)

5. Let G be a Lie group, $\mathfrak{g}_e$ its Lie algebra, endowed with a norm compatible with its topology. The function

$$F(t, \mathbf{u}, \mathbf{v}) = \log(\mathrm{Ad}(\exp_G(\mathbf{u}))\mathrm{Ad}(\exp_G(t\mathbf{v})))$$

with values in the Banach algebra $\mathrm{End}(\mathfrak{g}_e)$ (the logarithm being that of $\mathbf{GL}(\mathfrak{g}_e)$) is defined for $t \in [0, 1]$ and for $\mathbf{u}$, $\mathbf{v}$ in a sufficiently small neighborhood U of 0 in $\mathfrak{g}_e$. Show that for such values of t, $\mathbf{u}$, $\mathbf{v}$, the family of elements of $\mathfrak{g}_e$

$$(**) \qquad \frac{(-1)^{n+1}}{n(n+1)} \cdot \frac{((\mathrm{ad}\,\mathbf{u})^{p_1}(\mathrm{ad}\,t\mathbf{v})^{q_1}(\mathrm{ad}\,\mathbf{u})^{p_2}(\mathrm{ad}\,t\mathbf{v})^{q_2} \cdots (\mathrm{ad}\,\mathbf{u})^{p_n}) \cdot \mathbf{v}}{p_1! \cdots p_n! q_1! \cdots q_{n-1}!}$$

(where $n \geqq 1$, $p_i + q_i > 0$ for $1 \leqq i \leqq n-1$, and $p_n > 0$) is absolutely summable, and that its sum is equal to

$$\left(\frac{\partial}{\partial t} F(t, \mathbf{u}, \mathbf{v}) - I\right) \cdot \mathbf{v}.$$

(Use (19.16.5) and Problem 4.) Deduce that for $\mathbf{u}, \mathbf{v} \in$ U, the family

$$(***) \qquad \frac{(-1)^{n+1}}{n(n+1)} \cdot \frac{((\mathrm{ad}\,\mathbf{u})^{p_1}(\mathrm{ad}\,\mathbf{v})^{q_1}(\mathrm{ad}\,\mathbf{u})^{p_2}(\mathrm{ad}\,\mathbf{v})^{q_2} \cdots (\mathrm{ad}\,\mathbf{u})^{p_n}) \cdot \mathbf{v}}{p_1! \cdots p_n! q_1! \cdots q_{n-1}!(q_1 + \cdots + q_{n-1} + 1)}$$

is absolutely summable and that its sum is

$$\log_G(\exp_G(\mathbf{u})\exp_G(\mathbf{v})) - (\mathbf{u} + \mathbf{v})$$

(*Campbell–Hausdorff formula*). Hence give another proof of the result of Problem 1(b).

6. (a) Let A, B, C be three connected Lie groups immersed in a Lie group G, and let $\mathfrak{a}_e$, $\mathfrak{b}_e$, $\mathfrak{c}_e$ be their Lie algebras, which are subalgebras of $\mathfrak{g}_e = \mathrm{Lie}(G)$. Assume that $[\mathfrak{a}_e, \mathfrak{c}_e] \subset \mathfrak{c}_e$ and that $[\mathfrak{b}_e, \mathfrak{c}_e] \subset \mathfrak{c}_e$. Show that if $[\mathfrak{a}_e, \mathfrak{b}_e] \subset \mathfrak{c}_e$, then $(A, B) \subset C$. (By considering the Lie subalgebra $\mathfrak{a}_e + \mathfrak{b}_e + \mathfrak{c}_e$ of $\mathfrak{g}_e$, reduce to the case where $\mathfrak{c}_e$ is an ideal in $\mathfrak{g}_e$, and then use the Campbell–Hausdorff formula (Problem 5).) If $[\mathfrak{a}_e, \mathfrak{b}_e] = \mathfrak{c}_e$, show that $(A, B) = C$ (cf. Section 19.12, Problem 6).

(b) Deduce from (a) that if G is a connected Lie group, $\mathfrak{g}_e$ its Lie algebra, then the groups $\mathscr{C}^p(G)$ of the descending central series of G, defined by $\mathscr{C}^1(G) = G$, $\mathscr{C}^p(G) = (G, \mathscr{C}^{p-1}(G))$, are connected Lie groups immersed in G, whose Lie algebras are the $\mathfrak{C}^p(\mathfrak{g}_e)$.

7. (a) With the hypotheses of Problem 8(b) of Section **19.14**, show that the Lie subalgebra generated by the $\mathbf{u}_k$ such that $p \leq k \leq n$ has a multiplication table relative to this basis in which the structure constants are *rational* numbers. (Show that this is the case for the subalgebra $\mathfrak{m}_j$ generated by the $\mathbf{u}_k$ such that $k \geq j$, by descending induction on j; use Problems 8(a) and 6(a) of Section **19.14**, together with the Campbell–Hausdorff formula (Problem 5).)
(b) Conversely, suppose that the Lie algebra $\mathfrak{g}_e$ of a simply-connected nilpotent Lie group G has a basis relative to which the structure constants are rational numbers. Show that there exists a discrete subgroup D of G such that the homogeneous space G/D is compact. (Show first that there exists a basis $(\mathbf{u}_k)_{1 \leq k \leq n}$ of $\mathfrak{g}_e$ such that, if $\mathfrak{a}_j$ is the subspace of $\mathfrak{g}_e$ spanned by the $\mathbf{u}_k$ for which $k \geq j$, then the $\mathfrak{a}_j$ satisfy condition (v) of Section **19.12**, Problem 3, and the structure constants relative to the basis $(\mathbf{u}_k)$ are rational. Next observe that if

$$\exp(\xi_1 \mathbf{u}_1 + \xi_2 \mathbf{u}_2 + \cdots + \xi_n \mathbf{u}_n) = \exp(\zeta_1 \mathbf{u}_1) \exp(\zeta_2 \mathbf{u}_2) \cdots \exp(\zeta_n \mathbf{u}_n),$$

then the ζ_j are polynomials in $\xi_1, \ldots, \xi_n$ with rational coefficients. Complete the proof by induction on n.)
(c) With the hypotheses of (b) above, let D_0 be a discrete subgroup of G such that, in the description of D_0 given in Section **19.14**, Problem 8(b), the structure constants with respect to the basis $(\mathbf{u}_j)$ of $\mathfrak{g}_e$ are rational. Show that there exists an increasing sequence (D_m) of discrete subgroups of G containing D_0, such that $\bigcup_m D_m$ is dense in G. (With the notation of Section **19.14**, Problem 8(b), show that there exist arbitrarily large integers N with the property that the center Z of G contains an element b_n such that $b_n^N = a_n$; then proceed by induction on dim(G), by considering G/Z.)

8. (a) Let $\mathfrak{g}$ be a vector space of dimension 7 over **R**, and let $(\mathbf{e}_i)_{1 \leq i \leq 7}$ be a basis of $\mathfrak{g}$. Show that the formulas

$$[\mathbf{e}_i, \mathbf{e}_j] = \alpha_{ij} \mathbf{e}_{i+j} \qquad (1 \leq i < j \leq 7, i + j \leq 7),$$
$$[\mathbf{e}_i, \mathbf{e}_j] = 0 \qquad (i + j > 7)$$

define a Lie algebra structure on $\mathfrak{g}$, provided that the real numbers α_{ij} satisfy the conditions

$$\alpha_{23}\alpha_{15} - \alpha_{13}\alpha_{24} = 0,$$
$$\alpha_{12}\alpha_{34} - \alpha_{24}\alpha_{16} + \alpha_{14}\alpha_{25} = 0.$$

The Lie algebra $\mathfrak{g}$ so defined is nilpotent, and $\mathfrak{C}^j(\mathfrak{g})$ is spanned by the $\mathbf{e}_i$ such that $i \geq j + 1$ for $j = 2, \ldots, 6$; also, the centralizer $\mathfrak{h}$ of $\mathfrak{C}^5(\mathfrak{g})$ is spanned by $\mathbf{e}_2, \ldots, \mathbf{e}_7$. Deduce that if another basis $(\mathbf{e}'_i)_{1 \leq i \leq 7}$ of $\mathfrak{g}$ is such that $\mathfrak{C}^j(\mathfrak{g})$ is spanned by the $\mathbf{e}'_i$ such that $i \geq j + 1$ for $j = 2, \ldots, 6$, and such that $\mathfrak{h}$ is spanned by $\mathbf{e}'_2, \ldots, \mathbf{e}'_7$, then we must have

$$[\mathbf{e}'_i, \mathbf{e}'_j] = \alpha'_{ij} \mathbf{e}'_{i+j} + \mathbf{u}'_{ij},$$

where $\mathbf{u}'_{ij}$ is a linear combination of the $\mathbf{e}'_k$ with indices $k \geqq i+j+1$. Furthermore, we have

$$\alpha'_{14}\alpha'_{25}\alpha'^{-1}_{16}\alpha'^{-1}_{24} = \alpha_{14}\alpha_{25}\alpha^{-1}_{16}\alpha^{-1}_{24}.$$

(b) Deduce from (a) that for a suitable choice of the $\alpha_{ij} \in \mathbf{R}$ there exists *no* basis of $\mathfrak{g}$ for which the structure constants are rational. If G is the simply-connected nilpotent group having $\mathfrak{g}$ as Lie algebra, there exists *no* discrete subgroup D of G such that the homogeneous space G/D is compact.

(c) Let $\mathfrak{g}$ be a real vector space of dimension 8, $(\mathbf{e}_i)_{1 \leqq i \leqq 8}$ a basis of $\mathfrak{g}$. Show that the formulas

$$\begin{aligned}
&[\mathbf{e}_1, \mathbf{e}_2] = \mathbf{e}_3, \quad [\mathbf{e}_1, \mathbf{e}_3] = \mathbf{e}_4, \quad [\mathbf{e}_1, \mathbf{e}_4] = \mathbf{e}_5, \quad [\mathbf{e}_1, \mathbf{e}_5] = \mathbf{e}_6,\\
&[\mathbf{e}_1, \mathbf{e}_6] = [\mathbf{e}_1, \mathbf{e}_7] = \mathbf{e}_8, \quad [\mathbf{e}_2, \mathbf{e}_3] = \mathbf{e}_5, \quad [\mathbf{e}_2, \mathbf{e}_4] = \mathbf{e}_6, \quad [\mathbf{e}_2, \mathbf{e}_5] = \mathbf{e}_7,\\
&[\mathbf{e}_2, \mathbf{e}_6] = 2\mathbf{e}_8, \quad [\mathbf{e}_3, \mathbf{e}_4] = -\mathbf{e}_7 + \mathbf{e}_8, \quad [\mathbf{e}_3, \mathbf{e}_5] = -\mathbf{e}_8,\\
&[\mathbf{e}_i, \mathbf{e}_j] = 0 \qquad \text{for} \quad i+j > 8
\end{aligned}$$

define a Lie algebra structure on $\mathfrak{g}$. The Lie algebra $\mathfrak{g}$ so defined is nilpotent. By considering the $\mathfrak{C}^j(\mathfrak{g})$, the ideal $[\mathfrak{C}^2(\mathfrak{g}), \mathfrak{C}^2(\mathfrak{g})]$ and the transporter of $\mathfrak{C}^2(\mathfrak{g})$ into $\mathfrak{C}^4(\mathfrak{g})$, show that if $(\bar{\mathbf{e}}_i)$ is any other basis of $\mathfrak{g}$ for which the structure constants are rational, then there exists a third basis $(\mathbf{e}'_i)$ with the same property which is derived from $(\bar{\mathbf{e}}_i)$ by a transition matrix with rational entries, and is such that $\mathbf{e}'_i = \mu_i \mathbf{e}_i + \mathbf{u}'_i$, where $\mathbf{u}'_i$ is a linear combination of the $\mathbf{e}_k$ for which $k > i$, and the constants μ_i are *rational.*

(d) Let $\tilde{\text{G}}$ be a simply-connected nilpotent Lie group whose Lie algebra is isomorphic to the algebra described in (c) above. Show that there exists a discrete subgroup N of the center of $\tilde{\text{G}}$ such that the connected nilpotent group $\text{G} = \tilde{\text{G}}/\text{N}$ contains no discrete subgroup D such that G/D is compact.

9. Let G, G′ be two simply-connected nilpotent Lie groups, and let D (resp. D′) be a discrete subgroup of G (resp. G′); suppose that G/D is compact. Show that every homomorphism $f : \text{D} \to \text{D}'$ has a unique extension to a homomorphism $g : \text{G} \to \text{G}'$. (Use the description of D given in Section **19.14**, Problem 8 in terms of a basis $(\mathbf{u}_k)_{1 \leqq k \leqq n}$ of $\mathfrak{g}_e$; we may write $f(\exp_{\text{G}}(\mathbf{u}_k)) = \exp_{\text{G}'}(\mathbf{u}'_k)$ for each k, where $\mathbf{u}'_k \in \mathfrak{g}'_{e'}$ is well determined. To show that the linear mapping $\text{F} : \mathfrak{g}_e \to \mathfrak{g}'_{e'}$ defined by

$$\text{F}(\mathbf{u}_k) = \mathbf{u}'_k \qquad (1 \leqq k \leqq n),$$

is a homomorphism of Lie algebras, proceed by induction on the dimension of G, by introducing the subalgebras $\mathfrak{m}_j$ of $\mathfrak{g}_e$ (Section **19.14**, Problem 8). Use the Campbell–Hausdorff formula, Section **19.14**, Problem 6(a), and the formulas **(19.11.2.2)** and **(19.11.2.3)**.) If f is surjective (resp. bijective) and if G′/D′ is compact, then g is surjective (resp. bijective).

10. Let D be a countable discrete nilpotent group which admits a sequence

$$\text{D} = \text{D}_1 \supset \text{D}_2 \supset \text{D}_3 \supset \cdots \supset \text{D}_n \supset \text{D}_{n+1} = \{e\}$$

of normal subgroups such that $\text{D}_j/\text{D}_{j+1} \cong \mathbf{Z}$ for $1 \leqq j \leqq n$, and $(\text{D}, \text{D}_j) \subset \text{D}_{j+1}$. Show that there exists a simply-connected nilpotent Lie group G such that $\text{D} \subset \text{G}$ and G/D is compact. (Induction on n: if G_2 is a simply-connected nilpotent Lie group containing D_2 and such that G_2/D_2 is compact, and if $d \in D$ is such that the image of d in D/D_2 generates D/D_2, consider the automorphism $z \mapsto dzd^{-1}$ of D_2, and use Problem 9.)

11. Let G be a connected Lie group, H a connected Lie group immersed in G, and let $\mathfrak{g}_e$, $\mathfrak{h}_e$ be the Lie algebras of G, H respectively.

(a) Let $s \in G$. Then $sts^{-1}t^{-1} \in \mathscr{D}(H)$ for all $t \in H$ if and only if

$$\mathrm{Ad}(s) \cdot \mathbf{u} - \mathbf{u} \in \mathfrak{D}(\mathfrak{h}_e) = [\mathfrak{h}_e, \mathfrak{h}_e]$$

for all $\mathbf{u} \in \mathfrak{h}_e$. (Use the Campbell–Hausdorff formula to prove that the condition is sufficient.)

(b) Assume that H is dense in G. Show that $\mathscr{D}(H) = \mathscr{D}(G)$. (Use (a) to show that $(G, H) \subset \mathscr{D}(H)$, which implies that $[\mathfrak{g}_e, \mathfrak{h}_e] \subset [\mathfrak{h}_e, \mathfrak{h}_e]$; then show analogously that $(G, G) \subset \mathscr{N}(H)$.)

12. Show that if $\boldsymbol{\omega}$ is the canonical differential on a Lie group G, then $d\boldsymbol{\omega} + [\boldsymbol{\omega}, \boldsymbol{\omega}] = 0$ (*Maurer–Cartan equation*). (For each left-invariant scalar 1-form ω, evaluate

$$\langle \omega, X_{\mathbf{u}} \wedge X_{\mathbf{v}} \rangle$$

for any two elements $\mathbf{u}$, $\mathbf{v} \in \mathfrak{g}_e$, by using (17.15.3.6).) Consequently, if $(\mathbf{e}_i)_{1 \leq i \leq n}$ is a basis of $\mathfrak{g}_e$, and if we put

$$\boldsymbol{\omega} = \sum_i \omega_i \mathbf{e}_i, \qquad [\mathbf{e}_i, \mathbf{e}_j] = \sum_k a_{ijk} \mathbf{e}_k,$$

then we have

$$(1) \qquad d\omega_k + \tfrac{1}{2} \sum_{i,j} a_{ijk} \omega_j \wedge \omega_j = 0.$$

13. Let α be a left-invariant C^∞ differential p-form on a connected Lie group G. Show that α is right-invariant as well if and only if $d\alpha = 0$. (Observe that the condition of right-invariance, which is $\bigwedge^p ({}^t\mathrm{Ad}(s)) \cdot \alpha(e) = \alpha(e)$ for all $s \in G$, is equivalent to the condition

$$(2) \qquad \sum_{j=1}^{p} \langle \alpha(e), \mathbf{u}_1 \wedge \cdots \wedge \mathbf{u}_{j-1} \wedge [\mathbf{u}, \mathbf{u}_j] \wedge \mathbf{u}_{j+1} \wedge \cdots \wedge \mathbf{u}_p \rangle = 0$$

for all choices of $p + 1$ vectors $\mathbf{u}, \mathbf{u}_1, \ldots, \mathbf{u}_p$ in $\mathfrak{g}_e$; then use (17.15.3.5) for left-invariant vector fields.)

In particular, the dimension of the vector space of 1-forms on G which are both left- and right-invariant is equal to $\dim \mathfrak{g}_e - \dim[\mathfrak{g}_e, \mathfrak{g}_e]$.

14. For a connected Lie group G to be unimodular, it is necessary and sufficient that $\mathrm{Tr}(\mathrm{ad}(\mathbf{u})) = 0$ for all $\mathbf{u} \in \mathfrak{g}_e$. In particular, every connected nilpotent Lie group is unimodular.

15. Let G be a connected Lie group and H a connected closed subgroup of G; let $\mathfrak{g}_e$, $\mathfrak{h}_e$ be the Lie algebras of G and H, and let $\mathfrak{g}'_e$ be the Lie algebra of the closed subgroup $\overline{\mathscr{D}(G)}$ of G. Show that there exists on G/H a nonzero relatively G-invariant measure (Section 14.4, Problem 2) if and only if the kernel of the linear form $\mathbf{u} \mapsto \mathrm{Tr}(\mathrm{ad}_H(\mathbf{u}))$ on $\mathfrak{h}_e$ contains $\mathfrak{g}'_e \cap \mathfrak{h}_e$.

16. Let G be a connected Lie group of dimension n, let H be a connected closed subgroup of G of dimension $n-p$, and let $(\omega_i)_{1 \leq i \leq n}$ be a basis of the space of left-invariant differential 1-forms on G (19.16.3.1). Assume that the basis $(\mathbf{e}_i)_{1 \leq i \leq n}$ of $\mathfrak{g}_e$ has been chosen so that the inverse images ${}^t j(\omega_k)$ of the ω_k, $k \leq p$, under the canonical injection $j : \mathrm{H} \to \mathrm{G}$, are zero. If there exists a C^∞ differential q-form β on G/H, invariant under the action of G, then the q-form $\alpha = {}^t\pi(\beta)$ on G (where $\pi : \mathrm{G} \to \mathrm{G/H}$ is the canonical submersion) is a linear combination with constant coefficients of exterior products of the 1-forms $\omega_1, \ldots, \omega_p$, q at a time, and furthermore we must have $\overset{p}{\bigwedge}({}^t\mathrm{Ad}(s)) \cdot \alpha(e) = \alpha(e)$ for all $s \in \mathrm{H}$. Show that this condition is equivalent to

$$\sum_{j=1}^{q} \langle \alpha(e), \mathbf{u}_1 \wedge \cdots \wedge \mathbf{u}_{j-1} \wedge [\mathbf{u}, \mathbf{u}_j] \wedge \mathbf{u}_{j+1} \wedge \cdots \wedge \mathbf{u}_q \rangle = 0$$

for all $\mathbf{u} \in \mathfrak{h}_e$ and $\mathbf{u}_1, \ldots, \mathbf{u}_q \in \mathfrak{g}_e$.

In particular, there exists a nonzero G-*invariant measure* on G/H if and only if $d(\omega_1 \wedge \omega_2 \wedge \cdots \wedge \omega_p) = 0$. (Same method as in Problem 13.) This measure is unique up to a constant factor.

17. Deduce from Problem 16 that there exists no nonzero $\mathbf{SL}(n, \mathbf{R})$-invariant measure on the Grassmannian $\mathbf{G}_{n,p}$, where $1 \leq p \leq n-1$. (Observe that for the group $\mathbf{GL}(n, \mathbf{R})$ there exists a basis $(\omega_{ij})_{1 \leq i, j \leq n}$ of the space of left-invariant differential 1-forms for which the Maurer–Cartan equations are

$$d\omega_{ij} = -\sum_{k=1}^{q} \omega_{ik} \wedge \omega_{kj}$$

(cf. (19.4.2.2)).)

18. (a) Show that every Lie group which is a semidirect product $\mathrm{N} \times_\sigma \mathrm{L}$ of a commutative normal subgroup N and a compact Lie group L is unimodular (Section 14.3, Problem 5).

(b) Deduce from (a) that for every family $(\mathrm{V}_k)_{1 \leq k \leq r}$ of affine-linear subspaces of $\mathbf{R}^n$, if H denotes the closed subgroup of the group $\mathrm{I}(n)$ of isometries of $\mathbf{R}^n$ (Section 19.14, Problem 16) which stabilizes each of the V_k, then the homogeneous space $\mathrm{I}(n)/\mathrm{H}$ possesses a nonzero $\mathrm{I}(n)$-invariant measure.

19. Let G be a Lie group, $\mathfrak{g}_e$ its Lie algebra, and let $t \mapsto \mathbf{u}(t)$ be a continuous mapping of $\mathbf{R}$ into $\mathfrak{g}_e$. Define a continuous vector field X on the manifold $\mathrm{G} \times \mathbf{R}$ by the condition $X(x, t) = (\mathbf{u}(t) \cdot x, E(t))$ (in the notation of (18.1)).

(a) If $t \mapsto (v(t), t)$ is an integral curve of X defined in an open interval $\mathrm{I} \subset \mathbf{R}$, then for each $x \in \mathrm{G}$ the function $t \mapsto (v(t) \cdot x, t)$ is another integral curve. Deduce that the maximal integral curves of X are defined on the whole of $\mathbf{R}$. (In the notation of (18.2.2), observe that for each compact interval $\mathrm{K} \subset \mathbf{R}$, the functions $t^-(e, \xi)$ and $t^+(e, \xi)$ are bounded below by a number > 0 as ξ runs through K.)

(b) Deduce from (a) that there exists a unique C^1 mapping $v : \mathbf{R} \to \mathrm{G}$ such that $v'(t) = v(t) \cdot \mathbf{u}(t)$ for all $t \in \mathbf{R}$ (this is a generalization of (19.8.1)).

20. Let u be an *étale* (16.7.1) endomorphism of a Lie group G, with finite kernel. Show that u is proper and that, if β is a left Haar measure on G, we have

$$u(\beta) = \mathrm{Card}(\mathrm{Ker}(u)) \cdot |\det u_*^{-1}| \cdot (\beta \,|\, u(\mathrm{G})).$$

17. COMPLEX LIE GROUPS

(19.17.1) The definitions and results of Sections 19.1–19.9, 19.12, 19.15, and 19.16 (except for the discussion of Haar measure) apply equally well to complex Lie groups if throughout we replace C^∞-functions by *holomorphic* functions. Also, in 19.5, it should be noted that the Taylor expansions are automatically *convergent* in a suitable neighborhood of 0 (9.3.5.1). In 19.3, we have to consider left-invariant *complex* differential operators, which form an associative **C**-algebra $\mathfrak{G}$; it is this algebra which is called the *infinitesimal algebra* of the complex Lie group G. It may be identified with the algebra of *complex* distributions with support contained in $\{e\}$, which we denote by $\mathfrak{G}_e$, and just as in 19.3, we are led to define the (isomorphic) *complex* Lie algebras $\mathfrak{g}$ and $\mathfrak{g}_e$. The only simply-connected commutative complex Lie group of (complex) dimension n is $\mathbf{C}^n$, and every connected commutative complex Lie group is therefore isomorphic to $\mathbf{C}^n/D$, where D is a discrete subgroup of $\mathbf{C}^n$ (16.30.3); however, the classification up to isomorphism of these groups (as *complex* Lie groups) is a more difficult problem, which we shall not attempt.

(19.17.2) The relations between real Lie groups (resp. real Lie algebras) and complex Lie groups (resp. complex Lie algebras) are of two kinds. If G is a complex Lie group of (complex) dimension n, the *real*-analytic structure underlying the complex-analytic structure of G (16.1.6) is evidently compatible with the group-structure of G, and we denote by $G_{|\mathbf{R}}$ the real Lie group of dimension $2n$ so obtained. If $\mathfrak{g}_e$ is the (complex) Lie algebra of G, then the (real) Lie algebra of $G_{|\mathbf{R}}$ is the Lie algebra $\mathfrak{g}_{e|\mathbf{R}}$ obtained by restriction of scalars to **R** in the Lie algebra $\mathfrak{g}_e$. If $\mathbf{u}$ is a vector in $\mathfrak{g}_e$, the (real) one-parameter subgroup of $G_{|\mathbf{R}}$ corresponding to $\mathbf{u}$ is the restriction $t \mapsto \exp(t\mathbf{u})$ to **R** of the (complex) one-parameter subgroup $z \mapsto \exp(z\mathbf{u})$, where z runs through **C**; consequently the exponential mapping for $G_{|\mathbf{R}}$ is the same as for G.

The connected Lie groups immersed in $G_{|\mathbf{R}}$ are in one-to-one correspondence, as we have seen, with the (real) Lie subalgebras of $\mathfrak{g}_{e|\mathbf{R}}$. The connected (complex) Lie groups immersed in G are those which correspond to the Lie subalgebras of $\mathfrak{g}_e$, or equivalently to the Lie subalgebras of $\mathfrak{g}_{e|\mathbf{R}}$ which are also *complex* vector subspaces, i.e., which are *stable under the homothety* $\mathbf{u} \mapsto i\mathbf{u}$. A *closed* subgroup of G is therefore a Lie subgroup of $G_{|\mathbf{R}}$, but not in general a Lie subgroup of G: an obvious example is **R**, which is a closed subgroup of **C**.

These remarks enable us first of all to extend the results of (19.11) to complex Lie groups. So far as the adjoint representation is concerned (19.11.2), there is nothing to be changed. As to (19.11.3), if $\mathfrak{m}$ is a *complex* vector subspace of $\mathfrak{g}_e$, then the Lie algebra $\mathfrak{h}$ defined in (19.11.3(i)) is stable under $\mathbf{u} \mapsto i\mathbf{u}$, hence H is a *complex* subgroup of G; on the other hand, the Lie algebra $\mathfrak{k}$ defined in (19.11.3(ii)) is *always* stable under $\mathbf{u} \mapsto i\mathbf{u}$ when $\mathfrak{m}$ is a *real* vector subspace of $\mathfrak{g}_e$, and hence the subgroup K is always a *complex* subgroup of G. If H is a connected complex Lie group immersed in G, then $\mathscr{N}(\mathrm{H})$ is a *complex* closed subgroup of G; if H is a connected real Lie group immersed in $\mathrm{G}_{|\mathbf{R}}$, then $\mathscr{Z}(\mathrm{H})$ is a *complex* closed subgroup of G.

(19.17.3) Consider the automorphisms of a connected complex Lie group G. Such an automorphism is evidently also an automorphism of the real Lie group $\mathrm{G}_{|\mathbf{R}}$, and in view of (19.13.3) and (19.7.6) (the latter extended to complex Lie groups), the group Aut(G) of automorphisms of G may be identified with the subgroup of $\mathrm{Aut}(\mathrm{G}_{|\mathbf{R}})$ consisting of automorphisms u whose image u_* in $\mathrm{Aut}(\mathfrak{g}_{e|\mathbf{R}})$ *belongs to* $\mathrm{Aut}(\mathfrak{g}_e)$: in other words, if we identify $\mathrm{Aut}(\mathrm{G}_{|\mathbf{R}})$ with its image under $u \mapsto u_*$, we may write

$$\mathrm{Aut}(\mathrm{G}) = \mathrm{Aut}(\mathrm{G}_{|\mathbf{R}}) \cap \mathrm{Aut}(\mathfrak{g}_e).$$

If in addition G is simply-connected, Aut(G) may be canonically identified with $\mathrm{Aut}(\mathfrak{g}_e)$. In the general case, the characterization of Aut(G) as a subgroup of $\mathrm{Aut}(\tilde{\mathrm{G}})$, where $\tilde{\mathrm{G}}$ is the universal covering of G, is the same as in (19.13.3(iii)).

It follows from (19.13.8) and the remarks above in (19.17.2) that $\mathrm{Aut}(\tilde{\mathrm{G}})$, being isomorphic to $\mathrm{Aut}(\mathfrak{g}_e)$, is a *complex* Lie group; furthermore, the argument of (19.13.5(ii)) shows that the mapping $(u, x) \mapsto u(x)$ of $\mathrm{Aut}(\tilde{\mathrm{G}}) \times \tilde{\mathrm{G}}$ into $\tilde{\mathrm{G}}$ is *holomorphic*. We shall now show that Aut(G) is a *complex* Lie subgroup of $\mathrm{Aut}(\tilde{\mathrm{G}})$. For this purpose let $\mathrm{G} = \tilde{\mathrm{G}}/\mathrm{D}$, and for each $z \in \mathrm{D}$ let F_z be the set of $u \in \mathrm{Aut}(\tilde{\mathrm{G}})$ such that $u(z) \in \mathrm{D}$, so that Aut(G) is the intersection of the F_z for $z \in \mathrm{D}$. Let V be a connected open neighborhood of the identity element 1_{G} in the *real* Lie group Aut(G). By virtue of the continuity of the mapping $(u, x) \mapsto u(x)$, the image of V in D under the mapping $v \mapsto v(z)$ is connected, hence consists only of the point z because D is discrete. In other words, V is contained in the (closed) real Lie subgroup H_z of $\mathrm{Aut}(\tilde{\mathrm{G}})$ consisting of the automorphisms u such that $u(z) = z$; and conversely, the intersection of the H_z is a subgroup of Aut(G). Hence it is enough to show that each of the groups H_z is a complex Lie group, and for this it will be enough (19.17.2) to show that if $\mathbf{w} \in \mathrm{Lie}(\mathrm{H}_z)$, then also $\zeta\mathbf{w} \in \mathrm{Lie}(\mathrm{H}_z)$ for all complex ζ. Now, to say that $\mathbf{w} \in \mathrm{Lie}(\mathrm{H}_z) \subset \mathrm{Lie}(\mathrm{Aut}(\mathrm{G}))$ means that for all *real* ξ we have $\exp(\xi\mathbf{w}) \in \mathrm{H}_z$ (19.8.10), in other words $\exp(\xi\mathbf{w})(z) = z$. But if (U, φ, n) is a chart at the point z of the complex-analytic manifold $\tilde{\mathrm{G}}$, the function $\zeta \mapsto \varphi(\exp(\zeta\mathbf{w})(z))$ (with values in $\mathbf{C}^n$) is holomorphic in a neighborhood of 0 in $\mathbf{C}^n$; if it is constant for ζ real and close to 0, then it must also be constant for ζ *complex* and close to 0 (9.4.4), and this proves our assertion.

Now that we have established that Aut(G) is a complex Lie group, the same reasoning as in (19.13.5(ii)) shows once again that the mapping $(u, x) \mapsto u(x)$ is holomorphic on $\text{Aut(G)} \times \text{G}$. This being so, if we have two complex Lie groups N and L and a homomorphism of (complex) Lie groups $y \mapsto \sigma_y$ of L into Aut(N), then as in (19.14.5) we may form the semidirect product $\text{N} \times_\sigma \text{L}$, the structure of *complex*-analytic manifold of the product $\text{N} \times \text{L}$ being compatible with the group structure. The structure of the Lie algebra of this group is determined as in (19.14.7), and the result of (19.14.9) is valid without modification for complex Lie algebras and complex (simply-connected) Lie groups. From this it follows that (19.14.10) extends to solvable complex Lie groups, by replacing $\mathbf{R}$ by $\mathbf{C}$.

(19.17.4) The other type of relation between real and complex Lie groups arises from the possibility of constructing from an arbitrary *real* Lie algebra $\mathfrak{g}$ a *complex* Lie algebra $\mathfrak{g}_{(\mathbf{C})} = \mathfrak{g} \otimes_{\mathbf{R}} \mathbf{C}$ by extension of scalars: we recall that if $(\mathbf{u}_j)_{1 \leqq j \leqq n}$ is a basis of $\mathfrak{g}$, it can be identified canonically with a basis of $\mathfrak{g}_{(\mathbf{C})}$, with the *same* multiplication table (which therefore consists of *real* entries). The algebra $\mathfrak{g}_{(\mathbf{C})}$ is called the *complexification* of the real Lie algebra $\mathfrak{g}$. It should be remarked that not every complex Lie algebra is the complexification of a real Lie algebra (Problem 7). On the other hand, a given complex Lie algebra may be isomorphic to the complexification of several *nonisomorphic* real Lie algebras: these latter are called *real forms* of the given complex Lie algebra. We shall return to this question in Chapter XXI for the most important types of Lie algebras, namely the *semisimple* Lie algebras.

We shall also prove in Chapter XXI that every complex Lie algebra is the Lie algebra of a complex Lie group. The same result will then follow for a real Lie algebra $\mathfrak{g}$, for if $\mathfrak{h} = \mathfrak{g}_{(\mathbf{C})}$ is the complexification of $\mathfrak{g}$ and if H is a connected complex Lie group whose Lie algebra is isomorphic to $\mathfrak{h}$, then $\mathfrak{g}$ may be identified with a Lie subalgebra of $\mathfrak{h}_{|\mathbf{R}}$ and hence (19.7.4) is isomorphic to the Lie algebra of a connected Lie group G immersed in $\text{H}_{|\mathbf{R}}$. This will prove at the same time that on every (real) Lie group G there exists a structure of *real-analytic* manifold, compatible with the group structure of G and such that the underlying structure of differential manifold is the given one. For $\text{H}_{|\mathbf{R}}$ is endowed with such a structure, and therefore so also is every connected Lie group immersed in $\text{H}_{|\mathbf{R}}$. Moreover, this real-analytic structure on G is the unique one with these properties. This follows from the uniqueness assertion of (19.7.7), extended to real-analytic groups, which is proved in the same way as (19.7.7)—it depends on the uniqueness property of the maximal integral manifolds of an *analytic* field of directions (cf. Section 19.16, Problems 1 and 5).

PROBLEMS

1. Let G be a connected complex Lie group, and let H be a connected (real) Lie group immersed in $G_{|R}$. Show that there exists a smallest connected complex Lie group H^* immersed in G which contains H. Give an example where H is closed in $G_{|R}$, but H^* is not closed in G. (Take $G = \mathbf{C}^2/\mathbf{Z}^2$.)

2. Let A be a connected commutative complex Lie group, and let $A_{|R} = T \times V$, where T is the largest compact subgroup of $A_{|R}$ and V is isomorphic to $\mathbf{R}^n$ (19.7.9.2). Show that the smallest connected complex Lie group T^* immersed in A and containing T is closed (consider $A_{|R}/T$), and show that A is isomorphic to $T^* \times U$, where U is a closed Lie subgroup isomorphic to $\mathbf{C}^m$. (In Lie(A), consider a complex vector subspace supplementary to Lie(T^*), and observe that $A_{|R} = T^*_{|R} \times U_{|R}$, where U is the connected complex Lie group immersed in A which corresponds to this subspace.)

3. Let G be the connected commutative complex Lie group $\mathbf{C}^2/\mathbf{Z}^2$ and let $\pi : \mathbf{C}^2 \to G$ be the canonical homomorphism. Show that the image of $\mathbf{C}$ under the homomorphism $z \mapsto \pi((z, iz))$ is a closed complex one-parameter subgroup S of G, and that G/S is a compact complex Lie group, but that G is not isomorphic (as *complex* Lie group) to the product $S \times (G/S)$.

4. (a) Show that every *compact* connected complex Lie group G is necessarily commutative and hence is of the form $\mathbf{C}^n/D$, where D is a discrete subgroup of $\mathbf{C}^n$ of rank $2n$ over $\mathbf{Z}$. (Consider the adjoint representation of G, and use Section 16.3, Problem 3(b).)
(b) Show that every holomorphic differential 1-form ω on G is G-*invariant*. (The inverse image ${}^t\pi(\omega)$ of ω in $\mathbf{C}^n$ under the canonical homomorphism $\pi : \mathbf{C}^n \to G$ is of the form $\sum_{j=1}^{n} a_j(z)\, d\zeta^j$, where the a_j are holomorphic functions of $z = (\zeta^1, \ldots, \zeta^n) \in \mathbf{C}^n$, i.e., are entire functions: show that they are D-invariant, and hence constant (9.11.1).)
(c) Let $G' = \mathbf{C}^n/D'$ be another compact connected complex Lie group of complex dimension n. Show that every isomorphism of *complex manifolds* $u : G \to G'$ is of the form $s \mapsto v(s) + a'$, where v is an isomorphism of *complex Lie groups* and a' is a point of G'. (There exists an isomorphism of complex manifolds $\tilde{u} : \mathbf{C}^n \to \mathbf{C}^n$ such that $u \circ \pi = \pi' \circ \tilde{u}$, where $\pi' : \mathbf{C}^n \to G'$ is the canonical homomorphism (16.28.8). Observe that, for each holomorphic 1-form ω' on G', ${}^t\tilde{u}({}^t\pi'(\omega'))$ is a translation-invariant holomorphic 1-form on $\mathbf{C}^n$ (by (b)), and deduce that $\tilde{u}$ is an affine-linear mapping.)

5. (a) Every $\mathbf{R}$-isomorphism of vector spaces $\gamma : \mathbf{R}^{2n} \to \mathbf{C}^n$ defines on $\mathbf{R}^{2n}$ a structure of a complex vector space, by transporting the structure of $\mathbf{C}^n$ via γ^{-1} (that is, by defining $i \cdot \mathbf{x} = \gamma^{-1}(i\gamma(\mathbf{x}))$ for all $\mathbf{x} \in \mathbf{R}^{2n}$). Two such isomorphisms γ_1, γ_2 define the same complex vector space structure if and only if $\gamma_2 = s \circ \gamma_1$ for some $s \in \mathbf{GL}(n, \mathbf{C})$. For each $\mathbf{R}$-isomorphism $\gamma : \mathbf{R}^{2n} \to \mathbf{C}^n$, let $\gamma_{(\mathbf{C})} : \mathbf{C}^{2n} \to \mathbf{C}^n$ be the $\mathbf{C}$-homomorphism such that $\gamma_{(\mathbf{C})}(\mathbf{x} \otimes \zeta) = \zeta\gamma(\mathbf{x})$ for all $\mathbf{x} \in \mathbf{R}^{2n}$ and $\zeta \in \mathbf{C}$. Then we have $\dim_{\mathbf{C}}(\mathrm{Ker}(\gamma_{(\mathbf{C})})) = n$ and $\mathrm{Ker}(\gamma_{(\mathbf{C})}) \cap \mathbf{R}^{2n} = \{0\}$.
(b) Conversely, show that for each complex vector subspace V of $\mathbf{C}^{2n}$ of dimension n which is such that $V \cap \mathbf{R}^{2n} = \{0\}$, there exists an $\mathbf{R}$-isomorphism $\gamma : \mathbf{R}^{2n} \to \mathbf{C}^n$ such

that Ker $(\gamma_{(\mathbf{C})}) = \mathrm{V}$, and that if γ_1, γ_2 are two **R**-isomorphisms with this property, then $\gamma_2 = s \circ \gamma_1$ for a unique $s \in \mathbf{GL}(n, \mathbf{C})$. The set of **C**-vector space structures on $\mathbf{R}^{2n}$ for which the underlying **R**-vector space structure is the product structure on $\mathbf{R}^{2n}$ is therefore in canonical *bijective correspondence* with the set J_n of vector subspaces V of $\mathbf{C}^{2n}$ of dimension n over **C**, such that $\mathrm{V} \cap \mathbf{R}^{2n} = \{0\}$. Show that J_n is an open set in the Grassmannian $\mathbf{G}_{2n,n}(\mathbf{C})$ and is a homogeneous space for the action of $\mathbf{GL}(2n, \mathbf{R}) \subset \mathbf{GL}(n, \mathbf{C})$; also that J_n is diffeomorphic to $\mathbf{GL}(2n, \mathbf{R})/\mathbf{U}(n, \mathbf{C})$, hence to $\mathbf{R}^{n(2n+1)} \times (\mathbf{O}(2n, \mathbf{R})/\mathbf{U}(n, \mathbf{C}))$ (cf. Section **16.11**, Problem 5 and Section **16.10**, Problem 18).

(c) For each $\mathrm{V} \in \mathrm{J}_n$, if we endow $\mathbf{R}^{2n}$ with the corresponding structure of a complex vector space, we obtain on $\mathbf{R}^{2n}/\mathbf{Z}^{2n}$ a structure of complex manifold compatible with the group structure. Let G_V denote the compact complex Lie group so obtained. Show that for G_V and $\mathrm{G}_{\mathrm{V}'}$ to be isomorphic as *complex* Lie groups, it is necessary and sufficient that there should exist an automorphism $t \in \mathbf{GL}(2n, \mathbf{Z})$ of $\mathbf{R}^{2n}$ (which therefore leaves $\mathbf{Z}^{2n}$ invariant) such that $\mathrm{V}' = t(\mathrm{V})$. In other words, the isomorphism classes of compact connected complex Lie groups of complex dimension n are in a canonical bijective correspondence with the *orbits* of J_n under the action of $\mathbf{GL}(2n, \mathbf{Z})$. In particular, the set Θ_n of these classes is uncountable.

6. Let $\mathrm{U}_{2n,n}(\mathbf{C})$ be the canonical holomorphic vector bundle over the Grassmannian $\mathbf{G}_{2n,n}(\mathbf{C})$ consisting of the pairs $(\mathrm{V}, z) \in \mathbf{G}_{2n,n}(\mathbf{C}) \times \mathbf{C}^{2n}$ such that $z \in \mathrm{V}$ (Section **16.16**, Problem 1). Let N_n be the restriction of this bundle to the open set $\mathrm{J}_n \subset \mathbf{G}_{2n,n}(\mathbf{C})$ (Problem 5), so that N_n is a subbundle of the trivial bundle $\mathrm{J}_n \times \mathbf{C}^{2n}$. Let

$$\mathrm{M}_n = (\mathrm{J}_n \times \mathbf{C}^{2n})/\mathrm{N}_n$$

be the holomorphic quotient bundle, of (complex) rank n.

(a) Let $(\mathbf{e}_j)_{1 \leq j \leq 2n}$ be the canonical basis of $\mathbf{C}^{2n}$. For $1 \leq j \leq 2n$, let $\mathbf{s}_j$ be the holomorphic section of M_n which is the image under the canonical homomorphism $\mathrm{J}_n \times \mathbf{C}^{2n} \to \mathrm{M}_n$ of the section $\mathrm{V} \mapsto (\mathrm{V}, \mathbf{e}_j)$ of $\mathrm{J}_n \times \mathbf{C}^{2n}$. Show that the $\mathbf{s}_j$ form a framing of the real vector bundle over J_n which underlies M_n (observe that $\mathrm{V} \cap \mathbf{R}^{2n} = \{0\}$ for $\mathrm{V} \in \mathrm{J}_n$). This bundle is therefore trivializable.

(b) The group $\mathbf{Z}^{2n}$ acts freely and analytically on the complex manifold M_n by the rule

$$((m_1, \ldots, m_{2n}), \mathbf{u}_\mathrm{V}) \mapsto \mathbf{u}_\mathrm{V} + \sum_{j=1}^{2n} m_j \mathbf{s}_j(\mathrm{V})$$

for all $\mathrm{V} \in \mathrm{J}_n$. Show that the orbit manifold $\mathrm{T}_n = \mathrm{M}_n/\mathbf{Z}^{2n}$ exists (cf. Section **16.10**, Problem 1). On passing to the quotient, the canonical projection of M_n onto J_n induces a (holomorphic) submersion $\pi : \mathrm{T}_n \to \mathrm{J}_n$ such that, for all $\mathrm{V} \in \mathrm{J}_n$, the fiber $\pi^{-1}(\mathrm{V})$ is a complex manifold isomorphic to G_V (Problem 5). Deduce that $(\mathrm{T}_{n|\mathbf{R}}, \mathrm{J}_{n|\mathbf{R}}, \pi)$ is a trivializable fibration, so that $\mathrm{T}_{n|\mathbf{R}}$ is a real-analytic manifold isomorphic to $\mathrm{J}_{n|\mathbf{R}} \times \mathbf{T}^{2n}$, but that on the contrary $(\mathrm{T}_n, \mathrm{J}_n, \pi)$ *is not even a holomorphic fiber-bundle.*

7. Give an example of a nilpotent complex Lie algebra of dimension 7 which is not the complexification of a real Lie algebra. (Cf. Section **19.16**, Problem 8(a).)

8. Let G be the simply-connected real Lie group whose Lie algebra is $\mathfrak{sl}(2, \mathbf{R})$, so that G is the universal covering of $\mathbf{SL}(2, \mathbf{R})$. As a real-analytic manifold, G is isomorphic to $\mathbf{R}^3$, and the fundamental group $\pi_1(\mathbf{SL}(2, \mathbf{R}))$ is isomorphic to **Z**. Show that there exists no connected complex Lie group G′ with Lie algebra $\mathfrak{sl}(2, \mathbf{C})$ (the complexifica-

tion of $\mathfrak{sl}(2, \mathbf{R})$), such that G is the connected Lie group immersed in G′ which corresponds to the real Lie subalgebra $\mathfrak{sl}(2, \mathbf{R})$ of $\mathfrak{sl}(2, \mathbf{C})$. (First show that $\mathbf{SL}(2, \mathbf{C})$ is simply-connected, by using (16.30.6) and the fact that $\mathbf{SL}(2, \mathbf{C})$ is diffeomorphic to $\mathbf{SU}(2) \times \mathbf{R}^3$ (Section 11.5, Problem 15). Then observe that the center of $\mathbf{SL}(2, \mathbf{C})$ consists of two elements.)

9. Let G be a connected complex Lie group. If the mapping $\exp_G : \mathfrak{g}_e \to G$ is a local diffeomorphism at every point, show that G is nilpotent. (Use (19.16.6 and Section 19.12, Problem 3.)

CHAPTER XX

PRINCIPAL CONNECTIONS AND RIEMANNIAN GEOMETRY

The "naive" concept of differential geometry regards it as the study of curves and surfaces in physical space $\mathbf{R}^3$, and more generally of "manifolds" embedded in real space $\mathbf{R}^N$. For a long time the development of the subject centered on the metrical notions which were regarded as "natural" in $\mathbf{R}^N$, and which on "curved" submanifolds gave rise to the classical notions of length, area, volume, etc. This point of view emphasized the role of the group of Euclidean displacements,† under which these notions are invariant, and two submanifolds of $\mathbf{R}^N$ which are related by a displacement have always been regarded as intrinsically the same. From the inception of infinitesimal calculus, it was realized that one could attach to each point of a plane curve (for example) a number depending on the point which measured the "curvature" of the curve at that point, in the intuitive sense of the word; and it is easily shown that the knowledge of this number as a function of arc length determines the curve uniquely, up to a displacement. This is the starting point of the study, pursued unremittingly over two centuries, of the "differential invariants" of manifolds embedded in $\mathbf{R}^N$: a study which for a long time remained purely "local," but which from the beginning of this century came to include many "global" problems concurrently with the development of topology.

Superimposed on this line of development, since the time of Gauss, has been the "intrinsic" conception of manifolds equipped with an "infinitesimal" metric, independent of any embedding in real space $\mathbf{R}^N$. It required the efforts of Riemann and several subsequent generations of mathematicians to lay the foundations of this new theory, with the help of what was called "tensor calculus" (which is nothing more than "localized" multilinear algebra). Other efforts were necessary to extract and separate out the concept

† In the wide sense, i.e., with determinant ± 1.

of a differential manifold (not endowed with a metric) from that of a Riemannian manifold: on a Riemannian manifold, as was shown by Levi-Civita, the metric defines a unique "connection" (in the sense of (17.18)), which generalizes the intuitive idea of "parallelism" in $\mathbf{R}^N$ (20.9).

It appears at first sight, when one embarks on the study of general Riemannian geometry, that group theory ceases to play a part (in general, a Riemannian manifold possesses no metric-preserving diffeomorphism other than the identity). A fundamental advance, due to E. Cartan, was the recognition that Lie groups play as important a role in differential geometry as in classical geometry (in the sense of Klein). By refining and developing the "method of moving frames," which had proved its usefulness in the classical theory of surfaces, he showed that *vis-à-vis* an arbitrary Riemannian manifold M (or rather its tangent bundle T(M)), there is a principal fiber bundle, whose group is the orthogonal group, which plays a role entirely comparable to that played by a Lie group *vis-à-vis* its homogeneous spaces. Here again, the technique of "lifting everything up to the principal bundle of frames" reveals best the nature of problems in differential geometry (cf. Section 20.5).

This method need not be restricted to the orthogonal group, and according to the nature of the group G of the principal bundle under consideration, we obtain different geometries ("G-structures") in great variety. It has been impossible within the framework of this Treatise to do more than indicate the general principle which leads to these "geometries" (Sections 20.1 to 20.7), and we must refer the reader to specialized monographs for information on spaces endowed with an affine connection, a projective connection or a conformal connection, Finsler spaces, Hamiltonian structures (or contact structures), and their relations with dynamics, etc. (see [37, 54–56, 59, 61, 63, 70, 71]). The second part of the chapter is devoted to Riemannian manifolds; but here again the subject is so vast that we have regretfully had to pass by in silence many parts of it. The great theory of Riemannian symmetric spaces ([45, 62, 66]) is hardly touched on; the fundamental notion of the holonomy group is not mentioned, nor is the geometry of Hermitian manifolds and their most important particular case, the theory of Kähler manifolds ([18, 45]). Our study of geodesics does not go beyond the level of the most elementary theorems, and although this study constitutes one of the most attractive and complete chapters of the calculus of variations, this latter theory also remains beyond our horizon: both as regards the study in depth of extremal conditions in the case of simple integrals and the "multidimensional" general theory ([53, 58, 63, 65, 68]), and as regards the most beautiful application to analysis and topology of 1-dimensional calculus of variations, namely Morse theory ([63, 67]).

Although the techniques of analysis on manifolds which we have developed in Chapters XVI to XVIII underlie the results of this chapter, nevertheless it has a much more pronounced "geometrical" flavor than the other parts of this treatise, and corresponds to what used to be called "the application of analysis to geometry." However, in accordance with the spirit of modern mathematics, Geometry pays back in large measure the support it has drawn from Analysis, as our allusions above indicate, and as the reader will see for himself repeatedly in later chapters.

1. THE BUNDLE OF FRAMES OF A VECTOR BUNDLE

(20.1.1) Let B be a real differential manifold, E a real vector bundle over B and $p : E \to B$ the canonical projection. Let F be a real vector space of finite dimension, and consider the trivial bundle $B \times F$ over B, with fiber F. The general linear group $\mathbf{GL}(F)$ *acts on the right* on the vector bundle $\text{Hom}(B \times F, E)$ (16.16.3). Namely, the fiber of $\text{Hom}(B \times F, E)$ over a point $b \in B$ may be canonically identified with the vector space $\text{Hom}(F, E_b)$, because F_b is canonically identified with F by means of the projection $B \times F \to F$; the action of $\mathbf{GL}(F)$ on $\text{Hom}(F, E_b)$ is composition on the right $(u, s) \mapsto u \circ s$, where u is a linear mapping of F into E_b, and s is an automorphism of F. This action is differentiable, because by restricting attention to a sufficiently small neighborhood of a point of B, we may assume that E is trivial, of the form $B \times E_0$, where E_0 is a finite-dimensional real vector space; then $\text{Hom}(B \times F, E)$ is identified with the trivial bundle $B \times \text{Hom}(F, E_0)$, and the action of $\mathbf{GL}(F)$ is identified with the mapping $((b, v), s) \mapsto (b, v \circ s)$ of $(B \times \text{Hom}(F, E_0)) \times \mathbf{GL}(F)$ into $B \times \text{Hom}(F, E_0)$, which is of class C^∞.

Suppose now that B is a pure manifold and that the dimension of F is equal to the dimension of the fibers of E. Consider now the subset $\text{Isom}(B \times F, E)$ of $\text{Hom}(B \times F, E)$ defined by the condition that, for each $b \in B$, the intersection of this subset and the fiber $\text{Hom}(F, E_b) = (\text{Hom}(B \times F, E))_b$ is the set $\text{Isom}(F, E_b)$ of *isomorphisms* of the vector space F onto the vector space E_b. This set $\text{Isom}(B \times F, E)$ is a *dense open set* in $\text{Hom}(B \times F, E)$. To see this, again it is enough to consider the case where $E = B \times E_0$ is a trivial bundle; then $\text{Isom}(B \times F, E)$ may be identified with $B \times \text{Isom}(F, E_0)$, and $\text{Isom}(F, E_0)$ is a dense open subset of $\text{Hom}(F, E_0)$, because if u_0 is an isomorphism of F onto E_0, then $\text{Isom}(F, E_0)$ is the image of $\mathbf{GL}(F)$ under the mapping $s \mapsto u_0 \circ s$, and $\mathbf{GL}(F)$ is a dense open subset of $\text{End}(F)$ (for the polynomial $\det(x_{ij})$ in n^2 variables x_{ij} is not identically zero and therefore cannot vanish on a nonempty open set in $\mathbf{R}^{n^2}$ (9.4.2)). It is clear that $\text{Isom}(B \times F, E)$ is stable under the action of $\mathbf{GL}(F)$ on $\text{Hom}(B \times F, E)$, and $\mathbf{GL}(F)$ acts freely and transitively on each fiber $\text{Isom}(F, E_b)$. By virtue of (12.10.7), B may be identified with the orbit manifold $\mathbf{GL}(F)\backslash\text{Isom}(B \times F, E)$; consequently (16.14), for this action of GL(F) the space Isom(B × F, E) is a *principal bundle* over B. The C^∞ sections of this bundle over an open subset U of B are just the *framings* of E over U (16.15), and the principal bundle Isom(B × F, E) is called the *bundle of frames* of the vector bundle E.

If M is a pure differential manifold of dimension n, the fibers of T(M) are isomorphic to $\mathbf{R}^n$; the principal bundle Isom(M × $\mathbf{R}^n$, T(M)) with base M and group $\mathbf{GL}(n, \mathbf{R})$ is denoted by R(M) and is called the *bundle of frames* (or the *bundle of tangent frames*) *of the manifold* M. If $\Phi : \varphi(\mathrm{U}) \to \psi(\mathrm{U})$ is the transition diffeomorphism between two charts of M having the same domain of definition, then the transition diffeomorphism for the corresponding fibered charts of R(M) will be

(20.1.1.1) $$(x, S) \mapsto (\Phi(x), \mathrm{D}\Phi(x)S)$$

of $\varphi(\mathrm{U}) \times \mathbf{GL}(n, \mathbf{R})$ into $\psi(\mathrm{U}) \times \mathbf{GL}(n, \mathbf{R})$.

(20.1.2) We may identify F with $\mathbf{R}^\mathrm{N}$ by choosing a basis of F. Suppose that $\mathrm{F} = \mathbf{R}^\mathrm{N}$; then Isom($\mathbf{R}^\mathrm{N}$, E_b) may be identified with the set of sequences of N elements $(\mathbf{a}_1, \dots, \mathbf{a}_\mathrm{N})$ forming a *basis* of E_b, by identifying each isomorphism $v : \mathbf{R}^\mathrm{N} \to \mathrm{E}_b$ with the sequence of images $v(\mathbf{e}_j)$ of the vectors of the canonical basis of $\mathbf{R}^\mathrm{N}$. A *section* $\mathbf{s}$ of Isom(B × $\mathbf{R}^\mathrm{N}$, E) over an open subset U of B is therefore a mapping $y \mapsto (\mathbf{s}_1(y), \dots, \mathbf{s}_\mathrm{N}(y))$, where the $\mathbf{s}_j$ are sections of E over U with the property that, for each $y \in \mathrm{U}$, the vectors $\mathbf{s}_j(y)$ form a basis of E_y. Again, we may identify Isom(B × $\mathbf{R}^\mathrm{N}$, E) with a dense open subset of the direct sum $\mathrm{E}^\mathrm{N} = \mathrm{E} \oplus \mathrm{E} \oplus \cdots \oplus \mathrm{E}$ (N terms) (16.16.1); the action of a matrix $t = (t_{ij}) \in \mathbf{GL}(\mathrm{N}, \mathbf{R})$ on a fiber Isom($\mathbf{R}^\mathrm{N}$, E_b) is given by

$$(\mathbf{a}_1, \dots, \mathbf{a}_\mathrm{N}) \cdot t = (\mathbf{a}'_1, \dots, \mathbf{a}'_\mathrm{N}),$$

where

$$\mathbf{a}'_i = \sum_{j=1}^{\mathrm{N}} t_{ji} \mathbf{a}_j \qquad (1 \leqq i \leqq \mathrm{N}).$$

(20.1.3) Conversely, consider a principal bundle R with structure group G, base space B = G\R and projection π (where G acts on R on the right); consider also a *linear representation* $\rho : \mathrm{G} \to \mathbf{GL}(\mathrm{F})$ of G in a finite-dimensional vector space F, so that G acts differentiably on F (on the left) by the rule $(s, \mathbf{x}) \mapsto \rho(s) \cdot \mathbf{x}$ (which we denote by $s \cdot \mathbf{x}$) (16.10). Then, in the associated bundle $\mathrm{R} \times^\mathrm{G} \mathrm{F}$ with base B (16.14.7), we can define canonically on each fiber $\pi_\mathrm{F}^{-1}(b)$ a structure of a vector space isomorphic to F. Namely (with the notation of (16.14.7)), if U is a neighborhood of b over which R is trivializable, and if $\sigma : \mathrm{U} \to \pi^{-1}(\mathrm{U})$ is a C^∞ section of R over U, then we can transport the vector space structure of $\{b\} \times \mathrm{F}$ to $\pi_{\mathrm{F}}^{-1}(b)$ via the bijection $(b, \mathbf{y}) \mapsto \sigma(b) \cdot \mathbf{y}$. We must check that this structure does not depend on the choice of the section σ: now if σ' is another C^∞ section of R over U, then we can write $\sigma'(b) = \sigma(b) \cdot s$, where $s \in \mathrm{G}$, and since $\mathbf{y} \mapsto s \cdot \mathbf{y}$ is a linear bijection of F onto itself, this proves our assertion. It is clear that these vector space structures on the fibers of $\mathrm{R} \times^{\mathrm{G}} \mathrm{F}$ satisfy the condition (VB) of (16.15), and so we obtain a structure of *vector bundle* on $\mathrm{R} \times^{\mathrm{G}} \mathrm{F}$.

(20.1.4) Let us apply the construction of (20.1.3) to the case where $\mathrm{R} = \mathrm{Isom}(\mathrm{B} \times \mathrm{F}, \mathrm{E})$ (with the notation of (20.1.1)) and ρ is the identity mapping of **GL**(F). We have then a *canonical isomorphism* of E onto $\mathrm{R} \times^{\mathrm{G}} \mathrm{F}$. Indeed, for each open subset U of B over which E is trivializable, there exists a framing $\varphi : \mathrm{U} \times \mathrm{F} \to p^{-1}(\mathrm{U})$ (16.15) which may be identified with a section $b \mapsto \varphi(b, .)$ of $\mathrm{Isom}(\mathbf{B} \times \mathrm{F}, \mathrm{E})$ over U; on the other hand (16.14.7), this section defines an isomorphism $\psi : (b, \mathbf{z}) \mapsto \varphi(b, .) \cdot \mathbf{z}$ of $\mathrm{U} \times \mathrm{F}$ onto $\pi_{\mathrm{F}}^{-1}(\mathrm{U})$. It is sufficient to show that the isomorphism $\psi \circ \varphi^{-1} : p^{-1}(\mathrm{U}) \to \pi_{\mathrm{F}}^{-1}(\mathrm{U})$ does not depend on the framing φ chosen. But if $\varphi' : \mathrm{U} \times \mathrm{F} \to p^{-1}(\mathrm{U})$ is another framing, then the mapping $\varphi'^{-1} \circ \varphi$ is of the form $(b, \mathbf{z}) \mapsto (b, s(b) \cdot \mathbf{z})$, where s is a C^∞ mapping of U into **GL**(F). The mapping $\psi' : \mathrm{U} \times \mathrm{F} \to \pi_{\mathrm{F}}^{-1}(\mathrm{U})$ corresponding to φ' is

$$(b, \mathbf{z}) \mapsto \varphi'(b, .) \cdot \mathbf{z} = (\varphi(b, .) \circ s^{-1}) \cdot \mathbf{z} = \varphi(b, .) \cdot (s^{-1} \cdot \mathbf{z})$$

by (16.14.7). This signifies that $\psi' = \psi \circ (\varphi' \circ \varphi)^{-1}$, so that $\psi' \circ \varphi'^{-1} = \psi \circ \varphi^{-1}$.

With the same notation, let us remark now for example that **GL**(F) acts canonically on an arbitrary exterior power $\bigwedge^m \mathrm{F}$: in other words, we have a canonical linear representation $\rho : \mathbf{GL}(\mathrm{F}) \to \mathbf{GL}\left(\bigwedge^m \mathrm{F}\right)$, and with this representation there is associated by the mechanism of (20.1.3) a vector bundle $\mathrm{R} \times^{\mathrm{G}} \left(\bigwedge^m \mathrm{F}\right)$. This bundle is *isomorphic* to the vector bundle $\bigwedge^m \mathrm{E}$ defined in (16.16.2). For if $\mathbf{s}_1, \ldots, \mathbf{s}_m$ are m C^∞ sections of E over an open subset U of **B**, and if for brevity we put $\mathrm{F}' = \bigwedge^m \mathrm{F}$, then there exists in each fiber $\pi_{\mathrm{F}'}^{-1}(b)$, where $b \in \mathrm{U}$, a well-determined element $\mathbf{s}(b)$ such that, if $\mathbf{s}_j(b) = \varphi(b, \mathbf{y}_j)$, we have $\mathbf{s}(b) = \rho(\varphi(b, .)) \cdot (\mathbf{y}_1 \wedge \ldots \wedge \mathbf{y}_m)$ for every framing φ defined on $\mathrm{V} \times \mathrm{F}$, where V is a neighborhood of b. Moreover, this proves that $\mathbf{s}$ is a C^∞ section of $\mathrm{R} \times^{\mathrm{G}} \left(\bigwedge^m \mathrm{F}\right)$, and it suffices to apply the characterization given in (16.16.2) for $\bigwedge^m \mathrm{E}$. Likewise for tensor bundles over E, symmetric tensor bundles, etc.

PROBLEMS

1. Let B be a pure differential manifold and E a real vector bundle of rank n over B. For each $b \in B$, let $\mathrm{Aff}(\mathbf{R}^n, E_b)$ be the set of *affine-linear bijections* of $\mathbf{R}^n$ onto E_b, i.e., mappings of the form $\mathbf{x} \mapsto \mathbf{a} + v(\mathbf{x})$, where $v : \mathbf{R}^n \mapsto E_b$ is a linear bijection and $\mathbf{a} \in E_b$. Let $A = \mathrm{Aff}(B \times \mathbf{R}^n, E)$ be the disjoint union of the sets $\mathrm{Aff}(\mathbf{R}^n, E_b)$ for all $b \in B$ and let $\alpha : A \to B$ be the mapping which sends each element of $\mathrm{Aff}(\mathbf{R}^n, E_b)$ to b. Next, let $\mathbf{A}(n, \mathbf{R})$ be the Lie group of affine-linear bijections of $\mathbf{R}^n$ onto $\mathbf{R}^n$, identified with the subgroup of $\mathbf{GL}(n+1, \mathbf{R})$ consisting of the matrices

$$\begin{pmatrix} U & \mathbf{a} \\ 0 & 1 \end{pmatrix},$$

where $U \in \mathbf{GL}(n, \mathbf{R})$ and $\mathbf{a} \in \mathbf{R}^n$ (column-vector), the action of $\mathbf{A}(n, \mathbf{R})$ on $\mathbf{R}^n$ being

$$\begin{pmatrix} U & \mathbf{a} \\ 0 & 1 \end{pmatrix} \cdot \mathbf{x} = U \cdot \mathbf{x} + \mathbf{a}.$$

The group $\mathbf{A}(n, \mathbf{R})$ (called the *affine group of* $\mathbf{R}^n$) acts on A on the right by the rule $(u, s) \mapsto u \circ s$. Show that there exists on A a unique structure of principal bundle with base B, projection α, and group $\mathbf{A}(n, \mathbf{R})$ (cf. Section **16.14**, Problem 1). The set A, endowed with this structure, is called the *bundle of affine frames* of the vector bundle E.

If we put $G = \mathbf{A}(n, \mathbf{R})$, the left action of G on $\mathbf{R}^n$ defines a bundle of fiber-type $\mathbf{R}^n$ associated with A, namely $A \times^G \mathbf{R}^n$. Show that the fibration so defined is canonically B-isomorphic to that of E.

2. Let R be a principal fiber bundle with base B, group H and projection π. Suppose that H is a *connected Lie subgroup* of a Lie group G. Then H (as a subgroup of G) acts by left translation on the homogeneous space G/H, and we may therefore form the associated bundle $R \times^H (G/H) = X$ with base B and fibers diffeomorphic to G/H. Let $\dot{e}$ be the canonical image in G/H of the identity element e of G. Show that the mapping $r \mapsto r \cdot \dot{e}$ (**16.14.7**) factorizes into $r \mapsto \pi(r) \xrightarrow{\sigma} r \cdot \dot{e}$, where σ is a C^∞ *section* of X over B, called the *canonical* section; $B' = \sigma(B)$ is therefore a submanifold of X diffeomorphic to B. Let V(B′) be the vector bundle over B′ induced on the submanifold B′ by the vector bundle V(X) of vertical tangent vectors to X (**16.12.1**), which is a subbundle of T(X). Show that V(B′) is B′-isomorphic to the associated vector bundle $R \times^H (\mathfrak{g}_e/\mathfrak{h}_e)$, where $\mathfrak{g}_e$ and $\mathfrak{h}_e$ are the Lie algebras of G and H, respectively, and $t \in H$ acts on the left on $\mathfrak{g}_e/\mathfrak{h}_e$ by means of the tangent mapping at the point $\dot{e}$ of the mapping $\dot{s} \mapsto t \cdot \dot{s}$ (Section **19.7**, Problem 7).

Suppose now that dim(B) = dim(G/H). A *B′-isomorphism* of vector bundles of T(B′) onto V(B′) is called a *welding* of B and X (so that if such a welding exists, then the vector bundle T(B′) is isomorphic to V(B′), and consequently T(B) may be considered as the vector bundle associated with a principal bundle whose group H may be *larger* than the group $\mathbf{GL}(n, \mathbf{R})$(where $n = \dim(B)$)).

Suppose that there exists a welding of B and X, and let (σ^{-1}, α) be an isomorphism of V(B′) onto T(B). Identify $\mathfrak{g}_e/\mathfrak{h}_e$ with $\mathbf{R}^n$ by choosing a basis, and for each $t \in H$ let $\rho(t)$ be the image of t in $\mathbf{GL}(n, \mathbf{R})$, which is a (linear) operation of t on $\mathfrak{g}_e/\mathfrak{h}_e$. Let $\tilde{H} = \rho(H)$, which is a Lie group immersed in $\mathbf{GL}(n, \mathbf{R})$, isomorphic to a quotient of H. We may therefore form the principal bundle $R \times^{H} \tilde{H}$, where H acts on $\tilde{H}$ on the left by means of ρ. For each element $r'_b \cdot \rho(t)$ of this principal bundle, the mapping $x \mapsto \alpha(r'_b \cdot (\rho(t) \cdot x))$ is an isomorphism $u(r'_b \cdot \rho(t)) : x \mapsto r_b \cdot x$ of $\mathbf{R}^n$ onto $T_x(B)$, i.e., is an element of the fiber $R(B)_b$ of the bundle of frames R(B) of B. Show that in this way we obtain an injective B-*morphism* (u, j) of the principal bundle $R \times^{H} \tilde{H}$ into R(B), where j is the canonical injection of $\tilde{H}$ into $\mathbf{GL}(n, \mathbf{R})$. Conversely, such an injective B-morphism defines a well-determinded welding of B and X.

3. Let M be a pure differential manifold of dimension n.

(a) Show that the bundle of tangent frames R(M) of M may be canonically identified with the open submanifold of the manifold $J_0^1(\mathbf{R}^n, M)$ of jets of order 1 of $\mathbf{R}^n$ into M with source 0, consisting of the *invertible* jets (Section 16.5, Problem 9). For each $k \geqq 1$, the set $R_k(M)$ of invertible jets of order k of $\mathbf{R}^n$ into M, with source 0, is likewise an open submanifold of $J_0^k(\mathbf{R}^n, M)$ called the manifold of *frames of order* k of M.
(b) Let $G^k(n)$ be the group of invertible jets of order k of $\mathbf{R}^n$ into $\mathbf{R}^n$, with source and target 0 (Section 16.9, Problem 1): it acts on the right on $R_k(M)$ by the rule $(u, s) \mapsto u \circ s$. Show that, relative to this action, $R_k(M)$ is a principal bundle over M with group $G^k(n)$.
(c) For $h < k$, the restriction to $R_k(M)$ of the canonical mapping $J_0^k(\mathbf{R}^n, M) \to J_0^h(\mathbf{R}^n, M)$ is a submersion $p^{hk} : R_k(M) \to R_h(M)$. The canonical mapping $\rho^{hk} : G^k(n) \to G^h(n)$ is a surjective homomorphism of Lie groups, whose kernel $N^{hk}(n)$ is nilpotent and simply-connected (and commutative if $h = k - 1$); the pair (p^{hk}, ρ^{hk}) is a morphism of principal bundles (16.14.3), and $R_k(M)$ is a principal bundle over $R_h(M)$, with group $N^{hk}(n)$ and projection p^{hk}.
(d) Let M′ be a pure differential manifold. Show that the fibration $(J^k(M, M'), M, \pi)$ defined in Section 16.12, Problem 6 is isomorphic to the bundle associated to $R_k(M)$, of fiber-type $J_0^k(\mathbf{R}^n, M')$, on which $G^k(n)$ acts on the left by the rule $(s, j) \mapsto j \circ s^{-1}$. Show likewise how to define the fibrations

$$(J^k(M, M'), M', \pi') \quad \text{and} \quad (J^k(M, M'), M \times M', (\pi, \pi'))$$

as associated fiber bundles.

2. PRINCIPAL CONNECTIONS ON PRINCIPAL BUNDLES

(20.2.1) We return to the considerations of (17.16.1) which led us to the notion of a *connection* in a vector bundle E of rank p over a base-space B. We shall restrict our attention to the case envisaged in (17.16.1), where B is an open subset U of $\mathbf{R}^n$ and E is the trivial bundle $U \times \mathbf{R}^p$. The corresponding bundle of frames is then the trivial bundle $U \times \mathbf{GL}(p, \mathbf{R})$. The mapping $\mathbf{h} \mapsto F(\mathbf{h})$ considered in (17.16.1) is by hypothesis a mapping of a neighborhood V of 0 in $\mathbf{R}^n$, into $\mathbf{GL}(p, \mathbf{R})$ (if V is taken sufficiently small), and for each $\mathbf{h} \in \mathbf{z}$, $F(\mathbf{h})$ defines a mapping $(b, r) \mapsto (b + \mathbf{h}, F(\mathbf{h})^{-1} \circ r)$ of the fiber $\mathrm{Isom}(\mathbf{R}^p, E_b)$ of $\mathrm{Isom}(U \times \mathbf{R}^p, E)$ at the point b into the fiber $\mathrm{Isom}(\mathbf{R}^p, E_{b+\mathbf{h}})$

of this same bundle at the point $b + \mathbf{h}$. Hence we obtain an indefinitely differentiable mapping $(\mathbf{h}, r) \mapsto (b + \mathbf{h}, F(\mathbf{h})^{-1} \circ r)$ of $V \times \mathbf{GL}(p, \mathbf{R})$ into

$$U \times \mathbf{GL}(p, \mathbf{R})$$

(where $\mathbf{GL}(p, \mathbf{R})$ is considered as an open subset of $\mathrm{End}(\mathbf{R}^p) = \mathbf{R}^{p^2}$). Its *derivative* at the point $(0, r)$ is therefore $(\mathbf{k}, v) \mapsto (\mathbf{k}, v - (DF(0) \cdot \mathbf{k}) \circ r)$, a linear mapping of $\mathbf{R}^n \times \mathrm{End}(\mathbf{R}^p)$ into itself. Since the tangent space

$$T_{(b, r)}(\mathrm{Isom}(B \times \mathbf{R}^p, E))$$

may be identified with $T_b(U) \times T_r(\mathrm{End}(\mathbf{R}^p))$, the vectors $(\mathbf{k}, v)$ and

$$(\mathbf{k}, v - (DF(0) \cdot \mathbf{k}) \circ r)$$

should be considered as vectors in this tangent space, and hence also the vector

$$(\mathbf{k}, -(DF(0) \cdot \mathbf{k}) \circ r) = \mathbf{P}_b(\mathbf{k}, r).$$

On the other hand, $(\mathbf{k}, r)$ should be regarded as an element of the fiber over the point b of the fiber-product $T(B) \times_B \mathrm{Isom}(B \times \mathbf{R}^p, E)$ over B. Replacing the bundle of frames by an arbitrary principal bundle over B, we are led to make the following general and intrinsic definition:

(20.2.2) Given a *principal bundle* R with base B, projection π and group G, a *connection* (or *principal connection*) in R is defined to be a C^∞ mapping

(20.2.2.1) $$\mathbf{P} : T(B) \times_B R \to T(R),$$

with the following properties (where $b \in B$, $\mathbf{k}_b \in T_b(B)$, $r_b \in R_b$):

(20.2.2.2) $T(\pi)(\mathbf{P}_b(\mathbf{k}_b, r_b)) = \mathbf{k}_b, \qquad o_R(\mathbf{P}_b(\mathbf{k}_b, r_b)) = r_b;$

(20.2.2.3) $\mathbf{k}_b \mapsto \mathbf{P}_b(\mathbf{k}_b, r_b)$ is a *linear* mapping of $T_b(B)$ into $T_{r_b}(R)$;

(20.2.2.4) for each $s \in G$ we have $\mathbf{P}_b(\mathbf{k}_b, r_b \cdot s) = \mathbf{P}_b(\mathbf{k}_b, r_b) \cdot s$

(where on the right-hand side, $\mathbf{w} \mapsto \mathbf{w} \cdot s$ is the tangent linear mapping of $r \mapsto r \cdot s$ at the point r_b (16.10)).

The vector $\mathbf{P}_b(\mathbf{k}_b, r_b)$ is called the *horizontal lifting* of the vector $\mathbf{k}_b$ at the point r_b. If X is a vector field on B, then $r_b \mapsto \mathbf{P}_b(X(b), r_b)$ is a vector field on R, called the *horizontal lifting* of X.

It follows from (20.2.2.2) that the linear mapping $\mathbf{k}_b \mapsto \mathbf{P}_b(\mathbf{k}_b, r_b)$ is *injective* and that its image H_{r_b} is *supplementary* in the tangent space $T_{r_b}(R)$ to the *kernel* G_{r_b} of $T_{r_b}(\pi)$, which we have called the space of *vertical tangent vectors* (16.12.1) at the point $r_b \in R_b$. The space H_{r_b} is called the space of *horizontal tangent vectors* at the point r_b, relative to the connection $\mathbf{P}$. We have $H_{r_b \cdot s} = H_{r_b} \cdot s$ for all $s \in G$.

If R = B × G is trivial and B is an open set in $\mathbf{R}^n$ (so that T(B) is identified with B × $\mathbf{R}^n$), we may write

$$\mathbf{P}_b((b, \mathbf{k}), (b, s)) = ((b, \mathbf{k}), P(b, s) \cdot \mathbf{k})$$

for $(b, s) \in \mathrm{B} \times \mathrm{G}$ and $\mathbf{k} \in \mathbf{R}^n$, where $P(b, s)$ is a homomorphism of the vector space $\mathbf{R}^n$ into $\mathrm{T}_s(\mathrm{G})$ (we have identified $\mathrm{T}_{(b,s)}(\mathrm{B} \times \mathrm{G})$ with $\{b\} \times \mathbf{R}^n \times \mathrm{T}_s(\mathrm{G})$). For $s, t \in \mathrm{G}$, we derive from (20.2.2.4) the relation

(20.2.2.5) $$P(b, st) \cdot \mathbf{k} = (P(b, s) \cdot \mathbf{k}) \cdot t$$

(in the notation of (16.10)), whence, in particular,

(20.2.2.6) $$P(b, s) \cdot \mathbf{k} = (Q(b) \cdot \mathbf{k}) \cdot s,$$

where we have put $Q(b) = P(b, e)$, an element of $\mathrm{Hom}(\mathbf{R}^n, \mathfrak{g}_e)$. Conversely, given any C^∞ mapping Q of B into $\mathrm{Hom}(\mathbf{R}^n, \mathfrak{g}_e)$, we may define $P(b, s)$ by means of (20.2.2.6), then $\mathbf{P}_b((b, \mathbf{k}), (b, s))$ as above, and obtain a principal connection in B × G.

(20.2.3) For each $r_b \in \mathrm{R}_b$, there is a *projection* $p_{r_b} : \mathrm{T}_{r_b}(\mathrm{R}) \to \mathrm{T}_{r_b}(\mathrm{R})$ with image H_{r_b} and kernel G_{r_b}. For each tangent vector $\mathbf{h}_{r_b} \in \mathrm{T}_{r_b}(\mathrm{R})$, we have

(20.2.3.1) $$p_{r_b}(\mathbf{h}_{r_b}) = \mathbf{P}_b(T(\pi) \cdot \mathbf{h}_{r_b}, r_b),$$

which is called the *horizontal component* of $\mathbf{h}_{r_b}$, and is the *horizontal lifting* of the projection $\mathrm{T}(\pi) \cdot \mathbf{h}_{r_b}$ of $\mathbf{h}_{r_b}$ in $\mathrm{T}_b(\mathrm{B})$. It is clear that for all $s \in \mathrm{G}$ we have

(20.2.3.2) $$p_{r_b}(\mathbf{h}_{r_b} \cdot s) = (p_{r_b}(\mathbf{h}_{r_b})) \cdot s$$

with the notation of (16.10).

We remark that a horizontal vector $\mathbf{h} \in \mathrm{H}_{r_b}$ may always be considered as the value at the point r_b of a *horizontal vector field* Y of class C^∞. For there exists a C^∞ vector field Y_0 on R such that $Y_0(r_b) = \mathbf{h}$ (16.12.11), and we may define $Y(r')$ to be equal to $p_{r'}(Y_0(r'))$ for all $r' \in \mathrm{R}$.

The vector

$$\mathbf{h}_{r_b} - p_{r_b}(\mathbf{h}_{r_b}) = \mathbf{h}_{r_b} - \mathbf{P}_b(\mathrm{T}(\pi) \cdot \mathbf{h}_{r_b}, r_b)$$

is therefore *vertical*. Observe now that, for each $r_b \in \mathrm{R}_b$, the mapping $s \mapsto r_b \cdot s$ is a diffeomorphism of G onto R_b, and consequently

(20.2.3.3) $$\mathbf{t}_{r_b} : \mathbf{u} \mapsto r_b \cdot \mathbf{u}$$

(in the notation of (16.10)) is a *linear bijection of the Lie algebra* $\mathfrak{g}_e = \mathrm{T}_e(\mathrm{G})$ *onto the space of vertical tangent vectors* G_{r_b} at the point r_b. The mapping

(20.2.3.4) $$\boldsymbol{\omega}(r_b) : \mathbf{h}_{r_b} \mapsto \mathbf{t}_{r_b}^{-1}(\mathbf{h}_{r_b} - \mathbf{P}_b(\mathrm{T}(\pi) \cdot \mathbf{h}_{r_b}, r_b))$$

is therefore a *surjective linear mapping of* $T_{r_b}(R)$ *onto the Lie algebra* $\mathfrak{g}_e$: in other words, we have in this way defined a *vector-valued differential* 1-*form* ω *on* R, *with values in* $\mathfrak{g}_e$ (16.20.15). It is immediately verified that ω is of class C^∞: for we may assume that $R = B \times G$ is trivial, so that $\mathbf{t}_{r_b}^{-1}$, for $r_b = (b, s) \in B \times G$, is the composition of $\mathbf{h}_s \mapsto s^{-1} \cdot \mathbf{h}_s$ and the second projection $T_{(b,s)}(B \times G) \to T_s(G)$, from which the assertion follows immediately. The 1-form ω is called the *differential* 1-*form of the connection* **P**, or the *connection form* of **P**.

If $R = B \times G$ is trivial and B is an open set in $\mathbf{R}^n$, the mapping $\omega(b, e)$ may be identified (with the notation of (20.2.2)) with the projection

(20.2.3.5) $$\varpi_b : (\mathbf{k}, \mathbf{u}) \mapsto \mathbf{u} - Q(b) \cdot \mathbf{k}$$

of $\mathbf{R}^n \times \mathfrak{g}_e$ onto $\mathfrak{g}_e$, the mapping $b \mapsto \varpi_b$ of B into $\operatorname{End}(\mathbf{R}^n \times \mathfrak{g}_e)$ being of class C^∞, and $\omega(b, s)$ is identified with the linear mapping

(20.2.3.6)
$$((b, \mathbf{k}), (b, s \cdot \mathbf{u})) \mapsto \operatorname{Ad}(s^{-1}) \cdot (\mathbf{u} - Q(b) \cdot \mathbf{k}) = \operatorname{Ad}(s^{-1}) \cdot (\varpi_b \cdot (\mathbf{k}, \mathbf{u})).$$

(20.2.4) A differential q-form α on R, with values in a finite-dimensional vector space V, is said to be *vertical* (resp. *horizontal*) if

$$\alpha(r_b) \cdot (\mathbf{h}_1 \wedge \mathbf{h}_2 \wedge \cdots \wedge \mathbf{h}_q) = 0$$

whenever *one* of the tangent vectors $\mathbf{h}_j \in T_{r_b}(R)$ is *horizontal* (resp. *vertical*). Notice that the definition of a horizontal q-form does not depend on the presence of a connection in R, whereas the notion of a vertical q-form is meaningless except with respect to a connection. It is clear that the 1-form ω of the connection is *vertical*, and that the 2-form $[\omega, \omega]$ (16.20.15.8) is also vertical.

(20.2.5) *In order that a* C^∞ *vector-valued* 1-*form* ω *on* R, *with values in* $\mathfrak{g}_e$, *should be the* 1-*form of a connection in* R, *it is necessary and sufficient that it satisfy the following two conditions*:

(1) ω, *considered as a mapping of* T(R) *into* $\mathfrak{g}_e$, *is invariant* (19.1.4) *under the right action of* G *on* T(R) *induced* (19.1.4) *by the action of* G *on* R, *and the right action* $(\mathbf{u}, s) \mapsto \operatorname{Ad}(s^{-1}) \cdot \mathbf{u}$ *of* G *on* $\mathfrak{g}_e$ (19.2.1): *in other words, for* $\mathbf{h} \in T_{r_b}(R)$ *and* $s \in G$ *we have*

(20.2.5.1) $$\omega(r_b \cdot s) \cdot (\mathbf{h} \cdot s) = \operatorname{Ad}(s^{-1}) \cdot (\omega(r_b) \cdot \mathbf{h}).$$

(2) *For all* $\mathbf{u} \in \mathfrak{g}_e$, *if* $Z_{\mathbf{u}}$ *is the vertical field which is the transport of* $X_{\mathbf{u}}$ *by the action of* G (*the Killing field*) (19.3.7), *we have*

(20.2.5.2) $$\omega(r_b) \cdot Z_{\mathbf{u}}(r_b) = \mathbf{u}$$

for all $r_b \in R$. (In other words, the value of $\omega(r_b)$ on the *vertical* tangent vectors in $T_{r_b}(R)$ is determined *independently* of the connection $\mathbf{P}$ in R.)

Observe first that by definition $Z_{\mathbf{u}}(r_b) = r_b \cdot \mathbf{u} = \mathbf{t}_{r_b}(\mathbf{u})$, hence the property (20.2.5.2) of the 1-form ω of the connection $\mathbf{P}$ follows immediately from the definition (20.2.3.1). Next, since $p_{r_b \cdot s}(\mathbf{h} \cdot s) = (p_{r_b}(\mathbf{h})) \cdot s$, we have

$$\mathbf{h} \cdot s - p_{r_b \cdot s}(\mathbf{h} \cdot s) = (\mathbf{h} - p_{r_b}(\mathbf{h})) \cdot s,$$

and $(r_b \cdot s) \cdot \mathbf{u} = (r_b \cdot (s \cdot \mathbf{u} \cdot s^{-1})) \cdot s$ (16.10.1), or equivalently $\mathbf{t}_{r_b \cdot s}(\mathbf{u}) = (\mathbf{t}_{r_b}(\mathrm{Ad}(s) \cdot \mathbf{u})) \cdot s$, whence the formula (20.2.5.1) is a consequence of the definition (20.2.3.4).

Conversely, suppose that the 1-form ω satisfies the conditions of the proposition. Then (20.2.5.2) implies that $\mathbf{h} \mapsto \mathbf{t}_{r_b}(\omega(r_b) \cdot \mathbf{h})$ is a *projection* of $T_{r_b}(R)$ onto G_{r_b}. If we put $p_{r_b}(\mathbf{h}) = \mathbf{h} - \mathbf{t}_{r_b}(\omega(r_b) \cdot \mathbf{h})$, then p_{r_b} is a projection of $T_{r_b}(R)$ with kernel G_{r_b}, and the image H_{r_b} of p_{r_b} is therefore supplementary to G_{r_b}. It follows that the restriction of $T_{r_b}(\pi)$ to H_{r_b} is a bijection of H_{r_b} onto $T_b(B)$. If we denote the inverse bijection by $\mathbf{k}_b \mapsto \mathbf{P}_b(\mathbf{k}_b, r_b)$, then $\mathbf{P}$ is a mapping of $T(B) \times_B R$ into $T(R)$ which satisfies the conditions (20.2.2.2) and (20.2.2.3). From (20.2.5.1) and the relation $\mathbf{t}_{r_b \cdot s}(\mathbf{u}) = (\mathbf{t}_{r_b}(\mathrm{Ad}(s) \cdot \mathbf{u})) \cdot s$ it follows that $p_{r_b \cdot s}(\mathbf{h} \cdot s) = (p_{r_b}(\mathbf{h})) \cdot s$, and therefore $H_{r_b} \cdot s = H_{r_b \cdot s}$; and since $T_{r_b \cdot s}(\pi) \cdot (\mathbf{h} \cdot s) = T_{r_b}(\pi) \cdot \mathbf{h}$ for $\mathbf{h} \in T_{r_b}(R)$, it is clear that $\mathbf{P}$ also satisfies (20.2.2.4).

Finally, $\mathbf{P}$ is of class C^∞. For we may assume that $R = B \times G$ is trivial and that B is an open set in $\mathbf{R}^n$; and then, with the notation of (20.2.2) and (20.2.3), we have $Q(b) \cdot \mathbf{k} = -\varpi_b \cdot (\mathbf{k}, 0)$, and if $b \mapsto \varpi_b$ is of class C^∞, the same is true of $b \mapsto Q(b)$. Hence $\mathbf{P}$ is a principal connection, and ω is its connection form.

PROBLEMS

1. If $(\mathbf{P}_j)$ is a finite family of principal connections in a principal bundle R with base B, and (f_j) is a finite family of real-valued C^∞ functions defined on B, such that $\sum_j f_j = 1$, then the mapping $\sum_j f_j \mathbf{P}_j$ of $T(B) \times_B R$ into $T(R)$ is a principal connection. Deduce from this that every principal bundle carries a principal connection.

2. (a) Let R, R′ be two principal bundles over the same base B, with groups G, G′, respectively, and let (u, ρ) be a morphism of R into R′ corresponding to the identity mapping of B (16.14.3). Show that if $\mathbf{P}$ is a principal connection in R, then there exists a unique principal connection $\mathbf{P}'$ in R′ such that

$$\mathbf{P}'_b(\mathbf{k}_b, u(r_b)) = T_{r_b}(u) \cdot \mathbf{P}_b(\mathbf{k}_b, r_b)$$

for all $b \in B$, $\mathbf{k}_b \in T_b(B)$, $r_b \in R_b$.

If $\boldsymbol{\omega}$, $\boldsymbol{\omega}'$ are the 1-forms of the connections $\boldsymbol{P}$, $\boldsymbol{P}'$, respectively, then

$$\rho_*(\boldsymbol{\omega}(r_b)\cdot\mathbf{h}_{r_b}) = \boldsymbol{\omega}'(u(r_b))\cdot(\mathrm{T}_{r_b}(u)\cdot\mathbf{h}_{r_b})$$

with the notation of (20.2.3).

(b) Let R be a principal bundle with base B and group G, and let $R' = B' \times_B R$ be its inverse image under a mapping $f: B' \to B$ of class C^∞ (16.14.6). Show that if $\boldsymbol{P}$ is a principal connection in R, there exists a unique principal connection $\boldsymbol{P}'$ in R′ such that, if $u: R' \to R$ is the morphism corresponding to f, we have

$$\mathrm{T}_{r_{f(b')}}(u)\cdot(\boldsymbol{P}'_{b'}(\mathbf{k}_{b'}, r_{f(b')})) = \boldsymbol{P}_{f(b')}(\mathrm{T}_{b'}(f)\cdot\mathbf{k}_{b'}, r_{f(b')}).$$

If $\boldsymbol{\omega}$ is the 1-form of the connection $\boldsymbol{P}$, then the 1-form of $\boldsymbol{P}'$ is ${}^t u(\boldsymbol{\omega})$. The connection $\boldsymbol{P}'$ is called the *inverse image* of $\boldsymbol{P}$ by u.

(c) Define the notion of a product of two principal connections in principal bundles R, R′ with bases B, B′ and groups G, G′, respectively, as a principal connection in the product of the principal bundles R, R′ (Section 16.14, Problem 7).

3. Let $\boldsymbol{P}$ be a principal connection in a principal bundle X with structure group G, base B and projection π. Show that for each unending path $v: \mathbf{R} \to B$ of class C^∞ in B, and each point $x \in \pi^{-1}(v(0))$, there exists a unique unending path $w_x: \mathbf{R} \to X$ of class C^∞, called the *horizontal lifting* of v, such that $w_x(0) = x$, $\pi \circ w_x = v$, and such that the tangent vector $w_x'(t)$ is *horizontal* for all $t \in \mathbf{R}$. (Using the fact that the inverse image bundle ${}^t v(X)$ is trivializable (Section 16.26, Problem 7), show that there exists an unending path $u: \mathbf{R} \to X$ of class C^∞ such that $u(0) = x$ and which is a lifting of v. Then write $w_x(t) = u(t)\cdot f(t)$, where $f(t) \in G$, and use Section 19.16, Problem 19.)

Show that for each $s \in G$ we have $w_{x\cdot s}(t) = w_x(t)\cdot s$. The mapping $x \mapsto w_x(t)$, for a given $t \in \mathbf{R}$, is called the *parallel displacement* of the fiber $X_{v(0)}$ onto the fiber $X_{v(t)}$ along the path v. For each unending path u of class C^∞ in X such that $u(0) = x$ and which is a lifting of v, the path $t \mapsto g(t)$ in G, where $g(t)$ is the element of G such that $u(t) = w_x(t)\cdot g(t)$ in G, is called the *development* in G of the path u.

With this notation, show that $\boldsymbol{\omega}(u(t))\cdot u'(t) = -g'(t)\cdot g(t)^{-1}$ (in the notation of (18.1.2)), where $\boldsymbol{\omega}$ is the 1-form of the connection $\boldsymbol{P}$.

4. Let R be a principal bundle with base B, group G and projection π. If $\dim(B) = n$, show that for a C^∞ field of n-directions $r \mapsto H_r$ on R to consist of the spaces of horizontal tangent vectors of a principal connection in R, it is necessary and sufficient that $H_r \cdot s = H_{r\cdot s}$ for all $r \in R$ and $s \in G$, and that $\mathrm{T}(\pi)(H_r) = \mathrm{T}_{\pi(r)}(B)$.

3. COVARIANT EXTERIOR DIFFERENTIATION ATTACHED TO A PRINCIPAL CONNECTION. CURVATURE FORM OF A PRINCIPAL CONNECTION

(20.3.1) Let $\boldsymbol{P}$ be a principal connection in a principal bundle R with base B, projection π and group G. We retain the notion of (20.2.3). Given a vector-valued differential q-form $\boldsymbol{\alpha}$ on R, with values in a vector space V, for each $r_b \in R$ the mapping

$$(\mathbf{h}_1, \mathbf{h}_2, \ldots, \mathbf{h}_q) \mapsto \boldsymbol{\alpha}(r_b)\cdot(p_{r_b}(\mathbf{h}_1) \wedge p_{r_b}(\mathbf{h}_2) \wedge \cdots \wedge p_{r_b}(\mathbf{h}_q))$$

of $(T_{r_b}(R))^q$ into V is q-linear and alternating, hence can be written in the form

$$(\mathbf{h}_1, \ldots, \mathbf{h}_q) \mapsto \boldsymbol{\alpha}_{\mathbf{P}}(r_b) \cdot (\mathbf{h}_1 \wedge \mathbf{h}_2 \wedge \cdots \wedge \mathbf{h}_q),$$

where $\boldsymbol{\alpha}_{\mathbf{P}}$ is a differential q-form on R with values in V, which is evidently *horizontal* (20.2.4). Moreover, if $\boldsymbol{\alpha}$ is *invariant* (for the right action of G on $\bigwedge^q T(R)$ and a right action of G on V), then the same is true of $\boldsymbol{\alpha}_{\mathbf{P}}$, because

$$p_{r_b}(\mathbf{h}_1 \cdot s) \wedge \cdots \wedge p_{r_b}(\mathbf{h}_q \cdot s) = (p_{r_b}(\mathbf{h}_1) \wedge \cdots \wedge p_{r_b}(\mathbf{h}_q)) \cdot s.$$

It is immediately checked (by reduction to the case of a trivial principal bundle) that if $\boldsymbol{\alpha}$ is of class C^r then so also is $\boldsymbol{\alpha}_{\mathbf{P}}$.

(20.3.2) With the notation of (20.3.1), the *horizontal* differential $(q+1)$-form

(20.3.2.1) $$D\boldsymbol{\alpha} = (d\boldsymbol{\alpha})_{\mathbf{P}}$$

(also denoted by $D_{\mathbf{P}}\boldsymbol{\alpha}$) is called the *covariant exterior differential* of the vector-valued q-form $\boldsymbol{\alpha}$ (of class C^1), relative to the principal connection $\mathbf{P}$. It follows from (20.3.1) that if $\boldsymbol{\alpha}$ is *invariant* then so also is $D\boldsymbol{\alpha}$, because $d\boldsymbol{\alpha}$ is invariant (19.1.9.5).

If the q-form $\boldsymbol{\alpha}$ is *vertical*, and if $q \geqq 2$, then $D\boldsymbol{\alpha} = 0$. For it follows from the formula (17.15.3.5) that the value of $d\boldsymbol{\alpha}(r_b) \cdot (p_{r_b}(\mathbf{h}_0) \wedge \cdots \wedge p_{r_b}(\mathbf{h}_q))$ is a sum of terms of the form $\boldsymbol{\alpha}(r_b) \cdot (\mathbf{k}_1 \wedge \cdots \wedge \mathbf{k}_q)$, where $q-1$ of the vectors $\mathbf{k}_j$ are horizontal, and of terms which are values of Lie derivatives of the form $\boldsymbol{\alpha}_{\mathbf{P}}$, which is zero by hypothesis.

The *curvature form* of the connection $\mathbf{P}$ is defined to be the C^∞ vector-valued 2-form with values in $\mathfrak{g}_e$ which is the covariant exterior differential of the 1-form $\boldsymbol{\omega}$ of the connection:

(20.3.2.2) $$\boldsymbol{\Omega} = D\boldsymbol{\omega},$$

which is therefore *horizontal* and *invariant* (for the right action

$$(\mathbf{u}, s) \mapsto \mathrm{Ad}(s^{-1}) \cdot \mathbf{u}$$

of G on $\mathfrak{g}_e$).

(20.3.3) *The* 1-*form* $\boldsymbol{\omega}$ *of a principal connection in a principal bundle* R *satisfies the "structure equation"*

(20.3.3.1) $$d\boldsymbol{\omega} = -[\boldsymbol{\omega}, \boldsymbol{\omega}] + D\boldsymbol{\omega} = -[\boldsymbol{\omega}, \boldsymbol{\omega}] + \boldsymbol{\Omega}.$$

(In other words, for all tangent vectors $\mathbf{h}, \mathbf{k}$ at a point $r_b \in R$, we have

(20.3.3.2) $$d\boldsymbol{\omega}(r_b) \cdot (\mathbf{h} \wedge \mathbf{k}) = -[\boldsymbol{\omega}(r_b) \cdot \mathbf{h}, \boldsymbol{\omega}(r_b) \cdot \mathbf{k}] + D\boldsymbol{\omega}(r_b) \cdot (\mathbf{h} \wedge \mathbf{k}).)$$

Since both sides of (20.3.3.2) are bilinear in $\mathbf{h}$, $\mathbf{k}$, it is enough to verify this formula in the following three cases:

(i) $\mathbf{h}$ and $\mathbf{k}$ are horizontal. Since the form $\boldsymbol{\omega}$ is vertical, the relation (20.3.3.2) then reduces to the definition of $D\boldsymbol{\omega}$ (20.3.1).

(ii) $\mathbf{h}$ and $\mathbf{k}$ are vertical. We have then $\mathbf{h} = \mathbf{t}_{r_b}(\mathbf{u})$, $\mathbf{k} = \mathbf{t}_{r_b}(\mathbf{v})$, where $\mathbf{u}, \mathbf{v} \in \mathfrak{g}_e$, and $d\boldsymbol{\omega}(r_b) \cdot (\mathbf{h} \wedge \mathbf{k})$ is the value at the point r_b of the function $d\boldsymbol{\omega} \cdot (Z_{\mathbf{u}} \wedge Z_{\mathbf{v}})$. Now we have (17.15.8.1)

$$d\boldsymbol{\omega} \cdot (Z_{\mathbf{u}} \wedge Z_{\mathbf{v}}) = \theta_{Z_{\mathbf{u}}} \cdot (\boldsymbol{\omega} \cdot Z_{\mathbf{v}}) - \theta_{Z_{\mathbf{v}}} \cdot (\boldsymbol{\omega} \cdot Z_{\mathbf{u}}) - \boldsymbol{\omega} \cdot [Z_{\mathbf{u}}, Z_{\mathbf{v}}]. \tag{20.3.3.3}$$

But the function $\boldsymbol{\omega} \cdot Z_{\mathbf{u}}$ with values in $\mathfrak{g}_e$ is *constant* on R by virtue of the condition (20.2.5.2) characterizing a connection form. The right-hand side of (20.3.3.3) therefore reduces to $-\boldsymbol{\omega} \cdot [Z_{\mathbf{u}}, Z_{\mathbf{v}}]$. Now we have $[Z_{\mathbf{u}}, Z_{\mathbf{v}}] = Z_{[\mathbf{u}, \mathbf{v}]}$ (19.3.7.4), and therefore the value of $-\boldsymbol{\omega} \cdot [Z_{\mathbf{u}}, Z_{\mathbf{v}}]$ at the point r_b is indeed equal to the right-hand side of (20.3.3.2), bearing in mind (20.2.5.2) and the facts that $D\boldsymbol{\omega}$ is horizontal and $\mathbf{h}$, $\mathbf{k}$ vertical.

(iii) $\mathbf{h}$ is vertical, $\mathbf{k}$ is horizontal. The vector $\mathbf{h}$ is again the value at the point r_b of a vertical vector field of the form $Z_{\mathbf{u}}$, where $\mathbf{u} \in \mathfrak{g}_e$. On the other hand, we have seen (20.2.3) that there exists a *horizontal* vector field Y on R which takes the value $\mathbf{k}$ at the point r_b. The formula (17.15.8.1) now gives

$$d\boldsymbol{\omega} \cdot (Z_{\mathbf{u}} \wedge Y) = \theta_{Z_{\mathbf{u}}} \cdot (\boldsymbol{\omega} \cdot Y) - \theta_Y \cdot (\boldsymbol{\omega} \cdot Z_{\mathbf{u}}) - \boldsymbol{\omega} \cdot [Z_{\mathbf{u}}, Y]. \tag{20.3.3.4}$$

Since $\boldsymbol{\omega}$ is vertical, we have $\boldsymbol{\omega} \cdot Y = 0$. Also, as in (ii) above, we see that $\boldsymbol{\omega} \cdot Z_{\mathbf{u}}$ is constant, and therefore the right-hand side of (20.3.3.4) reduces to $-\boldsymbol{\omega} \cdot [Z_{\mathbf{u}}, Y]$. Since $\mathbf{h}$ is vertical and $D\boldsymbol{\omega}$ horizontal, the right-hand side of (20.3.3.2) is zero; hence we are reduced to proving the following lemma;

(20.3.3.5) *For each $\mathbf{u} \in \mathfrak{g}_e$ and each horizontal vector field Y on* R, *the vector field $[Z_{\mathbf{u}}, Y]$ on* R *is horizontal.*

It follows from the interpretation of the Lie bracket of two vector fields given in (18.2.14), and from the fact that the integral curves of the field $Z_{\mathbf{u}}$ are given by $t \mapsto r \cdot \exp(t\mathbf{u})$, that the value of $[Z_{\mathbf{u}}, Y]$ at a point $r \in \mathrm{R}$ is the limit, as $t \to 0$, of the tangent vector at the point r

$$\frac{1}{t}\,(Y(r \cdot \exp(t\mathbf{u})) \cdot \exp(-t\mathbf{u}) - Y(r)). \tag{20.3.3.6}$$

By virtue of (20.2.3.2), this vector is horizontal, and therefore so also is its limit.

When the base B is reduced to a point, so that R may be identified with the group G, there is only one connection on R, and the definition (20.2.3.4) shows that the form $\boldsymbol{\omega}$ is the *canonical* differential form defined in (19.16.1). Hence in this case we have

(20.3.3.7) $$d\boldsymbol{\omega} + [\boldsymbol{\omega}, \boldsymbol{\omega}] = 0$$

(*Maurer–Cartan equation*).

(20.3.4) *The curvature form* $\boldsymbol{\Omega}$ *satisfies the relation* $D\boldsymbol{\Omega} = 0$ (*Bianchi's identity*).

By virtue of the structure equation we have

$$D\boldsymbol{\Omega} = D(d\boldsymbol{\omega}) + D([\boldsymbol{\omega}, \boldsymbol{\omega}]).$$

By definition (20.3.2.1), $D(d\boldsymbol{\omega}) = 0$ because $d(d\boldsymbol{\omega}) = 0$; and since the 2-form $[\boldsymbol{\omega}, \boldsymbol{\omega}]$ is vertical, we have $D([\boldsymbol{\omega}, \boldsymbol{\omega}]) = 0$ by (20.3.2).

PROBLEMS

1. With the notation of (20.3.1), let $\rho : G \to GL(V)$ be the homomorphism such that $(s, v) \mapsto \rho(s^{-1}) \cdot v$ is the right action of G on V under consideration. For each differential q-form $\boldsymbol{\alpha}$ on R, with values in V, we denote by $\boldsymbol{\omega} \wedge_\rho \boldsymbol{\alpha}$ the differential $(q+1)$-form on R, with values in V, whose value is given by the formula

$$(\boldsymbol{\omega} \wedge_\rho \boldsymbol{\alpha})(r_b) \cdot (\mathbf{h}_1 \wedge \mathbf{h}_2 \wedge \cdots \wedge \mathbf{h}_{q+1})$$
$$= \frac{1}{q!} \boldsymbol{a} \cdot (\rho_*(\boldsymbol{\omega}(r_b) \cdot \mathbf{h}_1) \cdot (\boldsymbol{\alpha}(r_b) \cdot (\mathbf{h}_2 \wedge \mathbf{h}_3 \wedge \cdots \wedge \mathbf{h}_{q+1}))),$$

where $\boldsymbol{a}$ is the antisymmetrization operator (A.12.2). Show that if $\boldsymbol{\alpha}$ is *horizontal and invariant*, then we have

$$d\boldsymbol{\alpha} = -\boldsymbol{\omega} \wedge_\rho \boldsymbol{\alpha} + D\boldsymbol{\alpha}.$$

(Consider, as in (20.3.3), three cases: where all the $\mathbf{h}_j$ $(1 \leqq j \leqq q+1)$ are horizontal, where at least two of them are vertical, and where only one is vertical. In the last case, we may suppose that $\mathbf{h}_1$ is the value of a Killing field $Z_{\mathbf{u}}$, and that the $\mathbf{h}_j$ $(j \geqq 2)$ are horizontal and are the values of G-*invariant* fields; use (19.4.4.3) and (19.8.11).)

Deduce that

$$D(D\boldsymbol{\alpha}) = \boldsymbol{\Omega} \wedge_\rho \boldsymbol{\alpha},$$

the right-hand side being defined by antisymmetrization as above.

2. In the situation of Section 20.2, Problem 2(a), show that if $\boldsymbol{\Omega}$, $\boldsymbol{\Omega}'$ are the respective curvature forms of P, P', then we have

$$\rho_*(\boldsymbol{\Omega}(r_b) \cdot (\mathbf{h} \wedge \mathbf{k})) = \boldsymbol{\Omega}'(u(r_b)) \cdot ((T_{r_b}(u) \cdot \mathbf{h}) \wedge (T_{r_b}(u) \cdot \mathbf{k})).$$

In the situation of Section 20.2, Problem 2(b), show that $\boldsymbol{\Omega}' = {}^t u(\boldsymbol{\Omega})$.

3. Let R be a principal bundle with base B, group G and projection π. Let K be a Lie group acting on R on the *left*, such that $t \cdot (r \cdot s) = (t \cdot r) \cdot s$ for $s \in G$ and $t \in K$. Then for each $t \in K$ the mapping $r \mapsto t \cdot r$ is an *automorphism* of the principal bundle R. Let r_0 be a point of R, and let S denote the stabilizer of $\pi(r_0)$. For each $t \in S$, there exists a unique element $\rho(t) \in G$ such that $t \cdot r_0 = r_0 \cdot \rho(t)$, and ρ is a homomorphism of S into G.

Let $\mathfrak{g}_e$, $\mathfrak{k}_e$, $\mathfrak{s}_e$ be the respective Lie algebras of G, K, S. For each vector $\mathbf{w} \in \mathfrak{k}_e$, put $Y_{\mathbf{w}}(r) = \mathbf{w} \cdot r$ (19.3.7).

(a) Suppose that there exists a principal connection in R which is K-*invariant*. If $\boldsymbol{\omega}$ is the 1-form of the connection, then $\theta_{Y_{\mathbf{w}}} \cdot \boldsymbol{\omega} = 0$ (19.8.11.3). Deduce that, for each horizontal vector field X on R, and each $\mathbf{w} \in \mathfrak{k}_e$, the vector field $[Y_{\mathbf{w}}, X]$ is horizontal.
(b) For each vector $\mathbf{w} \in \mathfrak{k}_e$, put $\mathbf{f}(\mathbf{w}) = \boldsymbol{\omega}(r_0) \cdot Y_{\mathbf{w}}(r_0)$, so that $\mathbf{f}$ is a linear mapping of $\mathfrak{k}_e$ into $\mathfrak{g}_e$. Show that $\mathbf{f}(\mathbf{w}) = \rho_*(\mathbf{w})$ for all $\mathbf{w} \in \mathfrak{s}_e$, and that

$$\mathbf{f}(\mathrm{Ad}(t) \cdot \mathbf{w}) = \mathrm{Ad}(\rho(t)) \cdot \mathbf{f}(\mathbf{w})$$

for all $t \in S$. (Use (19.3.7.6) and (20.2.5.1).)
(c) If $\boldsymbol{\Omega}$ is the curvature form of the connection, show that

$$\boldsymbol{\Omega}(r_0) \cdot (Y_{\mathbf{v}}(r_0) \wedge Y_{\mathbf{w}}(r_0)) = [\mathbf{f}(\mathbf{v}), \mathbf{f}(\mathbf{w})] - \mathbf{f}([\mathbf{v}, \mathbf{w}]).$$

(Use the fact that $\theta_{Y_{\mathbf{w}}} \cdot \boldsymbol{\omega} = 0$.)
(d) Conversely, suppose that we are given a linear mapping $\mathbf{f} : \mathfrak{k}_e \to \mathfrak{g}_e$ satisfying the two conditions of (b) above, and suppose also that K acts *transitively* on B by the rule $t \cdot \pi(r) = \pi(t \cdot r)$. Show that there exists a unique K-invariant connection in R such that the form $\boldsymbol{\omega}$ of the connection satisfies the relation $\boldsymbol{\omega}(r_0) \cdot Y_{\mathbf{w}}(r_0) = \mathbf{f}(\mathbf{w})$ for all $\mathbf{w} \in \mathfrak{k}_e$. (The hypothesis of transitivity implies that for each point $r \in R$ and each tangent vector $\mathbf{h}_r \in T_r(R)$ there exist $t \in K$, $s \in G$, $\mathbf{w} \in \mathfrak{k}_e$, and $\mathbf{u} \in \mathfrak{g}_e$ such that $r_0 = t \cdot r \cdot s$ and $t \cdot \mathbf{h}_r \cdot s = \mathbf{w} \cdot r_0 + r_0 \cdot \mathbf{u}$. Show that the formula

$$\boldsymbol{\omega}(r) \cdot \mathbf{h}_r = \mathrm{Ad}(s) \cdot (\mathbf{f}(\mathbf{w}) + \mathbf{u})$$

defines a connection form with the required properties. The properties of $\mathbf{f}$ ensure that $\boldsymbol{\omega}(r)$ is independent, first of the choices of $\mathbf{w}$ and $\mathbf{u}$, and then also of the choices of t and s. To prove that $\boldsymbol{\omega}$ is of class C^∞, use a local section of K considered as a principal bundle over K/S in order to choose t, s, $\mathbf{w}$, and $\mathbf{u}$ in a neighborhood of a point of R.)
(e) The connection corresponding to $\mathbf{f}$ is flat (20.4.1) if and only if $\mathbf{f}$ is a Lie algebra homomorphism of $\mathfrak{k}_e$ into $\mathfrak{g}_e$.

4. EXAMPLES OF PRINCIPAL CONNECTIONS

(20.4.1) Given a trivial principal bundle $R = B \times G$, we define a principal connection $\mathbf{P}$ in R (called the *trivial* connection) by taking $\mathbf{P}_b(\mathbf{k}, (b, s))$ to be the image of $\mathbf{k}$ under the tangent linear mapping at b to the mapping $x \mapsto (x, s)$ of B into R. At a point $(b, s) \in R$, the tangent space is identified with $T_b(B) \times T_s(G)$, and therefore $\mathbf{P}_b(\mathbf{k}, (b, s)) = (\mathbf{k}, 0)$. If X is any vector field on B, its horizontal lifting $\tilde{X}$ is therefore given by $\tilde{X}(b, s) = (X(b), 0)$. It follows immediately that if X, Y are vector fields on B, then $[\tilde{X}, \tilde{Y}]$ is the horizontal lifting of $[X, Y]$. The formula (17.15.8.1) then shows that for two horizontal vectors $\mathbf{h}$, $\mathbf{k}$ in $T_{(b,s)}(R)$ we have $d\boldsymbol{\omega} \cdot (\mathbf{h} \wedge \mathbf{k}) = 0$, so that *the curvature form* $\boldsymbol{\Omega} = D\boldsymbol{\omega}$ *is zero.*

Conversely, suppose that $\boldsymbol{\Omega} = 0$ for a principal connection in a principal bundle R. If X, Y are two horizontal vector fields on R, we shall have $d\boldsymbol{\omega} \cdot (X \wedge Y) = 0$, and (17.15.8.1) then shows that $\boldsymbol{\omega} \cdot [X, Y] = 0$: in other words, $[X, Y]$ is also a horizontal vector field. This can also be interpreted by saying that the field of directions $r \mapsto H_r$ is *completely integrable* (18.14.5), and consequently (18.14.2) for each point $b \in B$ there exists an open neighborhood U of b in B and a C^∞ section $\mathbf{h}$ of R over U such that the tangent space at each point to the submanifold $\mathbf{h}(U)$ of R is horizontal. The same will be true for the translations $x \mapsto \mathbf{h}(x) \cdot s$ of the section $\mathbf{h}$ by elements $s \in G$, by virtue of (20.2.3). It follows that $\varphi : (x, s) \mapsto \mathbf{h}(x) \cdot s$ is a diffeomorphism of $U \times G$ onto $\pi^{-1}(U)$, and that the image under φ^{-1} of the given connection on $\pi^{-1}(U)$ is the *trivial* connection in $U \times G$. A principal connection is said to be *flat* if its curvature form is zero.

(20.4.2) Let G be a Lie group, H a Lie subgroup of G, and let $\mathfrak{g}_e$, $\mathfrak{h}_e$ be their Lie algebras. We may regard G as a principal bundle with group H (acting on G by right translations) and base G/H (16.14.2). We propose to consider when a principal connection in G (considered as a principal bundle over G/H) is G-*invariant* (for the action of G by left translations); in other words, if $\boldsymbol{\omega}_H$ is the 1-form of the connection, we must have

(20.4.2.1) $$\gamma(s)\boldsymbol{\omega}_H = \boldsymbol{\omega}_H \quad \text{for all} \quad s \in G$$

(where G is considered as acting trivially on $\mathfrak{g}_e$). Explicitly, we obtain (19.1.9.2)

(20.4.2.2) $$\boldsymbol{\omega}_H(x) \cdot \mathbf{h}_x = \boldsymbol{\omega}_H(sx) \cdot (s \cdot \mathbf{h}_x)$$

for all $s, x \in G$ and $\mathbf{h}_x \in T_x(G)$. Now consider the form $\boldsymbol{\omega}_H(e)$ which, by definition (20.2.5), is a *projection* of $\mathfrak{g}_e$ onto $\mathfrak{h}_e$; let $\mathfrak{m}$ be its kernel, which is a *supplementary subspace* for $\mathfrak{h}_e$ in $\mathfrak{g}_e$. It follows from (20.2.5.1) that for each vector $\mathbf{u} \in \mathfrak{g}_e$ and each $t \in H$ we must have

$$\boldsymbol{\omega}_H(t) \cdot (\mathbf{u} \cdot t) = \mathrm{Ad}(t^{-1}) \cdot (\boldsymbol{\omega}_H(e) \cdot \mathbf{u}),$$

which, in view of (20.4.2.2), can also be written as

$$\boldsymbol{\omega}_H(e) \cdot (t^{-1} \cdot \mathbf{u} \cdot t) = \mathrm{Ad}(t^{-1}) \cdot (\boldsymbol{\omega}_H(e) \cdot \mathbf{u})$$

or again as

(20.4.2.3) $$\boldsymbol{\omega}_H(e) \circ \mathrm{Ad}(t^{-1}) = \mathrm{Ad}(t^{-1}) \circ \boldsymbol{\omega}_H(e),$$

and implies in particular that

(20.4.2.4) $$\mathrm{Ad}(t) \cdot \mathfrak{m} \subset \mathfrak{m} \quad \text{for all} \quad t \in H.$$

Conversely, suppose that there exists a vector subspace $\mathfrak{m}$ of $\mathfrak{g}_e$ supplementary to $\mathfrak{h}_e$, such that (20.4.2.4) is satisfied for all $t \in \mathrm{H}$ (in which case the homogeneous space G/H is said to be *weakly reductive*). The projection P of $\mathfrak{g}_e$ onto $\mathfrak{h}_e$ with kernel $\mathfrak{m}$ then *commutes* with $\mathrm{Ad}(t)$, because $\mathrm{Ad}(t)$ is an endomorphism of the vector space $\mathfrak{g}_e$ which stabilizes $\mathfrak{h}_e$ and $\mathfrak{m}$. We define a 1-form ω_{H} on G with values in $\mathfrak{h}_e$ by the rule

$$\omega_{\mathrm{H}}(s) \cdot (s \cdot \mathbf{h}) = P \cdot \mathbf{h}$$

for all $\mathbf{h}$ in $\mathfrak{g}_e$ and s in G, and verify that ω_{H} satisfies the conditions (20.2.5.1) and (20.2.5.2), hence is the form of a G-invariant principal connection. It follows that there is a *canonical one-to-one correspondence* between the set of G-invariant principal connections and the set of supplementary subspaces $\mathfrak{m}$ of $\mathfrak{h}_e$ in $\mathfrak{g}_e$ satisfying (20.4.2.4). It is clear, by transport of structure, that the curvature form $\mathbf{\Omega}_{\mathrm{H}}$ is also invariant under G (acting by left-translations): it is given by

(20.4.2.5) $$\mathbf{\Omega}_{\mathrm{H}}(s) \cdot (s \cdot \mathbf{u} \wedge s \cdot \mathbf{v}) = -P \cdot [\mathbf{u}, \mathbf{v}]$$

for $\mathbf{u}, \mathbf{v} \in \mathfrak{g}_e$. For it is enough to calculate $d\omega_{\mathrm{H}} \cdot (X_{\mathbf{u}} \wedge X_{\mathbf{v}})$ for $\mathbf{u}, \mathbf{v} \in \mathfrak{m}$ by the formula (17.15.8.1), observing that by transport of structure $\omega_{\mathrm{H}} \cdot X_{\mathbf{u}}$ is a *constant* function on G.

When H is *connected*, the condition (20.4.2.4) is equivalent to

(20.4.2.6) $$[\mathfrak{h}_e, \mathfrak{m}] \subset \mathfrak{m}$$

by virtue of (19.11.2.4).

(20.4.3) An important particular case is that in which we are given an *involutory automorphism* $\sigma \neq 1_{\mathrm{G}}$ of a Lie group G, so that $\sigma^2 = 1_{\mathrm{G}}$. It is clear that the set H of points of G fixed by σ is a closed subgroup of G, hence a *Lie subgroup* (19.10.1). The derived homomorphism σ_* of the Lie algebra $\mathfrak{g}_e$ is an *involutory automorphism of the Lie algebra*. Since we have $\exp(t\sigma_*(\mathbf{u})) = \sigma(\exp(t\mathbf{u}))$ for all $t \in \mathbf{R}$ and $\mathbf{u} \in \mathfrak{g}_e$ (19.8.9), and since the relation

$$\exp(t\sigma_*(\mathbf{u})) = \exp(t\mathbf{u})$$

for all $t \in \mathbf{R}$ is equivalent to $\sigma_*(\mathbf{u}) = \mathbf{u}$ (19.8.6), it follows that the set of vectors in $\mathfrak{g}_e$ fixed by σ_* is the *Lie algebra* $\mathfrak{h}_e$ of H. For $r \in \mathrm{G}$ and $s \in \mathrm{H}$ we have, by definition,

$$\sigma(srs^{-1}) = s\sigma(r)s^{-1},$$

and by taking the derived homomorphisms of both sides (considered as functions of r) we obtain

(20.4.3.1) $$\sigma_* \circ \mathrm{Ad}(s) = \mathrm{Ad}(s) \circ \sigma_*.$$

Regarded as an involutory automorphism of the vector space $\mathfrak{g}_e$, σ_* has two eigenvalues, $+1$ and -1, and $\mathfrak{g}_e$ is the direct sum of $\mathfrak{h}_e$ and the eigenspace $\mathfrak{m}$ corresponds to the eigenvalue -1. The relation (20.4.3.1) then shows that $\mathrm{Ad}(s) \cdot \mathfrak{m} \subset \mathfrak{m}$ for all $s \in \mathrm{H}$. It follows that if H_0 is the identity component of H, then for each Lie subgroup H_1 of G such that $\mathrm{H}_0 \subset \mathrm{H}_1 \subset \mathrm{H}$ (i.e., each Lie subgroup of H having $\mathfrak{h}_e$ as Lie algebra) we can define in G/H_1 a principal connection **P** which is left-invariant under G, such that $\omega_{\mathrm{H}_1}(e)$ (for this connection) is the projection of $\mathfrak{g}_e$ onto $\mathfrak{h}_e$ with kernel $\mathfrak{m}$.

The principal connection **P** is the *unique principal connection in* G (considered as a principal bundle over G/H_1, with group H_1) *which is invariant under* G *and the automorphism* σ. For by virtue of (20.4.2) the 1-form of such a connection is determined by its value at the point e, which is a projection of $\mathfrak{g}_e$ onto $\mathfrak{h}_e$, with kernel $\mathfrak{m}'$, say. The image of the connection under σ is again G-invariant, hence (since $\sigma(\mathrm{H}_1) = \mathrm{H}_1$) corresponds, by transport of structure, to the projection of $\mathfrak{g}_e$ onto $\mathfrak{h}_e$ with kernel $\sigma_*(\mathfrak{m}')$. Since the only supplement $\mathfrak{m}'$ of $\mathfrak{h}_e$ in $\mathfrak{g}_e$ such that $\sigma_*(\mathfrak{m}') = \mathfrak{m}'$ is the subspace $\mathfrak{m}$, our assertion is proved.

We remark that in this case, since $\sigma_*(\mathbf{u}) = -\mathbf{u}$ for $\mathbf{u} \in \mathfrak{m}$, we have $\sigma_*([\mathbf{u}, \mathbf{v}]) = [\sigma_*(\mathbf{u}), \sigma_*(\mathbf{v})] = [\mathbf{u}, \mathbf{v}]$ for all $\mathbf{u}, \mathbf{v} \in \mathfrak{m}$, or in other words

(20.4.3.2) $$[\mathfrak{m}, \mathfrak{m}] \subset \mathfrak{h}_e$$

(cf. Problem 2). The curvature form at the point e is therefore given by

(20.4.3.3) $$\Omega_{\mathrm{H}}(e) \cdot (\mathbf{u} \wedge \mathbf{v}) = -[\mathbf{u}, \mathbf{v}]$$

for $\mathbf{u}, \mathbf{v} \in \mathfrak{m}$ (horizontal vectors at the point e).

The pair $(\mathrm{G}, \mathrm{H}_1)$ is said to be a *symmetric pair* if it arises from an involutory automorphism σ of G as described above. The homogeneous space G/H_1 is called a *symmetric homogeneous space* of G (defined by σ), and the connection **P** is called the *canonical principal connection* in G (considered as a principal bundle with group H_1 and base G/H_1).

PROBLEMS

1. Let R be a principal bundle over a simply-connected base space B. If R carries a flat connection, show that R is trivializable.

2. With the notation of (20.4.2), give an example in which the relation (20.4.2.6) is satisfied, but (20.4.3.2) is not. (Cf. Section 19.14, Problem 4.)

3. With the notation of Section 20.3, Problem 3, suppose that there exists a supplement $\mathfrak{m}$ of $\mathfrak{s}_e$ in the Lie algebra $\mathfrak{k}_e$ such that $\mathrm{Ad}(t) \cdot \mathfrak{m} \subset \mathfrak{m}$ for all $t \in \mathrm{S}$. Then there exists a canonical one-to-one correspondence between the set of K-invariant principal connections in R and the set of linear mappings $\mathbf{f}_{\mathfrak{m}} : \mathfrak{m} \to \mathfrak{g}_e$ such that $\mathbf{f}_{\mathfrak{m}}(\mathrm{Ad}(t) \cdot \mathbf{w}) = \mathrm{Ad}(\rho(t)) \cdot \mathbf{f}_{\mathfrak{m}}(\mathbf{w})$ for all $t \in \mathrm{S}$ and $\mathbf{w} \in \mathfrak{m}$: the mapping $\mathbf{f}$ considered in Section 20.3, Problem 3, has $\mathbf{f}_{\mathfrak{m}}$ as its restriction to $\mathfrak{m}$. The connection in R corresponding to $\mathbf{f}_{\mathfrak{m}} = 0$ is called *canonical* (relative to the given subspace $\mathfrak{m}$).

4. Problem 3 applies in particular when R is the *Stiefel manifold* $\mathbf{S}_{n,p}(\mathbf{R})$. Here we have $\mathrm{K} = \mathbf{O}(n)$, $\mathrm{S} = \mathbf{O}(n-p) \times \mathbf{O}(p)$, $\mathrm{G} = \mathbf{O}(p)$, and we may take $\mathfrak{m}$ to be the vector space of all $n \times n$ matrices of the form

$$\begin{pmatrix} 0 & -{}^tY \\ Y & 0 \end{pmatrix},$$

where Y is any $(n-p) \times p$ matrix. The point r_0 is chosen to be the matrix

$$E = \begin{pmatrix} I_p \\ 0 \end{pmatrix}$$

(see Section 19.7, Problem 9), the elements of $\mathbf{S}_{n,p}(\mathbf{R})$ being therefore them atrices $U = S \cdot E$, where $S \in \mathbf{O}(n)$. Equivalently, the matrices $U \in \mathbf{S}_{n,p}(\mathbf{R})$ are characterized in $\mathbf{R}^{np}$ by the relation ${}^tU \cdot U = I_p$. The tangent space to $\mathbf{S}_{n,p}(\mathbf{R})$ at the point $U = S \cdot E$ may be identified with the set of matrices $S \cdot \begin{pmatrix} X \\ Y \end{pmatrix}$, where $X \in \mathfrak{o}(p)$ is a skew-symmetric matrix and Y is any $(n-p) \times p$ matrix. If $\boldsymbol{\omega}$ is the 1-form of the canonical connection in $\mathbf{S}_{n,p}(\mathbf{R})$ (Problem 3), then $\boldsymbol{\omega}(U)$ maps the matrix $S \cdot \begin{pmatrix} X \\ Y \end{pmatrix}$ above to the matrix $X \in \mathfrak{o}(p)$. Equivalently, we may write $\boldsymbol{\omega} = {}^tU \cdot dU$, where by abuse of notation dU is the restriction to $\mathbf{S}_{n,p}(\mathbf{R})$ of the differential of the identity mapping $\mathbf{R}^{np} \to \mathbf{R}^{np}$, and $\mathfrak{o}(p)$ is canonically identified with the space of matrices $\begin{pmatrix} X \\ 0 \end{pmatrix} \in \mathbf{R}^{np}$ with X skew-symmetric.

Discuss in the same way the complex Stiefel manifolds $\mathbf{S}_{n,p}(\mathbf{C})$, where the orthogonal groups are replaced by unitary groups. Here the 1-form of the canonical connection is $\boldsymbol{\omega} = {}^t\overline{U} \cdot dU$, where $\overline{U}$ is the complex conjugate of U.

5. (a) Let Z be a trivializable principal bundle with base B, group $\mathbf{U}(p)$ and projection π, and let σ be a C^∞ section of Z over B. Let F_j ($1 \leq j \leq \mathrm{N}$) be functions on B with values in $\mathbf{M}_p(\mathbf{C})$ such that $\sum_{j=1}^{\mathrm{N}} {}^t\overline{F}_j \cdot F_j = I_p$ identically in B. Define a C^∞ mapping H of Z into the space $\mathbf{C}^{\mathrm{N}p^2} = \mathbf{M}_{\mathrm{N}p,p}(\mathbf{C})$ of $\mathrm{N}p \times p$ matrices by the formula

$$H(z) = \begin{pmatrix} F_1(\pi(z)) \\ \vdots \\ F_{\mathrm{N}}(\pi(z)) \end{pmatrix} \cdot V(z),$$

where $V(z) \in \mathbf{U}(p)$ is the unitary matrix such that $z = \sigma(\pi(z)) \cdot V(z)$. Show that ${}^t\overline{H}(z) \cdot H(z) = I_p$, and hence that H is a mapping of Z into the Stiefel manifold $\mathbf{S}_{\mathrm{N}p,p}(\mathbf{C})$; moreover, this mapping defines a morphism of principal bundles (with the same group $\mathbf{U}(p)$). Show that the inverse image ${}^t(H \circ \sigma)(\boldsymbol{\omega})$ of the 1-form of the canonical connection in $\mathbf{S}_{Np,p}(\mathbf{C})$ (Problem 4) is the vector-valued differential 1-form $\sum_{j=1}^{N} {}^t\bar{F}_j \cdot dF_j$ on B, with values in the Lie algebra $\mathfrak{u}(p)$ of $\mathbf{U}(p)$.

(b) Let V be an open subset of $\mathbf{R}^m$, let A be a positive-definite Hermitian $p \times p$ matrix, let f be a bounded C^∞ function, defined and >0 on V, and let c be a constant such that the Hermitian matrix $cI_p - f^2(x)A^2$ is >0 for all $x \in$ V. Let $F_2(x)$ be the Hermitian matrix >0 which is its square root (15.11.12). Let a be a real constant, and put $F_1(x) = f(x)e^{ai\xi^1}A^2$. Show that

$$ {}^t\bar{F}_1(x) \cdot F_1(x) + {}^t\bar{F}_2(x) \cdot F_2(x) = cI_p $$

and

$$ {}^t\bar{F}_1 \cdot dF_1 + {}^t\bar{F}_2 \cdot dF_2 = aif^2A^2\, d\xi^1, $$

a vector-valued differential 1-form with values in $\mathfrak{u}(p)$.

(c) Deduce from (b) that, for *every* bounded C^∞ vector-valued differential 1-form $\boldsymbol{\alpha}$ on V with values in $\mathfrak{u}(p)$, there exists an integer N depending only on the dimension m of V, and N mappings $F_j : \mathrm{V} \to \mathbf{M}_p(\mathbf{C})$ such that

$$ \sum_{j=1}^{N} {}^t\bar{F}_j \cdot F_j = I_p \qquad \text{and} \qquad \sum_{j=1}^{N} {}^t\bar{F}_j \cdot dF_j = \boldsymbol{\alpha}. $$

(Observe that the vector space $\mathfrak{u}(p)$ has a basis consisting of matrices iA_k, where the A_k are positive-definite Hermitian matrices.)

(d) Deduce from (a) and (c) that, for *every* principal bundle Z with group $\mathbf{U}(p)$ over a pure manifold B of dimension n, and for *every* principal connection $\mathbf{P}$ in Z, there exists an integer N depending only on n and p, and a morphism $(u, 1_{\mathbf{U}(p)})$ of Z into $\mathbf{S}_{N,p}(\mathbf{C})$ such that $\mathbf{P}$ is the inverse image under u (Section 20.2, Problem 2) of the canonical connection in $\mathbf{S}_{N,p}(\mathbf{C})$. (Use Section 16.25, Problems 10 and 11, and remark that in (16.4.1) the functions forming the partition of unity may be taken to be the squares of C^∞ functions.)

5. LINEAR CONNECTIONS ASSOCIATED WITH A PRINCIPAL CONNECTION

(20.5.1) Let R be a principal bundle with base M, group G, and projection π; also let F be a finite-dimensional real vector space, and ρ a *linear representation* of G in F, so that G acts on the left on F by $(s, \mathbf{y}) \mapsto \rho(s) \cdot \mathbf{y}$. We have seen (20.1.3) that the fiber-bundle $\mathrm{E} = \mathrm{R} \times^{\mathrm{G}} \mathrm{F}$ of fiber-type F associated with R is canonically endowed with a structure of a real vector bundle.

Suppose we are given a *principal connection* $\mathbf{P}$ in R. We shall show how to construct from it, in canonical fashion, a *linear connection* $\mathbf{C}$ (17.16.3) in E. For each point $x \in \mathrm{M}$ and each vector $\mathbf{u}_x \in \mathrm{E}_x$, there exists an element $r_x \in \mathrm{R}_x$ and $\mathbf{y} \in \mathrm{F}$ such that $\mathbf{u}_x = r_x \cdot \mathbf{y}$ (16.14.7). For each tangent vector $\mathbf{k}_x \in \mathrm{T}_x(\mathrm{M})$, we put

(20.5.1.1) $$ \mathbf{C}_x(\mathbf{k}_x, \mathbf{u}_x) = \mathbf{P}_x(\mathbf{k}_x, r_x) \cdot \mathbf{y} $$

with the notation of (16.14.7.3). First, it must be checked that this definition does not depend on the choice of the pair $(r_x, \mathbf{y})$ such that $r_x \cdot \mathbf{y} = \mathbf{u}_x$. However, any other choice is of the form $(r_x \cdot s, \rho(s^{-1}) \cdot \mathbf{y})$ for some $s \in G$, and by (20.2.2.4) and (16.14.7.4) we have

$$\begin{aligned}\mathbf{P}_x(\mathbf{k}_x, r_x \cdot s) \cdot (\rho(s^{-1}) \cdot \mathbf{y}) &= (\mathbf{P}_x(\mathbf{k}_x, r_x) \cdot s) \cdot (\rho(s^{-1}) \cdot \mathbf{y}) \\ &= \mathbf{P}_x(\mathbf{k}_x, r_x) \cdot \mathbf{y}.\end{aligned}$$

Second, if π_F is the canonical projection of E on M, we have $\pi_F(r \cdot \mathbf{y}) = \pi(r)$ for $r \in R$, and $\mathbf{y} \in F$, whence

$$T(\pi_F) \cdot \mathbf{C}_x(\mathbf{k}_x, \mathbf{u}_x) = T(\pi) \cdot \mathbf{P}_x(\mathbf{k}_x, r_x) = \mathbf{k}_x.$$

Next, if we put $m(r_x, \mathbf{y}) = r_x \cdot \mathbf{y}$, we have

$$m(., \alpha\mathbf{y} + \beta\mathbf{y}') = \alpha m(., \mathbf{y}) + \beta m(., \mathbf{y}');$$

taking the tangent linear mappings of these mappings of R into E, we see that, for each tangent vector $\mathbf{h} \in T_{r_x}(R)$, we have

$$\mathbf{h} \cdot (\alpha\mathbf{y} + \beta\mathbf{y}') = \alpha\mathbf{h} \cdot \mathbf{y} + \beta\mathbf{h} \cdot \mathbf{y}',$$

where the sum on the right-hand side is taken in the fiber $(T(E))_{T(\pi)\cdot\mathbf{h}}$ of T(E) considered as a vector bundle over T(M). This therefore shows that $\mathbf{u}_x \mapsto \mathbf{C}_x(\mathbf{k}_x, \mathbf{u}_x)$ is a linear mapping of E_x into $(T(E))_{\mathbf{k}_x}$. Since on the other hand $\mathbf{k}_x \mapsto \mathbf{P}_x(\mathbf{k}_x, r_x)$ is a linear mapping of $T_x(M)$ into $T_{r_x}(R)$, it follows that $\mathbf{k}_x \mapsto \mathbf{C}_x(\mathbf{k}_x, \mathbf{u}_x)$ is a linear mapping of $T_x(M)$ into $T_{\mathbf{u}_x}(E)$. Finally, to see that $\mathbf{C}$ is of class C^∞, we may assume that $R = M \times G$ is trivial and that M is an open set in $\mathbf{R}^n$; in that situation, E may be identified with the trivial bundle $M \times F$, and if $\mathbf{C}_x((x, \mathbf{k}), (x, \mathbf{y})) = ((x, \mathbf{y}), (\mathbf{k}, -\Gamma_x(\mathbf{k}, \mathbf{y})))$ is the local expression of $\mathbf{C}$, then with the notation of (20.2.2), and remembering that ρ_* is a homormorphism of $\mathfrak{g}_e$ into $\mathrm{End}(F) = \mathfrak{gl}(F)$, we have

$$\Gamma_x(\mathbf{k}, \mathbf{y}) = -\rho_*(Q(x) \cdot \mathbf{k})) \cdot \mathbf{y},$$

which proves our assertion.

(20.5.2) Conversely, if E is any vector bundle over a pure manifold M and if $\mathbf{C}$ is any linear connection in E, then $\mathbf{C}$ can be obtained by the construction of (20.5.1) from a well-determined principal connection $\mathbf{P}$ in the *bundle of frames* $R = \mathrm{Isom}(M \times F, E)$, E being identified with a bundle of fiber-type F associated with R (20.1.3). To see this, we take up again the procedure which was sketched in (20.2.1) in a nonintrinsic fashion. For this purpose, we remark that for each $r_x \in R_x = \mathrm{Isom}(F, E_x)$ and each tangent vector $\mathbf{h} \in T_{r_x}(R)$, the mapping $U(\mathbf{h}) : \mathbf{y} \mapsto \mathbf{h} \cdot \mathbf{y}$ (in the notation of (16.14.7.3)) is a linear mapping of F into $(T(E))_{\mathbf{k}_x}$, where $\mathbf{k}_x = T(\pi) \cdot \mathbf{h} = T(p) \cdot (\mathbf{h} \cdot \mathbf{y})$ (where π and p are the projections of the bundles R and E, respectively). For a fixed

$\mathbf{k}_x \in T_x(M)$, the mapping $\mathbf{h} \mapsto U(\mathbf{h})$ of $(T(R))_{\mathbf{k}_x}$ into $\mathrm{Hom}(F, (T(E))_{\mathbf{k}_x})$ is an *isomorphism*. To see this, we may assume that $E = M \times E_0$ is trivial and M is an open set in $\mathbf{R}^n$, so that $R = M \times \mathrm{Isom}(F, E_0)$. If $r_x = (x, u_0)$ and $\mathbf{k}_x = (x, \mathbf{k})$, the vectors $\mathbf{h} \in (T(R))_{\mathbf{k}_x}$ are the tangent vectors of the form $((x, u_0), (\mathbf{k}, u_0 \circ v))$, where $v \in \mathrm{End}(F)$, and we have

$$\mathbf{h} \cdot \mathbf{y} = ((x, u_0(\mathbf{y})), (\mathbf{k}, u_0(v(\mathbf{y})))),$$

from which the assertion follows since u_0 is bijective. Now the mapping $\mathbf{y} \mapsto \mathbf{C}_x(\mathbf{k}_x, r_x \cdot \mathbf{y})$ is a linear mapping of F into $(T(E))_{\mathbf{k}_x}$; hence, by what has been said above, there exists a unique tangent vector $\mathbf{P}_x(\mathbf{k}_x, r_x)$ in $(T(R))_{\mathbf{k}_x}$ satisfying (20.5.1.1) for all $\mathbf{y} \in F$. It is immediately verified that the mapping $\mathbf{P}$ thus defined satisfies (20.2.2.2) and (20.2.2.3). By reducing to the case where E is trivial, we see that $\mathbf{P}$ is of class C^∞. Finally, since $(r_x \cdot s) \cdot \mathbf{y} = r_x \cdot (s \cdot \mathbf{y})$ for all $s \in \mathbf{GL}(F)$, the mapping $\mathbf{P}$ satisfies (20.2.2.4) by virtue of its definition.

(20.5.3) We shall now show how this association of a principal connection with a given linear connection enables us to reduce the operations of *covariant differentiation of a section of* E (relative to $\mathbf{C}$), in the direction of a tangent vector of M (17.17.2.1), and of the *covariant exterior differential* (relative to $\mathbf{C}$) of a differential form on M with values in E (17.19.3), to much more elementary operations in the *principal bundle* R, namely, on the one hand the *Lie derivative* of a *function* on R with values in a vector space, and on the other hand *covariant exterior differentiation* (relative to $\mathbf{P}$) (20.3.2) of a vector-valued differential form on R.

Let then $\mathbf{s}$ be a C^∞ section of E over a neighborhood of x, and let r_x be any point in the fiber R_x. Since the question is local on M, we may assume that there exists a C^∞ section $y \mapsto \mathbf{R}(y)$ of R such that $\mathbf{R}(x) = r_x$. The section $\mathbf{s}$ can be written uniquely in the form (16.14.7.1)

(20.5.3.1) $$\mathbf{s}(y) = \mathbf{R}(y) \cdot \Phi(\mathbf{R}(y)),$$

where Φ is a C^∞ *mapping of a neighborhood of* r_x *in* R *into the vector space* F. Then we have the *first fundamental formula*, which gives $\nabla_{\mathbf{k}_x} \cdot \mathbf{s}$ for any tangent vector $\mathbf{h}_x \in T_x(M)$:

(20.5.3.2) $$\nabla_{\mathbf{h}_x} \cdot \mathbf{s} = r_x \cdot (\theta_{\mathbf{k}} \cdot \Phi + \tau_{\Phi(r_x)}((\omega(r_x) \cdot \mathbf{k}) \cdot \Phi(r_x)))$$

for all tangent vectors $\mathbf{k} \in T_{r_x}(R)$ *such that* $T(\pi) \cdot \mathbf{k} = \mathbf{h}_x$. We recall that $\theta_{\mathbf{k}} \cdot \Phi$, the Lie derivative of the function Φ in the direction of the vector $\mathbf{k}$, belongs to F (17.14.1), that $\omega(r_x) \cdot \mathbf{k} = \mathbf{u}$ is a vector in the Lie algebra $\mathfrak{gl}(F)$, and that the product $\mathbf{u} \cdot \mathbf{y} \in T_{\mathbf{y}}(F)$, for $\mathbf{y} \in F$, was defined in (16.10.1).

To prove (20.5.3.2), we start from the definition of $\nabla_{\mathbf{h}_x} \cdot \mathbf{s}$ (17.17.2.1):

$$\tau_{\mathbf{s}(x)}^{-1}(\nabla_{\mathbf{h}_x} \cdot \mathbf{s}) = \mathrm{T}_x(\mathbf{s}) \cdot \mathbf{h}_x - \mathbf{C}_x(\mathbf{h}_x, \mathbf{s}(x)) \in \mathrm{T}_{\mathbf{s}(x)}(\mathrm{E}),$$

and calculate $\mathrm{T}_x(\mathbf{s})$ by taking the tangent linear mappings of the two sides of (20.5.3.1) at the point x (16.14.7.5):

$$\mathrm{T}_x(\mathbf{s}) \cdot \mathbf{h}_x = (\mathrm{T}_x(\mathbf{R}) \cdot \mathbf{h}_x) \cdot \Phi(\mathbf{R}(x)) + \mathbf{R}(x) \cdot (\mathrm{T}_{\mathbf{R}(x)}(\Phi) \cdot (\mathrm{T}_x(\mathbf{R}) \cdot \mathbf{h}_x)).$$

Put $\mathbf{k} = \mathrm{T}_x(\mathbf{R}) \cdot \mathbf{h}_x \in \mathrm{T}_{r_x}(\mathrm{R})$, which is a vector such that $\mathrm{T}(\pi) \cdot \mathbf{k} = \mathbf{h}_x$; then $\mathrm{T}_{\mathbf{R}(x)}(\Phi) \cdot \mathbf{k}$ is a vector in $\mathrm{T}_{\Phi(r_x)}(\mathrm{F})$, equal to $\tau_{\Phi(r_x)}^{-1}(\theta_{\mathbf{k}} \cdot \Phi)$ (17.14.1); and since r_x is an isomorphism of F onto E_x, we have

$$r_x \cdot (\tau_{\Phi(r_x)}^{-1}(\theta_{\mathbf{k}} \cdot \Phi)) = \tau_{\mathbf{s}(x)}^{-1}(r_x \cdot (\theta_{\mathbf{k}} \cdot \Phi)).$$

Hence, bearing in mind (20.5.1.1), we may write

$$\tau_{\mathbf{s}(x)}^{-1}(\nabla_{\mathbf{h}_x} \cdot \mathbf{s}) = (\mathbf{k} - \mathbf{P}_x(\mathrm{T}(\pi) \cdot \mathbf{k}, r_x)) \cdot \Phi(r_x) + \tau_{\mathbf{s}(x)}^{-1}(r_x \cdot (\theta_{\mathbf{k}} \cdot \Phi)).$$

By the definition of the vector-valued form $\boldsymbol{\omega}$ (20.2.3.4) we have therefore

$$\tau_{\mathbf{s}(x)}^{-1}(\nabla_{\mathbf{h}_x} \cdot \mathbf{s}) = \tau_{\mathbf{s}(x)}^{-1}(r_x \cdot (\theta_{\mathbf{k}} \cdot \Phi)) + (\mathbf{t}_{r_x}(\boldsymbol{\omega}(r_x) \cdot \mathbf{k})) \cdot \Phi(r_x).$$

Now we have the following formula:

(20.5.3.3) $$\mathbf{t}_{r_x}(\mathbf{u}) \cdot \mathbf{y} = \tau_{r_x \cdot \mathbf{y}}^{-1}(r_x \cdot \tau_{\mathbf{y}}(\mathbf{u} \cdot \mathbf{y}))$$

for $\mathbf{u} \in \mathfrak{gl}(\mathrm{F})$ and $\mathbf{y} \in \mathrm{F}$. This formula is obtained by taking the tangent linear mappings at the point $e \in \mathbf{GL}(\mathrm{F})$ of the two functions $s \mapsto r_x \cdot (s \cdot \mathbf{y})$ and $s \mapsto (r_x \cdot s) \cdot \mathbf{y}$, which are equal, and using the definition of $\mathbf{t}_{r_x}$ (20.2.3.3). The formula (20.5.3.2) then follows from the fact that, for each $r_x \in \mathrm{R}_x$ and each tangent vector $\mathbf{k} \in \mathrm{T}_{r_x}(\mathrm{R})$ such that $\mathrm{T}(\pi) \cdot \mathbf{k} = \mathbf{h}_x$, there exists a section $\mathbf{R}$ of R over a neighborhood of x such that $\mathbf{R}(x) = r_x$ and $\mathrm{T}_x(\mathbf{R}) \cdot \mathbf{h}_x = \mathbf{k}$ (16.8.8).

In particular, let us take $\mathbf{k}$ to be the *horizontal lifting* $\mathrm{rel}(\mathbf{h}_x)$ of $\mathbf{h}_x$ at the point r_x (20.2.2). Then we have $\boldsymbol{\omega}(r_x) \cdot \mathbf{k} = 0$ because $\boldsymbol{\omega}$ is vertical, and we obtain

(20.5.3.4) $$\nabla_{\mathbf{h}_x} \cdot \mathbf{s} = r_x \cdot (\theta_{\mathrm{rel}(\mathbf{h}_x)} \cdot \Phi).$$

(20.5.4) Suppose now that the group $\mathbf{GL}(\mathrm{F})$ acts on three other vector spaces F_1, F_2, F_3 and that $\mathrm{B} : \mathrm{F}_1 \times \mathrm{F}_2 \to \mathrm{F}_3$ is a $\mathbf{GL}(\mathrm{F})$-*invariant* bilinear mapping, i.e., that

(20.5.4.1) $$\mathrm{B}(\rho_1(s) \cdot \mathbf{y}_1, \rho_2(s) \cdot \mathbf{y}_2) = \rho_3(s) \cdot \mathrm{B}(\mathbf{y}_1, \mathbf{y}_2)$$

for all $s \in \mathbf{GL}(\mathrm{F})$, where ρ_j ($j = 1, 2, 3$) is the linear representation of $\mathbf{GL}(\mathrm{F})$ in F_j.

Let E_1, E_2, E_3 be the vector bundles associated with these three representations, and endow each of them with the linear connection obtained by the procedure of (20.5.1) from the *same* principal connection $\boldsymbol{P}$ in the bundle R of frames of E. Let $\mathbf{s}_1$ (resp. $\mathbf{s}_2$) be a section of E_1 (resp. E_2) over a neighborhood of x. With the notation of (20.5.3), we may write $\mathbf{s}_j(y) = \boldsymbol{R}(y) \cdot \Phi_j(\boldsymbol{R}(y))$ $(j = 1, 2)$. Now define

$$\mathbf{s}_3(y) = \boldsymbol{R}(y) \cdot B(\Phi_1(\boldsymbol{R}(y)), \Phi_2(\boldsymbol{R}(y))),$$

which by virtue of (20.5.4.1) does not depend on the section $\boldsymbol{R}$ chosen. We write $\mathbf{s}_3 = B(\mathbf{s}_1, \mathbf{s}_2)$. Then, for each tangent vector $\mathbf{h}_x \in T_x(M)$, we have

$$\text{(20.5.4.2)} \qquad \nabla_{\mathbf{h}_x} \cdot B(\mathbf{s}_1, \mathbf{s}_2) = B(\nabla_{\mathbf{h}_x} \cdot \mathbf{s}_1, \mathbf{s}_2) + B(\mathbf{s}_1, \nabla_{\mathbf{h}_x} \cdot \mathbf{s}_2).$$

This is an immediate consequence of the definition of $B(\mathbf{s}_1, \mathbf{s}_2)$, the formula (20.5.3.4), and the fact that the point-distribution $\theta_{\mathrm{rel}(\mathbf{h}_x)}$ is a *derivation* (17.14.2.1).

If we take $E = T(M)$ and B to be a tensor product

$$\mathbf{T}^p_q(\mathbf{R}^n) \times \mathbf{T}^r_s(\mathbf{R}^n) \to \mathbf{T}^{p+r}_{q+s}(\mathbf{R}^n),$$

or the fundamental bilinear form on $\mathbf{R}^n \times (\mathbf{R}^n)^*$, the formula (20.5.4.2) shows that the canonical extension of the covariant differentiation in the direction of a tangent vector $\mathbf{h}_x$, to tensor fields on M (17.18.2), coincides with the covariant differentiation defined on each of the bundles $\mathbf{T}^p_q(M)$ from the connection $\boldsymbol{P}$ on R, by the formula (20.5.3.4).

(20.5.5) We shall now show how the calculation of the *covariant exterior differential* (relative to $\mathbf{C}$) of a *differential* 1*-form on* M *with values in* E (17.19.3) can be reduced to the calculation of the *covariant exterior differential* in the sense of (20.3.2.1) of a *vector-valued* 1*-form on* R, *with values in* F. Let ζ be a differential 1-form on M with values in E, i.e., an M-morphism of T(M) into E. By definition, the covariant exterior differential $\mathbf{d}\zeta$ is a differential 2-form on M with values in E (i.e., an M-morphism of $\bigwedge^2 T(M)$ into E) such that, for each pair of tangent vectors $\mathbf{h}_x$, $\mathbf{k}_x$ in $T_x(M)$, we have

(20.5.5.1)

$$\mathbf{d}\zeta(x) \cdot (\mathbf{h}_x \wedge \mathbf{k}_x) = \nabla_{\mathbf{h}_x} \cdot (\zeta \cdot Y) - \nabla_{\mathbf{k}_x} \cdot (\zeta \cdot X) - \zeta(x) \cdot ([X, Y](x)),$$

where X, Y are any two vector fields on M which take the values $\mathbf{h}_x$, $\mathbf{k}_x$, respectively, at the point x (17.19.3). With the notation of (20.5.3), since $\boldsymbol{R}$ is an immersion of a neighborhood V of x into R, there exists a section

$\mathbf{R}(y) \mapsto X'(\mathbf{R}(y))$ of T(R) over the submanifold $\mathbf{R}(V)$ such that $X'(\mathbf{R}(y)) = T_y(\mathbf{R}) \cdot X(y)$ for all $y \in V$, and X' can be extended to a vector field on R (16.12.11), which we shall also denote by X'. Likewise we define a vector field Y' on R, starting with Y.

We may always assume that $\mathbf{R}$ is such that $X'(\mathbf{R}(x)) = \text{rel}(\mathbf{h}_x)$ and $Y'(\mathbf{R}(x)) = \text{rel}(\mathbf{k}_x)$ (but, of course, at points y close to x, the vectors $X'(\mathbf{R}(y))$ and $Y'(\mathbf{R}(y))$ will not in general be horizontal).

This being so, we construct from ζ a *vector-valued differential* 1-*form* ξ *on* R, with values in F, by the formula

(20.5.5.2) $$\xi(r_x) \cdot \mathbf{z} = r_x^{-1} \cdot (\zeta(x) \cdot (T(\pi) \cdot \mathbf{z}))$$

for all tangent vectors $\mathbf{z} \in T_{r_x}(R)$. We shall now prove the *second fundamental formula*:

(20.5.5.3) $$\mathbf{d}\zeta(x) \cdot (\mathbf{h}_x \wedge \mathbf{k}_x) = r_x \cdot (D\xi(r_x) \cdot (\text{rel}(\mathbf{h}_x) \wedge \text{rel}(\mathbf{k}_x)))$$

for all tangent vectors $\mathbf{h}_x$, $\mathbf{k}_x$ at the point $x \in M$.

The section $\mathbf{s} = \zeta \cdot X$ of E may be written in the form $\mathbf{s}(y) = \mathbf{R}(y) \cdot \Phi(\mathbf{R}(y))$ with $\Phi = \xi \cdot X'$ by virtue of the above definitions. Applying the formula (20.5.3.4), we therefore obtain for the first two terms of the right-hand side of (20.5.5.1)

$$r_x \cdot (\theta_{\text{rel}(\mathbf{h}_x)} \cdot (\xi \cdot Y') - \theta_{\text{rel}(\mathbf{k}_x)} \cdot (\xi \cdot X'));$$

bearing in mind the definition of D (20.3.2.1) and the formula giving $d\xi$ (17.15.8.1), it is enough to show that we have

(20.5.5.4) $$\zeta(x) \cdot ([X, Y](x)) = \mathbf{R}(x) \cdot (\xi(\mathbf{R}(x)) \cdot [X', Y'](\mathbf{R}(x))).$$

However, since the vector fields X' and Y' are tangent to the submanifold $\mathbf{R}(V)$ of R at the points of this submanifold, and since $\mathbf{R}$ is an *isomorphism* of V onto $\mathbf{R}(V)$, the formula (20.5.5.4) follows simply by *transport of structure*, bearing in mind (20.5.5.2).

The formula corresponding to (20.5.5.3) when ζ is a *differential p-form on* M *with values in* E may be proved similarly, by using the general formula (17.15.3.5).

PROBLEMS

1. Let R be a principal fiber bundle with base M, group G and projection π. Let F be a differential manifold on which G acts differentiably on the left, and consider the associated bundle $E = R \times^G F$ with fiber-type F. If $\mathbf{P}$ is a principal connection in R, we may

again, for any element u_x of the fiber E_x of E over a point $x \in M$, write $u_x = r_x \cdot y$ for some $r_x \in R_x$ and $y \in F$, and we can therefore define a vector $\mathbf{C}_x(\mathbf{k}_x, u_x) \in T_{u_x}(E)$ by the formula (20.5.1.1); it does not depend on the choice of the pair (r_x, y) such that $r_x \cdot y = u_x$, and if π_F is the projection of E onto M, then $T(\pi_F) \cdot \mathbf{C}_x(\mathbf{k}_x, u_x) = \mathbf{k}_x$; finally, the mapping $\mathbf{k}_x \mapsto \mathbf{C}_x(\mathbf{k}_x, u_x)$ is linear, and its image is therefore supplementary in $T_{u_x}(E)$ to the subspace $V_{u_x}(E) = \operatorname{Ker}(T_{u_x}(\pi_F))$ of vertical tangent vectors. The vectors $\mathbf{C}_x(\mathbf{k}_x, u_x)$ are called the *horizontal* vectors in $T_{u_x}(E)$ corresponding to the connection $\boldsymbol{P}$.

Let $v : \mathbf{R} \to M$ be an unending path of class C^∞. For each $u_0 \in E_{v(0)}$, we may write $u_0 = r_0 \cdot y$ for some $y \in F$ and some $r_0 \in R_{v(0)}$. With the notation of Section 20.2, Problem 3, show that the unending path $t \mapsto G_{u_0}(t) = w_{r_0}(t) \cdot y$ is the unique unending path which lifts v to E and is such that $G_{u_0}(0) = u_0$ and such that, for each $t \in \mathbf{R}$, the vector $G'_{u_0}(t)$ is *horizontal* in the tangent space to E at the point $G_{u_0}(t)$. For each $t \in \mathbf{R}$, the mapping $u_0 \mapsto G_{u_0}(t)$ is a diffeomorphism φ_t of $E_{v(0)}$ onto $E_{v(t)}$. In the particular case of (20.5.1), where F is a vector space and G acts linearly on F, the diffeomorphism φ_t is a linear bijection; when $E = T(M)$, φ_t is the *parallel transport* along v, defined in (18.6.3).

Given an unending path $t \mapsto f(t)$ in E of class C^∞ which lifts v, then for each $t \in \mathbf{R}$ there exists a unique $u(t) \in E_{v(0)}$ such that $\varphi_t(u(t)) = f(t)$ (in other words, $G_{u(t)}(t) = f(t)$). The mapping $t \mapsto u(t)$ of $\mathbf{R}$ into $E_{v(0)}$ is of class C^∞ and is called the *development* of f in the fiber $E_{v(0)}$. Given an $r_0 \in R_{v(0)}$, we may write $u(t) = r_0 \cdot y(t)$, where $y(t) \in F$. Show that if G acts *transitively* on F, the vector $y'(t)$ is such that $w_{r_0}(t) \cdot y'(t)$ is the vertical component of $f'(t)$ in the decomposition of $T_{f(t)}(E)$ as the direct sum of the space of vertical vectors and the space of horizontal vectors. (Write $f(t)$ locally in the form $(w_{r_0}(t) \cdot g(t)) \cdot y_0$, where g is a C^∞ mapping of $\mathbf{R}$ into G.)

2. With the hypotheses and notation of Section 20.1, Problem 2 (so that in particular $\dim(B) = \dim(G/H) = n$), we may construct the principal bundle $R \times^H G$ with base B and group G (Section 16.14, Problem 17). The mapping $r \mapsto r \cdot e$ is an embedding of R into $R \times^H G$; also H acts on $R \times^H G$ on the right (by restriction of the action of G), and X may be canonically identified with the orbit-manifold $H \backslash (R \times^H G)$. A principal connection $\boldsymbol{P}$ in $R \times^H G$ is called a *Cartan connection for* R (relative to G) (or, by abuse of language, a Cartan connection *in* R) if, for all $r_b \in R_b$, the space H_{r_b} of horizontal tangent vectors (20.2.2), which is a subspace of $T_{r_b}(R \times^H G)$, is such that $H_{r_b} \cap T_{r_b}(R) = \{0\}$. An equivalent condition is that the *restriction* $\boldsymbol{\omega}_0$ to R of the 1-form $\boldsymbol{\omega}$ of the connection $\boldsymbol{P}$ is such that $\boldsymbol{\omega}_0(r_b)$ is an *injective* linear mapping (or, equivalently, *bijective* since $\dim(B) = \dim(G/H)$) of $T_{r_b}(R)$ into $\mathfrak{g}_e$.

(a) Conversely, let $\boldsymbol{\omega}_0$ be a differential 1-form on R with values *in* $\mathfrak{g}_e$ (not $\mathfrak{h}_e$) such that

(i) $\boldsymbol{\omega}_0(r_b \cdot t) \cdot (\mathbf{k} \cdot t) = \operatorname{Ad}(t^{-1}) \cdot (\boldsymbol{\omega}_0(r_b) \cdot \mathbf{k})$ for $\mathbf{k} \in T_{r_b}(R)$ and $t \in H$;
(ii) $\boldsymbol{\omega}_0(r_b) \cdot Z_{\mathbf{u}}(r_b) = \mathbf{u}$ for $\mathbf{u} \in \mathfrak{h}_e$;
(iii) for each $r_b \in R$, the mapping $\boldsymbol{\omega}_0(r_b) : T_{r_b}(R) \to \mathfrak{g}_e$ is in *injective* linear mapping.

Show that $\boldsymbol{\omega}_0$ has a unique extension to the 1-form on $R \times^H G$ of a Cartan connection for R.

(b) For each tangent vector $\mathbf{h}_b \in T_b(B)$, let $\beta_b(\mathbf{h}_b)$ be the vertical component of $T_b(\sigma) \cdot \mathbf{h}_b$ in the decomposition of $T_{\sigma(b)}(X)$ as a direct sum of the spaces of vertical and horizontal vectors (Problem 1). Show that β_b is bijective and that β is a *welding* of B and X, canonically associated with the Cartan connection $\boldsymbol{P}$.

For each unending path $t \mapsto f(t)$ in M, the *development* (Problem 1) in a fiber of X (isomorphic to G/H) of the path $t \mapsto \sigma(f(t))$ in X is also called the *development of f* in G/H.

(c) Conversely, show that if there exists a welding of B and X, then there exists on $R \times^H G$ a Cartan connection for R. (Argue as in Section 20.2, Problem 1.)

(d) Show that if there exists a Cartan connection for R (or, equivalently, a welding of B and X), then the manifold R is *parallelizable*.

3. Let G be a Lie group, H a Lie subgroup of G. Consider G as a principal bundle over G/H with group H, and let $\pi : G \to G/H$ denote the canonical projection. Show that the principal bundle $G \times^H G$ over G/H with group G is trivializable. (Define a canonical section of this bundle by noting that for $s \in G$, the product $s \cdot s^{-1}$ in $G \times^H G$ in the sense of (16.14.7) depends only on $\pi(s)$.) The canonical differential 1-form $\boldsymbol{\omega}_0$ on G (19.16.1) extends uniquely to the 1-form of a *Cartan connection* for G on $G \times^H G$, called the *canonical Cartan connection*. Show that this connection is flat (20.4.1).

4. Let X be a principal bundle with base B, stiucture group G, and projection π. Let H be a closed subgroup of G, and let $j : H \to G$ be the canonical injection; let Y be a principal bundle over B with group H, and let (u, j) be a morphism of Y into X, so that u is an embedding of Y in X (20.7.1). Suppose also that the Lie algebra $\mathfrak{g}_e$ of G contains a vector subspace $\mathfrak{m}$ supplementary to the Lie algebra $\mathfrak{h}_e$ of H, such that $\mathrm{Ad}(t) \cdot \mathfrak{m} \subset \mathfrak{m}$ for all $t \in H$.

(a) Let $\boldsymbol{P}$ be a principal connection in X, and let $\boldsymbol{\omega}$ be the 1-form of the connection. For each $y_b \in Y_b$ and each tangent vector $\mathbf{h}_{y_b} \in T_{y_b}(Y)$, let $\boldsymbol{\omega}_0(y_b) \cdot \mathbf{h}_{y_b}$ and $\boldsymbol{\varphi}(y_b) \cdot \mathbf{h}_{y_b}$ be the projections of $\boldsymbol{\omega}(y_b) \cdot \mathbf{h}_{y_b}$ onto $\mathfrak{h}_e$ and $\mathfrak{m}$, respectively. Show that $\boldsymbol{\omega}_0$ is the 1-form of a principal connection $\boldsymbol{P}_0$ in Y, and that $\boldsymbol{\varphi}$ is a vector-valued differential 1-form on Y, with values in $\mathfrak{m}$, which is *horizontal* and such that $\boldsymbol{\varphi}(y_b \cdot t) \cdot (\mathbf{h} \cdot t) = \mathrm{Ad}(t^{-1}) \cdot (\boldsymbol{\varphi}(y_b) \cdot \mathbf{h})$ for all $\mathbf{h} \in T_{y_b}(Y)$ and all $t \in H$.

Conversely, if we are given a principal connection $\boldsymbol{P}_0$ in Y, for which $\boldsymbol{\omega}_0$ is the connection form, and if $\boldsymbol{\varphi}$ is a horizontal differential 1-form on Y with values in $\mathfrak{m}$ which satisfies the above condition, then there exists a unique principal connection on X which gives rise to $\boldsymbol{\omega}_0$ and $\boldsymbol{\varphi}$ as above.

(b) If $X = Y \times^H G$ and $\dim(B) = \dim(G/H)$, then a connection in X with 1-form $\boldsymbol{\omega}$ is a Cartan connection if and only if, for each $y_b \in Y$, the mapping $\boldsymbol{\varphi}(y_b)$ of $T_{y_b}(Y)$ into $\mathfrak{m}$ is surjective.

(c) Suppose that G is a semidirect product $N \times_\sigma H$, so that $\mathfrak{m}$ may be taken to be the Lie algebra $\mathfrak{n}_e$ of N (19.14). Show that if N is *commutative*, the curvature forms $\boldsymbol{\Omega}$ and $\boldsymbol{\Omega}_0$ of the connections $\boldsymbol{P}$ and $\boldsymbol{P}_0$ are such that the restriction of $\boldsymbol{\Omega}$ to Y is equal to $\boldsymbol{\Omega}_0 + D\boldsymbol{\varphi}$ (covariant exterior differential relative to $\boldsymbol{P}_0$), and that

$$d\boldsymbol{\varphi} = -\boldsymbol{\omega}_0 \wedge \boldsymbol{\varphi} + D\boldsymbol{\varphi}.$$

(d) Suppose again that $G = N \times_\sigma H$ and $\mathfrak{m} = \mathfrak{n}_e$ (but not that N is necessarily commutative). Let $v : \mathbf{R} \to M$ be an unending path of class C^∞, and let y_0 be a point of $Y_{v(0)} \subset X_{v(0)}$. Denote by w and w_0 the unending paths in X and Y, respectively, which lift v and are such that their tangent vectors are *horizontal* relative to $\boldsymbol{P}$ and $\boldsymbol{P}_0$, respectively, and such that $w(0) = w_0(0) = y_0$ (Section 20.2, Problem 3). Show that $w(t) = w_0(t) \cdot h(t)$, where $h(t) \in N$, and that $\boldsymbol{\varphi}(w_0(t)) \cdot w_0'(t) = -h'(t) \cdot h(t)^{-1}$. (Use (20.2.5.1) and (20.2.5.2).)

6. THE METHOD OF MOVING FRAMES

(20.6.1) From now on we shall consider only vector bundles E which are *tangent bundles* T(M) of smooth manifolds M. In other words, we shall consider only *linear connections on a pure manifold* M of dimension n (17.18.1). These connections are in one-to-one correspondence (20.5) with the *principal connections* in the *bundle of frames* $R(M) = \text{Isom}(M \times \mathbf{R}^n, T(M))$ of M, and it is these latter that we shall consider first.

(20.6.2) The definition (20.1.1) of the bundle R(M) implies the existence of a *canonical vector-valued differential* 1-*form* on R(M) with values in $\mathbf{R}^n$ (independent of any connection in R(M)). Namely, for each $x \in M$ an element r_x of the fiber $R(M)_x$ is an isomorphism of $\mathbf{R}^n$ onto $T_x(M)$; if π is the projection of the bundle R(M), then the mapping

(20.6.2.1) $$\boldsymbol{\sigma}(r_x) : \mathbf{k}_{r_x} \mapsto r_x^{-1} \cdot (T(\pi) \cdot \mathbf{k}_{r_x})$$

is a surjective linear mapping of $T_{r_x}(R(M))$ onto $\mathbf{R}^n$. In other words, $\boldsymbol{\sigma}$ is a vector-valued differential 1-form on R(M), with values in $\mathbf{R}^n$, which is called the *canonical form* on R(M). This form is of class C^∞. For by trivializing T(M) and R(M) over an open set U by means of a chart of M, we may assume that M is an open set in $\mathbf{R}^n$, and hence $T(M) = M \times \mathbf{R}^n$, $R(M) = M \times \mathbf{GL}(n, \mathbf{R})$. A point r_x is then written as (x, U) with $U \in \mathbf{GL}(n, \mathbf{R})$, and a tangent vector $\mathbf{k}_{r_x}$ takes the form $((x, U), (\mathbf{v}, V))$ with $\mathbf{v} \in \mathbf{R}^n$ and $V \in \mathbf{M}_n(\mathbf{R})$. The vector $T(\pi) \cdot \mathbf{k}_{r_x}$ is then $(x, \mathbf{v})$, and hence $\boldsymbol{\sigma}(r_x)$ is the mapping

$$((x, U), (\mathbf{v}, V)) \mapsto U^{-1} \cdot \mathbf{v};$$

this proves our assertion.

It is clear that the canonical form $\boldsymbol{\sigma}$ is *horizontal* (20.2.4), from the definition of the vertical vectors of $T_{r_x}(R(M))$. Moreover, for each $s \in \mathbf{GL}(n, \mathbf{R})$ and $\mathbf{k} \in T_{r_x}(R(M))$, we have

(20.6.2.2) $$\boldsymbol{\sigma}(r_x \cdot s) \cdot (\mathbf{k} \cdot s) = s^{-1} \cdot (\boldsymbol{\sigma}(r_x) \cdot \mathbf{k}),$$

because $(r_x \cdot s)^{-1} = s^{-1} \circ r_x^{-1}$ and $T(\pi) \cdot (\mathbf{k} \cdot s) = T(\pi) \cdot \mathbf{k}$. In other words, $\boldsymbol{\sigma}$ is *invariant* (19.1) under the right action of $\mathbf{GL}(n, \mathbf{R})$ on R(M) and the canonical right action $(s, \mathbf{y}) \mapsto s^{-1} \cdot \mathbf{y}$ of $\mathbf{GL}(n, \mathbf{R})$ on $\mathbf{R}^n$.

Suppose now that we are given a principal connection $\boldsymbol{P}$ in R(M). Then the covariant exterior differential of the canonical form $\boldsymbol{\sigma}$ is a vector-valued 2-form $\boldsymbol{\Theta}$ *with values in* $\mathbf{R}^n$:

(20.6.2.3) $$\boldsymbol{\Theta} = D\boldsymbol{\sigma}$$

called the *torsion form* of $\boldsymbol{P}$. It is *horizontal* and *invariant* (for the action $(s, \mathbf{u}) \mapsto s^{-1} \cdot \mathbf{u}$ of $\mathbf{GL}(n, \mathbf{R})$ on $\mathbf{R}^n$).

(20.6.3) *The canonical form* $\boldsymbol{\sigma}$ *on* R(M) *satisfies the "structure equation"*

(20.6.3.1)

$$d\boldsymbol{\sigma}(r_x) \cdot (\mathbf{h} \wedge \mathbf{k})$$
$$= -(\boldsymbol{\omega}(r_x) \cdot \mathbf{h}) \cdot (\boldsymbol{\sigma}(r_x) \cdot \mathbf{k}) + (\boldsymbol{\omega}(r_x) \cdot \mathbf{k}) \cdot (\boldsymbol{\sigma}(r_x) \cdot \mathbf{h}) + \boldsymbol{\Theta}(r_x) \cdot (\mathbf{h} \wedge \mathbf{k})$$

for all tangent vectors **h**, **k** *at a point* $r_x \in$ R(M), *where* $\boldsymbol{\omega}$ *is the* 1*-form of the connection* **P**.

As in the proof of (20.3.3), we distinguish three cases:

(i) **h** and **k** are horizontal. Since the connection form $\boldsymbol{\omega}$ is vertical, the formula (20.6.3.1) reduces to the definition of $D\boldsymbol{\sigma}$ (20.3.2.1).

(ii) **h** and **k** are vertical. Then both sides of (20.6.3.1) are *zero*. This is clear as far as the right-hand side is concerned; on the other hand, if $\mathbf{u}, \mathbf{v} \in \mathfrak{gl}(n, \mathbf{R})$ are such that $\mathbf{h} = Z_{\mathbf{u}}(r_x)$, $\mathbf{k} = Z_{\mathbf{v}}(r_x)$, then we have

$$d\boldsymbol{\sigma} \cdot (Z_{\mathbf{u}} \wedge Z_{\mathbf{v}}) = \theta_{Z_{\mathbf{u}}} \cdot (\boldsymbol{\sigma} \cdot Z_{\mathbf{v}}) - \theta_{Z_{\mathbf{v}}} \cdot (\boldsymbol{\sigma} \cdot Z_{\mathbf{u}}) - \boldsymbol{\sigma} \cdot Z_{[\mathbf{u}, \mathbf{v}]},$$

and the right-hand side is 0 because $\boldsymbol{\sigma}$ is horizontal.

(iii) **h** is vertical; **k** is horizontal. Then **h** is the value at the point r_x of a Killing vector field $Z_{\mathbf{u}}$, where $\mathbf{u} \in \mathfrak{gl}(n, \mathbf{R})$. Also, there exists a vector field Y on R(M) which is *invariant* under the action of **GL**(n, **R**) and which takes the value **k** at the point r_x. For if Y_0 is a vector field on M whose value at the point x is the projection $\mathrm{T}(\pi) \cdot \mathbf{k}$ (16.12.11), then its *horizontal lifting* Y will have the required properties, because **k** is horizontal. We have then, by (17.15.8.1),

(20.6.3.2) $$d\boldsymbol{\sigma} \cdot (Z_{\mathbf{u}} \wedge Y) = \theta_{Z_{\mathbf{u}}} \cdot (\boldsymbol{\sigma} \cdot Y) - \theta_Y \cdot (\boldsymbol{\sigma} \cdot Z_{\mathbf{u}}) - \boldsymbol{\sigma} \cdot [Z_{\mathbf{u}}, Y].$$

Now, since $\boldsymbol{\sigma}$ is horizontal, we have $\boldsymbol{\sigma} \cdot Z_{\mathbf{u}} = 0$; also, because of the choice of Y, we have $[Z_{\mathbf{u}}, Y] = 0$ (19.8.11); finally, since $\boldsymbol{\sigma}$ and Y are invariant under the action of **GL**(n, **R**), the mapping $\boldsymbol{\sigma} \cdot Y$ of R(M) into $\mathbf{R}^n$ is also invariant under this action and under the action $(s, \mathbf{y}) \mapsto s^{-1} \cdot \mathbf{y}$ of **GL**(n, **R**) on $\mathbf{R}^n$. Hence (19.4.4.3) we have $\theta_{Z_{\mathbf{u}}} \cdot (\boldsymbol{\sigma} \cdot Y) = -\mathbf{u} \cdot (\boldsymbol{\sigma} \cdot Y)$ (recall that $\mathbf{u} \in \mathbf{M}_n(\mathbf{R})$). By definition, the value of this expression at the point r_x is

$$-(\boldsymbol{\omega}(r_x) \cdot \mathbf{h}) \cdot (\boldsymbol{\sigma}(r_x) \cdot \mathbf{k}).$$

On the other hand, $\boldsymbol{\omega}(r_x) \cdot \mathbf{k} = 0$ since **k** is horizontal, and $\boldsymbol{\Theta}(r_x) \cdot (\mathbf{h} \wedge \mathbf{k}) = 0$ since **h** is vertical. We have therefore verified that the two sides of (20.6.3.1) are equal.

In accordance with the conventions introduced in (16.20.15), we shall write this equation in abridged notation in the form

(20.6.3.3) $$d\boldsymbol{\sigma} = -\boldsymbol{\omega} \wedge \boldsymbol{\sigma} + \boldsymbol{\Theta}.$$

If the scalar-valued 1-forms ω_{ij} $(1 \leqq i, j \leqq n)$ and σ_i $(1 \leqq i \leqq n)$ are the components of $\boldsymbol{\omega}$ and $\boldsymbol{\sigma}$ with respect to the canonical bases of $\mathbf{M}_n(\mathbf{R})$ and $\mathbf{R}^n$, respectively, then the ith component $(1 \leqq i \leqq n)$ of $\boldsymbol{\omega} \wedge \boldsymbol{\sigma}$ is

(20.6.3.4) $$(\boldsymbol{\omega} \wedge \boldsymbol{\sigma})_i = \sum_{j=1}^{n} \omega_{ij} \wedge \sigma_j.$$

(20.6.4) *The curvature and torsion 2-forms of* **P** *satisfy the relation*

(20.6.4.1) $$D\boldsymbol{\Theta} = \boldsymbol{\Omega} \wedge \boldsymbol{\sigma}$$

(*Bianchi's identity*).

(The vector-valued 3-form on the right-hand side of (20.6.4.1) is defined by the formula

(20.6.4.2)
$$\begin{aligned}((\boldsymbol{\Omega} \wedge \boldsymbol{\sigma})(r_x)) \cdot (\mathbf{h}_1 \wedge \mathbf{h}_2 \wedge \mathbf{h}_3) \\ = (\boldsymbol{\Omega}(r_x) \cdot (\mathbf{h}_1 \wedge \mathbf{h}_2)) \cdot (\boldsymbol{\sigma}(r_x) \cdot \mathbf{h}_3) + (\boldsymbol{\Omega}(r_x) \cdot (\mathbf{h}_2 \wedge \mathbf{h}_3)) \cdot (\boldsymbol{\sigma}(r_x) \cdot \mathbf{h}_1) \\ + (\boldsymbol{\Omega}(r_x) \cdot (\mathbf{h}_3 \wedge \mathbf{h}_1)) \cdot (\boldsymbol{\sigma}(r_x) \cdot \mathbf{h}_2)\end{aligned}$$

for all $\mathbf{h}_1, \mathbf{h}_2, \mathbf{h}_3$ in $T_{r_x}(R(M))$.)

We obtain from (20.6.3.3), in view of the expressions (20.6.3.4) and the rules of calculation for exterior differentials (17.15.2.1), that

(20.6.4.3) $$d\boldsymbol{\Theta} = d\boldsymbol{\omega} \wedge \boldsymbol{\sigma} + \boldsymbol{\omega} \wedge d\boldsymbol{\sigma}.$$

To obtain the value of $D\boldsymbol{\Theta}$ at a trivector $\mathbf{h}_1 \wedge \mathbf{h}_2 \wedge \mathbf{h}_3$, we have therefore to evaluate the right-hand side of (20.6.4.3) at the trivector $\mathbf{h}'_1 \wedge \mathbf{h}'_2 \wedge \mathbf{h}'_3$, where $\mathbf{h}'_j$ is the *horizontal* component of $\mathbf{h}_j$ $(j = 1, 2, 3)$ (20.3.1). Now since $\boldsymbol{\omega}$ is a vertical form, the value of $\boldsymbol{\omega} \wedge d\boldsymbol{\sigma}$ at $\mathbf{h}'_1 \wedge \mathbf{h}'_2 \wedge \mathbf{h}'_3$ is zero, and the formula (20.6.4.1) follows.

(20.6.5) We shall now obtain expressions for the *covariant derivative*, the *curvature* and the *torsion* of the linear connection **C** on M defined (20.5.1) by the principal connection ***P*** in R(M). Since the question is local with respect to M, we shall suppose that there exists a C^∞ *section* of R(M) over M. Such a section $x \mapsto \mathbf{R}(x)$ is called a *moving frame* on M, and the existence of such a section is equivalent to that of n C^∞ vector fields $x \mapsto \mathbf{e}_i(x)$ on M, such that the vectors $\mathbf{e}_i(x)$ are linearly independent for all $x \in M$. By definition, we have

$$\mathbf{R}(x)^{-1} \cdot \mathbf{e}_i(x) = \mathbf{e}_i \qquad (1 \leqq i \leqq n), \tag{20.6.5.1}$$

where $(\mathbf{e}_i)$ is the canonical basis of $\mathbf{R}^n$.

We remark that the existence of a moving frame $\mathbf{R}$ on M implies that M is *orientable*, and defines an orientation on M for which $\mathbf{R}$ is *direct* at each point (16.21.2). If v_x is the n-covector at x that takes the value 1 at

$$\mathbf{e}_1(x) \wedge \cdots \wedge \mathbf{e}_n(x),$$

then $x \mapsto v_x$ is a C^∞ n-form on M which is nowhere zero.

To the vector-valued 1-forms $\boldsymbol{\sigma}$ and $\boldsymbol{\omega}$ on R(M) there correspond, under the mapping $\mathbf{R}$, their *inverse images* (16.20.15.4)

$$\boldsymbol{\sigma}^{(\mathbf{R})} = {}^t\mathbf{R}(\boldsymbol{\sigma}), \qquad \boldsymbol{\omega}^{(\mathbf{R})} = {}^t\mathbf{R}(\boldsymbol{\omega}), \tag{20.6.5.2}$$

which are vector-valued 1-forms *on* M, with values in $\mathbf{R}^n$ and $\mathbf{M}_n(\mathbf{R})$, respectively. We may therefore write

$$\boldsymbol{\sigma}^{(\mathbf{R})} = \sum_{i=1}^{n} \sigma_i^{(\mathbf{R})} \mathbf{e}_i, \qquad \boldsymbol{\omega}^{(\mathbf{R})} = \sum_{i,j} \omega_{ij}^{(\mathbf{R})} E_{ij}, \tag{20.6.5.3}$$

where the $\sigma_i^{(\mathbf{R})}$ and the $\omega_{ij}^{(\mathbf{R})}$ are *scalar* 1-forms on M. The forms $\boldsymbol{\sigma}^{(\mathbf{R})}$ and $\boldsymbol{\omega}^{(\mathbf{R})}$ (or their scalar components $\sigma_i^{(\mathbf{R})}$ and $\omega_{ij}^{(\mathbf{R})}$) are called the *canonical* and *connection forms* on M, corresponding to the moving frame $\mathbf{R}$ (or to the n vector fields $\mathbf{e}_i$).

Conversely, if we are given *any* 1-form $\boldsymbol{\varpi}$ on M with values in $\mathbf{M}_n(\mathbf{R})$, then there exists a unique connection form $\boldsymbol{\omega}$ on R(M) such that $\boldsymbol{\omega}^{(\mathbf{R})} = \boldsymbol{\varpi}$, since the value of $\boldsymbol{\omega}$ at one point of a fiber of R(M) determines its value at all other points (20.2.5.1). We have therefore another proof of the existence of a linear connection on M (17.16.8).

By definition, for each tangent vector $\mathbf{h}_x \in T_x(M)$, we have

$$\boldsymbol{\sigma}^{(\mathbf{R})}(x) \cdot \mathbf{h}_x = \boldsymbol{\sigma}(\mathbf{R}(x)) \cdot (T_x(\mathbf{R}) \cdot \mathbf{h}_x) \in \mathbf{R}^n,$$

and therefore, by virtue of the definition (20.6.2.1) of $\boldsymbol{\sigma}$, and of the fact that $T_{\mathbf{R}(x)}(\pi) \cdot (T_x(\mathbf{R}) \cdot \mathbf{h}_x) = \mathbf{h}_x$ (because $\mathbf{R}$ is a section), we obtain

$$\boldsymbol{\sigma}^{(\mathbf{R})}(x) \cdot \mathbf{h}_x = \mathbf{R}(x)^{-1} \cdot \mathbf{h}_x. \tag{20.6.5.4}$$

In other words, the $\sigma_i^{(\mathbf{R})}(x)$ are the n coordinate forms on $\mathrm{T}_x(\mathrm{M})$ relative to the basis $(\mathbf{e}_i(x))$, or equivalently,

(20.6.5.5) $$\mathbf{h}_x = \sum_i \langle \sigma_i^{(\mathbf{R})}(x), \mathbf{h}_x \rangle \mathbf{e}_i(x)$$

for all $\mathbf{h}_x \in \mathrm{T}_x(\mathrm{M})$.

We shall next express the *covariant derivative* $\nabla_{\mathbf{h}_x} \cdot Y$, for any vector $\mathbf{h}_x \in \mathrm{T}_x(\mathrm{M})$, of a vector field Y at the point x (17.18.1) in terms of the form $\boldsymbol{\omega}^{(\mathbf{R})}$. For this purpose it is enough to calculate the n vectors $\nabla_{\mathbf{h}_x} \cdot \mathbf{e}_i$ $(1 \leqq i \leqq n)$ since the $\mathbf{e}_i$ form a basis for the module of vector fields in a neighborhood of x. We apply the fundamental formula (20.5.3.2), taking $\mathbf{s} = \mathbf{e}_i$ and $\mathbf{k} = \mathrm{T}_x(\mathbf{R}) \cdot \mathbf{h}_x$. Now $\Phi(r_x) = \mathbf{e}_i$ is *constant* on $\mathrm{R}(\mathrm{M})$, and therefore the first term on the right-hand side of (20.5.3.2) is zero. Also, we have

$$\boldsymbol{\omega}(r_x) \cdot \mathbf{k} = \boldsymbol{\omega}^{(\mathbf{R})}(x) \cdot \mathbf{h}_x = \sum_{j,k} \langle \omega_{jk}^{(\mathbf{R})}(x), \mathbf{h}_x \rangle \cdot E_{jk},$$

and therefore we obtain

(20.6.5.6) $$\nabla_{\mathbf{h}_x} \cdot \mathbf{e}_i = \sum_j \langle \omega_{ji}^{(\mathbf{R})}(x), \mathbf{h}_x \rangle \cdot \mathbf{e}_j .$$

We shall usually omit the indication of the frame $\mathbf{R}$ in the notation for the components of the forms $\boldsymbol{\omega}^{(\mathbf{R})}$ and $\boldsymbol{\sigma}^{(\mathbf{R})}$. Introducing the covariant exterior differential $\boldsymbol{d}\mathbf{e}_i$, such that $\boldsymbol{d}\mathbf{e}_i \cdot \mathbf{h}_x = \nabla_{\mathbf{h}_x} \cdot \mathbf{e}_i$ (17.19.2), the formulas (20.6.5.5) and (20.6.5.6) take the forms

(20.6.5.7) $$1_{\mathrm{T}(\mathrm{M})} = \sum_{i=1}^{n} \sigma_i \otimes \mathbf{e}_i$$

or equivalently, $\langle \sigma_i, \mathbf{e}_j \rangle = \delta_{ij}$; and

(20.6.5.8) $$\boldsymbol{d}\mathbf{e}_i = \sum_{j=1}^{n} \omega_{ji} \otimes \mathbf{e}_j$$

(where $1_{\mathrm{T}(\mathrm{M})}$ is canonically identified with the Kronecker tensor field (16.18.3)); or, by abuse of notation,

(20.6.5.9) $$1_{\mathrm{T}(\mathrm{M})} = \sum_{i=1}^{n} \sigma_i \mathbf{e}_i ,$$

(20.6.5.10) $$\boldsymbol{d}\mathbf{e}_i = \sum_{j=1}^{n} \omega_{ji} \mathbf{e}_j .$$

From (20.6.5.9) it follows that, for each real-valued C^1 function f on M, we have

$$\langle df, \mathbf{h}_x\rangle = \sum_i \langle \sigma_i(x), \mathbf{h}_x\rangle\langle df, \mathbf{e}_i(x)\rangle,$$

which may also be written in the form

(20.6.5.11) $$df = \sum_i (\theta_{\mathbf{e}_i} \cdot f)\sigma_i = \sum_i (\nabla_{\mathbf{e}_i} \cdot f)\sigma_i .$$

(20.6.6) In the same way, to the vector-valued 2-forms $\mathbf{\Theta}$ and $\mathbf{\Omega}$ on R(M) there correspond, under the mapping $\mathbf{R}$, their inverse images

(20.6.6.1) $$\mathbf{\Theta}^{(\mathbf{R})} = {}^t\mathbf{R}(\mathbf{\Theta}), \qquad \mathbf{\Omega}^{(\mathbf{R})} = {}^t\mathbf{R}(\mathbf{\Omega}),$$

which are vector-valued 2-forms *on* M, with values in $\mathbf{R}^n$ and $\mathbf{M}_n(\mathbf{R})$, respectively. Put

(20.6.6.2) $$\mathbf{\Theta}^{(\mathbf{R})} = \sum_i \Theta_i^{(\mathbf{R})}\mathbf{e}_i, \qquad \mathbf{\Omega}^{(\mathbf{R})} = \sum_{i,j} \Omega_{ij}^{(\mathbf{R})}E_{ij},$$

where the $\Theta_i^{(\mathbf{R})}$ and $\Omega_{ij}^{(\mathbf{R})}$ are scalar-valued 2-forms on M, called respectively the *torsion* and *curvature 2-forms* of the linear connection $\mathbf{C}$ on M, corresponding to the frame $\mathbf{R}$ (or to the vector fields $\mathbf{e}_i$). In what follows, we shall generally omit $\mathbf{R}$ from the notation.

We have then the two *structure equations*

(20.6.6.3) $$d\sigma_i = -\sum_j \omega_{ij} \wedge \sigma_j + \Theta_i,$$

(20.6.6.4) $$d\omega_{ik} = -\sum_k \omega_{ik} \wedge \omega_{kj} + \Omega_{ij},$$

derived from (20.6.3.4) and (20.3.3.1) by taking inverse images with respect to $\mathbf{R}$ (17.15.3.2).

We shall show that the torsion and curvature of the connection $\mathbf{C}$, introduced in (17.19), are related to the torsion and curvature forms by the relations

(20.6.6.5) $$\boldsymbol{t} \cdot (X \wedge Y) = \sum_i \langle \Theta_i, X \wedge Y\rangle\mathbf{e}_i,$$

(20.6.6.6) $$(\boldsymbol{r} \cdot (X \wedge Y)) \cdot \mathbf{e}_i = \sum_j \langle \Omega_{ji}, X \wedge Y\rangle\mathbf{e}_j,$$

where X, Y are any two C^∞ vector fields on M.

By definition, we have $\boldsymbol{t} = \mathbf{d}(1_{T(M)})$, where $1_{T(M)}$ is considered as a differential 1-form on M with values in T(M). The differential 1-form on R, with values in $\mathbf{R}^n$, associated with $1_{T(M)}$ (20.5.5.2) is precisely the canonical form $\boldsymbol{\sigma}$ (20.6.2.1). If we replace ξ by $\boldsymbol{\sigma}$ in (20.5.5.3), we obtain

(20.6.6.7) $$\boldsymbol{t}(x) \cdot (\mathbf{h}_x \wedge \mathbf{k}_x) = r_x \cdot (\boldsymbol{\Theta}(r_x) \cdot (\mathrm{rel}(\mathbf{h}_x) \wedge \mathrm{rel}(\mathbf{k}_x))),$$

and (20.6.6.5) follows immediately.

Next, for each C^∞ vector field Z on M, let ζ_Z be the differential 1-form on M with values in T(M) defined by $\mathbf{h}_x \mapsto \nabla_{\mathbf{h}_x} \cdot Z$. By definition (17.19.3.1), we have

(20.6.6.8) $$(\boldsymbol{r} \cdot (X \wedge Y)) \cdot Z = \mathbf{d}\zeta_Z \cdot (X \wedge Y).$$

Since the value of the left-hand side of (20.6.6.8) at a point $x \in M$ depends only on the value of Z at this point, we may certainly assume that Z is a linear combination of the $\mathbf{e}_i$ with *constant* coefficients. Then, by virtue of (20.6.5.6) and (20.5.5.2), the vector-valued differential 1-form ξ on R with values in $\mathbf{R}^n$ corresponding to ζ_Z is given by

(20.6.6.9) $$\xi(r_x) : \mathbf{z} \mapsto (\boldsymbol{\omega}(r_x) \cdot \mathbf{z}) \cdot \mathbf{u},$$

where $\mathbf{u} \in \mathbf{R}^n$ is a vector such that $r_x \cdot \mathbf{u} = Z(x)$. Hence, from the definition of $\boldsymbol{\Omega} = D\boldsymbol{\omega}$ and (20.5.5.3), we have

(20.6.6.10) $$\mathbf{d}\zeta_Z(x) \cdot (\mathbf{h}_x \wedge \mathbf{k}_x) = r_x \cdot ((\boldsymbol{\Omega}(r_x) \cdot (\mathrm{rel}(\mathbf{h}_x) \wedge \mathrm{rel}(\mathbf{k}_x))) \cdot \mathbf{u})$$

and, by virtue of (20.6.6.8),

(20.6.6.11)
$$(\boldsymbol{r}(x) \cdot (X(x) \wedge Y(x))) \cdot Z(x) = r_x \cdot ((\boldsymbol{\Omega}^{(R)}(x) \cdot (X(x) \wedge Y(x))) \cdot (r_x^{-1} \cdot Z(x))),$$

from which the formula (20.6.6.6) follows immediately, if we bear in mind (20.6.6.2).

From the formulas (20.6.6.5) and (20.6.6.6) we obtain directly the expressions of the 2-forms Θ_i and Ω_{ji} in terms of the components of the torsion and curvature tensors relative to the basis $(\mathbf{e}_i)$: if

$$\mathbf{t} = \sum_{i,j,k} t^i_{jk} \sigma_j \otimes \sigma_k \otimes \mathbf{e}_i, \qquad \mathbf{r} = \sum_{i,j,k,l} r^l_{ijk} \sigma_i \otimes \sigma_j \otimes \sigma_k \otimes \mathbf{e}_l,$$

then we have

(20.6.6.12) $$\Theta_i = \sum_{j,k} t^i_{jk} \sigma_j \wedge \sigma_k,$$

(20.6.6.13) $$\Omega_{ji} = \sum_{h,k} r^j_{ihk} \sigma_h \wedge \sigma_k.$$

PROBLEMS

1. Let $\mathbf{P}$ be a principal connection in the bundle of frames R(M). For each vector $\mathbf{a} \in \mathbf{R}^n$, define a *horizontal* vector field $H_{\mathbf{a}}$ on R(M) by the condition

$$H_{\mathbf{a}}(r_x) = \mathbf{P}_x(r_x \cdot \mathbf{a}, r_x)$$

for all $r_x \in \mathrm{R(M)}$. Equivalently, $H_{\mathbf{a}}$ is defined by the conditions

$$\boldsymbol{\sigma} \cdot H_{\mathbf{a}} = \mathbf{a}, \qquad \boldsymbol{\omega} \cdot \mathrm{H}_{\mathbf{a}} = 0.$$

For each $s \in \mathbf{GL}(n, \mathbf{R})$, we have $H_{\mathbf{a}} \cdot s = H_{s^{-1} \cdot \mathbf{a}}$ If $(\mathbf{a}_i)_{1 \leqq i \leqq n}$ is a basis of $\mathbf{R}^n$ and $(\mathbf{u}_{ij})_{1 \leqq i,\, j \leqq n}$ a basis of $\mathfrak{gl}(n, \mathbf{R}) = \mathbf{M}_n(\mathbf{R})$, then the vector fields $H_{\mathbf{a}_i}$ and $Z_{\mathbf{u}_{ij}}$ form a *basis* of the $\mathscr{E}(\mathrm{R(M)})$-module of C^∞ vector fields on R(M).

(a) Show that, for all $\mathbf{a}, \mathbf{b} \in \mathbf{R}^n$ and $\mathbf{u} \in \mathfrak{gl}(n, \mathbf{R})$, we have

$$[Z_{\mathbf{u}}, H_{\mathbf{a}}] = H_{\mathbf{u} \cdot \mathbf{a}},$$

where $\mathbf{u} \cdot \mathbf{a}$ is the value at $\mathbf{a}$ of $\mathbf{u} \in \mathrm{End}(\mathbf{R}^n)$, and that

$$\boldsymbol{\sigma} \cdot [H_{\mathbf{a}}, H_{\mathbf{b}}] = -\boldsymbol{\Theta} \cdot (H_{\mathbf{a}} \wedge H_{\mathbf{b}}),$$

$$\boldsymbol{\omega} \cdot [H_{\mathbf{a}}, H_{\mathbf{b}}] = -\boldsymbol{\Omega} \cdot (H_{\mathbf{a}} \wedge H_{\mathbf{b}}).$$

(b) Deduce that the torsion and curvature of $\mathbf{P}$ are zero if and only if, for each $x \in \mathrm{M}$, there exists a chart $c = (\mathrm{U}, \varphi, n)$ of M at x such that if $\mathbf{R}$ is the moving frame defined by $\mathbf{R}^{-1} \cdot X_i = \mathbf{e}_i$ $(1 \leqq i \leqq n)$, where the X_i are the vector fields associated with the chart c (16.15.4.2), then the image of U under $\mathbf{R}$ is a submanifold all of whose tangent spaces are *horizontal*.

(c) The geodesics for the connection $\mathbf{C}$ defined by $\mathbf{P}$ are exactly the projections in M of the integral curves of the vector fields $H_{\mathbf{a}}$, for all $\mathbf{a} \in \mathbf{R}^n$.

(d) Deduce that for a curve $t \mapsto r(t)$ with values in R(M) to be such that $t \mapsto \pi(r(t))$ (where π is the canonical projection of R(M) onto M) is a geodesic for $\mathbf{C}$, the following condition is necessary and sufficient:

$$\frac{d}{dt}(\boldsymbol{\sigma}(r(t)) \cdot r'(t)) + (\boldsymbol{\omega}(r(t)) \cdot r'(t)) \cdot (\boldsymbol{\sigma}(r(t)) \cdot r'(t)) = 0.$$

(Argue as in Section 20.2. Problem 3, by using the structure equation (20.6.3.1).) Show that two principal connections $\mathbf{P}_1$, $\mathbf{P}_2$ in R(M) are such that the corresponding linear connections on M have the same geodesics if and only if

$$(\boldsymbol{\omega}_1(r) \cdot \mathbf{k}_r - \boldsymbol{\omega}_2(r) \cdot \mathbf{k}_r) \cdot (\boldsymbol{\sigma}(r), \mathbf{k}_r) = 0$$

for all $r \in \mathrm{R(M)}$ and all tangent vectors $\mathbf{k}_r$ at r, where $\boldsymbol{\omega}_1$, $\boldsymbol{\omega}_2$ are the 1-forms of the connections $\mathbf{P}_1$, $\mathbf{P}_2$.

2. Let M be a pure differential manifold of dimension n. For $0 \leq p \leq n$, $0 \leq q \leq n$, let

$$\mathbf{B}_q^p(\mathrm{M}) = \bigwedge^q (\mathrm{T}(\mathrm{M})^*) \otimes \bigwedge^p (\mathrm{T}(\mathrm{M}))$$

and define an M-morphism of multiplication

$$\mathbf{B}_q^p(\mathrm{M}) \otimes \mathbf{B}_s^r(\mathrm{M}) \to \mathbf{B}_{q+s}^{p+r}(\mathrm{M})$$

as follows: $(\alpha \otimes \mathbf{u}) \otimes (\beta \otimes \mathbf{v}) \mapsto (\alpha \wedge \beta) \otimes (\mathbf{u} \wedge \mathbf{v})$. The direct sum $\mathbf{B}(\mathrm{M})$ of the $\mathbf{B}_q^p(\mathrm{M})$ is a bundle of algebras over M. Let $\mathscr{B}_q^p(\mathrm{M})$ (resp. $\mathscr{B}(\mathrm{M})$) denote the $\mathscr{E}(\mathrm{M})$-module (resp. $\mathscr{E}(\mathrm{M})$-algebra) of C^∞ sections of $\mathbf{B}_q^p(\mathrm{M})$ (resp. $\mathbf{B}(\mathrm{M})$) over M. The elements of $\mathscr{B}_p^1(\mathrm{M})$ may be identified with the *differential p-forms on* M *with values in* T(M) (17.19.2).

(a) Let $\mathbf{C}$ be a linear connection on M, corresponding to a principal connection $\mathbf{P}$ in R(M). There corresponds therefore to $\mathbf{C}$ a covariant exterior differentiation operator $\boldsymbol{d}$ which maps each $\mathscr{B}_q^1(\mathrm{M})$ into $\mathscr{B}_{q+1}^1(\mathrm{M})$ (17.19.4). Show that there exists a unique differential operator $\mathbf{d} : \mathscr{B}(\mathrm{M}) \to \mathscr{B}(\mathrm{M})$, such that $\mathbf{d}(\mathscr{B}_q^p(\mathrm{M})) \subset \mathscr{B}_{q+1}^p(\mathrm{M})$, which agrees with the exterior differentiation d in each $\mathscr{B}_q^0(\mathrm{M}) = \mathscr{E}_q(\mathrm{M})$, and with the covariant exterior differentiation $\boldsymbol{d}$ in each $\mathscr{B}_q^1(\mathrm{M})$, and is such that

$$\mathbf{d}(\mathbf{vw}) = (\mathbf{dv})\mathbf{w} + (-1)^q \mathbf{v}(\mathbf{dw})$$

for all $\mathbf{v} \in \beta_q^p(\mathrm{M})$ and $\mathbf{w} \in \mathscr{B}_s^r(\mathrm{M})$. (The product in $\mathscr{B}(\mathrm{M})$ is denoted by juxtaposition.)

Extend the definition of $\mathbf{d}$ to matrices with entries in $\mathscr{B}(\mathrm{M})$: if $\mathbf{U} = (\mathbf{u}_{ij})$ is such a matrix, with a rows and b columns, put $\mathbf{dU} = (\mathbf{du}_{ij})$. Then we have, if $\mathbf{V}$ is a matrix with a rows and b columns and $\mathbf{W}$ is a matrix of b rows and c columns,

$$\mathbf{d}(\mathbf{VW}) = \mathbf{V}(\mathbf{dW}) + (-1)^q (\mathbf{dV})\mathbf{W}$$

if all the elements of $\mathbf{V}$ belong to $\mathscr{B}_q^p(\mathrm{M})$.

(b) Let $\mathbf{R}$ be a moving frame. With the notation of (20.6.5), identify $\mathbf{R}$ with the $n \times 1$ matrix formed by the $\mathbf{e}_i \in \mathscr{B}_0^1(\mathrm{M})$; $\boldsymbol{\sigma} = \boldsymbol{\sigma}^{(\mathbf{R})}$ with the $1 \times n$ matrix formed by the $\sigma_i \in \mathscr{B}_1^0(\mathrm{M})$; $\boldsymbol{\omega} = \boldsymbol{\omega}^{(\mathbf{R})}$ with the transpose of the $n \times n$ matrix formed by the $\omega_{ij} \in \mathscr{B}_1^0(\mathrm{M})$; $\boldsymbol{\Theta} = \boldsymbol{\Theta}^{(\mathbf{R})}$ with the $1 \times n$ matrix formed by the $\Theta_i \in \mathscr{B}_2^0(\mathrm{M})$; and $\boldsymbol{\Omega} = \boldsymbol{\Omega}^{(\mathbf{R})}$ with the transpose of the $n \times n$ matrix formed by the $\Omega_{ij} \in \mathscr{B}_2^0(\mathrm{M})$. Then the formulas (20.6.5.9) and (20.6.5.10) take the form

$$1_{\mathrm{T(M)}} = \boldsymbol{\sigma} \cdot \mathbf{R} \qquad (\in \mathscr{B}_1^1(\mathrm{M})),$$
$$\mathbf{dR} = \boldsymbol{\omega} \cdot \mathbf{R}.$$

The structure equations (20.6.6.3) and (20.6.6.4) take the form

$$\boldsymbol{\Theta} = d\boldsymbol{\sigma} - \boldsymbol{\sigma} \cdot \boldsymbol{\omega},$$
$$\boldsymbol{\Omega} = d\boldsymbol{\omega} - \boldsymbol{\omega}^2,$$

and by exterior differentiation we obtain Bianchi's identities (20.6.4) and (20.3.4) in the form

$$d\boldsymbol{\Theta} = -\boldsymbol{\Theta} \cdot \boldsymbol{\omega} + \boldsymbol{\sigma} \cdot \boldsymbol{\Omega},$$
$$d\boldsymbol{\Omega} = \boldsymbol{\omega} \cdot \boldsymbol{\Omega} - \boldsymbol{\Omega} \cdot \boldsymbol{\omega}.$$

(c) Prove the identities (for $r \geq 1$)

$$d(\boldsymbol{\Omega}^r) = \boldsymbol{\omega} \cdot \boldsymbol{\Omega}^r - \boldsymbol{\Omega}^r \cdot \boldsymbol{\omega},$$
$$d(\boldsymbol{\sigma} \cdot \boldsymbol{\Omega}^r) = \boldsymbol{\Theta} \cdot \boldsymbol{\Omega}^r + \boldsymbol{\sigma} \cdot \boldsymbol{\Omega}^r \cdot \boldsymbol{\omega},$$
$$d(\boldsymbol{\Theta} \cdot \boldsymbol{\Omega}^r) = \boldsymbol{\sigma} \cdot \boldsymbol{\Omega}^{r+1} - \boldsymbol{\Theta} \cdot \boldsymbol{\Omega}^r \cdot \boldsymbol{\omega}.$$

(d) Prove the identities (for $r \geqq 1$)

$$\mathbf{d}^{2r}(1_{T(M)}) = \boldsymbol{\sigma} \cdot \boldsymbol{\Omega}^r \cdot \mathbf{R}, \qquad \mathbf{d}^{2r+1}(1_{T(M)}) = \boldsymbol{\Theta} \cdot \boldsymbol{\Omega}^r \cdot \mathbf{R},$$
$$\mathbf{d}^{2r-1}\mathbf{R} = \boldsymbol{\omega} \cdot \boldsymbol{\Omega}^{r-1} \cdot \mathbf{R}, \qquad \mathbf{d}^{2r}\mathbf{R} = \boldsymbol{\Omega}^r \cdot \mathbf{R}.$$

(e) For each vector field $\mathbf{u} \in \mathscr{B}_0^1(M)$, prove that

$$\mathbf{d}((\mathbf{d}^{2r-1}\mathbf{u})^s) = s(\mathbf{d}^{2r}\mathbf{u})(\mathbf{d}^{2r-1}\mathbf{u})^{s-1},$$
$$\mathbf{d}((\mathbf{d}^{2r}\mathbf{u})^{2s-1}) = (\mathbf{d}^{2r+1}\mathbf{u})(\mathbf{d}^{2r}\mathbf{u})^{2s-2},$$
$$\mathbf{d}((\mathbf{d}^{2r}\mathbf{u})^{2s}) = 0.$$

for $r \geqq 1$ and $s \geqq 1$.

3. Deduce from Bianchi's identities (**20.6.4**) and (**20.3.4**) the following identities for the covariant derivation ∇, the torsion $\boldsymbol{t}$ and the curvature $\boldsymbol{r}$ of a linear connection $\mathbf{C}$:

$$\begin{aligned}(\boldsymbol{r} \cdot (X \wedge Y)) \cdot Z + (\boldsymbol{r} \cdot (Y \wedge Z)) \cdot X + (\boldsymbol{r} \cdot (Z \wedge X)) \cdot Y \\ = \boldsymbol{t} \cdot (\boldsymbol{t} \cdot (X \wedge Y) \wedge Z) + \boldsymbol{t} \cdot (\boldsymbol{t} \cdot (Y \wedge Z) \wedge X) + \boldsymbol{t} \cdot (\boldsymbol{t} \cdot (Z \wedge X) \wedge Y) \\ + (\nabla_X \cdot \boldsymbol{t}) \cdot (Y \wedge Z) + (\nabla_Y \cdot \boldsymbol{t}) \cdot (Z \wedge X) + (\nabla_Z \cdot \boldsymbol{t}) \cdot (X \wedge Y),\end{aligned}$$

$$\begin{aligned}(\nabla_X \cdot \boldsymbol{r}) \cdot (Y \wedge Z) + (\nabla_Y \cdot \boldsymbol{r}) \cdot (Z \wedge X) + (\nabla_Z \cdot \boldsymbol{r}) \cdot (X \wedge Y) \\ + \boldsymbol{r} \cdot (\boldsymbol{t} \cdot (X \wedge Y) \wedge Z) + \boldsymbol{r} \cdot (\boldsymbol{t} \cdot (Y \wedge Z) \wedge X) + \boldsymbol{r} \cdot (\boldsymbol{t} \cdot (Z \wedge X) \wedge Y) = 0\end{aligned}$$

for all C^∞ vector fields X, Y, Z on M. (Use the formula (**20.5.3.4**).)

4. Let f be a diffeomorphism of a pure differential manifold M onto a differential manifold M′. By transport of structure, f defines an isomorphism R(f) of the bundle of frames R(M) onto the bundle of frames R(M′): we have $R(f)(r_x) = T_x(f) \circ r_x$ for all $r_x \in R(M)_x$.

(a) Show that if $\boldsymbol{\sigma}$ and $\boldsymbol{\sigma}'$ are the canonical forms on R(M) and R(M′), respectively, then ${}^t R(f)(\boldsymbol{\sigma}') = \boldsymbol{\sigma}$. Conversely, every isomorphism F of the principal bundle R(M) onto the principal bundle R(M′) such that ${}^t F(\boldsymbol{\sigma}') = \boldsymbol{\sigma}$ is of the form R(f).
(b) Let $\mathbf{P}$ (resp. $\mathbf{P}'$) be a principal connection in R(M) (resp. R(M′)) and let $\boldsymbol{\omega}$ (resp. $\boldsymbol{\omega}'$) be the 1-form of the connection $\mathbf{P}$ (resp. $\mathbf{P}'$). Then $\mathbf{P}'$ is the image of $\mathbf{P}$ under R(f) if and only if ${}^t R(f)(\boldsymbol{\omega}') = \boldsymbol{\omega}$. In this case f is said to be an *isomorphism* of $\mathbf{P}$ onto $\mathbf{P}'$ (or of $\mathbf{C}$ onto $\mathbf{C}'$, where $\mathbf{C}$, $\mathbf{C}'$ denote the linear connections defined by $\mathbf{P}$, $\mathbf{P}'$).

5. Let M be a pure differential manifold, X a C^∞ vector field on M; let U be a relatively compact open subset of M, and $a > 0$ a number such that $U \times]-a, a[$ is contained in $\operatorname{dom}(F_X)$ (**18.2.8**). For $t \in]-a, a[$ and $x \in U$, put $g_t(x) = F_X(x, t)$, so that g_t is a diffeomorphism of U onto an open subset U_t of M. Then g_t defines an isomorphism $G_t = R(g_t)$ of the bundle of frames $\pi^{-1}(U)$ onto the bundle of frames $\pi^{-1}(U_t)$. The paths $t \mapsto G_t(x)$ are the integral curves of a vector field $\tilde{X}$ on R(M), called the *canonical lifting* of X. Show that $\tilde{X}$ is the unique vector field on R(M) which is invariant under $\mathbf{GL}(n, \mathbf{R})$, is such that $T(\pi) \cdot \tilde{X}(r_x) = X(x)$ for all $x \in M$ and all $r_x \in R(M)_x$, and for which $\theta_{\tilde{X}} \cdot \boldsymbol{\sigma} = 0$. The mapping $X \mapsto \tilde{X}$ is a bijective homomorphism of the Lie algebra $\mathscr{T}_0^1(M)$ onto a Lie subalgebra of $\mathscr{T}_0^1(R(M))$.

For each point $\mathbf{u}_x = r_x \cdot \mathbf{y}$ of $T_x(M)$, where $\mathbf{y} \in \mathbf{R}^n$ and $r_x \in R(M)_x$, the vector $\tilde{X}(r_x) \cdot \mathbf{y}$ depends only on $\mathbf{u}_x$. The vector field so defined *on* T(M) is precisely the *canonical lifting* of X (Section **18.6**, Problem 3).

6. (a) Using the notation of Problem 5, show that for a C^∞ vector field X on M and a principal connection $\mathbf{P}$ in R(M), the following conditions are equivalent:

(i) for each relatively compact open subset U of M, and each $a > 0$ such that $U \times]-a, a[\subset \text{dom}(F_X)$, and each $t \in]-a, a[$, the isomorphism G_t transforms the connection $\mathbf{P}|\pi^{-1}(U)$ into the connection $\mathbf{P}|\pi^{-1}(U_t)$;

(ii) the canonical lifting $\tilde{X}$ of X is such that $\theta_{\tilde{X}} \cdot \boldsymbol{\omega} = 0$, where $\boldsymbol{\omega}$ is the 1-form of the connection $\mathbf{P}$;

(iii) $[\tilde{X}, H_{\mathbf{a}}] = 0$ for each $\mathbf{a} \in \mathbf{R}^n$ (Problem 1);

(iv) $\theta_X \circ \nabla_Y - \nabla_Y \circ \theta_X = \nabla_{[X, Y]}$ for each vector field Y on M.

(Use (18.2.14).)

The field X is then said to be an *infinitesimal automorphism* of the connection $\mathbf{P}$ (or of M endowed with $\mathbf{P}$).

(b) For each C^∞ vector field X on M, the mapping $Y \mapsto \theta_X \cdot Y - \nabla_X \cdot Y$ of $\mathscr{T}_0^1(M)$ into itself is a differential operator of *order* 0, hence is of the form $Y \mapsto \mathbf{A}_X \cdot Y$, where $\mathbf{A}_X \in \mathscr{T}_1^1(M)$. Show that for X to be an infinitesimal automorphism of $\mathbf{P}$, it is necessary and sufficient that

$$\nabla_Y \cdot \mathbf{A}_X = \boldsymbol{r} \cdot (X \wedge Y) \tag{1}$$

for all C^∞ vector fields Y on M (cf. (17.20.4.1)). If the connection is torsion-free, then $\mathbf{A}_X \cdot Y = -\nabla_Y \cdot X$.

The set of infinitesimal automorphisms of $\mathbf{P}$ is a *Lie subalgebra* $\mathfrak{a}(\mathbf{P})$ of $\mathscr{T}_0^1(M)$. For two infinitesimal automorphisms X, Y of $\mathbf{P}$ we have

$$\mathbf{A}_{[X, Y]} = [\mathbf{A}_X, \mathbf{A}_Y] + \boldsymbol{r} \cdot (X \wedge Y).$$

(c) In the notation of (20.6.5), put

$$\nabla_{\mathbf{e}_i} \cdot \mathbf{e}_j = \sum_k \Gamma_{ij}^k \mathbf{e}_k$$

so that we have

$$\omega_{ji} = \sum_k \Gamma_{ki}^j \sigma_k$$

and

$$\nabla_{\mathbf{e}_i} \cdot \sigma_j = -\sum_k \Gamma_{ik}^j \sigma_k.$$

Deduce that the tensor $\mathbf{A}$ of type (1, 2) defined in (b) has components

$$A_{ij}^k = -(\Gamma_{ji}^k + t_{ij}^k)$$

in the notation of (20.6.6).

(d) Show that when the connection is torsion-free, the equation (1) is equivalent to the system of scalar equations

$$\theta_{\mathbf{e}_i} \cdot \langle \nabla_{\mathbf{e}_j} \cdot X, \sigma_h \rangle + \sum_k \Gamma_{ik}^h \langle \nabla_{\mathbf{e}_j} \cdot X, \sigma_k \rangle - \sum_k \Gamma_{ij}^h \langle \nabla_{\mathbf{e}_k} \cdot X, \sigma_h \rangle$$
$$= -\langle (\boldsymbol{r} \cdot (X \wedge \mathbf{e}_i)) \cdot \mathbf{e}_j, \sigma_h \rangle$$

in the notation of (20.6.5) and (a) above, where the indices i, j, h run independently from 1 to n.

7. (a) With the notation and hypotheses of Problem 5, let X be an infinitesimal automorphism of $\mathbf{P}$. Show that if M is connected and if $\tilde{X}$ vanishes at a point of R(M), then $X = 0$. (Using the invariance of $\tilde{X}$ and the fact that $[\tilde{X}, H_{\mathbf{a}}] = 0$, show that X is zero along each geodesic trajectory passing through a point x such that $\tilde{X}(r_x) = 0$ for some $r_x \in R(M)_x$, and use (18.4.6).)

Deduce that $\dim(\mathfrak{a}(\mathbf{P})) \leq n(n+1) = \dim \mathrm{R}(\mathrm{M})$.
(b) Show that if $\dim(\mathfrak{a}(\mathbf{P})) = n(n+1)$, then the curvature and torsion of $\mathbf{P}$ are zero. (Observe that for each point $r_x \in \mathrm{R}(\mathrm{M})$ and each $\mathbf{u} \in \mathfrak{gl}(n, \mathbf{R})$, there exists a unique element $X \in \mathfrak{a}(\mathbf{P})$ such that $\tilde{X}(r_x) = Z_{\mathbf{u}}(r_x)$. Using the relations $\theta_{\tilde{X}} \cdot \boldsymbol{\sigma} = 0$ and $[\tilde{X}, H_{\mathbf{a}}] = 0$, show that we have $\theta_{\tilde{X}} \cdot (\boldsymbol{\Theta} \cdot (H_{\mathbf{a}} \wedge H_{\mathbf{b}})) = 0$ in M, and in particular $\theta_{\tilde{X}(r_x)} \cdot (\boldsymbol{\Theta} \cdot (H_{\mathbf{a}} \wedge H_{\mathbf{b}})) = 0$, and therefore

$$\theta_{Z_{\mathbf{u}}(r_x)} \cdot (\boldsymbol{\Theta} \cdot (H_{\mathbf{a}} \wedge H_{\mathbf{b}})) = 0.$$

Next, show with the help of the structure equation that

$$\begin{aligned} &\theta_{Z_{\mathbf{u}}} \cdot (\boldsymbol{\Theta} \cdot (H_{\mathbf{a}} \wedge H_{\mathbf{b}})) \\ &\qquad = -\mathbf{u} \cdot (\boldsymbol{\Theta} \cdot (H_{\mathbf{a}} \wedge H_{\mathbf{b}})) + \boldsymbol{\Theta} \cdot ([Z_{\mathbf{u}}, H_{\mathbf{a}}] \wedge H_{\mathbf{b}}) + \boldsymbol{\Theta} \cdot (H_{\mathbf{a}} \wedge [Z_{\mathbf{u}}, H_{\mathbf{b}}]) \end{aligned}$$

and deduce that $\boldsymbol{\Theta}(r_x) \cdot (H_{\mathbf{a}}(r_x) \wedge H_{\mathbf{b}}(r_x)) = 0$, by taking $\mathbf{u}$ to be the identity element of $\mathrm{End}(\mathbf{R}^n)$. Repeat the argument, replacing $\boldsymbol{\sigma}$ by $\boldsymbol{\omega}$.)

8. A linear connection $\mathbf{C}$ on a connected differential manifold M (or the corresponding principal connection $\mathbf{P}$) is said to be *complete* if all geodesics for $\mathbf{C}$ are defined on the whole of $\mathbf{R}$; or, equivalently (Problem 1), if the integral curves of the fields $H_{\mathbf{a}}$ on R(M) are all defined on the whole of $\mathbf{R}$ (Section **20.2**, Problem 3). For each $\mathbf{a} \in \mathbf{R}^n$ and each $t \in \mathbf{R}$, let $h_{\mathbf{a}}(t)$ denote the diffeomorphism $r \mapsto \mathrm{F}_{H_{\mathbf{a}}}(r, t)$ of R(M) onto itself (**18.2.2**). Show that under these conditions, for every infinitesimal automorphism X of M, the integral curves of the field $\tilde{X}$ (and hence also those of X) are defined on the whole of $\mathbf{R}$. (Use Problem 6 to show that $\mathrm{F}_{\tilde{X}}(h_{\mathbf{a}}(t)(r), t') = h_{\mathbf{a}}(t)(\mathrm{F}_X(r, t'))$ whenever both sides are defined. Using (**18.4.6**) and the connectedness of M, show that for each $r_0 \in \mathrm{R}(\mathrm{M})$, each point $r \in \mathrm{R}(\mathrm{M})$ can be written in the form

$$r = (h_{\mathbf{a}_1}(t_1) \circ h_{\mathbf{a}_2}(t_2) \circ \cdots \circ h_{\mathbf{a}_k}(t_k))(r_0) \circ s$$

for suitable choices of $\mathbf{a}_j \in \mathbf{R}^n$, $t_j \in \mathbf{R}$, and $s \in \mathbf{GL}(n, \mathbf{R})$. Then show that if $\mathrm{F}_{\tilde{X}}(r_0, t)$ is defined for $|t| < \delta$, then the same is true of $\mathrm{F}_{\tilde{X}}(r, t)$, and that

$$\mathrm{F}_{\tilde{X}}(r, t) = (h_{\mathbf{a}_1}(t_1) \circ \cdots \circ h_{\mathbf{a}_k}(t_k))(\mathrm{F}_X(r_0, t) \circ s).$$

9. Let M, M′ be two pure differential manifolds and let $\mathbf{P}$ (resp. $\mathbf{P}'$) be a principal connection in R(M) (resp. R(M′)). A mapping $f: \mathrm{M} \to \mathrm{M}'$ is said to be a *local isomorphism* of $\mathbf{P}$ into $\mathbf{P}'$ (or of M into M′) if f is a local diffeomorphism and if, for each $x \in \mathrm{M}$, there exists an open neighborhood U of x in M such that f restricted to U is an isomorphism of the restriction of $\mathbf{P}$ to U onto the restriction of $\mathbf{P}'$ to $f(\mathrm{U})$. For this to be so it is necessary and sufficient that f should be a local diffeomorphism and that $\mathrm{T}(\mathrm{R}(f))$ should map each horizontal tangent vector to R(M) to a horizontal tangent vector to R(M′). We have then, for each $x \in \mathrm{M}$ and all sufficiently small vectors $\mathbf{u}_x \in \mathrm{T}_x(\mathrm{M})$, $f(\exp(\mathbf{u}_x)) = \exp(\mathrm{T}(f) \cdot \mathbf{u}_x)$.

(a) Suppose that M is connected, and let f, g be two local isomorphisms of M into M′. Show that if $f(x) = g(x)$ and $\mathrm{T}_x(f) = \mathrm{T}_x(g)$ for some point $x \in \mathrm{M}$, then $f = g$. (Consider the set of points at which f and g coincide, and use (**18.4.6**).)
(b) Suppose that M and M′ are real-analytic manifolds, and that the connections $\mathbf{P}$, $\mathbf{P}'$ are analytic. Let $f: \mathrm{M} \to \mathrm{M}'$ be an analytic mapping; suppose that M is connected and that f is a local diffeomorphism. If there exists a nonempty open subset U of M such that $f|\mathrm{U}$ is an isomorphism of $\mathbf{P}|\mathrm{U}$ onto $\mathbf{P}'|f(\mathrm{U})$, then f is a local isomorphism. (Use Section **18.8**, Problem 3.)

(c) Suppose that M and M′ are analytic, that the connections **P** and **P′** are analytic and that **P′** is complete. Let $x \in M$, $x' \in M'$, and let U be a neighborhood of x which is the bijective image of an open subset of $T_x(M)$ under the exponential mapping. If f is an analytic local isomorphism of a neighborhood $V \subset U$ of x in M, show that f has a unique extension to an analytic local isomorphism of U into M′, by considering $T_x(f)$ and using (b) above.
(d) Suppose that M and M′ are analytic, **P** and **P′** are analytic, **P′** complete, and M connected and simply-connected. Then every local isomorphism f of a nonempty connected open subset U of M into M′ has a unique extension to a local isomorphism of M into M′. (By using (c) above, extend f along a path with origin at a point $x_0 \in U$, and show that its value at the endpoint of the path depends only on this point, by arguing as in (9.6.3).)
(e) Suppose that M and M′ are analytic, connected and simply-connected, and that **P** and **P′** are analytic and complete. If f is an isomorphism of a nonempty connected open subset U of M onto an open subset of M′, then f has a unique extension to an isomorphism of M onto M′. (Consider the inverse g of the isomorphism f and apply (d) to f and g.) (Cf. Section 20.18, Problem 13.)

10. Show that the group A(**P**) of automorphisms of a principal connection **P** in R(M), where M is connected, is a Lie group acting differentiably on M. (Use Problem 4 above, and Section 19.10, Problem 6, applied to the parallelizable manifold R(M).) The Lie algebra of A(**P**) may be identified with a subalgebra of the Lie algebra $\mathfrak{a}(\mathbf{P})$ of infinitesimal automorphisms of **P** (Problem 6), and is equal to $\mathfrak{a}(\mathbf{P})$ if the connection **P** is complete (Problem 8).

11. Deduce from (19.4.4.3) that for the canonical form $\boldsymbol{\sigma}$ on R(M) and for any $\mathbf{u} \in \mathfrak{gl}(n, \mathbf{R})$, we have

$$i_{Z_\mathbf{u}} \cdot d\boldsymbol{\sigma} = -\mathbf{u} \circ \boldsymbol{\sigma},$$

u being considered as an element of $\mathrm{End}(\mathbf{R}^n)$.

12. Show that for each p-form α of class C^1 on M, we have (with the notation of (20.6.5))

$$d\alpha = \sum_i (\boldsymbol{\sigma}_i \wedge \nabla_{\mathbf{e}_i} \cdot \alpha) \tag{1}$$

if and only if the connection on M is torsion-free. (Prove that the right-hand side of (1) then satisfies the conditions of (17.15.2).) Deduce that when this is the case, the derivative $d\alpha$, considered as an antisymmetric covariant tensor of order $p+1$, may be identified with the antisymmetrization $(p!)^{-1}\boldsymbol{a}(\nabla\alpha)$.

13. If two moving frames $\mathbf{R}_1$, $\mathbf{R}_2$ on a manifold M endowed with a linear connection **C** are such that ${}^t\mathbf{R}_1(\boldsymbol{\sigma}) = {}^t\mathbf{R}_2(\boldsymbol{\sigma})$, show that $\mathbf{R}_1 = \mathbf{R}_2$.

14. Let M be a differential manifold, **C** a linear connection on M, and **P** the corresponding principal connection. Let $c = (U, \varphi, n)$ be a chart of M and let X_i $(1 \leq i \leq n)$ be the vector fields associated with this chart, forming a frame $\mathbf{R} = (X_i)$ over U. Writing σ_i and ω_{ij} in place of $\sigma_i^{(\mathbf{R})}$ and $\omega_{ij}^{(\mathbf{R})}$, the local expressions of the σ_i and the ω_{ij} in terms of the local expression (17.16.4.1) of the connection **C** are

$$\sigma_i = dx^i, \qquad \omega_{ij} = \sum_{k=1}^{n} \Gamma_{kj}^i \, dx^k.$$

Likewise, writing Θ_i and Ω_{ij} in place of $\Theta_i^{(\mathbf{R})}$ and $\Omega_{ij}^{(\mathbf{R})}$, the local expressions of the Θ_i and the Ω_{ij} in terms of the components t^i_{jk} and r^i_{ijk} of the torsion and curvature tensors (17.20) are

$$\Theta_i = \sum_{j<k} t^i_{jk}\, dx^j \wedge dx^k,$$

$$\Omega_{ij} = \sum_{k<l} r^i_{jkl}\, dx^k \wedge dx^l.$$

15. Let M be a differential manifold, **C** a linear connection on M and **P** the corresponding principal connection. Let U be a neighborhood of a point $x_0 \in$ M which is of the form exp(V), where V is a starlike neighborhood of $\mathbf{0}_{x_0}$ in $\mathrm{T}_{x_0}(\mathrm{M})$ in which the exponential mapping is a diffeomorphism onto U. If $(\mathbf{c}_j)_{1 \leq j \leq n}$ is a basis of $\mathrm{T}_{x_0}(\mathrm{M})$, the coordinates with respect to this basis of the point $\exp_{x_0}^{-1}(x)$ for $x \in$ U are called the *normal coordinates* of x with respect to $(\mathbf{c}_j)$.

(a) If v is the geodesic for **C** such that $v(0) = x_0$ and $v'(0) = \mathbf{h} = \sum_{j=1}^{n} h_j \mathbf{c}_j$, then the local expression of v in normal coordinates is

$$t \mapsto (h_i t)_{1 \leq i \leq n}$$

in a neighborhood of 0 in **R**. Deduce that the local expression of the connection **C**, for the normal coordinates in U, satisfies

$$\Gamma^i_{jk}(x_0) + \Gamma^i_{kj}(x_0) = 0$$

for all i, j, k. In particular, if **C** is torsion-free, then $\Gamma^i_{jk}(x_0) = 0$ for all i, j, k.
(b) Let r_0 be a frame at the point x_0, and identify r_0 with a basis $(\mathbf{c}_j)_{1 \leq j \leq n}$ of $\mathrm{T}_{x_0}(\mathrm{M})$. For each vector $\mathbf{h} \in$ V, if $t \in \mathbf{R}$ is such that $t\mathbf{h} \in$ V, then, putting $x = \exp(t\mathbf{h})$, the frame $\mathbf{R}(x)$ is defined to be the value at t of the integral curve of the field $H_{\mathbf{b}}$ (Problem 1) with origin r_0, where $r_0 \cdot \mathbf{b} = \mathbf{h}$ (in other words, the frame obtained by parallel transport of r_0 along the geodesic $\xi \mapsto \exp(\xi\mathbf{h})$ $(0 \leq \xi \leq t)$). The moving frame **R** is said to be *canonically associated* with V and r_0. Relative to the chart $c = (\mathrm{U}, \varphi, n)$ of M, where $\varphi = r_0 \circ \exp_{x_0}^{-1}$, the canonical form and the connection form corresponding to **R** have local expressions of the form

$$\sigma_i = \sum_j \mathrm{A}_{ij}\, dx^j, \qquad \omega_{ij} = \sum_k \mathrm{B}_{ijk}\, dx^k.$$

Show that, for each point $\mathbf{u} \in r_0^{-1}(\mathrm{V})$, we have

$$\sum_j \mathrm{A}_{ij}(t\mathbf{u})u^j = u^i \qquad \text{for} \quad 0 \leq t \leq 1 \quad \text{and} \quad 1 \leq i \leq n,$$

$$\sum_k \mathrm{B}_{ijk}(t\mathbf{u})u^k = 0 \qquad \text{for} \quad 0 \leq t \leq 1 \quad \text{and} \quad 1 \leq i, j \leq n.$$

(Use Problem 1.)
Relative to the same chart, let

$$\Theta_i = \sum_{j<k} \mathrm{T}^i_{jk}\, \sigma_j \wedge \sigma_k,$$

$$\Omega_{ij} = \sum_{k<l} \mathrm{R}^i_{jkl}\, \sigma_k \wedge \sigma_l.$$

Show that

$$\frac{d}{dt}(t\mathrm{A}_{ij}(t\mathbf{u})) = \delta_{ij} + t\sum_{l} \mathrm{B}_{ilj}(t\mathbf{u})u^l + t\sum_{l,m} \mathrm{T}^i_{lm}(t\mathbf{u})\mathrm{A}_{mj}(t\mathbf{u})u^l,$$

$$\frac{d}{dt}(t\mathrm{B}_{ijk}(t\mathbf{u})) = t\sum_{l,m} \mathrm{R}^i_{jkl}(t\mathbf{u})\mathrm{A}_{mk}(t\mathbf{u})u^l.$$

(Consider the inverse images of the forms σ_i, ω_{ij}, Θ_i, Ω_{ij} under the mapping $(t, u^1, \ldots, u^n) \mapsto (tu^1, \ldots, tu^n)$ of an open neighborhood of 0 in $\mathbf{R}^{n+1}$ into $\varphi(\mathrm{U})$, and write down the inverse images of the structure equations.)

With the same notation, show that the derivatives at the point $t = 0$ of the functions $t \mapsto \mathrm{T}^i_{jk}(t\mathbf{u})$ and $t \mapsto \mathrm{R}^i_{jkl}(t\mathbf{u})$ are completely determined by the values at the point x_0 of the covariant derivatives of all orders $\nabla^m \mathbf{t}$ and $\nabla^m \mathbf{r}$ of the torsion and curvature tensors. (Use (20.5.3.4).)

16. Let M, M′ be two real-analytic manifolds endowed with linear connections defined by analytic principal connections **P**, **P′**. Let ∇, ∇' denote the covariant derivatives corresponding to **P**, **P′**, respectively, and let **t**, **t′** be the torsion tensors and **r**, **r′** the curvature tensors of M, M′, respectively. Let $\mathrm{F} : \mathrm{T}_{x_0}(\mathrm{M}) \to \mathrm{T}_{y_0}(\mathrm{M}')$ be a bijective linear mapping such that the image of each of the tensors $\nabla^m \mathbf{t}(x_0)$, $\nabla^m \mathbf{r}(x_0)$ under the corresponding extensions of F is $\nabla'^m \mathbf{t}'(y_0)$, $\nabla'^m \mathbf{r}'(y_0)$, respectively. Then there exists a neighborhood U of x_0, a neighborhood V of y_0 and an analytic isomorphism f of **P**$|$U onto **P′**$|$V such that $f(x_0) = y_0$ and $\mathrm{T}_{x_0}(f) = \mathrm{F}$. (Use Problems 15 and 13.)

17. Let M, M′ be two differential manifolds endowed with linear connections corresponding to principal connections **P**, **P′**. Let ∇, ∇' be the corresponding covariant derivatives, **t**, **t′** the torsion tensors, and **r**, **r′** the curvature tensors. Suppose that $\nabla \mathbf{t} = 0$, $\nabla' \mathbf{t}' = 0$, $\nabla \mathbf{r} = 0$, $\nabla' \mathbf{r}' = 0$. Let $\mathrm{F} : \mathrm{T}_{x_0}(\mathrm{M}) \to \mathrm{T}_{y_0}(\mathrm{M}')$ be a bijective linear mapping such that the image of $\mathbf{t}(x_0)$ (resp. $\mathbf{r}(x_0)$) by the corresponding extension of F is $\mathbf{t}'(y_0)$ (resp. $\mathbf{r}'(y_0)$). Show that there exists a neighborhood U of x_0, a neighborhood V of y_0 and an isomorphism f of **P**$|$U onto **P′**$|$V such that $f(x_0) = y_0$ and $\mathrm{T}_{x_0}(f) = \mathrm{F}$. (Use Problems 13 and 15.)

18. Let M be a connected differential manifold. A linear connection **C** on M is said to be *invariant under parallelism* if, for any two points x, y of M and any C^∞ path γ from x to y in M (16.26.10), there exists a neighborhood U of x and an isomorphism f of U (for the connection **C**) onto a neighborhood of y, such that $f(x) = y$ and such that $\mathrm{T}_x(f)$ coincides with the parallel transport along γ (18.6.4). Show that for this to be the case it is necessary and sufficient that $\nabla \mathbf{t} = 0$ and $\nabla \mathbf{r} = 0$, where **t**, **r** are the torsion and curvature tensors of **C**. (Use Problem 17.)

19. Let M be a pure differential manifold of dimension n, and let R(M) be the bundle of tangent frames of M, with group $\mathbf{GL}(n, \mathbf{R})$. Let $\mathrm{A}(\mathrm{M}) = \mathrm{Aff}(\mathrm{M} \times \mathbf{R}^n, \mathrm{T}(\mathrm{M}))$ be the bundle of affine frames of T(M), which is a principal bundle with group $\mathbf{A}(n, \mathbf{R})$. Recall (Section 20.1, Problem 1) that $\mathbf{A}(n, \mathbf{R})$ is canonically isomorphic to the semi-direct product $\mathbf{R}^n \times_\rho \mathbf{GL}(n, \mathbf{R})$, where ρ is the natural action of $\mathbf{GL}(n, \mathbf{R})$ on $\mathbf{R}^n$.

(a) Show that there exists a canonical isomorphism of $\mathrm{R(M)} \times^{\mathbf{GL}(n,\mathbf{R})} \mathbf{A}(n, \mathbf{R})$ onto A(M), under which the product

$$r \cdot \begin{pmatrix} U & \mathbf{a} \\ 0 & 1 \end{pmatrix}, \qquad \text{where} \quad r \in \mathrm{R(M)}_x,$$

corresponds to the affine-linear bijection $\mathbf{y} \mapsto r(U \cdot \mathbf{y} + \mathbf{a})$ of $\mathbf{R}^n$ onto $\mathrm{T}_x(\mathrm{M})$.
(b) If we identify A(M) with the bundle $\mathrm{R(M)} \times^{\mathbf{GL}(n,\mathbf{R})} \mathbf{A}(n, \mathbf{R})$ obtained by extension of the structure group from $\mathbf{GL}(n, \mathbf{R})$ to $\mathbf{A}(n, \mathbf{R})$, the associated bundle

$$\mathrm{X} = \mathrm{R(M)} \times^{\mathbf{GL}(n,\mathbf{R})} (\mathbf{A}(n, \mathbf{R})/\mathbf{GL}(n, \mathbf{R})) = \mathrm{R(M)} \times^{\mathbf{GL}(n,\mathbf{R})} \mathbf{R}^n$$

is canonically identified with T(M): the product $r \cdot \mathbf{y}$ (where $r \in \mathrm{R(M)}$, $\mathbf{y} \in \mathbf{R}^n$) corresponds to the product given by the same notation in T(M). The canonical section of X (Section 20.1, Problem 2) is identified with the zero section of T(M). There exists a *canonical welding* $\mathbf{h}_x \mapsto r_x \cdot (r_x^{-1} \cdot \mathbf{h}_x)$ of M into X ($r_x^{-1} \cdot \mathbf{h}_x \in \mathbf{R}^n$, for $\mathbf{h}_x \in \mathrm{T}_x(\mathrm{M})$ and $r_x \in \mathrm{R(M)}_x$), which is identified with the identity mapping of T(M).
(c) For each principal connection $\boldsymbol{P}_0$ in R(M), show by using Section 20.5, Problem 4, that there exists a Cartan connection $\boldsymbol{P}$ on A(M) from which $\boldsymbol{P}_0$ is obtained by the procedure described in that Problem. The connection $\boldsymbol{P}$, for which the form $\boldsymbol{\varphi}$ is the canonical form $\boldsymbol{\sigma}$ of (20.6.2), corresponds to the canonical welding defined in (b) above. This connection on A(M) is called the *affine connection* on M associated with the principal connection $\boldsymbol{P}_0$ (or with the linear connection $\boldsymbol{C}_0$ corresponding to $\boldsymbol{P}_0$). The other principal connections $\boldsymbol{P}$ on A(M) which correspond to $\boldsymbol{P}_0$ are in one-to-one correspondence with the C^∞ mappings $r \mapsto u(r)$ of R(M) into $\mathrm{End}(\mathbf{R}^n) = \mathbf{M}_n(\mathbf{R})$ such that $u(r \cdot s) = s \circ u(r) \circ s^{-1}$ for all $s \in \mathbf{GL}(n, \mathbf{R})$.

Hence derive another proof of the structure equation (20.6.3.1).

20. (a) The manifold $\mathrm{R}_1(\mathbf{R}^n) = \mathrm{R}(\mathbf{R}^n)$ of frames of order 1 in $\mathbf{R}^n$ (Section 20.1, Problem 3) may be identified with the set of affine-linear mappings $\mathbf{x} \mapsto U \cdot \mathbf{x} + \mathbf{a}$ of $\mathbf{R}^n$ into $\mathbf{R}^n$, where $\mathbf{a} \in \mathbf{R}^n$ and $U \in \mathbf{GL}(n, \mathbf{R})$ (the jet of order 1 at the point 0 of a C^∞ mapping f of a neighborhood of 0 into $\mathbf{R}^n$ is identified with the affine-linear mapping

$$\mathbf{h} \mapsto \mathrm{D}f(0) \cdot \mathbf{h} + f(0)).$$

Consequently $\mathrm{R}_1(\mathbf{R}^n)$ may be canonically identified with the affine group $\mathbf{A}(n, \mathbf{R})$ (Section 20.1, Problem 1). Show that this identification is an *isomorphism* of principal bundles with structure group $\mathbf{GL}(n, \mathbf{R})$ (where $\mathbf{GL}(n, \mathbf{R})$, as a subgroup of $\mathbf{A}(n, \mathbf{R})$, acts on $\mathbf{A}(n, \mathbf{R})$ by right translation). From now on, we shall identify $\mathrm{R}_1(\mathbf{R}^n)$ with $\mathbf{A}(n, \mathbf{R})$ in this way: the identity element e of $\mathbf{A}(n, \mathbf{R})$ is therefore identified with the jet $\mathrm{J}_0^1(1_{\mathbf{R}^n})$, and the tangent space to $\mathrm{R}_1(\mathbf{R}^n)$ at the point e with the Lie algebra $\mathfrak{a}(n, \mathbf{R})$ of the group $\mathbf{A}(n, \mathbf{R})$, which can be canonically decomposed into $\mathbf{R}^n \oplus \mathfrak{gl}(n, \mathbf{R})$.
(b) Let M be a pure n-dimensional differential manifold. Let $p : \mathrm{R}_2(\mathrm{M}) \to \mathrm{R}_1(\mathrm{M}) = \mathrm{R(M)}$ and $\rho : \mathrm{G}^2(n) \to \mathrm{G}^1(n) = \mathbf{GL}(n, \mathbf{R})$ denote the canonical mappings (Section 20.1, Problem 3). Let f be a diffeomorphism of a neighborhood of 0 in $\mathbf{R}^n$ onto a neighborhood of $f(0)$ in M. By transport of structure, f defines a diffeomorphism $\mathrm{R}_1(f)$ of a neighborhood of e in $\mathrm{R}_1(\mathbf{R}^n)$ onto a neighborhood of $\mathrm{J}_0^1(f)$ in $\mathrm{R}_1(\mathrm{M})$, so that we have $\mathrm{R}_1(f)(u) = \mathrm{J}_0^1(f) \circ u$. Hence we obtain an isomorphism $\tilde{f} = \mathrm{T}_e(\mathrm{R}_1(f))$ of $\mathrm{T}_e(\mathrm{R}_1(\mathbf{R}^n)) = \mathfrak{a}(n, \mathbf{R})$ onto the tangent space $\mathrm{T}_{\mathrm{J}_0^1(f)}(\mathrm{R}_1(\mathrm{M}))$. For each tangent vector $\mathbf{h}$ to $\mathrm{R}_2(\mathrm{M})$ at the point $\mathrm{J}_0^2(f)$, the vector $\mathrm{T}(p) \cdot \mathbf{h}$ is a tangent vector to $\mathrm{R}_1(\mathrm{M})$ at the point $\mathrm{J}_0^1(f)$, and therefore $\tilde{f}^{-1}(\mathrm{T}(p) \cdot \mathbf{h})$ is a vector in the Lie algebra $\mathfrak{a}(n, \mathbf{R})$.

Show that it depends only on the point $J_0^2(f)$ and the tangent vector $\mathbf{h}$ at this point to $R_2(M)$, and therefore defines a *canonical vector-valued* 1-*form*

$$\tilde{\boldsymbol{\sigma}} : \mathbf{h} \mapsto \tilde{f}^{-1}(T(p) \cdot \mathbf{h})$$

on $R_2(M)$. The diagram

$$\begin{array}{ccc} T(R_2(M)) & \xrightarrow{T(p)} & T(R_1(M)) \\ \Big\downarrow{\scriptstyle \tilde{\boldsymbol{\sigma}}} & & \Big\downarrow{\scriptstyle \boldsymbol{\sigma}} \\ \mathfrak{a}(n, \mathbf{R}) & \xrightarrow[\mathrm{pr}_1]{} & \mathbf{R}^n \end{array}$$

where $\boldsymbol{\sigma}$ is the canonical form (20.6.2.1), is commutative.

(c) Consider the case where $M = \mathbf{R}^n$. The manifold of frames $R_2(\mathbf{R}^n)$ may be identified with the set of all mappings of $\mathbf{R}^n$ into $\mathbf{R}^n$ of the form $\mathbf{x} \mapsto \mathbf{a} + U \cdot \mathbf{x} + \mathbf{B} \cdot (\mathbf{x}, \mathbf{x})$, where $\mathbf{a} \in \mathbf{R}^n$, $U \in \mathbf{GL}(n, \mathbf{R})$ and $\mathbf{B} : (\mathbf{h}, \mathbf{k}) \mapsto \mathbf{B} \cdot (\mathbf{h}, \mathbf{k})$ is a *symmetric* bilinear mapping of $\mathbf{R}^n \times \mathbf{R}^n$ into $\mathbf{R}^n$ (the jet of order 2 of a C^∞ mapping f of a neighborhood of 0 into $\mathbf{R}^n$ is identified with the mapping $\mathbf{h} \mapsto f(0) + Df(0) \cdot \mathbf{h} + \frac{1}{2}D^2 f(0) \cdot (\mathbf{h}, \mathbf{h})$ (8.14.3)). For brevity we shall denote this mapping by $(\mathbf{a}, U, \mathbf{B})$. The projection $R_2(\mathbf{R}^n) \to \mathbf{R}^n$ is then the mapping $(\mathbf{a}, U, \mathbf{B}) \mapsto \mathbf{a}$; the group $G^2(n)$ is identified with the submanifold of $R_2(\mathbf{R}^n)$ defined by $\mathbf{a} = 0$; and the right action of $G^2(n)$ on $R_2(\mathbf{R}^n)$ is given by

$$(\mathbf{a}, U, \mathbf{B}) \cdot (U', \mathbf{B}') = (\mathbf{a}, UU', U \cdot \mathbf{B}' + \mathbf{B} \cdot (U', U')),$$

where UU' is the product of the matrices, $U \cdot \mathbf{B}'$ is the quadratic mapping $\mathbf{x} \mapsto U \cdot (\mathbf{B}' \cdot (\mathbf{x}, \mathbf{x}))$, and $\mathbf{B} \cdot (U', U')$ is the quadratic mapping $\mathbf{x} \mapsto \mathbf{B} \cdot (U' \cdot \mathbf{x}, U' \cdot \mathbf{x})$. If e denotes the jet $J_0^2(1_{\mathbf{R}^n})$, the tangent space $T_e(R_2(\mathbf{R}^n))$ may be identified with the vector space of all $(\mathbf{v}, V, \mathbf{W}))$, where $\mathbf{v} \in \mathbf{R}^n$, $V \in \mathrm{End}(\mathbf{R}^n) = M_n(\mathbf{R})$, and $\mathbf{W}$ runs through the subspace of the vector space $\mathscr{L}_2(\mathbf{R}^n, \mathbf{R}^n; \mathbf{R}^n)$ consisting of symmetric mappings. If f is a diffeomorphism of a neighborhood of 0 into $\mathbf{R}^n$, then f defines (again by transport of structure) a diffeomorphism $R_2(f)$ of a neighborhood of e in $R_2(\mathbf{R}^n)$ onto a neighborhood of $J_0^2(f) = (\mathbf{a}, U, \mathbf{B})$, by the formula $R_2(f)(u) = J_0^2(f) \circ u$. Show that the image under $T_e(R_2(f))$ of the tangent vector $(\mathbf{v}, V, \mathbf{W})$ is of the form

$$\mathbf{h} = (U \cdot \mathbf{v}, UV + \mathbf{B} \cdot ((\mathbf{v}, .) + (., \mathbf{v})), U \cdot \mathbf{W} + \mathbf{B} \cdot ((V, I) + (I, V)))$$

and consequently that

$$T(p) \cdot \mathbf{h} = (U \cdot \mathbf{v}, UV + \mathbf{B} \cdot ((\mathbf{v}, .) + (., \mathbf{v}))).$$

Changing the notation, deduce that the value at the point $(\mathbf{a}, U, \mathbf{B})$ of the canonical 1-form $\tilde{\boldsymbol{\sigma}}$ is identified with the mapping

$$(\mathbf{v}, V, \mathbf{W}) \mapsto (U^{-1} \cdot \mathbf{v}, U^{-1}V + U^{-1}(\mathbf{B} \cdot ((\mathbf{v}, .) + (., \mathbf{v})))).$$

(d) Suppose again that M is arbitrary. Show that for each element $\mathbf{w}$ of the Lie algebra $\mathfrak{g}^2(n)$ of $G^2(n)$, we have $\tilde{\boldsymbol{\sigma}} \cdot Z_{\mathbf{w}} = \rho_*(\mathbf{w})$, and that for each $s \in G^2(n)$,

$$\tilde{\boldsymbol{\sigma}}(u \cdot s) \cdot (\mathbf{h} \cdot s) = \mathrm{Ad}(\rho(s^{-1})) \cdot (\tilde{\boldsymbol{\sigma}}(u) \cdot \mathbf{h})$$

for all $u \in R_2(M)$ and all tangent vectors $\mathbf{h} \in T_u(R_2(M))$.

(e) Take as basis of $\mathfrak{a}(n, \mathbf{R})$ the union of the canonical basis $(\mathbf{e}_i)$ of $\mathbf{R}^n$ and the canonical basis (E_{ij}) of $\mathfrak{gl}(n, \mathbf{R}) = \mathbf{M}_n(\mathbf{R})$. Then the canonical 1-form $\tilde{\boldsymbol{\sigma}}$ on $\mathrm{R}_2(\mathrm{M})$ may be written as

$$\tilde{\boldsymbol{\sigma}} = \sum_i \sigma^i \mathbf{e}_i + \sum_{i, j} \sigma^i_j E_{ij},$$

where the σ_i and σ^i_j are scalar-valued differential 1-forms on $\mathrm{R}_2(\mathrm{M})$. Show that

$$d\sigma^i = -\sum_j \sigma^i_j \wedge \sigma^j.$$

(Reduce to the case $\mathrm{M} = \mathbf{R}^n$, and use (c).)

7. G-STRUCTURES

(20.7.1) Let X be a principal fiber bundle with base B, structure group G, and projection p. Let H be a *closed subgroup* of G, and $j : \mathrm{H} \to \mathrm{G}$ the canonical injection. If there exists a principal bundle Y with base B, structure group H and projection q, and a *morphism* (u, j) of Y into X such that u is a B-morphism of the fibration $(\mathrm{Y}, \mathrm{B}, q)$ into $(\mathrm{X}, \mathrm{B}, p)$ (16.14), then u is an *embedding* of Y into X, and the image of u is a closed submanifold of X. Since the question is local with respect to B, we may assume that $\mathrm{X} = \mathrm{B} \times \mathrm{G}$ and $\mathrm{Y} = \mathrm{B} \times \mathrm{H}$ are trivial; u is then of the form $(b, t) \mapsto (b, f(b, t))$, where f is a C^∞ mapping of $\mathrm{B} \times \mathrm{H}$ into G such that

$$f(b, tt') = f(b, t)j(t')$$

and consequently $f(b, t) = g(b)j(t)$, putting $g(b) = f(b, e)$. Since the mapping $v : (b, s) \mapsto (b, g(b)^{-1}s)$ is a diffeomorphism of $\mathrm{B} \times \mathrm{G}$ onto itself (16.12.2.1), and since $v \circ u$ is the canonical injection $(b, t) \mapsto (b, j(t))$, our assertion is proved.

A principal bundle Y and a morphism (u, j) with the properties above are said to constitute a *restriction* of the principal bundle X to the group H. We shall generally identify Y with its image under u (cf. Problem 1).

(20.7.2) Consider a pure differential manifold M of dimension n, and its bundle of frames R(M), with structural group $\mathbf{GL}(n, \mathbf{R})$. For a closed subgroup G of $\mathbf{GL}(n, \mathbf{R})$, a G-*structure on* M is by definition a *restriction* $\mathrm{S_G(M)}$ of the principal bundle R(M) to the group G ($\mathrm{S_G(M)}$ being identified with a closed submanifold of R(M)).

Let $u : \mathrm{M} \to \mathrm{M}'$ be a diffeomorphism. It gives rise canonically to an isomorphism of principal bundles (16.14) $\mathrm{R}(u) : \mathrm{R}(\mathrm{M}) \to \mathrm{R}(\mathrm{M}')$, which maps a frame $r \in \mathrm{Isom}(\mathrm{M} \times \mathbf{R}^n, \mathrm{T}(\mathrm{M}))$ to the frame

$$\mathrm{T}(u) \circ r \circ (u^{-1} \times 1) \in \mathrm{Isom}(\mathrm{M}' \times \mathbf{R}^n, \mathrm{T}(\mathrm{M}')).$$

Given a G-structure $S_G(M)$ on M and a G-structure $S'_G(M')$ on M', the diffeomorphism u is said to be an *isomorphism* of $S_G(M)$ onto $S'_G(M')$ if the image under $R(u)$ of $S_G(M)$ is $S'_G(M')$.

For each $s \in \mathbf{GL}(n, \mathbf{R})$, $S_G(M) \cdot s$ is an $(s^{-1}Gs)$-structure on M, because $S_G(M)$ is stable under the right action of G on R(M). The structure $S_G(M) \cdot s$ is said to be *conjugate* to the G-structure $S_G(M)$.

(20.7.3) If $S_G(M)$ is a G-structure on M, the reasoning of (20.1.4) applies without change and shows that T(M) is canonically isomorphic to the fiber bundle $S_G(M) \times^G \mathbf{R}^n$ with fiber-type $\mathbf{R}^n$, associated with $S_G(M)$. Given a principal connection $\boldsymbol{P}$ in $S_G(M)$, we may therefore construct canonically by the procedure of (20.5.1) a *linear connection* $\mathbf{C}$ on M. Such a connection is called a G-*connection associated with the* G-*structure* $S_G(M)$. A C^∞ section of $S_G(M)$ is called a *moving* G-*frame*. Everything in Section 20.6 remains valid for G-connections if we bear in mind that the connection form $\boldsymbol{\omega}$ and the curvature form $\boldsymbol{\Omega}$ *take their values in the Lie algebra* $\mathfrak{g}_e$ *of* G.

Conversely, we have seen that every linear connection $\mathbf{C}$ on M determines a unique principal connection $\boldsymbol{P}$ in R(M) (20.5.2). Since $\boldsymbol{P}$ is completely determined by knowledge of the space H_r of *horizontal* tangent vectors at each point r of R(M) (20.2.2), it follows from the definitions that for $\mathbf{C}$ to be a G-*connection*, associated with the G-structure $S_G(M)$, *it is necessary and sufficient that for each frame* $r \in S_G(M)$, *the space* H_r *should be contained in the tangent space to* $S_G(M)$ *at the point* r, for the restriction of $\boldsymbol{P}$ to $T(M) \times_M S_G(M)$ will then be a principal connection on $S_G(M)$ (20.2.2).

Examples of G-*structures*

(20.7.4) First take $G = \{e\}$: then a G-structure may be identified with the image of a C^∞ *section* of the principal bundle R(M) over M. The existence of such a section is equivalent to R(M) (and therefore also T(M)) being trivializable (16.14), in which case the manifold M is said to be *parallelizable*, and an $\{e\}$-structure on M is called a *parallelism* (or *total parallelism*). For example, a *Lie group* H is parallelizable: a C^∞ section $x \mapsto \boldsymbol{R}_g(x)$ of R(H) may be obtained by taking for $\boldsymbol{R}_g(x)$ the isomorphism $\mathbf{u} \mapsto x \cdot \mathbf{u}$ of the Lie algebra $\mathfrak{h}_e$ onto $T_x(H)$. Another section $\boldsymbol{R}_d$ may be obtained by taking for $\boldsymbol{R}_d(x)$ the isomorphism $\mathbf{u} \mapsto \mathbf{u} \cdot x$. There is here a unique $\{e\}$-connection, for which $\boldsymbol{\omega} = 0$ and therefore $\boldsymbol{\Omega} = 0$ and $\boldsymbol{\Theta} = d\boldsymbol{\sigma}$.

(20.7.5) Take $G = \mathbf{SL}(n, \mathbf{R})$: if $\mathbf{u}^*$ is the n-covector $\mathbf{e}_1^* \wedge \mathbf{e}_2^* \wedge \cdots \wedge \mathbf{e}_n^*$ on $E = \mathbf{R}^n$, then G may be defined as the subgroup of all $s \in \mathbf{GL}(n, \mathbf{R})$ such that $\bigwedge^n ({}^t s) \cdot \mathbf{u}^* = \mathbf{u}^*$. Let $S_G(M)$ be a G-structure on M; we shall show that there corresponds to it a C^∞ *differential n-form* υ on M such that $\upsilon(x) \neq 0$ for all $x \in M$ (a "volume form," cf. (16.21.1)). For this purpose, let $r_x \in S_G(M)_x$: since r_x is an isomorphism of $E = \mathbf{R}^n$ onto $T_x(M)$, it follows that $\bigwedge^n ({}^t r_x^{-1})$ is an isomorphism of $\bigwedge^n (E^*)$ onto $\bigwedge^n (T_x(M)^*)$, and we put $\upsilon(x) = \bigwedge^n ({}^t r_x^{-1}) \cdot \mathbf{u}^*$. This n-covector does not depend on the element r_x chosen in the fiber $S_G(M)_x$, because any other element of the fiber is of the form $r_x \cdot s$ with $s \in G$, and we have

$$\bigwedge^n({}^t(r_x \cdot s)^{-1}) \cdot \mathbf{u}^* = \left(\bigwedge^n({}^t r_x^{-1})\right) \cdot \left(\bigwedge^n({}^t s^{-1}) \cdot \mathbf{u}^*\right) = \upsilon(x)$$

by the definition of G. By reducing to the case where R(M) is trivial, it is immediately seen that the n-form υ is of class C^∞, and moreover it is clear that $\upsilon(x) \neq 0$ for all $x \in M$. *Conversely*, if such an n-form is given, we define $S_G(M)_x$ to be the set of all $r_x \in R(M)_x$ which satisfy the relation

$$\bigwedge^n ({}^t r_x^{-1}) \cdot \mathbf{u}^* = \upsilon(x);$$

it is immediately verified that the set $S_G(M)$ so defined is a restriction of R(M) to the group $G = \mathbf{SL}(n, \mathbf{R})$, and therefore we have established a one-to-one correspondence between $\mathbf{SL}(n, \mathbf{R})$-structures and volume forms on M. In particular, the existence of an $\mathbf{SL}(n, \mathbf{R})$-structure on M is equivalent to M being *orientable* (16.21.1).

(20.7.6) Let Φ be a *symmetric* or *alternating nondegenerate bilinear form* on $\mathbf{R}^n$, and take G to be the subgroup of $\mathbf{GL}(n, \mathbf{R})$, leaving this form *invariant* (16.11.2). If Φ is symmetric and of signature (p, q), to be given a G-structure on M is equivalent to being given a *symmetric covariant tensor field* $x \mapsto \mathbf{g}(x)$ of class C^∞ such that, for each $x \in M$, the symmetric bilinear form

$$(\mathbf{h}_1, \mathbf{h}_2) \mapsto \langle \mathbf{g}(x), \mathbf{h}_1 \otimes \mathbf{h}_2 \rangle$$

on $T_x(M)$ is nondegenerate and of signature (p, q). Namely, proceeding as in (20.7.5), we take an element $r_x \in S_G(M)_x$ and define $\mathbf{g}$ by the formula

$$\langle \mathbf{g}(x), \mathbf{h}_1 \otimes \mathbf{h}_2 \rangle = \Phi(r_x^{-1} \cdot \mathbf{h}_1, r_x^{-1} \cdot \mathbf{h}_2)$$

for $\mathbf{h}_1, \mathbf{h}_2 \in T_x(M)$: in other words, $\mathbf{g}(x) = \mathbf{T}_2^0(r_x) \cdot \Phi$. As in (20.7.5) it is easily checked that this does not depend on the choice of r_x in $S_G(M)_x$, and that $\mathbf{g}$ is a C^∞ tensor field of the type specified above. Conversely, we may define as in (20.7.5) a G-structure corresponding to such a tensor field.

The procedure is the same when Φ is alternating: the assignment of a G-structure on M is in this case equivalent to that of a C^∞ *differential* 2-*form* on M, nondegenerate at every point.

Up to equivalence (20.7.2), we may always assume that

$$\Phi(\mathbf{x}, \mathbf{y}) = \sum_{j=1}^{n} \varepsilon_j \xi_j \eta_j$$

with $\varepsilon_j = 1$ for $j \leqq p$ and $\varepsilon_j = -1$ for $j > p$, when Φ is symmetric, and

$$\Phi(\mathbf{x}, \mathbf{y}) = \sum_{j=1}^{m} (\xi_j \eta_{j+m} - \xi_{j+m} \eta_j)$$

when $n = 2m$ and Φ is alternating. In the first case, a G-structure on M is said to be *pseudo-Riemannian* of *signature* (p, q) (and *Riemannian* when all the ε_j are equal to $+1$, i.e., when the signature is $(n, 0)$); in the second case, the structure is said to be *almost-Hamiltonian*. The remainder of this chapter will be devoted to the study of pseudo-Riemannian and Riemannian structures.

We remark that on every pure manifold M of dimension n there exists a *canonical* almost-Hamiltonian structure on the cotangent bundle $T(M)^*$, defined by the canonical (nondegenerate) 2-form $-d\kappa_M$ (17.15.2.4).

(20.7.7) Suppose n even, say $n = 2m$. Identify $\mathbf{R}^{2m}$ with $\mathbf{C}^m$, and let J be the endomorphism $\mathbf{z} \mapsto i\mathbf{z}$ of $\mathbf{C}^m$, considered as an endomorphism of the real vector space $\mathbf{R}^{2m}$ (if $(\mathbf{e}'_k)_{1 \leqq k \leqq m}$ is the canonical basis of $\mathbf{C}^m$, and if we take $\mathbf{e}_k = \mathbf{e}'_k$, $\mathbf{e}_{m+k} = i\mathbf{e}'_k$, then J is the endomorphism defined by

$$J \cdot \mathbf{e}_k = \mathbf{e}_{k+m}, \; J \cdot \mathbf{e}_{k+m} = -\mathbf{e}_k \qquad (1 \leqq k \leqq m)).$$

We have $J^2 = -I$, where I is the identity automorphism of $\mathbf{R}^n$. The group $G = \mathbf{GL}(m, \mathbf{C})$ may be considered as the subgroup of $\mathbf{GL}(n, \mathbf{R})$ consisting of all $s \in \mathbf{GL}(n, \mathbf{R})$ which commute with J, and the assignment of a G-structure is equivalent to that of a *tensor field* $x \mapsto \mathbf{j}(x)$ in $\mathscr{T}_1^1(M)$, such that for each $x \in M$ the tensor $\mathbf{j}(x) \in T_x(M)^* \otimes T_x(M)$, regarded as an endomorphism J_x of $T_x(M)$, is such that $J_x^2 = -I_x$, where I_x is the identity automorphism of $T_x(M)$. Proceeding as before, we define J_x by the condition $J_x \cdot \mathbf{h}_x = r_x \cdot (J \cdot (r_x^{-1} \cdot \mathbf{h}_x))$ for some $r_x \in S_G(M)_x$ and all $\mathbf{h}_x \in T_x(M)$. The details are left to the reader. A G-structure for this group $G = \mathbf{GL}(m, \mathbf{C})$ is called an *almost-complex* structure on M. Clearly, the differential manifold underlying a pure *complex-analytic manifold* M of (complex) dimension m is canonically endowed with such a structure, for which J_x is the endomorphism $\mathbf{h}_x \mapsto i\mathbf{h}_x$ of the tangent space $T_x(M)$ (which is canonically endowed with a structure of a complex vector space of dimension m). But there exist almost-complex structures on real differential manifolds that do not arise in this way.

(20.7.8) Let F be a p-dimensional subspace of $\mathbf{R}^n$, and let G be the subgroup of $\mathbf{GL}(n, \mathbf{R})$ which stabilizes F. Suppose we are given a G-structure $S_G(M)$ on M; then for each $x \in M$ and $r_x \in S_G(M)_x$, $r_x \cdot F = L_x$ is a p-dimensional subspace of $T_x(M)$ which does not depend on the frame r_x chosen in $S_G(M)_x$. It is straightforward to verify that $x \mapsto L_x$ is a C^∞ *field of p-directions* on M (18.8) and that conversely such a field determines uniquely a G-structure on M.

(20.7.9) Let G be a Lie group, H a Lie subgroup of G, so that G is a principal bundle over G/H with structure group H (16.14.2). Let $\pi : G \mapsto G/H$ be the canonical projection, and put $x_0 = \pi(e)$. The tangent space $T_{x_0}(G/H)$ is the image under $T_e(\pi)$ of $T_e(G) = \mathfrak{g}_e$, and since the kernel of $T_e(\pi)$ is $\mathfrak{h}_e$, we may canonically identify $\mathfrak{g}_e/\mathfrak{h}_e$ with $T_{x_0}(G/H)$. Since G acts transitively on G/H by left translations, for each $s \in G$ there is a canonical bijection $r(s) : \mathbf{k} \mapsto s \cdot \mathbf{k}$ of $\mathfrak{g}_e/\mathfrak{h}_e$ onto $T_{s \cdot x_0}(G/H)$. In order that $r(s) = r(s')$, first of all it is necessary that $s' \cdot x_0 = s \cdot x_0$, i.e., that $s' = st$ for some $t \in H$; next, if we denote by $\rho(t)$ the automorphism $\mathbf{k} \mapsto t \cdot \mathbf{k}$ of $\mathfrak{g}_e/\mathfrak{h}_e$ for $t \in H$, it is necessary that $\rho(t) = 1$: in other words, t must belong to the *kernel* N of the homomorphism ρ of H into $\mathbf{GL}(\mathfrak{g}_e/\mathfrak{h}_e)$. If $\tilde{H} = \rho(H) \subset \mathbf{GL}(\mathfrak{g}_e/\mathfrak{h}_e)$, it is immediately verified that the set $S_{\tilde{H}}(G/H)$ of all frames $r(s)$, as s runs through G, is a *restriction* to $\tilde{H}$ of the frame bundle R(G/H), that is to say, it defines an $\tilde{H}$-*structure* on G/H.

(20.7.10) *Canonical linear connection on a symmetric homogeneous space.*

With the notation of (20.7.9), suppose that G is *connected* and that (G, H) is a *symmetric pair* (20.4.3) corresponding to an involutory automorphism σ of G. Suppose moreover that G acts *faithfully* on G/H, or equivalently, that the intersection of the stabilizers sHs^{-1} of the points of G/H is reduced to e, or equivalently again, that H *contains no normal subgroup of* G *other than* $\{e\}$.

Let $\mathfrak{m}$ be the set of all vectors $\mathbf{u} \in \mathfrak{g}_e$ such that $\sigma_*(\mathbf{u}) = -\mathbf{u}$, so that $\mathfrak{m}$ is a supplement of $\mathfrak{h}_e$ in $\mathfrak{g}_e$, and may be canonically identified with $\mathfrak{g}_e/\mathfrak{h}_e$, by projection parallel to $\mathfrak{h}_e$. Let us first show that, with this identification, the automorphism $\mathbf{k} \mapsto t \cdot \mathbf{k}$ of $\mathfrak{g}_e/\mathfrak{h}_e$, where $t \in H$, is identified with the automorphism $\mathbf{u} \mapsto \mathrm{Ad}(t) \cdot \mathbf{u}$ of $\mathfrak{m}$. Indeed, if $\mathbf{k}$ is the coset of $\mathbf{u}$ modulo $\mathfrak{h}_e$, then $t \cdot \mathbf{k}$ is the image under $T(\pi)$ of $t \cdot \mathbf{u} \in T_t(G)$, and this image is the same as that of $t \cdot \mathbf{u} \cdot t^{-1} = \mathrm{Ad}(t) \cdot \mathbf{u} \in \mathfrak{m}$. This being so, the fact that G acts faithfully on G/H implies that the homomorphism ρ of H into $\mathbf{GL}(\mathfrak{g}_e/\mathfrak{h}_e)$, defined in (20.7.9), is *injective*. For it may be identified with the homomorphism $t \mapsto \mathrm{Ad}(t)$ of H into $\mathbf{GL}(\mathfrak{m})$; if N is its kernel, then N centralizes $\exp(\mathfrak{m})$ (19.11.6); and since $\exp(\mathfrak{h}_e) \subset H$ normalizes N, it follows that $\exp(\mathfrak{h}_e)\exp(\mathfrak{m})$

normalizes N. Since $\exp(\mathfrak{h}_e)\exp(\mathfrak{m})$ generates the connected group G (19.9.14), we see that $N \subset H$ is normal in G, hence $N = \{e\}$ by hypothesis.

The mapping $s \mapsto r(s)$ of G into the bundle of frames R(G/H), defined in (20.7.9), is therefore an *isomorphism* of the principal bundle (G, G/H, π) onto ($S_{\tilde{H}}$(G/H), G/H, π_0), where π_0 denotes the restriction to $S_{\tilde{H}}$(G/H) of the projection of R(G/H) onto G/H. By virtue of the discussion above, we may therefore canonically identify T(G/H), considered as a vector bundle associated with $S_{\tilde{H}}$(G/H), with the vector bundle $G \times^H \mathfrak{m}$ associated with the principal bundle (G, G/H, π) by the action $(t, \mathbf{u}) \mapsto \mathrm{Ad}(t) \cdot \mathbf{u}$ of H on $\mathfrak{m}$. We may then construct, by the method of (20.5.1), from the *canonical principal connection* $\mathbf{P}$ in G (20.4.3) a linear $\tilde{H}$-connection $\mathbf{C}$ on G/H. This connection $\mathbf{C}$ is called the *canonical linear connection* on the symmetric homogeneous space G/H. It has the following remarkable properties:

(20.7.10.1) *The connection* $\mathbf{C}$ *is invariant under* G (acting by left translations on G/H) *and under the involutory diffeomorphism* σ_0 *of* G/H *onto itself defined by* $\sigma_0(\pi(s)) = \pi(\sigma(s))$ for $s \in G$ (recall that the elements of H are fixed by σ).

This follows from the definition of $\mathbf{C}$ (20.5.1.1) and the fact that $\mathbf{P}$ is invariant under G and under σ.

To avoid any confusion resulting from the identifications that have been made, if $\mathbf{u} \in \mathfrak{m}$ and $s \in G$, then $s \cdot \mathbf{u}$ shall denote the vector in $T_s(G)$ defined in (16.9.8), so that $T(\pi)(s \cdot \mathbf{u})$ is the vector $s \cdot \mathbf{k}$ in $T_{s \cdot x_0}(G/H)$ if $\mathbf{k}$ is the coset of $\mathbf{u}$ in $T_{x_0}(G/H)$. We have therefore, by (20.4.3),

(20.7.10.2) $$\mathbf{P}_{s \cdot x_0}(T(\pi) \cdot (s \cdot \mathbf{u}), s) = s \cdot \mathbf{u} \in T_s(G)$$

for $s \in G$ and $\mathbf{u} \in \mathfrak{m}$. From this we deduce (20.5.1.1)

(20.7.10.3) $$\mathbf{C}_{s \cdot x_0}(T(\pi) \cdot (s \cdot \mathbf{u}), T(\pi) \cdot (s \cdot \mathbf{y})) = (s \cdot \mathbf{u}) \cdot \mathbf{y}$$

for $s \in G$ and $\mathbf{u}, \mathbf{y} \in \mathfrak{m}$, the product on the right-hand side being that defined in (16.14.7.3) for the associated bundle $G \times^H \mathfrak{m}$.

(20.7.10.4) *The geodesic* v *for* C *such that* $v(0) = x_0$ *and* $v'(0) = \mathbf{u} \in \mathfrak{m}$ (18.6.1) *is given by* $v(t) = \pi(\exp(t\mathbf{u}))$ *and is defined for all* $t \in \mathbf{R}$. *The parallel transport along* v *of a vector* $\mathbf{y} \in \mathfrak{m}$ (18.6.4) *is given by*

$$\mathbf{w}(t) = T(\pi) \cdot (\exp(t\mathbf{u}) \cdot \mathbf{y}).$$

If $v(t) = \pi(\exp(t\mathbf{u}))$, then (19.8.1) $v'(t) = \mathrm{T}(\pi) \cdot (v(t) \cdot \mathbf{u})$. On the other hand, if φ is the mapping $s \mapsto \mathrm{T}(\pi) \cdot (s \cdot \mathbf{y})$ of G into $\mathrm{T}(\mathrm{G/H}) = \mathrm{G} \times^{\mathrm{H}} \mathfrak{m}$, then by definition (16.14.7.3) $(s \cdot \mathbf{u}) \cdot \mathbf{y} = \mathrm{T}(\varphi) \cdot \mathbf{u}$, and therefore $\mathbf{w}'(t) = (\exp(t\mathbf{u}) \cdot \mathbf{u}) \cdot \mathbf{y}$. These formulas and (20.7.10.3) show that

$$\mathbf{w}'(t) = \mathbf{C}_{v(t)}(v'(t), \mathbf{w}(t)),$$

and when $\mathbf{y} = \mathbf{u}$, we have $\mathbf{w}(t) = v''(t)$; hence $v''(t) = \mathbf{C}_{v(t)}(v'(t), v'(t))$, and the result follows ((18.6.1) and (18.6.4)).

(20.7.10.5) *The torsion of* $\mathbf{C}$ *is zero, and its* (G-*invariant*) *curvature is given by*

$$(\mathbf{r}(x_0) \cdot (\mathbf{u} \wedge \mathbf{v})) \cdot \mathbf{w} = -[[\mathbf{u}, \mathbf{v}], \mathbf{w}] \tag{20.7.10.6}$$

for $\mathbf{u}, \mathbf{v}, \mathbf{w}$ *in* $\mathfrak{m} = T_{x_0}(\mathrm{G/H})$.

The formula (20.7.10.6) follows immediately from (20.4.3.3) and (20.6.6.11), applied to the principal connection $\mathbf{P}$. Furthermore, the torsion $\mathbf{t}$ is invariant under σ_0, that is to say,

$$\mathbf{t}(x_0) \cdot (\mathrm{T}(\sigma_0) \cdot \mathbf{u} \wedge \mathrm{T}(\sigma_0) \cdot \mathbf{v}) = \mathrm{T}(\sigma_0) \cdot (\mathbf{t}(x_0) \cdot (\mathbf{u} \wedge \mathbf{v})).$$

Since $\mathrm{T}(\sigma_0) \cdot \mathbf{y} = -\mathbf{y}$ for all $\mathbf{y} \in \mathfrak{m}$, it follows that $\mathbf{t}(x_0) = 0$; hence $\mathbf{t} = 0$.

(20.7.10.7) *Every* G-*invariant tensor field* $\mathbf{Z}$ *on* G/H *satisfies* $\nabla \mathbf{Z} = 0$. *In particular, the curvature tensor of* $\mathbf{C}$ *satisfies* $\nabla \mathbf{r} = 0$.

If $\mathrm{E} = \mathbf{T}_q^p(\mathrm{G/H})$, then by (20.5.4) E may be identified with the vector bundle $\mathrm{G} \times^{\mathrm{H}} (\mathbf{T}_q^p(\mathfrak{m}))$ associated with G, and a tensor field $\mathbf{Z} \in \mathscr{T}_q^p(\mathrm{G/H})$ may then be written in the form $\mathbf{Z}(s \cdot x_0) = s \cdot \Phi(s)$, where Φ is a mapping of G into $\mathbf{T}_q^p(\mathfrak{m})$ (20.5.3). The G-invariance of $\mathbf{Z}$ is then expressed by the relation

$$s' \cdot \mathbf{Z}(s \cdot x_0) = \mathbf{Z}(s's \cdot x_0),$$

which implies that $\Phi(s's) = \Phi(s)$, so that Φ so *constant* on G. It now follows from (20.5.3.4) that $\nabla \mathbf{Z} = 0$.

Remark

(20.7.11) If $\mathrm{S_G(M)}$ is a G-structure on M, and if $\mathrm{G}' \supset \mathrm{G}$ is a subgroup of $\mathbf{GL}(n, \mathbf{R})$, then we may obtain canonically from $\mathrm{S_G(M)}$ a G′-structure $\mathrm{S_{G'}(M)}$ by taking the frames belonging to $\mathrm{S_{G'}(M)}$ to be the frames $r \cdot s'$, where $s' \in \mathrm{G}'$ and $r \in \mathrm{S_G(M)}$: it is immediately verified (by reduction to the case where R(M) is trivial) that we obtain in this way a restriction of R(M) to the group G′. A linear G-connection on M is also a G′-connection.

(20.7.12) We have seen in examples above that, for a given Lie subgroup G of $\mathbf{GL}(n, \mathbf{R})$, there need not exist a G-structure on M: the question depends in general on global topological properties of M. However, it is clear that if G is parallelizable (20.7.4), we can define a G-structure on M for *every* subgroup G of $\mathbf{GL}(n, \mathbf{R})$, by virtue of (20.7.11). Furthermore, in this case there always exist G-*connections* associated with a given G-structure. For if $\boldsymbol{\varpi}$ is an *arbitrary* differential 1-form on M, *with values in the Lie algebra* $\mathfrak{g}_e \subset \mathfrak{gl}(n, \mathbf{R})$, then there always exists a connection form $\boldsymbol{\omega}$ on $S_G(M)$ such that $\boldsymbol{\omega}^{(\mathbf{R})} = \boldsymbol{\varpi}$ for some section $\mathbf{R}$ of $S_G(M)$ over M.

Another important case in which no global topological condition is necessary is the case of Riemannian structures:

(20.7.13) *There exists a Riemannian structure on every pure differential manifold* M.

Consider a denumerable family of charts $c_\alpha = (U_\alpha, \varphi_\alpha, n)$ of M such that the U_α form a locally finite covering of M, and let $\psi_\alpha : U_\alpha \times \mathbf{R}^n \to o_M^{-1}(U_\alpha)$ be the framing of T(M) over U_α associated with the chart c_α. We define a tensor field $\mathbf{g}_\alpha$ on U_α which gives a Riemannian structure on U_α by the formula

$$\langle \mathbf{g}_\alpha(x), \psi_\alpha(x, \mathbf{u}) \otimes \psi_\alpha(x, \mathbf{v}) \rangle = (\mathbf{u} \mid \mathbf{v})$$

for all $x \in U_\alpha$ and $\mathbf{u}, \mathbf{v} \in \mathbf{R}^n$, where $(\mathbf{u} \mid \mathbf{v})$ is the Euclidean scalar product on $\mathbf{R}^n$. Let (h_α) be a partition of unity subordinate to (U_α), each h_α being a C^∞ mapping of M into [0, 1] (16.4.1), and put

$$\mathbf{g}(x) = \sum_\alpha h_\alpha(x) \mathbf{g}_\alpha(x)$$

for each $x \in M$ (with the convention that $h_\alpha(x)\mathbf{g}_\alpha(x) = 0$ for $x \notin U_\alpha$). Then $\mathbf{g}$ is a C^∞ tensor field on M, because every frontier point of U_α has a neighborhood on which $h_\alpha(x) = 0$. We assert that $\mathbf{g}$ defines a Riemannian structure on M. It is enough to show that if $x \in U_\alpha$, we have

$$\langle \mathbf{g}(x), \psi_\alpha(x, \mathbf{u}) \otimes \psi_\alpha(x, \mathbf{u}) \rangle > 0$$

for all $\mathbf{u} \neq 0$ in $\mathbf{R}^n$. Now if β is an index such that $h_\beta(x) \neq 0$, then we may write $\psi_\alpha(x, \mathbf{u}) = \psi_\beta(x, A_{\beta\alpha}(x) \cdot \mathbf{u})$, where $A_{\beta\alpha}(x) \in \mathbf{GL}(n, \mathbf{R})$ (16.15.1.1); consequently,

$$\begin{aligned}
&\langle \mathbf{g}(x), \psi_\alpha(x, \mathbf{u}) \otimes \psi_\alpha(x, \mathbf{u}) \rangle \\
&\quad = \sum_\beta h_\beta(x) \langle \mathbf{g}_\beta(x), \psi_\beta(x, A_{\beta\alpha}(x) \cdot \mathbf{u}) \otimes \psi_\beta(x, A_{\beta\alpha}(x) \cdot \mathbf{u}) \rangle \\
&\quad = \sum_\beta h_\beta(x) \| A_{\beta\alpha}(x) \cdot \mathbf{u} \|^2,
\end{aligned}$$

summed over all β such that $h_\beta(x) \neq 0$, the norm on $\mathbf{R}^n$ being the Euclidean norm. Since we have $\sum_\beta h_\beta(x) = 1$ and $\| A_{\beta\alpha}(x) \cdot \mathbf{u} \|^2 > 0$ for the indices β under consideration, our assertion is proved.

PROBLEMS

1. If Y is a restriction of a principal bundle X with group G to a closed subgroup H of G (20.7.1), then X is canonically isomorphic to the extension $Y \times^H G$ of Y (Section 16.14, Problem 17), and the associated bundle $X \times^G (G/H) = H\backslash X$ (16.14.8) is isomorphic to $Y \times^H (G/H)$ (Section 16.14, Problem 16), so that $X \times^G (G/H)$ admits a C^∞ *section* over B (Section 20.1, Problem 2). Conversely, if $X \times^G (G/H)$ admits a C^∞ section over B, then there exists a restriction of X to the subgroup H, and these restrictions are in one-to-one correspondence with the C^∞ sections of $X \times^G (G/H)$ over B. (If σ is such a section, consider the inverse image of $\sigma(B)$ under the mapping $x \mapsto x \cdot \dot{e}$ of X onto $X \times^G (G/H)$.)

   Consider in particular the case where G/H is diffeomorphic to $\mathbf{R}^n$, and hence deduce another proof of (20.7.13). (Cf. Section 11.5, Problem 15.)

2. Let K be a Lie group, H a Lie subgroup of K; let $M = K/H$ be the corresponding homogeneous space, $\pi : K \to M$ the canonical projection, and $x_0 = \pi(e)$. Suppose that the homomorphism ρ of H into $\mathbf{GL}(T_{x_0}(M))$ (20.7.9) is injective, so that K acts *freely* on the left on the bundle of frames R(M) (20.7.9): if $s \in K$ and $r \in R(M)$, so that r is a linear bijection of $T_{x_0}(M)$ (identified with $\mathbf{R}^n$) onto $T_x(M)$, the product $s \cdot r$ is the bijection $\mathbf{k} \mapsto s \cdot r(\mathbf{k})$ of $T_{x_0}(M)$ onto $T_{s \cdot x}(M)$, and we have $s \cdot (r \cdot \rho(t)) = (s \cdot r) \cdot \rho(t)$ for all $t \in H$. If G is a Lie subgroup of $\mathbf{GL}(n, \mathbf{R})$, a G-structure $S_G(M) \subset R(M)$ is said to be K-*invariant* if for all $r \in S_G(M)$, $s \in K$ and $t \in G$, we have $s \cdot (r \cdot t) = (s \cdot r) \cdot t$. Let r_0 be an element of $R(M)_{x_0}$; then for each $t \in H$ there exists a unique element $\lambda(t) \in G$ such that $t \cdot r_0 = r_0 \cdot \lambda(t)$, and from the definition of ρ we have $\lambda(t) = r_0^{-1} \circ \rho(t) \circ r_0$, so that λ is an injective homomorphism of H into G.

   (a) There exists a canonical one-to-one correspondence between the set of K-invariant principal connections in $S_G(M)$ and the set of linear mappings $\mathbf{f} : \mathfrak{k}_e \to \mathfrak{g}_e$ (the Lie algebras of K and G, respectively) such that: (i) $\mathbf{f}(\mathbf{w}) = \lambda_*(\mathbf{w})$ for $\mathbf{w} \in \mathfrak{h}_e$ (the Lie algebra of H); (ii) $\mathbf{f}(\mathrm{Ad}(t) \cdot \mathbf{w}) = \mathrm{Ad}(\lambda(t)) \cdot \mathbf{f}(\mathbf{w})$ for $\mathbf{w} \in \mathfrak{k}_e$ and $t \in H$. (Cf. Section 20.3, Problem 3(b).)
   (b) Show that the torsion and curvature of the linear connection $\mathbf{C}$ on M corresponding to the linear mapping $\mathbf{f}$ are given by the following formulas:

$$\mathbf{t} \cdot (Z_{\mathbf{u}}(x_0) \wedge Z_{\mathbf{v}}(x_0)) = r_0 \cdot (\mathbf{f}(\mathbf{u}) \cdot (r_0^{-1} \cdot Z_{\mathbf{v}}(x_0))) - r_0 \cdot (\mathbf{f}(\mathbf{v}) \cdot (r_0^{-1} \cdot Z_{\mathbf{u}}(x_0))) - Z_{[\mathbf{u},\mathbf{v}]}(x_0),$$
$$\mathbf{r} \cdot (Z_{\mathbf{u}}(x_0) \wedge Z_{\mathbf{v}}(x_0)) = r_0 \circ ([\mathbf{f}(\mathbf{u}), \mathbf{f}(\mathbf{v})] - \mathbf{f}([\mathbf{u}, \mathbf{v}])) \circ r_0^{-1}$$

   for any two vectors $\mathbf{u}, \mathbf{v} \in \mathfrak{k}_e$, where $Z_{\mathbf{u}}$ is the Killing field corresponding to $\mathbf{u}$ on $M = K/H$.
   (c) With the same notation, show that the covariant derivative (relative to $\mathbf{C}$) is given by the formula

$$\nabla_{Z_{\mathbf{v}}(x_0)} \cdot Z_{\mathbf{u}} = r_0 \cdot (\mathbf{f}(\mathbf{v}) \cdot (r_0^{-1} \cdot Z_{\mathbf{u}}(x_0))) + Z_{[\mathbf{u},\mathbf{v}]}(x_0).$$

3. The hypotheses are the same as in Problem 2. Suppose in addition that there exists a subspace $\mathfrak{m}$ of $\mathfrak{k}_e$ supplementary to $\mathfrak{h}_e$, such that $\mathrm{Ad}(t) \cdot \mathfrak{m} \subset \mathfrak{m}$ for all $t \in H$ (Section 20.4, Problem 3). Then the K-invariant principal connections in $S_G(M)$ correspond one-to-one to the linear mappings $\mathbf{f}_{\mathfrak{m}} : \mathfrak{m} \to \mathfrak{g}_e$ such that

$$\mathbf{f}_{\mathfrak{m}}(\mathrm{Ad}(t) \cdot \mathbf{w}) = \mathrm{Ad}(\lambda(t)) \cdot \mathbf{f}_{\mathfrak{m}}(\mathbf{w})$$

for $t \in \mathrm{H}$ and $\mathbf{w} \in \mathfrak{m}$ (*loc. cit.*). The formulas in Problem 2(b) then become, if we identify the vectors $\mathbf{u}$ *in* $\mathfrak{m}$ with the tangent vectors $Z_{\mathbf{u}}(x_0) \in \mathrm{T}_{x_0}(\mathrm{M})$,

$$\boldsymbol{t} \cdot (\mathbf{u} \wedge \mathbf{v}) = \mathbf{f}_{\mathfrak{m}}(\mathbf{u}) \cdot \mathbf{v} - \mathbf{f}_{\mathfrak{m}}(\mathbf{v}) \cdot \mathbf{u} - [\mathbf{u}, \mathbf{v}]_{\mathfrak{m}},$$

$$\boldsymbol{r} \cdot (\mathbf{u} \wedge \mathbf{v}) = [\mathbf{f}_{\mathfrak{m}}(\mathbf{u}), \mathbf{f}_{\mathfrak{m}}(\mathbf{v})] - \mathbf{f}_{\mathfrak{m}}([\mathbf{u}, \mathbf{v}]_{\mathfrak{m}}) - \lambda_*([\mathbf{u}, \mathbf{v}]_{\mathfrak{h}_e})$$

for $\mathbf{u}, \mathbf{v} \in \mathfrak{m}$, where $[\mathbf{u}, \mathbf{v}]_{\mathfrak{m}}$ and $[\mathbf{u}, \mathbf{v}]_{\mathfrak{h}_e}$ are the components of $[\mathbf{u}, \mathbf{v}]$ in $\mathfrak{m}$ and $\mathfrak{h}_e$, respectively, for the direct sum decomposition $\mathfrak{k}_e = \mathfrak{m} \oplus \mathfrak{h}_e$. The connection in $\mathrm{S_G(M)}$ corresponding to $\mathbf{f}_{\mathfrak{m}} = 0$ is called the *canonical* connection (for the choice of subspace $\mathfrak{m}$ supplementary to $\mathfrak{h}_e$).

For $\mathbf{u} \in \mathfrak{k}_e$, let $\tilde{Z}_{\mathbf{u}}$ denote the *canonical lifting* to $\mathrm{S_G(M)}$ of the vector field $Z_{\mathbf{u}}$ (Section 20.6, Problem 5). Then the relation $\mathbf{f}_{\mathfrak{m}}(\mathbf{u}) = 0$ for some $\mathbf{u} \in \mathfrak{m}$ signifies that the vector $\tilde{Z}_{\mathbf{u}}(x_0)$ is horizontal at the point r_0. Deduce that the canonical connection in $\mathrm{S_G(M)}$ is the only K-invariant connection such that for all $\mathbf{u} \in \mathfrak{m}$, if we put $g_t(x_0) = \exp(t\mathbf{u}) \cdot x_0$ and $\mathrm{G}_t = \mathrm{R}(g_t)$ (*loc. cit.*), the orbit $t \mapsto \mathrm{G}_t(r_0)$ is the horizontal lifting of $t \mapsto g_t(x_0)$ that passes through r_0 (Section 20.2, Problem 3). The paths $t \mapsto g_t(x_0)$ corresponding to the vectors $\mathbf{u} \in \mathfrak{m}$ are geodesics for the corresponding linear connection $\mathbf{C}$ on M; this connection is complete (Section 20.6, Problem 8), and we have $\nabla \boldsymbol{U} = 0$ for each K-invariant tensor field $\boldsymbol{U}$ on M.

Show that the K-invariant principal connections in $\mathrm{S_G(M)}$ for which the geodesics for the corresponding linear connection on M are the same as for the canonical connection, correspond to the mappings $\mathbf{f}_{\mathfrak{m}}$ such that $\mathbf{f}_{\mathfrak{m}}(\mathbf{u}) \cdot \mathbf{u} = 0$ for all $\mathbf{u} \in \mathfrak{m}$. In particular, there is just one of these linear connections which is torsion-free, and it corresponds to the mapping $\mathbf{f}_{\mathfrak{m}}$ defined by $\mathbf{f}_{\mathfrak{m}}(\mathbf{u}) \cdot \mathbf{v} = \frac{1}{2}[\mathbf{u}, \mathbf{v}]_{\mathfrak{m}}$ for $\mathbf{u}, \mathbf{v} \in \mathfrak{m}$.

Consider the case where $\mathrm{K} = \mathrm{L} \times \mathrm{L}$, L being a connected Lie group and K acting on L by the rule $((s, t), x) \mapsto sxt^{-1}$, so that L may be identified with K/H, where H is the diagonal of $\mathrm{L} \times \mathrm{L}$. We may then take $\mathfrak{m}$ to be any one of the subspaces $\{0\} \times \mathfrak{l}_e$, $\mathfrak{l}_e \times \{0\}$, or the image of $\mathfrak{l}_e$ under the mapping $\mathbf{u} \mapsto (\mathbf{u}, -\mathbf{u})$, where $\mathfrak{l}_e$ is the Lie algebra of L. Calculate the torsion and curvature of the canonical connections corresponding to these three choices of $\mathfrak{m}$.

4. Let M be the complement of the origin in $\mathbf{R}^n$. The group $\mathbf{GL}(n, \mathbf{R})$ acts transitively on M, so that M may be identified with the homogeneous space K/H, where $\mathrm{K} = \mathbf{GL}(n, \mathbf{R})$ and H is the subgroup which fixes some point $\neq 0$. The restriction to M of the canonical linear connection on $\mathbf{R}^n$ is K-invariant, but there exists no subspace $\mathfrak{m}$ supplementary to $\mathfrak{h}_e$ in $\mathfrak{k}_e$ such that $\mathrm{Ad}(t)\mathfrak{m} \subset \mathfrak{m}$ for all $t \in \mathrm{H}$.

5. Let M be a pure differential manifold of dimension n, let G be a Lie subgroup of $\mathbf{GL}(n, \mathbf{R})$, let $\mathfrak{g}$ be its Lie algebra, and let $\mathrm{S_G(M)}$ be a G-structure on M. We shall denote again by $\boldsymbol{\sigma}$ the restriction ${}^t j(\boldsymbol{\sigma})$ to $\mathrm{S_G(M)}$ of the canonical form $\boldsymbol{\sigma}$ on R(M) ($j : \mathrm{S_G(M)} \to \mathrm{R(M)}$ being the canonical injection); $\boldsymbol{\sigma}$ is therefore a vector-valued 1-form on $\mathrm{S_G(M)}$ with values in $\mathbf{R}^n$, which vanishes on vertical tangent vectors.

Let r be a point of $\mathrm{S_G(M)}$ and let H_1, H_2 be two subspaces of $\mathrm{T}_r(\mathrm{S_G(M)})$, both of which are supplementary to the space of vertical tangent vectors at the point r, so that the restrictions to H_1 and H_2 of $\mathrm{T}(\pi)$ (where $\pi : \mathrm{S_G(M)} \to \mathrm{M}$ is the canonical projection) are isomorphisms onto $\mathrm{T}_{\pi(r)}(\mathrm{M})$. For each vector $\mathbf{x} \in \mathbf{R}^n$, let $\mathbf{k}_1$, $\mathbf{k}_2$ be the vectors in H_1, H_2, respectively, such that $\mathrm{T}(\pi) \cdot \mathbf{k}_1 = \mathrm{T}(\pi) \cdot \mathbf{k}_2 = r \cdot \mathbf{x}$, or equivalently, such that $\boldsymbol{\sigma}(r) \cdot \mathbf{k}_1 = \boldsymbol{\sigma}(r) \cdot \mathbf{k}_2 = \mathbf{x}$. We may therefore write $\mathbf{k}_1 - \mathbf{k}_2 = Z_{T(\mathbf{x})}(r)$, where $T : \mathbf{R}^n \to \mathfrak{g}$ is a linear mapping uniquely determined by H_1 and H_2.

For $i = 1, 2$ and $\mathbf{x}, \mathbf{y} \in \mathbf{R}^n$, let

$$\mathbf{S}_i(\mathbf{x} \wedge \mathbf{y}) = d\sigma(r) \cdot (\mathbf{h}_i \wedge \mathbf{k}_i) \in \mathbf{R}^n,$$

where $\mathbf{h}_i, \mathbf{k}_i \in \mathrm{H}_i$ are such that $\sigma(r) \cdot \mathbf{h}_i = \mathbf{x}$, $\sigma(r) \cdot \mathbf{k}_i = \mathbf{y}$. Use Section 20.6, Problem 11 to show that

$$(*) \qquad \mathbf{S}_2(\mathbf{x} \wedge \mathbf{y}) - \mathbf{S}_1(\mathbf{x} \wedge \mathbf{y}) = T(\mathbf{x}) \cdot \mathbf{y} - T(\mathbf{y}) \cdot \mathbf{x}$$

(where $\mathfrak{g}$ is identified with a Lie subalgebra of $\mathfrak{gl}(n, \mathbf{R}) = \mathrm{End}(\mathbf{R}^n)$). Let

$$\partial : \mathrm{Hom}(\mathbf{R}^n, \mathfrak{g}) \to \mathrm{Hom}\left(\bigwedge^2 \mathbf{R}^n, \mathbf{R}^n\right)$$

be the linear mapping defined as follows: if $T \in \mathrm{Hom}(\mathbf{R}^n, \mathfrak{g})$, then ∂T is the linear mapping of $\bigwedge^2 \mathbf{R}^n$ into $\mathbf{R}^n$ induced by the alternating bilinear mapping

$$(\mathbf{x}, \mathbf{y}) \mapsto T(\mathbf{x}) \cdot \mathbf{y} - T(\mathbf{y}) \cdot \mathbf{x}.$$

It follows from $(*)$ that the class of $\mathbf{S}_1 \in \mathrm{Hom}\left(\bigwedge^2 \mathbf{R}^n, \mathbf{R}^n\right)$ modulo the subspace $\partial(\mathrm{Hom}(\mathbf{R}^n, \mathfrak{g}))$ depends only on the point $r \in \mathrm{S_G(M)}$ and not on the choice of the subspace H_1 supplementary to $\mathrm{Ker}(\mathrm{T}_r(\pi))$ in $\mathrm{T}_r(\mathrm{S_G(M)})$. This class $c(r)$ is the value at the point r of what is called the *first-order structure function* of the G-structure $\mathrm{S_G(M)}$.

The group $\mathrm{G} \subset \mathbf{GL}(n, \mathbf{R})$ acts in its natural way on $\mathbf{R}^n$, and on $\mathfrak{g}$ via the adjoint representation. Hence by transport of structure it acts on the left on $\mathrm{Hom}\left(\bigwedge^2 \mathbf{R}^n, \mathbf{R}^n\right)$ and on $\mathrm{Hom}(\mathbf{R}^n, \mathfrak{g})$. Show that, for each $s \in \mathrm{G}$ and each $T \in \mathrm{Hom}(\mathbf{R}^n, \mathfrak{g})$, we have $\partial(s \cdot T) = s \cdot \partial T$, so that G acts linearly on $\mathrm{Hom}\left(\bigwedge^2 \mathbf{R}^n, \mathbf{R}^n\right)/\partial(\mathrm{Hom}(\mathbf{R}^n, \mathfrak{g}))$. If ρ is the corresponding linear representation of G, show that $c(r \cdot s) = \rho(s^{-1}) \cdot c(r)$.

6. With the hypotheses and notation of Problem 5, there exists an $\mathrm{S_G(M)}$-morphism $h : (r, \mathbf{x}) \mapsto h_r \cdot \mathbf{x}$ of $\mathrm{S_G(M)} \times \mathbf{R}^n$ into the tangent bundle $\mathrm{T(S_G(M))}$ such that the image under h of $\{r\} \times \mathbf{R}^n$ is a subspace H_r of $\mathrm{T}_r(\mathrm{S_G(M)})$ supplementary to the subspace of vertical tangent vectors. If E is a subspace of $\mathrm{Hom}\left(\bigwedge^2 \mathbf{R}^n, \mathbf{R}^n\right)$ supplementary to $\partial(\mathrm{Hom}(\mathbf{R}^n, \mathfrak{g}))$, we may also assume that, for all $r \in \mathrm{S_G(M)}$, the element $\tilde{c}(r)$ of $\mathrm{Hom}\left(\bigwedge^2 \mathbf{R}^n, \mathbf{R}^n\right)$ defined by $\tilde{c}(r) \cdot (\mathbf{x} \wedge \mathbf{y}) = d\sigma(r) \cdot (\mathbf{h} \wedge \mathbf{k})$ (where $\mathbf{h}, \mathbf{k} \in \mathrm{H}_r$ are such that $\sigma(r) \cdot \mathbf{h} = \mathbf{x}$, $\sigma(r) \cdot \mathbf{k} = \mathbf{y}$) belongs to E (by adding to $\tilde{c}(r)$ if necessary an element $\partial T_r \in \partial(\mathrm{Hom}(\mathbf{R}^n, \mathfrak{g}))$, which is of class C^∞ as a function of r). We then identify the coset $c(r)$ with its representative $\tilde{c}(r) \in \mathrm{E}$.

Let $\mathrm{G}^{(1)}$ be the commutative subgroup of $\mathbf{GL}(\mathbf{R}^n \oplus \mathfrak{g})$ consisting of automorphisms of the form $(\mathbf{x}, \mathbf{u}) \mapsto (\mathbf{x}, T(\mathbf{x}) + \mathbf{u})$, where T runs through the subspace $\mathfrak{g}^{(1)} = \mathrm{Ker}(\partial)$ of $\mathrm{Hom}(\mathbf{R}^n, \mathfrak{g})$. For each frame $r \in \mathrm{S_G(M)}$, consider the frames in the tangent space to $\mathrm{S_G(M)}$ at the point r (i.e., isomorphisms of $\mathbf{R}^n \oplus \mathfrak{g}$ onto $\mathrm{T}_r(\mathrm{S_G(M)})$):

$$r^{(1)} : (\mathbf{x}, \mathbf{u}) \mapsto h_r \cdot \mathbf{x} + Z_{\mathbf{u}}(r) + Z_{T(\mathbf{x})}(r),$$

where T runs through the subspace $\mathfrak{g}^{(1)}$. These frames constitute a $\mathrm{G}^{(1)}$-*structure* $\mathrm{S}_{\mathrm{G}^{(1)}}(\mathrm{S_G(M)})$ on $\mathrm{S_G(M)}$, called the *first-order prolongation* of the G-structure $\mathrm{S_G(M)}$ on M. If we replace E by another supplement E′ of $\partial(\mathrm{Hom}(\mathbf{R}^n, \mathfrak{g}))$ in $\mathrm{Hom}\left(\bigwedge^2 \mathbf{R}^n, \mathbf{R}^n\right)$ and h by a morphism h' having the same properties relative to E′ as h has relative to E, then the $\mathrm{G}^{(1)}$-structure on $\mathrm{S_G(M)}$ is replaced by a *conjugate* structure having the same group. The Lie algebra of the commutative group $G^{(1)}$ may be identified with $\mathfrak{g}^{(1)}$, by identifying $T \in \mathrm{Hom}(\mathbf{R}^n, \mathfrak{g})$ with the endomorphism of $\mathbf{R}^n \oplus \mathfrak{g}$ which agrees with T on $\mathbf{R}^n$ and is zero on $\mathfrak{g}$. The structure function $c^{(1)}$ of the $G^{(1)}$-structure on $S_G(M)$ is called (by abuse of language) the *second-order structure function* of the G-structure $S_G(M)$ on M: its values lie in

$$\mathrm{Hom}\left(\bigwedge^2 (\mathbf{R}^n \oplus \mathfrak{g}), \mathbf{R}^n \oplus \mathfrak{g}\right)/\partial\mathrm{Hom}(\mathbf{R}^n \oplus \mathfrak{g}, \mathfrak{g}^{(1)}).$$

Let $\pi : S_G(M) \to M$, $\pi_1 : S_{G^{(1)}}(S_G(M)) \to S_G(M)$ be the canonical projections. Show that the canonical form $\boldsymbol{\sigma}^{(1)}$ on $S_{G^{(1)}}(S_G(M))$ may be written as $\boldsymbol{\sigma}^{(1)} = {}^t\pi_1(\boldsymbol{\sigma}) + \boldsymbol{\omega}_1$, where $\boldsymbol{\omega}_1$ is a vector-valued 1-form on $S_{G^{(1)}}(S_G(M))$ with values in the Lie algebra $\mathfrak{g}$.

The calculation of $c^{(1)}(r^{(1)})$ at a point of $S_{G^{(1)}}(S_G(M))$ is equivalent, once we have chosen a subspace $H^{(1)}$ of the tangent space to $S_{G^{(1)}}(S_G(M))$ at the point $r^{(1)}$, supplementary to the subspace of vertical tangent vectors, to the evaluation of

$$d\boldsymbol{\sigma}^{(1)}(r^{(1)}) \cdot (\mathbf{h}^{(1)} \wedge \mathbf{k}^{(1)}),$$

where $\mathbf{h}^{(1)}$, $\mathbf{k}^{(1)}$ are vectors in $H^{(1)}$. Putting $r = \pi_1(r^{(1)}) \in S_G(M)$, there are three cases to consider:

(i) $T(\pi_1) \cdot \mathbf{h}^{(1)} = Z_{\mathbf{u}}(r)$, $T(\pi_1) \cdot \mathbf{k}^{(1)} = Z_{\mathbf{v}}(r)$ for $\mathbf{u}, \mathbf{v} \in \mathfrak{g}$, vertical vectors in $T_r(S_G(M))$ such that $\boldsymbol{\omega}_1(r^{(1)}) \cdot \mathbf{h}^{(1)} = \mathbf{u}$, $\boldsymbol{\omega}_1(r^{(1)}) \cdot \mathbf{k}^{(1)} = \mathbf{v}$. We have then

$$d\boldsymbol{\sigma}^{(1)}(r^{(1)}) \cdot (\mathbf{h}^{(1)} \wedge \mathbf{k}^{(1)}) = [\mathbf{u}, \mathbf{v}] \in \mathfrak{g}.$$

(Extend $\mathbf{h}^{(1)}$ and $\mathbf{k}^{(1)}$ to $G^{(1)}$-invariant vector fields on a neighborhood of $r^{(1)}$, whose projections on $S_G(M)$ are $Z_{\mathbf{u}}$ and $Z_{\mathbf{v}}$.)

(ii) $T(\pi_1) \cdot \mathbf{h}^{(1)} = Z_{\mathbf{u}}(r), T(\pi_1) \cdot \mathbf{k}^{(1)} = h_r \cdot \mathbf{x}$ with $\mathbf{u} \in \mathfrak{g}, \mathbf{x} \in \mathbf{R}^n, \boldsymbol{\sigma}(r) \cdot (h_r \cdot \mathbf{x}) = \mathbf{x}$, $\boldsymbol{\omega}_1(r^{(1)}) \cdot \mathbf{h}^{(1)} = \mathbf{u}$, $\boldsymbol{\omega}_1(r^{(1)}) \cdot \mathbf{k}^{(1)} = 0$. Then (Section 20.6, Problem 11) we have

$$d\boldsymbol{\sigma}^{(1)}(r^{(1)}) \cdot (\mathbf{h}^{(1)} \wedge \mathbf{k}^{(1)}) = -\mathbf{u} \cdot \mathbf{x} + U_r(\mathbf{u}) \cdot \mathbf{x},$$

where $U_r \in \mathrm{Hom}(\mathfrak{g}, \mathrm{Hom}(\mathbf{R}^n, \mathfrak{g}))$. The group G acts by transport of structure on $R(S_G(M))$; if it leaves invariant the structure $S_{G^{(1)}}(S_G(M)) \subset R(S_G(M))$, then U_r takes its values in $\mathfrak{g}^{(1)}$. The converse is true if G is connected.

(iii) $T(\pi_1) \cdot \mathbf{h}^{(1)} = h_r \cdot \mathbf{x}$, $T(\pi_1) \cdot \mathbf{k}^{(1)} = h_r \cdot \mathbf{y}$, where $\mathbf{x}, \mathbf{y} \in \mathbf{R}^n$, $\boldsymbol{\sigma}(r) \cdot (h_r \cdot \mathbf{x}) = \mathbf{x}$, $\boldsymbol{\sigma}(r) \cdot (h_r \cdot \mathbf{y}) = \mathbf{y}$, $\boldsymbol{\omega}_1(r^{(1)}) \cdot \mathbf{h}^{(1)} = \boldsymbol{\omega}_1(r^{(1)}) \cdot \mathbf{k}^{(1)} = 0$. In this case we have

$$d\boldsymbol{\sigma}^{(1)}(r^{(1)}) \cdot (\mathbf{h}^{(1)} \wedge \mathbf{k}^{(1)}) = d\boldsymbol{\sigma}(r) \cdot (h_r \cdot \mathbf{x} \wedge h_r \cdot \mathbf{y}) + V_r \cdot (\mathbf{x} \wedge \mathbf{y}),$$

where $V_r \in \mathrm{Hom}\left(\bigwedge^2 \mathbf{R}^n, \mathfrak{g}\right)$. If $G^{(1)} = \{e\}$, we may identify $S_{G^{(1)}}(S_G(M))$ with $S_G(M)$; moreover, if G leaves invariant the $G^{(1)}$-structure on $S_G(M)$, then $\boldsymbol{\omega}_1$ is the 1-form of a principal connection in $S_G(M)$, and the vector-valued 2-form $\boldsymbol{\Omega}$ on $S_G(M)$ such that

$$\boldsymbol{\Omega}(r) \cdot (h_r \cdot \mathbf{x} \wedge h_r \cdot \mathbf{y}) = V_r \cdot (\mathbf{x} \wedge \mathbf{y}) \in \mathfrak{g}$$

is the curvature form of this connection. (For an example of this case, see 20.9.2.)

By induction we can define, for each integer $k > 1$, the kth-*order prolongation* of the G-structure $S_G(M)$ to be the first-order prolongation of the $G^{(k-1)}$-structure which is the $(k-1)$th order prolongation of $S_G(M)$. Let $G^{(k)}$ denote the corresponding group, $\mathfrak{g}^{(k)}$ its Lie algebra. The structure function of this $G^{(k)}$-structure is denoted by $c^{(k+1)}$ and is called the $(k+1)$th-*order structure function* of the G-structure $S_G(M)$.

7. For the manifold $M = \mathbf{R}^n$, the mapping $\boldsymbol{\tau}^{-1} : x \mapsto \boldsymbol{\tau}_x^{-1}$ (16.5.2) is a C^∞ section of the bundle of frames R(M). If G is any closed subgroup of $\mathbf{GL}(n, \mathbf{R})$, the union of the $\boldsymbol{\tau}_x^{-1} \cdot G$,

as x runs through M, is a G-structure $S_G(M)$ on M, called the *canonical flat* G-*structure* on $\mathbf{R}^n$. A G-structure on a differential manifold M′ of dimension n is said to be *flat* if, for each $x' \in M'$, there exists a diffeomorphism of a neighborhood U of x' onto $\mathbf{R}^n$ which (by transport of structure) transforms the induced G-structure on U into the canonical flat G-structure.

For the canonical flat G-structure, if $\pi : R(\mathbf{R}^n) \to \mathbf{R}^n$ is the canonical projection, the canonical form $\boldsymbol{\sigma}$ on $S_G(\mathbf{R}^n)$ is equal to ${}^t\pi(d(1_{\mathbf{R}^n})) = {}^t\pi(\boldsymbol{\tau})$, where $\boldsymbol{\tau}$ is considered as a vector-valued 1-form on $\mathbf{R}^n$, with values in $\mathbf{R}^n$ (16.20.15). Deduce that the first-order structure function of $S_G(\mathbf{R}^n)$ is *zero*.

With the notation of Problem 6, if we take $h(\boldsymbol{\tau}_x^{-1} \cdot s, \mathbf{y}) = T_x(\boldsymbol{\tau}_x^{-1} \cdot s) \cdot (\boldsymbol{\tau}_x^{-1} \cdot \mathbf{y})$, the subspaces H_r are the spaces of horizontal vectors for a principal connection in $S_G(\mathbf{R}^n)$ called the *canonical* connection. (The corresponding linear G-connection on $\mathbf{R}^n$ is also called the *canonical* linear G-connection.) Show that this connection is *flat* (20.4.1). If $\boldsymbol{\omega}$ is the differential 1-form of this connection, then (with the notation of Problem 6) $\boldsymbol{\omega}_1 = {}^t\pi_1(\boldsymbol{\omega})$. Show that the second-order structure function $c^{(1)}$ is constant, and generalize to structure functions of arbitrary order.

For every closed subgroup G of $\mathbf{GL}(n, \mathbf{R})$, the principal connection in $R(\mathbf{R}^n)$ which extends the canonical connection in $S_G(\mathbf{R}^n)$ (20.7.11) is the canonical connection. The Cartan connection corresponding to this in the bundle of affine frames $A(\mathbf{R}^n)$ (Section 20.6, Problem 19(c)) is identical with the canonical Cartan connection on the group $\mathbf{A}(n, \mathbf{R})$ relative to the subgroup $\mathbf{GL}(n, \mathbf{R})$ (Section 20.5, Problem 3).

8. Let M, M′ be two pure differential manifolds, and let $S_G(M)$ and $S_G(M')$ be G-structures on M, M′, respectively. A diffeomorphism f of M onto M′ is said (by abuse of language) to be an *isomorphism* of $S_G(M)$ onto $S_G(M')$ if the restriction $S_G(f)$ of $R(f)$ (Section 20.6, Problem 4) to $S_G(M)$ is a bijection of $S_G(M)$ onto $S_G(M')$.

(a) Let $F : S_G(M) \to S_G(M')$ be an isomorphism of principal bundles. In order that F should be of the form $S_G(f)$, where f is a diffeomorphism of M onto M′, it is necessary and sufficient that ${}^tF(\boldsymbol{\sigma}') = \boldsymbol{\sigma}$, where $\boldsymbol{\sigma}$ and $\boldsymbol{\sigma}'$ are the canonical 1-forms on $S_G(M)$ and $S_G(M')$, respectively.

(b) With the notation of Problem 6, show that if G is connected and if a diffeomorphism F of $S_G(M)$ onto $S_G(M')$ is an isomorphism of the $G^{(1)}$-structure $S_{G^{(1)}}(S_G(M))$ onto the $G^{(1)}$-structure $S_{G^{(1)}}(S_G(M'))$, then $F = S_G(f)$, where f is an isomorphism of M onto M′. (Use (a). If $Z_{\mathbf{u}}$ and $Z'_{\mathbf{u}}$ are the Killing fields on $S_G(M)$ and $S_G(M')$ corresponding to a vector $\mathbf{u} \in \mathfrak{g}_e$, begin by showing that F transforms $Z_{\mathbf{u}}$ into $Z'_{\mathbf{u}}$ for all $\mathbf{u} \in \mathfrak{g}_e$, and deduce that F is an isomorphism of principal bundles.)

9. Show that the group of automorphisms of a G-connection on a connected manifold M may be identified with a Lie group of dimension $\leq \dim(M) + \dim(G)$, and attains this maximum dimension only when M is the space $\mathbf{R}^n$ endowed with the canonical flat G-structure and the corresponding canonical G-connection (Problem 7). (Same method as in Section 20.6, Problem 7.)

10. With the notation of Problem 6, a G-structure $S_G(M)$ is said to be of *finite type* if there exists an index k such that $G^{(k)} = \{e\}$. Show that if a G-structure (for a connected group G) is of finite type, then its automorphism group is a Lie group. (Observe that to an $\{e\}$-structure there is intrinsically attached a principal connection, and use Problems 8 and 9.)

11. Let E, F be two finite-dimensional real vector spaces, let $\mathfrak{g}$ be a vector subspace of Hom(E, F), and let $\mathfrak{g}^{(1)}$ denote the subspace of Hom(E, $\mathfrak{g}$) consisting of all T such that $T(\mathbf{u}) \cdot \mathbf{v} = T(\mathbf{v}) \cdot \mathbf{u}$ for all $\mathbf{u}, \mathbf{v} \in$ E. For each integer $k > 1$ define $\mathfrak{g}^{(k)}$ inductively to be $(\mathfrak{g}^{(k-1)})^{(1)}$. The subspace $\mathfrak{g}$ is said to be of *finite type* if $\mathfrak{g}^{(k)} = \{0\}$ for some k. (Cf. Section 20.9, Problem 15.)

(a) If we identify Hom(E, F) with $E^* \otimes F$ (A.10.5.5), then Hom(E, $\mathfrak{g}$) is identified with a subspace of $E^* \otimes E^* \otimes F$, and $\mathfrak{g}^{(1)}$ is identified with the intersection $(E^* \otimes \mathfrak{g}) \cap (\mathbf{S}_2(E^*) \otimes F)$ (A.17). Deduce that, for each integer $k \geq 1$, $\mathfrak{g}^{(k)}$ is identified with the intersection

$$(\mathbf{S}_k(E^*) \otimes \mathfrak{g}) \cap (\mathbf{S}_{k+1}(E^*) \otimes F)$$

in the space $\mathbf{T}^{k+1}(E^*) \otimes F$.
(b) Deduce from (a) that if $\mathfrak{h}$ is a subspace of $\mathfrak{g}$, then $\mathfrak{h}^{(k)} \subset \mathfrak{g}^{(k)}$ for all k. If F is a subspace of a vector space F′, then $\mathfrak{g}^{(k)}$ is the same whether $\mathfrak{g}$ is regarded as a subspace of Hom(E, F) or of Hom(E, F′).
(c) If F = E and $\mathfrak{g}$ = End(E), then $\mathfrak{g}^{(k)} = \mathbf{S}_{k+1}(E^*) \otimes E$, and therefore $\mathfrak{g}$ is of infinite type. Deduce that every subspace of End(E) which contains an endomorphism of rank 1 is of infinite type. (Consider the subspace generated by such an endomorphism, and observe that it may be identified with End(**R**).)

12. Deduce from Problem 11 that if G = **GL**(n, **R**) or **SL**(n, **R**), all G-structures are of infinite type. Show that the same is true if G = **Sp**(Φ) is the symplectic group, leaving invariant a nondegenerate alternating bilinear form Φ on $\mathbf{R}^{2n}$. (Notice that Φ defines canonically an isomorphism of $E = \mathbf{R}^{2n}$ onto E*, and that under this isomorphism the Lie algebra $\mathfrak{g} = \mathfrak{sp}(\Phi)$ corresponds to $\mathbf{S}_2(E^*) \subset E^* \otimes E^* = \text{Hom}(E, E^*)$; deduce that $\mathfrak{g}^{(k)}$ may be identified with $\mathbf{S}_{k+2}(E^*)$.)

13. (a) Show that for G = **GL**(n, **R**) or **SL**(n, **R**), all G-structures are flat (Problem 7).
(b) If G is the subgroup of **GL**(n, **R**) which leaves invariant a subspace F of $\mathbf{R}^n$, and if $x \mapsto L_x$ is the corresponding field of directions on M (20.7.8), then a G-structure is flat if and only if the field $x \mapsto L_x$ is completely integrable. Show that this condition is equivalent to the vanishing of the first-order structure function. (Observe that $\partial(\text{Hom}(\mathbf{R}^n, \mathfrak{g}))$ is the kernel of the canonical mapping

$$\text{Hom}\left(\bigwedge^2 \mathbf{R}^n, \mathbf{R}^n\right) \to \text{Hom}\left(\bigwedge^2 F, \mathbf{R}^n/F\right)$$

which sends an alternating bilinear mapping $B : \mathbf{R}^n \times \mathbf{R}^n \to \mathbf{R}^n$ to the composition of the canonical mapping $\mathbf{R}^n \to \mathbf{R}^n/F$ with the restriction of B to F × F.)
(c) Take G to be the symplectic group **Sp**(n, **R**) ($n = 2m$). If

$$\Phi(\mathbf{x}, \mathbf{y}) = \sum_{j=1}^{m} (\xi_j \eta_{j+m} - \xi_{j+m} \eta_j),$$

so that a G-structure on M corresponds to a differential 2-form Ω on M such that $\Phi(r_x^{-1} \cdot \mathbf{h}_1, r_x^{-1} \cdot \mathbf{h}_2) = \langle \Omega(x), \mathbf{h}_1 \wedge \mathbf{h}_2 \rangle$ (20.7.6), show that ${}^t\pi(\Omega)$ (where $\pi : S_G(M) \to M$ is the canonical projection) is the 2-form $\Phi(\boldsymbol{\sigma}, \boldsymbol{\sigma})$ on $S_G(M)$ (16.20.15.8). Deduce that the following are equivalent: (i) the G-structure $S_G(M)$ is flat; (ii) $d\Omega = 0$; (iii) the first-order structure function is zero. (Observe that $d({}^t\pi(\Omega)) = \Phi(d\boldsymbol{\sigma}, \boldsymbol{\sigma}) - \Phi(\boldsymbol{\sigma}, d\boldsymbol{\sigma})$.) The G-structure is then called a *Hamiltonian* structure.
(d) Take G = **GL**(m, **C**), where $n = 2m$ (20.7.7). Then a G-structure is flat if and only if it comes from a complex-analytic structure on M.

When this is the case, the Nijenhuis torsion (Section **17.19**, Problem 3(f)) of the tensor $\boldsymbol{j}$ (**20.7.7**) is zero. (It can be shown that this necessary condition is also sufficient.) (Remark that if c is a chart of the *complex* manifold M, and if Z_k $(1 \leq k \leq m)$ are the holomorphic vector fields associated with this chart (**16.15.4.2**), then the fields Z_k and $\boldsymbol{j} \cdot Z_k$ are those associated with c considered as a chart of the underlying differential manifold M, and use the fact that the Nijenhuis torsion $\frac{1}{2}[\boldsymbol{u}, \boldsymbol{u}]$ is an M-morphism.)

14. Let $M = S_{n_1} \times S_{n_2} \times \cdots \times S_{n_r}$ be a product of spheres, where $r \geq 2$, $n_k \geq 1$ for all k and one of the n_k is *odd*. Show that M is *parallelizable* and hence that there exist G-structures on M for every Lie subgroup G of $\mathbf{GL}(n, \mathbf{R})$ (where $n = \dim(M)$). For this purpose, prove the following assertions:

(i) If $n = 2m - 1$ is odd, then S_n is embedded in $\mathbf{C}^m$ and there exists a vector field on S_n which vanishes nowhere, corresponding to the action on S_n of the group U of complex numbers of absolute value 1. Consequently, we have $T(S_n) = I \oplus E$, where I is the trivial line bundle (Section **16.15**, Problem 1).

(ii) For each n, $T(S_n) \oplus I$ is isomorphic to the trivial bundle $(n + 1)I$ (Section **16.20**, Problem 3).

15. Let M be a pure differential manifold of dimension n, and $R_2(M)$ be the principal bundle of frames of order 2 on M (Section **20.1**, Problem 3). For each Lie subgroup G of $G^2(n)$, a restriction $S_G^2(M)$ of the bundle $R_2(M)$ to the group G is called a *second-order G-structure* on M.

Consider in particular the case where $G = G^1(n) = \mathbf{GL}(n, \mathbf{R})$. Let $\boldsymbol{C}$ be a *torsion-free* linear connection on M, and (U, φ, n) a chart of M. Consider the local expression $(\mathbf{k}, \mathbf{u}) \mapsto (\mathbf{k}, -\Gamma_x(\mathbf{k}, \mathbf{u}))$ of $\boldsymbol{C}$, where for each point $x \in U$ the mapping Γ_x is a *symmetric* bilinear mapping of $\mathbf{R}^n \times \mathbf{R}^n$ into $\mathbf{R}^n$; and consider, for each $x \in U$, the set of elements of the fiber $R_2(M)_x$ whose local expression, relative to the chart (U, φ, n), is $(\varphi(x), V, -\Gamma_x \cdot (V, V))$, where V runs through $\mathbf{GL}(n, \mathbf{R})$ (Section **20.6**, Problem 19). Show that this set is independent of the choice of chart, and that the union of these sets is a *second-order* $\mathbf{GL}(n, \mathbf{R})$-*structure* on M. In this way we obtain a *canonical one-to-one correspondence* between torsion-free linear connections on M and second-order $\mathbf{GL}(n, \mathbf{R})$-structures on M. For each of these structures $S^2_{\mathbf{GL}(n, \mathbf{R})}(M)$, there exists a canonical M-isomorphism $h : R(M) \to S^2_{\mathbf{GL}(n, \mathbf{R})}(M)$ of principal bundles whose local expression, relative to a chart (U, φ, n) as above, is

$$(\varphi(x), V) \mapsto (\varphi(x), V, -\Gamma_x \cdot (V, V)).$$

Show that ${}^t h\left(\sum_i \sigma^i \mathbf{e}_i\right)$ is the canonical form and ${}^t h\left(\sum_{i,j} \sigma^i_j E_{ij}\right)$ the connection form of the principal connection $\boldsymbol{P}$ corresponding to the linear connection $\boldsymbol{C}$ (in the notation of Section **20.6**, Problem 19).

If H is a Lie subgroup of $G^2(n)$ containing $G^1(n) = \mathbf{GL}(n, \mathbf{R})$, every second-order $\mathbf{GL}(n, \mathbf{R})$-structure is contained in a unique second-order H-structure.

16. The projective group $\mathbf{PGL}(n + 1, \mathbf{R})$ is the quotient of $\mathbf{GL}(n + 1, \mathbf{R})$ by its center Z, consisting of the scalar multiples of the identity. This group acts transitively on the projective space $\mathbf{P}_n(\mathbf{R})$. If H_0 is the stabilizer of the point $x_0 \in \mathbf{P}_n(\mathbf{R})$ with homogeneous coordinates $(1, 0, \ldots, 0)$, there exists a neighborhood V of the identity element of H_0 and a neighborhood W of x_0 in $\mathbf{R}^n$ (identified with the subspace of

$\mathbf{P}_n(\mathbf{R})$ consisting of all points with first homogeneous coordinate equal to 1) such that the action of V on W may be written in the form

(1) $$((U, \mathbf{b}^*), \mathbf{x}) \mapsto (1 + \langle \mathbf{b}^*, \mathbf{x} \rangle)^{-1}(U \cdot \mathbf{x}),$$

V being identified with a neighborhood of $(I, 0)$ in $\mathbf{M}_n(\mathbf{R}) \times (\mathbf{R}^n)^*$. The mapping which sends (1) to its jet of order 2 at the point $((I, 0), 0)$ defines an isomorphism of H_0 onto the subgroup H of $\mathrm{G}^2(n)$ consisting of all pairs $(U, \mathbf{B})$, where $\mathbf{B} : \mathbf{R}^n \times \mathbf{R}^n \to \mathbf{R}^n$ is a bilinear mapping of the form $(\mathbf{h}, \mathbf{k}) \mapsto \langle \mathbf{b}^*, \mathbf{h} \rangle \mathbf{k} + \langle \mathbf{b}^*, \mathbf{k} \rangle \mathbf{h}$ for some linear form $\mathbf{b}^*$ on $\mathbf{R}^n$. By abuse of language, a *second-order* H-*structure* on M (Problem 15) is called a *projective structure* on the differential manifold M.

Deduce from Problem 15 that there exist projective structures on every differential manifold M, and, more precisely, that to each torsion-free linear connection on M there corresponds a well-determined projective structure on M. The projective structures corresponding to two torsion-free linear connections $\mathbf{C}$, $\mathbf{C}'$ on M are the same if and only if the difference $\mathrm{B} = \mathbf{C} - \mathbf{C}'$ (17.16.6) is a bilinear morphism such that

$$\mathrm{B}_x(\mathbf{h}, \mathbf{k}) = \langle \omega(x), \mathbf{h} \rangle \mathbf{k} + \langle \omega(x), \mathbf{k} \rangle \mathbf{h}$$

for all $x \in \mathrm{M}$, where ω is a C^∞ differential 1-form on M.

8. GENERALITIES ON PSEUDO-RIEMANNIAN MANIFOLDS

(20.8.1) In this section we shall suppose that M is a pure differential manifold of dimension n, endowed with a pseudo-Riemannian structure of signature (p, q) (with $p + q = n$); M is said to be a *pseudo-Riemannian manifold* of signature (p, q), and a *Riemannian manifold* when $p = n$ and $q = 0$. The symmetric covariant tensor field $\mathbf{g}$ which defines the pseudo-Riemannian structure of M is called the *pseudo-Riemannian* (or *Riemannian* when $p = n$) *metric tensor*. A *real-analytic manifold* M endowed with a pseudo-Riemannian metric tensor $\mathbf{g}$ is called an *analytic pseudo-Riemannian manifold* if the tensor field $\mathbf{g}$ is *analytic*. In the same way we may define the notion of a *complex-analytic Riemannian manifold* (but in this case the signature is no longer meaningful).

Let u be a local diffeomorphism of a pure manifold M into a pure manifold M_1. If M_1 carries a pseudo-Riemannian metric tensor $\mathbf{g}_1$, then the inverse image (16.20.9) ${}^t u(\mathbf{g}_1)$ is a pseudo-Riemannian metric tensor of the same signature on M. If M is endowed with a pseudo-Riemannian metric tensor $\mathbf{g}$, the mapping u is said to be a *local isometry* of the pseudo-Riemannian manifold M into the pseudo-Riemannian manifold M_1 if ${}^t u(\mathbf{g}_1) = \mathbf{g}$; if moreover u is a diffeomorphism of M onto M_1, then u is said to be an *isometry* of M onto M_1. If there exists a function $m(x) > 0$ on M such that ${}^t u(\mathbf{g}_1)(x) = m(x)\mathbf{g}(x)$ for all $x \in \mathrm{M}$, then u is said to be a *conformal* mapping of M into M_1.

Let M, M_1 be two pseudo-Riemannian manifolds and let $a \in M$, $a_1 \in M_1$. We say that M and M_1 are *locally isometric* at the points a and a_1 if there exists an open neighborhood V (resp. V_1) of a (resp. a_1) and an isometry of V onto V_1 which sends a to a_1.

If M is a *covering* of M_1 and $\pi : M \to M_1$ is the canonical projection, then for any pseudo-Riemannian metric tensor $\mathbf{g}_1$ on M_1, the pseudo-Riemannian metric tensor $\mathbf{g} = {}^t\pi(\mathbf{g}_1)$ on M is said to be *canonically induced* by $\mathbf{g}_1$, and the manifold M, endowed with $\mathbf{g}$, is said to be a *pseudo-Riemannian covering* of M_1 (a *Riemannian covering* if $p = n$).

In particular, if M is the *universal covering* of M_1, then whenever we refer to M as a Riemannian manifold it is always the canonically induced metric tensor on M that is meant. If the fundamental group $\pi_1(M_1)$ is considered as acting on the left on M (16.29.2), it is then a group of *isometries* of M onto itself.

(20.8.2) Let M be a pseudo-Riemannian manifold and $\mathbf{g}$ its metric tensor. Where there is no risk of confusion, for each pair of tangent vectors $\mathbf{h}_x$, $\mathbf{k}_x$ in $T_x(M)$, we shall write

(20.8.2.1) $$(\mathbf{h}_x | \mathbf{k}_x)_{\mathbf{g}} = \langle \mathbf{g}(x), \mathbf{h}_x \otimes \mathbf{k}_x \rangle.$$

This is called the *scalar product* (relative to $\mathbf{g}$) of $\mathbf{h}_x$ and $\mathbf{k}_x$, and is also denoted simply by $(\mathbf{h}_x | \mathbf{k}_x)$. When the metric tensor is *Riemannian* we also put

(20.8.2.2) $$\|\mathbf{h}_x\| = (\mathbf{h}_x | \mathbf{h}_x)^{1/2},$$

which is called the *length* of the tangent vector $\mathbf{h}_x$.

Any angle φ such that $\cos \varphi = (\mathbf{h}_x | \mathbf{k}_x)/(\|\mathbf{h}_x\| \cdot \|\mathbf{k}_x\|)$ is called an *angle between the tangent vectors* $\mathbf{h}_x$, $\mathbf{k}_x$.

For two vector fields X, Y on M, the function $x \mapsto (X(x) | Y(x))$ is denoted by $(X | Y)$ (and $(X | X)$ is denoted by $\|X\|^2$). A vector field X is said to be a *unit vector field* if $\|X\| = 1$.

(20.8.2.3) Let $\mathbf{R} = (\mathbf{e}_1, \mathbf{e}_2, \ldots, \mathbf{e}_n)$ be a moving frame on M, and put

$$g_{ij} = \langle \mathbf{g}, \mathbf{e}_i \otimes \mathbf{e}_j \rangle = \langle \Phi, \mathbf{R}^{-1} \cdot \mathbf{e}_i \otimes \mathbf{R}^{-1} \cdot \mathbf{e}_j \rangle,$$

where Φ is the symmetric bilinear form on $\mathbf{R}^n$ defining the pseudo-Riemannian structure of M (20.7.6). If for each $x \in M$ the basis of $T_x(M)^*$ dual to $(\mathbf{e}_1(x), \ldots, \mathbf{e}_n(x))$ is $(\sigma_1(x), \ldots, \sigma_n(x))$, then the σ_j are the canonical 1-forms corresponding to $\mathbf{R}$ (20.6.5), and we may write

(20.8.2.4) $$\mathbf{g} = \sum_{i,j} g_{ij} \sigma_i \otimes \sigma_j .$$

In particular, if $\mathbf{R}$ is the frame $(X_i)_{1 \leqq i \leqq n}$ associated with a chart (U, φ, n) of M (16.15.4), then $\sigma_j = d\varphi^j$ and therefore

(20.8.2.5) $$\mathbf{g} = \sum_{i,j} g_{ij}\, d\varphi^i \otimes d\varphi^j.$$

Conversely, the assignment of a C^∞ function $x \mapsto (g_{ij}(x))$ on U, with values in the space of matrices of symmetric bilinear forms of signature $(p, n-p)$, defines by (20.8.2.5) a pseudo-Riemannian structure on U.

For each point $x_0 \in \mathrm{M}$, there exists an open neighborhood U of x_0 and a *moving frame* $(\mathbf{e}_1, \mathbf{e}_2, \ldots, \mathbf{e}_n)$ (20.6.5) defined on U such that

(20.8.2.6) $$\begin{cases} (\mathbf{e}_i(x) \mid \mathbf{e}_j(x)) = 0 & \text{for } i \neq j, \\ (\mathbf{e}_i(x) \mid \mathbf{e}_i(x)) = \varepsilon_i, & \text{where } \varepsilon_i = 1 \text{ for } 1 \leqq i \leqq p, \\ & \qquad \varepsilon_i = -1 \text{ for } p+1 \leqq i \leqq p+q = n. \end{cases}$$

First, there exist n vectors $\mathbf{c}_i \in \mathrm{T}_{x_0}(\mathrm{M})$ satisfying the conditions (20.8.2.6); next, there exist n C^∞ vector fields $\boldsymbol{a}_i$ defined on a neighborhood U_0 of x_0, such that $\boldsymbol{a}_i(x_0) = \mathbf{c}_i$ for $1 \leqq i \leqq n$ (16.12.11). There exists a neighborhood $\mathrm{U}_1 \subset \mathrm{U}_0$ of x_0 such that $(\boldsymbol{a}_1(x) \mid \boldsymbol{a}_1(x))$ has the sign of ε_1 for all $x \in \mathrm{U}_1$. Put $\mathbf{e}_1(x) = \boldsymbol{a}_1(x)/|(\boldsymbol{a}_1(x) \mid \boldsymbol{a}_1(x))|^{1/2}$, so that $(\mathbf{e}_1(x) \mid \mathbf{e}_1(x)) = \varepsilon_1$ for $x \in \mathrm{U}_1$. We now proceed by induction, assuming that $\mathbf{e}_1, \ldots, \mathbf{e}_k$ are defined on a neighborhood $\mathrm{U}_k \subset \mathrm{U}_0$ of x_0 and satisfy the conditions (20.8.2.6) for $i \leqq k$ and $j \leqq k$, and such that $\mathbf{e}_i(x_0) = \mathbf{c}_i$ for $i \leqq k$. Now form the vector field on U_k

$$\boldsymbol{b}_{k+1} = \boldsymbol{a}_{k+1} - \sum_{j=1}^{k} \varepsilon_j (\boldsymbol{a}_{k+1} \mid \mathbf{e}_j)\mathbf{e}_j;$$

we have $(\boldsymbol{b}_{k+1} \mid \mathbf{e}_j) = 0$ for $j \leqq k$, and $\boldsymbol{b}_{k+1}(x_0) = \mathbf{c}_{k+1}$. There exists a neighborhood $\mathrm{U}_{k+1} \subset \mathrm{U}_k$ of x_0 in which $(\boldsymbol{b}_{k+1}(x) \mid \boldsymbol{b}_{k+1}(x))$ has the same sign as ε_{k+1}; by putting $\mathbf{e}_{k+1}(x) = \boldsymbol{b}_{k+1}(x)/|(\boldsymbol{b}_{k+1}(x) \mid \boldsymbol{b}_{k+1}(x))|^{1/2}$, the induction step is completed, and when $k = n$ gives us the required moving frame.

The formula (20.8.2.4) therefore now takes the form

(20.8.2.7) $$\mathbf{g} = \sum_i \varepsilon_i \sigma_i \otimes \sigma_i.$$

When M is a Riemannian manifold, so that $\varepsilon_i = 1$ for each i, $(\mathbf{e}_1, \ldots, \mathbf{e}_n)$ is called an *orthonormal moving frame.*

(20.8.3) To the covariant tensor $\mathbf{g}(x)$ of order 2 there corresponds canonically an isomorphism G_x of the tangent vector space $\mathrm{T}_x(\mathrm{M})$ onto its dual $\mathrm{T}_x(\mathrm{M})^*$, such that

(20.8.3.1) $$(\mathbf{h}_x \mid \mathbf{k}_x) = \langle G_x \cdot \mathbf{h}_x, \mathbf{k}_x \rangle.$$

One sees immediately that $G : \mathbf{h}_x \mapsto G_x \cdot \mathbf{h}_x$ is an M-*isomorphism* of the vector bundle T(M) onto the vector bundle T(M)*. This isomorphism and its inverse enable us to define isomorphisms of *tensor bundles* on M: if $r \geqq 1$, $s \geqq 1$ and if $1 \leqq i \leqq r$, $1 \leqq j \leqq s$, we define an isomorphism G_j^i of $\mathbf{T}_s^r(M)$ onto $\mathbf{T}_{s+1}^{r-1}(M)$ by the formula

$$G_j^i(\mathbf{h}_1^* \otimes \cdots \otimes \mathbf{h}_s^* \otimes \mathbf{k}_1 \otimes \cdots \otimes \mathbf{k}_r)$$
$$= \mathbf{h}_1^* \otimes \cdots \otimes \mathbf{h}_{j-1}^* \otimes (G \cdot \mathbf{k}_i) \otimes \cdots \otimes \mathbf{h}_s^* \otimes \mathbf{k}_1 \otimes \cdots \otimes \widehat{\mathbf{k}_i} \otimes \cdots \otimes \mathbf{k}_r.$$

G_j^i (resp. $(G_j^i)^{-1}$) is referred to as *lowering the* ith *contravariant index to the* jth *place* (resp. *raising the* jth *covariant index to the* ith *place*).

These isomorphisms give rise canonically to isomorphisms of the corresponding spaces of tensor fields. In particular, if f is a C^∞ real-valued function on M, so that df is a differential 1-form on M, then $G^{-1} \cdot df$ is a *vector field* on M, called the *gradient* of f and denoted by grad(f). We have therefore, for any C^∞ vector field X on M

(20.8.3.2) $$(\operatorname{grad}(f) \mid X) = \langle df, X \rangle = \theta_X \cdot f.$$

(20.8.4) For each $x \in M$, the nondegenerate symmetric bilinear form $\mathbf{g}(x)$ on $T_x(M)$ induces canonically a nondegenerate symmetric bilinear form $\mathbf{g}_r(x)$ on the space of tangent r-vectors $\bigwedge^r T_x(M)$, defined by

(20.8.4.1) $$\langle \mathbf{g}_r(x), (\mathbf{h}_1 \wedge \cdots \wedge \mathbf{h}_r) \otimes (\mathbf{k}_1 \wedge \cdots \wedge \mathbf{k}_r) \rangle = \det((\mathbf{h}_i \mid \mathbf{k}_j)_{\mathbf{g}})$$

for all $\mathbf{h}_i, \mathbf{k}_j \in T_x(M)$. This form is positive definite if $\mathbf{g}(x)$ is positive definite; in this case, we put

(20.8.4.2) $$\|\mathbf{h}_1 \wedge \cdots \wedge \mathbf{h}_r\| = (\det(\mathbf{h}_i \mid \mathbf{h}_j))^{1/2}.$$

This nonnegative real number is called the *r-dimensional area* of the tangent r-vector $\mathbf{h}_1 \wedge \mathbf{h}_2 \wedge \cdots \wedge \mathbf{h}_r$. (For $r = 2$, it is called the *area* of $\mathbf{h}_1 \wedge \mathbf{h}_2$.)

Next, $\bigwedge^r G^{-1}$ is an isomorphism of $\bigwedge^r T(M)^*$ onto $\bigwedge^r T(M)$. Hence, for each $x \in M$, we may derive from the form $\mathbf{g}_r(x)$ on $\bigwedge^r T_x(M)$ a nondegenerate symmetric bilinear form $\mathbf{g}_r^*(x)$ on the space of tangent r-covectors $\bigwedge^r T_x(M)^*$, as follows:

(20.8.4.3) $$\mathbf{g}_r^*(\alpha, \beta) = \mathbf{g}_r\left(\left(\bigwedge^r G^{-1}\right) \cdot \alpha, \left(\bigwedge^r G^{-1}\right) \cdot \beta\right)$$

for any two differential r-forms α, β on M. If $\mathbf{g}(x)$ is positive definite, then so is $\mathbf{g}_r^*(x)$.

(20.8.5) When $r = n$, the n-dimensional area is also called the *volume* of a tangent n-vector at x to a *Riemannian* manifold. If we denote the volume by $V_x(\mathbf{h}_1 \wedge \mathbf{h}_2 \wedge \cdots \wedge \mathbf{h}_n)$, then we have

$$V_x(\lambda(\mathbf{h}_1 \wedge \cdots \wedge \mathbf{h}_n)) = |\lambda| V_x(\mathbf{h}_1 \wedge \cdots \wedge \mathbf{h}_n)$$

for all $\lambda \in \mathbf{R}$. From this we shall deduce the existence of a *positive Lebesgue measure* (16.22) on M canonically associated with the Riemannian structure of M, called the *Riemannian volume* on M and denoted by $v_{\mathbf{g}}$ or v or $\mathrm{vol}_{\mathbf{g}}$ or vol. (When $n = 1$ (resp. 2), it is called *length* (resp. *area*).)

It is clearly sufficient (13.1.9) to define this measure on the domain of an arbitrary *chart* (U, φ, n) of M, and to show that it is independent of the chart chosen. Let $f \in \mathscr{K}(U)$; then the local expression $f \circ \varphi^{-1}$ is a function belonging to $\mathscr{K}(\varphi(U))$. The local expression of the Riemannian metric tensor $\mathbf{g}$ is $x \mapsto (x, \Phi_x)$, where $x \mapsto \Phi_x$ is a C^∞ mapping of $\varphi(U)$ into the space $\mathscr{H}_{n,0}(\mathbf{R}^n)$ of positive definite symmetric bilinear forms on $\mathbf{R}^n$ (16.11.1). By definition (20.8.4), $V_x(\mathbf{e}_1 \wedge \mathbf{e}_2 \wedge \cdots \wedge \mathbf{e}_n) = (g(x))^{1/2}$, where $g(x) > 0$ is the *discriminant* of the form Φ_x relative to the canonical basis of $\mathbf{R}^n$; since this function is of class C^∞ on $\varphi(U)$, it is clear that the linear form

(20.8.5.1)
$$f \mapsto \int_{\varphi(U)} f(\varphi^{-1}(x))(g(x))^{1/2}\, d\lambda(x)$$

(λ being Lebesgue measure) is a *positive Lebesgue measure* on U. Consider now another chart (U, ψ, n) with the same domain of definition, and let $u = \varphi \circ \psi^{-1} : \psi(U) \to \varphi(U)$ be the transition diffeomorphism. The local expression of $\mathbf{g}$ relative to this second chart is $x \mapsto (x, \Psi_x)$, where the form Ψ_x is given by

$$\Psi_x(\mathbf{h}, \mathbf{k}) = \Phi_{u(x)}(Du(x) \cdot \mathbf{h}, Du(x) \cdot \mathbf{k})$$

for all $x \in \psi(U)$. If $J(x)$ is the *Jacobian* of u at the point x (8.10), the discriminant $g_1(x)$ of Ψ_x is given by

$$g_1(x) = J(x)^2 g(u(x)).$$

Since the measure on U corresponding to the chart (U, ψ, n) is given by

$$f \mapsto \int_{\psi(U)} f(\psi^{-1}(x))(g_1(x))^{1/2}\, d\lambda(x),$$

and since this integral can be written in the form

$$\int_{\psi(U)} f(\varphi^{-1}(u(x)))\, |J(x)|\, g(u(x))^{1/2}\, d\lambda(x),$$

it follows from the formula for change of variables (16.22.1) that it is equal to the right-hand side of (20.8.5.1), as required.

(20.8.6) Suppose now that the Riemannian manifold M is *oriented*, and let υ_0 be a differential n-form on M belonging to the orientation of M (16.21.2). We can then associate with $\mathbf{g}$ a *positive n-form* $\upsilon_{\mathbf{g}}$ on M such that

(20.8.6.1) $$|\langle \upsilon_{\mathbf{g}}(x), \mathbf{h}_1 \wedge \cdots \wedge \mathbf{h}_n \rangle| = \mathrm{V}_x(\mathbf{h}_1 \wedge \cdots \wedge \mathbf{h}_n).$$

For if the local expression of υ_0, relative to a chart (U, φ, n) of M, is

$$\upsilon_0(\varphi^{-1}(x)) = w(\xi^1, \ldots, \xi^n)\, d\xi^1 \wedge d\xi^2 \wedge \cdots \wedge d\xi^n,$$

we put

(20.8.6.2) $$\upsilon_{\mathbf{g}}(\varphi^{-1}(x)) = \pm\,(g(x))^{1/2}\, d\xi^1 \wedge d\xi^2 \wedge \cdots \wedge d\xi^n,$$

the sign being that of the function w. It is immediately verified that this form does not depend on the choice of local expression. The form $\upsilon_{\mathbf{g}}$ (also denoted by υ) is called the *canonical volume form* on the oriented Riemannian manifold M. For each $v_{\mathbf{g}}$-integrable function f on M, we have therefore

(20.8.6.3) $$\int f\, dv_{\mathbf{g}} = \int f\upsilon_{\mathbf{g}}.$$

Let $\mathbf{R} = (\mathbf{e}_1, \ldots, \mathbf{e}_n)$ be a direct orthonormal moving frame (20.8.2) on an open set U in M, and let σ_j $(1 \leqq j \leqq n)$ be the corresponding canonical forms. Then the form $\upsilon_{\mathbf{g}}$ is given in U by

(20.8.6.4) $$\upsilon_{\mathbf{g}} = \sigma_1 \wedge \sigma_2 \wedge \cdots \wedge \sigma_n.$$

For at each point $x \in \mathrm{U}$ we can choose a chart such that the covectors $\sigma_j(x)$ have for local expressions the values of the $d\xi^j$ at the point corresponding to x; the formula (20.8.6.4) then follows from (20.8.6.2) and the fact that the discriminant of $\mathbf{g}(x)$ relative to the basis $(\mathbf{e}_j(x))$ is equal to 1.

(20.8.7) Let M be an oriented Riemannian manifold of dimension n, and let υ be the canonical volume form on M. From exterior algebra it follows that, for each differential r-form α on M, there exists a differential $(n-r)$-form on M, denoted by $*\alpha$, such that

(20.8.7.1) $$\beta \wedge (*\alpha) = \mathbf{g}_r^*(\alpha, \beta)\upsilon$$

for all differential r-forms β (A.15.3). The form $*\alpha$ is called the *adjoint* of the form α. We have the following formulas:

(20.8.7.2)
$$*1 = \upsilon,$$

(20.8.7.3)
$$*(*\alpha) = (-1)^{r(n-r)}\alpha,$$

(20.8.7.4)
$$\mathbf{g}_r^*(\alpha, \beta) = \mathbf{g}_{n-r}^*(*\alpha, *\beta).$$

PROBLEMS

1. Let M_1, M_2 be two pseudo-Riemannian manifolds and $\mathbf{g}_1$, $\mathbf{g}_2$ their respective metric tensors. On the product manifold $M = M_1 \times M_2$, show that $\mathbf{g} = {}^t\mathrm{pr}_1(\mathbf{g}_1) + {}^t\mathrm{pr}_2(\mathbf{g}_2)$ is a pseudo-Riemannian metric tensor of signature $(p_1 + p_2, q_1 + q_2)$, where (p_j, q_j) is the signature of $\mathbf{g}_j$ $(j = 1, 2)$. This metric tensor is called the *product* of $\mathbf{g}_1$ and $\mathbf{g}_2$, and the manifold M endowed with this metric tensor is called the *product of the pseudo-Riemannian manifolds* M_1, M_2. At each point $(x_1, x_2) \in M_1 \times M_2$, the tangent spaces to the submanifolds $M_1 \times \{x_2\}$ and $\{x_1\} \times M_2$ are totally orthogonal. If M_1, M_2 are Riemannian manifolds, the Riemannian volume $v_{\mathbf{g}}$ is equal to $v_{\mathbf{g}_1} \otimes v_{\mathbf{g}_2}$. If N_1 (resp. N_2) is a pseudo-Riemannian covering of M_1 (resp. M_2), then the product $N_1 \times N_2$ is a pseudo-Riemannian covering of $M_1 \times M_2$.

2. Let M be a compact Riemannian manifold, N a Riemannian k-sheeted covering of M, where k is finite. Show that the volume of N is k times that of M.

3. Let M be a pseudo-Riemannian manifold, $(\mathbf{h} | \mathbf{k})$ the corresponding scalar product in T(M). Let $E(\mathbf{h})$ denote $(\mathbf{h} | \mathbf{h})$.

(a) Show that for every vector $\mathbf{h} \in T(M)$ and every *vertical* tangent vector $\mathbf{v} \in T_{\mathbf{h}}(T(M))$, we have
$$\theta_{\mathbf{v}} \cdot E = 2(\mathbf{h} | \tau_{\mathbf{h}}(\mathbf{v})).$$

(b) With the notation of (20.8.3), let α_M or α denote the differential 1-form ${}^t G(\kappa_M)$, κ_M being the canonical 1-form on $T(M)^*$ (16.20.6). Show that for every vector $\mathbf{h} \in T(M)$ and every tangent vector $\mathbf{k_h} \in T_{\mathbf{h}}(T(M))$, we have
$$\langle \alpha(\mathbf{h}), \mathbf{k_h} \rangle = (\mathbf{h} | T(o_M) \cdot \mathbf{k_h}).$$

(c) Show that the differential 2-form $d\alpha$ on T(M) is nondegenerate. For each vector $\mathbf{h} \in T(M)$ and each pair of vectors $\mathbf{k_h}$, $\mathbf{v_h} \in T_{\mathbf{h}}(T(M))$ with $\mathbf{v_h}$ vertical, we have
$$\langle d\alpha(\mathbf{h}), \mathbf{k_h} \wedge \mathbf{v_h} \rangle = -(T(o_M) \cdot \mathbf{k_h} | \tau_{\mathbf{h}}(\mathbf{v_h})).$$

(d) Relative to a moving frame satisfying (20.8.2.6), let $\tilde{\sigma}_i$ denote the *function* $\mathbf{h}_x \mapsto \langle \sigma_i(x), \mathbf{h}_x \rangle$ on T(M). Show that

$$\alpha = \sum_i \varepsilon_i \tilde{\sigma}_i \cdot ({}^t\mathrm{T}(o_{\mathrm{M}})(\sigma_i)).$$

Let σ_i^* denote ${}^t\mathrm{T}(o_{\mathrm{M}})(\sigma_i)$, for brevity. Then the forms $d\tilde{\sigma}_i$ and σ_i^* $(1 \leqq i \leqq n)$ form a basis for the module of differential 1-forms on T(U), where U is the open set on which the moving frame is defined, and we have

$$d\alpha = \sum_i \varepsilon_i (d\tilde{\sigma}_i) \wedge \sigma_i^* + \sum_i \varepsilon_i \tilde{\sigma}_i \cdot (d\sigma_i^*).$$

4. (a) Let M be a pure manifold of dimension n and let $f : \mathrm{M} \to \mathbf{R}$ be a submersion. Suppose M to be endowed with a Riemannian structure (20.7.13). For each $t \in \mathbf{R}$, $f^{-1}(t)$ is a submanifold of M of dimension $n-1$. For each point $x \in f^{-1}(t)$, let $X(x)$ denote the tangent vector to M which is orthogonal in $\mathrm{T}_x(\mathrm{M})$ to the tangent space $\mathrm{Ker}(\mathrm{T}_x(f))$ to $f^{-1}(t)$ and is such that $\mathrm{T}_x(f) \cdot X(x) = E(t)$, in the notation of (18.1). Show that if f is *proper* (17.3.7), the integral curves of the field X are defined on the whole of **R**. (Use (18.2.10).) Deduce that for each $t \in \mathbf{R}$ the mapping $x \mapsto \mathrm{F}_X(x, t)$ is a diffeomorphism of $f^{-1}(0)$ onto $f^{-1}(t)$.
(b) Deduce from (a) that if $f : \mathrm{X} \to \mathrm{Y}$ is a *proper submersion* and if Y is connected, then $(\mathrm{X}, \mathrm{Y}, f)$ is a *fibration* (16.12.1) (*Ehresmann's theorem*). (Induction on dim Y.)
(c) Show that every submersion $f : \mathrm{X} \to \mathrm{Y}$, such that every fiber $f^{-1}(y)$ is *compact* and *connected*, is proper (and therefore a fibration if Y is connected). (Show by contradiction that for each compact neighborhood V of $f^{-1}(y_0)$ in X there exists a neighborhood W of y_0 in Y such that $f^{-1}(\mathrm{W}) \subset \mathrm{V}$, by considering the intersections of the fibers $f^{-1}(y)$ for $y \in \mathrm{W}$ with the frontier of V in X.)

5. Let M be a differential manifold, f a C^∞ real-valued function on M, and $a < b$ two real numbers. Suppose that the inverse image $f^{-1}([a, b])$ is compact and contains no critical points of f. Endow M with a structure of a Riemannian manifold; then the hypothesis on f signifies that $(\mathrm{grad}\, f)(x) \neq 0$ for $x \in f^{-1}([a, b])$. Then there exists a C^∞ vector field X on M with compact support, such that $X = \rho \cdot \mathrm{grad}(f)$, where ρ is a C^∞ function, everywhere $\geqq 0$, which is chosen so that $\|X\| \leqq 1$ throughout M and $\|X\| = 1$ throughout $f^{-1}([a, b])$. If we now put $h_t(x) = \mathrm{F}_X(x, t)$, the h_t form a one-parameter group of diffeomorphisms of M (18.2.11). For each $t \in \mathbf{R}$, let M_t denote the open set of all $x \in \mathrm{M}$ such that $f(x) < t$; show that for $a < c < d < b$, h_{d-c} is a diffeomorphism of M_c onto M_d. (Consider the function $t \mapsto f(h_t(x))$.) Deduce that there exists a C^∞ isotopy (16.26.7) of the identity mapping of M_c onto h_{d-c}.

6. Let M be a pure *compact* differential manifold of dimension n, and let f be a C^∞ real-valued function on M which has exactly *two* critical points in M, both of which are nondegenerate (Section 16.5, Problem 4). Show that M is homeomorphic to $\mathbf{S}_n$ (*Reeb's theorem*). (The two critical points are the point at which f attains its minimum a, and the point at which f attains its maximum b, and we have $a < b$. Using Section 16.5, Problem 4, show that M_c (in the notation of Problem 5 above) is homeomorphic to $\mathbf{R}^n$ for $a < c < b$ and c sufficiently close to a. Then use Problem 5 and Morton Brown's theorem (Section 16.2, Problem 5).)

9. THE LEVI–CIVITA CONNECTION

(20.9.1) Let M be a pure differential manifold of dimension n, and $\mathbf{g}$ a pseudo-Riemannian metric tensor of signature (p, q). Let Φ be a nondegenerate symmetric bilinear form of signature (p, q) on $\mathbf{R}^n$, and let $\mathbf{O}(\Phi) = \mathrm{G}$ be the subgroup of $\mathbf{GL}(n, \mathbf{R})$ which leaves Φ invariant. We recall that G is a Lie group of dimension $\frac{1}{2}n(n-1)$ (16.11.1). If we write $(\mathbf{u}|\mathbf{v}) = \Phi(\mathbf{u}, \mathbf{v})$, then $(\mathbf{u}|\mathbf{v}) = (\mathbf{v}|\mathbf{u})$, and the Lie algebra $\mathfrak{g}_e$ of G is the set of endomorphisms S of $\mathbf{R}^n$ such that

(20.9.1.1) $$(S \cdot \mathbf{u}|\mathbf{v}) + (\mathbf{u}|S \cdot \mathbf{v}) = 0$$

identically in $\mathbf{u}, \mathbf{v} \in \mathbf{R}^n$ (19.4.3.2). We have seen in (20.7.6) that there corresponds canonically to the pseudo-Riemannian metric tensor $\mathbf{g}$ a principal bundle $\mathrm{S_G(M)}$ over M with G as group. We propose now to determine *all principal connections* in $\mathrm{S_G(M)}$. We begin with two lemmas.

(20.9.2.) *Let T be a linear mapping of $\mathbf{R}^n$ into $\mathfrak{g}_e \subset \mathrm{End}(\mathbf{R}^n)$ such that*

(20.9.2.1) $$(T \cdot \mathbf{u}) \cdot \mathbf{v} = (T \cdot \mathbf{v}) \cdot \mathbf{u}$$

for all $\mathbf{u}, \mathbf{v} \in \mathbf{R}^n$. Then $T = 0$.

Consider the scalar product $((T \cdot \mathbf{u}) \cdot \mathbf{v}|\mathbf{w})$ for an arbitrary $\mathbf{w} \in \mathbf{R}^n$. By virtue of the symmetry of $(\mathbf{u}|\mathbf{v})$, the hypothesis (20.9.2.1), and the relation (20.9.1.1), in which we may replace S by the image under T of any vector in $\mathbf{R}^n$, we calculate:

$$\begin{aligned}((T \cdot \mathbf{u}) \cdot \mathbf{v}|\mathbf{w}) &= ((T \cdot \mathbf{v}) \cdot \mathbf{u}|\mathbf{w}) = -(\mathbf{u}|(T \cdot \mathbf{v}) \cdot \mathbf{w}) \\ &= -(\mathbf{u}|(T \cdot \mathbf{w}) \cdot \mathbf{v}) = ((T \cdot \mathbf{w}) \cdot \mathbf{u}|\mathbf{v}) = ((T \cdot \mathbf{u}) \cdot \mathbf{w}|\mathbf{v}) \\ &= (\mathbf{v}|(T \cdot \mathbf{u}) \cdot \mathbf{w}) = -((T \cdot \mathbf{u}) \cdot \mathbf{v}|\mathbf{w})\end{aligned}$$

and therefore $((T \cdot \mathbf{u}) \cdot \mathbf{v}|\mathbf{w}) = 0$. Since Φ is nondegenerate, it follows that $(T \cdot \mathbf{u}) \cdot \mathbf{v} = 0$ for all $\mathbf{u}, \mathbf{v}$ and hence $T = 0$.

(20.9.3) *Let $\boldsymbol{\sigma}$ be the canonical form on* $\mathrm{S_G(M)}$ (20.6.2). *With the notation of* (20.6.3), *the mapping*

(20.9.3.1) $$\boldsymbol{\varpi} \mapsto \boldsymbol{\varpi} \wedge \boldsymbol{\sigma}$$

is a bijection of the vector space of vector-valued 1*-forms* $\boldsymbol{\varpi}$ *on* $S_G(M)$ *with values in* $\mathfrak{g}_e$ *which are horizontal and* G*-invariant* (*for the action*

$$(s, \mathbf{u}) \mapsto \mathrm{Ad}(s^{-1}) \cdot \mathbf{u}$$

of G *on* $\mathfrak{g}_e$), *onto the vector space of vector-valued* 2*-forms on* $S_G(M)$ *with values in* $\mathbf{R}^n$ *which are horizontal and* G*-invariant.*

Let $r \in S_G(M)$ and let H be a supplement to the subspace of vertical vectors in the tangent space $T_r(S_G(M))$. If $\boldsymbol{\varpi}$ is horizontal, then clearly so is $\boldsymbol{\varpi} \wedge \boldsymbol{\sigma}$, and its values on $\bigwedge^2 (T_r(S_G(M)))$ are determined by its values at bivectors $\mathbf{u} \wedge \mathbf{v}$, where $\mathbf{u}$, $\mathbf{v}$ are in H:

$$(\boldsymbol{\varpi} \wedge \boldsymbol{\sigma})(r) \cdot (\mathbf{u} \wedge \mathbf{v}) = (\boldsymbol{\varpi}(r) \cdot \mathbf{u})(\boldsymbol{\sigma}(r) \cdot \mathbf{v}) - (\boldsymbol{\varpi}(r) \cdot \mathbf{v})(\boldsymbol{\sigma}(r) \cdot \mathbf{u}).$$

Now $\boldsymbol{\sigma}(r)$ is a *bijection* of H onto $\mathbf{R}^n$. If we put $T = \boldsymbol{\varpi}(r) \circ (\boldsymbol{\sigma}(r))^{-1}$, the relation $(\boldsymbol{\varpi} \wedge \boldsymbol{\sigma})(r) = 0$ signifies that the linear mapping T of $\mathbf{R}^n$ into $\mathfrak{g}_e$ satisfies the relation (20.9.2.1), and therefore $T = 0$ by (20.9.2); in other words, $\boldsymbol{\varpi}(r) = 0$. This proves that the mapping (20.9.3.1) is *injective*. On the other hand, the vector space $V(r)$ of the $\boldsymbol{\varpi}(r)$ (for fixed r) is isomorphic to $\mathrm{Hom}_{\mathbf{R}}(H, \mathfrak{g}_e)$, hence has dimension $\frac{1}{2}n^2(n-1)$. The vector space $W(r)$ of the $\boldsymbol{\Xi}(r)$, where $\boldsymbol{\Xi}$ runs through the space of horizontal vector-valued 2-forms with values in $\mathbf{R}^n$, is likewise isomorphic to $\mathrm{Hom}_{\mathbf{R}}\left(\bigwedge^2 H, \mathbf{R}^n\right)$, hence is also of dimension $\frac{1}{2}n^2(n-1)$. It follows (A.4.19) that, for all $r \in S_G(M)$, the mapping $\boldsymbol{\varpi}(r) \mapsto (\boldsymbol{\varpi} \wedge \boldsymbol{\sigma})(r)$ is a *bijection* of $V(r)$ onto $W(r)$. The coordinates of the solution $\boldsymbol{\varpi}(r)$ of the equation $((\boldsymbol{\varpi} \wedge \boldsymbol{\sigma})(r) = \boldsymbol{\Xi}(r)$ relative to a basis of $V(r)$ are given by Cramer's formulas; from this it follows that, for each C^∞ horizontal 2-form $\boldsymbol{\Xi}$ on $S_G(M)$ with values in $\mathbf{R}^n$, the unique horizontal 1-form $\boldsymbol{\varpi}$ on $S_G(M)$ with values in $\mathfrak{g}_e$ which satisfies the equation $\boldsymbol{\varpi} \wedge \boldsymbol{\sigma} = \boldsymbol{\Xi}$ is of class C^∞. Finally, it is immediately seen (by virtue of the uniqueness of the solution and the invariance of $\boldsymbol{\sigma}$) that if $\boldsymbol{\Xi}$ is invariant, then so is $\boldsymbol{\varpi}$, and vice versa.

(20.9.4) (Levi-Civita) *The mapping which associates to each principal connection in* $S_G(M)$ *its torsion form is a bijection of the set of principal connections onto the set of horizontal, invariant, vector-valued* 2*-forms on* $S_G(M)$ *with values in* $\mathbf{R}^n$. *In particular, there exists a unique torsion-free principal connection in* $S_G(M)$.

Suppose first of all that M is parallelizable, so that (20.7.11) there exists at least one principal connection in $S_G(M)$. Let $\boldsymbol{\omega}$ be the 1-form of the connection. Every other principal connection in $S_G(M)$ has a connection form which can be expressed uniquely as $\boldsymbol{\omega} + \boldsymbol{\varpi}$, where $\boldsymbol{\varpi}$ is a vector-valued 1-form on $S_G(M)$ which is *horizontal* (by virtue of (20.2.5.2)) and G-invariant.

If $\boldsymbol{\Theta}$ and $\boldsymbol{\Theta} + \boldsymbol{\Xi}$ are the torsion forms of these two connections, it follows from the structure equation (20.6.3.3) that $\boldsymbol{\Xi} = \boldsymbol{\varpi} \wedge \boldsymbol{\sigma}$. Conversely, if $\boldsymbol{\varpi}$ is any invariant horizontal 1-form on $S_G(M)$, it follows from (20.2.5) that $\boldsymbol{\omega} + \boldsymbol{\varpi}$ is a connection form on $S_G(M)$; and since the torsion 2-forms are horizontal and invariant, the proposition follows from (20.9.3). In particular, there is a unique invariant horizontal 1-form $\boldsymbol{\varpi}$ such that $\boldsymbol{\Theta} + \boldsymbol{\Xi} = 0$.

Consider now the general case, and cover M by a family of open sets U_α which are domains of charts of M. From what has just been proved, there is a unique torsion-free principal connection in each of the $\pi^{-1}(U_\alpha)$ (where π is the projection $S_G(M) \to M$); and for each pair of indices α, β, the connections induced in $\pi^{-1}(U_\alpha) \cap \pi^{-1}(U_\beta)$ by those in $\pi^{-1}(U_\alpha)$ and $\pi^{-1}(U_\beta)$ are the same, by virtue of the uniqueness already established. We have therefore defined a torsion-free principal connection in $S_G(M)$; and now the argument of the previous paragraph, applied to the corresponding connection form on $S_G(M)$, completes the proof.

The torsion-free principal connection in $S_G(M)$ that has just been defined is called the *Levi–Civita (principal) connection* in $S_G(M)$ corresponding to the pseudo-Riemannian metric tensor $\mathbf{g}$. The corresponding linear G-connection on M (20.7.3) is also called the Levi–Civita (linear) connection defined by $\mathbf{g}$. Throughout the remainder of this chapter, whenever a connection on a pseudo-Riemannian manifold is referred to, explicitly or implicitly, *it is always the Levi–Civita connection that is meant*, unless the contrary is expressly stated. In particular, a *geodesic* (resp. a *geodesic trajectory*, a *geodesic arc*) *in a pseudo-Riemannian manifold* is to be understood as a geodesic (resp. a geodesic trajectory, a geodesic arc) relative to the Levi–Civita connection.

Next, we have the following *characterization* of G-connections on M:

(20.9.5) *A linear connection on a pseudo-Riemannian manifold* M *with metric tensor* $\mathbf{g}$ *is a* G-*connection if and only if*

(20.9.5.1) $$\nabla \mathbf{g} = 0.$$

Let $r_x \in S_G(M) \subset R(M)$. In a neighborhood V of r_x *in* R(M) we may write $\mathbf{g}(y) = \mathbf{T}_2^0(r_y) \cdot \Psi(r_y)$, where Ψ is a C^∞ mapping of V into the space $\mathscr{H}_{p,q}(\mathbf{R}^n)$ of symmetric bilinear forms of signature (p, q) on $\mathbf{R}^n$. To say that $r_y \in S_G(M)$ means by definition (20.7.6) that $\Psi(r_y) = \Phi$, so that Ψ is *constant* on the neighborhood $V \cap S_G(M)$ of r_x in $S_G(M)$. If H is the subspace of horizontal vectors at the point r_x for a principal connection *in* $S_G(M)$, we have therefore $\theta_{\mathbf{k}} \cdot \Psi = 0$ for all vectors $\mathbf{k} \in H$; hence for the corresponding G-connection $\mathbf{C}$ on M we have (20.5.3.4) $\nabla_{\mathbf{h}_x} \cdot \mathbf{g} = 0$ for all tangent vectors $\mathbf{h}_x \in T_x(M)$.

Conversely, suppose we are given a linear connection $\mathbf{C}'$ on M satisfying (20.9.5.1); let H′ be the subspace of horizontal vectors at the point r_x for the corresponding principal connection in R(M) (20.5.2). The formula (20.5.3.4) then shows that $\theta_{\mathbf{k}'} \cdot \Psi = 0$ for all $\mathbf{k}' \in \mathrm{H}'$. Now both H and H′ are supplementary to the subspace L of vertical tangent vectors to R(M) at the point r_x (20.2.2). If $\mathbf{k}$ is the projection of $\mathbf{k}' \in \mathrm{H}'$ on H parallel to L, we have seen that $\theta_{\mathbf{k}} \cdot \Psi = 0$, and therefore the vector $\mathbf{v} = \mathbf{k}' - \mathbf{k} \in \mathrm{L}$ is such that $\theta_{\mathbf{v}} \cdot \Psi = 0$. Now we may write $\mathbf{v} = r_x \cdot \mathbf{u}$, where $\mathbf{u} \in \mathfrak{gl}(n, \mathbf{R})$; if A is the matrix of the form Φ, then the matrix of the form $\Psi(r_x \cdot s)$ for $s \in \mathbf{GL}(n, \mathbf{R})$ is ${}^t s \cdot A \cdot s$; the relation $\theta_{\mathbf{v}} \cdot \Psi = 0$ therefore signifies that $\mathbf{u}$ annihilates the derivative of the mapping $s \mapsto {}^t s \cdot A \cdot s$ of $\mathbf{GL}(n, \mathbf{R})$ into $\mathbf{R}^{n^2}$; in other words (19.4.3), that $\mathbf{u} \in \mathfrak{g}_e$. Consequently, the vector $\mathbf{v}$ is *tangent* to $\mathrm{S_G(M)}$ at the point r_x; therefore, so is each vector $\mathbf{k}' \in \mathrm{H}'$, and the proof is complete (20.7.3).

In view of the properties of the covariant derivative (17.18.2), the result of (20.9.5) may be expressed as follows: for any three C^∞ vector fields X, Y, Z on M, and *any* G-connection on M, we have

(20.9.5.2) $$\theta_Z \cdot \langle \mathbf{g}, X \otimes Y \rangle = \langle \mathbf{g}, \nabla_Z \cdot X \otimes Y \rangle + \langle \mathbf{g}, X \otimes \nabla_Z \cdot Y \rangle$$

or, with the notation of (20.8.2),

(20.9.5.3) $$\theta_Z \cdot (X \mid Y) = (\nabla_Z \cdot X \mid Y) + (X \mid \nabla_Z \cdot Y).$$

Furthermore (17.18.2), these formulas remain valid in the more general situation where we have a mapping $f: \mathrm{N} \to \mathrm{M}$ of class C^r ($r \geqq 1$) of a differential manifold N into M, a vector field Z *on* N and two C^r *liftings* X, Y of f to T(M). In particular, taking $\mathrm{N} = \mathbf{R}$, if $\mathbf{w}_1, \mathbf{w}_2$ are two liftings to T(M) of a mapping f of an open interval of $\mathbf{R}$ into M, then we have

(20.9.5.4) $$\frac{d}{dt}(\mathbf{w}_1 \mid \mathbf{w}_2) = (\nabla_t \cdot \mathbf{w}_1 \mid \mathbf{w}_2) + (\mathbf{w}_1 \mid \nabla_t \cdot \mathbf{w}_2),$$

where $\nabla_t \cdot \mathbf{w}_j$ means $\nabla_{E(t)} \cdot \mathbf{w}_j$, in the notation of (18.1.1).

It follows that if $\mathbf{w}_1, \mathbf{w}_2$ are *parallel transports* along the curve f, the scalar product $(\mathbf{w}_1 \mid \mathbf{w}_2)$ is *constant.*

(20.9.6) The formula (20.9.5.2) enables us to calculate explicitly the Levi–Civita connection when we are given a moving frame $\mathbf{R} = (\mathbf{e}_1, \mathbf{e}_2, \ldots, \mathbf{e}_n)$

on M. We have the functions g_{ij} on M defined by (20.8.2.3), and we define functions Γ^i_{kj} by the formula

$$\nabla_{\mathbf{e}_k} \cdot \mathbf{e}_j = \sum_{i=1}^{n} \Gamma^i_{kj}\, \mathbf{e}_i \tag{20.9.6.1}$$

We then obtain from (20.9.5.2), for all i, j, k,

$$\theta_{\mathbf{e}_k} \cdot g_{ij} = \sum_l (g_{lj} \Gamma^l_{ki} + g_{il} \Gamma^l_{kj}) = \Gamma_{kji} + \Gamma_{kij}, \tag{20.9.6.2}$$

where

$$\Gamma_{kji} = \sum_l g_{lj} \Gamma^l_{ki}. \tag{20.9.6.3}$$

Next, expressing that the torsion is zero, we have for all j, k

$$\nabla_{\mathbf{e}_k} \cdot \mathbf{e}_j - \nabla_{\mathbf{e}_j} \cdot \mathbf{e}_k = [\mathbf{e}_k, \mathbf{e}_j] \tag{20.9.6.4}$$

and these relations, together with (20.9.6.2), determine the Γ^i_{kj}. In the particular case where the $\mathbf{e}_i$ are the vector fields X_i associated with a chart of M (16.15.4.2), we have

$$\theta_{X_k} \cdot g_{ij} = \frac{\partial \tilde{g}_{ij}}{\partial \xi^k},$$

where $\tilde{g}_{ij}$ is the local expression of g_{ij} relative to this chart. The formula (20.9.6.2) then takes the form

$$\frac{\partial \tilde{g}_{ij}}{\partial \xi^k} = \tilde{\Gamma}_{kji} + \tilde{\Gamma}_{kij}, \tag{20.9.6.5}$$

where $\tilde{\Gamma}_{kji}$ is the local expression of Γ_{kji}. Since $[X_k, X_j] = 0$, we have $\Gamma^i_{kj} = \Gamma^i_{jk}$ and hence, by (20.9.6.3), $\tilde{\Gamma}_{kji} = \tilde{\Gamma}_{ijk}$. By cyclic permutation of the indices in (20.9.6.5), we therefore obtain for the local expressions of the Γ_{ijk}

$$\tilde{\Gamma}_{ijk} = \frac{1}{2}\left(\frac{\partial \tilde{g}_{ij}}{\partial \xi^k} + \frac{\partial \tilde{g}_{jk}}{\partial \xi^i} - \frac{\partial \tilde{g}_{ik}}{\partial \xi^j}\right). \tag{20.9.6.6}$$

We remark that, by virtue of (20.9.6.1) and (20.6.5.6), the connection 1-forms ω_{ij} of the Levi-Civita connection (relative to the frame **R**) are given by

$$\omega_{ij} = \sum_{k=1}^{n} \Gamma^i_{kj}\, \sigma_k. \tag{20.9.6.7}$$

(20.9.7) A consequence of the fundamental formula (20.9.5.1) is that, for the covariant tensor $\mathbf{g}_r$ of order $2r$ defined in (20.8.4.1), we have

$$\nabla \mathbf{g}_r = 0. \tag{20.9.7.1}$$

For if Y_i and Z_i $(1 \leqq i \leqq r)$ are $2r$ C^∞ vector fields on M, then it follows immediately, from the fact that the Lie derivative θ_X is a derivation, that $\theta_X \cdot \det((Y_i|Z_j))$ is the sum of the determinants of the r matrices obtained by replacing the entries in the jth column of the matrix $((Y_i|Z_j))$ by $\theta_X \cdot (Y_i|Z_j)$. Hence, by virtue of (20.8.4.1), (20.9.5.3), and the definition of the covariant derivative (17.18.2.2), we have

$$\langle \nabla \mathbf{g}_r, (Y_1 \wedge \cdots \wedge Y_r) \otimes (Z_1 \wedge \cdots \wedge Z_r) \rangle = 0.$$

Since the Y_i and the Z_i were arbitrarily chosen, this establishes (20.9.7.1). If M is an oriented Riemannian manifold, we have likewise

$$\nabla v_{\mathbf{g}} = 0, \tag{20.9.7.2}$$

$v_{\mathbf{g}}$ being considered as a covariant tensor of order n. The proof is analogous and rests on the relation

$$\det((Y_i|Y_j)) = (\langle v_{\mathbf{g}}, Y_1 \wedge \cdots \wedge Y_n \rangle)^2.$$

(20.9.8) If $f: \mathrm{M}' \to \mathrm{M}$ is a *local isometry* of pseudo-Riemannian manifolds, then it is immediate that the inverse image under f of the Levi-Civita connection on M (17.18.4) is the Levi-Civita connection on M′.

PROBLEMS

1. Show that, for the Levi-Civita connection on a pseudo-Riemannian manifold, if X, Y, Z are C^∞ vector fields, we have

$$\begin{aligned} 2(\nabla_X \cdot Y|Z) = {} & \theta_X \cdot (Y|Z) + \theta_Y \cdot (X|Z) - \theta_Z \cdot (X|Y) \\ & + ([X, Y]|Z) + ([Z, X]|Y) + (X|[Z, Y]). \end{aligned}$$

2. The notation is that of Section 20.8, Problem 3. If $\mathbf{C}$ is the Levi-Civita connection on M, let $S(\mathbf{h}_x) = \mathbf{C}_x(\mathbf{h}_x, \mathbf{h}_x)$ be the corresponding geodesic field on T(M) (18.6.1).

   (a) Show that

$$\langle \alpha, S \rangle = \mathrm{E}, \qquad i_S \cdot d\alpha = -\tfrac{1}{2}\, d\mathrm{E},$$
$$\theta_S \cdot \alpha = \tfrac{1}{2}\, d\mathrm{E}, \qquad \theta_S \cdot d\alpha = 0.$$

Deduce that, for each $\mathbf{h} \in \mathrm{T(M)}$ and each $t \in \mathrm{J}(\mathbf{h})$ (in the notation of (18.2.2))

$$\mathrm{E}(\mathrm{F_S}(\mathbf{h}, t)) = \mathrm{E}(\mathbf{h}), \qquad {}^t\mathrm{F_S}(\,.\,, t)(\alpha) - \alpha = \tfrac{1}{2}\, d\mathrm{E}.$$

(b) For each $\mathbf{h}_x \in \mathrm{T(M)}$ and each $\mathbf{k}_{\mathbf{h}_x} \in \mathrm{T}_{\mathbf{h}_x}(\mathrm{T(M)})$, put

$$Q_{\mathbf{h}_x} \cdot \mathbf{k}_{\mathbf{h}_x} = \boldsymbol{\tau}_{\mathbf{h}_x}(\mathbf{k}_{\mathbf{h}_x} - \mathbf{C}_x(\mathbf{h}_x, \mathrm{T}(o_\mathrm{M}) \cdot \mathbf{k}_{\mathbf{h}_x})) \in \mathrm{T}_x(\mathrm{M}).$$

Show that we have

$$\langle d\alpha(\mathbf{h}_x), \mathbf{k}'_{\mathbf{h}_x} \wedge \mathbf{k}''_{\mathbf{h}_x}\rangle = (Q_{\mathbf{h}_x} \cdot \mathbf{k}'_{\mathbf{h}_x} | \mathrm{T}(o_\mathrm{M}) \cdot \mathbf{k}''_{\mathbf{h}_x}) - (Q_{\mathbf{h}_x} \cdot \mathbf{k}''_{\mathbf{h}_x} | \mathrm{T}(o_\mathrm{M}) \cdot \mathbf{k}'_{\mathbf{h}_x})$$

for any two vectors $\mathbf{k}'_{\mathbf{h}_x}$, $\mathbf{k}''_{\mathbf{h}_x}$ in $\mathrm{T}_{\mathbf{h}_x}(\mathrm{T(M)})$. Hence there exists a covariant tensor $\boldsymbol{B}$ of order 2 on T(M) such that $d\alpha$ is the antisymmetrization of $\boldsymbol{B}$ and such that

$$\langle \boldsymbol{B}(\mathbf{h}_x), \mathbf{k}'_{\mathbf{h}_x} \otimes \mathbf{k}''_{\mathbf{h}_x}\rangle = (Q_{\mathbf{h}_x} \cdot \mathbf{k}'_{\mathbf{h}_x} | \mathrm{T}(o_\mathrm{M}) \cdot \mathbf{k}''_{\mathbf{h}_x}).$$

Show that the covariant tensor $\theta_S \cdot \boldsymbol{B}$ of order 2 is symmetric and that it is given by

$$\begin{aligned}\langle(\theta_S \cdot \boldsymbol{B})(\mathbf{h}_x), \mathbf{k}'_{\mathbf{h}_x} \otimes \mathbf{k}''_{\mathbf{h}_x}\rangle &= ((\boldsymbol{r} \cdot (\mathbf{h}_x \wedge \mathrm{T}(o_\mathrm{M}) \cdot \mathbf{k}'_{\mathbf{h}_x})) \cdot \mathbf{h}_x | \mathrm{T}(o_\mathrm{M}) \cdot \mathbf{k}''_{\mathbf{h}_x}) \\ &\quad + (Q_{\mathbf{h}_x} \cdot \mathbf{k}'_{\mathbf{h}_x} | Q_{\mathbf{h}_x} \cdot \mathbf{k}''_{\mathbf{h}_x}).\end{aligned}$$

(Consider the geodesic v such that $v(0) = x$ and having $\mathbf{h}_x$ as tangent vector at this point, and the liftings $\mathbf{w}_1$, $\mathbf{w}_2$ of v' to T(T(M)) such that $\theta_S \cdot \mathbf{w}_1 = \theta_S \cdot \mathbf{w}_2 = 0$ and $\mathbf{w}_1(0) = \mathbf{k}'_{\mathbf{h}_x}$, $\mathbf{w}_2(0) = \mathbf{k}''_{\mathbf{h}_x}$; and calculate $\langle \theta_S \cdot \boldsymbol{B}, \mathbf{w}_1 \otimes \mathbf{w}_2\rangle$.)

3. Let M be a pure pseudo-Riemannian manifold of dimension n, with metric tensor $\mathbf{g}$.

(a) With the notation of Problem 2, show that the following formula defines a pseudo-Riemannian metric tensor $\mathbf{g}_\mathrm{T}$ on the manifold T(M):

$$\langle \mathbf{g}_\mathrm{T}(\mathbf{h}), \mathbf{k}'_\mathbf{h} \otimes \mathbf{k}''_\mathbf{h}\rangle = (Q_\mathbf{h} \cdot \mathbf{k}'_\mathbf{h} | Q_\mathbf{h} \cdot \mathbf{k}''_\mathbf{h}) + (\mathrm{T}(o_\mathrm{M}) \cdot \mathbf{k}'_\mathbf{h} | \mathrm{T}(o_\mathrm{M}) \cdot \mathbf{k}''_\mathbf{h})$$

for $\mathbf{k}'_\mathbf{h}$, $\mathbf{k}''_\mathbf{h}$ in $\mathrm{T}_\mathbf{h}(\mathrm{T(M)})$. If $\mathbf{g}$ is of signature (p, q), then $\mathbf{g}_\mathrm{T}$ is of signature $(2p, 2q)$. Relative to $\mathbf{g}_\mathrm{T}$, every vertical tangent vector at a point $\mathbf{h} \in \mathrm{T(M)}$ is orthogonal to every horizontal tangent vector at $\mathbf{h}$.

(b) Show that for the metric tensor $\mathbf{g}_\mathrm{T}$ on T(M) we have

$$\mathrm{grad}(\mathrm{E})(\mathbf{h}) = 2\boldsymbol{\tau}_\mathbf{h}^{-1}(\mathbf{h}).$$

(c) On T(M) the $2n$-form $(d\alpha)^{\wedge n}$ vanishes nowhere and hence defines a canonical orientation of T(M) (Section 16.21, Problem 4). Assume for the rest of this problem that M is a Riemannian manifold. Then, for the canonical orientation on T(M), the canonical volume form for $\mathbf{g}_\mathrm{T}$ is given by

$$v_{\mathbf{g}_\mathrm{T}} = (-1)^{n(n-1)/2}(n!)^{-1}(d\alpha)^{\wedge n}.$$

(Calculate the value of $(d\alpha)^{\wedge n}$ at the tangent $2n$-vector defined as follows: let $x \in \mathrm{M}$ and $\mathbf{h} \in \mathrm{T}_x(\mathrm{M})$; let $(\mathbf{h}_1, \ldots, \mathbf{h}_n)$ be an orthonormal basis of $\mathrm{T}_x(\mathrm{M})$, and let $\mathbf{k}'_j = \boldsymbol{\tau}_\mathbf{h}^{-1}(\mathbf{h}_j)$, $\mathbf{k}''_j = \mathbf{C}_x(\mathbf{h}, \mathbf{h}_j)$; then form the $2n$-vector $\mathbf{k}'_1 \wedge \mathbf{k}''_1 \wedge \mathbf{k}'_2 \wedge \mathbf{k}''_2 \wedge \cdots \wedge \mathbf{k}'_n \wedge \mathbf{k}''_n$.)

(d) If the volume $\mathrm{vol}_\mathbf{g}(\mathrm{M})$ is finite, show that

$$\mathrm{vol}_{\mathbf{g}_\mathrm{T}}(\mathrm{B(M)}) = \mathrm{V}_n \, \mathrm{vol}_\mathbf{g}(\mathrm{M}),$$

where V_n is the Lebesgue measure of the unit ball $\mathbf{B}_n$ (14.3.11.3) and B(M) is the set of all $\mathbf{h} \in \mathrm{T(M)}$ such that $\mathrm{E}(\mathbf{h}) \leq 1$. (Consider first the case where M is oriented, and use (16.24.8); for the general case, use a suitable partition of unity on M.)

(e) The set S(M) of vectors $\mathbf{h} \in T(M)$ satisfying $E(\mathbf{h}) = 1$ is a closed submanifold of T(M) of dimension $2n - 1$, whose tangent space at each point $\mathbf{h}$ is orthogonal to $\operatorname{grad}(E)(\mathbf{h})$. Let $\mathbf{g}_S$ denote the restriction to S(M) of the Riemannian metric tensor $\mathbf{g}_T$. Show that on S(M), canonically oriented (16.21.9.2), the canonical volume form for $\mathbf{g}_S$ is given by

$$v_{\mathbf{g}_S} = (-1)^{n(n-1)/2}((n-1)!)^{-1}\alpha \wedge (d\alpha)^{\wedge(n-1)}$$

and, if $\operatorname{vol}_{\mathbf{g}}(M)$ is finite,

$$\operatorname{vol}_{\mathbf{g}_S}(S(M)) = \Omega_n \operatorname{vol}_{\mathbf{g}}(M),$$

where $\Omega_n = nV_n$ is the superficial measure of S_{n-1} (16.24.9.3).

4. Show that the Levi-Civita principal connection on the product of two pseudo-Riemannian manifolds (Section 20.8, Problem 1) is the product of the Levi-Civita principal connections on the two manifolds (Section 20.2, Problem 2).

5. Let M be a pseudo-Riemannian manifold, $\mathbf{g}$ its metric tensor. If a diffeomorphism f of M onto M is both an automorphism of the Levi-Civita connection on M and a conformal mapping (20.8.1), show that ${}^t f(\mathbf{g}) = c \cdot \mathbf{g}$, where c is a *constant* $\neq 0$. (Use (20.9.5).) Such a mapping f is called a *homometry* of M.

6. Let M, M′ be two pure pseudo-Riemannian manifolds whose metric tensors have the same signature: we may therefore assume that the group G of the corresponding G-structures (20.7.6) is the same for both manifolds. Show that for a diffeomorphism f of M onto M′ to be an isometry, it is necessary and sufficient that the isomorphism $R(f)$ of R(M) onto R(M′) (Section 20.6, Problem 4) should transform $S_G(M)$ into $S_G(M')$.

Show that the group I(M) of isometries of M is a Lie group (Section 20.6, Problem 10).

7. Let M be a pure pseudo-Riemannian manifold of dimension n and metric tensor $\mathbf{g}$. With the notation of Section 20.6, Problem 5, the vector field X is said to be an *infinitesimal isometry* of M if, for each relatively compact open subset U of M and each $a > 0$ such that $U \times]-a, a[\subset \operatorname{dom}(F_X)$, g_t is an isometry of U onto U_t for all $t \in]-a, a[$.

Show that the following conditions are equivalent:

(i) X is an infinitesimal isometry.

(ii) The canonical lifting $\tilde{X}$ of X to R(M) (Section 20.6, Problem 5) is tangent to $S_G(M)$ (in the notation of Problem 6) at all points of this subbundle.

(iii) $\theta_X \cdot \mathbf{g} = 0$ (cf. (18.2.14.6)).

(iv) With the notation of Section 20.6, Problem 6,

$$(\mathbf{A}_X \cdot Y \mid Z) = -(\mathbf{A}_X \cdot Z \mid Y) \tag{1}$$

for all C^∞ vector fields Y, Z on M. (Use the fact that $\mathbf{A}_X$ is a derivation.)

The set of all infinitesimal isometries of M is a Lie subalgebra i(M) of $\mathscr{T}_1^0(M)$, of dimension $\leq \frac{1}{2}n(n+1)$ (Section 20.6, Problem 7). The Lie algebra of the group I(M) of isometries of M may be identified with a Lie subalgebra of i(M).

8. Let M, M_1 be two connected pseudo-Riemannian manifolds with metric tensors $\mathbf{g}$, $\mathbf{g}_1$, respectively, and let f be an isomorphism of the Levi–Civita connection of M onto that of M_1. For f to be an isometry of M onto M_1, it is necessary and sufficient that at a point $x \in M$, the image under $T_x(f)$ of the form $\mathbf{g}(x)$ should be the form $\mathbf{g}_1(f(x))$. (Consider the parallel transports along a path γ through x, and their images under $T(f)$ along the path $f \circ \gamma$, and use (20.9.5.4).) Hence generalize the results of Section 20.6, Problem 9, by replacing isomorphisms of connections by isometries.

9. Let M be a pseudo-Riemannian manifold. With the notation of (20.8.3) show that, for the Levi–Civita connection, if $\mathbf{Z} \in \mathscr{T}_s^r(M)$ and $X \in \mathscr{T}_1^0(M)$, then

$$\nabla_X \cdot (G_j^i \cdot \mathbf{Z}) = G_j^i(\nabla_X \cdot \mathbf{Z})$$

for $1 \leqq i \leqq r$ and $1 \leqq j \leqq s$.

10. Let M be an oriented pure Riemannian manifold of dimension n, and let v be the canonical volume form on M. For every C^r differential p-form α on M, let

$$\partial\alpha = (-1)^{np+p+1} * (d(*\alpha)),$$

which is a differential $(p-1)$-form of class C^{r-1}. With the notation of (20.6.5), show that

$$\partial\alpha = -\sum_j i(G^{-1} \cdot \sigma_j) \cdot (\nabla_{\mathbf{e}_j} \cdot \alpha).$$

(Use Problem 9.)

11. Let $\mathbf{f} = (f_1, \ldots, f_n)$ be a diffeomorphism of an open subset U of $\mathbf{R}^n$ onto an open set $V = \mathbf{f}(U)$, and suppose that $\mathbf{f}$ is a *conformal* mapping of U onto V, relative to the canonical Riemannian structure on $\mathbf{R}^n$ (20.11.2). Then $(V, \mathbf{f}^{-1}, n)$ is a chart of V, relative to which the local expression of the metric tensor $\mathbf{g}$ on V is

$$\frac{1}{\rho^2}((d\xi^1)^2 + (d\xi^2)^2 + \cdots + (d\xi^n)^2),$$

where ρ is a C^∞ function on U which does not take the value 0. Use the formulas (20.9.6.6) and (20.9.6.7) to show that the connection 1-forms on V are given by

$$\omega_{ij} = \frac{1}{\rho}\left(\frac{\partial\rho}{\partial\xi^j}\, d\xi^i - \frac{\partial\rho}{\partial\xi^i}\, d\xi^j\right) \quad \text{if } i \neq j,$$

$$\omega_{ii} = -\frac{d\rho}{\rho}.$$

By writing down the structure equations (20.6.6.4) with $\Omega_{ij} = 0$, deduce that if $n \geqq 3$, we have

$$\frac{\partial^2\rho}{\partial\xi^i\, \partial\xi^j} = 0 \quad \text{if } i \neq j,$$

$$\frac{\partial^2\rho}{(\partial\xi^j)^2} = c,$$

where $c \in \mathbf{R}$ is independent of j.

Consider the two cases $c = 0$ and $c \neq 0$, and show that $\mathbf{f}$ is the restriction to U of a transformation belonging to the *conformal group*, generated by similitudes and the inversion $\mathbf{x} \mapsto \mathbf{x}/\|\mathbf{x}\|^2$† (*Liouville's theorem*). Consider the analogous result when $\mathbf{R}^n$ is replaced by $\mathbf{C}^n$. What can be said when $n = 2$?

12. (a) Identity $\mathbf{R}^n$ with the subspace of $\mathbf{P}_n(\mathbf{R})$ consisting of the points with first homogeneous coordinate equal to 1. Then there exists a neighborhood V of the identity element of $\mathbf{PGL}(n+1, \mathbf{R})$ and a neighborhood W of 0 in $\mathbf{R}^n$ such that the action of V on W may be written in the form

$$((\mathbf{a}, U, \mathbf{b}^*), \mathbf{x}) \mapsto (1 + \langle \mathbf{b}^*, \mathbf{x} \rangle)^{-1}(U \cdot \mathbf{x} + \mathbf{a}),$$

V being identified with a neighborhood of $(0, I, 0)$ in $\mathbf{R}^n \times \mathbf{M}_n(\mathbf{R}) \times (\mathbf{R}^n)^*$. Deduce that the Lie algebra $\mathfrak{pgl}(n+1, \mathbf{R})$ of $\mathbf{PGL}(n+1, \mathbf{R})$ may be identified with the direct sum $\mathfrak{n} \oplus \mathfrak{gl}(n, \mathbf{R}) \oplus \mathfrak{n}^*$, where $\mathfrak{n} = \mathbf{R}^n$ is identified with the space of $n \times 1$ matrices, and its dual $\mathfrak{n}^*$ with the space of $1 \times n$ matrices. The Lie bracket in $\mathfrak{pgl}(n+1, \mathbf{R})$ induces on $\mathfrak{gl}(n, \mathbf{R})$ its usual Lie algebra structure, and for $\mathbf{u}, \mathbf{v} \in \mathfrak{n}$, $\mathbf{u}^*, \mathbf{v}^* \in \mathfrak{n}^*$ and $U \in \mathfrak{gl}(n, \mathbf{R}) = \mathbf{M}_n(\mathbf{R})$, we have the formulas

$$[\mathbf{u}, \mathbf{v}] = 0, \qquad [\mathbf{u}^*, \mathbf{v}^*] = 0, \qquad [\mathbf{u}, \mathbf{u}^*] = \mathbf{u} \cdot \mathbf{u}^* + (\mathbf{u}^* \cdot \mathbf{u}) \cdot I_n,$$
$$[U, \mathbf{u}] = U \cdot \mathbf{u}, \qquad [\mathbf{u}^*, U] = \mathbf{u}^* \cdot U.$$

Let $(\mathbf{e}_i)_{1 \leq i \leq n}$ denote the canonical basis of $\mathfrak{n} = \mathbf{R}^n$, $(\mathbf{e}_i^*)_{1 \leq i \leq n}$ the dual basis of $\mathfrak{n}^*$, and $(E_{ij})_{1 \leq i, j \leq n}$ the canonical basis of $\mathbf{M}_n(\mathbf{R})$.

(b) Write $\mathrm{G} = \mathbf{PGL}(n+1, \mathbf{R})$ and consider the connected Lie subgroup H_0 of G which fixes the origin in $\mathbf{R}^n$, so that G/H_0 may be identified with $\mathbf{P}_n(\mathbf{R})$. Let M be a pure manifold of dimension n, let R be a principal bundle with base M, projection π, and group H_0, and let $\boldsymbol{\omega}_0$ be a differential 1-form on R with values in the Lie algebra $\mathfrak{pgl}(n+1, \mathbf{R})$ which satisfies the conditions of Problem 2(a) of Section **20.5**, so that $\boldsymbol{\omega}_0$ has a unique extension to the connection form $\boldsymbol{\omega}$ on $\mathrm{R} \times^{\mathrm{H}_0} \mathrm{G}$ of a Cartan connection for R.

We may then write

$$(1) \qquad \boldsymbol{\omega}_0 = \sum_i \omega^i \mathbf{e}_i + \sum_{i,j} \omega^i_j E_{ij} + \sum_i \omega_i \mathbf{e}_i^*,$$

where ω^i, ω^i_j, and ω_i are $n^2 + 2n$ differential 1-forms which form a *basis* of the $\mathscr{E}(\mathrm{R})$-module $\mathscr{T}^0_1(\mathrm{R})$ of C^∞ differential 1-forms on R. Condition (i) of Section **20.5**, Problem 2(a) implies that the real vector space F of dimension $n^2 + 2n$ spanned by the ω^i, ω^i_j, and ω_i is *stable* under the action of H_0 on $\mathscr{T}^0_1(\mathrm{R})$ (induced canonically by its right action on R). More precisely, by calculating the endomorphism $\mathrm{Ad}(t^{-1})$ of $\mathfrak{pgl}(n+1, \mathbf{R})$, where $t = (0, U, \mathbf{b}^*) \in \mathrm{H}_0$, show that:

(i) The subspace F_1 of F spanned by the n forms ω^i $(1 \leq i \leq n)$ is stable under H_0, and with the previous notation, we have

$$\begin{pmatrix} \omega^1(r \cdot t) \\ \omega^2(r \cdot t) \\ \vdots \\ \omega^n(r \cdot t) \end{pmatrix} = U^{-1} \cdot \begin{pmatrix} \omega^1(r) \\ \omega^2(r) \\ \vdots \\ \omega^n(r) \end{pmatrix}.$$

† See J. Dieudonné, "Linear Algebra and Geometry," Appendix III, pp.159–164, Houghton-Mifflin, Boston, 1969.

(ii) the subspace F_2 of F spanned by F_1 and the n^2 forms ω_j^i ($1 \leqq i, j \leqq n$) is stable under H_0, and we have

$$(\omega_j^i(r \cdot t)) = U^{-1} \cdot (\omega_j^i(r)) \cdot U + U^{-1} \cdot (\omega^i(r)) \cdot \mathbf{b}^* + (\mathbf{b}^* \cdot U^{-1})(\omega^i(r)) I_n .$$

(iii) $(\omega_1(r \cdot t), \ldots, \omega_n(r \cdot t)) \equiv (\omega_1(r), \ldots, \omega_n(r)) \cdot U$ (mod. F_2).

Condition (ii) of Section 20.5, Problem 2(a) signifies that, for each vector

$$\mathbf{x} = \sum_i b_i \mathbf{e}_i^* + \sum_{i,j} u_j^i E_{ij}$$

in the Lie algebra $\mathfrak{h}_0 = \mathfrak{n}^* + \mathfrak{gl}(n, \mathbf{R})$ of H_0, we have

(iv) $\langle \omega^i, Z_{\mathbf{x}} \rangle = 0, \qquad \langle \omega_j^i, Z_{\mathbf{x}} \rangle = u_j^i \qquad (1 \leqq i, j \leqq n)$

and

(v) $\langle \omega_i, Z_{\mathbf{x}} \rangle = b_i \qquad (1 \leqq i \leqq n).$

Finally, condition (iii) of Section 20.5, Problem 2(a), signifies that

(vi) the vector subspace of $T_r(R)$ defined by the n equations $\langle \omega^i(r), \mathbf{h}_r \rangle = 0$ ($1 \leqq i \leqq n$) is the subspace $\operatorname{Ker}(T_r(\pi))$ of vertical tangent vectors at the point r (spanned by the $Z_{\mathbf{x}}(r)$ for $\mathbf{x} \in \mathfrak{h}_0$).

(c) *Conversely*, suppose we are given a system of $n^2 + n$ C^∞ differential 1-forms ω^i, ω_j^i on R whose values at each point are linearly independent, and which satisfy conditions (i), (ii), (iv), and (vi) of (b) above. Show that there exists at least one system $(\omega_i)_{1 \leqq i \leqq n}$ of n C^∞-differential 1-forms on R such that the formula (1) defines a Cartan connection for R. (Reduce to the case where the principal bundle R is trivial.) Furthermore, every other such system $(\omega_i')_{1 \leqq i \leqq n}$ of differential 1-forms with this property is given by the formulas

$$\omega_i' - \omega_i = \sum_j A_{ij} \omega^j, \tag{2}$$

where (A_{ij}) is an $n \times n$ matrix of C^∞ functions on R such that (with the notation of (b) above)

$$(A_{ij}(r \cdot t)) = {}^t U \cdot (A_{ij}(r)) \cdot U. \tag{3}$$

13. With the notation of Problem 12, suppose that $\boldsymbol{\omega}_0$ defines a Cartan connection for R.

(a) By writing the structure equation (20.3.3.1) for the restriction to R of the connection 1-form $\boldsymbol{\omega}$, show that

$$d\omega^i = -\sum_k \omega_k^i \wedge \omega^k + \Omega^i,$$
$$d\omega_j^i = -\sum_k \omega_k^i \wedge \omega_j^k - \omega^i \wedge \omega_j + \delta_{ij} \sum_k (\omega_k \wedge \omega^k) + \Omega_j^i,$$
$$d\omega_j = -\sum_k \omega_k \wedge \omega_j^k + \Omega_j,$$

where Ω^i, Ω_j^i, Ω_j are the components of the restriction to R of the curvature 2-form $\boldsymbol{\Omega}$ of $\boldsymbol{\omega}$.

Using the fact that $\mathbf{\Omega}$ is horizontal (20.3.2), show that

$$\Omega^i = \tfrac{1}{2} \sum_{k,l} \mathrm{K}^i_{kl}\, \omega^k \wedge \omega^l,$$

$$\Omega^i_j = \tfrac{1}{2} \sum_{k,l} \mathrm{K}^i_{jkl}\, \omega^k \wedge \omega^l,$$

$$\Omega_j = \tfrac{1}{2} \sum_{k,l} \mathrm{K}_{jkl}\, \omega^k \wedge \omega^l,$$

where the coefficients K^i_{kl}, K^i_{jkl}, and K_{jkl} are C^∞ functions on R.

(b) Show that conditions (i), (ii), and (iii) of Problem 12(b) remain satisfied when ω^i, ω^i_j, and ω_j are replaced by Ω^i, Ω^i_j, and Ω_j, respectively.

(c) Prove Bianchi's identities

$$d\Omega^i = \sum_k \Omega^i_k \wedge \omega^k,$$

$$d\Omega^i_j = \sum_k \Omega^i_k \wedge \omega^k_j - \sum_k \omega^i_k \wedge \Omega^k_j + \Omega^i \wedge \omega_j - \omega^i \wedge \Omega_j + \delta_{ij} \sum_k \omega_k \wedge \Omega^k - \delta_{ij} \sum_k \Omega_k \wedge \omega^k,$$

$$d\Omega_j = \sum_k \Omega_k \wedge \omega^k_j - \sum_k \omega_k \wedge \Omega^k_j.$$

(d) The mapping $(0, U, \mathbf{b}^*) \mapsto U$ of H_0 onto $\mathbf{GL}(n, \mathbf{R})$ is a homomorphism whose kernel N^* is the normal subgroup of H_0 with Lie algebra $\mathfrak{n}^*$. Deduce that R is a principal bundle over $\mathrm{R}_1 = \mathrm{N}^* \backslash \mathrm{R}$ with group N^*, and that R_1 is a principal bundle over M with group $\mathbf{GL}(n, \mathbf{R})$ (16.14.8). Let $p: \mathrm{R} \to \mathrm{R}_1$ be the canonical projection. Show that the forms Ω^i may be written as ${}^t p(\Theta^i)$, where the Θ^i are differential 2-forms on R_1. The 2-form $\mathbf{\Theta} = (\Theta^i)$ with values in $\mathbf{R}^n$ is invariant under the action $(U, \mathbf{x}) \mapsto U^{-1} \cdot \mathbf{x}$ of $\mathbf{GL}(n, \mathbf{R})$ on $\mathbf{R}^n$ and the action of $\mathbf{GL}(n, \mathbf{R})$ on R_1. (Show that for each $\mathbf{u} \in \mathfrak{n}^*$ we have $i_{Z\mathbf{u}} \cdot \Omega^i = 0$ and $\theta_{Z\mathbf{u}} \cdot \Omega^i = 0$ by using (c) above, and then apply the argument of (18.16.8).)

(e) The forms Ω^i are called the *torsion forms* of the Cartan connection defined by $\boldsymbol{\omega}_0$, and the connection is said to be *torsion-free* if the Ω^i are zero. Show that if the Cartan connection defined by $\boldsymbol{\omega}_0$ is torsion-free, then the Ω^i_j may be written in the form ${}^t p(\Upsilon^i_j)$, where the Υ^i_j are 2-forms on R_1. The 2-form $\mathbf{\Upsilon} = (\Upsilon^i_j)$ with values in $\mathfrak{gl}(n, \mathbf{R})$ is invariant under the action of $\mathbf{GL}(n, \mathbf{R})$ on R_1, and the right action $(U, V) \mapsto \mathrm{Ad}(U^{-1}) \cdot V = U^{-1} V U$ of $\mathbf{GL}(n, \mathbf{R})$ on $\mathfrak{gl}(n, \mathbf{R})$. (Same method.)

(f) Suppose again that the Cartan connection defined by $\boldsymbol{\omega}_0$ is *torsion-free*. For each $r \in \mathrm{R}$, let $\mathbf{K}(r) \in \mathbf{T}^1_3(\mathbf{R}^n)$ be the tensor whose components relative to the canonical basis are the $\mathrm{K}^i_{jkl}(r)$. With the notation of Problem 12, show that

$$\mathbf{K}(r \cdot t) = \mathbf{T}^1_3(U^{-1}) \cdot \mathbf{K}(r)$$

for $t \in \mathrm{H}_0$. The contracted tensors $\mathbf{B}(r) = c^1_1(\mathbf{K}(r))$ and $\mathbf{C}(r) = c^1_2(\mathbf{K}(r))$ then belong to $\mathbf{T}^0_2(\mathbf{R}^n)$ and may be identified with matrices $(\mathrm{B}_{ij}(r))$ and $(\mathrm{C}_{ij}(r))$, where

$$\mathrm{B}_{ij} = \sum_k \mathrm{K}^k_{kij}, \qquad \mathrm{C}_{ij} = \sum_k \mathrm{K}^k_{ikj}.$$

Show that these matrices satisfy the relation (3) of Problem 12. Deduce that if ω^i, ω^i_j are $n^2 + n$ differential 1-forms on R, whose values at each point are linearly independent, which satisfy conditions (i), (ii), (iv), and (vi) of Problem 12(b), and which also satisfy

$$d\omega^i = -\sum_k \omega^i_k \wedge \omega^k,$$

then there exists a *unique Cartan connection* for R defined by a form $\boldsymbol{\omega}_0$ whose first n^2+n components are the ω^i and ω^i_j (and which is therefore torsion-free) and for which

$$c^1_1(\mathbf{K}) = c^1_2(\mathbf{K}) = 0 \tag{1}$$

(which implies that also $c^1_3(\mathbf{K}) = 0$, since $\mathrm{K}^i_{jkl} = -\mathrm{K}^i_{jlk}$). The first of these conditions is equivalent to $\sum_i \Omega^i_i = 0$. (Use Problem 11(c) and the structure equations of (a) above.)

(g) Show that if the form $\boldsymbol{\omega}_0$ defines a torsion-free Cartan connection satisfying (1), then we have

$$\sum_j \Omega^i_j \wedge \omega^j = 0, \qquad \sum_j \Omega_j \wedge \omega^j = 0.$$

Deduce that if $n \geqq 3$ and if the Ω^i_j are all zero, then the Ω_j are also zero. (Use (b).)

14. We shall apply the results of Problem 13 in the case where the principal bundle R is a *projective structure* P on the manifold M (Section 20.7, Problem 16); P may be considered as a principal bundle with group H_0. The subgroup N^* is then identified with the intersection of $\mathrm{H} \subset \mathrm{G}^2(n)$ and the kernel of the canonical homomorphism

$$\rho : \mathrm{G}^2(n) \to \mathrm{G}^1(n)$$

(Section 20.6, Problem 19), and consequently $\mathrm{N}^*\backslash\mathrm{P} = \mathrm{R}_1$ is identified with the frame bundle $\mathrm{R(M)} = \mathrm{R}_1(\mathrm{M})$. A Cartan connection for P is called (by abuse of language) a *projective connection on* M if it is defined by a form $\boldsymbol{\omega}_0$ such that ω^i is the restriction to P of σ^i, and ω^i_j the restriction to P of σ^i_j for each pair of indices i, j, where the σ^i and σ^i_j are the canonical differential forms on $\mathrm{R}_2(\mathrm{M})$ (*loc. cit*).

(a) Deduce from Problem 13(f) that for each projective structure P on M there exists a unique torsion-free projective connection that satisfies the relations (1) of Problem 13. This connection is called the *normal projective connection* corresponding to P. The 2-form Υ on R(M) is then called the *Weyl projective curvature form* of the projective structure P. By the formula analogous to (20.6.6.6) in which Ω_{ji} is replaced by Υ^i_j, we obtain a morphism $\boldsymbol{w}$ of $\bigwedge^2 \mathrm{T(M)}$ into $\mathrm{T(M)}^* \otimes \mathrm{T(M)}$, and hence, just as in (17.20.5), a tensor $\mathbf{w}$ of type (1, 3) on M, called the *Weyl projective curvature tensor* of the projective structure P.

(b) To each torsion-free linear connection $\mathbf{C}$ on M there corresponds canonically a projective structure P on M, and hence a normal projective connection on M. For each vector $\mathbf{a} = \sum_i a^i \mathbf{e}_i \in \mathfrak{n}$, there exists a unique C^∞ vector field $H'_{\mathbf{a}}$ on P such that

$$\langle \omega^i, H'_{\mathbf{a}} \rangle = a^i, \qquad \langle \omega^i_j, H'_{\mathbf{a}} \rangle = 0, \qquad \langle \omega_j, H'_{\mathbf{a}} \rangle = 0$$

for all i, j. Show that the projections in M of the integral curves of the fields $H'_{\mathbf{a}}$ are the geodesics of the connection $\mathbf{C}$. (Use Section 20.6, Problem 1.)

(c) If f is a diffeomorphism of a pure manifold M onto a pure manifold M′, then f gives rise by transport of structure to an isomorphism $\mathrm{R}_2(f)$ of the principal bundle $\mathrm{R}_2(\mathrm{M})$ onto the principal bundle $\mathrm{R}_2(\mathrm{M}')$. If P (resp. P′) is a projective structure on M (resp. M′), then f is said to be an *isomorphism* of P onto P′ if the restriction of $\mathrm{R}_2(f)$ to P is a bijection of P onto P′. Show that the group of automorphisms of a projective structure on M is a Lie group. (The argument is the same as in Problem 6, using the normal projective connection.)

(d) Let X be a C^∞ vector field on M. As in Section 20.6, Problem 5, we define a *canonical lifting* $\tilde{\tilde{X}}$ of X to $R_2(M)$, by replacing $R(g_t)$ by $R_2(g_t)$ in the definition; $\tilde{\tilde{X}}$ is a lifting of $\tilde{X}$, relative to the canonical projection $R_2(M) \to R_1(M) = R(M)$.

If P is a projective structure on M, the field X is said to be an *infinitesimal automorphism* of P if the canonical lifting $\tilde{\tilde{X}}$ is tangent to the submanifold P of $R_2(M)$ at each of its points.

Show that the set of infinitesimal automorphisms of P is a Lie subalgebra $\mathfrak{p}(P)$ of $\mathcal{T}_0^1(M)$, of dimension $\leq n^2 + 2n$. If $\dim(\mathfrak{p}(P)) = n^2 + 2n$, then the forms Ω_j^i and Ω_j for the normal projective connection corresponding to P are zero. (Argue as in Section 20.6, Problems 6 and 7.)

(e) If $M = \mathbf{P}_n(\mathbf{R})$, the set of jets of order 2 at the point 0 of mappings of $\mathbf{R}^n$ (canonically identified with an open subset of $\mathbf{P}_n(\mathbf{R})$, as in Problem 9) into $\mathbf{P}_n(\mathbf{R})$, which are restrictions of mappings belonging to $\mathbf{PGL}(n+1, \mathbf{R})$, is a projective structure called the *canonical* projective structure on M. This principal bundle may be canonically identified with $\mathbf{PGL}(n+1, \mathbf{R})$ considered as a principal bundle with base $M = \mathbf{PGL}(n+1, \mathbf{R})/H_0$ and group H_0, and the normal projective connection is then identified with the canonical Cartan connection of Section 20.5, Problem 3. The forms Ω_j^i and Ω_j are zero for this connection.

(f) A projective structure P on a pure manifold M of dimension n is said to be *flat* if, for each point $x_0 \in M$, there exists an open neighborhood U of x_0 and an isomorphism of the projective structure induced by P on U, onto the projective structure induced by the canonical structure on an open subset of $\mathbf{P}_n(\mathbf{R})$. Show that P is flat if and only if, for the corresponding normal projective connection, the forms Ω_j^i and Ω_j are zero.

15. Let $\mathbf{GO}(n, \mathbf{R})$ be the subgroup of $\mathbf{GL}(n, \mathbf{R})$ consisting of automorphisms U of the vector space $\mathbf{R}^n$ which are similitudes for the Euclidean scalar product $(\mathbf{x}|\mathbf{y})$, i.e., such that $(U \cdot \mathbf{x} | U \cdot \mathbf{y}) = \mu(U)(\mathbf{x}|\mathbf{y})$, where $\mu(U)$ is a real number > 0. The Lie algebra $\mathfrak{go}(n, \mathbf{R})$ of this group consists of the endomorphisms $V \in \mathrm{End}(\mathbf{R}^n)$ such that

$$(V \cdot \mathbf{x} | \mathbf{y}) + (\mathbf{x} | V \cdot \mathbf{y}) = \lambda(V)(\mathbf{x}|\mathbf{y}),$$

with $\lambda(V) \in \mathbf{R}$.

(a) If we take $\mathfrak{g} = \mathfrak{go}(n, \mathbf{R})$, then the Lie algebra $\mathfrak{g}^{(1)}$ (Section 20.7, Problem 6) is the set of mappings $\mathbf{u} \mapsto T_\mathbf{u}$ of $\mathbf{R}^n$ into $\mathrm{End}(\mathbf{R}^n)$ such that $(T_\mathbf{u} \cdot \mathbf{x} | \mathbf{y}) + (\mathbf{x} | T_\mathbf{u} \cdot \mathbf{y}) = \lambda_\mathbf{u}(\mathbf{x}|\mathbf{y})$, where $\mathbf{u} \mapsto \lambda_\mathbf{u}$ is a linear form on $E = \mathbf{R}^n$, so that $\lambda_\mathbf{u} = \langle \mathbf{u}, \rho(T) \rangle$, where $\rho(T) \in E^*$. Show that ρ is a bijection of $\mathfrak{g}^{(1)}$ onto E^*. (The injectivity of ρ follows from (20.9.2). To prove surjectivity, observe that if G is the canonical isomorphism of E onto E^*, so that $\langle \mathbf{x}, G \cdot \mathbf{y} \rangle = (\mathbf{x}|\mathbf{y})$, then for each $\mathbf{a}^* \in E^*$, if we put

$$(\gamma_{\mathbf{a}^*})_\mathbf{u} \cdot \mathbf{x} = \langle \mathbf{u}, \mathbf{a}^* \rangle \mathbf{x} + \langle \mathbf{x}, \mathbf{a}^* \rangle \mathbf{u} - (\mathbf{u}|\mathbf{x}) G^{-1} \cdot \mathbf{a}^*,$$

the mapping $\mathbf{u} \mapsto (\gamma_{\mathbf{a}^*})_\mathbf{u}$ belongs to $\mathfrak{g}^{(1)}$, and $\rho(T) = 2\mathbf{a}^*$.)

(b) The Lie algebra $\mathfrak{g}^{(2)}$ (Section 20.7, Problems 6 and 11) may be identified with the set of symmetric bilinear mappings $(\mathbf{u}, \mathbf{v}) \mapsto T_{\mathbf{u},\mathbf{v}}$ of $\mathbf{R}^n \times \mathbf{R}^n$ into $\mathrm{End}(\mathbf{R}^n)$ such that

$$(T_{\mathbf{u},\mathbf{v}} \cdot \mathbf{x} | \mathbf{y}) + (\mathbf{x} | T_{\mathbf{u},\mathbf{v}} \cdot \mathbf{y}) = \lambda_{\mathbf{u},\mathbf{v}}(\mathbf{x}|\mathbf{y}),$$

where λ is a symmetric bilinear form on $\mathbf{R}^n \times \mathbf{R}^n$. Show that if $(\mathbf{u}|\mathbf{v}) = 0$, then

$$\lambda_{\mathbf{u},\mathbf{u}}(\mathbf{v}|\mathbf{v}) = -\lambda_{\mathbf{v},\mathbf{v}}(\mathbf{u}|\mathbf{u}),$$

and deduce that for $n \geq 3$ we have $\lambda_{\mathbf{u},\mathbf{u}} = 0$ for all $\mathbf{u} \in \mathbf{R}^n$; hence $\lambda = 0$ and therefore $\mathfrak{g}^{(2)} = \{0\}$, by (20.9.2). What can be said when $n = 2$?

10. THE RIEMANN–CHRISTOFFEL TENSOR

In this section, "vector field" will always mean "C^∞ vector field."

(20.10.1) *On a pseudo-Riemannian manifold* M, *the curvature* $\boldsymbol{r}$ (17.20.4) *satisfies the identity*

(20.10.1.1) $(\boldsymbol{r}\cdot(X\wedge Y))\cdot Z+(\boldsymbol{r}\cdot(Z\wedge X))\cdot Y+(\boldsymbol{r}\cdot(Y\wedge Z))\cdot X=0.$

Since the torsion of the Levi–Civita connection is zero, Bianchi's identity (20.6.4.1) takes the form

$$\boldsymbol{\Omega}\wedge\boldsymbol{\sigma}=0.$$

In view of the formulas (20.6.4.2) and (20.6.6.11) and the definition of $\boldsymbol{\sigma}$, this gives (20.10.1.1) immediately.

If we lower the contravariant index (20.8.3) of the curvature tensor of the Levi–Civita connection on M, we obtain a *covariant tensor* $\mathbf{K}$ of order 4, called the *Riemann–Christoffel tensor* of M: it is defined by the condition

(20.10.2) $\langle\mathbf{K}, X\otimes Y\otimes Z\otimes W\rangle=(X|(\boldsymbol{r}\cdot(Z\wedge W))\cdot Y)_{\boldsymbol{g}}$

for any four vector fields X, Y, Z, W on M.

(20.10.3) *The Riemann–Christoffel tensor satisfies the identities*

(20.10.3.1) $\langle\mathbf{K}, X\otimes Y\otimes Z\otimes W+X\otimes Z\otimes W\otimes Y+X\otimes W\otimes Y\otimes Z\rangle=0.$

(20.10.3.2) $\langle\mathbf{K}, X\otimes Y\otimes Z\otimes W\rangle=-\langle\mathbf{K}, Y\otimes X\otimes Z\otimes W\rangle.$

(20.10.3.3) $\langle\mathbf{K}, X\otimes Y\otimes Z\otimes W\rangle=-\langle\mathbf{K}, X\otimes Y\otimes W\otimes Z\rangle.$

(20.10.3.4) $\langle\mathbf{K}, X\otimes Y\otimes Z\otimes W\rangle=\langle\mathbf{K}, Z\otimes W\otimes X\otimes Y\rangle.$

If $\mathbf{K}'$ *is another covariant tensor of order* 4 *on M which satisfies these four identities and is such that*

(20.10.3.5) $\langle\mathbf{K}, X\otimes Y\otimes X\otimes Y\rangle=\langle\mathbf{K}', X\otimes Y\otimes X\otimes Y\rangle$

for all vector fields X, Y, *on* M, *then* $\mathbf{K}'$ *is equal to* $\mathbf{K}$.

The identity (20.10.3.1) is an immediate consequence of the definition of **K** and of (20.10.1.1). Next, (20.10.3.3) is trivial. As to (20.10.3.2), it is enough to show that for each pair of indices i, j we have

(20.10.3.6) $$\Big(\sum_k \Omega_{ki}\,\mathbf{e}_k \,\Big|\, \mathbf{e}_j\Big) = -\Big(\sum_k \Omega_{kj}\,\mathbf{e}_k \,\Big|\, \mathbf{e}_i\Big),$$

or equivalently (20.6.6.1)

$$(\mathbf{\Omega}^{(\mathbf{R})} \cdot \mathbf{e}_i \,|\, \mathbf{e}_j) + (\mathbf{e}_i \,|\, \mathbf{\Omega}^{(\mathbf{R})} \cdot \mathbf{e}_j) = 0.$$

But this follows from the fact that the vector-valued 2-form $\mathbf{\Omega}^{(\mathbf{R})}$ takes its values in the Lie algebra of the group $\mathbf{O}(\Phi)$, characterized by the relation (20.9.1.1). Next, the identity (20.10.3.4) is a consequence of the three preceding ones. For if we temporarily abbreviate the expression $\langle \mathbf{K}, X \otimes Y \otimes Z \otimes W\rangle$ to $(XYZW)$, then it follows from (20.10.3.1), (20.10.3.2), and (20.10.3.3) that

$$\begin{aligned}(XYZW) &= -(XZWY) - (XWYZ) = (ZXWY) + (WXZY)\\ &= -(ZWYX) - (ZYXW) - (WYZX) - (WZXY)\\ &= 2(ZWXY) + (YZXW) + (YWZX)\\ &= 2(ZWXY) - (YXWZ)\\ &= 2(ZWXY) - (XYZW),\end{aligned}$$

from which (20.10.3.4) follows.

As to the last assertion of the proposition, by replacing **K** by $\mathbf{K}' - \mathbf{K}$ we may assume that $\langle \mathbf{K}, X \otimes Y \otimes X \otimes Y\rangle = 0$ identically, and we now have to prove that $\mathbf{K} = 0$. Using the same abridged notation as above, if we replace X by $X + Z$ in $(XYXY) = 0$ we obtain

$$(XYXY) + (ZYZY) + (ZYXY) + (XYZY) = 0,$$

which gives $(XYZY) + (ZYXY) = 0$ and therefore, using (20.10.3.4), $(XYZY) = 0$. Now replace Y by $Y + W$ in this relation, and in the same way we obtain

$$(XYZW) + (XWZY) = 0;$$

hence, using (20.10.3.3), $(XYZW) = (XWYZ)$, and so by cyclic permutation of Y, Z, W also $(XYZW) = (XZWY)$. But now, using (20.10.3.1), we have $3(XYZW) = 0$, and the proof is complete.

Given a moving frame $\mathbf{R} = (\mathbf{e}_1, \mathbf{e}_2, \ldots, \mathbf{e}_n)$ on M, let

(20.10.4) $$\mathrm{K}_{hijk} = \langle \mathbf{K}, \mathbf{e}_h \otimes \mathbf{e}_i \otimes \mathbf{e}_j \otimes \mathbf{e}_k\rangle.$$

Then the relations proved in (20.10.3) are equivalent, by linearity, to the following:

(20.10.4.1) $$\mathrm{K}_{hijk}+\mathrm{K}_{hjki}+\mathrm{K}_{hkij}=0,$$

(20.10.4.2) $$\mathrm{K}_{hijk}=-\mathrm{K}_{ihjk},$$

(20.10.4.3) $$\mathrm{K}_{hijk}=-\mathrm{K}_{hikj},$$

(20.10.4.4) $$\mathrm{K}_{hijk}=\mathrm{K}_{jkhi},$$

for each quadruple of indices h, i, j, k.

If **R** is an *orthogonal* moving frame, such that $(\mathbf{e}_i|\mathbf{e}_i)=\varepsilon_i$, where $\varepsilon_i=1$ for $i\leq p$ and $\varepsilon_i=-1$ for $i>p$ (20.8.2), the formulas (20.10.2) and (20.6.6.6) enable us to express the K_{hijk} in terms of the Ω_{ij}:

(20.10.4.5) $$\mathrm{K}_{hijk}=\varepsilon_h\langle\Omega_{hi},\mathbf{e}_j\wedge\mathbf{e}_k\rangle,$$

or equivalently,

(20.10.4.6) $$\Omega_{hi}=\sum_{j,k}\varepsilon_h\mathrm{K}_{hijk}\sigma_j\wedge\sigma_k.$$

Hence, by virtue of (20.10.4.2), we have

(20.10.4.7) $$\varepsilon_j\Omega_{ji}=-\varepsilon_i\Omega_{ij}.$$

(20.10.5) Given any two vector fields X, Y on a pseudo-Riemannian manifold M, we can use the curvature $\boldsymbol{r}$ of M to define a linear M-morphism of T(M) into itself: this morphism is denoted by $\boldsymbol{r}'(X, Y)$ and is defined by the condition that

(20.10.5.1) $$\boldsymbol{r}'(X,Y)\cdot Z=(\boldsymbol{r}\cdot(Z\wedge X))\cdot Y$$

for all vector fields Z on M. In particular, for each vector field X on M, we put

(20.10.5.2) $$\boldsymbol{r}''(X)=\boldsymbol{r}'(X,X).$$

For each $x\in\mathrm{M}$, the restriction $\boldsymbol{r}''(X)_x$ of $\boldsymbol{r}''(X)$ to $\mathrm{T}_x(\mathrm{M})$ is a *self-adjoint* endomorphism of $\mathrm{T}_x(\mathrm{M})$, relative to the bilinear form $\mathbf{g}(x)$ on this vector space. For if Z, W are any two vector fields on M, we have

$$\begin{aligned}(W|\boldsymbol{r}''(X)\cdot Z)=\langle\mathbf{K},W\otimes X\otimes Z\otimes X\rangle&=\langle\mathbf{K},Z\otimes X\otimes W\otimes X\rangle\\&=(Z|\boldsymbol{r}''(X)\cdot W)\end{aligned}$$

by virtue of (20.10.3.4).

(20.10.6) The *Ricci tensor* of M is by definition the covariant tensor $\mathbf{K}'$ of order 2 defined by the condition

(20.10.6.1) $$\langle \mathbf{K}', X \otimes Y \rangle = \operatorname{Tr}(\boldsymbol{r}'(X, Y))$$

for any two vector fields X, Y on M. This tensor is *symmetric*. For if

$$(\mathbf{e}_1, \mathbf{e}_2, \ldots, \mathbf{e}_n)$$

is an orthogonal moving frame such that $(\mathbf{e}_i | \mathbf{e}_i) = \varepsilon_i$, where $\varepsilon_i = 1$ for $i \leq p$ and $\varepsilon_i = -1$ for $i > p$ (20.8.2), then by definition we have

$$\begin{aligned}\operatorname{Tr}(\boldsymbol{r}'(X, Y)) &= \sum_{i=1}^{n} \varepsilon_i (\mathbf{e}_i | \boldsymbol{r}'(X, Y) \cdot \mathbf{e}_i) \\ &= \sum_{i=1}^{n} \varepsilon_i \langle \mathbf{K}, \mathbf{e}_i \otimes Y \otimes \mathbf{e}_i \otimes X \rangle \\ &= \sum_{i=1}^{n} \varepsilon_i \langle \mathbf{K}, \mathbf{e}_i \otimes X \otimes \mathbf{e}_i \otimes Y \rangle \\ &= \operatorname{Tr}(\boldsymbol{r}'(Y, X))\end{aligned}$$

by virtue of (20.10.3). An equivalent definition of $\mathbf{K}'$ is that it is obtained by contraction of the contravariant index and the first covariant index in the curvature tensor of the Levi–Civita connection (17.20.5). Relative to the same moving frame, if we put

(20.10.6.2) $$\mathrm{K}'_{jk} = \langle \mathbf{K}', \mathbf{e}_j \otimes \mathbf{e}_k \rangle,$$

then we have

(20.10.6.3) $$\mathrm{K}'_{jk} = \sum_{i=1}^{n} \varepsilon_i \mathrm{K}_{jiki}$$

with the notation of (20.10.4).

(20.10.7) For each tangent vector $\mathbf{h}_x \in \mathrm{T}_x(\mathrm{M})$, the number

(20.10.7.1) $$\operatorname{Ric}(\mathbf{h}_x) = \langle \mathbf{K}'(x), \mathbf{h}_x \otimes \mathbf{h}_x \rangle = \operatorname{Tr}(\boldsymbol{r}''(X)_x)$$

(where $X(x) = \mathbf{h}_x$) is called the *Ricci curvature* of M in the direction of the tangent vector $\mathbf{h}_x$. If $(\mathbf{e}_1, \ldots, \mathbf{e}_n)$ is a moving frame satisfying the same conditions as in (20.10.6), and if $\mathbf{h}_x = \sum_{j=1}^{n} \xi^j \mathbf{e}_j(x)$, then we have

(20.10.7.2) $$\operatorname{Ric}(\mathbf{h}_x) = \sum_{j,k} \mathrm{K}'_{jk}(x) \xi^j \xi^k.$$

(20.10.8) Since $\mathbf{h}_x \mapsto \operatorname{Ric}(\mathbf{h}_x)$ is a quadratic form on $T_x(M)$, it determines a *self-adjoint* endomorphism $\tilde{r}_x$ of $T_x(M)$ by the rule

(20.10.8.1) $$(\tilde{r}_x(\mathbf{h}_x) \mid \mathbf{k}_x) = \langle \mathbf{K}'(x), \mathbf{h}_x \otimes \mathbf{k}_x \rangle.$$

The number

(20.10.8.2) $$S(x) = \operatorname{Tr}(\tilde{r}_x)$$

is called the *scalar curvature* of M at the point x. If the moving frame

$$(\mathbf{e}_1, \dots, \mathbf{e}_n)$$

satisfies the same conditions as in (20.10.6), then with the same notation we have

(20.10.8.3) $$S = \sum_{j=1}^{n} K'_{jj} = \sum_{i,j} \varepsilon_i K_{jiji},$$

the second equality following from (20.10.4.3).

(20.10.9) *The cases* $n = 1, 2$.

On a manifold of dimension 1 all 2-forms are zero; hence the curvature of a Riemannian manifold of dimension 1 is zero.

Next suppose $n = 2$, and consider first the case of a Riemannian manifold M. Let $(\mathbf{e}_1, \mathbf{e}_2)$ be an *orthonormal* moving frame; then if we put

(20.10.9.1) $$K_{1212} = K$$

relative to this frame, the only nonzero components K_{hijk} are

(20.10.9.2) $$K_{1212} = K_{2121} = -K_{1221} = -K_{2112} = K$$

by virtue of (20.10.4). We have therefore for the Ricci tensor

(20.10.9.3) $$K'_{11} = K'_{22} = K, \qquad K'_{12} = 0,$$

so that the Ricci curvature is given by

(20.10.9.4) $$\operatorname{Ric}(\mathbf{h}_x) = K(x)((\xi^1)^2 + (\xi^2)^2).$$

Finally, the scalar curvature is $S = 2K$. The number $K(x)$ is called the *Gaussian curvature* of M at the point x. Relative to the orthonormal frame we have chosen, by virtue of (20.10.4.5) we have $K = \langle \Omega_{12}, \mathbf{e}_1 \wedge \mathbf{e}_2 \rangle$, so that $\Omega_{12} = K\sigma_1 \wedge \sigma_2$, and the formulas (20.6.5.10), (20.6.6.3), and (20.6.6.4) become in this case

(20.10.9.5) $$d\mathbf{e}_1 = \omega_{12}\,\mathbf{e}_2, \qquad d\mathbf{e}_2 = -\omega_{12}\mathbf{e}_1,$$

(20.10.9.6) $$d\sigma_1 = -\omega_{12} \wedge \sigma_2, \qquad d\sigma_2 = \omega_{12} \wedge \sigma_1,$$

(20.10.9.7) $$d\omega_{12} = K\sigma_1 \wedge \sigma_2.$$

If $n = 2$ and M is a pseudo-Riemannian manifold of signature (1, 1), we choose an orthogonal moving frame $(\mathbf{e}_1, \mathbf{e}_2)$ such that $(\mathbf{e}_1 | \mathbf{e}_1) = 1$, $(\mathbf{e}_2 | \mathbf{e}_2) = -1$. With the same notation (20.10.9.1), the relations (20.10.9.2) remain valid, but this time $K'_{11} = -K$, $K'_{22} = K$, $K'_{12} = 0$, and the scalar curvature S is zero.

PROBLEMS

1. In the notation of (20.10.4), show that Bianchi's identity (20.3.4) takes the form
$$\nabla_{\mathbf{e}_m} \cdot K_{ijkl} + \nabla_{\mathbf{e}_k} \cdot K_{ijlm} + \nabla_{\mathbf{e}_l} \cdot K_{ijkm} = 0$$
for all indices i, j, k, l, m. (Cf. Section 20.6, Problem 3.)

2. A Riemannian manifold M is said to be an *Einstein manifold* if there exists a C^∞ real-valued function λ on M such that the Ricci tensor $\mathbf{K}'$ is equal to $\lambda\mathbf{g}$. Show that if M is a connected Einstein manifold of dimension $n \geqq 3$, then the function λ is necessarily *constant*. (Using the formula (20.10.6.3) and Problem 1, show that we have
$$(n-2)\,\nabla_{\mathbf{e}_m} \cdot \lambda = 0$$
for all indices m.)

3. (a) Let M be an oriented differential manifold of dimension n, and let v be a C^∞ differential n-form in the orientation of M. For each C^∞ vector field X on M, $\theta_X \cdot v$ is a C^∞ n-form on M; hence can be written uniquely as $(\operatorname{div} X)v$, where $\operatorname{div} X$ is a C^∞ scalar-valued function on M, called the *divergence* of the vector field X. If X has compact support, show that
$$\int (\operatorname{div} X)v = 0.$$
(Use (17.15.3.4) and (17.15.5.1).)

(b) Suppose now that M carries a linear connection **C** such that $\nabla v = 0$. If the tensor field $\mathbf{A}_X$ is defined as in Section 20.6, Problem 6, show that for each C^∞ vector field X on M, we have

$$\operatorname{div} X = -\operatorname{Tr}(\mathbf{A}_X).$$

In particular, if the connection **C** is torsion-free, then

$$\operatorname{div} X = c_1^1(\nabla X). \tag{1}$$

In the notation of (20.6.5), this takes the form

$$\operatorname{div} X = \sum_{i=1}^{n} \langle \nabla_{\mathbf{e}_i} \cdot X, \sigma_i \rangle. \tag{2}$$

If f is any C^∞ real-valued function on M, then we have

$$\operatorname{div}(fX) = f \cdot \operatorname{div} X + \theta_X \cdot f.$$

(c) On a Riemannian manifold M, the definition (1) of the divergence coincides with that given in (a) by taking v to be the canonical volume form (20.8.5). If X, Y are two C^∞ vector fields on M, show that

$$\operatorname{div}(\mathbf{A}_X \cdot Y + (\operatorname{div} X)Y) = (\operatorname{div} X)(\operatorname{div} Y) - \operatorname{Tr}(\mathbf{A}_X \mathbf{A}_Y) - \langle \mathbf{K}', X \otimes Y \rangle \tag{3}$$

(*Yano's formula*). (Use the formula (2), with the moving frame $(\mathbf{e}_i)$ chosen so that $[\mathbf{e}_i, \mathbf{e}_j] = 0$ for all i, j.)

(d) Deduce from Section 20.6, Problem 6(c), that if X is an infinitesimal automorphism of the Levi–Civita connection, then div X is a *constant*.

4. Let M be a Riemannian manifold, X an infinitesimal isometry of M (Section 20.9, Problem 7).

(a) Show that for each tangent vector $\mathbf{h}_x \in T_x(M)$

$$\theta_{\mathbf{h}_x} \cdot \|X\|^2 = 2(\mathbf{h}_x \mid \mathbf{A}_X(x) \cdot X(x)).$$

(b) Let v be the geodesic (for the Levi–Civita connection) defined on a neighborhood of 0 in **R**, such that $v(0) = x$ and $v'(0) = \mathbf{h}_x$. Put $f(t) = \|(X \cdot v)(t)\|^2$. Deduce from (a) that

$$f''(0) = (\mathbf{h}_x \mid \nabla_{\mathbf{h}_x} \cdot (\mathbf{A}_X \cdot X)).$$

(Use the fact that $\nabla_{v'} \cdot v' = 0$, and Section 20.9, Problem 7.)

(c) Show that at a point x_0 where the function $\|X\|^2$ attains a relative maximum (resp. minimum), we have $\operatorname{div}(\mathbf{A}_X \cdot X) \leq 0$ (resp. ≥ 0). (Use (b), replacing $\mathbf{h}_x$ by the values at x_0 of an orthonormal moving frame, together with Problem 3(b).)

5. Let M be a connected Riemannian manifold whose Ricci tensor $\mathbf{K}'$ is such that at each point $x \in M$ the quadratic form $\mathbf{h}_x \mapsto \langle \mathbf{K}'(x), \mathbf{h}_x \otimes \mathbf{h}_x \rangle$ is *negative definite*. Let X be a vector field which is an infinitesimal isometry of M. Show that if the function $\|X\|$ attains a relative maximum at some point of M, then $X = 0$. (Use the hypothesis and Section 20.9, Problem 7, to show that at such a point x we have

$$\langle \mathbf{K}'(x), X(x) \otimes X(x) \rangle = 0$$

and therefore $X(x) = 0$; then use Section 20.6, Problem 7.)

6. (a) Let M be a compact Riemannian manifold whose Ricci tensor is such that the quadratic form $\mathbf{h}_x \mapsto \langle \mathbf{K}'(x), \mathbf{h}_x \otimes \mathbf{h}_x \rangle$ is negative definite at each point $x \in \mathrm{M}$. Show that the group of isometries of M is *discrete* (and in fact *finite*, cf. Section 20.16, Problem 11). (By replacing M if necessary by a two-sheeted covering, reduce to the case where M is oriented; deduce from Problem 3(c) and (d) that, if X is an infinitesimal isometry, we have

$$\int (\langle \mathbf{K}', X \otimes X \rangle + \mathrm{Tr}(\mathbf{A}_X \circ \mathbf{A}_X))\, dv = 0$$

and use Section 20.9, Problem 7.)

(b) Let M be a compact Riemannian manifold whose Ricci tensor is identically zero. Show that $\nabla X = 0$ for every infinitesimal isometry X. (Same method.)

7. (a) Let M be a pure Riemannian manifold of dimension n with metric tensor $\mathbf{g}$. Let a be a C^∞ function on M, which is everywhere strictly positive, and let $\mathbf{g}' = a^2\mathbf{g}$, thus defining a second Riemannian structure on M. Let $(\mathbf{e}_i)_{1 \leq i \leq n}$ be an orthonormal moving frame relative to $\mathbf{g}$, and let σ_i, ω_{ij} be the corresponding canonical and connection forms; then $(a^{-1}\mathbf{e}_i)_{1 \leq i \leq n}$ is an orthonormal moving frame relative to $\mathbf{g}'$, and the $a\sigma_i$ are the corresponding canonical forms. Let ω'_{ij} be the connection forms for $\mathbf{g}'$; show that

$$\omega'_{ij} - \omega_{ij} = b_j \sigma_i - b_i \sigma_j,$$

where the b_i are defined by

$$a^{-1}\, da = \sum_i b_i \sigma_i.$$

(Use (20.6.6.3).)

(b) Deduce from (a) that if S, S′ are the scalar curvatures for $\mathbf{g}$, $\mathbf{g}'$, respectively, then

$$\mathrm{S} - \mathrm{S}'a^2 = 2(n-1)\sum_i b_{ii}, \tag{1}$$

where

$$b_{ii} = \theta_{\mathbf{e}_i} \cdot b_i + \tfrac{1}{2}\left(\sum_{j \neq i} b_j^2 - b_i^2\right).$$

(Use (20.6.6.4).)

Show that the formula (1) can also be written in the form

$$(\mathrm{S} - \mathrm{S}'a^2)a^{(n-2)/2} = 2(n-1)\,\mathrm{div}(\mathrm{grad}(c)), \tag{2}$$

where $c = 2a^{(n-2)/2}/(n-2)$ if $n > 2$ and $c = \log a$ if $n = 2$.

(c) Suppose that M is compact and oriented. If v, v' are the Riemannian volumes for $\mathbf{g}$, $\mathbf{g}'$, respectively, show that $v' = a^n \cdot v$. Deduce from (2) and Problem 3(a) that

$$\int \mathrm{S}a^{(n-2)/2}\, dv = \int \mathrm{S}'a^{-(n-2)/2}\, dv'.$$

(d) Suppose again that M is compact and oriented, and moreover that the scalar curvatures S and S′ are *constant*. Show that there exists a C^∞ function f on $\mathbf{R}$ such that $f'(c) = \mathrm{S} - \mathrm{S}'a^2$, in the notation of (b). If $\mathrm{S} \leq 0$ (and therefore also $\mathrm{S}' \leq 0$, by (c)), deduce that

$$f'(c)\,\mathrm{div}(\mathrm{grad}(c)) \geq 0, \qquad f''(c) \geq 0.$$

By calculating div(grad($f(c)$)) and using Problem 3(a), prove that c (and hence also a) is necessarily *constant* on M.

(e) Deduce from (d) and (c) that if an oriented compact Riemannian manifold has *constant* scalar curvature ≤ 0, then every diffeomorphism of M onto itself which is *conformal* is an isometry.

(f) Let M be a Riemannian manifold, $\mathbf{g}$ its metric tensor, G a group of isometries of M, and ρ a C^∞ function on M which is everywhere strictly positive and which is invariant under no element $\neq e$ of G. Let $\mathbf{g}' = \rho^2 \mathbf{g}$. If M is endowed with the metric tensor $\mathbf{g}'$, then G is a group of conformal diffeomorphisms of M, none of which (other than the identity mapping) is an isometry.

8. (a) Let M be a Riemannian manifold, $(\mathbf{e}_i)$ an orthonormal moving frame defined on an open subset U of M. Show that for each C^∞ vector field X on U, we have

$$\tfrac{1}{2}\operatorname{div}(\operatorname{grad}\|X\|^2) = \sum_i (\nabla_{\mathbf{e}_i} \cdot (\nabla_{\mathbf{e}_i} \cdot X) | X) + \operatorname{Tr}(\mathbf{A}_X \mathbf{A}_X^*),$$

where, for each $x \in M$, $\mathbf{A}_X^*(x)$ is the *adjoint* of the endomorphism $\mathbf{A}_X(x)$ relative to the scalar product $\mathbf{g}(x)$ ($\mathbf{g}$ being the metric tensor on M) (11.5.1). (Use (20.9.5.3).)

(b) Suppose that X is defined on all of M and is an infinitesimal automorphism of the Levi–Civita connection. Show that in U we have

$$\sum_i (\nabla_{\mathbf{e}_i} \cdot (\nabla_{\mathbf{e}_i} \cdot X) | X) = -\langle \mathbf{K}', X \otimes X \rangle.$$

(Use formula (1) of Section 20.6, Problem 6, and the fact that we may assume that $\nabla_{\mathbf{e}_i(x)} \cdot \mathbf{e}_i = 0$ for all i at a given point $x \in M$ (Section 20.6, Problem 15).)

(c) Deduce from (a) and (b) and formula (3) of Problem 3 that if X is any infinitesimal automorphism of the Levi–Civita connection, then

$$(\operatorname{div} X)^2 - \operatorname{Tr}(\mathbf{A}_X^2) - \operatorname{Tr}(\mathbf{A}_X \mathbf{A}_X^*) = \operatorname{div}(\mathbf{A}_X \cdot X + (\operatorname{div} X)X + \tfrac{1}{2}\operatorname{grad}(\|X\|^2)).$$

(d) Assume now that M is *compact*, connected, and oriented. Show that every infinitesimal automorphism X of the Levi–Civita connection is an *infinitesimal isometry* (*Bochner–Yano theorem*). (Using Problems 3(a) and 3(d), show first that div $X = 0$. Then use (c) and Problem 3(a), and observe that if U is an endomorphism of a finite-dimensional Hilbert space E, then

$$\operatorname{Tr}((U + U^*)^2) = 2(\operatorname{Tr}(U^2) + \operatorname{Tr}(UU^*)) \geqq 0,$$

with equality only if $U + U^* = 0$ (11.5.7); finally, use Section 20.9, Problem 7.)

(e) Let M_1 be a compact connected Riemannian manifold, $\mathbf{g}_1$ its metric tensor. Let c be a positive constant $\neq 1$ and let M_2 be the Riemannian manifold defined by the metric tensor $\mathbf{g}_2 = c\mathbf{g}_1$ on M_1. Let $M = M_1 \times M_2$ be the product Riemannian manifold (Section 20.7, Problem 1). Show that the mapping $(x_1, x_2) \mapsto (x_2, x_1)$ is a diffeomorphism of M which is an automorphism of the Levi–Civita connection, but is not an isometry.

9. With the notation of (20.10.7) and (20.10.8), show that

$$\nabla S = 2c_2^1(\nabla((G_1^1)^{-1} \cdot \mathbf{K}')).$$

(Use Bianchi's identity (20.3.4) in the form given in Problem 1.)

11. EXAMPLES OF RIEMANNIAN AND PSEUDO-RIEMANNIAN MANIFOLDS

(20.11.1) With the notation of (20.7.10), consider a *symmetric pair* (G, H) relative to an involutory automorphism σ of G, where G is connected and acts faithfully on G/H. We propose to find all G-*invariant* pseudo-Riemannian metric tensors $\mathbf{g}$ on G/H. For such a tensor $\mathbf{g}$, if $x_0 = \pi(e)$, $\mathbf{g}(x_0)$ is a nondegenerate symmetric bilinear form on the tangent space $T_{x_0}(G/H)$, which may be identified with the subspace $\mathfrak{m}$ of $\mathfrak{g}_e$ (20.7.10). The action of H on $T_{x_0}(G/H)$ is then identified with the action $(t, \mathbf{u}) \mapsto \operatorname{Ad}(t) \cdot \mathbf{u}$ of H on $\mathfrak{m}$, and therefore $\Phi = \mathbf{g}(x_0)$ *must be invariant under the operators* $\operatorname{Ad}(t)$ *for all* $t \in H$. This necessary condition is also sufficient; for if we define

$$\mathbf{g}(s \cdot x_0) = s \cdot \mathbf{g}(x_0) \quad (= \mathbf{T}_2^0(s) \cdot \mathbf{g}(x_0))$$

for all $s \in G$, then the form $\mathbf{g}(s \cdot x_0)$ will depend only on the point $s \cdot x_0$, because $t \cdot \mathbf{g}(x_0) = \mathbf{g}(x_0)$ for all $t \in H$. We remark also that the metric tensor $\mathbf{g}$ is automatically *invariant under* σ_0, since $T_{x_0}(\sigma_0)$ is the symmetry $\mathbf{u} \mapsto -\mathbf{u}$ in $\mathfrak{m}$, which leaves invariant every bilinear form on $\mathfrak{m}$.

For *every* nondegenerate symmetric bilinear form Φ satisfying the above condition, the Levi-Civita connection corresponding to $\mathbf{g}$ is always equal to the *canonical linear connection* $\mathbf{C}$ on G/H defined in (20.7.10). For since $\mathbf{g}$ is a G-invariant tensor, we have $\nabla \mathbf{g} = 0$ relative to $\mathbf{C}$ (20.7.10.7); hence $\mathbf{C}$ is an $\mathbf{O}(\Phi)$-connection (20.9.5); and since moreover its torsion is zero (20.7.10.5), $\mathbf{C}$ is the Levi-Civita connection by virtue of (20.9.4).

(20.11.2) As a first example we take $G = \mathbf{R}^n$, $H = \{0\}$. This is clearly a symmetric pair relative to the symmetry $\sigma : \mathbf{x} \mapsto -\mathbf{x}$. Every nondegenerate symmetric bilinear form Φ on $\mathfrak{m} = \mathfrak{g}_e = \mathbf{R}^n$ trivially satisfies the condition of (20.11.1); hence defines a translation-invariant pseudo-Riemannian structure on $\mathbf{R}^n$; the corresponding connection has zero curvature, and corresponds to the trivial connection (20.4.1) in the (trivial) principal frame bundle. The geodesic trajectories are *straight lines*. If we take Φ to be the Euclidean scalar product on $\mathbf{R}^n$, the corresponding Riemannian structure on $\mathbf{R}^n$ is called *canonical*. Whenever we refer to $\mathbf{R}^n$ as a Riemannian manifold, it is always this canonical structure that is meant, unless the contrary is expressly stated.

It is immediately seen that for this structure *the canonical volume form* (20.8.6) for the canonical orientation of $\mathbf{R}^n$ is $d\xi^1 \wedge d\xi^2 \wedge \cdots \wedge d\xi^n$. All Euclidean displacements (with determinant ± 1) are evidently isometries of $\mathbf{R}^n$, and it can be shown that they are the only ones (Problem 1). By virtue of the invariance of the metric on $\mathbf{R}^n$ under translations, for each *discrete* subgroup D of $\mathbf{R}^n$ there exists on $\mathbf{R}^n/D$ a unique Riemannian structure for which $\mathbf{R}^n$ is a *Riemannian covering* of $\mathbf{R}^n/\mathrm{D}$. All these manifolds have *zero curvature*, and their geodesic trajectories are the canonical images of the straight lines in $\mathbf{R}^n$. The surface $\mathbf{R}^2/(\mathbf{Z} \times \{0\})$ is called the *circular cylinder*, and the surface $\mathbf{R}^2/(\mathbf{Z} \times \mathbf{Z})$, diffeomorphic to $\mathbf{T}^2$, is called the *flat torus*.

(20.11.3) Consider now a symmetric pair (G, H), where H is *compact* (cf. Section 20.16, Problem 11). In this case, there always exists a *Riemannian metric tensor* on G/H, invariant under G (and σ_0). By virtue of (20.11.1), this is a consequence of the following proposition:

(20.11.3.1) *Let* E *be a finite-dimensional real vector space and* K *a compact subgroup of* **GL**(E). *Then there exists a positive-definite symmetric bilinear form on* E *which is* K-*invariant.*

Let Φ be any positive-definite symmetric bilinear form on E. For all $x, y \in \mathrm{E}$, the function $s \mapsto \Phi(s \cdot x, s \cdot y)$ is continuous on K. If μ is a Haar measure on K (14.1.5), define

(20.11.3.2)
$$\Psi(x, y) = \int_{\mathrm{K}} \Phi(s \cdot x, s \cdot y)\, d\mu(s).$$

Clearly Ψ is a symmetric bilinear form on E. For each $t \in \mathrm{K}$ we have

$$\begin{aligned}\Psi(t \cdot x, t \cdot y) &= \int_{\mathrm{K}} \Phi(st \cdot x, st \cdot y)\, d\mu(s)\\ &= \int_{\mathrm{K}} \Phi(s \cdot x, s \cdot y)\, d\mu(s) = \Psi(x, y)\end{aligned}$$

by virtue of the invariance of μ. Finally, we have

$$\Psi(x, x) = \int_{\mathrm{K}} \Phi(s \cdot x, s \cdot x)\, d\mu(s),$$

and if $x \neq 0$, the continuous function $s \mapsto \Phi(s \cdot x, s \cdot x)$ is nonnegative for all $s \in \mathrm{K}$ and strictly positive at the point e. Since the support of μ is K (14.1), we have $\Psi(x, x) > 0$ (13.9), and the proof is complete.

The space G/H, endowed with a G-invariant Riemannian metric, is called a *Riemannian symmetric space*.

Remark

(20.11.3.3) If E is a finite-dimensional *complex* vector space and K is a compact subgroup of **GL**(E), the same proof as in (20.11.3.1) shows that there exists a K-invariant *positive-definite Hermitian form* on E.

(20.11.4) As a first example of (20.11.3), we shall show that the *sphere* S_n $(n \geqq 1)$ is a Riemannian symmetric space G/H, with H compact. From (16.11.5), S_n is canonically diffeomorphic to $\mathbf{SO}(n+1)/\mathbf{SO}(n)$ if $n \geqq 1$. If $(\mathbf{e}_i)_{1 \leqq i \leqq n+1}$ is the canonical basis of $\mathbf{R}^{n+1}$, and $s \in \mathbf{O}(n+1)$ the reflection in the hyperplane perpendicular to $\mathbf{e}_{n+1}$ (identified with $\mathbf{R}^n$), so that $s(\mathbf{e}_j) = \mathbf{e}_j$ for $j \leqq n$, and $s(\mathbf{e}_{n+1}) = -\mathbf{e}_{n+1}$, then the fact that s is involutory implies that $\sigma : t \mapsto sts^{-1}$ is an involutory automorphism of $\mathbf{SO}(n+1)$. Moreover it is immediately seen that the only elements of $\mathbf{SO}(n+1)$ that commute with s are those which map $\mathbf{e}_{n+1}$ to $\pm\mathbf{e}_{n+1}$, so that the subgroup of σ-invariant elements of $\mathbf{SO}(n+1)$ is $\mathbf{O}(n)$, of which $\mathrm{H} = \mathbf{SO}(n)$ is the identity component. Hence $(\mathbf{SO}(n+1), \mathbf{SO}(n))$ is a symmetric pair relative to σ, and H contains no normal subgroup $\neq \{e\}$. Now (19.11.2.4) the derived homomorphism σ_* is $\mathrm{Ad}(s) : \mathbf{u} \mapsto s\mathbf{u}s^{-1}$, acting on the space $\mathfrak{so}(n+1)$ of skew-symmetric endomorphisms of $\mathbf{R}^{n+1}$ (relative to the Euclidean scalar product $(\mathbf{x} \mid \mathbf{y})$). Let us determine the subspace $\mathfrak{m}$ of $\mathfrak{so}(n+1)$ consisting of the $\mathbf{u}$ such that $\sigma_*(\mathbf{u}) = -\mathbf{u}$. For this purpose, for each vector $\mathbf{z} \in \mathbf{R}^n$ consider the skew-symmetric endomorphism $a_\mathbf{z}$ of $\mathbf{R}^{n+1}$ whose matrix relative to the canonical basis is

$$\begin{pmatrix} 0_n & \mathbf{z} \\ -{}^t\mathbf{z} & 0 \end{pmatrix} \tag{20.11.4.1}$$

($\mathbf{z}$ being identified with a column matrix).

If $\mathbf{z} \neq 0$, it is easily checked, by taking an orthonormal basis of $\mathbf{R}^{n+1}$ containing $\mathbf{z}/\|\mathbf{z}\|$ and $\mathbf{e}_{n+1}$, that $a_\mathbf{z}$ leaves invariant the vectors in $\mathbf{R}^n$ perpendicular to $\mathbf{z}$, and that its restriction to the plane $\mathbf{Re}_{n+1} \oplus \mathbf{Rz}$, oriented by taking $(\mathbf{e}_{n+1}, \mathbf{z})$ as a positive basis, is a rotation of $\frac{1}{2}\pi$ followed by a homothety of ratio $\|\mathbf{z}\|$. It is immediate that $a_\mathbf{z} \in \mathfrak{m}$; hence $\mathbf{z} \mapsto a_\mathbf{z}$ is an injective linear mapping of $\mathbf{R}^n$ into $\mathfrak{m}$; and since

$$\dim(\mathfrak{m}) = \dim(\mathfrak{so}(n+1)) - \dim(\mathfrak{so}(n)) = n$$

(16.11.2), it follows (A.4.19) that the $a_\mathbf{z}$ are the whole of $\mathfrak{m}$. For each element $t \in \mathrm{H} = \mathbf{SO}(n)$, $\mathrm{Ad}(t)$ is the mapping $\mathbf{u} \mapsto t\mathbf{u}t^{-1}$(19.11.2.4), hence

$$\mathrm{Ad}(t) \cdot a_\mathbf{z} = a_{t \cdot \mathbf{z}},$$

and therefore H acts transitively on the lines and on the planes in $\mathfrak{m}$ (passing through 0). We may choose the Riemannian metric tensor $\mathbf{g}$ so that $\mathbf{g}(\pi(e))$ is the image under $\mathbf{z} \mapsto a_\mathbf{z}$ of the Euclidean scalar product on $\mathbf{R}^n$. The Riemannian structure so defined on S_n is called *canonical*, and whenever we refer to S_n as a Riemannian manifold it will always be this structure that is meant, unless the contrary is expressly stated.

By virtue of (20.7.10.4), the geodesics with origin $x_0 = \pi(e)$, identified with $\mathbf{e}_{n+1}$, are the curves $\xi \mapsto \exp(\xi a_{\mathbf{z}}) \cdot \mathbf{e}_{n+1}$, i.e. (if $\mathbf{z} \neq 0$), taking the same orthonormal basis as before,

(20.11.4.2) $$\xi \mapsto \cos(\|\mathbf{z}\|\xi)\mathbf{e}_{n+1} + \sin(\|\mathbf{z}\|\xi)\frac{\mathbf{z}}{\|\mathbf{z}\|}.$$

The geodesic trajectories are therefore the "great circles" on $\mathbf{S}_n$.

(20.11.5) If we take (for even n) $\mathrm{H} = \mathbf{O}(n)$ instead of $\mathbf{SO}(n)$ in $\mathrm{G} = \mathbf{SO}(n+1)$, we obtain the projective space $\mathbf{P}_n(\mathbf{R})$ as a Riemannian symmetric space (16.11.8 and 16.14.9) (for odd n, we must replace $\mathbf{O}(n)$ and $\mathbf{SO}(n+1)$ by their quotients by their common center $\{\pm 1\}$). The space $\mathfrak{m}$ is the same as in (20.11.4), and we take the $\mathbf{SO}(n+1)$-invariant Riemannian metric tensor on $\mathbf{P}_n(\mathbf{R})$ to be that which on $\mathfrak{m}$ is the image under (20.11.4.1) of the Euclidean scalar product. Then $\mathbf{S}_n$ is a two-sheeted *Riemannian covering* of $\mathbf{P}_n(\mathbf{R})$, and the geodesic trajectories in $\mathbf{P}_n(\mathbf{R})$ are the *projective lines*, the canonical images of the curves (20.11.4.2).

(20.11.6) In (20.11.4) let us now replace G by $\mathbf{SU}(n+1)$, the space $\mathbf{R}^{n+1}$ being replaced by $\mathbf{C}^{n+1}$, and $(\mathbf{x}|\mathbf{y})$ now denoting the canonical Hermitian scalar product. With the same definitions of s and σ, the elements of $\mathbf{SU}(n+1)$ which commute with s are now those which map $\mathbf{e}_{n+1}$ to $\zeta\mathbf{e}_{n+1}$, where $|\zeta| = 1$, and hence the subgroup H of σ-invariant elements may be identified with $\mathbf{U}(n)$ (which is connected), so that the symmetric homogeneous space G/H is canonically diffeomorphic to *complex projective space* $\mathbf{P}_n(\mathbf{C})$ ($n \geqq 1$); moreover, H contains no nontrivial normal subgroup of G. One shows as in (20.11.4) that $\mathfrak{m}$ consists of the skew-Hermitian endomorphisms $a_{\mathbf{z}}$ whose matrices are of the form

(20.11.6.1) $$\begin{pmatrix} O_n & \mathbf{z} \\ -{}^t\bar{\mathbf{z}} & 0 \end{pmatrix},$$

where $\mathbf{z} \in \mathbf{C}^n$ and $\bar{\mathbf{z}}$ is the vector whose components are the complex conjugates of the components of $\mathbf{z}$. The group H again acts via the adjoint representation, and we have $\mathrm{Ad}(t) \cdot a_{\mathbf{z}} = a_{t \cdot \mathbf{z}}$ for $t \in \mathbf{U}(n)$ and $\mathbf{z} \in \mathbf{C}^n$; this time H acts transitively on the real lines in $\mathfrak{m}$, but not on the real planes. We may take as nondegenerate symmetric bilinear form on $\mathfrak{m}$ the image under the mapping $\mathbf{z} \mapsto a_{\mathbf{z}}$ of the real part $\mathscr{R}(\mathbf{x}|\mathbf{y})$ of the canonical Hermitian scalar product on $\mathbf{C}^n$. The geodesic trajectories passing through $x_0 = \pi(e)$, the canonical image of $\mathbf{e}_{n+1}$ in $\mathbf{P}_n(\mathbf{C})$, are again the canonical images of the curves (20.11.4.2), hence are diffeomorphic to circles.

We may also replace G in (20.11.4) by the quaternionic unitary group $\mathbf{U}(n+1, \mathbf{H})$, and $\mathbf{R}^{n+1}$ by $\mathbf{H}^{n+1}$. This time the subgroup H may be identified with $\mathbf{U}(n, \mathbf{H})$, and G/H with *quaternionic projective space* $\mathbf{P}_n(\mathbf{H})$ (16.11.9). Everything said above extends immediately to this case; in particular, the geodesic trajectories are again diffeomorphic to circles.

(20.11.7) Now let G be the identity component of the group $\mathbf{O}(\Psi)$, where Ψ is the form $\sum_{j=1}^{n} \xi_j \eta_j - \xi_{n+1}\eta_{n+1}$ on $\mathbf{R}^{n+1}$ (16.11.1). If s and σ are defined as in (20.11.4), the identity component H of the subgroup of σ-invariant elements is again $\mathbf{SO}(n)$. The subspace $\mathfrak{m}$ of $\mathfrak{g}_e$ is now the image of $\mathbf{R}^n$ under the mapping $\mathbf{z} \mapsto a_{\mathbf{z}}$, where this time $a_{\mathbf{z}}$ is the symmetric endomorphism of $\mathbf{R}^n$ whose matrix relative to the canonical basis is

(20.11.7.1)
$$\begin{pmatrix} O_n & \mathbf{z} \\ {}^t\mathbf{z} & 0 \end{pmatrix}.$$

The group H acts on $\mathfrak{m}$ via the adjoint representation, and we have $\mathrm{Ad}(t) \cdot a_{\mathbf{z}} = a_{t \cdot \mathbf{z}}$. It acts transitively on the lines and the planes in $\mathfrak{m}$.

Since the elements of G have determinant equal to 1, the group H may be identified with the stabilizer of the point $\mathbf{e}_{n+1}$, for the canonical action of G on $\mathbf{R}^{n+1}$; consequently G/H is canonically diffeomorphic to the connected component of $\mathbf{e}_{n+1}$ in the $\mathbf{O}(\Psi)$-orbit of $\mathbf{e}_{n+1}$ (16.10.8), i.e., the *demi-quadric* Q_+ in $\mathbf{R}^{n+1}$ defined by the relations

(20.11.7.2)
$$\xi_{n+1} \geqq 0, \qquad \xi_1^2 + \xi_2^2 + \cdots + \xi_n^2 - \xi_{n+1}^2 = 1.$$

We may again take the form Φ on $\mathfrak{m}$ to be that which corresponds to the Euclidean scalar product on $\mathbf{R}^n$ under the isomorphism $\mathbf{z} \mapsto a_{\mathbf{z}}$. The geodesics with origin $\mathbf{e}_{n+1}$ are the curves

$$\xi \mapsto \cosh(\|\mathbf{z}\|\xi)\mathbf{e}_{n+1} + \sinh(\|\mathbf{z}\|\xi)\frac{\mathbf{z}}{\|\mathbf{z}\|};$$

hence their images are the intersections of Q_+ with the planes passing through 0 and $\mathbf{e}_{n+1}$. It follows that we obtain all geodesic trajectories by intersecting Q_+ with planes through 0, and these trajectories are therefore branches of hyperbolas with center at 0.

Since Q_+ may be considered as the graph of the function

$$(\xi_1, \ldots, \xi_n) \mapsto (1 - \xi_1^2 - \cdots - \xi_n^2)^{1/2},$$

it follows from (16.8.13) that the projection of Q_+ on $\mathbf{R}^n$ is a diffeomorphism; by transporting the Riemannian structure of Q_+ to $\mathbf{R}^n$ by means of this diffeomorphism, we obtain a structure of Riemannian manifold on $\mathbf{R}^n$. The space $\mathbf{R}^n$ with this Riemannian structure is called *hyperbolic n-space* and is denoted by $\mathbf{Y}_n$.

(20.11.8) Let G be a connected Lie group *whose center is* $\{e\}$. We shall show that G may be regarded as a *symmetric homogeneous space.* Consider the product group $G \times G$, and let σ be the involutory automorphism $(s, t) \mapsto (t, s)$. The set of σ-invariant elements is then the diagonal subgroup D of all elements (s, s), which contains no nontrivial normal subgroup of $G \times G$, because $(s, t)(x, x)(s^{-1}, t^{-1}) = (sxs^{-1}, txt^{-1})$ belongs to D for all $(s, t) \in G \times G$ only if x is in the center of G. The subspace $\mathfrak{m}$ of $\mathfrak{g}_e \times \mathfrak{g}_e$ formed by the eigenvectors of σ_* for the eigenvalue -1 is clearly the space of all vectors $(\mathbf{u}, -\mathbf{u})$, where $\mathbf{u} \in \mathfrak{g}_e$, and we have $\mathrm{Ad}(s, s) \cdot (\mathbf{u}, -\mathbf{u}) = (\mathrm{Ad}(s) \cdot \mathbf{u}, -\mathrm{Ad}(s) \cdot \mathbf{u})$ for $s \in G$. In order that we should be able to apply the method of (20.11.1) to the symmetric pair $(G \times G, D)$, it is therefore necessary and sufficient that there should exist on $\mathfrak{g}_e$ a *nondegenerate symmetric bilinear form* Φ *which is invariant under the operators* $\mathrm{Ad}(s)$ *for all* $s \in G$.

Let G_0 be the subgroup $\{e\} \times G$ of $G \times G$. The mapping

$$\varphi : (x, y) \mapsto ((e, yx^{-1}), (x, x))$$

is a diffeomorphism of $G \times G$ onto $G_0 \times D$, and the inverse diffeomorphism is $\varphi^{-1} : ((e, t), (s, s)) \mapsto (s, ts)$. We have $\varphi(xs, ys) = ((e, yx^{-1}), (xs, xs))$; hence if $\pi : G \times G \to (G \times G)/D$ is the canonical projection, there exists a diffeomorphism $\psi : (G \times G)/D \to G_0$ such that $\psi(\pi(x, y)) = (e, yx^{-1})$ (16.10.4), and φ is therefore an *isomorphism* of the principal bundle

$$(G \times G, (G \times G)/D, \pi)$$

onto the trivial principal bundle $(G_0 \times D, G_0, \mathrm{pr}_1)$ with the same group D. The tangent linear mapping $T_{(e, e)}(\varphi)$ is $(\mathbf{u}, \mathbf{v}) \mapsto ((0, \mathbf{v} - \mathbf{u}), (\mathbf{u}, \mathbf{u}))$ (16.9.9), and therefore defines, by restriction to $\mathfrak{m}$ and canonical identification of $\mathrm{Lie}(G_0)$ with $\mathfrak{g}_e$, an isomorphism $\rho : (\mathbf{u}, -\mathbf{u}) \mapsto -2\mathbf{u}$ of $\mathfrak{m}$ onto $\mathfrak{g}_e$. We see therefore that there exists a unique pseudo-Riemannian structure on G which is invariant under *left translations* by the elements of G and whose metric tensor $\mathbf{g}$ is such that $\mathbf{g}(e) = \Phi$. The invariance of Φ under the adjoint representation shows moreover that $\mathbf{g}$ is also invariant under *right translations*, since $\mathbf{g}(x) \cdot s = x \cdot \mathbf{g}(e) \cdot s = (xs) \cdot (s^{-1} \cdot \mathbf{g}(e) \cdot s) = (xs) \cdot \mathbf{g}(e) = \mathbf{g}(xs)$. Finally, the formula (20.7.10.6) giving the curvature of the canonical connection on $(G \times G)/D$ at the point $\pi(e)$ also gives, by transport of structure via φ and ρ, the curvature of the connection on the pseudo-Riemannian manifold G at the point e:

(20.11.8.1) $$(r(e) \cdot (\mathbf{u} \wedge \mathbf{v})) \cdot \mathbf{w} = -\tfrac{1}{4}[[\mathbf{u}, \mathbf{v}], \mathbf{w}]$$

for $\mathbf{u}, \mathbf{v}, \mathbf{w} \in \mathfrak{g}_e$ (19.8.4.2). We recall that the geodesic trajectories are the *translates of the one-parameter subgroups* of G and that, for $\mathbf{u}, \mathbf{v} \in \mathfrak{g}_e$, we have

(20.11.8.2) $$\nabla_{X_{\mathbf{u}}} \cdot X_{\mathbf{v}} = \tfrac{1}{2} X_{[\mathbf{u}, \mathbf{v}]} = \tfrac{1}{2}[X_{\mathbf{u}}, X_{\mathbf{v}}].$$

Moreover, since the vector fields $X_{\mathbf{u}}$ and $X_{\mathbf{v}}$ are by definition invariant under left translations, the function $(X_{\mathbf{u}} | X_{\mathbf{v}})$ is *constant* on G. Hence, for each vector field Y on G we have, by (20.9.5.3),

$$0 = \theta_Y \cdot (X_{\mathbf{u}} | X_{\mathbf{v}}) = (\nabla_Y \cdot X_{\mathbf{u}} | X_{\mathbf{v}}) + (X_{\mathbf{u}} | \nabla_Y \cdot X_{\mathbf{v}});$$

taking in particular $Y = X_{\mathbf{w}}$, where $\mathbf{w} \in \mathfrak{g}_e$, we obtain by virtue of (20.11.8.2)

(20.11.8.3) $$([X_{\mathbf{u}}, X_{\mathbf{v}}] | X_{\mathbf{w}}) = (X_{\mathbf{u}} | [X_{\mathbf{v}}, X_{\mathbf{w}}]).$$

We remark that, when G is *compact*, it follows from above and from (20.11.3.1) that there exists a *Riemannian* structure on G, *invariant under left and right translations.*

Remark

(20.11.9) With the notation of (20.11.1), if H acts *transitively* via the adjoint representation on the lines in $\mathfrak{m}$, then the values of $\mathbf{g}(x_0)$ in $\mathfrak{m}$ are determined by the value at *one* nonzero vector; hence there exists (up to a constant factor) only *one* G-invariant pseudo-Riemannian metric tensor on G/H.

PROBLEMS

1. Show that Euclidean displacements (with determinant ± 1) are the only isometries of $\mathbf{R}^n$ for its canonical Riemannian structure, by using Section 20.6, Problem 9(a).

2. (a) The linear connection on $\mathbf{R}^n$ corresponding to the trivial connection (20.4.1) in the space of frames $R(\mathbf{R}^n)$ (endowed with the canonical trivialization $r_x \mapsto (x, \tau_x \circ r_x)$) has the property that, relative to the canonical moving frame $x \mapsto \tau_x^{-1}$, all the connection forms ω_{ij} are identically zero. For brevity, we shall call this connection the *canonical* linear connection on $\mathbf{R}^n$. The pseudo-Riemannian structures whose Levi–Civita connection is the canonical linear connectton are the canonical structures defined in (20.11.2): relative to the canonical moving frame, the components g_{ij} of the metric tensor $\mathbf{g}$ are *constants* (cf. (20.9.6.5)).
(b) Let G be a discrete group acting properly and freely on $\mathbf{R}^n$; then $\mathbf{R}^n$ is a covering of $M = \mathbf{R}^n/G$ (Section 16.28, Problem 4). There exists a unique linear connection on M whose inverse image under the projection $p : \mathbf{R}^n \mapsto M$ is the canonical linear connection. In order that this connection should be the Levi–Civita connection of a pseudo-Riemannian structure on M, it is necessary and sufficient that the group G should be a group of isometries for one of the canonical structures on $\mathbf{R}^n$ defined in (20.11.2).
(c) Take $n = 2$ in (b), and take G to be the group of affine-linear mappings $(x, y) \mapsto (x + ny + m, y + n)$, where $m, n \in \mathbf{Z}$. Show that $\mathbf{R}^2/G$ is diffeomorphic to

the torus $\mathbf{T}^2$, but that the (flat) linear connection induced by the canonical linear connection on $\mathbf{R}^2$ is not the Levi–Civita connection for any pseudo-Riemannian structure on M.

3. Let K be a connected Lie group, H a connected Lie subgroup of K. Let $\mathfrak{k}_e$, $\mathfrak{h}_e$ be the Lie algebras of K, H, respectively, and suppose that $\mathfrak{k}_e = \mathfrak{h}_e \oplus \mathfrak{m}$, where $\mathfrak{m}$ is a vector subspace of $\mathfrak{k}_e$ such that $\mathrm{Ad}(t) \cdot \mathfrak{m} \subset \mathfrak{m}$ for all $t \in \mathrm{H}$. Let $\pi : \mathrm{K} \to \mathrm{K/H}$ be the canonical projection, and put $x_0 = \pi(e)$.

(a) Show that the K-invariant pseudo-Riemannian structures on K/H are in one-to-one correspondence with the nondegenerate symmetric bilinear forms Φ on $\mathfrak{m}$ (canonically identified with $\mathrm{T}_{x_0}(\mathrm{K/H})$) which are invariant under the adjoint action of H, or equivalently, satisfy $\Phi([\mathbf{w}, \mathbf{u}], \mathbf{v}) + \Phi(\mathbf{u}, [\mathbf{w}, \mathbf{v}]) = 0$ for all $\mathbf{u}, \mathbf{v} \in \mathfrak{m}$ and $\mathbf{w} \in \mathfrak{h}_e$.
(b) Let Φ be a nondegenerate symmetric bilinear form on $\mathfrak{m}$ which satisfies this condition, and let G be the subgroup of $\mathbf{GL}(n, \mathbf{R})$ which leaves Φ invariant ($\mathfrak{m}$ being identified with $\mathbf{R}^n$), so that the pseudo-Riemannian structure on K/H corresponding to Φ is a G-structure. With the notation of Section 20.7, Problem 3, the K-invariant linear connections on K/H are in one-to-one correspondence with the linear mappings $\mathbf{f}_{\mathfrak{m}} : \mathfrak{m} \mapsto \mathfrak{g}_e$ such that $\mathbf{f}_{\mathfrak{m}}(\mathrm{Ad}(t) \cdot \mathbf{w}) = \mathrm{Ad}(\lambda(t)) \cdot \mathbf{f}_{\mathfrak{m}}(\mathbf{w})$ for all $t \in \mathrm{H}$ and $\mathbf{w} \in \mathfrak{m}$. Show that the Levi–Civita connection on K/H corresponds to the linear mapping $\mathbf{f}_{\mathfrak{m}}$ given by

$$\mathbf{f}_{\mathfrak{m}}(\mathbf{u}) \cdot \mathbf{v} = \tfrac{1}{2}[\mathbf{u}, \mathbf{v}]_{\mathfrak{m}} + \mathbf{B}(\mathbf{u}, \mathbf{v})$$

for $\mathbf{u}, \mathbf{v} \in \mathfrak{m}$, where $\mathbf{B} : \mathfrak{m} \times \mathfrak{m} \to \mathfrak{m}$ is the symmetric bilinear mapping defined by the relation

$$2\Phi(\mathbf{B}(\mathbf{u}, \mathbf{v}), \mathbf{w}) = \Phi(\mathbf{u}, [\mathbf{w}, \mathbf{v}]_{\mathfrak{m}}) + \Phi([\mathbf{w}, \mathbf{u}]_{\mathfrak{m}}, \mathbf{v})$$

for $\mathbf{u}, \mathbf{v}, \mathbf{w} \in \mathfrak{m}$. (Use the fact that $Z_{\mathbf{u}}$ is an infinitesimal isometry for all $\mathbf{u} \in \mathfrak{m}$; equation (1) of Section 20.9, Problem 7; and Section 20.7, Problem 3.) The Levi–Civita connection corresponding to Φ coincides with the torsion-free connection corresponding to the canonical connection in $\mathrm{S_G}(\mathrm{K/H})$ (Section 20.7, Problem 3) if and only if

$$\Phi(\mathbf{u}, [\mathbf{w}, \mathbf{v}]_{\mathfrak{m}}) + \Phi([\mathbf{w}, \mathbf{u}]_{\mathfrak{m}}, \mathbf{v}) = 0$$

for all $\mathbf{u}, \mathbf{v}, \mathbf{w} \in \mathfrak{m}$.

The curvature of the Levi–Civita connection corresponding to Φ then satisfies the relation

$$\langle \mathbf{g}(x_0), \mathbf{u} \otimes (\boldsymbol{r} \cdot (\mathbf{u} \wedge \mathbf{v})) \cdot \mathbf{v} \rangle = \tfrac{1}{4}\Phi([\mathbf{u}, \mathbf{v}]_{\mathfrak{m}}, [\mathbf{u}, \mathbf{v}]_{\mathfrak{m}}) - \Phi([[\mathbf{u}, \mathbf{v}]_{\mathfrak{h}_e}, \mathbf{v}], \mathbf{u})$$

for all $\mathbf{u}, \mathbf{v} \in \mathfrak{m}$, where $\mathbf{g}$ is the pseudo-Riemannian metric tensor on K/H.

4. Generalize the results of (20.11.4) to the spaces $\mathbf{G}'_{n,p}(\mathbf{R})$ (Section 16.21, Problem 1) by considering the symmetry $s \in \mathbf{O}(n)$ such that $s(\mathbf{e}_j) = -\mathbf{e}_j$ for $j \leqq p$ and $s(\mathbf{e}_j) = \mathbf{e}_j$ for $p + 1 \leqq j \leqq n$, and the involutory automorphism $\sigma : t \mapsto sts^{-1}$ of $\mathbf{SO}(n)$. The space $\mathfrak{m}$ may here be identified with the space of real matrices

$$a_X = \begin{pmatrix} O_p & -{}^tX \\ X & O_{n-p} \end{pmatrix},$$

where X is any $(n-p) \times p$ matrix. We may take the metric tensor $\mathbf{g}$ to be that whose restriction to $\mathfrak{m}$ is $(a_X | a_Y) = -\tfrac{1}{2}\,\mathrm{Tr}(a_X a_Y)$.

5. In complex projective space $\mathbf{P}_n(\mathbf{C})$, let $Q_{n-1}(\mathbf{C})$ denote the *complex quadric* defined by the homogeneous equation

$$(z^0)^2 + (z^1)^2 + \cdots + (z^n)^2 = 0,$$

where (z^j) are the homogeneous coordinates of a point in $\mathbf{P}_n(\mathbf{C})$. Show that the subgroup $\mathbf{SO}(n+1)$ of $\mathbf{SU}(n+1)$, which acts on $\mathbf{P}_n(\mathbf{C})$ by restriction of the canonical action of $\mathbf{SU}(n+1)$, acts transitively on $Q_{n-1}(\mathbf{C})$, and that $Q_{n-1}(\mathbf{C})$ is diffeomorphic to $\mathbf{G}'_{n+1,2}(\mathbf{R})$, and hence is endowed with a structure of Riemannian symmetric space (Problem 4).

6. Consider the Hermitian form on $\mathbf{C}^n$ of signature $(p, n-p)$

$$\Psi(x, y) = \sum_{j=1}^{p} \xi_j \bar{\eta}_j - \sum_{j=p+1}^{n} \xi_j \bar{\eta}_j$$

and the unitary group $\mathbf{U}(\Psi)$, which is connected. The space $R_{n,p}$ of sequences $(\mathbf{x}_k)_{1 \leq k \leq p}$ of p vectors in $\mathbf{C}^n$ which are *orthonormal* relative to Ψ may be identified with the homogeneous space $\mathbf{U}(\Psi)/\mathbf{U}(p)$ (argue as in (**16.11.4**)). If $P_{n,p}$ is the subspace of $G_{n,p}(\mathbf{C})$ consisting of the p-dimensional subspaces of $\mathbf{C}^n$ on which the restriction of Ψ is positive definite, then $P_{n,p}$ may be identified with $R_{n,p}/\mathbf{U}(n-p)$ and hence with $\mathbf{U}(\Psi)/(\mathbf{U}(p) \times \mathbf{U}(n-p))$. Define a structure of Riemannian symmetric space on $P_{n,p}$ by proceeding as in Problem 4. Show that $P_{n,p}$ may be canonically identified with the space of $(n-p) \times p$ matrices Z with complex entries such that the Hermitian matrix $I_p - {}^t\bar{Z} \cdot Z$ is positive definite. (An element of $R_{n,p}$ may be identified with an $n \times p$ matrix of the form $\begin{pmatrix} X \\ Y \end{pmatrix}$, where X is a $p \times p$ matrix, such that ${}^t\bar{X} \cdot X - {}^t\bar{Y} \cdot Y = I_p$. To this matrix corresponds $Z = YX^{-1}$.)

7. (a) Let M be a connected differential manifold of dimension n, and $\mathbf{C}$ a linear connection on M. For each $x \in$ M, let U be a symmetric neighborhood of $\mathbf{0}_x$ in $T_x(M)$ contained in the domain of definition of the exponential mapping $\exp_x$, and such that $\exp_x | U$ is a diffeomorphism onto an open neighborhood V of x in M. Then there exists a unique diffeomorphism s_x of V onto itself such that $s_x(\exp_x(\mathbf{u})) = \exp_x(-\mathbf{u})$ for all $\mathbf{u} \in$ U, and we have $s_x \circ s_x = 1_V$. The diffeomorphism s_x is called the *symmetry* with center x in V. If U′ is another symmetric neighborhood of $\mathbf{0}_x$ having the same properties as U, and if V′ is its image under $\exp_x$, then the symmetries of V and V′ with center x coincide on V ∩ V′. For every tensor field $\mathbf{Z} \in \mathcal{T}^p_q(M)$ we have $\mathrm{T}^p_q(\mathrm{T}_x(s_x)) \cdot \mathbf{Z}(x) = (-1)^{p+q}\mathbf{Z}(x)$.
(b) For each $x \in$ M the symmetry s_x, defined on a symmetric open neighborhood V_x of x in M, is an automorphism of the connection induced by $\mathbf{C}$ on V_x if and only if the torsion tensor $\mathbf{t}$ is zero and the curvature tensor $\mathbf{r}$ satisfies $\nabla \mathbf{r} = 0$. (To show that the condition is sufficient, use Section **20.6**, Problems 17 and 9(a).) The connection $\mathbf{C}$ is then said to be *locally symmetric*.

8. With the notation and hypotheses of Problem 7, the connection $\mathbf{C}$ is said to be *symmetric* if, for each $x \in$ M, the symmetry s_x is the restriction of an automorphism of $\mathbf{C}$ (which is unique by Section **20.6**, Problem 9(a)). We denote this automorphism also by s_x. For the rest of this Problem, assume that $\mathbf{C}$ is symmetric.

(a) Let $v : \mathrm{I} \to \mathrm{M}$ be a geodesic for $\mathbf{C}$ defined on an open interval $\mathrm{I} \subset \mathbf{R}$. Show that for each $t_0 \in$ I, if $x = v(t_0)$, the mapping $u : t \mapsto s_x(v(t))$ of I into M is a geodesic for $\mathbf{C}$ such that $u(t_0) = v(t_0)$ and $u'(t_0) = -v'(t_0)$.

(b) Deduce from (a) that the connection **C** is complete (Section 20.6, Problem 8).
(c) Deduce from (a) that if $x, y \in M$ are the endpoints of a geodesic arc, then there exists $z \in M$ such that $s_z(x) = y$. Deduce that the group G of automorphisms of **C** is *transitive* on M. (Use (20.17.5).)
(d) In general, if a Lie group G acts differentiably and transitively on a connected differential manifold M, then its identity component G_0 also acts transitively on M. (Observe that the orbits of G_0 in M are open sets.) Hence deduce from (c) that if G is the identity component of the Lie group of automorphisms of **C** (Section 20.6, Problem 10), M may be identified with G/H_x, where H_x is the stabilizer of a point $x \in M$. Furthermore, if σ is the involutory automorphism $t \mapsto s_x \circ t \circ s_x^{-1}$ and if H is the subgroup of σ-invariant elements of G, then H_x is contained in H and contains the identity component of H, so that (G, H_x) is a symmetric pair (20.4.3), and **C** is the canonical connection on G/H_x. (To show that H_x contains the identity component of H, observe that the orbit of x for a one-parameter subgroup of H consists of points invariant under s_x.)
(e) Show that the Riemannian symmetric spaces are the Riemannian manifolds whose Levi-Civita connection is symmetric. (Use Section 20.9, Problem 8.)

9. Let (G, H) be a symmetric pair and let G′ be a σ-stable connected Lie group immersed in G; then $H' = G' \cap H$ is closed in G′ for the proper topology of G′. Show that the canonical mapping of G′/H′ into G/H is bijective if and only if the Lie algebra $\mathfrak{g}_e'$ of G′ contains $\mathfrak{m}$. The subspace $\mathfrak{m} + [\mathfrak{m}, \mathfrak{m}]$ is an ideal in $\mathfrak{g}_e$, and the Lie algebras $\mathfrak{g}_e'$ which contain $\mathfrak{m}$ are those which contain this ideal.

Hence construct an example of a Riemannian symmetric space G/H such that G is not the identity component of the group of isometries of G/H.

10. Let M be a connected differential manifold endowed with a symmetric linear connection, and let Γ be a discrete group acting properly and freely on M, so that M is a covering of $M' = M/\Gamma$ (Section 16.28, Problem 4). Suppose moreover that Γ leaves invariant the connection **C**, so that M is canonically endowed with a linear connection **C**′, the canonical image of **C** under the projection $\pi : M \to M'$ (17.18.6). The connection **C**′ is then locally symmetric (Problem 7); for it to be symmetric, it is necessary and sufficient that, for each $x \in M$, the image of any orbit of Γ under the symmetry s_x should be an orbit of Γ.

Take M to be the sphere S_3, considered as the submanifold of $\mathbf{C}^2$ defined by the equation $|z^1|^2 + |z^2|^2 = 1$; let p, q be coprime integers, and take Γ to be the cyclic group of order p generated by the orthogonal transformation

$$(z^1, z^2) \mapsto (z^1 \exp(2\pi i/p), z^2 \exp(2\pi iq/p)).$$

Show that the manifold M/Γ is not symmetric.

12. RIEMANNIAN STRUCTURE INDUCED ON A SUBMANIFOLD

(20.12.1) Let M be a pure pseudo-Riemannian manifold of dimension n, **g** its metric tensor. Let M′ be a pure submanifold of M, of dimension $n' < n$, and let $f: M' \to M$ be the canonical injection. Consider the inverse image $\mathbf{g}' = {}^t f(\mathbf{g})$ of the covariant tensor **g** on M, which is a symmetric covariant tensor of order 2 on M′. At a point $x \in M'$, the symmetric bilinear form $(\mathbf{h}_x, \mathbf{k}_x) \mapsto \langle \mathbf{g}'(x), \mathbf{h}_x \otimes \mathbf{k}_x \rangle = \langle \mathbf{g}(x), \mathbf{h}_x \otimes \mathbf{k}_x \rangle$ on $T_x(M') \subset T_x(M)$ is nondegenerate if and only if $T_x(M')$ is not an isotropic subspace of $T_x(M)$, in which case we shall say that M′ is *nonisotropic* (relative to M) at the point x. Since the nondegeneracy of $\mathbf{g}'(x)$ may be expressed by the nonvanishing of its discriminant, relative to a basis of $T_x(M')$, it follows that the set of points $x \in M'$ at which M′ is nonisotropic is *open* in M′. It is equal to M′ if M is a *Riemannian manifold.*

From now on in this section we shall consider only the case where M is a Riemannian manifold. The reader may verify for himself that the results generalize easily (at the cost of some notational complication) to pseudo-Riemannian manifolds, provided that we restrict consideration to submanifolds which are nonisotropic at all points (cf. Section 20.13, Problem 5).

Example

(20.12.1.1) If we equip the sphere $\mathbf{S}_n$ with the Riemannian structure induced by the canonical structure on $\mathbf{R}^{n+1}$, we obtain the canonical structure on $\mathbf{S}_n$ defined in (20.11.4). For since both $\mathbf{S}_n$ and the canonical structure on $\mathbf{R}^{n+1}$ are invariant under $\mathbf{SO}(n+1)$, it is enough to verify that the two Riemannian metric tensors agree at the point $\mathbf{e}_{n+1}$, and this is an immediate consequence of the definitions.

(20.12.2) With the notation of (20.12.1), suppose that M (and therefore also M′) is Riemannian. Let x be a point of M′, and let $\boldsymbol{R}' = (\mathbf{e}'_1, \dots, \mathbf{e}'_{n'})$ be an *orthonormal moving frame* for the Riemannian manifold M′ (20.8.2) defined on an open neighborhood U′ of x in M′. We shall show that there exists an open neighborhood U of x *in* M and an orthonormal moving frame $\boldsymbol{R} = (\mathbf{e}_1, \dots, \mathbf{e}_n)$ for the Riemannian manifold M such that $\mathbf{e}_j | M' = \mathbf{e}'_j$ for $1 \leqq j \leqq n'$ on a neighborhood V of x contained in $U \cap U'$. For this purpose, extend the sections $\mathbf{e}_j$ $(1 \leqq j \leqq n')$ of $T(M') \subset T(M)$ over U′ to C^∞ sections $\boldsymbol{a}_j$ $(1 \leqq j \leqq n')$ of T(M) over a neighborhood of x in M (16.12.11); next, choose $n - n'$ vectors $\mathbf{c}_k$ $(n' + 1 \leqq k \leqq n)$ in $T_x(M)$ which together with the $\mathbf{e}'_j(x)$ $(1 \leqq j \leqq n')$ form an orthonormal basis of $T_x(M)$, and extend the $\mathbf{c}_k$ to C^∞ sections $\boldsymbol{a}_k$ $(n' + 1 \leqq k \leqq n)$ of T(M) over a neighborhood of x in M. Then, for $y \in M$ sufficiently close to x, the $\boldsymbol{a}_i(y)$ $(1 \leqq i \leqq n)$ will form a basis of $T_y(M)$, and we can apply to the $\boldsymbol{a}_i$ the method described in (20.8.2) to obtain the frame $\boldsymbol{R}$.

At each point $y \in M'$, let N_y be the subspace of $T_y(M)$ orthogonal to $T_y(M')$: at each point $y \in V \cap M'$, the vectors $\mathbf{e}_k(y)$ $(n' + 1 \leqq k \leqq n)$ form an orthonormal basis of N_y. Hence (16.17.1) the union N of the N_y is a *vector subbundle* of $f^*(\mathrm{T}(\mathrm{M}))$, and we have $f^*(\mathrm{T}(\mathrm{M})) = \mathrm{T}(\mathrm{M}') \oplus \mathrm{N}$. The bundle N is therefore canonically isomorphic to $f^*(\mathrm{T}(\mathrm{M}))/\mathrm{T}(\mathrm{M}')$, which we have called the *normal bundle* of M′ in M (16.19.2). We shall usually identify these two bundles.

(20.12.3) The fundamental formulas of (20.6.5) and (20.6.6) for the Levi–Civita connection **C** on M and the moving frame **R** are:

(20.12.3.1) $$1_{\mathrm{T}(\mathrm{M})} = \sum_{i=1}^{n} \sigma_i \, \mathbf{e}_i,$$

(20.12.3.2) $$\mathbf{de}_i = \sum_{j=1}^{n} \omega_{ji} \, \mathbf{e}_j,$$

(20.12.3.3) $$d\sigma_i = -\sum_{j=1}^{n} \omega_{ij} \wedge \sigma_j,$$

(20.12.3.4) $$d\omega_{ij} = -\sum_{k=1}^{n} \omega_{ik} \wedge \omega_{kj} + \Omega_{ij}.$$

Since here the basis $(\mathbf{e}_i(y))$ of $\mathrm{T}_y(\mathrm{M})$ is orthonormal, we must have

(20.12.3.5) $$\omega_{ji} = -\omega_{ij} \qquad (1 \leqq i, j \leqq n),$$

so that the matrix $(\omega_{ij}(y))$ belongs to the Lie algebra $\mathfrak{o}(n, \mathbf{R})$ for all $y \in \mathrm{V}$.

Let $\sigma_i' = {}^t f(\sigma_i)$. Then, by (20.12.3.1), for each $y \in \mathrm{V} \cap \mathrm{M}'$ and each tangent vector $\mathbf{h}_y' \in \mathrm{T}_y(\mathrm{M}')$, we have

$$\mathbf{h}_y' = \sum_{i=1}^{n} \langle \sigma_i'(y), \mathbf{h}_y' \rangle \mathbf{e}_i(y).$$

From the definition of the frame **R**, this implies first that

(20.12.3.6) $$\sigma_\alpha' = 0 \qquad \text{for} \quad n' + 1 \leqq \alpha \leqq n$$

and second that the σ_j' for $1 \leqq j \leqq n'$ are the *canonical forms* corresponding to the given moving frame **R**′.

We have, therefore,

(20.12.3.7) $$\mathbf{g}' = \sum_{i=1}^{n'} \sigma_i' \otimes \sigma_i';$$

the tensor field **g**′ is also called the *first fundamental form* on M′.

Next, let $\omega'_{ij} = {}^t f(\omega_{ij})$. Then, first of all,

(20.12.3.8) $$\omega'_{ji} = -\omega'_{ij} \quad \text{for} \quad 1 \leqq i, j \leqq n;$$

this shows that for all $y \in V \cap M'$ the matrix $(\omega'_{ij}(y))_{1 \leqq i, j \leqq n'}$ belongs to the Lie algebra $\mathfrak{o}(n', \mathbf{R})$, and consequently the ω'_{ij} define an $\mathbf{O}(n', \mathbf{R})$-connection $\mathbf{C}'$ (20.6.5) on the Riemannian manifold $V \cap M'$. Moreover, by virtue of (20.12.3.3) and (20.12.3.6), we have

(20.12.3.9) $$d\sigma'_i = -\sum_{j=1}^{n'} \omega'_{ij} \wedge \sigma'_j$$

for $1 \leqq i \leqq n'$, which by comparison with (20.6.6.3) shows that the torsion of $\mathbf{C}'$ is zero. In other words, $\mathbf{C}'$ is the *Levi–Civita connection* on M′, and the ω'_{ij} $(1 \leqq i, j \leqq n')$ are the *connection forms* of $\mathbf{C}'$ corresponding to the frame $\mathbf{R}'$.

(20.12.4) Let ∇ (resp. ∇') denote covariant differentiation relative to the connection $\mathbf{C}$ (resp. $\mathbf{C}'$). Since a vector field X' on M′ can be considered as a mapping of M′ into T(M), the covariant derivative $\nabla_{\mathbf{h}'_y} \cdot X'$ is defined for all vectors $\mathbf{h}'_y \in T_y(M')$ (17.17.2). Comparison of (20.12.3.2) and the analogous formula for the connection $\mathbf{C}'$ shows that we have

(20.12.4.1) $$\nabla_{\mathbf{e}'_j} \cdot \mathbf{e}'_i = \nabla'_{\mathbf{e}'_j} \cdot \mathbf{e}'_i + \sum_{\alpha=n'+1}^{n} \langle \omega'_{\alpha i}, \mathbf{e}'_j \rangle \mathbf{e}_\alpha$$

for $1 \leqq i, j \leqq n'$. Let us write

$$l_{\alpha ij} = \langle \omega'_{\alpha i}, \mathbf{e}'_j \rangle,$$

which is a C^∞ function on $V \cap M'$; then we have

(20.12.4.2) $$l_{\alpha ji} = l_{\alpha ij}$$

for $1 \leqq i, j \leqq n'$ and $n' + 1 \leqq \alpha \leqq n$. For the formulas (20.12.3.3) and (20.12.3.6) give, for $n' + 1 \leqq \alpha \leqq n$,

(20.12.4.3) $$\sum_{j=1}^{n'} \omega'_{\alpha j} \wedge \sigma'_j = 0$$

and hence, evaluating the left-hand side at $\mathbf{e}'_i \wedge \mathbf{e}'_k$,

$$\sum_{j=1}^{n'} (\langle \omega'_{\alpha j}, \mathbf{e}'_i \rangle \langle \sigma'_j, \mathbf{e}'_k \rangle - \langle \omega'_{\alpha j}, \mathbf{e}'_k \rangle \langle \sigma'_j, \mathbf{e}'_i \rangle) = 0;$$

since $\langle \sigma'_i, \mathbf{e}'_k \rangle = \delta_{ik}$, the relation (20.12.4.2) follows. For each $\alpha > n'$ there is therefore a *symmetric covariant tensor field* $\boldsymbol{l}_\alpha$ *of order* 2 on M′ such that

$$\langle \boldsymbol{l}_\alpha, \mathbf{e}'_i \otimes \mathbf{e}'_j \rangle = l_{\alpha ij}$$

for $1 \leqq i, j \leqq n'$. Equivalently, we have

(20.12.4.4) $$\boldsymbol{l}_\alpha = \sum_{i=1}^{n'} \omega'_{\alpha i} \otimes \sigma'_i$$

by virtue of (20.12.4.2).

The $\boldsymbol{l}_\alpha$ $(n' + 1 \leqq \alpha \leqq n)$ are called the *second fundamental forms* on M′; their assignment is equivalent to that of the $n'(n - n')$ differential forms $\omega'_{\alpha i}$. It follows immediately from (20.12.4.1) and (17.17.3.4) that if X', Y' are any two vector fields *on* M′, then we have

(20.12.4.5) $$\nabla_{X'} \cdot Y' - \nabla'_{X'} \cdot Y' = \sum_{\alpha > n'} \langle \boldsymbol{l}_\alpha, X' \otimes Y' \rangle \mathbf{e}_\alpha,$$

which shows that the $\boldsymbol{l}_\alpha$ are independent of the choice of the moving frame $\boldsymbol{R}'$; they do depend on the choice of the $\mathbf{e}_\alpha$ (the orthonormal frame of the normal bundle N), but the sum

(20.12.4.6) $$\boldsymbol{l} = \sum_{\alpha > n'} \boldsymbol{l}_\alpha \mathbf{e}_\alpha$$

is independent of this choice, by virtue of (20.12.4.5): it is an M′-morphism of $\mathbf{T}_0^2(\mathrm{M}')$ into N, and is called the (*vector-valued*) *second fundamental form* on M′.

We see therefore that *the Levi–Civita connection on* M′ *completely determines the vector-valued second fundamental form on* M′, *and conversely.* We remark that the value of $\nabla'_{X'} \cdot Y'$ at each point $x \in \mathrm{M}'$ belongs to $\mathrm{T}_x(\mathrm{M}')$, and the value of the right-hand side of (20.12.4.5) belongs to N_x; hence (20.12.4.5) gives at each point $x \in \mathrm{M}'$ the canonical decomposition of the value of $\nabla_{X'} \cdot Y'$ into its components in the two orthogonal subspaces $\mathrm{T}_x(\mathrm{M}')$ and N_x. These components are called respectively the *tangential* and *normal components*.

(20.12.5) Let Ω'_{ij} denote the *curvature* 2-*forms* of $\mathbf{C}'$, relative to the frame $\boldsymbol{R}'$ $(1 \leqq i, j \leqq n')$, and let $\tilde{\Omega}_{ij} = {}^t f(\Omega_{ij})$ for $1 \leqq i, j \leqq n$. Then we deduce from (20.12.3.4)

(20.12.5.1) $$d\omega'_{ij} = -\sum_{k=1}^{n} \omega'_{ik} \wedge \omega'_{kj} + \tilde{\Omega}_{ij} \qquad (1 \leqq i, j \leqq n)$$

and hence, by comparison with (20.6.6.4),

(20.12.5.2) $$\tilde{\Omega}_{ij} - \Omega'_{ij} = \sum_{\alpha > n'} \omega'_{i\alpha} \wedge \omega'_{\alpha j} \qquad (1 \leqq i, j \leqq n').$$

(20.12.6) Consider in particular the case where $M = \mathbf{R}^n$, endowed with its canonical structure (20.11.2). We have then $\tilde{\Omega}_{ij} = 0$ for all i, j, and consequently the forms ω'_{ij} satisfy the structure equations

(20.12.6.1) $$d\omega'_{ij} = -\sum_{k=1}^{n} \omega'_{ik} \wedge \omega'_{kj} \qquad (1 \leqq i, j \leqq n).$$

We shall show that the assignment of the $n' + n^2$ differential forms σ'_i, ω'_{ij} *determines* the submanifold M′ of $\mathbf{R}^n$ (locally) *up to a Euclidean displacement.* More precisely, we have the following existence and uniqueness theorem:

(20.12.7) *Let z_0 be a point in $\mathbf{R}^{n'}$, let U be an open neighborhood of z_0, and let n be an integer $> n'$. Let ρ_i $(1 \leqq i \leqq n')$, ϖ_{ij} $(1 \leqq i, j \leqq n)$ be a system of $n' + n^2$ differential 1-forms on U, such that the ρ_i are linearly independent, and satisfying the following relations:*

(20.12.7.1) $$\varpi_{ji} = -\varpi_{ij} \qquad (1 \leqq i, j \leqq n),$$

(20.12.7.2) $$d\varpi_{ij} = -\sum_{k=1}^{n} \varpi_{ik} \wedge \varpi_{kj} \qquad (1 \leqq i, j \leqq n),$$

(20.12.7.3) $$d\rho_i = -\sum_{j=1}^{n'} \varpi_{ij} \wedge \rho_j \qquad (1 \leqq i \leqq n'),$$

(20.12.7.4) $$0 = \sum_{j=1}^{n'} \varpi_{\alpha j} \wedge \rho_j \qquad (n' + 1 \leqq \alpha \leqq n).$$

Let $\mathbf{x}_0$ be a point of $\mathbf{R}^n$ and let $(\mathbf{b}_i)_{1 \leqq i \leqq n}$ be an orthonormal basis of $\mathbf{R}^n$. Then there exists a connected open neighborhood V of z_0 contained in U, and an embedding (16.8.4) $\mathbf{F}$ *of* V *into* $\mathbf{R}^n$ *having the following properties:*

(i) $\mathbf{F}(z_0) = \mathbf{x}_0$ *and* $D_i \mathbf{F}(z_0) = \mathbf{b}_i$ $(1 \leqq i \leqq n')$.

(ii) *The forms* $\sigma'_i = {}^t\mathbf{F}^{-1}(\rho_i)$ $(1 \leqq i \leqq n')$, $\omega'_{ij} = {}^t\mathbf{F}^{-1}(\varpi_{ij})$ $(1 \leqq i, j \leqq n)$ *on the submanifold* $M' = \mathbf{F}(V)$ of $\mathbf{R}^n$ *are the differential forms induced on* M′ *by the first n' canonical forms and the connection forms of an orthonormal moving frame* $\mathbf{R}$ *on* $\mathbf{R}^n$ *whose first n' vectors are tangent to* M′ *at each point and which is equal to* $(\tau_{x_0}(\mathbf{b}_i))$ *at the point* x_0.

Moreover, if there exists another neighborhood $V_1 \subset U$ *of* z_0 *and an embedding* $\mathbf{F}_1$ *of* V_1 *into* $\mathbf{R}^n$ *with these two properties, then* $\mathbf{F}$ *and* $\mathbf{F}_1$ *coincide on the connected component of* z_0 *in* $V \cap V_1$.

Suppose that there exists an embedding $\mathbf{F}$ and a frame $\mathbf{R} = (\mathbf{e}_i)_{1 \leqq i \leqq n}$ with the desired properties. With the notation of (16.5.2), put

$$\mathbf{v}_i(z) = \tau_{\mathbf{F}(z)}^{-1}(\mathbf{e}_i(\mathbf{F}(z))) \qquad (1 \leqq i \leqq n), \tag{20.12.7.5}$$

so that the $\mathbf{v}_i$ are n C^∞ mappings of V into $\mathbf{R}^n$. Bearing in mind that $\mathbf{R}^n$ is flat (20.4.1), the relations (20.12.3.2) at the point $\mathbf{F}(z)$ of M′ give us

$$d\mathbf{v}_i = \sum_{j=1}^{n} \varpi_{ji} \mathbf{v}_j \qquad (1 \leqq i \leqq n). \tag{20.12.7.6}$$

Next, from the relation (20.12.3.1) at the point $\mathbf{F}(z)$, we have for each vector $\mathbf{h}_z \in T_z(\mathbf{R}^{n'})$,

$$T_z(\mathbf{F}) \cdot \mathbf{h}_z = \sum_{i=1}^{n'} \langle \sigma_i(\mathbf{F}(z)), T_z(\mathbf{F}) \cdot \mathbf{h}_z \rangle \mathbf{e}_i(\mathbf{F}(z)),$$

from which we derive (16.5.7)

$$d\mathbf{F} = \sum_{i=1}^{n} \rho_i \mathbf{v}_i. \tag{20.12.7.7}$$

Conversely, we shall first show that in a neighborhood of z_0 there exist mappings $\mathbf{v}_i$ $(1 \leqq i \leqq n)$ into $\mathbf{R}^n$ satisfying (20.12.7.6) and such that $\mathbf{v}_i(z_0) = \mathbf{b}_i$ for $1 \leqq i \leqq n$. If v_{ih} $(1 \leqq h \leqq n)$ are the components of $\mathbf{v}_i$, the equations (20.12.7.6) form a Pfaffian system of n^2 equations

$$dv_{ih} - \sum_{j=1}^{n} \varpi_{ji} v_{jh} = 0 \qquad (1 \leqq i, h \leqq n). \tag{20.12.7.8}$$

This system is *completely integrable*. For, by virtue of (20.12.7.2), we have

$$\begin{aligned} d\left(\sum_{j=1}^{n} \varpi_{ji} v_{jh}\right) &= \sum_{j=1}^{n} (v_{jh}\, d\varpi_{ji} + dv_{jh} \wedge \varpi_{ji}) \\ &= -\sum_{j=1}^{n} v_{jh}\left(\sum_{k=1}^{n} \varpi_{jk} \wedge \varpi_{ki}\right) + \sum_{k=1}^{n} dv_{kh} \wedge \varpi_{ki} \\ &= \sum_{k=1}^{n} \left(dv_{kh} - \sum_{j=1}^{n} \varpi_{jk} v_{jh}\right) \wedge \varpi_{ki}, \end{aligned}$$

from which our assertion follows (18.14.3). The desired functions $\mathbf{v}_i$ therefore exist in a neighborhood of z_0 and are uniquely determined (18.14.3).

We seek now a mapping $\mathbf{F}$ of a neighborhood of z_0 into $\mathbf{R}^n$ which satisfies the equation (20.12.7.7) for the functions $\mathbf{v}_i$ just determined, and which is such that $\mathbf{F}(z_0) = \mathbf{x}_0$. If F_h $(1 \leqq h \leqq n)$ are the components of $\mathbf{F}$, the equation (20.12.7.7) is equivalent to the Pfaffian system of n equations

(20.12.7.9)
$$d\mathrm{F}_h = \sum_{i=1}^{n'} \rho_i v_{ih} \qquad (1 \leqq h \leqq n).$$

Again, this system is *completely integrable.* For by virtue of (20.12.7.3) and (20.12.7.8), we have

$$\begin{aligned} d\left(\sum_{i=1}^{n'} \rho_i v_{ih}\right) &= \sum_{i=1}^{n'} (v_{ih}\, d\rho_i - \rho_i \wedge dv_{ih}) \\ &= -\sum_{i=1}^{n'} v_{ih}\left(\sum_{j=1}^{n'} \varpi_{ij} \wedge \rho_j\right) - \sum_{i=1}^{n'} \rho_i \wedge \left(\sum_{j=1}^{n} \varpi_{ji} v_{jh}\right). \end{aligned}$$

In this last expression, it is clear that the coefficient of v_{ih} is zero for $1 \leqq i \leqq n'$, and the coefficient of $v_{\alpha h}$ for $\alpha > n'$ is zero by virtue of (20.12.7.4). From this follow the existence and uniqueness of the function $\mathbf{F}$ in a neighborhood of z_0 (18.14.3).

We shall next show that $(\mathbf{v}_i | \mathbf{v}_j) = \delta_{ij}$ for $1 \leqq i, j \leqq n$. The functions $w_{ij} = (\mathbf{v}_i | \mathbf{v}_j)$ satisfy the relations

(20.12.7.10)
$$\begin{aligned} dw_{ij} &= (d\mathbf{v}_i | \mathbf{v}_j) + (\mathbf{v}_i | d\mathbf{v}_j) \\ &= \sum_{k=1}^{n} \varpi_{ki} w_{kj} + \sum_{k=1}^{n} \varpi_{kj} w_{ik} \qquad (1 \leqq i, j \leqq n). \end{aligned}$$

These n^2 equations again form a *completely integrable* Pfaffian system: This is verified as above, using (20.12.7.2). But *by virtue of* (20.12.7.1) the constant functions δ_{ij} satisfy this system of equations; and since at the point z_0 we have $w_{ij}(z_0) = \delta_{ij}$, by hypothesis, it follows that $w_{ij} = \delta_{ij}$ throughout V, since V is connected.

Since the $\mathbf{v}_i(z)$ form a basis of $\mathbf{R}^n$ for all z, and since the ρ_i are linearly independent, $d\mathbf{F}(z)$ is of rank n' at every point of V by (20.12.7.7); hence, replacing V if necessary by a smaller neighborhood, it follows that $\mathbf{F}$ is an *embedding* (16.8.8). We may then at each point of $\mathrm{M}' = \mathbf{F}(\mathrm{V})$ define the n vectors $\mathbf{e}_i(\mathbf{F}(z))$ by the formula (20.12.7.5), and it is clear that the embedding $\mathbf{F}$ and the frame $\mathbf{R} = (\mathbf{e}_i)$ (extended to a neighborhood of M′ by the method of (20.8.2)) have the required properties. The uniqueness follows from the uniqueness of the solutions $\mathbf{v}_i$ and $\mathbf{F}$ of the Pfaffian systems (20.12.7.5) and (20.12.7.7), and the remarks made at the beginning of the proof. Q.E.D.

Theorem (20.12.7) reduces many problems of the determination of Riemannian submanifolds of $\mathbf{R}^n$ satisfying given conditions to the integration of Pfaffian systems in which the unknowns are the differential forms ρ_i ($1 \leqq i \leqq n'$) and ϖ_{ij} ($1 \leqq i, j \leqq n$), and which are obtained by adjoining to the relations in (20.12.7) additional relations expressing the conditions of the given problem. We shall now give some examples of this (see also (20.15.1)).

Examples

(20.12.8) Let us determine all connected submanifolds M′ of $\mathbf{R}^n$ of dimension $n' < n$ whose *second fundamental form is identically zero* (cf. (20.13.7) and (20.23.6)). In the notation of (20.12.4), this means that $\omega'_{\alpha i} = 0$ for $1 \leqq i \leqq n'$ and $n' + 1 \leqq \alpha \leqq n$. It follows immediately (in the notation of (20.12.7)) that *each* of the two Pfaffian systems

$$d\mathbf{v}_i = \sum_{j=1}^{n'} \varpi_{ji} \mathbf{v}_j \qquad (1 \leqq i \leqq n'),$$

$$d\mathbf{v}_\alpha = \sum_{\beta > n'} \varpi_{\beta\alpha} \mathbf{v}_\beta \qquad (\alpha > n')$$

is completely integrable. Now the initial value of $(\mathbf{v}_i | \mathbf{b}_\alpha)$ is zero for $i \leqq n'$ and $\alpha > n'$, and the $(\mathbf{v}_i | \mathbf{b}_\alpha)$ ($1 \leqq i \leqq n'$) are solutions of a completely integrable Pfaffian system which admits also the solution 0; hence $(\mathbf{v}_i | \mathbf{b}_\alpha) = 0$; in other words, the vector subspace E of $\mathbf{R}^n$ spanned by the $\mathbf{v}_i$ ($1 \leqq i \leqq n'$) is fixed. But then the projection $\mathbf{G}$ of $\mathbf{F}$ on the subspace orthogonal to E satisfies $d\mathbf{G} = 0$ by virtue of (20.12.7.7), and consequently is *constant*. It follows that M′ is a *connected open subset of a linear subvariety of dimension* n'.

(20.12.9) A point $x \in$ M′ is called an *umbilic* if at this point all the second fundamental forms $\mathbf{l}_\alpha(x)$ are scalar multiples of the first fundamental form $\mathbf{g}'(x)$. We shall determine all connected submanifolds M′ of $\mathbf{R}^n$, of dimension $n' \geqq 2$, such that *every point of* M′ *is an umbilic*. Leaving aside the case dealt with in (20.12.8), let us assume that the $\mathbf{l}_\alpha$ are not all identically zero. By changing the moving frame $\mathbf{R}$, we may assume that $\mathbf{l}_\alpha = 0$ for

$$n' + 1 \leqq \alpha \leqq n - 1$$

and that $\mathbf{l}_n \neq 0$. For at each point $y \in$ M′ we may write

$$\sum_\alpha \mathbf{l}_\alpha(y)\mathbf{e}_\alpha(y) = \mathbf{g}'(y)\mathbf{a}_n(y),$$

where $\mathbf{a}_n(y) \neq 0$ in the space N_y; now choose a moving frame in N whose first vector at each point is $\mathbf{a}_n(y)$, and orthonormalize it by the method of

(20.8.2), beginning with $\boldsymbol{a}_n$; by virtue of (20.12.4.5), the new frame so obtained will have the desired properties. Since the moving frame $\boldsymbol{R}$ is orthonormal, the hypothesis $\boldsymbol{l}_n(y) = \mathrm{A}(y)\mathbf{g}'(y)$ with $\mathrm{A}(y) \neq 0$ for all y may be written in the form

(20.12.9.1) $$\omega'_{ni} = \mathrm{A}\sigma'_i \qquad (1 \leqq i \leqq n'),$$

and likewise we have

(20.12.9.2) $$\omega'_{\alpha i} = 0 \qquad (1 \leqq i \leqq n', \quad n' + 1 \leqq \alpha \leqq n - 1).$$

From the structure equations (20.12.6.1), bearing in mind (20.12.3.8) and (20.12.3.9), we now obtain

(20.12.9.3) $$0 = d\omega'_{\alpha i} = \mathrm{A}\omega'_{n\alpha} \wedge \sigma'_i \qquad (1 \leqq i \leqq n', \quad n' + 1 \leqq \alpha \leqq n - 1),$$

$$d\omega'_{ni} = \mathrm{A}\; d\sigma'_i + d\mathrm{A} \wedge \sigma'_i = -\sum_{k=1}^{n} \omega'_{nk} \wedge \omega'_{ki} = \mathrm{A} \sum_{k=1}^{n'} \sigma'_k \wedge \omega'_{ik},$$

and therefore

(20.12.9.4) $$d\mathrm{A} \wedge \sigma'_i = 0 \qquad (1 \leqq i \leqq n').$$

These relations imply that $d\mathrm{A} = 0$ (A.13.3.1), so that A is a *constant* $\neq 0$ on M′. Hence, from (20.12.9.3), we have

$$\omega'_{n\alpha} \wedge \sigma'_i = 0$$

for $1 \leqq i \leqq n'$ and $n' + 1 \leqq \alpha \leqq n - 1$, so that $\omega'_{n\alpha} = 0$ for these values of α. With the notation of (20.12.7), it follows that the Pfaffian system

$$dv_i = \sum_{j=1}^{n'} \varpi_{ji} v_j + \varpi_{ni} v_n \qquad (1 \leqq i \leqq n'),$$

$$dv_n = \sum_{j=1}^{n'} \varpi_{jn} v_j$$

is completely integrable, and as in (20.12.8) that the subspace of $\mathbf{R}^n$ of dimension $n' + 1$ generated by $\mathbf{v}_1, \ldots, \mathbf{v}_{n'}$ and $\mathbf{v}_n$ is *fixed*. Moreover, from (20.12.9.1) we obtain

$$d\mathbf{v}_n = \sum_{j=1}^{n'} \varpi_{jn} \mathbf{v}_j = -\mathrm{A} \sum_{j=1}^{n'} \sigma_j \mathbf{v}_j = -\mathrm{A}\; d\mathbf{F},$$

so that the point $\mathbf{v}_n + \mathrm{A}\mathbf{F}$ is *fixed* in $\mathbf{R}^n$: we may assume that it is the origin, i.e., that $\mathbf{F} = -\mathrm{A}^{-1}\mathbf{v}_n$. Thus, finally, we see that M′ is a *connected open subset of a sphere* $\mathrm{A}^{-1}\mathrm{S}_{n'}$, up to a Euclidean displacement.

PROBLEMS

1. Let M be a pure submanifold of $\mathbf{R}^N$ of dimension n.

(a) Consider a chart (U, φ, n) of M, and put $\mathbf{f} = \varphi^{-1}$ in $\varphi(U) \subset \mathbf{R}^n$. Then the local expression of the first fundamental form on $\varphi(U)$ is, if the local coordinates are $u^1, \ldots, u^n$,

$$\mathbf{g} = \sum_{i,j} g_{ij}\, du^i \otimes du^j,$$

where

$$g_{ij} = \left(\frac{\partial \mathbf{f}}{\partial u^i} \,\middle|\, \frac{\partial \mathbf{f}}{\partial u^j} \right)$$

and $(\mathbf{x} | \mathbf{y})$ is the Euclidean scalar product. The local expressions of the second fundamental forms $\boldsymbol{l}_\alpha$ are

$$\boldsymbol{l}_\alpha = \sum_{i,j} \left(\mathbf{e}_\alpha \,\middle|\, \frac{\partial^2 \mathbf{f}}{\partial u^i\, \partial u^j} \right) du^i \otimes du^j.$$

(Observe that $(\mathbf{e}_\alpha | \, \partial \mathbf{f}/\partial u^i) = 0$.)

(b) If we identify the tangent bundle $T(\mathbf{R}^N)$ canonically with $\mathbf{R}^{2N}$, the normal bundle P of M is an N-dimensional submanifold of $\mathbf{R}^{2N}$. For each vector $\mathbf{n}_x \in P$ with origin $x \in M \subset \mathbf{R}^N$, let $\mathbf{p}(\mathbf{n}_x) = x + \boldsymbol{\tau}_x(\mathbf{n}_x) \in \mathbf{R}^N$. The images under $\mathbf{p}$ of the critical points of $\mathbf{p}$ (16.23) are called the *focal points* of M, and $\mathbf{n}_x$ is a *focal vector of multiplicity* μ if the rank of $\mathbf{p}$ at the point $\mathbf{n}_x$ is $N - \mu$. The set of focal points of M has measure zero in $\mathbf{R}^N$.

For each *unit* vector $\mathbf{n}_x \in P$ normal to M at the point x, let $S_{\mathbf{n}_x}$ denote the endomorphism of $T_x(M)$ defined by the relation

$$(\boldsymbol{l}(x) \cdot (\mathbf{h}_x \otimes \mathbf{k}_x) | \mathbf{n}_x) = (S_{\mathbf{n}_x} \cdot \mathbf{h}_x | \, \mathbf{k}_x),$$

which is Hermitian relative to the scalar product in $T_x(M)$. The eigenvalues $K_j(\mathbf{n}_x)$ $(1 \leq j \leq n)$ of this endomorphism, each counted according to its multiplicity, are called the *principal curvatures of* M *in the direction* $\mathbf{n}_x$. Show that the focal points of M on the line $t \mapsto x + t\boldsymbol{\tau}_x(\mathbf{n}_x)$ are the points corresponding to $t = K_j(\mathbf{n}_x)^{-1}$ for the values of j such that $K_j(\mathbf{n}_x) \neq 0$, and that the multiplicity of the focal vector $K_j(\mathbf{n}_x)^{-1}\mathbf{n}_x$ is equal to the multiplicity of the eigenvalue $K_j(\mathbf{n}_x)$. (Take a chart of M and compute the square of the determinant of the Jacobian matrix of the mapping

$$(u^1, \ldots, u^n, t^1, \ldots, t^{N-n}) \mapsto \mathbf{f}(u^1, \ldots, u^n) + \sum_\alpha t^\alpha \mathbf{w}_\alpha(u^1, \ldots, u^n),$$

where $\mathbf{w}_\alpha(u^1, \ldots, u^n)$ is the local expression of $\boldsymbol{\tau}_x(\mathbf{e}_\alpha(x))$; use (a) and observe that we may assume that at the point x the matrix (g_{ij}) is the unit matrix.)

2. With the hypotheses and notation of Problem 1, for each point $y \in \mathbf{R}^N$ let $E_y : M \to \mathbf{R}$ be the function defined by $E_y(x) = \|x - y\|^2$.

(a) Show that x is a critical point of E_y if and only if $y = \mathbf{p}(\mathbf{n}_x)$ for some vector $\mathbf{n}_x$ normal to M at the point x; that x is a degenerate critical point of E_y if and only if y is a focal point of M; and that if μ is the multiplicity of the focal vector $\mathbf{n}_x$, then the rank of the Hessian of E_y at the point x is $n - \mu$. Deduce that for almost all points $y \in \mathbf{R}^N$ the critical points of E_y are nondegenerate.

(b) If x is a nondegenerate critical point of E_y, the Morse index of E_y at the point x (Section 16.5, Problem 3) is equal to the number of focal points (counted according to multiplicity) contained in the segment with endpoints x and y.
(c) Suppose that $N = 2m$, so that $\mathbf{R}^N$ may be identified with $\mathbf{C}^m$, and that M is a *complex-analytic* submanifold of $\mathbf{C}^m$ of complex dimension $\frac{1}{2}n$. Show that at each nondegenerate critical point x of E_y, the Morse index of E_y (Section 16.5, Problem 3) at the point x is $\leqq \frac{1}{2}n$. (Observe that if Q is a complex quadratic form on $\mathbf{C}^q$ and $Q' = \mathscr{R}(Q)$ is the real part of Q, which is a real quadratic form on $\mathbf{R}^{2q}$, then for each eigenvalue λ of the Hermitian endomorphism corresponding to Q', $-\lambda$ is also an eigenvalue, with the same multiplicity.)

3. Let M be a differential manifold. Show that in the Fréchet space $\mathscr{E}_{\mathbf{R}}(M)$ of C^∞ real-valued functions on M (17.1), the set of functions having no degenerate critical points is *dense*. (If $f \in \mathscr{E}_{\mathbf{R}}(M)$, there exists an embedding $\mathbf{h} = (h_1, h_2, \ldots, h_N)$ of M into $\mathbf{R}^N$ such that $h_1 = f$ (Section 16.25, Problem 2). Choose a point $y \in \mathbf{R}^N$ of the form $(-c, \varepsilon_2, \ldots, \varepsilon_N)$, where c is large and positive and the ε_j are small, such that E_y has no degenerate critical points (Problem 2(a)), and consider the real valued function $g(x) = (E_y(x) - c^2)/2c$ on M.)

4. With the hypotheses and notation of Problem 1, consider the submanifold Q (of dimension $N - 1$) of P consisting of the *unit* vectors, and the mapping $\mathbf{q} : Q \mapsto S_{N-1}$ defined by $\mathbf{q}(\mathbf{n}_x) = \boldsymbol{\tau}_x(\mathbf{n}_x)$.

(a) Let $y \in \mathbf{R}^N$, $y \neq 0$. Show that the point $\mathbf{n}_x \in Q$ is a critical point of the function $\mathbf{n}_x \mapsto (y|\mathbf{q}(\mathbf{n}_x))$ if and only if $\mathbf{q}(\mathbf{n}_x)$ is collinear with y, and that this critical point is nondegenerate if and only if $\mathbf{n}_x$ is a critical point of the mapping $\mathbf{q}$ (by using the same method as in Problem 1(b)). These latter points are also the points $\mathbf{n}_x \in Q$ such that

$$K(\mathbf{n}_x) = \det(S_{\mathbf{n}_x}) = \prod_{j=1}^{n} K_j(\mathbf{n}_x) = 0$$

in the notation of Problem 1(b). The number $K(\mathbf{n}_x)$ is called the *total curvature of* M *in the direction* $\mathbf{n}_x$. The image F under $\mathbf{q}$ of the set of critical points of $\mathbf{q}$ is a set of measure zero in S_{N-1}. For each point $y \in S_{N-1} - F$, the set $\mathbf{q}^{-1}(y)$ is the *discrete* closed set of critical points of the function $\mathbf{n}_x \mapsto (y|\mathbf{q}(\mathbf{n}_x))$ on Q (Section 16.5, Problem 4).
(b) The submanifold Q of $\mathbf{R}^{2N}$ is canonically endowed with a structure of a Riemannian manifold, hence also with a Lebesgue measure, namely, its Riemannian volume v_Q. If v is the inverse image ${}^t\mathbf{q}(\sigma^{(N-1)})$ of the solid angle form $\sigma^{(N-1)}$ on S_{N-1} (16.21.10) and if μ_v is the corresponding Lebesgue measure (16.24), show that $\mu_v = |K| \cdot v_Q$. (Take a chart of M.)
(c) Suppose that the manifold M is *compact* and of dimension $n \geqq 1$. Then the set F is *closed* in S_{N-1}; the mapping $\mathbf{q} : Q \to S_{N-1}$ is *surjective* (consider, for each $y \in S_{N-1}$, the function $\mathbf{n}_x \mapsto (y|\mathbf{q}(\mathbf{n}_x))$ on the compact manifold Q, and use (a)); and the restriction of $\mathbf{q}$ to the open set $\mathbf{q}^{-1}(S_{N-1} - F)$ is a surjective *local diffeomorphism* of this open set onto $S_{N-1} - F$.

The *integral curvature* of M is the number

$$\kappa(M) = \frac{1}{\Omega_N} \int_Q |K(\mathbf{n}_x)| \, dv_Q(\mathbf{n}_x).$$

For every C^∞ real-valued function f on M which has only *nondegenerate* critical points (and therefore cannot be locally constant anywhere) let $\beta(M, f)$ denote the (finite) number of critical points of f, and let $\beta(M)$ denote the infimum of $\beta(M, f)$ for all such functions f. Then $\beta(M) \geqq 2$. Show that

$$\kappa(M) = \frac{1}{\Omega_N} \int_{S_{N-1}-F} \beta(M, (y|\mathbf{q}))\, d\nu_{N-1}(y),$$

where ν_{N-1} is the canonical Riemannian volume on S_{N-1} induced by the solid angle form $\sigma^{(N-1)}$. (Use (16.24.8).) Deduce that

$$\kappa(M) \geqq \beta(M) \geqq 2$$

(*Chern–Lashof theorem*).

(d) With the hypotheses of (c), show that if $\beta(M) < 3$, then M is homeomorphic to S_n. (Observe that there exists a point $y \in S_{N-1}$ such that the function $(y|\mathbf{q})$ has only two critical points in M, both of which are nondegenerate, and apply Reeb's theorem (Section 20.8, Problem 6).)

5. Let $\mathbf{f}$ be a C^∞ mapping of a differential manifold N into $\mathbf{R}^n$, and let $\mathbf{G}$ be a C^∞ lifting of $\mathbf{f}$ to $T(\mathbf{R}^n)$. If $\mathbf{R}^n$ is endowed with its canonical connection (20.11.2), we have $\tau_{\mathbf{f}(z)}(d\mathbf{G}(z)) = d_z\,\mathbf{G}$, where $\mathbf{G}(z) = \tau_{\mathbf{f}(z)}(\mathbf{G}(z))$. We may therefore *identify* $\mathbf{G}$ with $\mathbf{G}$, and then the covariant exterior differential $\mathbf{d}\mathbf{G}$ is identified with the differential $d\mathbf{G}$ of the vector-valued function $\mathbf{G}$ (16.20.15).

If N is a submanifold M′ of $\mathbf{R}^n$ and $\mathbf{f}$ is the canonical injection, we write $d\mathbf{x}$ for the differential of $\mathbf{f} : M' \to \mathbf{R}^n$. If $(\mathbf{e}_1, \ldots, \mathbf{e}_n)$ is a moving frame having the properties of (20.12.2), then with the preceding conventions the formulas (20.12.3.1) and (20.12.3.2) may be written in the form

$$d\mathbf{x} = \sum_{i=1}^{n} \sigma_i \mathbf{e}_i,$$

$$d\mathbf{e}_i = \sum_{j=1}^{n} \omega_{ji} \mathbf{e}_j,$$

and the structure equations (20.12.3.3) and (20.12.3.4) are obtained simply by writing down the relations $d(d\mathbf{x}) = 0$ and $d(d\mathbf{e}_i) = 0$ for the exterior differentials (17.15.3.1). The relations (20.12.3.5) are obtained by remarking that $(\mathbf{e}_i | \mathbf{e}_j) = \delta_{ij}$ and hence, taking exterior differentials, that $(d\mathbf{e}_i | \mathbf{e}_j) + (\mathbf{e}_i | d\mathbf{e}_j) = 0$. Since the Riemannian metric tensor on $\mathbf{R}^n$ may be identified with the mapping $(\mathbf{u}, \mathbf{v}) \mapsto (\mathbf{u} | \mathbf{v})$ of $\mathbf{R}^{2n}$ into $\mathbf{R}^n$, the first fundamental form of M′ may be written as $(d\mathbf{x} | d\mathbf{x})$, and each of the second fundamental forms as $-(d\mathbf{e}_\alpha | d\mathbf{x})$. Finally, the curvature form $\mathbf{\Omega}'$ of M′ may be written as

$$\mathbf{\Omega}' = \sum_{\alpha > n'} (d\mathbf{e}_\alpha \cdot d\mathbf{e}_\alpha),$$

the product being that defined in Section 20.6, Problem 2.

13. CURVES IN RIEMANNIAN MANIFOLDS

(20.13.1) We recall that a *curve* C in a Riemannian manifold M is a *one-dimensional submanifold* of M; we regard C as endowed with the Riemannian metric tensor induced by the Riemannian metric tensor $\mathbf{g}$ on M. We shall study the properties of C in relation to M, *in a sufficiently small neighborhood* of a point $c \in C$. We suppose, therefore, that there exists an *embedding* u of an open interval $I \subset \mathbf{R}$ into M such that $u(I) = C$, that $0 \in I$ and that $u(0) = c$. The hypothesis that u is an embedding implies that the vector $u'(t) \in T_{u(t)}(C)$ (defined in (18.1.2.3)) is always $\neq 0$. By transport of structure, the image under u of the canonical orientation of $\mathbf{R}$ is an orientation of C, called the *orientation defined by u*. If $\alpha, \beta \in I$ are such that $\alpha < \beta$, and if $a = u(\alpha)$, $b = u(\beta)$ are the corresponding points of C, we shall sometimes say that C, endowed with this orientation, is *oriented in the direction from a to b*.

Let l be the *length* on C, namely, the positive Lebesgue measure defined by the Riemannian metric tensor induced by $\mathbf{g}$ (20.8.5). For each interval $[\alpha, \beta] \subset I$, we have therefore by definition (20.8.5.1)

$$l(u([\alpha, \beta])) = \int_{\alpha}^{\beta} \|u'(\xi)\| \, d\xi, \tag{20.13.1.1}$$

where $\|u'(\xi)\| = \langle \mathbf{g}(u(\xi)), u'(\xi) \otimes u'(\xi) \rangle$. For each $t \in I$, let

$$\varphi(t) = \int_{0}^{t} \|u'(\xi)\| \, d\xi, \tag{20.13.1.2}$$

which is therefore the length of the arc $u([0, t])$ if $t \geqq 0$, and the negative of the length of $u([t, 0])$ if $t \leqq 0$. This number *depends only on* $u(t)$ *and the choice of c and the orientation of* C. For if $f: I_1 \to I$ is a diffeomorphism of an open interval $I_1 \subset \mathbf{R}$ containing 0 onto the open interval I, such that f preserves the orientation and $f(0) = 0$, then $f'(\zeta) > 0$ for all $\zeta \in I_1$. Now put $u_1 = u \circ f$, so that (16.5.4.1) $u_1'(\zeta) = f'(\zeta)u'(f(\zeta))$ and $\|u_1'(\zeta)\| = f'(\zeta)\|u'(f(\zeta))\|$; if $t = f(t_1)$, our assertion follows from the formula (8.7.4) for change of variable in an integral:

$$\int_{0}^{t_1} \|u_1'(\zeta)\| \, d\zeta = \int_{0}^{t_1} \|u'(f(\zeta))\| f'(\zeta) \, d\zeta$$
$$= \int_{0}^{t} \|u'(\xi)\| \, d\xi.$$

The number $\varphi(t)$ is called the *curvilinear coordinate of the point* $x = u(t)$ *on* C, *endowed with the orientation defined by u, with respect to the origin c*. If we denote it by $\psi(x)$, then ψ is a *chart* of C on the interval $\varphi(I) = J$ (an interval which is therefore entirely determined by the curve C, the point c, and the chosen orientation), and we have $\varphi(t) = \psi(u(t))$. The mapping

$$s \mapsto v(s) = u(\varphi^{-1}(s))$$

of J into M is called the *parametrization of* C *by arc length* (relative to the given choices of c and the orientation).

We have by definition

$$\langle d\psi(u(t)), u'(t)\rangle^2 = (\varphi'(t))^2 = \|u'(t)\|^2 = \langle \mathbf{g}(u(t)), u'(t) \otimes u'(t)\rangle;$$

since C is 1-dimensional, this may also be written in the form $d\psi \otimes d\psi = {}^t h(\mathbf{g})$, where $h : \mathrm{C} \to \mathrm{M}$ is the canonical injection. In other words, $d\psi \otimes d\psi$ is the metric tensor induced on C. If (U, ζ, n) is a chart of a neighborhood of c in M, then writing the metric tensor $\mathbf{g}$ in the form (20.8.2.5), we have on C,

$$d\psi \otimes d\psi = \sum_{i,j} g_{ij}\, d(\zeta^i \circ h) \otimes d(\zeta^j \circ h);$$

or, equivalently, on I,

$$d\varphi \otimes d\varphi = \sum_{i,j} (g_{ij} \circ u)\, du^i \otimes du^j,$$

where we have put $u^i = \zeta^i \circ u$ $(1 \leqq i \leqq n)$. By abuse of notation, we shall sometimes write ds in place of $d\varphi$, and abbreviate the above formula to

$$ds^2 = \sum_{i,j} g_{ij}\, du^i\, du^j.$$

Because of this formula, the metric tensor $\mathbf{g}$ on M is sometimes referred to as "ds^2 on M."

(20.13.2) In this section we shall assume that the point c and the orientation of C have been chosen once for all, and we retain the previous notation. By definition, we have $\|v'(s)\| = 1$ for $s \in \mathrm{J}$; the vector $v'(s)$ is called the *unit tangent vector* to C at the point $v(s)$. It depends only on the choice of orientation (and changes sign if the opposite orientation is chosen).

We recall that if a mapping $\mathbf{w}$ of J into T(M) is a lifting of v, the covariant exterior differential $\mathbf{dw}$ is the differential 1-form on J, with values in T(M), such that

$$\mathbf{dw}(s) \cdot E(s) = \nabla_{E(s)} \cdot \mathbf{w}$$

for all $s \in \mathrm{J}$ (17.19.4.2), where E is the field of unit vectors on $\mathbf{R}$ (18.1.1). Recall also that, by abuse of notation, we write $\nabla_s \cdot \mathbf{w}$ in place of $\nabla_E \cdot \mathbf{w}$ (18.7.2).

With this notation established, a *Frenet frame* of C is by definition a C^∞ mapping $s \mapsto (\mathbf{f}_1(s), \ldots, \mathbf{f}_n(s))$ of J into the frame bundle R(M) of M, such that the $\mathbf{f}_j(s)$ belong to $\mathrm{T}_{v(s)}(\mathrm{M})$ and form an *orthonormal basis* (relative to $\mathbf{g}$)

of this space, and such that the following relations (*Frenet's formulas*) hold in J:

$$(20.13.2.1)\qquad \begin{cases} \mathbf{f}_1 = v' \\ d\mathbf{f}_1 = (k_1 \cdot ds)\mathbf{f}_2, \\ d\mathbf{f}_2 = -(k_1 \cdot ds)\mathbf{f}_1 + (k_2 \cdot ds)\mathbf{f}_3, \\ d\mathbf{f}_3 = -(k_2 \cdot ds)\mathbf{f}_2 + (k_3 \cdot ds)\mathbf{f}_4, \\ \vdots \\ d\mathbf{f}_{n-1} = -(k_{n-2} \cdot ds)\mathbf{f}_{n-2} + (k_{n-1} \cdot ds)\mathbf{f}_n, \\ d\mathbf{f}_n = -(k_{n-1} \cdot ds)\mathbf{f}_{n-1}, \end{cases}$$

where the k_j are C^∞ real-valued functions on J, everywhere strictly positive.

(20.13.3) We shall show that *in general* there exists a *unique Frenet frame* in a sufficiently small open subset of C: the meaning of the words "in general" will be made clear by the proof. From the relation $(\mathbf{f}_1 \mid \mathbf{f}_1) = 1$, by taking the covariant derivatives of both sides and using (20.9.5.4), we obtain $2(\nabla_s \cdot \mathbf{f}_1 \mid \mathbf{f}_1) = 0$. We recall that $\nabla_s \cdot \mathbf{f}_1 = 0$ at all points of C if and only if v is a *geodesic* of M (18.6.1.2). Assume that this is not so (the "general case"); then there exists an open set U_1 in C in which $\nabla_s \cdot \mathbf{f}_1$ does not vanish, and we may therefore write $\nabla_s \cdot \mathbf{f}_1 = k_1 \mathbf{f}_2$, where $\mathbf{f}_2$ is *orthogonal* to $\mathbf{f}_1$, of length 1, and $k_1 > 0$ at all points of U_1.

Suppose that we have determined a decreasing sequence $(U_j)_{1 \leqq j \leqq i-1}$ of open sets in C such that in U_{i-1} the first i of the formulas (20.13.2.1) hold, the $\mathbf{f}_j$ for $j \leqq i$ being mutually orthogonal and of length 1, and the functions k_j $(j \leqq i-1)$ strictly positive at each point of U_{i-1}. By taking the covariant derivatives of the relations $(\mathbf{f}_j \mid \mathbf{f}_i) = 0$ for $j < i$ and $(\mathbf{f}_i \mid \mathbf{f}_i) = 1$, we obtain immediately

$$(\nabla_s \cdot \mathbf{f}_i \mid \mathbf{f}_j) = 0 \qquad (1 \leqq j \leqq i-2),$$
$$(\nabla_s \cdot \mathbf{f}_i \mid \mathbf{f}_{i-1}) + k_{i-1} = 0,$$
$$(\nabla_s \cdot \mathbf{f}_i \mid \mathbf{f}_i) = 0,$$

which shows that $\nabla_s \cdot \mathbf{f}_i + k_{i-1}\mathbf{f}_{i-1}$ is orthogonal to $\mathbf{f}_1, \ldots, \mathbf{f}_{i-1}, \mathbf{f}_i$ at each point of U_{i-1}. If $i < n$, it may happen that $\nabla_s \cdot \mathbf{f}_i + k_{i-1}\mathbf{f}_{i-1}$ vanishes identically on U_{i-1}. Assume that this is not so (the "general case"); then there exists an open subset $U_i \subset U_{i-1}$ of C on which this vector field does not vanish, and then we may write $\nabla_s \cdot \mathbf{f}_i + k_{i-1}\mathbf{f}_{i-1} = k_i \mathbf{f}_{i+1}$, where $\mathbf{f}_{i+1}$ is orthogonal to $\mathbf{f}_1, \ldots, \mathbf{f}_i$, of length 1, and $k_i > 0$ at all points of U_i. If $i = n$, we necessarily have $\nabla_s \cdot \mathbf{f}_n + k_{n-1}\mathbf{f}_{n-1} = 0$.

If there exists a Frenet frame of C, the number $k_j(s) > 0$ is called the jth *curvature* of C at the point $x = v(s)$, and the number $1/k_j(s)$ the jth *radius of curvature*; the vector $\mathbf{f}_{j+1}(s)$ is called the jth *unit normal vector* to C at x, the line $\mathbf{R}\mathbf{f}_{j+1}(s)$ in $T_x(M)$ the jth *normal*, and the plane $\mathbf{R}\mathbf{f}_1(s) + \mathbf{R}\mathbf{f}_{j+1}(s)$ the jth *osculating plane*. When $n = 2$, k_1 is simply called the *curvature*, and the first normal is called the *normal*. When $n = 3$, k_1 and k_2 are called, respectively, the *curvature* and the *torsion*; the first and second normals are called, respectively, the *principal normal* and the *binormal*, and the first and second osculating planes are called, respectively, the *osculating plane* and the *rectifying plane*.

When $M = \mathbf{R}^n$, this terminology is commonly taken to refer, not to the lines and planes in $T_x(\mathbf{R}^n)$ just defined, but to their images under the composition of the translation $\mathbf{h} \mapsto \mathbf{h} + x$ and the canonical bijection τ_x, so that they are taken to be affine lines and planes contained in $\mathbf{R}^n$ (cf. 16.8.6).

(20.13.4) When $M = \mathbf{R}^n$ and the curve C possesses a Frenet frame, the $n - 1$ functions k_j $(1 \leqq j \leqq n - 1)$ *determine the curve* C *up to a Euclidean displacement*. Precisely, we have:

(20.13.5) *Let* $J \subset \mathbf{R}$ *be an interval containing* 0, *and let* k_j $(1 \leqq j \leqq n - 1)$ *be* C^∞ *functions on* J *which are everywhere* >0. *Let* c *be a point of* $\mathbf{R}^n$ *and let* $\mathbf{b}_i$ $(1 \leqq i \leqq n)$ *be* n *vectors in* $T_c(\mathbf{R}^n)$ *forming an orthonormal basis. Then there exists an interval* $I \subset J$ *containing* 0, *a mapping* $v : I \to \mathbf{R}^n$, *and* n *mappings* $\mathbf{f}_i : I \to T(\mathbf{R}^n)$ $(1 \leqq i \leqq n)$, *such that*:

(i) v *is an embedding of* I *in* $\mathbf{R}^n$, $s \mapsto v(s)$ *is the arc length parametrization of* $C = v(I)$ (*for the orientation which is the image of the canonical orientation of* $\mathbf{R}$), *and* $(\mathbf{f}_i)_{1 \leqq i \leqq n}$ *is a Frenet frame of* C *satisfying the formulas* (20.13.2.1) *for the given functions* k_j;

(ii) $v(0) = c$ *and* $\mathbf{f}_i(0) = \mathbf{b}_i$ $(1 \leqq i \leqq n)$.

Furthermore, if I_1 *is an interval containing* 0 *and contained in* J, *and if* $v^{(1)}$, $\mathbf{f}_i^{(1)}$ $(1 \leqq i \leqq n)$ *are functions satisfying the above conditions, then* $v = v^{(1)}$ *and* $\mathbf{f}_i = \mathbf{f}_i^{(1)}$ $(1 \leqq i \leqq n)$ *in* $I \cap I_1$.

This is a particular case of (20.12.7), with $n' = 1$. In this case there are no compatibility conditions, nor is it necessary to invoke Frobenius' theorem since we are dealing with ordinary differential equations.

Henceforth, whenever we consider a curve (other than a geodesic trajectory) in a Riemannian manifold, we shall always assume that it possesses a Frenet frame, unless the contrary is expressly stated.

Remark

(20.13.5) When the manifold M is *oriented*, it is often convenient to modify the definition of a Frenet frame, replacing $\mathbf{f}_n$ by $\varepsilon \mathbf{f}_n$, where $\varepsilon = \pm 1$ is chosen so that $(\mathbf{f}_1, \ldots, \mathbf{f}_{n-1}, \varepsilon \mathbf{f}_n)$ is a *direct* frame (16.21.2). The Frenet formulas remain valid provided that we replace k_{n-1} by εk_{n-1} (which therefore can be of either sign).

(20.13.6) Let M be a Riemannian manifold of dimension n, and M′ a submanifold of dimension n', where $1 < n' < n$. Let C be a curve *in* M′. Applying the formula (20.12.4.4) with $X' = Y' = \mathbf{f}_1 \circ v^{-1}$ (in the notation of (20.13.2)), we obtain

(20.13.6.1) $$\nabla_s \cdot \mathbf{f}_1 = \nabla'_s \cdot \mathbf{f}_1 + \sum_\alpha \langle \mathbf{l}_\alpha, \mathbf{f}_1 \otimes \mathbf{f}_1 \rangle \mathbf{e}_\alpha .$$

Since the two terms on the right-hand side are the components of the left-hand side in $\mathrm{T}_x(\mathrm{M}')$ and the normal space N_x, we deduce immediately:

(20.13.6.2) (i) *Every curve* C *in* M′ *which is a geodesic trajectory of* M *is also a geodesic trajectory of* M′.

(ii) *Let* C *be a curve in* M′ *which is not a geodesic trajectory of* M; *in order that* C *should be a geodesic trajectory of* M′, *it is necessary and sufficient that its first normal* (*relative to* M) *should be normal to* M′.

(20.13.6.3) *The normal component of* $\nabla_s \cdot \mathbf{f}_1$ (called the *normal curvature vector of* C) *at a point x is the same for all curves in* M′ *having the same tangent at the point x* (*Meusnier's theorem*). *The first curvature at x is the same for all curves in* M *having the same tangent and the same first normal at x.*

The latter assertion is a consequence of the following formula, which comes from (20.13.6.1) and (20.13.2.1):

(20.13.6.4) $$k_1(\mathbf{f}_2 \mid \mathbf{e}_\alpha) = \langle \mathbf{l}_\alpha, \mathbf{f}_1 \otimes \mathbf{f}_1 \rangle \qquad (n' + 1 \leqq \alpha \leqq n).$$

A curve $\mathrm{C} \subset \mathrm{M}'$ is called an *asymptotic line* of M′ if its normal curvature vector is zero, or equivalently, if $\langle \mathbf{l}_\alpha, \mathbf{f}_1 \otimes \mathbf{f}_1 \rangle = 0$ for $n' + 1 \leqq \alpha \leqq n$. If C is not a geodesic trajectory of M, another equivalent definition is that the first osculating plane of C should be *tangent* to M′.

(20.13.7) A submanifold M′ of M is said to be *geodesic at the point x* if *all* the geodesic trajectories *of* M′ passing through x are also geodesic trajectories *of* M. The formula (20.13.6.1) shows that this condition signifies that *all the second fundamental forms of* M′ *are zero at the point x*. The submanifold M′ is said to be *totally geodesic* if it is geodesic at *every* point, or equivalently if all the second fundamental forms of M′ are *identically zero*.

PROBLEMS

1. Let C be a curve in $\mathbf{R}^n$. With the notation of (20.13.3), suppose that we have the first i formulas (20.13.2.1), but that $\nabla_s \cdot f_i + k_{i-1} f_{i-1} = 0$ identically in U_{i-1}. Show that in this case U_{i-1} is contained in an affine-linear subspace of $\mathbf{R}^n$ of dimension i. (Observe that, for a fixed vector $\mathbf{a} \in \mathbf{R}^n$, the i functions $(\mathbf{a}|f_j)$ for $j \leq i$ satisfy a homogeneous linear system of differential equations.)

Generalize to the case where C is a curve in an arbitrary Riemannian manifold: With the same hypotheses, show that the i-vector which is the product of the first i unit normal vectors at any point of C is obtained from its value at a point $x_0 \in C$ by parallel transport along C.

2. Show that if there exists a Frenet frame for a curve C defined by an embedding $u : t \mapsto u(t)$ ($t \in I$ being any parameter), then the jth curvature of C is the following function of t:

$$k_j = \frac{\|\nabla_s \cdot u \wedge \cdots \wedge \nabla_s^{(j+1)} \cdot u\| \cdot \|\nabla_s \cdot u \wedge \cdots \wedge \nabla_s^{(j-1)} \cdot u\|}{\|\nabla_s \cdot u\| \cdot \|\nabla_s \cdot u \wedge \cdots \wedge \nabla_s^{(j)} \cdot u\|^2},$$

where $\nabla_s \cdot u = u'$ and $\nabla_s^{(k)} \cdot u = \nabla_s \cdot (\nabla_s^{(k-1)} \cdot u)$, the norms being the "$k$-dimensional areas" defined in (20.8.4).

3. (a) The second fundamental forms (20.12.4) of a curve C, relative to the Frenet frame of C, are given by

$$l_{211} = k_1, \qquad l_{\alpha 11} = 0 \qquad \text{for} \quad \alpha \geq 3.$$

(b) Show that for a compact curve C in $\mathbf{R}^N$, the integral curvature (Section 20.12, Problem 4) is given by

$$\frac{2\Omega_{N-2}}{(N-1)\Omega_{N-1}} \int_0^L k_1(s)\, ds,$$

where L is the length of the curve. Deduce that

$$\int_0^L k_1(s)\, ds \geq (N-1)\Omega_{N-1}/\Omega_{N-2}$$

($= 2\pi$ if $N = 3$) (*Fenchel's inequality*).

(c) A compact connected curve C in $\mathbf{R}^3$ is said to be a *knot* if there exists no homeomorphism f of the disk $\|\mathbf{x}\| \leq 1$ in $\mathbf{R}^2$ onto a subspace of $\mathbf{R}^3$ such that the image under f of the circle $\|\mathbf{x}\| = 1$ is C. Show that if C is a knot, we have

$$\int_0^L k_1(s)\, ds \geq 4\pi.$$

(Supposing that $\int_0^L k_1(s)\, ds < 4\pi$, deduce as in Section 20.12, Problem 4(d), that there exists a unit vector $\mathbf{y} \in S_2$ such that the function $s \mapsto (\mathbf{y}|\mathbf{v}(s))$ has a derivative which vanishes at only two points of the interval [0, L[, which we may assume to be 0 and

$L' < L$. Then define a homeomorphism f with the properties described above.) (*Fary–Milnor theorem*).

4. A curve C in the plane $\mathbf{R}^2$ is said to be *strictly convex* (by abuse of language, cf. Section 8.5, Problem 8) if C meets any line in at most two distinct points.

(a) Let C be a connected, strictly convex curve. Show that for any point $x \in C$ the curve C lies entirely on one side of the tangent to C at the point x. (Prove first that this is the case for a sufficiently small neighborhood of x in C, arguing by contradition.) Deduce that if C is compact, then C is the frontier of the convex set D which is the intersection of the closed half-planes containing C, or equivalently, is the convex hull of C (Section 12.14, Problem 13).
(b) Let C be a compact, strictly convex curve in $\mathbf{R}^2$ (identified with $\mathbf{C}$); let L be its length and $s \mapsto v(s)$ its arc length parametrization $(0 \leq s < L)$ for some choice of orientation. Show that $v'(s) = e^{i\theta(s)}$, where θ is a strictly monotone mapping of $[0, L[$ onto $[0, 2\pi[$ or $]-2\pi, 0]$. Conversely, if θ is a C^∞ mapping satisfying this condition, and such that $\int_0^L e^{i\theta(s)}\,ds = 0$, then the image of $[0, L[$ under the mapping $s \mapsto \int_0^s e^{i\theta(t)}\,dt$ is a compact strictly convex curve.
(c) Show that on a compact, strictly convex curve C there exist at least four points at which the curvature (considered as a function $s \mapsto k(s)$ of the arc length) is *stationary* (*Mukhopadhyaya's theorem*). (Assume that the result is false and show that, up to a displacement, the curve is the image of a mapping $s \mapsto (\xi^1(s), \xi^2(s))$ of $[0, L[$ into $\mathbf{R}^2$ such that, for some $s_0 \in]0, L[$, the function k is strictly decreasing on $[0, s_0]$, strictly increasing on $[s_0, L[$, and such that $\xi^2(s) > 0$ for $s \in]0, s_0[$ and $\xi^2(s) < 0$ for $s \in]s_0, L[$. Deduce from Frenet's formulas that

$$\int_0^L k(s)\frac{d\xi^2}{ds}\,ds = 0;$$

split this integral into $\int_0^{s_0} + \int_{s_0}^L$ and integrate each of these by parts; then use the mean-value theorem to obtain a contradiction.) Consider in particular the case where C is an ellipse.
(d) Give an example of an *immersion* u of an interval $I \subset \mathbf{R}$ into $\mathbf{R}^2$ which is not injective but is such that each point $t_0 \in I$ has a neighborhood $J \subset I$ such that the restriction of u to J is an embedding and $u(J)$ is strictly convex.

5. Let h be a C^∞ function on $\mathbf{R}$, periodic with period 2π, and taking only values >0. Show that there exists a compact strictly convex curve C in $\mathbf{R}^2$ (Problem 4) such that the origin is an interior point of the convex hull D of C and such that the function of support of D (Section 16.5, Problem 7) is given by $H(re^{i\theta}) = rh(\theta)$. (Assuming that the problem has been solved and that C is defined as the image $\theta \mapsto (x(\theta), y(\theta))$ of $[0, 2\pi[$ under a C^∞ mapping, express that the tangent at the point with parameter θ is orthogonal to the vector $e^{i\theta}$.)

Let U be the open subset of $\mathbf{S}_1 \times \mathbf{S}_1$ consisting of pairs (ζ_1, ζ_2) such that $\zeta_2 \neq \pm\zeta_1$. For each point $(e^{i\theta_1}, e^{i\theta_2}) \in U$, let $P(\theta_1, \theta_2)$ be the intersection of the lines of support (Section 5.8, Problem 3) of D with equations $(z|e^{i\theta_1}) = h(\theta_1)$, $(z|e^{i\theta_2}) = h(\theta_2)$. Show that this mapping P makes U a two-sheeted covering of $\mathbf{R}^2 - D$, and that its Jacobian at the point $(e^{i\theta_1}, e^{i\theta_2})$ is equal in absolute value to $t_1(P)t_2(P)/|\sin \omega(P)|$, where $t_1(P)$ and

$t_2(P)$ are the distances from P to the points of contact of the two tangents to C which pass through P, and $\omega(P)$ is an angle between these two tangents. Deduce that

$$\int_{\mathbf{R}^2 - D} \frac{|\sin \omega(P)|}{t_1(P)t_2(P)} \, dx \, dy = 2\pi^2$$

(*Crofton's formula*).

6. (a) Let G be the group of displacements of $\mathbf{R}^2$ with determinant $+1$, so that G is the semidirect product of the group $\mathbf{R}^2$ and $\mathbf{SO}(2, \mathbf{R})$. Any element $\sigma \in G$ is of the form

$$(x, y) \mapsto (u + x \cos \theta - y \sin \theta, v + x \sin \theta + y \cos \theta).$$

The forms

$$\omega_1 = \cos \theta \, du + \sin \theta \, dv, \qquad \omega_2 = -\sin \theta \, du + \cos \theta \, dv, \qquad \omega_3 = d\theta$$

constitute a basis of the space of left-invariant differential forms on G, so that $v_G = \omega_1 \wedge \omega_2 \wedge \omega_3 = du \wedge dv \wedge d\theta$ is a left- and right-invariant 3-form on G, and the corresponding measure μ is a Haar measure on G.

Consider two curves C_1, C_2 in the plane $\mathbf{R}^2$, of finite lengths l_1, l_2, parametrized by their curvilinear coordinates s_1, s_2. For each pair (s_1, s_2) of parameters and each angle $\varphi \in [0, 2\pi[$, let $\sigma = g(s_1, s_2, \varphi)$ be the element of G which maps the point M_2 of C_2 with parameter s_2 to the point M_1 of C_1 with parameter s_1, and the unit tangent vector at M_2 to C_2 to a vector making the angle φ with the unit tangent vector at M_1 to C_1. Show that ${}^t g(v_G) = \pm \sin \varphi \, ds_1 \wedge ds_2 \wedge d\varphi$. Using Sard's theorem (**16.23.1**), deduce that for almost all $\sigma \in G$ (relative to the measure μ) the set $C_1 \cap (\sigma \cdot C_2)$ is finite, and that if $n(\sigma)$ is the number of elements in this set, then

$$\int_G n(\sigma) \, d\mu(\sigma) = 4 l_1 l_2$$

(*Poincaré's formula*).

(b) Let C be a compact strictly convex curve in $\mathbf{R}^2$, of length L; let D be its convex hull, and $V_r(D)$ the set of points of $\mathbf{R}^2$ whose distance from D is $\leqq r$, for some $r > 0$. For each integer $k > 0$, let m_k be the Lebesgue measure of the set of points $(x, y) \in \mathbf{R}^2$ such that the circle with center (x, y) and radius r meets C in k points. Show that $m_k = 0$ if k is odd. By applying Poincaré's formula with $C_1 = C$ and C_2 a circle of radius r, show that

$$m_2 + 2m_4 + 3m_6 + \cdots = 2rL. \tag{1}$$

Let A be the Lebesgue measure of D. Using the formula of Steiner–Minkowski to evaluate the measure of $V_r(D)$ (Section **16.24**, Problem 7(b)), prove that

$$m_0' + rL - A - \pi r^2 = m_4 + 2m_6 + 3m_8 + \cdots, \tag{2}$$

where m_0' is the measure of the set of points of $V_r(D)$ such that the circle with center at the point and radius r does not intersect C. Let r_i be the maximum of the radii of closed disks contained in D, and r_e the minimum of the radii of the closed disks containing D. Deduce from (2) that

$$L^2 - 4\pi A \geqq \pi^2 (r_e - r_i)^2$$

(*Bonnesen's inequality*). (Observe that $m_0' = 0$ if $r_i \leqq r \leqq r_e$.)

7. (a) With the notation of Problem 6, let H be the subgroup of G consisting of the displacements which leave the line $y = 0$ invariant as a whole, so that H consists of the $\sigma \in G$ for which $v = 0$ and $\theta = 0$ or π. The homogeneous space G/H may be identified with the space of affine lines in $\mathbf{R}^2$ (a submanifold of the Grassmannian $\mathbf{G}_{3,2}$). If $\pi : G \to G/H$ is the canonical submersion, show that there exists a G-invariant differential 2-form $v_{G/H}$ on G/H such that ${}^t\pi(v_{G/H}) = \omega_2 \wedge \omega_3$ (Section 19.16, Problem 16). Let ν be the corresponding Lebesgue measure on G/H (16.24.1).

Consider a curve C in the plane $\mathbf{R}^2$, of finite length l, parametrized by its arc length s. For each parameter s and angle $\varphi \in [0, \pi[$, let $\Delta(s, \varphi)$ be the line through the point M of C with parameter s, inclined at the angle φ to the tangent at M to C. Show that ${}^t\Delta(v_{G/H}) = \pm \sin \varphi \, ds \wedge d\varphi$. Deduce that for almost all lines γ in $\mathbf{R}^2$ the set $\gamma \cap C$ is finite, and that if $n(\gamma)$ is the number of elements in this set, then

$$\int_{G/H} n(\gamma) \, d\nu(\gamma) = 2l$$

(*Crofton's formula*).

(b) Deduce from (a) that if C is contained in the convex hull D_1 of a compact strictly convex curve C_1 of length l_1, then there exists at least one line γ such that the number of points of $\gamma \cap C$ is $\geq 2l/l_1$ (Apply Crofton's formula to C and C_1).

(c) Generalize Crofton's formula to curves in $\mathbf{R}^n$ and their intersections with hyperplanes in $\mathbf{R}^n$.

8. (a) The existence "in general" of a Frenet frame for a curve C may be expressed in the following terms. Let $S_{O(n)}(M)$ be the bundle of orthonormal frames of M, π its projection on M_1 and σ_i, ω_{ij} the canonical and connection forms on $S_{O(n)}(M)$. Then there exist two sections of $S_{O(n)}(M)$ over C (corresponding to the two orientations of C) such that the $(n+1) \times n$ matrix of the forms σ_i', ω_{ij}' which are the inverse images of σ_i, ω_{ij}, respectively, under one of these sections is of the form

$$(1) \qquad \begin{pmatrix} \sigma_1' & 0 & 0 & \cdots & 0 & 0 \\ 0 & -\omega_{21}' & 0 & \cdots & 0 & 0 \\ \omega_{21}' & 0 & -\omega_{32}' & \cdots & 0 & 0 \\ \vdots & \vdots & \vdots & \vdots & \vdots & \vdots \\ 0 & 0 & 0 & \cdots & \omega_{n,n-1}' & 0 \end{pmatrix},$$

where $\omega_{j+1,j}' = k_j \sigma_1'$, the k_j being functions >0. These frames may be determined by the following systematic inductive procedure: the principal bundle P_0 over $\pi^{-1}(C)$ induced by $S_{O(n)}(M)$, with group $\mathbf{O}(n)$, has dimension $\frac{1}{2}n(n-1)+1$, hence for each $r \in \pi^{-1}(C)$ the kernel $K_0(r)$ of the canonical surjective mapping

$${}^tT_r(j_0) : T_r(S_{O(n)}(M))^* \to T_r(P_0)^*,$$

corresponding to the canonical injection $j_0 : P_0 \to S_{O(n)}(M)$, has dimension $n-1$. The frames $r \in P_0$ for which $K_0(r)$ is the vector subspace spanned by the covectors $\sigma_2(r), \ldots, \sigma_n(r)$ form a *principal bundle* P_1 over C with group $\mathbf{O}(n-1) \times \{\pm 1\}$, hence of dimension $\frac{1}{2}(n-1)(n-2)+1$. For each $r \in P_0$, the kernel $K_1(r)$ of the canonical surjective mapping

$${}^tT_r(j_1) : T_r(P_0)^* \to T_r(P_1)^*,$$

corresponding to the canonical injection $j_1 : P_1 \to P_0$, has dimension $n - 1$. The frames $r \in P_1$ for which $K_1(r)$ is spanned by the covectors $\omega_{j1}(r)$ for $j \geqq 3$ and $\omega_{21}(r) - k_1\sigma_1(r)$, for some number k_1 depending only on $\pi(r) \in C$, form a *principal bundle* P_2 over C with group $\mathbf{O}(n-2) \times \{\pm 1\}^2$. If j_2 is the canonical injection $P_2 \to P_1$, we consider at the next stage the kernel of ${}^t T_r(j_2)$, of dimension $n - 2$; and so on.

(b) Define in the same way the Frenet frame for a curve C in a pseudo-Riemannian manifold M, assuming that C is not isotropic at any point.

(c) Let M be a pseudo-Riemannian manifold of signature $(n-1, 1)$, so that the corresponding group G (20.7.6) is the group $\mathbf{O}(\Phi)$ of a symmetric bilinear form Φ on $\mathbf{R}^n$ such that $\Phi(e_1, e_1) = \Phi(e_2, e_2) = 0$, $\Phi(e_1, e_2) = \Phi(e_2, e_1) = 1$, $\Phi(e_i, e_j) = \delta_{ij}$ for $i, j \geqq 2$. A curve $C \subset M$ is said to be a *curve of zero length* if it is isotropic at every point. Prove the existence "in general" of a "Frenet frame" for C in the following sense: If $S_G(M)$ is the bundle of frames of the G-structure of M, then the connection forms satisfy the relations

$$\omega_{11} + \omega_{22} = 0, \qquad \omega_{21} = \omega_{12} = 0, \qquad \omega_{1j} + \omega_{j2} = 0 \quad \text{for } j \geqq 3,$$
$$\omega_{j1} + \omega_{2j} = 0 \quad \text{for } j \geqq 3, \qquad \omega_{ij} + \omega_{ji} = 0 \quad \text{for } i, j \geqq 3.$$

There exist a *finite* number of sections of $S_G(M)$ over C, such that the $(n+1) \times n$ matrix formed by the inverse images of σ_i, ω_{ij} under one of these sections is of the form

$$\begin{pmatrix}
\sigma_1' & 0 & 0 & 0 & 0 & \cdots & 0 & 0 \\
\sigma_1' & 0 & -\omega_{32}' & 0 & 0 & \cdots & 0 & 0 \\
0 & -\sigma_1' & 0 & 0 & 0 & \cdots & 0 & 0 \\
0 & \omega_{32}' & 0 & -\omega_{43}' & 0 & \cdots & 0 & 0 \\
0 & 0 & \omega_{43}' & 0 & -\omega_{54}' & \cdots & 0 & 0 \\
\vdots & \vdots & \vdots & \vdots & \vdots & \cdots & \vdots & \vdots \\
0 & 0 & 0 & 0 & 0 & \cdots & \omega_{n,n-1}' & 0
\end{pmatrix}$$

(reduce to the case $n = 3$). Generalize to the case of a pseudo-Riemannian metric tensor of arbitrary signature. Calculate the number of sections having the above property.

Suppose that $M = \mathbf{R}^n$ and that there exists an embedding u of an open interval $I \subset \mathbf{R}$ (containing 0) into $\mathbf{R}^n$ such that $u(I) = C$, thereby defining on C an orientation and an origin $c = u(0)$. The *pseudo-arc-length* of a point $x = u(t)$ on C is then defined to be the value of the integral over C of the form $f_x \sigma_1'$ (16.24.2), where f_x is the characteristic function of the set $u([0, t])$ if $t \geqq 0$, and the negative of the characteristic function of the set $u([t, 0])$ if $t \leqq 0$. If s is the real-valued function on C so defined, then $\sigma_1' = ds$, and we may write $\omega_{j,j-1}' = K_{j-2}(s)\, ds$ for $3 \leqq j \leqq n$. Show that the assignment of $n - 2$ functions $K_{j-2}(s)$ of class C^∞ and everywhere >0 on an open interval of $\mathbf{R}$ defines a curve of zero length in $\mathbf{R}^n$ satisfying the relations above, which is uniquely determined up to a displacement.

9. Endow $\mathbf{R}^n$ with the canonical flat G-structure, where $G = \mathbf{SL}(n, \mathbf{R})$, and with the corresponding canonical connection (Section 20.7, Problem 7). Let σ_i and ω_{ij} be the canonical and connection forms on $S_G(M)$ (which satisfy the relation

$$\omega_{11} + \omega_{22} + \cdots + \omega_{nn} = 0).$$

Show that for a curve $C \subset \mathbf{R}^n$ there exist in general a finite number of sections of $S_G(M)$ over C such that the $(n+1) \times n$ matrix formed by the inverse images of the σ_i and ω_{ij} under one of these sections is of the form

$$\begin{pmatrix} \sigma'_1 & 0 & 0 & \cdots & 0 & 0 \\ 0 & \sigma'_1 & 0 & \cdots & 0 & 0 \\ 0 & 0 & \sigma'_1 & \cdots & 0 & 0 \\ 0 & 0 & 0 & \cdots & 0 & \sigma'_1 \\ \omega'_{n1} & \omega'_{n2} & \omega'_{n3} & \cdots & \omega'_{n,n-1} & 0 \end{pmatrix}.$$

(The notational conventions of Section 20.12, Problem 5, will be found useful.) Define as in Problem 8 a pseudo-arc-length s on C, and show that if we put $\omega'_{nj} = K_j(s)\, ds$, then the assignment of the $n-1$ functions K_j in an open interval of $\mathbf{R}$ defines a curve C satisfying the relations above; moreover, every other curve satisfying these relations is obtained from C by a translation followed by an automorphism belonging to $\mathbf{SL}(n, \mathbf{R})$.

14. HYPERSURFACES IN RIEMANNIAN MANIFOLDS

(20.14.1) We recall that a *hypersurface* in a Riemannian manifold M of dimension n is a submanifold V of dimension $n-1$. Suppose that M is *oriented* (which involves no loss of generality if we are concerned only with local properties). Since locally M is diffeomorphic to the product $V \times \mathbf{R}$, an orthonormal moving frame $\mathbf{R}' = (\mathbf{e}_1, \ldots, \mathbf{e}_{n-1})$ on the Riemannian manifold V determines uniquely a *direct* orthonormal frame $(\mathbf{e}_1, \ldots, \mathbf{e}_{n-1}, \mathbf{e}_n)$ for M. The vector field $\mathbf{n} = \mathbf{e}_n$ is then called the *oriented unit normal vector field* of V, relative to the orientations of M and V (the latter relative to $\mathbf{R}'$). There is only one *second fundamental form* $\mathbf{l} = \mathbf{l}_n$ on V, which is uniquely determined up to sign (the sign depending on the orientations of M and V) and is given by the formula

(20.14.2)
$$\mathbf{l} = \sum_{i=1}^{n-1} \omega'_{ni} \otimes \sigma'_i .$$

As soon as the orientation of M is fixed, that of V is fixed by the choice of $\mathbf{n}$, and the form $\mathbf{l}$ changes sign with $\mathbf{n}$. The $n-1$ eigenvalues of $\mathbf{l}(x)$ (counted according to multiplicity) with respect to the positive definite form $\mathbf{g}'(x)$ are real (of either sign), and are called the *principal curvatures* of V at the point x; they change sign with $\mathbf{n}$. Denoting them by

$$\rho_1(x) \leqq \rho_2(x) \leqq \cdots \leqq \rho_{n-1}(x),$$

the numbers

$$H(x) = \frac{1}{n-1}(\rho_1(x) + \cdots + \rho_{n-1}(x))$$

and

$$K(x) = \rho_1(x)\rho_2(x) \cdots \rho_{n-1}(x)$$

are called, respectively, the *mean curvature* and the *total curvature* of V at the point x.

The inverses $(\rho_j(x))^{-1}$ for those indices j such that $\rho_j(x) \neq 0$ are called the *principal radii of curvature* of V at the point x. When $\mathrm{M} = \mathbf{R}^n$, the points $x + \rho_j(x)^{-1}\tau_x(\mathbf{n}(x))$ (for $\rho_j(x) \neq 0$) are called the *principal centers of curvature* of V at the point x; they do not depend on the sign of $\mathbf{n}$.

Suppose that the principal curvatures $\rho_j(x)$ are all *distinct*. Since they are the roots of the equation $\det(l_{ij}(x) - \delta_{ij}t) = 0$, and these roots are all simple, it follows that the derivative of this polynomial with respect to t is nonzero at each root; hence the ρ_j $(1 \leqq j \leqq n-1)$ are well-defined functions of class C^∞ in a sufficiently small neighborhood of x (10.2.3) and their values at each point of this neighborhood are all distinct. It follows therefore from Cramer's formulas that there exist $n-1$ fields of unit vectors $\mathbf{c}_j$ $(1 \leqq j \leqq n-1)$ of class C^∞ on V such that at each point y the lines $\mathbf{R}\mathbf{c}_j(y)$ are the *principal axes* of $\mathbf{l}(y)$ with respect to $\mathbf{g}'(y)$. These lines are called the *principal directions* of V at the point y. A curve in V which at each of its points is tangent to a principal direction is called a *line of curvature* of V.

We may always assume that the $\mathbf{c}_j$ (which are determined only up to sign) have been chosen so that $(\mathbf{c}_1, \ldots, \mathbf{c}_{n-1}, \mathbf{n})$ is a *direct* orthonormal frame. Relative to this frame, the second fundamental form has the expression

(20.14.2.1) $$\mathbf{l} = \sum_{j=1}^{n-1} \rho_j \sigma'_j \otimes \sigma'_j .$$

For any curve C in V, the number $k_1(\mathbf{f}_2 | \mathbf{n})$ (in the notation of (20.13.2)) is called the *normal curvature* of C (relative to the chosen normal vector $\mathbf{n}$ to V). It follows from (20.14.2.1) and (20.13.6.4) that we have

(20.14.3) $$k_1(\mathbf{f}_2 | \mathbf{n}) = \sum_{j=1}^{n-1} \rho_j(\mathbf{f}_1 | \mathbf{c}_j)^2 \qquad (\textit{Euler's formula}).$$

The relation (20.14.2.1) also signifies that relative to the frame

$$(\mathbf{c}_1, \ldots, \mathbf{c}_{n-1}, \mathbf{n}),$$

we have, with the notation of (20.12.3),

(20.14.4) $$\omega'_{ni} = \rho_i \sigma'_i \qquad (1 \leqq i \leqq n-1),$$

and hence the formula (20.12.5.2) may be written as

(20.14.5) $$\tilde{\Omega}_{ij} - \Omega'_{ij} = \rho_i \rho_j \sigma'_j \wedge \sigma'_i.$$

In particular, when $n = 3$ and $M = \mathbf{R}^3$, we have (assuming that $\rho_1 \neq \rho_2$)

(20.14.6) $$K = \rho_1 \rho_2,$$

i.e., *the product of the principal curvatures is equal to the Gaussian curvature* (and therefore depends only on the induced Riemannian metric tensor $\mathbf{g}'$) (*Gauss's theorema egregium*).

It should be remarked that this result remains true when $\rho_1 = \rho_2$. This is clear by continuity at an umbilic which is a limit of nonumbilical points; the only other cases are the sphere and the plane, for which the theorem is immediately verified (20.12.9).

(20.14.7) In general, for $n \geqq 4$, hypersurfaces in $\mathbf{R}^n$ are *rigid*: the second fundamental form is in general *uniquely determined by the Riemannian metric tensor* $\mathbf{g}'$. To be precise:

(20.14.8) *Let* V, V_1 *be two connected hypersurfaces in* $\mathbf{R}^n$, *where* $n \geqq 4$, *and suppose that the principal curvatures of* V *are distinct and nonzero at every point of* V. *Then every isometry* f *of* V *onto* V_1 *is a Euclidean displacement* (*Beez's theorem*).

Let us retain the notation introduced above for the second fundamental form of V, and let $(l_{ij}(x))$ be the matrix at the point x, relative to the frame $(\mathbf{c}_i(x))_{1 \leqq i \leqq n-1}$, of the inverse image under f of the second fundamental form of V_1. Since the curvature of the connection on $\mathbf{R}^n$ is zero, the hypothesis that f is an isometry implies, by virtue of (20.14.5) and (20.12.5.2), that

(20.14.8.1) $$\langle -\Omega'_{ij}, \mathbf{c}_r \wedge \mathbf{c}_s \rangle = \rho_i \rho_j (\delta_{jr} \delta_{is} - \delta_{ir} \delta_{js}) = l_{jr} l_{is} - l_{ir} l_{js}$$

for all indices i, j, r, s. This signifies that if u, v are the endomorphisms of $\mathbf{R}^{n-1}$ whose matrices are $(l_{ij}(x))$ and $\operatorname{diag}(\rho_1(x), \ldots, \rho_{n-1}(x))$, respectively, then $\bigwedge^2 u = \bigwedge^2 v$; we have to show that $u = \pm v$. Since by hypothesis v is invertible, it is enough to show that $v^{-1}(u(\mathbf{z}))$ is a scalar multiple of $\mathbf{z}$, for all $\mathbf{z} \in \mathbf{R}^{n-1}$. Suppose not; then there exists $\mathbf{z} \in \mathbf{R}^{n-1}$ such that $u(\mathbf{z})$ and $v(\mathbf{z})$ are linearly independent. Since $n - 1 \geqq 3$, there exists a vector $\mathbf{y}$ such that $u(\mathbf{z})$, $v(\mathbf{y})$, and

$v(\mathbf{z})$ are linearly independent, and therefore $u(\mathbf{z}) \wedge v(\mathbf{y}) \wedge v(\mathbf{z}) \neq 0$; but by hypothesis we have $v(\mathbf{y}) \wedge v(\mathbf{z}) = u(\mathbf{y}) \wedge u(\mathbf{z})$, which leads to a contradiction. Hence $u = \alpha v$ for some scalar α; since $\bigwedge^2 u = \bigwedge^2 v$, this implies that $\alpha^2 = 1$ and completes the proof, in view of (20.12.7).

By contrast, in $\mathbf{R}^3$ there exist surfaces whose principal curvatures are distinct and $\neq 0$, and for which there exist isometric surfaces not obtainable by any Euclidean displacement (Problem 8).

PROBLEMS

1. With the hypotheses and notation of (20.14.1), for each $x \in \mathrm{V}$ let $S(x)$ denote the self-adjoint endomorphism of $\mathrm{T}_x(\mathrm{V})$ associated (Section 11.5, Problem 3) with the symmetric bilinear form $\mathbf{l}(x)$, i.e., such that $(S(x) \cdot \mathbf{h}_x | \mathbf{k}_x) = \langle \mathbf{l}(x), \mathbf{h}_x \otimes \mathbf{k}_x \rangle$ (cf. Section 20.12, Problem 1(b)). The eigenvalues of $S(x)$ are therefore the principal curvatures $\rho_j(x)$. The rank of $S(x)$ is equal to the rank at the point x of the normal mapping $\mathbf{n}$ of V into $\mathbf{S}_{n-1}$ (Section 20.12, Problem 5).

   Suppose that $\mathrm{M} = \mathbf{R}^n$ and that V (of dimension $n-1$) is the image of an open set $\mathrm{D} \subset \mathbf{R}^{n-1}$ under a C^∞ mapping

$$\mathbf{f} : (x^1, \ldots, x^{n-1}) \mapsto (x^1, \ldots, x^{n-1}, u(x^1, \ldots, x^{n-1})).$$

   Orient V by taking the oriented unit normal vector at the point $x = \mathbf{f}(x^1, \ldots, x^{n-1})$ to be $\mathbf{n}(x) = \|\mathbf{w}(x)\|^{-1}\mathbf{w}(x)$, where

$$\mathbf{w} = (-p_1, \ldots, -p_{n-1}, 1)$$

   and $p_j = \partial u/\partial x^j$. Show that, relative to the basis of $\mathrm{T}_x(\mathrm{V})$ formed by the vectors $\tau_x^{-1}(\partial \mathbf{f}/\partial x^j)$, the matrix of $S(x)$ is

$$-\frac{1}{\|\mathbf{w}\|} R \cdot \left(I - \frac{1}{\|\mathbf{w}\|^2} \mathbf{p} \cdot {}^t\mathbf{p} \right),$$

   where

$$R = (r_{ij}), \qquad r_{ij} = \partial^2 u/\partial x^i\, \partial x^j,$$

   and $\mathbf{p}$ is the column vector with entries $p_1, p_2, \ldots, p_{n-1}$.

   Deduce that the mean curvature $\mathrm{H}(x)$ and the total curvature $\mathrm{K}(x)$ of V at the point x are given by the formulas

$$\mathrm{H} = -\frac{1}{n\|\mathbf{w}\|} \left(\sum_i r_{ii} - \frac{1}{\|\mathbf{w}\|^2} \sum_{i,j} r_{ij} p_i p_j \right),$$

$$\mathrm{K} = \frac{(-1)^{n-1}}{\|\mathbf{w}\|^{n+1}} \det(R).$$

If ω denotes the differential $(n-2)$-form on V given by

$$\omega = \sum_i (-1)^{i-1} \frac{p_i}{\|\mathbf{w}\|} dx^1 \wedge \cdots \wedge \widehat{dx^i} \wedge \cdots \wedge dx^{n-1},$$

show that

$$d\omega = -n\mathrm{H}\, dx^1 \wedge dx^2 \wedge \cdots \wedge dx^{n-1}.$$

2. With the notation of Problem 1, suppose that D is the open ball with center 0 and radius R in $\mathbf{R}^{n-1}$. If there exists a number $a > 0$ such that $\mathrm{H}(x) \geqq a$ for all $x \in \mathrm{V}$, show that $a\mathrm{R} \leqq 1$ (*Heinz's inequality*). (If $0 < r < \mathrm{R}$ and if B_r (resp. S_r) denotes the open ball (resp. the sphere) with center 0 and radius r, then by virtue of the elementary version of Stokes's formula (**16.24.11**) we have

$$\int_{\mathrm{S}_r} \omega = \int_{\mathrm{B}_r} d\omega.$$

Obtain a lower bound for the absolute value of the integral on the right by using the hypothesis and Problem 1; then obtain an upper bound for the integral on the left by using the solid angle measure (**16.24.9**) and the Cauchy–Schwarz inequality.)

3. Generalizing the notion of a strictly convex curve (Section **20.13**, Problem 4), a hypersurface V in $\mathbf{R}^n$ is said to be *strictly convex* if it meets each line in at most two distinct points.

(a) Show that a connected hypersurface V is strictly convex provided that at each point $x \in \mathrm{V}$ the principal curvatures $\rho_j(x)$ $(1 \leqq j \leqq n-1)$ are nonzero and all of the same sign. At each point x of a hypersurface V at which this property is satisfied, there exists an open neighborhood U of x in $\mathbf{R}^n$ such that $\mathrm{U} \cap \mathrm{V}$ is a strictly convex hypersurface, and V is said to be *strictly convex at the point x*.
(b) Show that if V is a compact, strictly convex hypersurface, then V is the frontier of its convex hull.
(c) Show that if V is any compact hypersurface in $\mathbf{R}^n$, then there exists at least one point of V at which V is strictly convex. (Consider a point at which the distance from a fixed point in $\mathbf{R}^n$ attains its maximum.)
(d) Let V be a compact connected hypersurface in $\mathbf{R}^n$. Show that the following conditions are equivalent:

(α) For each $x \in \mathrm{V}$, there exists no isotropic vector $\neq 0$ for the second fundamental form $\mathit{I\!I}(x)$ ($\mathit{I\!I}$ being defined up to sign in a neighborhood of x).

(β) V is orientable, and for a choice of orientation on V the normal mapping $x \mapsto \mathbf{n}(x)$ (notation of (**20.14.1**)) is a diffeomorphism of V onto $\mathbf{S}_{n-1}$.

(γ) The total curvature of V is never zero.

(To show that (α) implies (β), use Section **20.12**, Problem 4 and Section **16.12**, Problem 1. To show that (γ) implies (α), use (c) above.)

If these conditions are satisfied, V is strictly convex.

(e) Give an example of a connected quadric which is neither compact nor strictly convex in $\mathbf{R}^3$, and whose principal curvatures are never zero.

4. Let V be a hypersurface in $\mathbf{R}^n$.

(a) Suppose that, at all points x of an orientable open subset U of V, the rank of the endomorphism $S(x)$ (Problem 1)—or, equivalently, the rank at the point x of the normal mapping $\boldsymbol{n}$ of U into S_{n-1} (for any choice of orientation on U)—is constant and equal to $n-k-1$, where $k \geqq 1$.

Show that for each point $x_0 \in U$ there exists a chart $(U_0, \varphi, n-1)$ of V at the point x_0, with $U_0 \subset U$ and $\varphi(U_0) = W \times T$, where $W \subset \mathbf{R}^{n-k-1}$ and $T \subset \mathbf{R}^k$ are two open sets, such that for each point $w \in W$ the set $L_w = \varphi^{-1}(\{w\} \times T)$ is the intersection of U_0 and an affine-linear subspace of dimension k in $\mathbf{R}^n$; moreover the normal vector $\boldsymbol{n}$ is *constant* on each L_w. (Choose local coordinates $u^1, \ldots, u^{n-1}$ such that $\partial\boldsymbol{n}/\partial u^j = 0$ in $\varphi(U_0)$ for $1 \leqq j \leqq k$, and such that the vectors $\partial\boldsymbol{n}/\partial u^h$ for $k < h \leqq n-1$ are linearly independent; then differentiate the relations $(\boldsymbol{n} \mid \partial\mathbf{f}/\partial u^j) = 0$ for $1 \leqq j \leqq n-1$, where $\mathbf{f} = \varphi^{-1}$.)

(b) Let y be a point in the closure of U and in the closure of one of the sets L_w defined above. Show that the rank $\text{rk}_y(\boldsymbol{n})$ is still equal to $n-k-1$. (There exists a moving frame $(\mathbf{e}_1, \ldots, \mathbf{e}_{n-1}, \boldsymbol{n})$ defined in a neighborhood U_0 of y such that the vectors $\mathbf{e}_j(z)$ are parallel to L_w for $1 \leqq j \leqq k$ and $z \in U_0 \cap L_w$; we have then $\omega'_{jn} = 0$ for $1 \leqq j \leqq k$ in $U_0 \cap L_w$, so that if we write

$$\omega'_{jn} = \sum_{h=1}^{n-1} A_{jh}\sigma'_h,$$

then $A_{jh} = 0$ in $U_0 \cap L_w$ for $1 \leqq j \leqq k$, and the determinant Δ of the matrix $(A_{jh})_{n-k-1 \leqq j,h \leqq n-1}$ is invertible at the points of $U_0 \cap L_w$. By differentiating the relation

$$\omega'_{k+1,n} \wedge \omega'_{k+2,n} \wedge \cdots \wedge \omega'_{n-1,n} = \Delta\sigma'_{k+1} \wedge \sigma'_{k+2} \wedge \cdots \wedge \sigma'_{n-1},$$

show that if we put

$$\omega'_{jh} = \sum_{l=1}^{n-1} a_{jhl}\sigma'_l$$

$(1 \leqq j \leqq k,\quad k+1 \leqq h \leqq n-1)$ then in $U_0 \cap L_w$ we have $a_{jhl} = 0$ for $1 \leqq l \leqq k$ and

$$d\Delta + \Delta\left(\sum_{h=k+1}^{n-1} \sum_{j=1}^{k} a_{jhh}\sigma'_h\right) = 0.$$

Deduce that as $z \to y$ in $U_0 \cap L_w$, $\Delta(z)$ tends to a limit $\neq 0$, and therefore the matrix (A_{jh}) remains invertible at the point y).

(c) For each integer k such that $0 \leqq k \leqq n-1$, let V_k denote the set of points $x \in V$ at which $\text{rk}_x(\boldsymbol{n}) = n-k-1$. Show that for each point $x \in V$ and each neighborhood T of x in V, there exists an integer $m \leqq n-1$ and a nonempty open set $T_1 \subset T$ contained in V_m (cf. (**16.5.3**)). Let $y \in T_1$ and let H be the affine hyperplane tangent to V at the point y (in the sense of (**16.8.6**)). Deduce from (a) that there exists in H an affine-linear subspace L of dimension m such that the interior E of $L \cap V$ *with respect to* L contains y, and such that H is the tangent affine hyperplane to V at all points of E. Furthermore, if $E \neq L$, every frontier point z of E in L also belongs to V_m by (b) above, but is not an interior point of V_m. Deduce that in every neighborhood of z in V there exist points belonging to $V_{m'}$ for some $m' < m$.

(d) In the plane $\mathbf{R}^2$, identified with C, consider three vectors $\mathbf{a}_1 = 1$, $\mathbf{a}_2 = j$ (a complex cube root of unity), $\mathbf{a}_3 = \bar{j} = j^2$. Let D be the open set in $\mathbf{R}^2$ which is the union of the three "angles" A_{jk} defined by $(\mathbf{x} \mid \mathbf{a}_j) < 1$ and $(\mathbf{x} \mid \mathbf{a}_k) < 1$. In the angle A_{23}, define a function z_1 to be equal to 0 if $(\mathbf{x} \mid \mathbf{a}_1) \leqq 1$, and equal to $\exp(-((\mathbf{x} \mid \mathbf{a}_1) - 1)^{-2})$ if $(\mathbf{x} \mid \mathbf{a}_1) > 1$; define z_2 in A_{31} and z_3 in A_{12} analogously. Then these three functions coincide in the triangle D′ which is the intersection of any two of A_{23}, A_{31}, A_{12}, and therefore define a C^∞ function z in D. Let C be the hypersurface in $\mathbf{R}^3$ which is the graph of the function z. With the notation of (c) above, show that $V_0 = \varnothing$, $V_1 = V - \bar{D}'$, $V_2 = \bar{D}'$, and that through a point of D' there passes no line wholly contained in V.

(e) For each point $x \in V$, let $\mathscr{H}(x)$ denote the affine tangent hyperplane to V. If $\mathbf{R}^n$ is identified with an open subset of $\mathbf{P}_n(\mathbf{R})$, then $\mathscr{H}(x)$ is an element of the Grassmannian $\mathbf{G}_{n+1,n}$. The mapping $x \mapsto \mathscr{H}(x)$ of V into $\mathbf{G}_{n+1,n}$ is of class C^∞ (it is called the *tangential image* of V). With the notation of (c) show that if V is *compact* then $\mathscr{H}(V_0)$ is dense in $\mathscr{H}(V)$.

5. Let V be a compact hypersurface in $\mathbf{R}^n$. Then the integral curvature of V (Section **20.12**, Problem 4) is equal to

$$\kappa(V) = \frac{2}{\Omega_n} \int_V |K(x)| \, dv(x),$$

where $K(x)$ is the total curvature at the point $x \in V$ and v is the Riemannian volume of V.

(a) Suppose that $\kappa(V) = 2$ (cf. *loc. cit.*). Show that if a point $x \in V$ belongs to V_0 (notation of Problem 4(c)), then V lies entirely to one side of the tangent hyperplane to V at the point x. (Observe that $\boldsymbol{n}$ is a local diffeomorphism at the point x: if V were not to one side of H, there would exist a neighborhood of $\boldsymbol{n}(x)$ in $\mathbf{S}_{n-1}$ such that, for all points $\mathbf{z}$ in this neighborhood, there would be at least three distinct points of V at which $\boldsymbol{n}$ took the value $\mathbf{z}$. Then use Section **20.12**, Problem 4.)

(b) Deduce that if $\kappa(V) = 2$, then V is the frontier of a convex body in $\mathbf{R}^n$. (Use Problem 4(c).)

6. Let M be a submanifold of $\mathbf{R}^n$, let $\boldsymbol{r}$ be its curvature morphism and $\boldsymbol{l}$ the second (vector-valued) fundamental form on M.

(a) Show that for each $x \in M$ and each system of tangent vectors $\mathbf{u}$, $\mathbf{v}$, $\mathbf{w}$, $\mathbf{t}$ in $T_x(M)$, we have

$$(1) \qquad (\mathbf{t} \mid (\boldsymbol{r}(x) \cdot (\mathbf{u} \wedge \mathbf{v})) \cdot \mathbf{w}) = (\langle \boldsymbol{l}(x), \mathbf{u} \otimes \mathbf{w} \rangle \mid \langle \boldsymbol{l}(x), \mathbf{v} \otimes \mathbf{t} \rangle) - (\langle \boldsymbol{l}(x), \mathbf{v} \otimes \mathbf{w} \rangle \mid \langle \boldsymbol{l}(x), \mathbf{u} \otimes \mathbf{t} \rangle).$$

(Apply (**20.12.5.2**) and (**20.10.4.6**).)

(b) Let $T'_x(M)$ denote the vector subspace of $T_x(M)$ consisting of the vectors $\mathbf{u}$ such that $\boldsymbol{r}(x) \cdot (\mathbf{u} \wedge \mathbf{v}) = 0$ for *all* $\mathbf{v} \in T_x(M)$, and let $T''_x(M)$ denote the vector subspace of $T_x(M)$ consisting of the vectors $\mathbf{u}$ such that $\langle \boldsymbol{l}(x), \mathbf{u} \otimes \mathbf{v} \rangle = 0$ for *all* $\mathbf{v} \in T_x(M)$. Show that $T''_x(M) \subset T'_x(M)$.

(c) Use (a) to show that every vector $\mathbf{u} \in T'_x(M)$ such that $\langle \boldsymbol{l}(x), \mathbf{u} \otimes \mathbf{u} \rangle = 0$ belongs to $T''_x(M)$, and deduce that

$$\dim T'_x(M) - \dim T''_x(M) \leq n - \dim(M).$$

(Use the following algebraic lemma: If E, F are two finite-dimensional real vector spaces and $u : E \times E \to F$ is a symmetric bilinear mapping, then if $\dim F < \dim E$, there exist two nonzero vectors $\mathbf{x}, \mathbf{y} \in E$ such that $u(\mathbf{x}, \mathbf{x}) = u(\mathbf{y}, \mathbf{y})$ and $u(\mathbf{x}, \mathbf{y}) = 0$. To prove this lemma, consider the symmetric bilinear mapping $u_{(\mathbf{C})} : E_{(\mathbf{C})} \times E_{(\mathbf{C})} \to F_{(\mathbf{C})}$ (the complexification of u) and show that there exists $\mathbf{z} \neq 0$ in $E_{(\mathbf{C})}$ such that $u_{(\mathbf{C})}(\mathbf{z}, \mathbf{z}) = 0$, by using the fact that in $\mathbf{C}^m$ a system of $m - 1$ homogeneous polynomials always has a common zero other than the origin.)

7. Let V be a hypersurface in $\mathbf{R}^n$, and let $\boldsymbol{r}$ be its curvature morphism.

(a) Show that $\text{rk}_x(\boldsymbol{n}) = 0$ or 1 if and only if $\boldsymbol{r}(x) = 0$.
(b) With the notation of Problem 6, show that if $\text{rk}_x(\boldsymbol{n}) \geqq 2$, then $\text{T}'_x(\text{V}) = \text{T}''_x(\text{V})$. (With the notation of Problem 1, show that

$$(1) \qquad (\boldsymbol{r}(x) \cdot (\mathbf{u} \wedge \mathbf{v})) \cdot \mathbf{w} = (S(x) \cdot \mathbf{v} | \mathbf{w})(S(x) \cdot \mathbf{u}) - (S(x) \cdot \mathbf{u} | \mathbf{w})(S(x) \cdot \mathbf{v})$$

for $\mathbf{u}, \mathbf{v}, \mathbf{w} \in \text{T}_x(\text{V})$. Let N be the orthogonal supplement of $\text{T}''_x(\text{V})$ in $\text{T}_x(\text{V})$; show that $\text{N} \cap \text{T}'_x(\text{V}) = \{0\}$ by remarking that $\dim(\text{N}) \geqq 2$ and that if $\mathbf{u}, \mathbf{v} \in \text{N}$ are linearly independent, then so are $S(x) \cdot \mathbf{u}$ and $S(x) \cdot \mathbf{v}$. Now use (1) to obtain a contradiction.)
(c) Generalize the result of (20.14.8) by replacing the hypothesis that the principal curvatures of V at each point are distinct and nonzero by the hypothesis that $\text{rk}_x(\boldsymbol{n}) \geqq 3$ for all $x \in \text{V}$. (Observe that the orthogonal supplement N of $\text{T}'_x(\text{V})$ depends only on the Riemannian metric tensor of V and is stable under $S(x)$ and the analogous endomorphism $S_1(x)$ derived from the inverse image of the second fundamental form of V_1.)

8. Let G be a hypersurface in $\mathbf{R}^n$ with equation $\xi^n = \zeta(y)$, where y runs through an open subset D of $\mathbf{R}^{n-1}$ and ζ is a C^∞ function on D which is everywhere >0. The set V of points $(y, \zeta(y)x)$ in $\mathbf{R}^{n+m} = \mathbf{R}^{n-1} \times \mathbf{R}^{m+1}$, where $y \in \text{D}$ and $x \in \mathbf{S}_m$, is called the *hypersurface of revolution with axis* $\mathbf{R}^{n-1}$ *generated by* G. The hypersurface V is canonically diffeomorphic to $\text{G} \times \mathbf{S}_m$, and G may be identified with the submanifold of points of V for which $x = \mathbf{e}_n$. If G_x is the set of points of V for which x is fixed, then G_x is obtained from G by any orthogonal transformation $s \in \mathbf{O}_{m+1}(\mathbf{R})$ which sends $\mathbf{e}_n$ (identified with the first vector in the canonical basis of $\mathbf{R}^{m+1}$) to x. The tangent space to V at the point $(y, \zeta(y)x)$ is the direct sum of the subspaces $s \cdot \text{T}_y(\text{G})$ and $\zeta(y)\text{T}_x(\text{M})$, which are orthogonal. If $\boldsymbol{l}$ (resp. $\mathbf{L}$) is the second fundamental form of G (resp. V), the restriction of $\mathbf{L}$ to $s \cdot \text{T}_y(\text{G})$ is $s \cdot \boldsymbol{l}$. The principal centers of curvature at the point $(y, \zeta(y)x)$ are the images under s of the principal centers of curvature of G at the point y and the point where the normal to G at y meets the axis $\mathbf{R}^{n-1}$. The submanifolds G_x of V are totally geodesic.

If $n = 1$ and G is the semicircle $\xi^2 = (1 - (\xi^1)^2)^{1/2}$, D being the open interval $]-1, 1[$, then V is the sphere $\mathbf{S}_{m+1}$ with the poles $\pm\mathbf{e}_1$ removed.

Let D be an open subset of the plane $\mathbf{R}^2$, let σ_1, σ_2 be two differential 1-forms on D which are linearly independent at each point of D, and let ρ_1, ρ_2 be two C^∞ functions defined on D.

Put $d\sigma_1 = -c_1\sigma_1 \wedge \sigma_2$, $d\sigma_2 = -c_2\sigma_1 \wedge \sigma_2$, and for each C^∞ function f on D put $df = f_1\sigma_1 + f_2\sigma_2$. In order that the forms $\sigma_1, \sigma_2, \omega_{12} = c_1\sigma_1 + c_2\sigma_2, \omega_{31} = \rho_1\sigma_1$, $\omega_{32} = \rho_2\sigma_2$ should satisfy the conditions of (20.12.7) (and hence define a surface in $\mathbf{R}^3$, unique up to a displacement, for which the principal curvatures are ρ_1, ρ_2 and the lines of curvature are given by the equations $\sigma_1 = 0$ and $\sigma_2 = 0$) it is necessary and sufficient that ρ_1 and ρ_2 satisfy the following relations:

$$\rho_1\rho_2 + (c_1)^2 + (c_2)^2 = c_{21} - c_{12},$$
$$\rho_{12} = (\rho_2 - \rho_1)c_1,$$
$$\rho_{21} = (\rho_2 - \rho_1)c_2.$$

Take $\sigma_1 = d\varphi$, $\sigma_2 = \cos\varphi\, d\theta$, so that $c_1 = 0$, $c_2 = \tan\varphi$. Then a solution of these equations is $\rho_1 = \rho_2 = 1$, which corresponds to the sphere $\mathbf{S}_2$ with the poles $\pm\mathbf{e}_1$ removed. Show that there exist infinitely many other solutions, depending on a parameter.

9. (a) On a differential manifold M, let $\mathbf{u}_j$ $(1 \leqq j \leqq p)$ be p continuous functions with values in $\mathbf{R}^n$, and let $\boldsymbol{\omega}_k$ $(1 \leqq k \leqq n-p)$ be $n-p$ differential 1-forms with values in $\mathbf{R}^n$. Define a differential $(n-p)$-form α on M by the condition

$$\langle \alpha(x), \mathbf{h}_1 \wedge \mathbf{h}_2 \wedge \cdots \wedge \mathbf{h}_{n-p} \rangle = \sum_{\pi \in \mathfrak{S}_{n-p}} \varepsilon_\pi \det(\mathbf{u}_1(x), \ldots, \mathbf{u}_p(x), \boldsymbol{\omega}_1(x) \cdot \mathbf{h}_{\pi(1)}, \ldots, \boldsymbol{\omega}_{n-p}(x) \cdot \mathbf{h}_{\pi(n-p)})$$

for each system of vectors $\mathbf{h}_1, \ldots, \mathbf{h}_{n-p}$ in $T_x(M)$, the determinant being the determinant of the matrix whose jth column is $\mathbf{u}_j(x)$ (relative to the canonical basis of $\mathbf{R}^n$) if $1 \leqq j \leqq p$, and $\boldsymbol{\omega}_k(x) \cdot \mathbf{h}_{\pi(k)}$ if $j = p + k$, $1 \leqq k \leqq n-p$. We write $\alpha = [\mathbf{u}_1, \ldots, \mathbf{u}_p, \boldsymbol{\omega}_1, \ldots, \boldsymbol{\omega}_{n-p}]$. If $d\boldsymbol{\omega}_k = 0$ for $1 \leqq k \leqq n-p$, show that

(1) $$d\alpha = \sum_{i=1}^{p} [\mathbf{u}_1, \ldots, d\mathbf{u}_i, \ldots, \mathbf{u}_p, \boldsymbol{\omega}_1, \ldots, \boldsymbol{\omega}_{n-p}],$$

where

$$[\mathbf{u}_1, \ldots, d\mathbf{u}_i, \ldots, \mathbf{u}_p, \boldsymbol{\omega}_1, \ldots, \boldsymbol{\omega}_{n-p}] = (-1)^{p-i}[\mathbf{u}_1, \ldots, \hat{\mathbf{u}}_i, \ldots, \mathbf{u}_p, d\mathbf{u}_i, \boldsymbol{\omega}_1, \ldots, \boldsymbol{\omega}_{n-p}].$$

(b) Let V be a surface in $\mathbf{R}^3$. Using the identifications of Section 20.12, Problem 5, and the operations defined in (a), establish the following formulas for the canonical "area" form v on V, the mean curvature H and the total curvature K:

$$[\mathbf{n}, d\mathbf{x}, d\mathbf{x}] = 2v,$$
$$[\mathbf{n}, d\mathbf{x}, d\mathbf{n}] = -2\mathrm{H}v,$$
$$[\mathbf{n}, d\mathbf{n}, d\mathbf{n}] = 2\mathrm{K}v.$$

Generalize to hypersurfaces in $\mathbf{R}^n$.

(c) Suppose that V is *compact*. If A is the area of V and if $p = (\mathbf{x} \mid \mathbf{n})$, show that

$$\mathrm{A} = \int p\mathrm{H}v,$$
$$\int \mathrm{H}v = \int p\mathrm{K}v$$

(*Minkowski's formulas*). (Apply the formula (1) to $[\mathbf{x}, \mathbf{n}, d\mathbf{x}]$ and $[\mathbf{x}, \mathbf{n}, d\mathbf{n}]$.)

(d) Let $\mathrm{V}^{(1)}$, $\mathrm{V}^{(2)}$ be two compact surfaces in $\mathbf{R}^3$, with nonvanishing total curvature at each point. Then there exists a diffeomorphism $\mathbf{z} \mapsto \mathbf{f}^{(i)}(\mathbf{z})$ $(i = 1, 2)$ of S_2 onto $\mathrm{V}^{(i)}$ such that $\mathbf{n}(\mathbf{f}^{(i)}(\mathbf{z})) = \mathbf{z}$ for $i = 1, 2$ (Problem 3(d)). Let $\mathrm{R}_1^{(i)}(\mathbf{z})$, $\mathrm{R}_2^{(i)}(\mathbf{z})$ be the principal radii of curvature (which are >0) at the point $\mathbf{f}^{(i)}(\mathbf{z})$, and let

$$\mathrm{P}_1^{(i)} = \tfrac{1}{2}(\mathrm{R}_1^{(i)} + \mathrm{R}_2^{(i)}), \qquad \mathrm{P}_2^{(i)} = \mathrm{R}_1^{(i)}\mathrm{R}_2^{(i)}, \qquad \mathrm{Q} = \tfrac{1}{2}(\mathrm{R}_1^{(1)}\mathrm{R}_2^{(2)} + \mathrm{R}_2^{(1)}\mathrm{R}_1^{(2)}).$$

If $p^{(i)}(\mathbf{z}) = (\mathbf{z} \mid \mathbf{f}^{(i)}(\mathbf{z}))$, prove the formulas

$$\int_{\mathbf{S}_2} (p^{(i)}\mathrm{P}_1^{(i)} - \mathrm{P}_2^{(i)})\alpha = 0,$$
$$\int_{\mathbf{S}_2} (p^{(1)}\mathrm{P}_1^{(2)} - \mathrm{Q})\alpha = 0,$$
$$\int_{\mathbf{S}_2} (p^{(1)}\mathrm{P}_2^{(2)} - p^{(2)}\mathrm{Q})\alpha = 0,$$

α being the solid angle form on $\mathbf{S}_2$. (The first formula is a restatement of Minkowski's second formula. For the other two formulas, consider the forms $[\mathbf{f}^{(1)}(\mathbf{z}), \mathbf{z}, d\mathbf{f}^{(2)}(\mathbf{z})]$ and $[\mathbf{f}^{(1)}(\mathbf{z}), \mathbf{f}^{(2)}(\mathbf{z}), d\mathbf{f}^{(2)}(\mathbf{z})]$ on $\mathbf{S}_2$.)

(e) With the hypotheses and notation of (d), prove that if $P_1^{(1)} = P_1^{(2)}$ or if $P_2^{(1)} = P_2^{(2)}$ on S_2, then $V^{(2)}$ is obtained from $V^{(1)}$ by a translation in $\mathbf{R}^3$. (In the first case, use (a) to prove that if $F = (R_1^{(1)} - R_1^{(2)})(R_2^{(1)} - R_2^{(2)})$, then we have $\int_{S_2} F\alpha = 0$. In the second case, observe that the relation $P_2^{(1)} = P_2^{(2)}$ implies $Q \geqq P_2^{(1)}$, with equality if and only if $R_1^{(1)} = R_1^{(2)}$ and $R_2^{(1)} = R_2^{(2)}$; prove also that

$$\int_{S_2} p^{(1)}(P_2^{(2)} - Q)\alpha = 0.$$

In both cases, use (20.12.7).)
(f) Deduce from (e) that if $V^{(1)}$ and $V^{(2)}$ are two compact surfaces in $\mathbf{R}^3$, with total curvature $\neq 0$ at each point, which are *isometric*, then they are related by a Euclidean displacement of $\mathbf{R}^3$ (*Cohn–Vossen's theorem*).

10. Let V be a hypersurface in $\mathbf{R}^n$.

(a) Show that the Ricci tensor on V is given by the formula

$$\langle \boldsymbol{K}', X \otimes Y \rangle = (S \cdot X \mid Y)\operatorname{Tr}(S) - (S \cdot X \mid S \cdot Y),$$

where $S(x)$ is the endomorphism of $T_x(V)$ defined in Problem 1. (Use the formula (1) of Problem 7.)
(b) If $\boldsymbol{K}' = \lambda \mathbf{g}$, where the function λ is everywhere >0 on V (Section 20.10, Problem 2), show that all points of V are umbilics. (Deduce from (a) that the principal curvatures satisfy the relations

$$\left(\sum_{j=1}^{n-1} \rho_j\right)\rho_k - \rho_k^2 = \lambda$$

for $1 \leqq k \leqq n-1$.)

11. Let M be a connected differential manifold of dimension n, endowed with a torsion-free linear connection $\mathbf{C}$, arising from a principal connection $\boldsymbol{P}$ in the bundle of frames R(M). Let σ_i, ω_{ij} denote the canonical 1-forms and connection 1-forms on R(M), and let $\pi : R(M) \to M$ be the projection. Let V be a hypersurface in M.

(a) Show that in $\pi^{-1}(V)$ the set P_1 of frames r such that the inverse image of the form σ_n under the canonical injection $\pi^{-1}(V) \to R(M)$ vanishes at r, is a submanifold of codimension 1 in $\pi^{-1}(V)$. Using the same notation σ_i, ω_{ij} to denote the inverse images of these forms under the canonical injection $P_1 \to R(M)$, show that on P_1 we have

$$\omega_{nj} = \sum_{k=1}^{n-1} m_{jk}\sigma_k \qquad (1 \leqq j \leqq n-1),$$

where the m_{jk} are C^∞ functions on P_1, forming a *symmetric* matrix $M = (m_{jk})$ (use (20.6.3.3)). Let r' be another frame in P_1, lying in the fiber $\pi^{-1}(\pi(r))$, so that $r' = r \cdot s$, where $s \in \mathbf{GL}(n, \mathbf{R})$: we identify s with the transpose of the $n \times n$ matrix

$$S = \begin{pmatrix} & A & & s_{1n} \\ & & & \vdots \\ & & & s_{n-1,n} \\ s_{n1} & \cdots & s_{n,n-1} & s_{nn} \end{pmatrix}.$$

Show that $s_{nj}=0$ for $1 \leq j \leq n-1$ and that $s_{nn}M(r') = A \cdot M(r) \cdot {}^tA$. (Use the calculations of Section 20.6, Problem 2, by considering r as the value at the point $x = \pi(r)$ of a moving frame **R**, so that r' is the value at x of the moving frame $S \cdot \mathbf{R}$, in the notation of *loc. cit.*)

(b) Suppose that, for each $r \in P_1$, the matrix $M(r)$ is invertible. Suppose also that, at a point $r_0 \in P_1$, the matrix $M(r_0)$ has signature $(p, n-1-p)$. Show that P_1 is the union of four connected open sets, on each of which $M(r)$ has signature $(p, n-1-p)$ or $(n-1-p, p)$.

Assume for simplicity that $p = n-1$; then the set of frames $r \in P_1$ such that $M(r) = I_{n-1}$ is a submanifold P_2 of P_1.

Let P_3 be the set of frames $r \in P_2$ such that the inverse image, under the canonical injection $P_3 \to \pi^{-1}(V)$, of the form

$$\tfrac{1}{2}(n-1)\omega_{nn} - \sum_{j=1}^{n-1} \omega_{jj}$$

vanishes at the point r. Show that P_3 is a submanifold of P_2. (Use the same method as in (a), taking S this time to be a function of x.) Show that if $r' = r \cdot s$ is a frame in P_3 belong to the same fiber as r, then with the same notation as in (a) we have $A \cdot {}^tA = s_{nn} I_{n-1}$ and $s_{jn} = 0$ for $1 \leq j \leq n-1$. This latter relation signifies that if

$$r = (\mathbf{a}_1, \ldots, \mathbf{a}_{n-1}, \mathbf{a}_n)$$

and

$$r' = (\mathbf{a}'_1, \ldots, \mathbf{a}'_{n-1}, \mathbf{a}'_n)$$

are considered as bases of $T_x(M)$ (20.1.2), then the vectors $\mathbf{a}_n$ and $\mathbf{a}'_n$ are *proportional*. The vectors $\mathbf{a}_j$ and $\mathbf{a}'_j$ for $1 \leq j \leq n-1$ lie in the tangent space $T_x(V)$, by the definition of P_1.) The line $\mathbf{R}\mathbf{a}_n$ is therefore intrinsically determined by V and **C**, and is called the *pseudo-normal* to V at the point $x \in V$ (relative to the connection **C**). For every automorphism u of the connection **C** which leaves V stable, the image under $T_x(u)$ of the pseudo-normal to V at a point $x \in V$ is the pseudo-normal to V at the point $u(x)$.

Show that P_3 is a G-structure $S_G(V)$ on V for the group $G = \mathbf{GO}(n-1, \mathbf{R})$ (Section 20.9, Problem 15).

(c) Let $\mathbf{R} = (\boldsymbol{a}_1, \ldots, \boldsymbol{a}_n)$ be a moving frame defined on a neighborhood U in M of a point of V, and such that its restriction to $U \cap V$ is a section of P_3. Let σ'_i $(1 \leq i \leq n-1)$ and ω'_{ij} $(1 \leq i, j \leq n-1)$ be the inverse images of σ_i, ω_{ij}, respectively, under the restriction of **R** to $U \cap V$. Show that if $\mathbf{R}'$ is the moving frame on V defined by the first $n-1$ vector fields of **R** (which take their values in T(V)), then σ'_i and ω'_{ij} are the canonical and connection forms, relative to $\mathbf{R}'$, for a linear connection $\mathbf{C}'_{U \cap V}$ on $U \cap V$ which is *independent* of the choice of the moving frame **R** (subject to the above condition). (Use (20.5.5.6).) The linear connections $\mathbf{C}'_{U \cap V}$ (for variable U) are restrictions of one and the same linear connection $\mathbf{C}'$ on V, said to be *induced* by **C**. If we put $\boldsymbol{l} = \sum_{i=1}^{n-1} \sigma'_i \otimes \sigma'_i$, and if we denote by ∇ and ∇' the covariant differentiations relative to **C** and $\mathbf{C}'$, respectively, then we have

$$\nabla_{X'} \cdot Y' - \nabla'_{X'} \cdot Y' = \langle \boldsymbol{l}, X' \otimes Y' \rangle \boldsymbol{a}_n$$

for any two vector fields X', Y' on V, which shows that $\boldsymbol{l}\boldsymbol{a}_n$ does not depend on the choice of the frame **R**. The connection $\mathbf{C}'$ is torsion-free. If Ω'_{ij} $(1 \leq i, j \leq n-1)$ are the curvature 2-forms of $\mathbf{C}'$ relative to $\mathbf{R}'$, and $\tilde{\Omega}_{ij}$ ($1 \leq i, j \leq n-1$) the inverse images of the curvature 2-forms Ω_{ij} of $\mathbf{C}$ under the restriction of $\mathbf{R}$ to $U \cap V$, then we have

$$\tilde{\Omega}_{ij} - \Omega'_{ij} = \omega'_{in} \wedge \omega'_{nj} \qquad \text{for} \quad 1 \leq i, j \leq n-1.$$

(d) Generalize the preceding results to the case where the signature $(p, n-1-p)$ is arbitrary. Extend the theory to the case where the hypersurface V is replaced by a connected submanifold of M.

12. With the notation of Problem 11, suppose that $M = \mathbf{R}^n$ and that $\mathbf{C}$ is the canonical $\mathbf{SL}(n, \mathbf{R})$ connection (Section 20.7, Problem 7), which is flat. Assuming that $p = n-1$ as in Problem 11(b), we have now additionally $\omega_{nn} = 0$ and $\sum_{j=1}^{n-1} \omega_{jj} = 0$ in P_3, so that the vector field $\mathbf{a}_n$ is completely determined, and the connection $\mathbf{C}'$ is an $\mathbf{SL}(n-1, \mathbf{R})$-connection on V. Show moreover that in P_3 we have $\omega_{ij} + \omega_{ji} = \sum_{k=1}^{n-1} c_{ijk}\, \sigma_k$, where the coefficients c_{ijk} are invariant under all permutations of the indices i, j, k. (Use the structure equations.)

Show that the symmetric covariant tensors

$$\mathbf{l} = \sum_{i=1}^{n-1} \sigma'_i \otimes \sigma'_i, \qquad \mathbf{c} = \sum_{i,j,k} c_{ijk}\, \sigma'_i \otimes \sigma'_j \otimes \sigma'_k$$

on V are independent of the choice of the moving frame $\mathbf{R}$. If V_1 is another hypersurface in $\mathbf{R}^n$ such that there exists a diffeomorphism of V onto V_1 which transforms the tensor fields $\mathbf{l}$ and $\mathbf{c}$ into the analogous tensor fields $\mathbf{l}^{(1)}$ and $\mathbf{c}^{(1)}$ on V_1, show that there exists an *affine-linear* bijection of $\mathbf{R}^n$ onto itself (Section 20.1, Problem 1) which sends V onto V_1. (Use (20.9.3) for the forms $\omega_{ji} - \omega_{ij}$.)

Show that the covariant tensor field

$$\mathbf{k} = \sum_{i=1}^{n-1} \omega'_{in} \otimes \sigma'_i$$

on V is symmetric and that it is uniquely determined by the tensors $\mathbf{l}$ and $\mathbf{c}$ (use the structure equations). The $n-1$ eigenvalues of $\mathbf{k}(x)$ relative to the positive-definite form $\mathbf{l}(x)$ are called the *principal affine curvatures* of V at the point x.

15. THE IMMERSION PROBLEM

As an example of the use of the method of moving frames, we shall prove the following theorem:

(20.15.1) (E. Cartan) *Let* M *be an analytic Riemannian manifold* (20.8.1) *of dimension n. For each point* $x_0 \in M$, *there exists a neighborhood* U *of* x_0 *in* M *and an analytic embedding of* U *into* $\mathbf{R}^N$, *where* $N = \frac{1}{2}n(n+1)$, *which is an isometry of* U *onto a Riemannian submanifold of* $\mathbf{R}^N$.

Let $\mathbf{R} = (\mathbf{e}_i)_{1 \leq i \leq n}$ be an analytic orthonormal moving frame on M defined on a neighborhood of x_0, and let (σ_i), (ω_{ij}), and (Ω_{ij}) be the corresponding canonical, connection, and curvature forms, satisfying the relations (20.12.3.1)–(20.12.3.4). Suppose that the problem has been solved, and let $f\colon \mathrm{U} \to \mathbf{R}^{\mathrm{N}}$ be an isometry of U onto $f(\mathrm{U})$; then it follows from (20.12.2) and (20.12.3) that there exists an orthonormal moving frame $\mathbf{R}_0$ defined on a neighborhood of $f(\mathrm{U})$ in $\mathbf{R}^{\mathrm{N}}$ such that, if $(\rho'_h)_{1 \leq h \leq \mathrm{N}}$ and $(\varpi'_{hk})_{1 \leq h, k \leq \mathrm{N}}$ are the corresponding canonical and connection forms, then we have

$$\begin{aligned} {}^t f(\rho'_i) &= \sigma_i && (1 \leq i \leq n), \\ {}^t f(\rho'_\alpha) &= 0 && (n+1 \leq \alpha \leq \mathrm{N}), \\ {}^t f(\varpi'_{ij}) &= \omega_{ij} && (1 \leq i, j \leq n). \end{aligned}$$

These relations may be interpreted as follows. Recall that $\rho'_h = {}^t\mathbf{R}_0(\rho_h)$ and $\varpi'_{hk} = {}^t\mathbf{R}_0(\varpi_{hk})$, where (ρ_h) and (ϖ_{hk}) are the canonical and connection forms on the *principal bundle* P *of orthonormal frames* on $\mathbf{R}^{\mathrm{N}}$, of dimension $\frac{1}{2}\mathrm{N}(\mathrm{N}+1)$, identified with $\mathbf{R}^{\mathrm{N}} \times \mathbf{O}(\mathrm{N}, \mathbf{R})$ (20.6.5). If we put

$$\mathrm{F} = \mathbf{R}_0 \circ f\colon \mathrm{U} \to \mathrm{P},$$

the relations above are equivalent to

$$\begin{aligned} {}^t\mathrm{F}(\rho_i) &= \sigma_i && (1 \leq i \leq n), \\ {}^t\mathrm{F}(\rho_\alpha) &= 0 && (n+1 \leq \alpha \leq \mathrm{N}), \\ {}^t\mathrm{F}(\varpi_{ij}) &= \omega_{ij} && (1 \leq i, j \leq n). \end{aligned}$$

On the product manifold M × P, by abuse of notation we shall write σ_i and ω_{ij} in place of ${}^t\mathrm{pr}_1(\sigma_i)$ and ${}^t\mathrm{pr}_1(\omega_{ij})$, and ρ_h, ϖ_{hk} in place of ${}^t\mathrm{pr}_2(\rho_h)$, ${}^t\mathrm{pr}_2(\varpi_{hk})$. Then the preceding relations are equivalent to the fact that the *graph* Γ_F of F is an *n-dimensional integral manifold of the Pfaffian system of* $\mathrm{N} + \frac{1}{2}n(n-1)$ *equations*

(20.15.1.1)
$$\begin{cases} \rho_i - \sigma_i = 0 & (1 \leq i \leq n), \\ \rho_\alpha = 0 & (n+1 \leq \alpha \leq \mathrm{N}), \\ \varpi_{ij} - \omega_{ij} = 0 & (1 \leq i, j \leq n). \end{cases}$$

Conversely, suppose that we have established the existence, at a point $z_0 = (x_0, r_0) \in \mathrm{M} \times \mathrm{P}$, of an *integral element* L_{z_0} (18.10.2) of dimension n of this system, such that the projection of L_{z_0} onto $\mathrm{T}_{x_0}(\mathrm{M})$ is *bijective* and such that L_{z_0} contains a *regular* integral element (18.10.4) of dimension $n-1$. Then the Cartan–Kähler theorem (which is applicable by virtue of the hypotheses of analyticity) proves (18.13.9) the existence of an integral manifold G of the system (20.15.1.1), of dimension n, containing z_0 and such that $\mathrm{T}_{z_0}(\mathrm{G}) = \mathrm{L}_{z_0}$. Furthermore, on restriction to a neighborhood of z_0 if necessary, G is the graph of an analytic mapping F of a neighborhood U of x_0 into P, and the first

n equations (20.15.1.1) show that F and $f = \pi \circ F$ (where π is the canonical projection of P on $\mathbf{R}^N$) are embeddings, and that $F = \boldsymbol{R}'_0 \circ f$, where $\boldsymbol{R}'_0$ is a well-defined analytic section of P over the submanifold $f(U)$ of $\mathbf{R}^N$. Since $\boldsymbol{R}'_0$ can be extended to a C^∞ section of P over a neighborhood of $f(U)$ in $\mathbf{R}^N$, it follows that f is an *isometry* onto $f(U)$.

(20.15.1.2) To establish the existence of L_{z_0} with the desired properties, we must first consider the differential ideal generated by the forms on the left-hand sides of the equations (20.15.1.1). Bearing in mind the structure equations for M and for P, this ideal is generated by the left-hand sides of (20.15.1.1) and the 2-forms

(20.15.1.3)
$$\sum_{j=1}^{n} \varpi_{\alpha j} \wedge \rho_j \qquad (n + 1 \leqq \alpha \leqq N),$$

(20.15.1.4)
$$\sum_{\alpha=n+1}^{N} \varpi_{\alpha i} \wedge \varpi_{\alpha j} - \Omega_{ij} \qquad (1 \leqq i < j \leqq n)$$

(since $\Omega_{ji} = -\Omega_{ij}$).

(20.15.1.5) We shall now apply the definitions of (18.10). An essential remark is that at each point $r \in P$, the $N + \frac{1}{2}N(N-1)$ covectors $\rho_i(r)$ $(1 \leqq i \leqq N)$ and $\varpi_{ij}(r)$ $(1 \leqq i \leqq j \leqq N)$ are always *linearly independent*, the canonical and connection forms of a principal connection being surjective at every point ((20.2.3) and (20.6.2)). A vector $\mathbf{v} \in T_r(P)$ is therefore completely determined by the numbers $\langle \rho_i(r), \mathbf{v} \rangle$ and $\langle \varpi_{ij}(r), \mathbf{v} \rangle$, which may be arbitrarily prescribed.

(20.15.1.6) We shall show by induction on p that there exist vectors $\mathbf{v}_p$ $(1 \leqq p \leqq n)$ in $T_{r_0}(P)$ such that, for each p, the p vectors

$$\mathbf{u}_j = (\mathbf{e}_j(x_0), \mathbf{v}_j) \qquad (1 \leqq j \leqq p)$$

generate an *integral element* in $T_{z_0}(M \times P) = T_{x_0}(M) \times T_{r_0}(P)$ which is *regular for* $p \leqq n - 1$; for $p = n$, this integral element L_{z_0} will satisfy the conditions stated above, and the theorem will then be proved.

In the first place, for $z = (x, r)$, a nonzero vector $\mathbf{u}' = (\mathbf{c}', \mathbf{v}')$ in

$$T_x(M) \times T_r(P)$$

generates an integral element if and only if it satisfies the relations

(20.15.1.7)
$$\begin{cases} \langle \rho_i(r), \mathbf{v}' \rangle = \langle \sigma_i(x), \mathbf{c}' \rangle & (1 \leqq i \leqq n), \\ \langle \rho_\alpha(r), \mathbf{v}' \rangle = 0 & (n + 1 \leqq \alpha \leqq N), \\ \langle \varpi_{ij}(r), \mathbf{v}' \rangle = \langle \omega_{ij}(x), \mathbf{c}' \rangle & (1 \leqq i < j \leqq n). \end{cases}$$

Since, for a given $\mathbf{c}' \in T_x(M)$, the left-hand sides of these linear equations in $\mathbf{v}'$ are *linearly independent* (20.15.1.5), the rank $s_0(z)$ (18.10.3) is *constant* in a neighborhood of z_0.

(20.15.1.8) We shall find it convenient to associate with a vector $\mathbf{v} \in T_r(P)$ a system of n vectors $\mathbf{w}_j(\mathbf{v})$ $(1 \leqq j \leqq n)$ in $\mathbf{R}^{n(n-1)/2}$, given by

$$\mathbf{w}_j(\mathbf{v}) = (\langle \varpi_{\alpha j}(r), \mathbf{v} \rangle)_{n+1 \leqq \alpha \leqq N} \qquad (1 \leqq j \leqq n). \tag{20.15.1.9}$$

For each system of n vectors $\mathbf{b}_j$ $(1 \leqq j \leqq n)$ in $\mathbf{R}^{n(n-1)/2}$, there exist vectors $\mathbf{v} \in T_r(P)$ satisfying (20.15.1.7) and such that $\mathbf{w}_j(\mathbf{v}) = \mathbf{b}_j$ for $1 \leqq j \leqq n$.

In what follows, the scalar product for vectors of the form $\mathbf{w}_j(\mathbf{v})$ will be the Euclidean scalar product in $\mathbf{R}^{n(n-1)/2}$.

(20.15.1.10) The vectors $\mathbf{u}_j$ $(1 \leqq j \leqq p)$ will be chosen in such a way that the number $s_p(z_0, \mathbf{u}_1, \ldots, \mathbf{u}_p)$ is *maximal* for $p \leqq n-1$; by virtue of the remark (18.10.4.1), this will imply the assertion of (20.15.1.6). We must therefore compare this number with the number $s_p(z, \mathbf{u}'_1, \ldots, \mathbf{u}'_p)$ for $(z, \mathbf{u}'_1, \ldots, \mathbf{u}'_p)$ near to $(z_0, \mathbf{u}_1, \ldots, \mathbf{u}_p)$. Now the equations (18.10.3.3) with $(i_1 \ldots, i_{r-1}) = (1, 2, \ldots, p)$ involve only the p-vector $\mathbf{u}_1 \wedge \mathbf{u}_2 \wedge \cdots \wedge \mathbf{u}_p$, and therefore we may restrict our attention to systems $(z, \mathbf{u}'_1, \ldots, \mathbf{u}'_p)$ near $(z_0, \mathbf{u}_1, \ldots, \mathbf{u}_p)$ for which $\mathbf{u}'_j = (\mathbf{c}'_j(x), \mathbf{v}_j)$, where the $\mathbf{c}'_j(x)$ form an *orthonormal system* in $T_x(M)$.

(20.15.1.11) Consider first the case $p = 2$. We begin with a vector $\mathbf{u}_1 = (\mathbf{e}_1(x_0), \mathbf{v}_1)$ satisfying (20.15.1.7). Let $\mathbf{u}'_1 = (\mathbf{c}'_1(x), \mathbf{v}'_1)$ be a vector satisfying (20.15.1.7), with $\mathbf{c}'_1(x)$ a unit vector. In order that a vector $\mathbf{u}'_2 = (\mathbf{c}'_2(x), \mathbf{v}'_2)$ be such that $\mathbf{c}'_1(x)$ and $\mathbf{c}'_2(x)$ form an orthonormal system and the vectors $\mathbf{u}'_1$, $\mathbf{u}'_2$ determine an integral element, it is necessary and sufficient that the 2-forms (20.15.1.3) and (20.15.1.4) vanish at the bivector

$$(\mathbf{c}'_1(x), \mathbf{v}'_1) \wedge (\mathbf{c}'_2(x), \mathbf{v}'_2).$$

Let $\boldsymbol{R}'$ be an orthonormal moving frame on M defined in a neighborhood of x_0, whose first two vectors at the point x are $\mathbf{c}'_1(x)$, $\mathbf{c}'_2(x)$. If $K'_{hijk}(x)$ denote the corresponding components (20.10.4) of the Riemann–Christoffel tensor $\boldsymbol{K}$ at the point x, then it is immediately verified (bearing in mind (20.15.1.7) and (20.10.4.5)) that the conditions imposed on $\mathbf{v}'_2$ may be expressed as

$$\mathbf{w}_1(\mathbf{v}'_2) = \mathbf{w}_2(\mathbf{v}'_1) \tag{20.15.1.12}$$

(which is equivalent to $\frac{1}{2}n(n-1)$ scalar equations in $\mathbf{v}_2'$) and

(20.15.1.13)

$$(\mathbf{w}_j(\mathbf{v}_2') \mid \mathbf{w}_i(\mathbf{v}_1')) - (\mathbf{w}_i(\mathbf{v}_2') \mid \mathbf{w}_j(\mathbf{v}_1')) - K'_{ij12}(x) = 0 \qquad (1 \leqq i < j \leqq n).$$

We may assume that at $z = z_0$ the vector $\mathbf{v}_1 \in T_{r_0}(P)$ has been chosen so that the $n-1$ vectors

$$\mathbf{w}_1(\mathbf{v}_1), \ldots, \mathbf{w}_{n-1}(\mathbf{v}_1)$$

are *linearly independent* in $\mathbf{R}^{n(n-1)/2}$. In a neighborhood of $(z_0, \mathbf{u}_1)$, the $n-1$ vectors $\mathbf{w}_1(\mathbf{v}_1'), \ldots, \mathbf{w}_{n-1}(\mathbf{v}_1')$ may therefore also be assumed to be linearly independent. Let us now show that the left-hand sides of the scalar equations (20.15.1.12) and (20.15.1.13) determining $\mathbf{v}_2'$ are *linearly independent* forms: this will imply that $s_1(z_0, \mathbf{u}_1)$ is a maximum, and therefore that the integral element generated by $\mathbf{u}_1$ is *regular*. Now, the vector $\mathbf{w}_1(\mathbf{v}_2')$ is completely determined by (20.15.1.12). For $i=1$, $j=2$, the equation (20.15.1.13) signifies that the vector $\mathbf{w}_2(\mathbf{v}_2')$ lies in a well-determined hyperplane

$$E_1 \subset \mathbf{R}^{n(n-1)/2},$$

since $\mathbf{w}_1(\mathbf{v}_1') \neq 0$. Next, the equations (20.15.1.13) for $i=1$, $j=3$ and for $i=2$, $j=3$ signify that $\mathbf{w}_3(\mathbf{v}_2')$ belongs to a well-determined affine-linear subspace $E_2 \subset \mathbf{R}^{n(n-1)/2}$ of codimension 2, since $\mathbf{w}_1(\mathbf{v}_1')$ and $\mathbf{w}_2(\mathbf{v}_1')$ are linearly independent. Hence, proceeding by induction on j, it is clear that the equations (20.15.1.13) for $i = 1, 2, \ldots, j-1$ signify that $\mathbf{w}_j(\mathbf{v}_2')$ belongs to a well-determined affine-linear subspace $E_{j-1} \subset \mathbf{R}^{n(n-1)/2}$ of codimension $j-1$. For $n > 2$ we have $\frac{1}{2}n(n-1) > n-1$; hence none of the subspaces E_j $(1 \leqq j \leqq n-1)$ is empty, and in view of (20.15.1.8) this establishes our assertion. Furthermore, for $n > 3$, the inequality $\frac{1}{2}n(n-1) > (n-1) + (n-2)$ shows that $\mathbf{v}_2 \in T_{r_0}(P)$ may be chosen so that the vectors

$$\mathbf{w}_1(\mathbf{v}_1), \ldots, \mathbf{w}_{n-1}(\mathbf{v}_1), \mathbf{w}_2(\mathbf{v}_2), \ldots, \mathbf{w}_{n-1}(\mathbf{v}_2)$$

are *linearly independent.*

(20.15.1.14) We now assume as inductive hypothesis that, for some $p \leqq n$, we have determined vectors $\mathbf{v}_1, \ldots, \mathbf{v}_{p-1}$ in $T_{r_0}(P)$ such that

(i) if $\mathbf{u}_j = (\mathbf{e}_j(x_0), \mathbf{v}_j)$ for $1 \leqq j \leqq p-1$, the $\mathbf{u}_j$ generate an integral element of dimension $p-1$, and the $\mathbf{u}_j$ with index $\leqq p-2$ a *regular* integral element;

(ii) the $(n-1) + (n-2) + \cdots + (n-p+1)$ vectors

$$\mathbf{w}_k(\mathbf{v}_j) \qquad (1 \leqq k \leqq n-1, \quad 1 \leqq j \leqq p-1, \quad j \leqq k)$$

are linearly independent in $\mathbf{R}^{n(n-1)/2}$.

Let $\mathbf{u}'_j = (\mathbf{c}'_j(x), \mathbf{v}'_j)$ be a system of $p-1$ vectors such that the $\mathbf{c}'_j(x)$ $(1 \leqq j \leqq p-1)$ form an orthonormal system and such that $(z, \mathbf{u}'_1, \ldots, \mathbf{u}'_{p-1})$ lies in $(\mathrm{M} \times \mathrm{P})_{p-1}$ (notation of (18.10.4)). Since we have only to consider a neighborhood of $(z_0, \mathbf{u}_1, \ldots, \mathbf{u}_{p-1})$ in $(\mathrm{M} \times \mathrm{P})_{p-1}$, we may assume that the vectors $\mathbf{w}_k(\mathbf{v}'_j)$ $(1 \leqq k \leqq n-1,\ 1 \leqq j \leqq p-1,\ j \leqq k)$ remain *linearly independent*. In order that a vector $\mathbf{u}'_p = (\mathbf{c}'_p(x), \mathbf{v}'_p)$ should be such that the $\mathbf{c}'_j(x)$ $(1 \leqq j \leqq p)$ form an orthonormal system and that $\mathbf{u}'_p$ determines together with $\mathbf{u}'_1, \ldots, \mathbf{u}'_{p-1}$ an integral element, it is necessary and sufficient that the 2-forms (20.15.1.3) and (20.15.1.4) should vanish at all the bivectors $(\mathbf{c}'_k(x), \mathbf{v}'_k) \wedge (\mathbf{c}'_p(x), \mathbf{v}'_p)$ $(1 \leqq k \leqq p-1)$. Let $\mathbf{R}'$ be an orthonormal moving frame on M, defined in a neighborhood of x_0, and such that its first p vectors at the point x are the $\mathbf{c}'_j(x)$ $(1 \leqq j \leqq p)$. Denoting again by $\mathrm{K}'_{hijk}(x)$ the corresponding components of the Riemann–Christoffel tensor $\mathbf{K}$ at the point x, the conditions which $\mathbf{v}'_p$ has to satisfy may be expressed as

(20.15.1.15) $$\mathbf{w}_k(\mathbf{v}'_p) = \mathbf{w}_p(\mathbf{v}'_k) \qquad (1 \leqq k \leqq p-1),$$

(20.15.1.16) $$(\mathbf{w}_j(\mathbf{v}'_p) \mid \mathbf{w}_i(\mathbf{v}'_k)) - (\mathbf{w}_i(\mathbf{v}'_p) \mid \mathbf{w}_j(\mathbf{v}'_k)) - \mathrm{K}'_{ijkp}(x) = 0$$
$$(1 \leqq i < j \leqq n, \quad 1 \leqq k \leqq p-1).$$

For brevity we shall put

(20.15.1.17) $$(ijkh) = (\mathbf{w}_j(\mathbf{v}'_h) \mid \mathbf{w}_i(\mathbf{v}'_k)) - (\mathbf{w}_i(\mathbf{v}'_h) \mid \mathbf{w}_j(\mathbf{v}'_k)) - \mathrm{K}'_{ijkh}(x)$$

for $1 \leqq i \leqq n$, $1 \leqq j \leqq n$, $1 \leqq k \leqq p$, $1 \leqq h \leqq p$.

The $\mathbf{w}_k(\mathbf{v}'_p)$ are completely determined by (20.15.1.15) *for* $k \leqq p-1$. We shall first show that *for* $i \leqq p-1$ *and* $j \leqq p-1$, the equations (20.15.1.16) are identically satisfied by these values of $\mathbf{w}_i(\mathbf{v}'_p)$ and $\mathbf{w}_j(\mathbf{v}'_p)$. Namely, we have $\mathbf{w}_i(\mathbf{v}'_k) = \mathbf{w}_k(\mathbf{v}'_i)$ and $\mathbf{w}_j(\mathbf{v}'_k) = \mathbf{w}_k(\mathbf{v}'_j)$ for these values of i, j and for $k \leqq p-1$, by virtue of the inductive hypothesis; bearing in mind the identity $\mathrm{K}'_{ijkp} = \mathrm{K}'_{kpij}$ (20.10.4.4), it is therefore evident that $(ijkp) = (kpij) = 0$ by virtue of the inductive hypothesis. We need therefore to consider only those of the equations (20.15.1.16) in which *one of the indices* i, j *is* $\geqq p$. Suppose first that $i \leqq p-1$ and $j \geqq p$; from (20.10.4) we have the identity $\mathrm{K}'_{ijkp} - \mathrm{K}'_{kjip} = \mathrm{K}'_{jpik}$, from which we deduce immediately that $(ijkp) - (kjip) = (jpik) = 0$ by the inductive hypothesis, because $i \leqq p-1$ and $k \leqq p-1$. Consequently, in order to determine the $\mathbf{w}_j(\mathbf{v}'_p)$ for $j \geqq p$ we need retain only those of the equations (20.15.1.16) for which $p \leqq j \leqq n$, $k \leqq p-1$ and $1 \leqq i < j$. This system splits up into $n-p+1$ partial systems which determine successively $\mathbf{w}_p(\mathbf{v}'_p)$, then $\mathbf{w}_{p+1}(\mathbf{v}'_p)$ as a function of $\mathbf{w}_p(\mathbf{v}'_p)$, then $\mathbf{w}_{p+2}(\mathbf{v}'_p)$ as a function of $\mathbf{w}_p(\mathbf{v}'_p)$ and $\mathbf{w}_{p+1}(\mathbf{v}'_p)$, and so on.

Now, each of these systems

(20.15.1.18) $$(\mathbf{w}_j(\mathbf{v}'_p) \mid \mathbf{w}_i(\mathbf{v}'_k)) = (\mathbf{w}_i(\mathbf{v}'_p) \mid \mathbf{w}_j(\mathbf{v}'_k)) + \mathrm{K}'_{ijkp}(x)$$

(j fixed, $k \leqq p - 1$, $k \leqq i < j$) is a system of nonhomogeneous linear equations, whose left-hand sides are *linearly independent* forms by virtue of the *linear independence* of the $\mathbf{w}_i(\mathbf{v}'_k)$ for $k \leqq p - 1$ and $k \leqq i$. This proves that in $(\mathrm{M} \times \mathrm{P})_{p-1}$ the number $s_{p-1}(z_0, \mathbf{u}_1, \ldots, \mathbf{u}_{p-1})$ takes its *maximum* value and hence that the integral element generated by $\mathbf{u}_1, \ldots, \mathbf{u}_{p-1}$ is *regular*. Furthermore, the equations determining $\mathbf{v}'_p$ have at least one solution; hence we can choose a vector $\mathbf{u}_p = (\mathbf{e}_p(x_0), \mathbf{v}_p)$ which together with $\mathbf{u}_1, \ldots, \mathbf{u}_{p-1}$ determines an integral element of dimension p. Finally, if $p < n$, we have

$$\tfrac{1}{2}n(n-1) > (n-1) + (n-2) + \cdots + (n-p),$$

and moreover we can choose $\mathbf{v}_p$ such that the $\mathbf{w}_k(\mathbf{v}_j)$ are again *linearly independent* for $1 \leqq k \leqq n-1$, $1 \leqq j \leqq p$ and $j \leqq k$; hence the induction can continue, and the proof of the theorem (20.15.1) is complete.

(20.15.2) By using much deeper results from the theory of partial differential equations, it can be shown that every Riemannian manifold of dimension n is isometric to a submanifold of $\mathbf{R}^{\mathrm{N}}$, for some larger value of N.

16. THE METRIC SPACE STRUCTURE OF A RIEMANNIAN MANIFOLD: LOCAL PROPERTIES

If M is a Riemannian manifold, a *path* (9.6, 16.7) $\gamma : [a, b] \to \mathrm{M}$ is said to be *piecewise of class* C^r or *piecewise*-C^r (r an integer $\geqq 1$, or $+\infty$) if there exists a strictly increasing sequence $a = a_0 < a_1 < \cdots < a_m = b$ of points in $[a, b]$ such that the restriction of γ to each interval $[a_{j-1}, a_j]$ ($1 \leqq j \leqq m$) is equal to the restriction to this interval of a C^r mapping of an *open* interval (*containing* $[a_{j-1}, a_j]$) into M.

If γ is a piecewise-C^1 path, the vector $\gamma'(t) \in \mathrm{T(M)}$ (18.1.2.3) is therefore defined on each of the open intervals $]a_{j-1}, a_j[$, on which γ is of class C^1, and has a right (resp. left) limit at the point a_{j-1} (resp. a_j). Hence, assigning arbitrary values to the function γ' at the subdivision points a_j, the function $t \mapsto \gamma'(t)$ is a *regulated* function on $[a, b]$. Generalizing (20.13.1.1), the *length of the path* γ in M is defined to be the number

(20.16.1) $$\mathrm{L}(\gamma) = \int_a^b \|\gamma'(t)\|\, dt.$$

It is immediately obvious that $L(\gamma^0) = L(\gamma)$, where γ^0 is the *opposite* of the path γ, and that for a *juxtaposition* (9.6) of two piecewise-C^1 paths γ_1, γ_2, we have $L(\gamma_1 \vee \gamma_2) = L(\gamma_1) + L(\gamma_2)$.

The length $L(\gamma)$ of the path γ is also called the *length* of the image C of the path in M, and is denoted also by L(C).

(20.16.2) Let M be a *connected* Riemannian manifold of dimension $\geqq 1$, and x, y two points of M. By (16.26.10) there exists a C^∞ path in M with endpoints x and y. The number

(20.16.2.1)
$$d(x, y) = \inf_\gamma L(\gamma),$$

where γ runs through the *set of piecewise-*C^1 *paths with endpoints* x, y, is therefore finite and $\geqq 0$. We shall prove the following proposition:

(20.16.3) *The function d defined by* (20.16.2.1) *is a distance defining the topology of* M.

Clearly $d(x, x) = 0$, and by virtue of (20.16.1) we have $d(y, x) = d(x, y)$ and $d(x, z) \leqq d(x, y) + d(y, z)$ for $x, y, z \in M$. It remains to show that $d(x, y) > 0$ if $x \neq y$, and that the open balls in M with center x (relative to d) form a fundamental system of neighborhoods of x in M. We shall obtain these results by showing that, for $y \in M$ sufficiently close to x, the lower bound *$d(x, y)$ is attained by a geodesic path* (i.e., the restriction of a geodesic to a compact interval contained in its interval of definition) *with endpoints x and y*.

In what follows, we shall denote by exp the exponential mapping defined by the geodesic field of the Levi–Civita connection, by $\Omega \subset T(M)$ its domain of definition (18.4.3 and 18.6.1), and by $\exp_x$ its restriction to $\Omega \cap T_x(M)$, so that for $x \in M$ and $\mathbf{h}_x \neq 0$ in $T_x(M)$, the unending path $t \mapsto \exp(t\mathbf{h}_x) = \exp_x(t\mathbf{h}_x)$ is a geodesic through x, tangent to the vector $\mathbf{h}_x$ at x, and defined on the open interval consisting of all $t \in \mathbf{R}$ such that $t\mathbf{h}_x \in \Omega \cap T_x(M)$. We shall first prove the following lemmas:

(20.16.3.1) *Let x be a point of* M, *let* $\mathbf{h}_x$, $\mathbf{k}_x$ *be two tangent vectors in* $T_x(M)$ *such that* $\mathbf{h}_x \in \Omega$, *and let* $\mathbf{h}'_x$, $\mathbf{k}'_x$ *be their images in* $T_{\mathbf{h}_x}(T_x(M))$ *under the canonical bijection* $\tau_{\mathbf{h}_x}^{-1}$ (16.5.2).

(i) *We have*

(20.16.3.2)
$$\|T_{\mathbf{h}_x}(\exp_x) \cdot \mathbf{h}'_x\| = \|\mathbf{h}_x\|.$$

(ii) *If* $\mathbf{h}_x$, $\mathbf{k}_x$ *are orthogonal in* $T_x(M)$, *then the vectors* $T_{\mathbf{h}_x}(\exp_x) \cdot \mathbf{h}'_x$ *and* $T_{\mathbf{h}_x}(\exp_x) \cdot \mathbf{k}'_x$ *are orthogonal in* $T_{\exp(\mathbf{h}_x)}(M)$ *(Gauss's lemma).*

Consider the family of geodesics $t \mapsto f(t, \xi) = \exp(t(\mathbf{h}_x + \xi\mathbf{k}_x))$ (18.7.9), and the corresponding Jacobi field $t \mapsto \mathbf{w}(t) = f'_\xi(t, 0)$. If $v(t) = f(t, 0)$, we have, therefore, for all $t \in \mathbf{R}$ such that $t\mathbf{h}_x \in \Omega$,

$$v'(t) = T_{t\mathbf{h}_x}(\exp_x) \cdot (\tau_{t\mathbf{h}_x}^{-1}(\mathbf{h}_x)), \qquad \mathbf{w}(t) = t(T_{t\mathbf{h}_x}(\exp_x) \cdot (\tau_{t\mathbf{h}_x}^{-1}(\mathbf{k}_x))).$$

To prove (20.16.3.2) it is enough to show that the function $(v' \mid v')$ is *constant*: and this is immediate since by (20.9.5.4), the derivative $D(v' \mid v') = 2(\nabla_t \cdot v' \mid v')$ is zero because v is a geodesic (notation of (18.7.1)). Again, to prove the second assertion of the proposition, it is enough to show that the function $(v' \mid \mathbf{w})$ is identically zero. Now, by (20.9.5.4), we have

$$D(v' \mid \mathbf{w}) = (\nabla_t \cdot v' \mid \mathbf{w}) + (v' \mid \nabla_t \cdot \mathbf{w}) = (v' \mid \nabla_t \cdot \mathbf{w})$$

because v is a geodesic; hence

$$\begin{aligned} D^2(v' \mid \mathbf{w}) &= (\nabla_t \cdot v' \mid \nabla_t \cdot \mathbf{w}) + (v' \mid \nabla_t \cdot (\nabla_t \cdot \mathbf{w})) \\ &= (v' \mid \nabla_t \cdot (\nabla_t \cdot \mathbf{w})) \\ &= (v' \mid (\mathbf{r} \cdot (v' \wedge \mathbf{w})) \cdot v'), \end{aligned}$$

because $\mathbf{w}$ is a Jacobi field (18.7.5.1). This can be written in the form

$$D^2(v' \mid \mathbf{w}) = \langle \mathbf{K}, v' \otimes v' \otimes \mathbf{w} \otimes v' \rangle = 0$$

by virtue of (20.10.2) and (20.10.3.2). The function $D(v' \mid \mathbf{w})$ is therefore constant. Since its value at $t = 0$ is $(\mathbf{h}_x \mid \mathbf{k}_x) = 0$ by hypothesis (by virtue of (18.7.9)), it follows that $(v' \mid \mathbf{w})$ is constant; but it too is zero for $t = 0$, and the proof is complete.

(20.16.3.3) *Let* $x \in M$ *and let* $\gamma : [a, b] \to \Omega \cap T_x(M)$ *be a piecewise-*C^1 *path. Then*

(20.16.3.4) $$L(\exp_x \circ \gamma) \geqq \|\gamma(b)\| - \|\gamma(a)\|.$$

Suppose in addition that $\exp_x$ *is a local diffeomorphism at all points of* $\gamma([a, b])$. *Then the two sides of* (20.16.3.4) *are equal if and only if* $\gamma(t) = \rho(t)\mathbf{h}_x$, *where* $\mathbf{h}_x$ *is a fixed unit vector in* $T_x(M)$, *and* $t \mapsto \rho(t)$ *is an increasing, piecewise-*C^1 *function on* $[a, b]$.

Replacing γ by its opposite γ^0 if necessary, we may assume that

$$\|\gamma(b)\| \geqq \|\gamma(a)\|$$

and $\|\gamma(b)\| > 0$. Let F be the closed set of points $t \in [a, b]$ such that $\gamma(t) = \mathbf{0}_x$. It is clear that $L(\exp_x \circ \gamma)$ is greater than or equal to the sum of lengths of the restrictions of $\exp_x \circ \gamma$ to the intervals which are the connected components of $\complement F$ (3.19.6). On the other hand, if F is not empty and if $c < b$ is the largest element of F, we have $\gamma(t) \neq \mathbf{0}_x$ for all $t \in]c, b]$, and if F is empty we have $\gamma(t) \neq \mathbf{0}_x$ for all $t \in]a, b]$. To prove (20.16.3.4), we may therefore assume that $\gamma(t) \neq \mathbf{0}_x$ for all $t \in]a, b]$. By definition we have

$$(\exp_x \circ \gamma)'(t) = T_{\gamma(t)}(\exp_x) \cdot (\tau_{\gamma(t)}^{-1}(D\gamma(t))).$$

Since $\gamma(t) \neq \mathbf{0}_x$ in $]a, b]$, we may put $\gamma(t) = \rho(t)\mathbf{u}(t)$ in this interval, where $\rho(t) = \|\gamma(t)\| > 0$ is piecewise of class C^1, $\mathbf{u}(t)$ a unit vector in $T_x(M)$, and $t \mapsto \mathbf{u}(t)$ piecewise of class C^1. We have then

$$D\gamma(t) = (D\rho(t))\mathbf{u}(t) + \rho(t)D\mathbf{u}(t)$$

at all points where $D\gamma$ is defined; on the other hand, by differentiating the relation $(\mathbf{u}(t) | \mathbf{u}(t)) = 1$ we have $2(\mathbf{u}(t) | D\mathbf{u}(t)) = 0$ (8.1.4) at all points where $D\gamma$ is defined. Consequently

$$(\exp_x \circ \gamma)'(t) = (D\rho(t) \cdot (\rho(t))^{-1})T_{\mathbf{h}_x}(\exp_x) \cdot \mathbf{h}'_x + T_{\mathbf{h}_x}(\exp_x) \cdot \mathbf{k}'_x$$

in the notation of (20.16.3.2), with $\mathbf{h}_x = \gamma(t)$ and $\mathbf{k}_x = \rho(t)D\mathbf{u}(t)$. Since the vectors $\mathbf{h}_x$ and $\mathbf{k}_x$ are orthogonal, we can apply (20.16.3.1) and obtain

(20.16.3.5) $$\|(\exp_x \circ \gamma)'(t)\|^2 = |D\rho(t)|^2 + \|T_{\mathbf{h}_x}(\exp_x) \cdot \mathbf{k}'_x\|^2$$

and consequently $\|(\exp_x \circ \gamma)'(t)\| \geqq |D\rho(t)|$. The inequality (20.16.3.4) now follows from the mean value theorem (8.7.7).

Since any connected component of $\complement F$ contains a nonempty open interval in which $\|D\gamma(t)\| > 0$, the same argument shows that equality cannot occur in (20.16.3.4) unless $F = \varnothing$ or $\{a\}$. Moreover, since both sides of (20.16.3.5) are regulated functions of t, it follows from (8.5.3) that equality can only occur in (20.16.3.4) if $T_{\mathbf{h}_x}(\exp_x) \cdot \mathbf{k}_x = 0$ except at points where $D\gamma$ is undefined; but since by hypothesis $T_{\mathbf{h}_x}(\exp_x)$ is a bijection, this relation is equivalent to $D\mathbf{u}(t) = 0$ except at the point a and the points where $D\gamma$ is undefined. By virtue of (8.5.3), this completes the proof of the lemma (20.16.3.3).

(20.16.3.6) We are now in a position to complete the proof of (20.16.3). For each point $x \in M$ and each neighborhood V of x in M, there exists (18.4.6) in $T_x(M)$ an open ball $B : \|\mathbf{h}_x\| < r$ with center $\mathbf{0}_x$ such that $B \subset \Omega$ and the restriction of $\exp_x$ to B is a *diffeomorphism* of B onto an open neighborhood $\exp_x(B) \subset V$ of x in M. We shall show that for each $\mathbf{h}_x \in B$ we have

(20.16.3.7) $$d(x, \exp_x(\mathbf{h}_x)) = \|\mathbf{h}_x\|.$$

For this purpose let $\varphi : [a, b] \to \mathrm{M}$ be a piecewise-C^1 path in M, with origin x and endpoint $\exp_x(\mathbf{h}_x)$. For each r_1 such that $0 < r_1 < \|\mathbf{h}_x\|$, let $\mathrm{B}_1 \subset \mathrm{B}$ be the open ball with center $\mathbf{0}_x$ and radius r_1 in the normed space $\mathrm{T}_x(\mathrm{M})$, and let t_0 be the least number in $[a, b]$ such that $\varphi(t_0) \in \mathrm{M} - \exp_x(\mathrm{B}_1)$. Then there exists a unique piecewise-C^1 mapping

$$\gamma : [a, t_0] \to \overline{\mathrm{B}_1} \subset \Omega \cap \mathrm{T}_x(\mathrm{M})$$

such that $\varphi(t) = \exp_x(\gamma(t))$ for all $t \in [a, t_0]$. Clearly we have $\mathrm{L}(\varphi) \geqq \mathrm{L}(\exp_x \circ \gamma)$ and hence, by (20.16.3.3),

$$\mathrm{L}(\varphi) \geqq \|\gamma(t_0)\| - \|\gamma(a)\| = r_1;$$

this shows that $d(x, \exp_x(\mathbf{h}_x)) \geqq \|\mathbf{h}_x\|$. On the other hand, the path

$$t \mapsto \exp(t\mathbf{h}_x)$$

is defined on $[0, 1]$, of class C^∞, with origin a and endpoint $\exp_x(\mathbf{h}_x)$; by virtue of (20.16.3.2), its length is $\|\mathbf{h}_x\|$, which proves (20.16.3.7).

With the same notation (r_1 being any number such that $0 < r_1 < r$) we remark that, since $\overline{\mathrm{B}_1} - \mathrm{B}_1$ is the frontier of B_1 in B, therefore $\exp_x(\overline{\mathrm{B}_1} - \mathrm{B}_1)$ is the frontier of $\exp_x(\mathrm{B}_1)$ *in* $\exp_x(\mathrm{B})$; but since $\exp_x(\overline{\mathrm{B}_1})$ is compact and therefore closed in M (3.17.2), $\exp_x(\overline{\mathrm{B}_1} - \mathrm{B}_1)$ is also the frontier of $\exp_x(\mathrm{B}_1)$ *in* M. For each frontier point z of $\exp_x(\mathrm{B}_1)$ in M, we have therefore $d(x, z) = r_1$ by (20.16.3.7). On the other hand, if $y \in \mathrm{M} - \exp_x(\mathrm{B}_1)$, then for each piecewise-C^1 path $\varphi : [a, b] \to \mathrm{M}$ with origin x and endpoint y, the closed set of points $t \in [a, b]$ such that $\varphi(t) \in \exp_x(\overline{\mathrm{B}_1} - \mathrm{B}_1)$ is not empty (3.19.9); if t_0 is the smallest element of this set, and φ_0 the restriction of φ to $[a, t_0]$, we have $\mathrm{L}(\varphi) \geqq \mathrm{L}(\varphi_0) \geqq r_1$ from what has already been proved, and therefore $d(x, y) \geqq r_1$. We have thus shown that $\exp_x(\mathrm{B}_1)$ is exactly the set of points $y \in \mathrm{M}$ such that $d(x, y) < r_1$; and this, together with (20.16.3.7), completes the proof of (20.16.3).

The function d defined by (20.16.2.1) is called the *Riemannian distance* on the Riemannian manifold M. Whenever we refer to M as a metric space, it is always to be understood that the distance is the Riemannian distance, unless the contrary is expressly stated.

(20.16.4) *For each* $x \in \mathrm{M}$ *and each* $r > 0$ *such that the open ball*

$$\mathrm{B}(\mathbf{0}_x; r) : \|\mathbf{h}_x\| < r$$

in $\mathrm{T}_x(\mathrm{M})$ *is contained in* Ω *and such that* $\exp_x$ *is a diffeomorphism of this ball onto an open subset of* M (18.4.6), *this open set* $\exp_x(\mathrm{B}(\mathbf{0}_x; r))$ *is the open ball* $\mathrm{B}(x; r)$ *relative to the Riemannian distance on* M. *Moreover, in order that a piecewise-*C^1 *path* $\varphi : [a, b] \to \mathrm{M}$ *with origin* x *and endpoint* $y = \exp_x(\mathbf{h}_x) \in \mathrm{B}(x; r)$

should satisfy $d(x, y) = \mathrm{L}(\varphi)$, *it is necessary and sufficient that there should exist a piecewise-*C^1, *surjective, increasing mapping* $\rho: [a, b] \to [0, 1]$, *such that* $\varphi(t) = \exp_x(\rho(t)\mathbf{h}_x)$.

The first assertion follows from the proof of (20.16.3). If $\mathrm{L}(\varphi) = d(x, y)$, then $\varphi([a, b])$ is wholly contained in $\mathrm{B}(x; r)$: because otherwise it would contain frontier points of $\mathrm{B}(x; r)$, hence points $y' \in \mathrm{B}(x; r)$, such that

$$d(x, y') > d(x, y),$$

and *a fortiori* we should have $\mathrm{L}(\varphi) > d(x, y)$. We can therefore put $\varphi = \exp_x \circ \gamma$, where $\gamma : [a, b] \to \mathrm{B}(\mathbf{0}_x; r)$ is a piecewise-C^1 path with origin $\mathbf{0}_x$ and endpoint $\mathbf{h}_x$, and the result now follows from (20.16.3.3).

A path $\varphi : [a, b] \to \mathrm{M}$ of the form $t \mapsto \exp_x(\rho(t)\mathbf{h}_x)$, where ρ is a piecewise-C^1 increasing function, is called a *rectilinear path* in M. We have

(20.16.4.1) $$\mathrm{L}(\varphi) = (\rho(b) - \rho(a))\|\mathbf{h}_x\|$$

by virtue of (20.16.3.4).

(20.16.5) Under the hypotheses of (20.16.4), the diffeomorphism $\exp_x^{-1}$ of $\mathrm{B}(x; r)$ onto $\mathrm{B}(\mathbf{0}_x; r)$, followed by a linear bijection of $\mathrm{T}_x(\mathrm{M})$ onto $\mathbf{R}^n$, defines a *chart* of $\mathrm{B}(x; r)$ for which the images of the geodesic arcs with origin x are straight-line segments with origin 0 in $\mathbf{R}^n$. The local coordinates corresponding to such a chart are called *normal coordinates* at x.

Remark

(20.16.6) Given a piecewise-C^r path $\varphi : [a, b] \to \mathrm{M}$, for each $\varepsilon > 0$ there exists a C^r-path $\varphi_1 : [a, b] \to \mathrm{M}$ that *coincides* with φ except in the intervals $[a_j - \varepsilon, a_j + \varepsilon]$ (where the $a_j \in [a, b]$ are the points at which the rth derivative of φ is undefined), is such that $d(\varphi(t), \varphi_1(t)) \leqq \varepsilon$ for all $t \in [a, b]$, and for which $|\mathrm{L}(\varphi_1) - \mathrm{L}(\varphi)| \leqq \mathrm{C}\varepsilon$. where C is a constant independent of ε. It is immediately seen that it is sufficient to consider the case in which M is an open subset of $\mathbf{R}^n$, the interval of definition of φ is an open interval containing $[-a, a]$, and the origin is the only point at which $\mathrm{D}^r\varphi$ is undefined: we then have to find φ_1 which is equal to φ except in the interval $[-\varepsilon, \varepsilon]$, such that $\|\varphi_1 - \varphi\| \leqq \varepsilon$ in this interval and such that $\|\mathrm{D}\varphi_1\|$ is bounded in the interval $[-\varepsilon, \varepsilon]$ by a number independent of ε. For this purpose, let h be a C^∞ function on $\mathbf{R}$, with support contained in the interval $[-1, 1]$, with values in

$[0, 1]$, and such that $h(t) = 1$ for $|t| \leqq \frac{1}{2}$ (16.4.1); if $\int_{-\infty}^{+\infty} h(t)\,dt = c$, put $\rho(t) = h(t)/c$, so that $\int_{-\infty}^{+\infty} \rho(t)\,dt = 1$. Then the function ψ_m defined by

$$\psi_m(t) = m \int_{-\infty}^{+\infty} \varphi(s)\rho(m(t-s))\,ds$$

is of class C^∞ in an open interval containing $[-a, a]$ (17.12.2) and converges uniformly to φ as $m \to \infty$ (14.11.2); hence by taking m large enough we shall have $\|\psi_m - \varphi\| \leqq \varepsilon$ in $[-a, a]$. Moreover (17.11.11), in the intervals $[-a, 0[$ and $]0, a]$ we have

$$D\psi_m(t) = m \int_{-\infty}^{+\infty} D\varphi(t-s)\rho(ms)\,ds,$$

and therefore there exists a constant A independent of m, such that

$$\|D\psi_m\| \leqq A$$

for all m and $\|D\varphi\| \leqq A$. Now take

$$\varphi_1(t) = \psi_m(t) - (1 - h(t/\varepsilon))(\psi_m(t) - \varphi(t))$$

so that $\varphi_1(t) = \varphi(t)$ for $|t| > \varepsilon$, and

$$\|\varphi_1(t) - \varphi(t)\| = h(t/\varepsilon)\|\psi_m(t) - \varphi(t)\| \leqq \varepsilon$$

for $t \in [-\varepsilon, \varepsilon]$, and $\varphi_1(t) = \psi_m(t)$ for $t \in [-\frac{1}{2}\varepsilon, \frac{1}{2}\varepsilon]$, so that φ_1 is of class C^r. Moreover, if B is an upper bound of $|h'(t)|$, one sees immediately that $\|D\varphi_1(t)\| \leqq 3A + B$. This completes the proof.

PROBLEMS

1. With the hypotheses and notation of (20.16.4), for each r_1 such that $0 < r_1 < r$, let $S(x; r_1)$ denote the sphere with center x and radius r_1 relative to the distance function d; it is a hypersurface in M, diffeomorphic to $\mathbf{S}_{n-1}$. Show that for each point $y = \exp(\mathbf{h}_x)$ of $S(x; r_1)$, the normal to $S(x; r_1)$ at the point y coincides with the tangent to the geodesic $t \mapsto \exp(t\mathbf{h}_x)$ at the point y. (Use Gauss's Lemma.)

2. Under the hypotheses of (20.16.4), let $(\mathbf{c}_j)_{1 \leqq j \leqq n}$ be an *orthonormal* basis of $T_x(M)$. The normal coordinates of a point $y \in B(x; r)$ relative to $(\mathbf{c}_j)$ may be written as $y^i = p^i t$, where $(p^i)_{1 \leqq i \leqq n}$ is a point of $\mathbf{S}_{n-1}$ and $0 \leqq t < r$. Choose as moving frame on $B(x; r)$ the frame $\boldsymbol{R}$ *canonically associated* with this neighborhood and the basis $(\mathbf{c}_j)$ (Section 20.6, Problem 15). Show that for this frame the canonical and connection forms are such that
   (i) $\sigma_i = p^i\,dt + \zeta_i$, where ζ_i does not contain the dp^i;
   (ii) ω_{ij} contains only the dp^i, not dt (*loc. cit.*)

Furthermore, the forms ζ_i and ω_{ij} vanish at $t=0$. Show that

$$d\zeta_i = -\Big(dp^i + \sum_j \omega_{ij} p^j\Big) \wedge dt + \alpha_i,$$

$$d\omega_{ij} = -\sum_{k,l} r^i_{jkl} p^k \zeta_l \wedge dt + \beta_{ij},$$

where the r^i_{jkl} are the components of the curvature tensor relative to the moving frame $\mathbf{R}$, and α_i, β_{ij} are 2-forms containing only the dp^i.

Finally, the metric tensor, relative to the frame $\mathbf{R}$, is given by

$$\mathbf{g} = dt \otimes dt + \sum_i \zeta_i \otimes \zeta_i.$$

(Use Problem 1.)

3. We propose to calculate the Riemannian volume of the sphere $S(x; r_1)$ of dimension $n-1$, relative to the Riemannian distance on M (Problem 1). With the notation of Problem 2, consider the diffeomorphism f of $\mathbf{S}_{n-1}$ onto $S(x; r_1)$ which sends each point $\mathbf{p} = (p^i)$ of $\mathbf{S}_{n-1}$ to the point $y = f(\mathbf{p}) = \exp_x\Big(r_1 \cdot \sum_{j=1}^{n} p^j \mathbf{c}_j\Big)$. The inverse image ${}^t f(v)$ of the canonical volume form v on $S(x; r_1)$ (oriented by transport of the canonical orientation of $\mathbf{S}_{n-1}$ by f) may be written as $F\sigma$, where σ is the solid angle form on $\mathbf{S}_{n-1}$ (**16.24.7**) and F is a C^∞ function.

(a) Let $(\mathbf{p}_1, \ldots, \mathbf{p}_n)$ be an orthonormal basis of $\mathbf{R}^n$ such that $\mathbf{p}_1 = \mathbf{p}$, and let $(\mathbf{h}_1, \ldots, \mathbf{h}_n)$ be its image in $T_x(M)$ under the linear mapping $(p^j) \mapsto \sum_j p^j \mathbf{c}_j$. The vectors

$$T_{r_1\mathbf{h}}(\exp_x) \cdot (\tau^{-1}_{r_1\mathbf{h}}(\mathbf{h}_j)) \qquad (2 \leq j \leq n),$$

where $\mathbf{h} = \mathbf{h}_1$, form a basis of $T_y(S(x; r_1))$. Using the fact that

$$\langle \sigma(\mathbf{p}), \tau^{-1}_{\mathbf{p}}(\mathbf{p}_2) \wedge \cdots \wedge \tau^{-1}_{\mathbf{p}}(\mathbf{p}_n)\rangle = 1,$$

show that

$$F(y) = \langle v(y), T_{r_1\mathbf{h}}(\exp_x) \cdot (\tau^{-1}_{r_1\mathbf{h}}(\mathbf{h}_2)) \wedge \cdots \wedge T_{r_1\mathbf{h}}(\exp_x) \cdot (\tau^{-1}_{r_1\mathbf{h}}(\mathbf{h}_n))\rangle.$$

Let $\mathbf{w}_j$ $(2 \leq j \leq n)$ be the Jacobi field along the geodesic $t \mapsto \exp_x(t\mathbf{h})$ such that $\mathbf{w}_j(0) = 0$ and $(\nabla_t \cdot \mathbf{w}_j)(0) = \mathbf{h}_j$, and let $\tilde{\mathbf{w}}_j(r_1)$ be the vector in $T_x(M)$ obtained by parallel transport along this geodesic from the vector $\mathbf{w}_j(r_1) \in T_y(M)$. Show that

$$F(y) = \|\tilde{\mathbf{w}}_2(r_1) \wedge \cdots \wedge \tilde{\mathbf{w}}_n(r_1)\|.$$

(b) Show that, as $r_1 \to 0$, we have

$$F(y) = r_1^{n-1}(1 - \tfrac{1}{6} r_1^2 \operatorname{Ric}(\mathbf{h}) + o_2(r_1)),$$

where $o_2(r_1)/r_1^2 \to 0$. (Use Section **18.7**, Problem 5.)

(c) Show that, as $r_1 \to 0$, we have

$$\operatorname{vol}(S(x; r_1)) = \Omega_n r_1^{n-1}\Big(1 - \frac{1}{6n} S(x) r_1^2 + o_2(r_1)\Big),$$

where $S(x)$ is the scalar curvature of M at the point x (**20.10.8**).

4. On a Riemannian manifold M, let f be a C^∞ function, Z the vector field grad(f). Suppose that $\|Z\|$ is *constant*. Show that for every C^∞ vector field X on M, we have

$$\theta_Z(X \mid Z) = (X \mid \nabla_Z \cdot Z) - ([X, Z] \mid Z).$$

Deduce that

$$\nabla_Z \cdot Z = 0$$

and hence that the integral curves of the field Z are geodesics. (Consider two cases, according as $X = Z$ or $(X \mid Z) = 0$; in the second case, observe that $\theta_X \cdot f = (X \mid Z) = 0$ and deduce that $([X, Z] \mid Z) = \theta_{[X, Z]} \cdot f = 0$.

5. Let X be an infinitesimal automorphism (Section **20.6**, Problem 6) of the Levi–Civita connection of a Riemannian manifold M. Show that along each geodesic v in M, the vector field $t \mapsto X(v(t))$ is a Jacobi field. (With the notation of *loc. cit.*, observe that for each C^∞ vector field Y on M we have

$$\nabla_Y \cdot (\nabla_Y \cdot X) + (\boldsymbol{r} \cdot (X \wedge Y)) \cdot Y + \mathbf{A}_X \cdot (\nabla_Y \cdot Y) = 0.)$$

6. Let v be a geodesic in a Riemannian manifold M. Then the liftings $\mathbf{w}_1$ and $\mathbf{w}_2$ of v defined by $\mathbf{w}_1(t) = v'(t)$ and $\mathbf{w}_2(t) = tv'(t)$ are Jacobi fields along v. Show that every Jacobi field along v can be written uniquely in the form

$$\mathbf{w} = a\mathbf{w}_1 + b\mathbf{w}_2 + \mathbf{z},$$

where a, b are constants and $\mathbf{z}$ is a Jacobi field such that $(\mathbf{z} \mid v') = 0$. Deduce that if a Jacobi field $\mathbf{w}$ along v is such that $(\mathbf{w} \mid v')$ vanishes at two distinct points, then this scalar product is identically zero.

7. Let v be a geodesic in a Riemannian manifold M, and let $\mathbf{w}$ be a Jacobi field along v. Show that

$$\frac{d^2}{dt^2}(\|\mathbf{w}\|^2) = 2\|\nabla_t \cdot \mathbf{w}\|^2 + 2\langle \mathbf{K}, \mathbf{w} \otimes v' \otimes v' \otimes \mathbf{w}\rangle.$$

If $t = 0$ belongs to the interval of definition of v, and if $v(0) = x$, $\mathbf{w}(0) = \mathbf{0}_x$ and $(\nabla_t \cdot \mathbf{w})(0) = \mathbf{h}$, then for t small and positive we have

$$\|\mathbf{w}(t)\| = t + \tfrac{1}{6}t^3\langle \mathbf{K}(x), \mathbf{h} \otimes v'(0) \otimes v'(0) \otimes \mathbf{h}\rangle + o_3(t),$$

where $o_3(t)/t^3 \to 0$ as $t \to 0$ (cf. Section **18.7**, Problem 5).

8. Let $\gamma : [a, b] \to$ M be a C^∞ path in a Riemannian manifold M. Show that

$$\text{(1)} \qquad \mathrm{L}(\gamma) = \sup \sum_{i=0}^{p-1} d(\gamma(t_i), \gamma(t_{i+1})),$$

where d is the Riemannian distance, and the supremum is taken over all finite sequences such that

$$a = t_0 < t_1 < \cdots < t_p = b.$$

(If the right-hand side of (1) is denoted by $\mathrm{L}_1(\gamma)$, then clearly $\mathrm{L}(\gamma) \geqq \mathrm{L}_1(\gamma)$. To prove the reverse inequality, argue by contradiction : suppose that $\mathrm{L}(\gamma) = \mathrm{L}_1(\gamma) + k$, where $k > 0$, and show that there would exist a decreasing sequence of closed intervals

$[a_n, b_n]$ in $[a, b]$ such that $b_n - a_n = 2^{-n}(b-a)$ and such that, if γ_n denotes the restriction of γ to $[a_n, b_n]$, we have $L(\gamma_n) \geqq L_1(\gamma_n) + k/2^n$. We may assume that $t = 0$ is the common point of the intervals $[a_n, b_n]$, and then we have

$$\|\gamma'(0)\| \geqq \frac{k}{b-a} + \limsup_{n\to\infty} \frac{d(\gamma(a_n), \gamma(0)) + d(\gamma(0), \gamma(b_n))}{b_n - a_n}.$$

Observe that $d(\gamma(t), \gamma(0))/|t| \to \|\gamma'(0)\|$ as $t \to 0$, and hence derive a contradiction.)

9. Show that if d is the Riemannian distance on a Riemannian manifold M, the function $(x, y) \mapsto (d(x, y))^2$ is of class C^∞ in a sufficiently small neighborhood of the diagonal in $M \times M$ (cf. Section 20.18, Problem 5).

10. In a Riemannian manifold M, let v_1, v_2 be two geodesics passing through the same point x, parametrized by their curvilinear coordinates with origin x. If $(v_1'(0) \mid v_2'(0)) = \cos\alpha$ with $0 < \alpha < \pi$, show that

$$\sin\tfrac{1}{2}\alpha = \lim_{s\to 0} \frac{1}{2s} d(v_1(s), v_2(s)),$$

where d is the Riemannian distance on M. (Using Problem 7, show that for each $\varepsilon > 0$ there exists a neighborhood U of $\mathbf{0}_x$ in $T_x(M)$ such that, for all $\mathbf{h} \in U$ and all nonzero tangent vectors $\mathbf{k_h} \in T_{\mathbf{h}}(T_x(M))$, we have

$$1 - \varepsilon \leqq \|T_{\mathbf{h}}(\exp_x) \cdot (\tau_{\mathbf{h}}^{-1}(\mathbf{k_h}))\| / \|\tau_{\mathbf{h}}^{-1}(\mathbf{k_h})\| \leqq 1 + \varepsilon.)$$

11. If M is a compact, connected Riemannian manifold, then the group of isometries of M is compact. For every connected Riemannian manifold M, the subgroup of the group of isometries of M which fixes a given point is compact. (Observe that an isometry leaves invariant the Riemannian distance function, and use Ascoli's theorem, Section 7.5, Problem 10, and Section 20.6, Problem 9(a).) In particular, if G/H is a Riemannian symmetric space (20.11.3), then H must be *compact*.

17. STRICTLY GEODESICALLY CONVEX BALLS

A Riemannian manifold is said to be *geodesically convex* if it is *convex relative to the geodesic spray* of the Levi-Civita connection (18.5.1).

(20.17.1) *If* M *is a geodesically convex Riemannian manifold, then for each point* $x \in M$ *and each vector* $\mathbf{h}_x \in \Omega \cap T_x(M)$, *we have* $d(x, \exp_x(\mathbf{h}_x)) = \|\mathbf{h}_x\|$.

The length of the geodesic path $t \mapsto \exp_x(t\mathbf{h}_x)$, with origin x, defined on $[0, 1]$, is $\|\mathbf{h}_x\|$ by (20.16.4.1), and hence $d(x, \exp_x(\mathbf{h}_x)) \leqq \|\mathbf{h}_x\|$. On the other hand, let $\varphi : [a, b] \to M$ be a piecewise-C^1 path with origin x and endpoint $\exp_x(\mathbf{h}_x)$. Since M is geodesically convex, we have $\varphi = \exp_x \circ \gamma$, where

$\gamma : [a, b] \to \Omega \cap T_x(M)$ is a piecewise-C^1 path with origin $\mathbf{0}_x$ and endpoint $\mathbf{h}_x$ (18.5.2). Consequently we have $L(\varphi) \geqq \|\mathbf{h}_x\|$ (20.16.3.4), whence

$$d(x, \exp_x(\mathbf{h}_x)) \geqq \|\mathbf{h}_x\|.$$

(20.17.2) Let M be a connected Riemannian manifold, and let d be the Riemannian distance on M. For each connected open set $U \subset M$, let d_U be the Riemannian distance on U endowed with the Riemannian structure induced by that of M. It follows immediately from the definitions that $d(x, y) \leqq d_U(x, y)$ for all $x, y \in U$. The open set U is said to be *strictly geodesically convex* if it is geodesically convex and if moreover we have $d = d_U$ on $U \times U$. For each pair of points x, y in U, there then exists a *unique geodesic arc contained in* U with endpoints x, y (18.5.2), and $d(x, y)$ is equal to the *length* of this arc, by virtue of (20.17.1). Furthermore, it can be shown that there exists no other geodesic arc with endpoints x, y (not contained in U) of length $d(x, y)$ (Section 20.18, Problem 1).

Examples

(20.17.3) In $\mathbf{R}^n$, every convex open set is strictly geodesically convex. On the other hand, in the cylinder $M = \mathbf{R}^2/(\mathbf{Z} \times \{0\})$ (20.11.2), the canonical image of a strip $I \times \mathbf{R}$, where I is an open interval in $\mathbf{R}$, is geodesically convex if the length of I is <1, but is not strictly geodesically convex unless this length is $<\frac{1}{2}$.

(20.17.4) *Let* M *be a connected Riemannian manifold,* U *a geodesically convex open subset of* M. *Let* V *be a geodesically convex open set contained in* U *such that, for each pair* x, y *of points of* V, *we have*

$$d_U(x, y) \leqq d(x, M - U) + d(y, M - U).$$

Then V *is strictly geodesically convex.*

For points x, y *in* V, every path φ in M, with endpoints x and y, which is not contained in U, has length $\geqq d_U(x, y)$; for if $z \in M - U$ is in the image of φ, then clearly $L(\varphi) \geqq d(x, z) + d(z, y) \geqq d_U(x, y)$ by hypothesis. Consequently $d(x, y) = d_U(x, y)$. Furthermore, there exists a unique $\mathbf{h}_x \in T_x(M) \cap \Omega$ such that the geodesic path $t \mapsto \exp(t\mathbf{h}_x)$ defined in $[0, 1]$ is contained in V and has endpoints x, y (18.5.2). Since U and V are geodesically convex, it follows from (18.5.2) and (20.17.1) that

$$d_U(x, y) = d_V(x, y) = \|\mathbf{h}_x\|,$$

whence the result.

(20.17.5) *Let* M *be a connected Riemannian manifold. For each point* $a \in M$, *there exists a neighborhood* W *of* a *in* M *and a real number* $r > 0$ *such that, for all* $b \in W$ *and all* ρ *such that* $0 < \rho < r$, *the open ball with center* b *and radius* ρ *is strictly geodesically convex.*

Let us first show that (18.5.3) may be applied to the geodesic field of M by taking $Q(x, y) = (d(x, y))^2$. In normal coordinates in a neighborhood of the point a (20.16.5), we have $Q_a(x) = \|\exp_a^{-1}(x)\|^2$: in other words, the local expression of Q_a is the quadratic form $\mathbf{u} \mapsto (\mathbf{u} | \mathbf{u})$ on $T_x(M)$, which is positive definite, stationary at the origin $\mathbf{0}_a$, and has second derivative $(\mathbf{u}, \mathbf{v}) \mapsto 2(\mathbf{u} | \mathbf{v})$ ((8.1.4) and (8.2.1)). The conditions of (18.5.3) are therefore satisfied. Let W be a compact neighborhood of a in M and r a positive real number such that the assertion of (18.5.3) is true for W and r. Since Ω is open in T(M) and W is compact, there exists $r' > 0$ such that, for each $x \in W$, the ball $\|\mathbf{h}_x\| < r'$ in $T_x(M)$ is contained in Ω (3.17.10). It now follows from (20.16.4) that, for each $x \in W$, the ball $B(x; \rho)$ is connected for all $\rho < r'$. We conclude therefore from (18.5.3) that for $x \in W$ and $\rho < \inf(r, r')$, the ball $U = B(x; \rho)$ is *geodesically convex*. From this it follows that $V = B(x; \frac{1}{3}\rho)$ is *strictly geodesically convex*; for if $y, z \in V$, we have

$$d_U(y, z) \leqq d_U(x, y) + d_U(x, z) = d(x, y) + d(y, z) \leqq \tfrac{2}{3}\rho$$

by (20.16.3.7) and (20.17.1); but on the other hand, $d(y, M - U) \geqq \frac{1}{3}\rho$ and $d(z, M - U) \geqq \frac{1}{3}\rho$; hence the result follows from (20.17.4).

PROBLEMS

1. Let M be a Riemannian manifold and U a geodesically convex open subset of M such that $\bar{U}$ is contained in a geodesically convex open set and $\bar{U} - U$ is a hypersurface V. If x_1, x_2 are any two points of V, the geodesic arc with endpoints x_1, x_2 is then contained in $\bar{U}$. Assume that if $x_1 \neq x_2$, the points of this arc, other than the endpoints, belong to U, and conversely, that every geodesic trajectory which meets U meets V in two distinct points. The set G of geodesic trajectories which meet U may then be endowed with a structure of a differential manifold by identifying G with the set of orbits of the group of two elements acting on $V \times V - \Delta$ (where Δ is the diagonal of $V \times V$) by permutation of the factors.

   In the manifold S(M) of unit tangent vectors to M (Section 20.9, Problem 3(e)) let W be the submanifold of dimension $2n - 2$ consisting of the vectors $\mathbf{h}_x$ with origin $x \in V$ such that, if v is the geodesic such that $v(0) = x$ and $v'(0) = \mathbf{h}_x$, then $v(t) \in U$ for all sufficiently small $t > 0$. If $\gamma(\mathbf{h}_x)$ is the corresponding geodesic trajectory, then γ makes W into a two-sheeted covering of G. Consider on W the differential $(2n - 2)$-form which is the inverse image of $((n - 1)!)^{-1}(d\alpha)^{\wedge(n-1)}$ under the canonical injection $W \to T(M)$, in the notation of Section 20.9, Problem 3, and let μ be the Lebesgue measure on G which is the image under γ of the measure on W corresponding to this volume form.

Now let V′ be a hypersurface contained in U, with finite $(n-1)$-dimensional area A. Show that, for almost all geodesic trajectories g which meet U, the set $V' \cap g$ is finite, and that if $n(g)$ is the number of elements in this set, then

$$\int_G n(g)\, d\mu(g) = V_{n-1} A,$$

where V_{n-1} is the volume of the unit ball $\mathbf{B}_{n-1}$. (Argue as in Section 20.13, Problem 7.)

2. Let M be a Riemannian manifold, U an open set in M with nonempty frontier. Show that there exists a geodesic v whose interval of definition contains [0, 1], such that $v(t) \in U$ for $0 \leq t < 1$, and such that $v(1)$ is a frontier point of U. (Consider a geodesically convex ball with center at a frontier point a of U, a geodesic arc joining a to a point of U and contained in this ball, and the intersection of this arc with Fr(U).)

18. THE METRIC SPACE STRUCTURE OF A RIEMANNIAN MANIFOLD: GLOBAL PROPERTIES. COMPLETE RIEMANNIAN MANIFOLDS

(20.18.1) *Let* M *be a connected Riemannian manifold, d the Riemannian distance on* M. *For each $a \in M$ and $r > 0$, the open ball $B(a; r)$ in* M *is connected, and its closure in* M *is the closed ball $B'(a; r)$ with same center and radius.*

Since $B'(a; r)$ is closed in M, we have $\overline{B(a; r)} \subset B'(a; r)$; hence the second assertion will be proved if we can show that the *sphere* $S(a; r)$ with center a and radius r is contained in $\overline{B(a; r)}$. Now, if $x \in S(a; r)$, then for each ε such that $0 < \varepsilon < r$, there exists a path $\varphi : [\alpha, \beta] \to M$ with origin a and endpoint x such that $L(\varphi) \leq r + \varepsilon$. Since $t \mapsto d(a, \varphi(t))$ is continuous on $[\alpha, \beta]$ and varies from 0 to r, there exists $t_0 \in]\alpha, \beta[$ such that $d(a, \varphi(t_0)) = r - \varepsilon$ (3.19.8). If φ_1, φ_2 are the restrictions of φ to the intervals $[\alpha, t_0]$ and $[t_0, \beta]$, respectively, we have $L(\varphi) = L(\varphi_1) + L(\varphi_2)$, and $d(a, \varphi(t_0)) \leq L(\varphi_1)$; consequently

$$d(\varphi(t_0), x) \leq L(\varphi_2) = L(\varphi) - L(\varphi_1) \leq r + \varepsilon - (r - \varepsilon) = 2\varepsilon,$$

from which the result follows. The fact that for each $x \in B(a; r)$ there exists a path φ with origin a and endpoint x, of length $L(\varphi) < r$, implies that this path is contained in $B(a; r)$, and hence that $B(a; r)$ is connected (3.19.3).

(20.18.2) *Let* M *be a connected Riemannian manifold, d the Riemannian distance on* M, *and a, b two distinct points of* M. *If $\varphi : [\alpha, \beta] \to M$ is a piecewise-C^1 path with origin a and endpoint b such that $L(\varphi) = d(a, b) = l$, then φ is a rectilinear path $t \mapsto \exp_a(\rho(t)\mathbf{h}_a)$* (20.16.4) *with $\|\mathbf{h}_a\| = 1$, $\rho(\alpha) = 0$, $\rho(\beta) = l$, and $l\mathbf{h}_a \in \Omega$. Furthermore, there exists an open interval $I \supset [0, l]$ in* **R** *such that $v : s \mapsto \exp_a(s\mathbf{h})$ is an embedding of* I *in* M, *where s is the curvilinear coordinate* *of $v(s)$ on the curve $C = v(I)$ with respect to the origin a and the orientation defined by v.*

If $\lambda < \mu$ are any two points of $[\alpha, \beta]$, let $p = \varphi(\lambda), q = \varphi(\mu)$ be their images; if $p \neq q$, then we must have $L(\varphi_{\lambda,\mu}) = d(p, q)$, where $\varphi_{\lambda,\mu}$ denotes the restriction of φ to $[\lambda, \mu]$; for otherwise we should have $d(p, q) < L(\varphi_{\lambda,\mu})$, and therefore by definition there would exist a path $\psi : [\lambda, \mu] \to M$ with origin p and endpoint q such that $d(p, q) \leqq L(\psi) < L(\varphi_{\lambda,\mu})$, and then the path which is equal to φ on $[\alpha, \lambda]$ and $[\mu, \beta]$ and to ψ on $[\lambda, \mu]$ would have length $< d(a, b)$ which is absurd.

In particular, consider a ball $B(a; r)$ with center a which is strictly geodesically convex (20.17.5), and choose $\lambda > \alpha$ such that $p = \varphi(\lambda)$ is distinct from a and contained in $B(a; r)$. From the previous paragraph and from (20.16.4), there exists a unit vector $\mathbf{h}_a \in T_a(M)$ such that $\varphi(t) = \exp_a(\rho(t)\mathbf{h}_a)$ in $[\alpha, \lambda]$, where ρ is increasing and piecewise of class C^1; furthermore, $\mathbf{h}_a$ does not depend on the choice of p in $B(a; r)$. Let t_0 denote the least upper bound of the numbers $\mu \in [\alpha, \beta]$ such that $\varphi(t) = \exp_a(\rho(t)\mathbf{h}_a)$ for all $t \in [\alpha, \mu]$; since $\rho(t) = d(a, \varphi(t))$ by the preceding remarks and (20.16.4.1), the function ρ is well-defined. We have to show that $t_0 = \beta$. Suppose not, and put $x_0 = \varphi(t_0)$; also let $r' > 0$ be such that every open ball of radius r' with center lying in $B(x_0; r')$ is strictly geodesically convex (20.17.5). If we put $l_0 = d(a, x_0)$, then we have $s\mathbf{h}_a \in \Omega$ for $0 \leqq s < l_0$, and $\varphi(t) = v_0(\rho(t))$ for $\alpha \leqq t < t_0$, where $v_0(s) = \exp_a(s\mathbf{h}_a)$ for $0 \leqq s < l_0$. Let $t_1 < t_0 < t_2$ be two values of t such that $x_1 = \varphi(t_1)$ and $x_2 = \varphi(t_2)$ are distinct from x_0 and contained in $B(x_0; \frac{1}{2}r')$; put $l_1 = \rho(t_1) = d(a, x_1) < l_0$, and $l_2 = d(a, x_2) > l_0$. Also let $\mathbf{h}_{x_1} = v_0'(l_1)$, which is a unit vector in $T_{x_1}(M)$ (20.16.3.2). By virtue of the choice of t_1, t_2 and the remark at the beginning, the path φ_{t_1,t_2} is rectilinear, of the form $t \mapsto \exp_{x_1}(\rho_1(t)\mathbf{h}_{x_1}) = v_1(s)$, where $v_1(s) = \exp_{x_1}(s\mathbf{h}_{x_1})$ is defined for

$$0 \leqq s \leqq l_2 - l_1$$

(which implies that $s\mathbf{h}_{x_1} \in \Omega$ for these values of s) (20.16.4). But if

$$0 \leqq s < l_0 - l_1,$$

we have $v_1(s) = v_0(l_1 + s)$ by virtue of (18.2.3.2); hence (18.2.2) we have $s\mathbf{h}_a \in \Omega$ for $0 \leqq s < l_2$, and the function $s \mapsto \exp_a(s\mathbf{h}_a)$ is defined in this interval. Putting $\rho(t) = l_1 + \rho_1(t)$ for $t_1 \leqq t \leqq t_2$, we have therefore $\varphi(t) = \exp_a(\rho(t)\mathbf{h}_a)$ for $\alpha \leqq t < t_2$, contradicting the definition of t_0.

To complete the proof, it is enough to show that $s \mapsto v(s) = \exp_a(s\mathbf{h}_a)$ is *injective* in $[0, l]$. Now, if $v(s_1) = v(s_2)$ for $0 \leqq s_1 < s_2 \leqq l$, the juxtaposition of the paths $s \mapsto v(s)$ for $0 \leqq s \leqq s_1$ and $s \mapsto v(s + s_2, s_1)$ for $s_1 \leqq s \leqq s_1 + l - s_2$ would be a piecewise-C^1 path with endpoints a, b and length $< d(a, b)$, which is absurd.

(20.18.3) *Let a, b, c be three distinct points of* M, *and let* C_1 (resp. C_2) *be a geodesic arc with endpoints a, b* (resp. b, c). *Suppose that* $d(a, b) = L(C_1)$ *and* $d(b, c) = L(C_2)$. *If, when* C_1 (resp. C_2) *is oriented positively from a to b* (resp. *from b to c*), *the unit tangent vectors to* C_1 *and* C_2 *at the point b are distinct, then we have*

(20.18.3.1) $$d(a, c) < d(a, b) + d(b, c)$$

(*strict triangle inequality*).

Let $v_1 : [0, l_1] \to M$ and $v_2 : [0, l_2] \to M$ be the geodesic paths with images C_1, C_2, respectively. If the result were false, the path $v : [0, l_1 + l_2] \to M$ such that $v(s) = v_1(s)$ for $0 \leqq s \leqq l_1$ and $v(s) = v_2(s - l_1)$ for $l_1 \leqq s \leqq l_1 + l_2$ would have endpoints a and c and length $L(v) = d(a, c)$; hence it would be a rectilinear path by (20.18.2), contrary to hypothesis.

(20.18.4) In general, given two points a, b of a connected Riemannian manifold M, there need not exist a piecewise-C^1 path φ with endpoints a, b such that $L(\varphi) = d(a, b)$; consider for example $M = \mathbf{R}^2 - \{0\}$ and two points a, $-a$. It is remarkable that there is a condition involving only the *metric space* structure of M which implies the existence of "shortest paths" between any two points:

(20.18.5) (Hopf–Rinow theorem) *Let* M *be a connected Riemannian manifold, d the Riemannian distance on* M. *Then the following properties are equivalent*:

(a) *Every closed bounded* (*relative to d*) *subset of* M *is compact.*
(b) *The metric space* M *is complete* (*relative to d*).
(c) $\Omega = T(M)$.
(d) *There exists a point* $a \in M$ *such that* $T_a(M) \subset \Omega$.

Furthermore, these properties imply that, for each pair x, y of points of M, *there exists a geodesic path* $t \mapsto \exp_x(t\mathbf{h}_x)$ *defined on* $[0, 1]$, *with origin x and endpoint* $y = \exp_x(\mathbf{h}_x)$, *of length* $d(x, y) = \|\mathbf{h}_x\|$; *and moreover we have*

(20.18.5.1) $$B(x; r) = \exp_x(B(\mathbf{0}_x; r)), \qquad B'(x; r) = \exp_x(B'(\mathbf{0}_x; r))$$

for all $x \in M$ *and all* $r > 0$.

We shall prove that (a) $\Rightarrow$ (b) $\Rightarrow$ (c) $\Rightarrow$ (d) $\Rightarrow$ (a). Since every Cauchy sequence is bounded, the implication (a) $\Rightarrow$ (b) is immediate (3.16.1), and so also is (c) $\Rightarrow$ (d). The remaining implications and assertions of the theorem will result from the following lemmas.

(20.18.5.2) *Let a, b be two distinct points of a connected Riemannian manifold* M. *Then there exists a vector* $\mathbf{h}_a \neq 0$ *in* $\Omega \cap \mathrm{T}_a(\mathrm{M})$ *such that*

$$\|\mathbf{h}_a\| = d(a, \exp_a(\mathbf{h}_a)),$$
$$\|\mathbf{h}_a\| + d(\exp_a(\mathbf{h}_a), b) = d(a, b).$$

Let $r > 0$ be sufficiently small so that $\exp_a$ is a diffeomorphism of $\mathrm{B}(\mathbf{0}_a; r)$ onto $\mathrm{B}(a; r)$ (20.16.4), and choose $\rho \in]0, r[$ sufficiently small so that $b \notin \mathrm{B}(a; \rho)$. Let S denote the sphere with center $\mathbf{0}_a$ and radius ρ in $\mathrm{T}_a(\mathrm{M})$, so that $\exp_a(\mathrm{S})$ is the sphere with center a and radius ρ in M. Since $\exp_a(\mathrm{S})$ is compact, there exists $c \in \exp_a(\mathrm{S})$ such that $d(b, c) = d(b, \exp_a(\mathrm{S}))$ (3.17.10). Let

$$\mathbf{h}_a \in \mathrm{T}_a(\mathrm{M}) \cap \Omega$$

be such that $c = \exp_a(\mathbf{h}_a)$; then we have

$$d(a, b) \leqq d(a, c) + d(c, b) = \|\mathbf{h}_a\| + d(b, c)$$

by (20.16.4). On the other hand, S being the frontier of $\mathrm{B}(\mathbf{0}_a; \rho)$, $\exp_a(\mathrm{S})$ is the frontier of $\mathrm{B}(a; \rho)$ (20.16.3.6); a path γ with endpoints a and b must meet $\exp_a(\mathrm{S})$ (3.19.10); hence there exists $c' \in \exp_a(\mathrm{S})$ such that

$$\begin{aligned} \mathrm{L}(\gamma) &\geqq d(a, c') + d(b, c') \\ &\geqq d(a, \exp_a(\mathrm{S})) + d(b, \exp_a(\mathrm{S})) \\ &= d(a, \exp_a(\mathrm{S})) + d(b, c) \\ &= \rho + d(b, c) \\ &= \|\mathbf{h}_a\| + d(b, c). \end{aligned}$$

Hence, by definition, $d(a, b) \geqq \|\mathbf{h}_a\| + d(b, c)$, which proves the lemma.

(20.18.5.3) *Let a, b be two distinct points in a connected Riemannian manifold* M. *If* $\mathbf{h}_a \in \Omega \cap \mathrm{T}_a(\mathrm{M})$ *is a nonzero vector such that*

$$\|\mathbf{h}_a\| + d(\exp_a(\mathbf{h}_a), b) = d(a, b),$$

then for each $t \in \mathbf{R}$ *such that* $t\mathbf{h}_a \in \Omega$ *and* $0 \leqq \|t\mathbf{h}_a\| \leqq d(a, b)$, *we have*

$$\|t\mathbf{h}_a\| + d(\exp_a(t\mathbf{h}_a), b) = d(a, b).$$

First observe that if $t \geqq 0$ and $t\mathbf{h}_a \in \Omega$, then $\exp_a(t\mathbf{h}_a)$ is defined and $d(a, \exp(t\mathbf{h}_a)) \leqq \|t\mathbf{h}_a\|$ by (20.16.4.1), so that

$$\begin{aligned} \|t\mathbf{h}_a\| + d(\exp_a(t\mathbf{h}_a), b) &\geqq d(a, \exp_a(t\mathbf{h}_a)) + d(\exp_a(t\mathbf{h}_a), b) \\ &\geqq d(a, b). \end{aligned}$$

Hence, if $\|t\mathbf{h}_a\| + d(\exp_a(t\mathbf{h}_a), b) = d(a, b)$, it follows that for $0 \leqq t' \leqq t$ we have

$$\begin{aligned}\|t'\mathbf{h}_a\| + d(\exp_a(t'\mathbf{h}_a), b) &\leqq \|t'\mathbf{h}_a\| + d(\exp_a(t'\mathbf{h}_a), \exp_a(t\mathbf{h}_a)) + d(\exp_a(t\mathbf{h}_a), b)\\ &\leqq \|t'\mathbf{h}_a\| + \|(t - t')\mathbf{h}_a\| + d(a, b) - \|t\mathbf{h}_a\|\\ &= d(a, b)\end{aligned}$$

by (20.16.4.1), and therefore from above

$$\|t'\mathbf{h}_a\| + d(\exp_a(t'\mathbf{h}_a), b) = d(a, b).$$

In other terms, the set I of numbers $t \geqq 0$ such that $t\mathbf{h}_a \in \Omega \cap \mathrm{T}_a(\mathrm{M})$ and such that

(20.18.5.4) $$\|t\mathbf{h}_a\| + d(\exp_a(t\mathbf{h}_a), b) = d(a, b)$$

is a *bounded interval* in $\mathbf{R}$, with endpoints 0 and $t_0 \geqq 1$. The lemma will be established if we can show that it is not possible that $t_0\,\mathbf{h}_a \in \Omega$ and

$$\|t_0\,\mathbf{h}_a\| < d(a, b)$$

simultaneously. Assume the contrary; then by continuity it follows from (20.18.5.4) that $t_0 \in \mathrm{I}$. Since $d(a, b) > \|t_0\mathbf{h}_a\| \geqq d(a, c)$, where $c = \exp_a(t_0\,\mathbf{h}_a)$, we have $b \neq c$. Lemma (20.18.5.2) then shows that there exists a vector $\mathbf{h}_c \neq 0$ in $\Omega \cap \mathrm{T}_c(\mathrm{M})$ such that

$$\|\mathbf{h}_c\| = d(c, \exp_c(\mathbf{h}_c)), \qquad \|\mathbf{h}_c\| + d(\exp_c(\mathbf{h}_c), b) = d(c, b),$$

and therefore

$$\|t_0\,\mathbf{h}_a\| + \|\mathbf{h}_c\| + d(\exp_c(\mathbf{h}_c), b) = d(a, b).$$

By (20.16.3.3) and the triangle inequality, this implies that

$$\|t_0\,\mathbf{h}_a\| + \|\mathbf{h}_c\| = d(a, \exp_c(\mathbf{h}_c)).$$

It follows now from (20.16.3.3) and (20.18.2), applied to the juxtaposition of the geodesic paths $t \mapsto \exp_a(t\mathbf{h}_a)$ $(0 \leqq t \leqq t_0)$ and $t \mapsto \exp_c((t - t_0)\mathbf{h}_c)$ $(0 \leqq t - t_0 \leqq 1)$, that $\exp_c(\mathbf{h}_c) = \exp_a(t'\,\mathbf{h}_a)$, where $t' = t_0 + \lambda t_0$ and

$$\lambda = \|\mathbf{h}_c\|/\|t_0\,\mathbf{h}_a\|.$$

Consequently, we have

$$\|t'\mathbf{h}_a\| + d(\exp_a(t'\mathbf{h}_a), b) = d(a, b)$$

and $t'\mathbf{h}_a \in \Omega$, contrary to the definition of t_0.

(20.18.5.5) *Let* M *be a connected Riemannian manifold, a a point of* M. *If the closed ball* $\mathrm{B}'(\mathbf{0}_a; r)$ *in* $\mathrm{T}_a(\mathrm{M})$ *is contained in* Ω, *then the closed ball* $\mathrm{B}'(a; r)$ (resp. *the open ball* $\mathrm{B}(a; r)$) *in* M *is equal to* $\exp_a(\mathrm{B}'(\mathbf{0}_a; r))$ (resp. $\exp_a(\mathrm{B}(\mathbf{0}_a; r))$).

Clearly we have $B'(a; r) \supset \exp_a(B'(\mathbf{0}_a; r))$, by (20.16.4.1). Since on the other hand $B(a; r)$ is the union of the $B'(a; r')$ for $r' < r$, it is enough to show that $B'(a; r)$ is contained in $\exp_a(B'(\mathbf{0}_a; r))$. Let b be a point of M such that $b \notin \exp_a(B'(\mathbf{0}_a; r))$; we shall show that $d(a, b) > r$. By virtue of (20.18.5.2), there exists a vector $\mathbf{h}_a \neq 0$ in $\Omega \cap T_a(M)$ such that

$$d(a, b) = \|\mathbf{h}_a\| + d(\exp_a(\mathbf{h}_a), b).$$

Let $t = \inf(r/\|\mathbf{h}_a\|, d(a, b)/\|\mathbf{h}_a\|)$; then the hypotheses imply that $t\mathbf{h}_a \in \Omega$ and $0 < \|t\mathbf{h}_a\| \leq d(a, b)$, and by virtue of (20.18.5.3) we have

$$d(a, b) = \|t\mathbf{h}_a\| + d(\exp_a(t\mathbf{h}_a), b);$$

but since $b \notin \exp_a(B'(\mathbf{0}_a; r))$ and this set is compact, we have

$$d(\exp_a(t\mathbf{h}_a), b) > 0$$

(3.17.10), hence $\|t\mathbf{h}_a\| < d(a, b)$. This implies that $\|t\mathbf{h}_a\| = r$; hence $d(a, b) = r + d(\exp_a(t\mathbf{h}_a), b) > r$, which proves the lemma.

(20.18.5.6) *Let* M *be a connected Riemannian manifold, and let a be a point of* M *such that* $T_a(M) \subset \Omega$. *Then for each* $x \in M$ *there exists a vector* $\mathbf{h}_a \in T_a(M)$ *such that* $\|\mathbf{h}_a\| = d(a, x)$ *and* $x = \exp_a(\mathbf{h}_a)$, *and for each* $r > 0$ *we have* $B(a; r) = \exp_a(B(\mathbf{0}_a; r))$ *and* $B'(a; r) = \exp_a(B'(\mathbf{0}_a; r))$.

We may assume that $x \neq a$; then, by virtue of (20.18.5.2), there exists a vector $\mathbf{h}_a \neq 0$ in $T_a(M) \subset \Omega$ such that

$$\|\mathbf{h}_a\| + d(\exp_a(\mathbf{h}_a), x) = d(a, x).$$

But then, by virtue of the hypothesis and (20.18.5.3), if $t > 0$ is such that $\|t\mathbf{h}_a\| = d(a, x)$, we have

$$\|t\mathbf{h}_a\| + d(\exp_a(t\mathbf{h}_a), x) = d(a, x),$$

which is possible only if $x = \exp_a(t\mathbf{h}_a)$. The remaining assertions follow from (20.18.5.5).

(20.18.5.7) These lemmas already show that (d) $\Rightarrow$ (a) in (20.18.5) and establish the last two assertions of the theorem. It remains to prove that (b) implies (c). Let $a \in M$ and $\mathbf{h}_a \in T_a(M)$, and let t_0 be the least upper bound of the $t \in \mathbf{R}$ such that $t\mathbf{h}_a \in \Omega$; we have to show that $t_0 = +\infty$. Suppose not, and put $v(t) = \exp_a(t\mathbf{h}_a)$. The sequence $(v(t_0 - (1/n)))_{n>1}$ is a Cauchy sequence in M, because

$$d\left(v\left(t_0 - \frac{1}{n}\right), v\left(t_0 - \frac{1}{m}\right)\right) \leq \left|\frac{1}{n} - \frac{1}{m}\right| \|\mathbf{h}_a\|$$

by virtue of (20.16.4.1). Let b be the limit of this sequence in M. There exists a neighborhood W of b in M and a real number $r > 0$ such that Ω contains the closed ball $B'(\mathbf{0}_x; r)$ for all $x \in W$. Choose n sufficiently large so that $c = v(t_0 - (1/n)) \in W$ and $r/\|\mathbf{h}_a\| > 1/n$. Then the vector $\mathbf{h}_c = v'(t_0 - (1/n))$ belongs to $T_c(M)$, and its norm is equal to $\|\mathbf{h}_a\|$ (20.16.3.2), and therefore we have

$$\frac{r}{\|\mathbf{h}_a\|} \cdot \mathbf{h}_c \in \Omega.$$

By virtue of (18.2.3.1) applied to the geodesic field, it follows that

$$\left(t_0 - \frac{1}{n} + \frac{r}{\|\mathbf{h}_a\|}\right)\mathbf{h}_a \in \Omega,$$

which contradicts the definition of t_0. This completes the proof of (20.18.5).

A connected Riemannian manifold M is said to be *complete* if it satisfies the equivalent conditions of (20.18.5).

Examples

(20.18.6) Every *compact* connected Riemannian manifold is obviously complete. Real n-space $\mathbf{R}^n$, endowed with its canonical Riemannian structure, is a complete Riemannian manifold. If (G, H) is a symmetric pair with H *compact*, then *the Riemannian symmetric space* G/H (20.11.3) *is a complete Riemannian manifold*: for it follows from the properties of the canonical linear connection on G/H (20.7.10.4) that a geodesic with origin $x_0 = \pi(e)$ is defined on the whole of $\mathbf{R}$, hence condition (d) of (20.18.5) is satisfied. In a Riemannian manifold M, an open subset U, which is a complete Riemannian manifold for the structure induced by that of M, is closed in M, as follows from condition (a) of (20.18.5), which shows that the frontier of U is empty.

(20.18.7) Let M_1 be a complete connected Riemannian manifold, M a connected *covering* of M_1, and $\pi : M \to M_1$ the canonical projection. If we endow M with the Riemannian structure canonically induced from that of M_1 (20.8.1), then M is also a *complete* Riemannian manifold. For since π is a local isometry, each geodesic $t \mapsto v(t)$ of M determines a geodesic $t \mapsto \pi(v(t))$ of M_1 (by transport of structure), and conversely the lifting to M (16.28.1) of a geodesic in M_1 is a geodesic in M. Since the geodesics in M_1 are defined on the whole of $\mathbf{R}$, the same is true of the geodesics in M_1, by reason of the uniqueness of liftings (16.28.1).

Conversely, we have the following proposition:

(20.18.8) *Let* M *be a complete connected Riemannian manifold, and let f be a local isometry of* M *into a connected Riemannian manifold* M_1. *Then* M *is a Riemannian covering of* M_1, *with f as projection, and* M_1 *is complete.*

The set $f(M)$ is open in M_1. It is enough (20.18.6) to prove (20.18.8) with M_1 replaced by $f(M)$. Let a_1 be a point of $f(M)$, and V_1 a geodesically convex (20.17) neighborhood of a_1. By (18.5.3), there exists a C^∞ mapping $(x_1, y_1) \mapsto \mathbf{s}(x_1, y_1)$ of $V_1 \times V_1$ into $T(M_1)$ such that $o_{M_1}(\mathbf{s}(x_1, y_1)) = x_1$ and $\exp_{x_1}(\mathbf{s}(x_1, y_1)) = y_1$. Let u denote the mapping $y_1 \mapsto \mathbf{s}(a_1, y_1)$ of V_1 into $T_{a_1}(M_1)$, which is a diffeomorphism of V_1 onto an open neighborhood of $\mathbf{0}_{a_1}$ in $T_{a_1}(M_1)$, the inverse of which is the restriction of $\exp_{a_1}$ to this open set (18.5.2). For each point $a \in f^{-1}(a_1)$, the linear mapping $T_a(f)$ is by hypothesis an isomorphism of $T_a(M)$ onto $T_{a_1}(M_1)$. Let v_a denote the inverse of this isomorphism. Since M is complete, the mapping $g_a = \exp_a \circ v_a \circ u$ is defined and of class C^∞ on V_1. On the other hand, for each $\mathbf{h}_a \in T_a(M)$ and each $t \in \mathbf{R}$, $\exp_{a_1}(t(T_a(f) \cdot \mathbf{h}_a))$ is defined, and we have

(20.18.8.1) $$f(\exp_a(t\mathbf{h}_a)) = \exp_{a_1}(t(T_a(f) \cdot \mathbf{h}_a)).$$

For, since f is a local isometry, it is clear that $t \mapsto f(\exp_a(t\mathbf{h}_a))$ is (by transport of structure) a geodesic in M_1 defined on the whole of $\mathbf{R}$, whence our assertion follows. We have therefore, in V_1,

$$f \circ g_a = \exp_{a_1} \circ T_a(f) \circ v_a \circ u = \exp_{a_1} \circ u = 1_{V_1},$$

from which it follows (16.8.8) that g_a is a diffeomorphism of V_1 onto an open set $g_a(V_1)$, the inverse mapping being the restriction of f to $g_a(V_1)$. Let a' be a point of $f^{-1}(a_1)$, other than a; then we have $g_a(V_1) \cap g_{a'}(V_1) = \varnothing$. For if $g_a(x_1) = g_{a'}(x_1)$ for some x_1, then also $g_a(y_1) = g_{a'}(y_1)$ for all y_1 sufficiently close to x_1, because the restriction of f to a neighborhood of $g_a(x_1)$ is injective. The set of points $x_1 \in V_1$ such that $g_a(x_1) = g_{a'}(x_1)$ is therefore both open and closed (3.15.1), hence is either the whole of V_1 or the empty set, because V_1 is connected. Since $g_a(a_1) \neq g_{a'}(a_1)$, it is the second alternative which holds. Again, if $x_1 \neq y_1$, then we have $g_a(x_1) \neq g_{a'}(y_1)$, because $f(g_a(x_1)) = x_1$ and $f(g_{a'}(y_1)) = y_1$. Consequently $g_a(V_1) \cap g_{a'}(V_1) = \varnothing$, as asserted. To complete the proof, it is enough to show that $f^{-1}(V_1)$ is the union of the sets $g_a(V_1)$ as a runs through $f^{-1}(a_1)$ (16.12.4.1). Let $x \in f^{-1}(V_1)$, $x_1 = f(x)$, and consider the tangent vector $\mathbf{h}_{x_1} = \mathbf{s}(x_1, a_1)$ at the point x_1; there exists a unique tangent vector $\mathbf{h}_x \in T_x(M)$ such that $T_x(f) \cdot \mathbf{h}_x = \mathbf{h}_{x_1}$. By definition, we have $\exp_{x_1}(\mathbf{h}_{x_1}) = a_1$, and the formula (20.18.8.1) shows that $f(\exp_x(\mathbf{h}_x)) = a_1$, so that $\exp_x(\mathbf{h}_x) = a \in f^{-1}(a_1)$. The definition of g_a now shows that $x = g_a(x_1)$.

The fact that $f(M)$ is complete has been established during the course of the proof, because we have shown that $f(M)$ satisfies condition (d) of (20.18.5).

PROBLEMS

1. In a Riemannian manifold M, let U be a strictly geodesically convex open set. If x, y are any two distinct points of U, show that the unique geodesic arc with endpoints x, y contained in U is the *only* geodesic arc *in* M with endpoints x, y, of length $d(x, y)$.

(Assume the result false, and use the strict triangle inequality (20.18.3) to obtain a contradiction.)

Deduce that every finite intersection of strictly geodesically convex open sets is strictly geodesically convex.

2. Give an example of a noncomplete Riemannian manifold with the property that through any two distinct points there passes a unique geodesic trajectory.

3. Give an example of an unbounded connected open set U in $\mathbf{R}^2$ such that no two points $x, y \in U$ such that $d(x, y) > 1$ can be joined by a geodesic arc contained in U.

4. Show that the product of two complete Riemannian manifolds (Section 20.8, Problem 1) is a complete Riemannian manifold.

5. Let M be a complete, connected, non-simply-connected Riemannian manifold. Show that the function $(x, y) \mapsto (d(x, y))^2$ on $M \times M$ cannot be of class C^∞. (Consider the simply-connected Riemannian covering $\tilde{M}$ of M, and two distinct points $a, b \in \tilde{M}$ which project to the same point x of M. If $d(a, b) = r$ in $\tilde{M}$, consider the sphere with center x and radius $\frac{1}{2}r$ in M.)

6. Let M be a noncomplete connected Riemannian manifold, $\mathbf{g}$ its metric tensor. For each $x \in M$, let $\rho(x)$ be the least upper bound of the real numbers r such that the closed ball $B'(x; r)$ is compact. The hypothesis on M implies that $0 < \rho(x) < +\infty$ for each $x \in M$.

(a) Show that $\rho(y) \leqq \rho(x) + d(x, y)$ for all x, y in M, and hence that ρ is a continuous function on M.

(b) Let f be a C^∞ function on M such that $f(x) > 1/\rho(x)$ for all $x \in M$, and consider the metric tensor $\mathbf{g}_1 = f^2\mathbf{g}$. Let $\gamma : I \to M$ be a piecewise-C^∞ path in M. If L, L_1 are its lengths relative to $\mathbf{g}$, $\mathbf{g}_1$, respectively, then

$$L_1 \geqq L \Big/ \Big(\sup_{x \in \gamma(I)} \rho(x) \Big).$$

Deduce that if d, d_1 are the Riemannian distances corresponding to $\mathbf{g}$, $\mathbf{g}_1$, respectively, and if the endpoints a, b of γ satisfy $d(a, b) \geqq \frac{1}{2}\rho(a)$, then we have $L_1 \geqq \frac{1}{3}$. (Use (a).)

(c) Deduce from (b) that the closed ball with center a and radius $\frac{1}{6}$ (relative to the distance d_1) is contained in the closed ball with center a and radius $\frac{1}{2}\rho(a)$ (relative to the distance d). Consequently, the manifold M is *complete* relative to the metric tensor $\mathbf{g}_1$.

7. Let M be a complete, noncompact, connected Riemannian manifold, and let $\mathbf{g}$ be its metric tensor. Given a point $a \in M$, let h be a C^∞ function on M such that $h(x) > d(a, x)$ for all $x \in M$, where d is the Riemannian distance. Show that for the Riemannian metric tensor $\mathbf{g}_2 = e^{-2h}\mathbf{g}$ on M, the diameter of M is $\leqq 1$, and hence that M is not complete relative to $\mathbf{g}_2$.

8. Let M be a complete Riemannian manifold, d the Riemannian distance on M, and let M′ be a submanifold on M, endowed with the Riemannian structure induced by that of M. If d' is the Riemannian distance on M′, then $d'(x, y) \geqq d(x, y)$ for all $x, y \in M'$, and consequently every Cauchy sequence in M′ (relative to d') converges to a point of M (relative to d). Suppose that every point of M has a neighborhood U such that every connected component of $U \cap M'$ is closed in U. Show that M′ is then complete. (Argue by contradiction, by supposing that there exists a geodesic in M′ which is not defined on the whole of **R**.) Give an example in which the condition above is satisfied, but the distances d and d' are not uniformly equivalent (3.14) on M′.

9. Let M be a connected Riemannian manifold. For each $x \in M$, let $a(x)$ denote the supremum of the radii of open balls with center $\mathbf{0}_x$ contained in $T_x(M) \cap \Omega$, and by $r(x)$ the supremum of the real numbers r such that (relative to the Riemannian distance d on M) every open ball contained in the ball with center x and radius r is strictly geodesically convex. Then $r(x) \leqq a(x)$.

(a) Show that if $r(x_0) = +\infty$ for some $x_0 \in M$, then $r(x) = +\infty$ for all $x \in M$. If $r(x)$ is finite for all $x \in M$, then $|r(x) - r(y)| \leqq d(x, y)$, and therefore $r(x)$ is a continuous function of x on M.
(b) We have $a(x_0) = +\infty$ if and only if M is complete; hence $a(x_0) = +\infty$ implies that $a(x) = +\infty$ for all $x \in M$. If $a(x)$ is finite for all $x \in M$, show that

$$|a(x) - a(y)| \leqq d(x, y),$$

and hence $a(x)$ is a continuous function of x on M. (Argue by contradiction, using the fact that for $r < a(x)$ the closed ball $B(x; r)$ is compact, and proceeding as in the proof of (b) ⇒ (c) in (20.18.5).)

10. In $\mathbf{R}^2$, consider the connected Riemannian submanifold $\mathbf{R}^2 - \{0\}$, and its simply-connected universal Riemannian covering M, which is not complete. Show that there exists no connected Riemannian manifold N containing M, in which M is a proper open subset of N. (Argue by contradiction. Let a be a frontier point of M in N, and U a strictly geodesically convex neighborhood of a in M. Observe that through each point $x \in M$ there passes only one geodesic not defined on the whole of **R**, and deduce that the complement of $U \cap M$ in U consists of the single point a. Then consider in U the set $S(a; \rho)$ of points whose distance from a is constant and equal to ρ, where ρ is sufficiently small, and obtain a contradiction by observing that this set is homeomorphic to a circle, and must be contained in the set of points of M which project onto the circle with center 0 and radius ρ in $\mathbf{R}^2 - \{0\}$.)

11. Consider on **R** the Riemannian metric tensor $\mathbf{g} = e^x\, dx \otimes dx$, which is not complete. The mapping $x \mapsto x + 1$ is a homometry (Section 20.9, Problem 5) and hence an automorphism for the Levi–Civita connection. Hence construct an example of a linear connection on $\mathbf{T} = \mathbf{R}/\mathbf{Z}$, which is not complete in the sense of Section 20.6, Problem 8, and by extension (17.18.5) a noncomplete linear connection on $\mathbf{S}_2$.

12. (a) Let M be a connected Riemannian manifold, d the Riemannian distance on M. Let $\gamma : I \to M$ be a path of class C^0 such that for all $t_1 < t_2 < t_3$ in I, we have

$$d(\gamma(t_1), \gamma(t_3)) = d(\gamma(t_1), \gamma(t_2)) + d(\gamma(t_2), \gamma(t_3)).$$

Show that $\gamma(I)$ is a geodesic arc. (By considering a strictly geodesically convex neighborhood of $\gamma(t)$, show that there exists $\varepsilon > 0$ such that $\gamma([t - \varepsilon, t + \varepsilon])$ is a geodesic arc.)
(b) Let M, M′ be two connected Riemannian manifolds and d, d' the Riemannian distances on M, M′, respectively. Let f be a mapping of M onto M′ such that $d'(f(x), f(y)) = d(x, y)$ for all $x, y \in M$. Show that f is an isometry in the sense of (20.8.1). (Using (a), prove first that for each geodesic $\gamma : I \to M$ in M, the path

$f \circ \gamma : \mathrm{I} \to \mathrm{M}'$ is a geodesic in M′. Deduce that for each $x \in \mathrm{M}$ there exists a bijection F_x of $\mathrm{T}_x(\mathrm{M})$ onto $\mathrm{T}_{f(x)}(\mathrm{M}')$ such that $\mathrm{F}_x(c\mathbf{h}_x) = c\mathrm{F}_x(\mathbf{h}_x)$ for all $c \in \mathbf{R}$ and $\mathbf{h}_x \in \mathrm{T}_x(\mathrm{M})$, and such that $f \circ \exp_x = \exp_{f(x)} \circ \mathrm{F}_x$ in a neighborhood of $\mathbf{0}_x$ in $\mathrm{T}_x(\mathrm{M})$. Finally show that $(\mathrm{F}_x(\mathbf{h}_x) | \mathrm{F}_x(\mathbf{k}_x)) = (\mathbf{h}_x | \mathbf{k}_x)$ for all $\mathbf{h}_x, \mathbf{k}_x \in \mathrm{T}_x(\mathrm{M})$, by using Section **20.16**, Problem 10.)

13. Consider a curve C in $\mathbf{R}^3$ defined by $\xi^2 = 0$, $\xi^1 \in \mathrm{I}$, where I is an open interval in **R**, and $\xi^3 = f(\xi^1)$, where f is a C^∞ function on I which is everywhere >0. If α is an endpoint of I and $f(t) \to 0$, $f'(t) \to \pm\infty$ as $t \to \alpha$, show that if S is the surface of revolution with axis $\mathbf{R}\mathbf{e}_1$ generated by C (Section **20.14**, Problem 7), then $\mathrm{S} \cup \{\alpha\mathbf{e}_1\}$ is a differential manifold.

Show that there exists a function f (of class C^∞) defined on the interval $\mathrm{I} =]-\infty, 1[$ such that $f(\xi^1) = 1$ for $\xi^1 \leqq 0$, $f(\xi^1) > 0$ throughout I, and $f''(\xi^1) < 0$ in $]0, 1[$, and such that $f(\xi^1) \to 0$ and $f'(\xi^1) \to -\infty$ as $\xi^1 \to 1$. Let V be the corresponding surface, which is closed in $\mathbf{R}^3$.

Let V′ be the union of the set of points of V such that $\xi^2 \geqq 0$ and the mirror-image of this set with respect to the plane $\xi^2 = 0$. Show that V′ is also a differential manifold and that there exists an isometry of a neighborhood of $\mathbf{e}_1$ in V onto a neighborhood of $\mathbf{e}_1$ in V′, which cannot be extended in an isometry of V onto V′, although V and V′ are complete and simply-connected. (Compare with Section **20.6**, Problem 9(e) and Section **20.9**, Problem 8.)

14. Let M, N be two connected Riemannian manifolds of the same dimension. A C^∞ mapping $f: \mathrm{M} \to \mathrm{N}$ is said to be *complete* if there exists a continuous positive-valued function λ on N such that, for all $y \in \mathrm{N}$, all $x \in f^{-1}(y)$ and all tangent vectors $\mathbf{h}_x \in \mathrm{T}_x(\mathrm{M})$, we have $\|\mathrm{T}(f) \cdot \mathbf{h}_x\| \geqq \lambda(y)\|\mathbf{h}_x\|$. This condition implies that f is a local diffeomorphism.

(a) Show that if f is complete and if the manifold M is complete, then $f(\mathrm{M}) = \mathrm{N}$ and f makes M a covering of N. (For the first assertion, argue by contradiction, by supposing that there exists a frontier point y_0 of $f(\mathrm{M})$ in N. Deduce that there exists a frontier point y of $f(\mathrm{M})$ and a geodesic path $v : [0, 1] \to \mathrm{N}$ with origin $b \in f(\mathrm{M})$ and endpoint y, such that $v(t) \in f(\mathrm{M})$ for $0 \leqq t < 1$. Show that there exists a C^∞ mapping $u : [0, 1[\to \mathrm{M}$ which lifts v, such that $u(t)$ tends to a limit as $t \to 1$, by using the completeness of f and M. To show that M is a covering of N, use Section **16.29**, Problem 5.)

(b) The mapping f is said to be *uniformly complete* if $\lambda(x)$ is bounded in every bounded subset of N (relative to the Riemannian distance). Show that if f is uniformly complete and M is complete, then N is complete. (Consider a Cauchy sequence in a strictly geodesically convex ball in N, and show that it can be lifted to a Cauchy sequence in M.) Consider the case where f is a local isometry.

(c) Under the hypotheses of (a), if we suppose in addition that the fundamental group of N is *finite*, then f is proper and N is complete (cf. Section **16.12**, Problem 1).

19. PERIODIC GEODESICS

(20.19.1) *Let* M *be a complete connected Riemannian manifold, a and b two points of* M, *and* $\gamma : [\alpha, \beta] \to \mathrm{M}$ *a piecewise-*C^1 *path with origin a and endpoint b. Then there exists a piecewise-*C^1 *path* $\gamma_0 : [\alpha, \beta] \to \mathrm{M}$, *with origin a and endpoint b, of length d(a, b) (so that* γ_0 *is rectilinear* **(20.18.2)**) *which is homotopic to* γ *under a homotopy leaving a and b fixed.*

Let M′ be the universal covering of M, endowed with the Riemannian structure canonically induced by that of M (20.8.2), so that M′ is a Riemannian covering of M. Let γ' be a lifting of γ to M′ (16.28.1) with origin a' and endpoint b'. Since M′ is complete (20.18.7), there exists a rectilinear path γ_0' in M′ with origin a' and endpoint b', whose length is equal to the distance from a' to b'. The projection γ_0 of γ_0' has the desired properties since the length of a path in M is equal to that of any lifting of the path to M′, and since the paths with origin a' and endpoint b' are precisely the liftings of paths with origin a and endpoint b which are homotopic to γ under a homotopy leaving a and b fixed (16.27).

(20.19.2) In a Riemannian manifold M, a *periodic geodesic* is by definition a geodesic $t \mapsto \varphi(t)$ in M, defined on the whole of **R**, not reduced to a single point, such that φ is a periodic mapping with period $\neq 0$. If τ is the smallest period of φ, and if φ restricted to $[0, \tau[$ is injective, then the image of φ is diffeomorphic to the circle S_1.

(20.19.3) *Let* M *be a compact connected Riemannian manifold, and let* γ *be a piecewise-*C^1 *loop in* M *which is not homotopic (as a loop) to a point. Then (after a change of parameters) there exists a loop* γ_0 *in* M *which is loop-homotopic to* γ *and which is the restriction of a periodic geodesic (to an interval of length equal to a period); moreover,* $L(\gamma_0)$ *is the minimum of the lengths of piecewise-*C^1 *loops which are loop-homotopic to* γ.

Let x_0 be the origin of γ. For each $x \in M$ there exists a piecewise-C^1 loop with origin x which is loop-homotopic to γ (16.27.3.1). Let $H_{\gamma, x}$ denote the set of all piecewise-C^1 loops with origin x which are loop-homotopic to γ, and let $\lambda(x)$ denote the *greatest lower bound of the lengths* of the loops belonging to $H_{\gamma, x}$.

Let M′ be the Riemannian universal covering of M, d, d' the Riemannian distances on M, M′, respectively, and $p : M' \to M$ the projection. The loops in $H_{\gamma, x}$ are precisely the projections of the piecewise-C^1 paths in M′ with endpoints x' and x'', where x' is any point of $p^{-1}(x)$, and x'' is the image of x' under the element s of $\pi_1(M)$ which is the class of the loop γ (16.29.2). Since the length of a path in M′ is equal to that of its projection by p, it follows that $\lambda(x) = d'(x', s \cdot x')$. Hence $\lambda(x)$ is a *continuous* function of x on M: for if x_1 is any point of M, there exists a neighborhood V′ of $x_1' \in p^{-1}(x_1)$ such that the restriction of p to V′ is a diffeomorphism of V′ onto a neighborhood V of x_1 in M; if q is the inverse diffeomorphism, we have $\lambda(x) = d(q(x), s \cdot p(x))$ for $x \in V_1$, and our assertion is now obvious because s acts continuously on M′. Since M is compact, there exists a point $a \in M$ at which λ attains its minimum l, which is $\neq 0$ because γ is not homotopic to a point. Moreover, since M′ is complete, l is the length of a loop γ_0 with origin a which is a geodesic path and the projection of a geodesic path with endpoints a' and $s \cdot a'$ in M′ (where $a' \in p^{-1}(a)$) of length $d'(a', s \cdot a')$. We shall show that γ_0 has the required properties. Let $t \mapsto \gamma_0(t)$ be the arc length parametrization of γ_0; we have to show that $\gamma_0'(0) = \gamma_0'(l)$. Suppose that this is not the case, and let B be a strictly geodesically convex open ball with center a (20.17.5). Choose $\varepsilon > 0$ sufficiently small so that $\gamma_0(\varepsilon)$ and $\gamma_0(l - \varepsilon)$ both belong to B. Then there exists a rectilinear path $w : [l - \varepsilon, l + \varepsilon] \to M$ contained in B, with origin $\gamma_0(l - \varepsilon)$ and endpoint $\gamma_0(\varepsilon)$, and length equal to $d(\gamma_0(l - \varepsilon), \gamma_0(\varepsilon))$. Now the hypothesis implies that

$$d(\gamma_0(l - \varepsilon), \gamma_0(\varepsilon)) < d(\gamma_0(l - \varepsilon), a) + d(a, \gamma_0(\varepsilon))$$

by virtue of (20.18.3). The loop $\gamma_1 : [l - \varepsilon, 2l - \varepsilon] \to M$ which is equal to $w(t)$ in $[l - \varepsilon, l + \varepsilon]$ and to $\gamma_0(t - l)$ in $[l + \varepsilon, 2l - \varepsilon]$ then has length $< l$; but since B is simply-connected (16.27.7), the path w is homotopic to the path with the same endpoints which is equal to $\gamma_0(t)$ for $l - \varepsilon \leqq t \leqq l$ and to $\gamma_0(t - l)$ for $l \leqq t \leqq l + \varepsilon$ (under a homotopy leaving the endpoints fixed). It follows immediately that γ_1 is homotopic to γ, which contradicts the definition of l.

20. FIRST AND SECOND VARIATION OF ARC LENGTH. JACOBI FIELDS ON A RIEMANNIAN MANIFOLD

Let M be a connected Riemannian manifold and let v be a geodesic in M, defined on an open interval $I \subset \mathbf{R}$ containing $[a, b]$. Assume that v is parametrized by the curvilinear coordinate with origin $v(a)$. We propose to give conditions under which the length of the geodesic path obtained by restricting v to $[a, b]$ is *minimal among "neighboring" paths with the same endpoints.*

(20.20.1) We shall therefore consider a *one-parameter family f of curves* (18.7.1), defined on $I \times J$, where J is an open interval in $\mathbf{R}$ containing 0. Suppose that f is of class C^r, where $r \geqq 3$, and use the notation of (18.7). For each $\xi \in J$,

(20.20.1.1) $$L(\xi) = \int_a^b \|f_t'(t, \xi)\| \, dt$$

is the length of the path $C_\xi : t \mapsto f(t, \xi)$ defined on $[a, b]$. The integral

(20.20.1.2) $$E(\xi) = \int_a^b \|f_t'(t, \xi)\|^2 \, dt$$

is called the *energy* of this path. By the Cauchy–Schwarz inequality (6.2.1), we have

(20.20.1.3) $$(L(\xi))^2 \leqq (b - a)E(\xi),$$

with equality when $t \mapsto \|f_t'(t, \xi)\|$ is *constant* on $[a, b]$, in other words, when the parameter t is proportional to the *curvilinear coordinate* of the path C_ξ, with origin $f(a, \xi)$.

By virtue of our hypothesis on f, we may calculate the first and second derivatives of $E(\xi)$ by Leibniz's formula (8.11.2). First of all we have, by virtue of (20.9.5.4) and (18.7.4.2),

(20.20.1.4) $$\frac{1}{2}\frac{d}{d\xi}(\|f_t'\|^2) = (\nabla_\xi \cdot f_t' | f_t') = (\nabla_t \cdot f_\xi' \mid f_t')$$
$$= \frac{d}{dt}(f_\xi' | f_t') - (f_\xi' | \nabla_t \cdot f_t'),$$

and then, bearing in mind (18.7.2.1) and the preceding formulas,

(20.20.1.5) $$\frac{1}{2}\frac{d^2}{d\xi^2}(\|f_t'\|^2) = (\nabla_\xi \cdot (\nabla_t \cdot f_\xi') | f_t') + (\nabla_t \cdot f_\xi' \mid \nabla_\xi \cdot f_t')$$
$$= (\nabla_t \cdot (\nabla_\xi \cdot f_\xi') | f_t') + ((\mathbf{r} \cdot (f_\xi' \wedge f_t')) \cdot f_\xi' | f_t') + \|\nabla_t \cdot f_\xi'\|^2$$
$$= \frac{d}{dt}(\nabla_\xi \cdot f_\xi' | f_t') - (\nabla_\xi \cdot f_\xi' | \nabla_t \cdot f_t') + \|\nabla_t \cdot f_\xi'\|^2$$
$$+ \langle \mathbf{K}, f_t' \otimes f_\xi' \otimes f_\xi' \otimes f_t' \rangle$$

by introducing the Riemann–Christoffel tensor (20.10.2); and finally, by virtue of (20.10.3),

(20.20.1.6) $$\frac{1}{2}\frac{d^2}{d\xi^2}(\|f_t'\|)^2 = \frac{d}{dt}(\nabla_\xi \cdot f_\xi' | f_t') - (\nabla_\xi \cdot f_\xi' | \nabla_t \cdot f_t')$$
$$+ \|\nabla_t \cdot f_\xi'\|^2 + ((\mathbf{r} \cdot (f_t' \wedge f_\xi')) \cdot f_t' | f_\xi').$$

Integrating with respect to t from a to b, we obtain the formulas

(20.20.1.7) $$\tfrac{1}{2}\mathrm{E}'(\xi) = (f'_\xi \mid f'_t)\Big|_{(a,\,\xi)}^{(b,\,\xi)} - \int_a^b (\nabla_t \cdot f'_t \mid f'_\xi)\,dt,$$

(20.20.1.8) $$\tfrac{1}{2}\mathrm{E}''(\xi) = (\nabla_\xi \cdot f'_\xi \mid f'_t)\Big|_{(a,\,\xi)}^{(b,\,\xi)} - \int_a^b (\nabla_\xi \cdot f'_\xi \mid \nabla_t \cdot f'_t)\,dt$$
$$+ \int_a^b (((\boldsymbol{r} \cdot (f'_t \wedge f'_\xi)) \cdot f'_t \mid f'_\xi) + \|\nabla_t \cdot f'_\xi\|^2)\,dt.$$

(20.20.2) Suppose now that $v(t) = f(t, 0)$ is the restriction to $[a, b]$ of a *geodesic*, so that $\nabla_t \cdot f'_t = \nabla_t \cdot v' = 0$ in $[a, b]$. Then the formulas (20.20.1.7) and (20.20.1.8) for $\xi = 0$ simplify:

(20.20.2.1) $$\tfrac{1}{2}\mathrm{E}'(0) = (f'_\xi(\,.\,, 0) \mid v')\Big|_{(a,\,0)}^{(b,\,0)},$$

(20.20.2.2) $$\tfrac{1}{2}\mathrm{E}''(0) = ((\nabla_\xi \cdot f'_\xi)(\,.\,, 0) \mid v')\Big|_{(a,\,0)}^{(b,\,0)} + \mathrm{I}(f'_\xi(\,.\,, 0), f'_\xi(\,.\,, 0)),$$

where, for two *piecewise-*C^1 *liftings* $\mathbf{w}_1$, $\mathbf{w}_2$ of v to T(M), we put

(20.20.2.3)
$$\mathrm{I}(\mathbf{w}_1, \mathbf{w}_2) = \int_a^b (((\boldsymbol{r} \cdot (v' \wedge \mathbf{w}_1)) \cdot v' \mid \mathbf{w}_2) + (\nabla_t \cdot \mathbf{w}_1 \mid \nabla_t \cdot \mathbf{w}_2))\,dt.$$

This is a symmetric bilinear form on the vector space $\mathscr{R}^1(v)$ of these liftings, called the *index form*.

(20.20.3) *Let f be a one-parameter family of curves defined on* I × J, *such that $t \mapsto f(t, 0) = v(t)$ is a geodesic, and let* $\mathbf{w}$ *be a piecewise-*C^1 *lifting of v to* T(M), *defined on* $[a, b]$, *and such that on the interval* $[a, c]$ *for some* $c \in \,]a, b[$, $\mathbf{w}$ *is a Jacobi field along v, with* $\mathbf{w}(a) = \mathbf{0}_{v(a)}$. *If there exists no point on v conjugate to $v(a)$* (18.7.12) *and if* $\mathbf{z}$ *denotes the unique Jacobi field along v such that* $\mathbf{z}(a) = \mathbf{0}_{v(a)}$ *and* $\mathbf{z}(b) = \mathbf{w}(b)$ (18.7.15), *then we have*

(20.20.3.1) $$\mathrm{I}(\mathbf{w}, \mathbf{w}) > \mathrm{I}(\mathbf{z}, \mathbf{z})$$

unless $\mathbf{w} = \mathbf{z}$.

Let $(\mathbf{u}_j)_{1 \leq j \leq n}$ be a basis of $\mathrm{T}_{v(a)}(\mathrm{M})$, and for each j let $\mathbf{z}_j$ be the Jacobi field along v such that $\mathbf{z}_j(a) = \mathbf{0}_a$ and $(\nabla_t \cdot \mathbf{z}_j)(a) = \mathbf{u}_j$ (18.7.6). The hypotheses imply that for each $t \in \,]a, b]$ the n vectors $\mathbf{z}_j(t)$ $(1 \leq j \leq n)$ form a *basis* of $T_{v(t)}(M)$; for if $\sum_j \lambda_j \mathbf{z}_j(t) = 0$ with scalars λ_j not all zero, then the Jacobi field $\sum_j \lambda_j \mathbf{z}_j$ is not identically zero but vanishes at two distinct points a and t, which is impossible (18.7.15). We may therefore write, for each $t \in]a, b]$,

(20.20.3.2)
$$\mathbf{w}(t) = \sum_{j=1}^{n} h_j(t)\mathbf{z}_j(t),$$

where the h_j are piecewise of class C^1 on $]a, b]$ and are *constant* on $]a, c]$ (18.7.7). It follows (17.17.3.4) that

$$\nabla_t \cdot \mathbf{w} = \sum_j h'_j \mathbf{z}_j + \sum_j h_j(\nabla_t \cdot \mathbf{z}_j).$$

Hence, putting

(20.20.3.3)
$$\mathbf{f} = \sum_j h'_j \mathbf{z}_j, \qquad \mathbf{u} = \sum_j h_j(\nabla_t \cdot \mathbf{z}_j),$$

we have

(20.20.3.4)
$$\|\nabla_t \cdot \mathbf{w}\|^2 = \|\mathbf{f}\|^2 + 2(\mathbf{f} \mid \mathbf{u}) + \|\mathbf{u}\|^2.$$

On the other hand, we have

$$((\boldsymbol{r} \cdot (v' \wedge \mathbf{w})) \cdot v' \mid \mathbf{w}) = \sum_j h_j((\boldsymbol{r} \cdot (v' \wedge \mathbf{z}_j)) \cdot v' \mid \mathbf{w})$$

and since the $\mathbf{z}_j$ are Jacobi fields and therefore satisfy (18.7.5.1), we have

$$\begin{aligned}((\boldsymbol{r} \cdot (v' \wedge \mathbf{w})) \cdot v' \mid \mathbf{w}) &= \sum_j h_j(\nabla_t \cdot (\nabla_t \cdot \mathbf{z}_j) \mid \mathbf{w}) \\ &= (\nabla_t \cdot \mathbf{u} \mid \mathbf{w}) - \sum_j h'_j(\nabla_t \cdot \mathbf{z}_j \mid \mathbf{w})\end{aligned}$$

and therefore, using (20.9.5.4),

(20.20.3.5)
$$\begin{aligned}&((\boldsymbol{r} \cdot (v' \wedge \mathbf{w})) \cdot v' \mid \mathbf{w}) \\ &\quad = \frac{d}{dt}(\mathbf{u} \mid \mathbf{w}) - (\mathbf{u} \mid \nabla_t \cdot \mathbf{w}) - \sum_{j,k} h'_j h_k(\nabla_t \cdot \mathbf{z}_j \mid \mathbf{z}_k) \\ &\quad = \frac{d}{dt}(\mathbf{u} \mid \mathbf{w}) - (\mathbf{u} \mid \mathbf{f} + \mathbf{u}) - \sum_{j,k} h'_j h_k(\nabla_t \cdot \mathbf{z}_j \mid \mathbf{z}_k).\end{aligned}$$

But since the $\mathbf{z}_j$ are Jacobi fields, we have by (20.9.5.4) and (18.7.5.1),

$$\begin{aligned}\frac{d}{dt}((\nabla_t \cdot \mathbf{z}_j | \mathbf{z}_k) - (\mathbf{z}_j | \nabla_t \cdot \mathbf{z}_k)) &= (\nabla_t \cdot (\nabla_t \cdot \mathbf{z}_j) | \mathbf{z}_k) - (\mathbf{z}_j | \nabla_t \cdot (\nabla_t \cdot \mathbf{z}_k)) \\ &= ((\boldsymbol{r} \cdot (v' \wedge \mathbf{z}_j)) \cdot v' | \mathbf{z}_k) - ((\boldsymbol{r} \cdot (v' \wedge \mathbf{z}_k)) \cdot v' | \mathbf{z}_j) \\ &= 0,\end{aligned}$$

by virtue of the symmetry properties (20.10.3.2), (20.10.3.3), and (20.10.3.4) of the Riemann–Christoffel tensor. The function $(\nabla_t \cdot \mathbf{z}_j | \mathbf{z}_k) - (\mathbf{z}_j | \nabla_t \cdot \mathbf{z}_k)$ is therefore *constant* on $[a, b]$, and since it vanishes at the point a, it follows that

(20.20.3.6) $$(\nabla_t \cdot \mathbf{z}_j | \mathbf{z}_k) - (\mathbf{z}_j | \nabla_t \cdot \mathbf{z}_k) = 0.$$

Consequently,

$$\sum_{j,k} h'_j h_k (\nabla_t \cdot \mathbf{z}_j | \mathbf{z}_k) = \sum_{j,k} h'_j h_k (\nabla_t \cdot \mathbf{z}_k | \mathbf{z}_j) = (\mathbf{f} | \mathbf{u})$$

and finally the relations (20.20.3.4) and (20.20.3.5) give

$$\|\nabla_t \cdot \mathbf{w}\|^2 + ((\boldsymbol{r} \cdot (v' \wedge \mathbf{w})) \cdot v' | \mathbf{w}) = \frac{d}{dt}(\mathbf{u} | \mathbf{w}) + \|\mathbf{f}\|^2.$$

Now $\mathbf{f}$ and $\mathbf{u}$ both tend to limits at the point a, by virtue of the hypothesis on $\mathbf{w}$, and therefore we may integrate both sides of this relation from a to b, thus obtaining the following expression for the index form:

(20.20.3.7) $$\mathrm{I}(\mathbf{w}, \mathbf{w}) = \int_a^b \|\mathbf{f}\|^2 \, dt + (\mathbf{u}(b) | \mathbf{w}(b)).$$

We may repeat the same calculation with $\mathbf{w}$ replaced by $\mathbf{z}$. This time the h_j are constants *in* $]a, b]$ (18.7.7), and $\mathbf{f}$ is replaced by 0. Since $\mathbf{w}(b) = \mathbf{z}(b)$, we obtain

(20.20.3.8) $$\mathrm{I}(\mathbf{w}, \mathbf{w}) - \mathrm{I}(\mathbf{z}, \mathbf{z}) = \int_a^b \|\mathbf{f}\|^2 \, dt.$$

Since the right-hand side vanishes only when the h_j are constants (8.5.3), the proof is complete.

Remark

(20.20.3.9) The assertion of (20.20.3) remains valid if we assume only that $\mathbf{w}(a) = \mathbf{0}_{v(a)}$. Let $\mathbf{s}_j$ be the parallel transport of $\mathbf{u}_j$ along the path v (18.6.4)

$(1 \leqq j \leqq n)$; then the vectors $\mathbf{s}_j(t)$ form a basis of $T_{v(t)}(M)$ for all $t \in [a, b]$, and we may write

$$\mathbf{z}_j = \sum_{k=1}^{n} z_{jk}\mathbf{s}_k, \qquad \mathbf{w} = \sum_{j=1}^{n} w_j \mathbf{s}_j;$$

since $\nabla_t \cdot \mathbf{s}_j = 0$ (18.6.3.1), we have

$$\nabla_t \cdot \mathbf{z}_j = \sum_{k=1}^{n} z'_{jk}\mathbf{s}_k, \qquad \nabla_t \cdot \mathbf{w} = \sum_{j=1}^{n} w'_j \mathbf{s}_j.$$

By hypothesis, the functions z'_{jk} and w'_j are continuous at the point a, and the matrix $(z'_{jk}(t))$ tends to the unit matrix I_n as $t \to a$. Since $z_{jk}(a) = 0$ for all j, k, the matrix $((t-a)^{-1}z_{jk}(t))$ also tends to I_n as $t \to a$, and the same argument shows that the functions $(t-a)^{-1}w_j(t)$ tend to finite limits as $t \to a$. This being so, we have

$$\sum_{k=1}^{n} h_k(t)\left(\frac{1}{t-a} z_{kj}(t)\right) = \frac{1}{t-a} w_j(t),$$

and Cramer's formulas show that the $h_k(t)$ tend to *finite limits* as $t \to a$. It follows that the limit of $(\mathbf{u}(t) \mid \mathbf{w}(t))$, as $t \to a$, exists and is finite; the formula (20.20.3.7) is therefore still valid, and the integral $\int_a^b \|\mathbf{f}\|^2\, dt$ is finite (13.8.1), hence the conclusion is unaltered.

(20.20.4) *With the hypotheses and notation of* (20.20.3), *suppose that the parameter t is the curvilinear coordinate of the geodesic $t \mapsto v(t) = f(t, 0)$, with origin $v(a)$, and that $f(a, \xi) = v(a)$ and $f(b, \xi) = v(b)$ for all $\xi \in J$. Then, if there exists no point on v conjugate to $v(a)$* (18.7.12) *and if $f'_\xi(t, 0)$ is not identically zero in $[a, b]$, we have*

$$L(\xi) > L(0) = b - a$$

for all sufficiently small $\xi \neq 0$.

Since $\|f'_t(t, 0)\| = 1$ for all $t \in [a, b]$, we may assume that $f'_t(t, \xi) \neq 0$ for all $(t, \xi) \in I \times J$, and the change of parameter $(t, \xi) \mapsto (s(t, \xi), \xi)$, where $s(u, \xi) = \int_0^u \|f'_t(t, \xi)\|\, dt$ for $a \leqq u \leqq b$, allows us to assume that for all $\xi \in J$ the parameter t is proportional to the curvilinear coordinate of the curve $t \mapsto f(t, \xi)$. Since $f'_\xi(a, \xi)$ and $f'_\xi(b, \xi)$ are by hypothesis *zero* for all $\xi \in J$, the formula (20.20.2.2) becomes here

$$\tfrac{1}{2}E''(0) = I(f'_\xi(\cdot, 0), f'_\xi(\cdot, 0)).$$

Now the hypotheses imply that the only Jacobi field $\mathbf{z}$ along v such that $\mathbf{z}(a) = \mathbf{0}_a$ and $\mathbf{z}(b) = \mathbf{0}_b$ is the *zero* field (18.7.11). Since $f'_\xi(t, 0)$ is not identically zero in $[a, b]$, we have $\mathrm{I}(f'_\xi(.\,, 0), f'_\xi(.\,, 0)) > 0$ by (20.20.3.9), hence $\mathrm{E}''(0) > 0$; and since $\mathrm{E}'(0) = 0$ by (20.20.2.1), we have $\mathrm{E}(\xi) > \mathrm{E}(0)$ for all sufficiently small $\xi \neq 0$, by virtue of Taylor's formula (8.14.2). Since

$$\mathrm{E}(\xi) = (\mathrm{L}(\xi))^2/(b - a),$$

by virtue of our choice of the parameter t, the result is proved.

(20.20.5) *Let v be a geodesic in* M *defined on an open interval* $\mathrm{I} \subset \mathbf{R}$ *containing* $[a, b]$, *and suppose that $v(b)$ is the first point on v conjugate to $v(a)$. Then, for all sufficiently small $\varepsilon > 0$, there exists a* C^∞ *path defined on* $[a, b + \varepsilon]$ *whose distance from v (relative to the topology of uniform convergence and the Riemannian distance d on* M) *is* $\leqq \varepsilon$, *with endpoints $v(a)$ and $v(b + \varepsilon)$, and of length* strictly less *than that of the restriction of v to* $[a, b + \varepsilon]$.

In view of (20.16.6), it will suffice to establish the existence of a *piecewise-*C^∞ path with the properties asserted. We shall first show that there exists a piecewise-C^∞ lifting $\mathbf{w}$ of v (restricted to $[a, b + \varepsilon]$) such that

$$\mathrm{I}(\mathbf{w}, \mathbf{w}) < 0, \ \mathbf{w}(a) = \mathbf{0}_a$$

and $\mathbf{w}(b + \varepsilon) = \mathbf{0}_{b+\varepsilon}$. For each interval $[\alpha, \beta] \subset \mathrm{I}$, let $\mathrm{I}_{\alpha, \beta}(\mathbf{w}, \mathbf{w})$ denote the value of the index form relative to the restrictions of v and $\mathbf{w}$ to $[\alpha, \beta]$. Let ρ be the radius of a geodesically convex ball with center $v(b)$ (20.17.5), and let $\varepsilon > 0$ be sufficiently small so that $v(t)$ lies in this ball for $b - \varepsilon \leqq t \leqq b + \varepsilon$. By hypothesis, there exists a Jacobi field $\mathbf{z}_1$ along v which is not identically zero on $[a, b]$ and is such that $\mathbf{z}_1(a) = \mathbf{0}_a$ and $\mathbf{z}_1(b) = \mathbf{0}_b$ (18.7.11). On the other hand, the choice of ε implies that there exists no point conjugate to $v(b + \varepsilon)$ on the restriction of v to $[b - \varepsilon, b + \varepsilon]$ (18.5.2); consequently, there exists a Jacobi field $\mathbf{z}_2$ along this restriction of v such that $\mathbf{z}_2(b + \varepsilon) = \mathbf{0}_{b+\varepsilon}$ and $\mathbf{z}_2(b - \varepsilon) = \mathbf{z}_1(b - \varepsilon)$ (18.7.11). Put $\mathbf{w}(t) = \mathbf{z}_1(t)$ for $a \leqq t \leqq b - \varepsilon$, and $\mathbf{w}(t) = \mathbf{z}_2(t)$ for $b - \varepsilon \leqq t \leqq b + \varepsilon$; also let $\mathbf{u}$ denote the lifting of v such that $\mathbf{u}(t) = \mathbf{w}(t)$ for $a \leqq t \leqq b$ and $\mathbf{u}(t) = 0$ for $b \leqq t \leqq b + \varepsilon$. The formula (20.20.3.7) applied to $\mathbf{z}_1$ on an interval $[a, b']$ with $b' < b$ gives $\mathrm{I}_{a, b}(\mathbf{z}_1, \mathbf{z}_1) = 0$ on letting b' tend to b, and this may also be written as

$$\mathrm{I}_{a, b+\varepsilon}(\mathbf{u}, \mathbf{u}) = \mathrm{I}_{a, b}(\mathbf{z}_1, \mathbf{z}_1) = 0.$$

But also we have

$$\begin{aligned}\mathrm{I}_{a, b+\varepsilon}(\mathbf{u}, \mathbf{u}) &= \mathrm{I}_{a, b-\varepsilon}(\mathbf{z}_1, \mathbf{z}_1) + \mathrm{I}_{b-\varepsilon, b+\varepsilon}(\mathbf{u}, \mathbf{u}) \\ &= \mathrm{I}_{a, b-\varepsilon}(\mathbf{w}, \mathbf{w}) + \mathrm{I}_{b-\varepsilon, b+\varepsilon}(\mathbf{z}, \mathbf{z}).\end{aligned}$$

Finally, we may apply (20.20.3) in the interval $[b-\varepsilon, b+\varepsilon]$ to the Jacobi field $\mathbf{z}_2$ and the lifting $\boldsymbol{u}$ (since the latter is zero in a neighborhood of $b+\varepsilon$), and hence we obtain

$$\mathrm{I}_{b-\varepsilon, b+\varepsilon}(\boldsymbol{u}, \boldsymbol{u}) > \mathrm{I}_{b-\varepsilon, b+\varepsilon}(\mathbf{z}_2, \mathbf{z}_2) = \mathrm{I}_{b-\varepsilon, b+\varepsilon}(\mathbf{w}, \mathbf{w}),$$

and thus $0 = \mathrm{I}_{a, b+\varepsilon}(\boldsymbol{u}, \boldsymbol{u}) > \mathrm{I}_{a, b+\varepsilon}(\mathbf{w}, \mathbf{w})$.

The conclusion of (20.20.5) will therefore result from the following proposition (in which the notation has been slightly altered):

(20.20.5.1) *Let v be a geodesic defined on an open interval $\mathrm{I} \supset [a, b]$, and let $\mathbf{w}$ be a piecewise-C^3 lifting of v to* T(M), *defined on $[a, b]$ and such that $\mathbf{w}(a) = \mathbf{0}_a$, $\mathbf{w}(b) = \mathbf{0}_b$, and $\mathrm{I}_{a, b}(\mathbf{w}, \mathbf{w}) < 0$. Then for each $\varepsilon > 0$ there exists a piecewise-C^3 path $\varphi : [a, b] \to \mathrm{M}$ such that*

$$\varphi(a) = v(a), \qquad \varphi(b) = v(b), \qquad d(\varphi(t), v(t)) \leqq \varepsilon$$

for all $t \in [a, b]$, and $\mathrm{L}(\varphi) < \mathrm{L}(v)$.

There exists a real number $\delta > 0$ such that $\delta\mathbf{w}(t) \in \Omega$ for all $t \in [a, b]$, so that the function $f(t, \xi) = \exp_{v(t)}(\xi\mathbf{w}(t))$ is defined and piecewise of class C^3 in $[a, b] \times]-\delta, \delta[$; moreover f'_ξ is continuous on this set, and for each $t \in [a, b]$ the function $\xi \mapsto f'_\xi(t, \xi)$ is of class C^∞ and satisfies the equation $(\nabla_\xi \cdot f'_\xi)(t, \xi) = 0$ and the boundary condition $f'_\xi(t, 0) = \mathbf{w}(t)$. Applying the formulas (20.20.2.1) and (20.20.2.2) in each of the intervals $[a_j, a_{j+1}]$ on which $\mathbf{w}$ is of class C^3, we obtain

$$\mathrm{E}'(0) = 0, \qquad \mathrm{E}''(0) = 2\mathrm{I}(\mathbf{w}, \mathbf{w}) < 0.$$

For sufficiently small $\xi > 0$ we have therefore $d(f(t, \xi), v(t)) \leqq \varepsilon$ for $t \in [a, b]$, and $\mathrm{E}(\xi) < \mathrm{E}(0)$, by Taylor's formula. By virtue of (20.20.1.3), this implies $\mathrm{L}(\xi) < \mathrm{L}(0)$, whence the result.

(20.20.6) It should be carefully noted that the length of a geodesic arc with endpoints p, q in M may well be $> d(p, q)$, even if there exists *no point conjugate* to p or q on the arc, as the example of the cylinder (20.17.3) shows. The property of minimizing the length of an arc with the same endpoints, when the geodesic arc under consideration contains no point conjugate to the endpoints, holds only for "neighboring" arcs.

(20.20.7) *Let* M *be a complete connected Riemannian manifold. If a point $a \in \mathrm{M}$ is such that no geodesic with origin a contains a point conjugate to a, then* $(\mathrm{T}_a(\mathrm{M}), \mathrm{M}, \exp_a)$ *is the universal covering of* M.

The hypothesis that M is complete implies that $\exp_a$ is a surjection of $T_a(M)$ onto M (20.18.5), and the hypothesis that no geodesic contains any point conjugate to a implies that $\exp_a$ is a local diffeomorphism (18.7.12). If $\mathbf{g}$ is the Riemannian metric tensor on M, consider the Riemannian metric tensor ${}^t\exp_a(\mathbf{g}) = \mathbf{g}_1$ on $T_a(M)$, relative to which $\exp_a$ is a local isometry. Since for each $\mathbf{h}_a \in T_a(M)$ the curve $t \mapsto \exp_a(t\mathbf{h}_a)$ is a geodesic in M defined on the whole of $\mathbf{R}$, it follows that $t \mapsto t\mathbf{h}_a$ is a geodesic in $T_a(M)$ relative to the metric tensor $\mathbf{g}_1$, defined on the whole of $\mathbf{R}$; by virtue of (20.18.5), $T_a(M)$ is *complete* relative to $\mathbf{g}_1$ and the result follows from (20.18.8).

(20.20.8) *Let* M *be a connected Riemannian manifold of dimension n, let a be a point of* M, *and let* $B(\mathbf{0}_a; r)$ *be an open ball in* $T_a(M) \cap \Omega$ *on which* $\exp_a$ *is injective. Then* $\exp_a$ *is a diffeomorphism of* $B(\mathbf{0}_a; r)$ *onto the open ball* $B(a; r)$ *in* M, *such that* $d(a, \exp_a(\mathbf{h}_a)) = \|\mathbf{h}_a\|$ *for* $\|\mathbf{h}_a\| < r$.

Suppose that there exists a point $\mathbf{h}_a \in B(\mathbf{0}_a; r)$ at which the rank of $\exp_a$ is $<n$. Then, by virtue of (20.20.5), there exists a point $t\mathbf{h}_a \in B(\mathbf{0}_a; r)$ such that $d(a, \exp_a(t\mathbf{h}_a)) < \|t\mathbf{h}_a\|$. If r' is such that $d(a, \exp_a(t\mathbf{h}_a)) < r' < \|t\mathbf{h}_a\|$, then it follows from the fact that $B(a; r')$ is the image under $\exp_a$ of $B(\mathbf{0}_a; r)$ (20.18.5.5) that there exists a vector $\mathbf{h}'_a \in T_a(M)$ such that $\exp_a(\mathbf{h}'_a) = \exp_a(t\mathbf{h}_a)$ and $\|\mathbf{h}'_a\| < r' < \|t\mathbf{h}_a\|$, contrary to the hypothesis that $\exp_a$ is injective on $B(\mathbf{0}_a; r)$. Since $\exp_a$, restricted to $B(\mathbf{0}_a; r)$, is therefore a bijective local diffeomorphism of this ball onto $B(a; r)$, it follows (16.8.8) that it is a diffeomorphism.

(20.20.9) *Let* M *be a complete connected Riemannian manifold and a, a point of* M. *For* $\exp_a$ *to be injective on* $T_a(M)$, *it is necessary and sufficient that* M *should be simply-connected and that no geodesic with origin a should contain any point conjugate to a. The mapping* $\exp_a$ *is then a diffeomorphism of* $T_a(M)$ *onto* M.

The necessity of the condition follows from (20.20.8) and (18.7.12), and the sufficiency from (20.20.7), since M is simply-connected.

PROBLEMS

1. Let $\mathbf{P}_n(K)$ be projective n-space over $K = \mathbf{R}$, $\mathbf{C}$, or $\mathbf{H}$, endowed with the Riemannian structure defined in (20.11.5) and (20.11.6). All geodesics in $\mathbf{P}_n(K)$ are periodic with period π (relative to the curvilinear coordinate). If two geodesics v_1, v_2 have the same origin $x_0 \in \mathbf{P}_n(K)$ and if (relative to the curvilinear coordinate with origin x_0) we put $\mathbf{h}_1 = v'_1(0)$ and $\mathbf{h}_2 = v'_2(0)$, then these two geodesics have no common point $\neq x_0$ unless the vectors $\mathbf{h}_1$, $\mathbf{h}_2$ are linearly dependent for the K-*vector-space* structure on

$T_{x_0}(\mathbf{P}_n(K))$; in which case they have a second common point, with curvilinear coordinate $\frac{1}{2}\pi$. The maximum radius $d(x_0)$ of a ball $B(\mathbf{0}_{x_0}; r)$, such that $\exp_{x_0}$ is injective when restricted to this ball, is $\frac{1}{2}\pi$, and the ball $B(x_0; \frac{1}{2}\pi)$ is strictly geodesically convex.

2. Let $M = G/H$ be a Riemannian symmetric space (**20.11.3**), with H compact, and let $v : \mathbf{R} \to M$ be a geodesic parametrized by the curvilinear coordinate with origin $x_0 = v(0)$. Put $v'(0) = \mathbf{h}$ (so that $v(t) = \exp_{x_0}(t\mathbf{h})$), and let $R(\mathbf{h})$ denote the endomorphism

$$\mathbf{k} \mapsto (r(x_0) \cdot (\mathbf{k} \wedge \mathbf{h})) \cdot \mathbf{h}$$

of $T_{x_0}(M)$, which is self-adjoint (**20.10.5**). If, for every Jacobi field $\mathbf{w}$ along v, we denote by $\tilde{\mathbf{w}}(t)$, the vector in $T_{x_0}(M)$ obtained by parallel transport of $\mathbf{w}(t)$ along v, show that the mappings $\tilde{\mathbf{w}}$ of $\mathbf{R}$ into $T_{x_0}(M)$ are the solutions of the equation $\tilde{\mathbf{w}}'' = R(\mathbf{h}) \cdot \tilde{\mathbf{w}}$. (Use the fact that the connection on M is G-invariant.) There exists an orthonormal basis $(\mathbf{h}_i)_{1 \le i \le n}$ of $T_{x_0}(M)$ consisting of eigenvectors of $R(\mathbf{h})$, with $\mathbf{h}_1 = \mathbf{h}$; if λ_i is the eigenvalue corresponding to $\mathbf{h}_i$, then $\lambda_1 = 0$. Show that the Jacobi fields along v which are zero at the point x_0 are linear combinations of the n fields $\mathbf{w}_i$ given by the conditions $\mathbf{w}_i(0) = 0$, $(\nabla_t \cdot \mathbf{w}_i)(0) = \mathbf{h}_i$ $(1 \le i \le n)$. We have

$$\mathbf{w}_i(t) = (-\lambda_i)^{-1/2} \sin((-\lambda_i)^{1/2} t)\mathbf{h}_i \quad \text{if } \lambda_i < 0,$$
$$\mathbf{w}_i(t) = t\mathbf{h}_i \quad \text{if } \lambda_i = 0,$$
$$\mathbf{w}_i(t) = \lambda_i^{-1/2} \sinh(\lambda_i^{1/2} t)\mathbf{h}_i \quad \text{if } \lambda_i > 0.$$

3. (a) In Problem 2, take M to be one of the projective spaces $\mathbf{P}_n(K)$ of Problem 1. Use Problem 1 to show that the eigenvalues λ_i are equal to -1 or -4.
(b) With the notation of Problem 2, consider the family of geodesics

$$(t, \alpha) \mapsto \exp_{x_0}(t(\cos\alpha \cdot \mathbf{h} + \sin\alpha \cdot \mathbf{h}_i)) = f(t, \alpha),$$

so that $f(t, 0) = v(t)$. Show that $f'_\alpha(\frac{1}{2}\pi, \alpha) = \mathbf{w}_\alpha(\frac{1}{2}\pi)$, where $\mathbf{w}_\alpha$ is the Jacobi field along v such that $\mathbf{w}_\alpha(0) = \mathbf{0}_{x_0}$ and $(\nabla_t \cdot \mathbf{w}_\alpha)(0) = -\sin\alpha \cdot \mathbf{h} + \cos\alpha \cdot \mathbf{h}_i$.
(c) Deduce from (b) and Problem 1 that $\lambda_i = -4$ if and only if $\mathbf{h}$ and $\mathbf{h}_i$ are linearly dependent over the field K. (To show that the condition is necessary, remark that $-\sin\alpha \cdot \mathbf{h} + \cos\alpha \cdot \mathbf{h}_i$ is an eigenvector of $R(\cos\alpha \cdot \mathbf{h} + \sin\alpha \cdot \mathbf{h}_i)$ (cf. (**20.21.2**)) and deduce that if $\lambda_i = -4$ we must have $f'_\alpha(\frac{1}{2}\pi, \alpha) = 0$ for all α.) Hence find the values of the λ_i in each of the three cases $K = \mathbf{R}$, $K = \mathbf{C}$, and $K = \mathbf{H}$.

4. Let M be a Riemannian manifold all of whose geodesics are *periodic with the same minimum period l* (when parametrized by arc length).

(a) Show that M is a complete manifold and that for all $x \in M$ we have $M = \exp_x(B'(\mathbf{0}_x; \frac{1}{2}l))$, and hence that M is compact.
(b) Show that two geodesic paths with origin x and length l (and therefore with endpoint x) are loop-homotopic in M. (If v_1, v_2 are the paths, defined on $[0, l]$, consider a path in the unit sphere with center $\mathbf{0}_x$, with endpoints $v_1'(0)$ and $v_2'(0)$.) Show that the fundamental group $\pi_1(M)$ has one or two elements. (Use (**20.19.1**) for the Riemannian universal covering of M.)

5. Let f be a C^∞ function defined on $]0, 1[$. Let $S \subset \mathbf{R}^3$ be the surface of revolution with axis $\mathbf{R}\mathbf{e}_3$, given by the equation

$$\xi^3 = f(((\xi^1)^2 + (\xi^2)^2)^{1/2})$$

(Section 20.18, Problem 13). The mapping $(r, \varphi) \mapsto (r \cos \varphi, r \sin \varphi, f(r))$ is a diffeomorphism of the open subset or $\mathbf{R}^2$ defined by $0 < r < 1$, $0 < \varphi < 2\pi$ onto a dense open subset U of S. Show that, relative to the corresponding chart of S with domain U, we have

$$\mathbf{g} = (1 + (f'(r))^2)\, dr \otimes dr + r^2\, d\varphi \otimes d\varphi.$$

Deduce that, for each geodesic in U, parametrized by arc-length with a suitable orientation, there exists a constant $a \geqq 0$ such that

$$r^2 \frac{d\varphi}{ds} = a$$

and

$$\left(\frac{d\varphi}{dr}\right)^2 = \frac{a^2}{r^2} \cdot \frac{1 + f'(r)^2}{r^2 - a^2},$$

implying that $r \geqq a$ at all points of this geodesic.

6. Let f_1, f_2 be two C^∞ functions defined on $[0, 1[$, such that $f_i(1) = 0, f_i'(0) = 0, f_i'(t) \leqq 0$ for $0 \leqq t < 1$ and $\lim_{t \to 1} f_i'(t) = -\infty$ $(i = 1, 2)$. Consider the two surfaces of revolution S_1, S_2 obtained by the procedure of Problem 5, by taking $f = f_1$ and $f = -f_2$, respectively.

(a) Let $\lambda(t)$ be a polynomial such that $\lambda(0) = 0$ and

$$(1 - t^2)^{-1/2} \pm \lambda(t) \geqq 1$$

for $0 \leqq t < 1$. Choose f_1 and f_2 so that

$$(1 + f_1'^2)^{1/2} = (1 - t^2)^{-1/2} + \lambda(t),$$
$$(1 + f_2'^2)^{1/2} = (1 - t^2)^{-1/2} - \lambda(t).$$

Show that the closure S of $S_1 \cup S_2$ in $\mathbf{R}^3$ is a compact *analytic* surface diffeomorphic to S_2 (*Zoll's surface*). (If $u = (1 - r^2)^{1/2}$ on S_1, $u = -(1 - r^2)^{1/2}$ on S_2, show that $du/d\xi^3 = F(u)$ in a neighborhood of $\xi^3 = 0$ in S_1 and in S_2, for the *same* analytic function F.)
(b) Show that all the geodesics of S are periodic and have the same minimum period. (If a is the minimum value of r on a geodesic, calculate the variation of φ as ξ^3 varies from $-f_2(a)$ to $f_1(a)$.)
(c) When $\lambda(t) = \frac{3}{2}t^4$, show that the total curvature of S takes opposite signs (cf. Section 20.14, Problem 7).

7. Let M be a complete connected Riemannian manifold, and let x be a point of M. Consider a geodesic $t \mapsto v(t) = \exp_x(t\mathbf{h})$, parametrized by arc length (where $\mathbf{h} \in T_x(M)$ and $\|\mathbf{h}\| = 1$).

(a) Show that the set I of numbers $s > 0$ such that $d(x, v(s)) = s$ is an interval, either $]0, +\infty[$ or $]0, r]$ with r finite. In the latter case, the point $v(r)$ is called the *cut-point* on the positive geodesic ray with origin x defined by $\mathbf{h}$.

(b) If there exists a cut-point $v(r)$ on v, show that one of the following two alternatives holds:

(α) $v(r)$ is the first point on v conjugate to x (18.7.12);

(β) There exists at least one geodesic $v_1 : t \mapsto \exp_x(t\mathbf{h}_1)$ through x with $\mathbf{h}_1 \neq \mathbf{h}$, $v_1(r) = v(r)$, and $d(x, v_1(r)) = r$.

(Let (a_n) be a decreasing sequence of numbers $> r$ and tending to r, and for each n let $t \mapsto \exp_x(t\mathbf{h}_n)$ be a geodesic parametrized by arc length and such that

$$d(x, \exp_x(b_n \mathbf{h}_n)) = d(x, \exp_x(a_n \mathbf{h})) < a_n.$$

We may assume that the sequence $(b_n \mathbf{h}_n)$ tends to a limit $\mathbf{k}$ in $T_x(M)$. Examine the two possibilities $\mathbf{k} \neq r\mathbf{h}$ and $\mathbf{k} = r\mathbf{h}$; in the second case, argue by contradiction, observing that $\exp_x$ cannot be a local diffeomorphism at the point $\mathbf{k}$.)

(c) Show that if $y = v(r)$ is the cut-point on the positive geodesic ray $t \mapsto v(t)$ with origin x, then x is the cut-point on the positive geodesic ray $t \mapsto v(r-t)$ with origin y. (Use (b).)

(d) Let S_x be the unit sphere $\|\mathbf{h}_x\| = 1$ in $T_x(M)$. For each $\mathbf{h}_x \in S_x$, let $\mu(\mathbf{h}_x)$ be the number r if $\exp_x(r\mathbf{h}_x)$ is the cut-point on the positive geodesic ray with origin x defined by $\mathbf{h}_x$, and let $\mu(\mathbf{h}_x) = +\infty$ if there is no cut-point on this geodesic ray. Show that the mapping μ of S_x into $[0, +\infty] \subset \bar{\mathbf{R}}$ is continuous. (Argue by contradiction, by considering a sequence $(\mathbf{h}_n)$ of points of S_x with limit $\mathbf{h}$ such that the sequence $(\mu(\mathbf{h}_n))$ tends to a limit $c \neq \mu(\mathbf{h})$; distinguish two cases according as $c < \mu(\mathbf{h})$ or $c > \mu(\mathbf{h})$. In the first case, observe that $\exp_x(c\mathbf{h})$ is not conjugate to x, and hence that $\exp_x$ is a diffeomorphism of a neighborhood of $c\mathbf{h}$ onto a neighborhood of $\exp_x(c\mathbf{h})$; now use (b) to obtain a contradiction. In the second case, we may assume that $\mu(\mathbf{h}_n) \geq \mu(\mathbf{h}) + b$ for some $b > 0$; observe that there exists a vector $\mathbf{h}' \neq \mathbf{h}$ such that $\exp_x((\mu(\mathbf{h}) + b)\mathbf{h}) = \exp_x((\mu(\mathbf{h}) + b')\mathbf{h}')$ for some $b' < b$. Consider the path obtained by juxtaposition of the path $t \mapsto \exp_x(t\mathbf{h}')$ for $0 \leq t \leq \mu(\mathbf{h}) + b'$ and a geodesic path from $\exp_x((\mu(\mathbf{h}) + b')\mathbf{h}')$ to $\exp_x((\mu(\mathbf{h}) + b)\mathbf{h}_n)$, of length equal to the distance between these two points, and hence arrive at a contradiction.)

(e) The set $P(x)$ of points $\exp_x(\mu(\mathbf{h}_x)\mathbf{h}_x)$, where $\mathbf{h}_x$ runs through the set of points of S_x such that $\mu(\mathbf{h}_x) < +\infty$, is called the *cut locus* of x. If E_x is the set of all $t\mathbf{h}_x \in T_x(M)$ such that $t < \mu(\mathbf{h}_x)$, show that E_x is homeomorphic to $T_x(M)$, that $\exp_x$ is a diffeomorphism of E_x onto an open subset $\exp_x(E_x)$ of M, and that M is the disjoint union of $\exp_x(E_x)$ and $P(x)$. (Use (a) and the Hopf–Rinow theorem.)

8. Determine the cut locus of a point when M is a projective space $\mathbf{P}_n(K)$ (Problem 1) or the flat torus $\mathbf{T}^2$.

9. (a) Let M be a complete connected Riemannian manifold and x_0 a point of M. Show that M is compact if and only if there is a cut-point on each geodesic ray with origin x_0. (Use Problem 7(d) to show that the condition is sufficient.)

(b) Let M be a complete connected Riemannian manifold whose universal covering is not compact. Show that for each point $x \in M$ there exists a geodesic ray $t \mapsto \exp_x(t\mathbf{h})$ $(t \geq 0)$ with origin x which contains no point conjugate to x. (Reduce to the case where M is simply-connected, and then use (a).) If M is simply-connected, we may assume that $d(x, \exp_x(t\mathbf{h})) = t$ along a geodesic ray.

10. Let M be a complete connected Riemannian manifold. For each $x \in M$ let $d(x)$ denote the radius of the largest open ball $B(\mathbf{0}_x; r)$ on which $\exp_x$ is injective.

(a) Show that $d(x)$ is also the radius of the largest open ball with center $\mathbf{0}_x$ contained in the set E_x (Problem 7(e)). (Observe that the proof of Problem 7(d) shows that the function μ is continuous, not only on each S_x, but on the submanifold U(M) of T(M) which is the union of the S_x.)
(b) Deduce from (a) that the set of $x \in M$ such that no geodesic through x contains a point conjugate to x is closed in M. (Reduce to the case where M is simply-connected, and then use (a) and (20.20.7).)
(c) Take M to be the surface given by the equation $\xi^3 = \frac{1}{2}((\xi^1)^2 + (\xi^2)^2)$ in $\mathbf{R}^3$. Show that the origin is the only point x of M such that no geodesic through x contains a point conjugate to x. (Use Problem 5.)

11. Let M be a complete pure Riemannian manifold, and let X be an infinitesimal isometry of M (Section 20.9, Problem 7). Let S be the set of points $x \in M$ such that $X(x) = 0$.

(a) The flow F_X of X has $M \times \mathbf{R}$ as domain (Section 20.6, Problem 8); if we put $\varphi_t(x) = F_X(x, t)$, then φ_t is an isometry of M onto itself for all $t \in \mathbf{R}$, and leaves fixed the points of S. For each $x \in S$, the mapping $t \mapsto T_x(\varphi_t)$ is a homomorphism of $\mathbf{R}$ into the orthogonal group relative to the scalar product $(\mathbf{u}|\mathbf{v})$ defined on $T_x(M)$ by the metric tensor of M, and hence $T_x(M)$ is the Hilbert sum of suspaces E_j $(1 \leq j \leq r)$ of dimension 2 and a subspace N of dimension $\dim(M) - 2r$, stable under $T_x(\varphi_t)$ for all $t \in \mathbf{R}$ (cf. (21.8.1)). Show that the geodesics through x all of whose points are fixed by the isometries φ_t are those whose tangent vector at x lies in N, and that the union of these geodesics is a totally geodesic submanifold of M.
(b) Deduce from (a) that the connected components of S are totally geodesic submanifolds of M. If V_1, V_2 are two distinct components of S, show that there exists an infinite number of distinct geodesic trajectories of length $d(x_1, x_2)$, with endpoints x_1 and x_2, for each point $x_1 \in V_1$ and each point $x_2 \in V_2$.
In particular, if for each point $x \in M$ and each plane $P_x \subset T_x(M)$ we have $A(P_x) \leq 0$ (20.22.1), then S must be connected.

21. SECTIONAL CURVATURE

(20.21.1) Let M be a Riemannian manifold, $\mathbf{K}$ its Riemann–Christoffel tensor (20.10.2). It follows immediately from the symmetry properties of $\mathbf{K}$ (20.10.3) that, for any two vectors $\mathbf{h}_x$, $\mathbf{k}_x$ in $T_x(M)$, the number

$$\langle \mathbf{K}(x), \mathbf{h}_x \otimes \mathbf{k}_x \otimes \mathbf{h}_x \otimes \mathbf{k}_x \rangle$$

depends only on the *bivector* $\mathbf{h}_x \wedge \mathbf{k}_x$, and is multiplied by λ^2 when this bivector is multiplied by a scalar λ. If $\mathbf{h}_x \wedge \mathbf{k}_x \neq 0$, i.e., if the two vectors $\mathbf{h}_x$, $\mathbf{k}_x$ are linearly independent, the number

(20.21.1.1) $$A(\mathbf{h}_x, \mathbf{k}_x) = \langle \mathbf{K}(x), \mathbf{h}_x \otimes \mathbf{k}_x \otimes \mathbf{h}_x \otimes \mathbf{k}_x \rangle / \|\mathbf{h}_x \wedge \mathbf{k}_x\|^2$$

therefore depends only on the *plane* P_x spanned by $\mathbf{h}_x$ and $\mathbf{k}_x$. This number is called the *sectional curvature* (or *Riemannian curvature*) of M for the plane

P_x, and is denoted by $A(P_x)$. We have seen in (20.10.3) that knowledge of $\mathbf{g}(x)$ and $A(P_x)$ for all planes P_x in $T_x(M)$ completely determines the Riemann–Christoffel tensor $\mathbf{K}(x)$. For an orthonormal moving frame $(\mathbf{e}_1, \dots, \mathbf{e}_n)$, with the notation of (20.10.4) we have

(20.21.1.2) $$A(\mathbf{e}_i(x), \mathbf{e}_j(x)) = K_{ijij}(x) = \langle \Omega_{ij}(x), \mathbf{e}_i(x) \wedge \mathbf{e}_j(x) \rangle$$

for $i \neq j$, since by definition $\|\mathbf{e}_i \wedge \mathbf{e}_j\| = 1$ (20.8.4.2). When $n = 2$, $T_x(M)$ is itself a plane, and $A(T_x(M))$ is the *Gaussian curvature* of M at the point x. The formulas (20.10.6.3) and (20.10.7.2) give the values of the Ricci curvature in the directions of the vectors $\mathbf{e}_i(x)$ in terms of the sectional curvature:

(20.21.1.3) $$\operatorname{Ric}(\mathbf{e}_i(x)) = \sum_{j \neq i} A(\mathbf{e}_i(x), \mathbf{e}_j(x))$$

and likewise for the scalar curvature we have

(20.21.1.4) $$S(x) = \sum_{i \neq j} A(\mathbf{e}_i(x), \mathbf{e}_j(x)).$$

Examples

(20.21.2) The sectional curvature of a flat Riemannian manifold is evidently zero, and conversely. The formula (20.7.10.6) giving the curvature of the canonical linear connection on a Riemannian symmetric space G/H also enables us to compute the sectional curvatures for such a space. In view of the invariance of the metric tensor under G, it is enough to perform the calculation at the point $x_0 = \pi(e)$; identifying $T_{x_0}(G/H)$ with $\mathfrak{m}$, we have then

(20.21.2.1) $$A(\mathbf{u}, \mathbf{v}) = ([[\mathbf{u}, \mathbf{v}], \mathbf{u}] \mid \mathbf{v}) / \|\mathbf{u} \wedge \mathbf{v}\|^2$$

for all $\mathbf{u}, \mathbf{v} \in \mathfrak{m}$. For example, consider the complex projective space $\mathbf{P}_n(\mathbf{C})$, identified with $\mathbf{SU}(n+1)/\mathbf{U}(n)$; take for $\mathbf{u}$ and $\mathbf{v}$ endomorphisms whose matrices are of the type (20.11.6.1); since $H = \mathbf{U}(n)$ acts transitively on the real lines in $\mathfrak{m}$, we may assume that $\mathbf{u} = a_{\mathbf{e}_1}$. If

$$\mathbf{z} = \begin{pmatrix} \xi_1 + i\eta_1 \\ \xi_2 + i\eta_2 \\ \vdots \\ \xi_n + i\eta_n \end{pmatrix},$$

then an elementary calculation with matrices (bearing in mind that $\mathfrak{m}$ is to be considered as a real vector space, so that $\|\mathbf{e}_1 \wedge (i\mathbf{e}_1)\| = 1$) gives the result

(20.21.2.2)
$$\mathrm{A}(a_{\mathbf{e}_1}, a_{\mathbf{z}}) = \frac{4\eta_1^2 + \sum_{j=2}^{n} (\xi_j^2 + \eta_j^2)}{\eta_1^2 + \sum_{j=2}^{n} (\xi_j^2 + \eta_j^2)}.$$

For the sphere $\mathbf{S}_n$, the imaginary parts η_j are replaced by zero, and we obtain the *constant* 1. (The fact that $\mathrm{A}(\mathbf{u}, \mathbf{v})$ is here independent of $\mathbf{u}$ and $\mathbf{v}$ could have been foreseen without calculation from the fact that in this case $\mathrm{H} = \mathbf{SO}(n)$ acts transitively on the *planes* in $\mathfrak{m}$.) Likewise, for hyperbolic space $\mathbf{Y}_n$ (20.11.7), we obtain for $\mathrm{A}(\mathbf{u}, \mathbf{v})$ the *constant* value -1.

Finally, for a *compact Lie group* G with center $\{e\}$, the formulas (20.11.8.1) and (20.11.8.3) give

(20.21.2.3)
$$\mathrm{A}(\mathbf{u}, \mathbf{v}) = \tfrac{1}{4}\|[\mathbf{u}, \mathbf{v}]\|^2 / \|\mathbf{u} \wedge \mathbf{v}\|^2$$

for all $\mathbf{u}$, $\mathbf{v}$ in $\mathfrak{g}_e$.

(20.21.3) Let M be a pure Riemannian manifold of dimension n, and let M′ be a pure submanifold of M of dimension n'. We wish to express the sectional curvature $\mathrm{A}'(\mathrm{P}_x)$ of M′ for a plane $\mathrm{P}_x \subset \mathrm{T}_x(\mathrm{M}')$ in terms of the sectional curvature $\mathrm{A}(\mathrm{P}_x)$ of M for P_x, and the second fundamental forms. We shall use the notation of Section 20.12. By a suitable choice of the moving frame $\boldsymbol{R}'$, we may suppose that P_x is spanned by the vectors $\mathbf{e}_i(x)$, $\mathbf{e}_j(x)$ $(i \neq j)$. It then follows from the formulas (20.21.1.2), (20.10.4.5), and (20.12.5.2) that

$$\begin{aligned}\mathrm{A}(\mathrm{P}_x) - \mathrm{A}'(\mathrm{P}_x) = -\sum_{\alpha} (&\langle \omega'_{\alpha i}(x), \mathbf{e}_i(x)\rangle\langle \omega'_{\alpha j}(x), \mathbf{e}_j(x)\rangle \\ &- \langle \omega'_{\alpha i}(x), \mathbf{e}_j(x)\rangle\langle \omega'_{\alpha j}(x), \mathbf{e}_i(x)\rangle)\end{aligned}$$

or

(20.21.3.1)
$$\mathrm{A}(\mathrm{P}_x) - \mathrm{A}'(\mathrm{P}_x) = -\sum_{\alpha} (l_{\alpha ii}(x) l_{\alpha jj}(x) - (l_{\alpha ij}(x))^2)$$

(*Gauss's formula*). From this formula we can derive a simple *geometrical interpretation* of the sectional curvature. Consider an open ball $\mathrm{B}(\mathbf{0}_x; r)$ contained in $\Omega \cap \mathrm{T}_x(\mathrm{M})$, and take M′ to be the *surface* $\exp_x(\mathrm{P}_x \cap \mathrm{B}(\mathbf{0}_x; r))$ which is the union of the *geodesic trajectories* passing through $x \in \mathrm{M}$ whose tangent vectors at x lie in P_x. This signifies that the values $\langle \boldsymbol{l}_\alpha(x), \mathbf{h}_x \otimes \mathbf{h}_x\rangle$

are zero for all $\mathbf{h}_x \in P_x$ (20.13.6.2), which is possible only if the restriction $(\mathbf{h}_x, \mathbf{k}_x) \mapsto \langle \mathbf{l}_\alpha(x), \mathbf{h}_x \otimes \mathbf{k}_x \rangle$ of the symmetric bilinear form $\mathbf{l}_\alpha(x)$ to P_x is *identically zero* for all α. The right-hand side of (20.21.3.1) therefore vanishes; bearing in mind (20.21.1.2), it follows that $A(P_x)$ is the *Gaussian curvature of the surface* M'.

The formula (20.21.3.1) also has as a consequence the following proposition:

(20.21.4) *Let* M' *be a submanifold of* M *and suppose that* M' *contains a geodesic trajectory* C *of* M. *Then for each* $x \in C$ *and each plane* $P_x \subset T_x(M)$ *containing the tangent vector to* C *at* x *we have*

(20.21.4.1) $$A'(P_x) \leqq A(P_x)$$

(*Synge's lemma*).

We may assume that the frame $\boldsymbol{R}'$ has been chosen such that $\mathbf{e}_1(x)$ is tangent to C and that P_x is spanned by $\mathbf{e}_1(x)$ and $\mathbf{e}_2(x)$. Then we have $l_{\alpha 11}(x) = 0$ for all α (20.13.6.2), and consequently

$$A(P_x) - A'(P_x) = \sum_\alpha (l_{\alpha 12}(x))^2$$

giving the result.

22. MANIFOLDS WITH POSITIVE SECTIONAL CURVATURE OR NEGATIVE SECTIONAL CURVATURE

(20.22.1) *Let* M *be a Riemannian manifold and* v *a geodesic in* M. *Suppose that at each point* x *of this geodesic and each plane* $P_x \subset T_x(M)$ *we have* $A(P_x) \leqq 0$. *Then there exists no pair of conjugate points on* v.

Let $\mathbf{w}$ be a Jacobi field along v, not identically zero. Then for each t in the interval of definition I of v, we have (18.7.5)

$$\nabla_t \cdot (\nabla_t \cdot \mathbf{w}) = (\boldsymbol{r} \cdot (v' \wedge \mathbf{w})) \cdot v'.$$

We shall see that $\mathbf{w}$ cannot vanish at two distinct points of I. First let us show that the function $t \mapsto (\mathbf{w}(t) | \nabla_t \cdot \mathbf{w})$ is *increasing*. By (20.9.5.4), we have

(20.22.1.1) $$\frac{d}{dt}(\mathbf{w} | \nabla_t \cdot \mathbf{w}) = \|\nabla_t \cdot \mathbf{w}\|^2 + (\mathbf{w} | (\boldsymbol{r} \cdot (v' \wedge \mathbf{w})) \cdot v')$$
$$= \|\nabla_t \cdot \mathbf{w}\|^2 - A(v', \mathbf{w}) \|v' \wedge \mathbf{w}\|^2 \geqq 0$$

by virtue of the hypothesis, which proves our assertion. Moreover, if

$$(\mathbf{w} \mid \nabla_t \cdot \mathbf{w})$$

were to vanish on an interval $J \subset I$ of length >0, then we should have $\nabla_t \cdot \mathbf{w} = 0$ on this interval by the formula above, and it follows immediately from (18.7.6) that if $\mathbf{w}$ vanished at a point in J, then $\mathbf{w}$ would vanish identically on I. This proves the proposition.

(20.22.2) (Hadamard–Cartan theorem) *Let* M *be a complete connected Riemannian manifold of dimension n, such that* $A(P_x) \leqq 0$ *for each* $x \in M$ *and each plane* $P_x \subset T_x(M)$. *Then the universal covering of* M *is diffeomorphic to* $\mathbf{R}^n$, *and if* M *is simply-connected then* M *is strictly geodesically convex.*

This follows immediately from (20.22.1) and (20.20.7).

(20.22.3) (Myers's theorem) *Let* M *be a complete connected Riemannian manifold of dimension* $n \geqq 2$. *If there exists a number* $c > 0$ *such that the Ricci curvature of* M (20.10.7) *satisfies the inequality* $\operatorname{Ric}(\mathbf{h}_x) \geqq c\|\mathbf{h}_x\|^2$ *for all* $\mathbf{h}_x \in T(M)$, *then* M *is compact, the diameter of* M *satisfies the inequality*

(20.22.3.1)
$$\delta(M) \leqq \pi \sqrt{\frac{n-1}{c}},$$

and the fundamental group $\pi_1(M)$ *is finite.*

Let a, b be two distinct points of M, and put $l = d(a, b)$. Since M is complete, there exists a geodesic path $v : t \mapsto \exp_a(t\mathbf{h}_a)$ defined on $[0, 1]$, with origin a and endpoint b, such that $\|\mathbf{h}_a\| = l$ (20.18.5). Let $(\mathbf{e}_i)_{1 \leqq i \leqq n}$ be an orthonormal basis of $T_a(M)$ such that $\mathbf{h}_a = l\mathbf{e}_n$, and let $\mathbf{u}_j$ denote the parallel transport of $\mathbf{e}_j$ along v (18.6.4), so that $l\mathbf{u}_n = v'$ and $(\mathbf{u}_j \mid \mathbf{u}_k) = \delta_{jk}$ for all pairs of indices j, k (20.9.5.4). Put $\mathbf{w}_j(t) = \mathbf{u}_j(t) \sin \pi t$ for $1 \leqq j \leqq n$ and $0 \leqq t \leqq 1$, and let us calculate the index form $I(\mathbf{w}_j, \mathbf{w}_j)$ (20.20.2.3). We have

$$\nabla_t \cdot \mathbf{w}_j = (\nabla_t \cdot \mathbf{u}_j) \sin \pi t + \pi \mathbf{u}_j(t) \cos \pi t = \pi \mathbf{u}_j(t) \cos \pi t$$

by the definition of a parallel transport; since $\|\mathbf{u}_j(t)\| = 1$, we obtain

(20.22.3.2)
$$I(\mathbf{w}_j, \mathbf{w}_j) = \int_0^1 (\pi^2 + ((\mathbf{r} \cdot (v' \wedge \mathbf{u}_j)) \cdot v' \mid \mathbf{u}_j)) \sin^2 \pi t \, dt.$$

By definition (20.10.7.2), we have

$$\operatorname{Ric}(v') = -\sum_{j=1}^{n} ((\mathbf{r}' \cdot (v' \wedge \mathbf{u}_j)) \cdot v' \mid \mathbf{u}_j) = -\sum_{j=1}^{n-1} ((\mathbf{r} \cdot (v' \wedge \mathbf{u}_j)) \cdot v' \mid \mathbf{u}_j)$$

since $l\mathbf{u}_n = v'$; hence we obtain from (20.22.3.2) the formula

$$\sum_{j=1}^{n-1} \mathrm{I}(\mathbf{w}_j, \mathbf{w}_j) = \int_0^1 ((n-1)\pi^2 - \mathrm{Ric}(v'(t))) \sin^2 \pi t \, dt$$

and consequently the inequality

$$\sum_{j=1}^{n-1} \mathrm{I}(\mathbf{w}_j, \mathbf{w}_j) \leqq \int_0^1 ((n-1)\pi^2 - cl^2) \sin^2 \pi t \, dt. \tag{20.22.3.3}$$

This being so, if $l^2 > (n-1)\pi^2/c$, at least one of the numbers

$$\mathrm{I}(\mathbf{w}_j, \mathbf{w}_j) \qquad (1 \leqq j \leqq n-1)$$

would be negative, and it would follow from (20.20.5.1) that $d(a, b) < l$, which is absurd. This establishes the inequality (20.22.3.1), and since M is complete it follows that M, being equal to a closed ball, is compact (20.18.5). Now let $\tilde{\mathrm{M}}$ be the universal covering of M, endowed with the Riemannian metric canonically induced by that of M (20.8.1). Clearly $\tilde{\mathrm{M}}$ satisfies the same hypotheses as M (20.18.7), hence $\tilde{\mathrm{M}}$ is compact. Every fiber in $\tilde{\mathrm{M}}$ over a point of M is therefore compact and discrete, hence finite (3.16.3); but by (16.28.3) the fibers of the covering $\tilde{\mathrm{M}}$ are each in one-to-one correspondence with $\pi_1(\mathrm{M})$, and therefore $\pi_1(\mathrm{M})$ is finite.

(20.22.4) *Let* M *be a complete connected Riemannian manifold of dimension* $n \geqq 2$. *If there exists* $c > 0$ *such that* $\mathrm{A}(\mathrm{P}_x) \geqq c$ *for all* $x \in \mathrm{M}$ *and all planes* $\mathrm{P}_x \subset \mathrm{T}_x(\mathrm{M})$, *then* M *is compact, its diameter satisfies the inequality*

$$\delta(\mathrm{M}) \leqq \frac{\pi}{c^{1/2}}, \tag{20.22.4.1}$$

and the fundamental group $\pi_1(\mathrm{M})$ *is finite.*

For it follows from (20.21.1.3) that $\mathrm{Ric}(\mathbf{h}_x) \geqq (n-1)c\|\mathbf{h}_x\|^2$ for all $\mathbf{h}_x \in \mathrm{T}(\mathrm{M})$, and the result therefore follows from (20.22.3).

(20.22.5) (H. Weyl's theorem) *If* G *is a compact connected Lie group with discrete* (or equivalently (3.16.3) *finite*) *center, then the universal covering of* G *is compact.*

We shall begin by reduction to the case where the center of G is $\{e\}$. For this purpose we need the following topological lemma:

(20.22.5.1) *Let* G *be a connected Hausdorff topological group,* D *a discrete subgroup of the center* Z *of* G. *Then the center of* G/D *is* Z/D.

Let $\pi : G \to G/D$ be the canonical mapping. If $s \in G$ is such that $\pi(s)$ is in the center of G/D, we have $\pi(sxs^{-1}x^{-1}) = \pi(e)$ for all $x \in G$, so that $sxs^{-1}x^{-1} \in D$. Now the mapping $x \mapsto sxs^{-1}x^{-1}$ is continuous, and since G is connected and D is discrete, it follows that $sxs^{-1}x^{-1}$ takes the same value for all $x \in G$ (3.19.7). Taking $x = e$, we see that $sx = xs$ for all $x \in G$, and therefore $s \in Z$.

This lemma being established, let $\tilde{G}$ be the universal covering of G, and let Z_0 be the center of $\tilde{G}$, wihch is *discrete* because $\tilde{G}$ is locally isomorphic with G, so that its Lie algebra has a trivial center (19.11.7). The group G is isomorphic to $\tilde{G}/D$, where D is a subgroup of Z_0 (16.30.2); by virtue of (20.22.5.1), the center Z of G is Z_0/D, and G/Z is isomorphic to $\tilde{G}/Z_0$ (16.10.8). Since G is compact and connected, the same is true of G/Z, and the center of G/Z is trivial by virtue of (20.22.5.1); but $\tilde{G}$ is the universal covering of $\tilde{G}/Z_0$, and we may therefore assume henceforth that the center of G is trivial.

The group G being endowed with a Riemannian structure as in (20.11.8), if $(\mathbf{e}_i)_{1 \leqq i \leqq n}$ is an orthonormal basis of $\mathfrak{g}_e$, it cannot be the case that $[\mathbf{e}_i, \mathbf{e}_j] = 0$ for all $j \neq i$, since this would imply that $\mathbf{e}_i$ belonged to the center of $\mathfrak{g}_e$, which by hypothesis is trivial (19.11.7). The formulas (20.21.1.3) and (20.21.2.3) therefore show that $\text{Ric}(\mathbf{e}_i) > 0$; but since $\mathbf{e}_i$ can be any unit vector in $\mathfrak{g}_e$, and since the unit sphere in $\mathfrak{g}_e$ is compact and the function $\mathbf{u} \mapsto \text{Ric}(\mathbf{u})$ is continuous on this sphere, we see (3.17.10) that there exists $c > 0$ such that $\text{Ric}(\mathbf{h}_x) \geqq c\|\mathbf{h}_x\|^2$ for all $\mathbf{h}_x \in T(G)$, having regard to the fact that the Riemannian metric is translation-invariant. The result now follows from (20.22.3).

(20.22.6) (Synge's theorem) *Let* M *be an orientable, compact, connected Riemannian manifold of even dimension. If* $A(P_x) > 0$ *for all* $x \in M$ *and all planes* $P_x \subset T_x(M)$, *then* M *is simply-connected.*

Suppose that $\pi_1(M)$ is not the trivial group. Then there exists (20.19.3) a periodic geodesic $v : t \mapsto \exp(t\mathbf{h}_a)$ of period 1, so that $v(0) = v(1) = a$ and $v'(0) = v'(1) = \mathbf{h}_a$; moreover, $l = \|\mathbf{h}_a\|$ is the *smallest* of the lengths of loops homotopic to v. As in the proof of (20.22.3), we define a sequence $(\boldsymbol{u}_j)_{1 \leqq j \leqq n}$ of parallel transports along v, such that $v' = l\boldsymbol{u}_n$ and $(\boldsymbol{u}_j | \boldsymbol{u}_k) = \delta_{jk}$. Since $v(1) = v(0) = a$, the sequence $(\boldsymbol{u}_j(1))_{1 \leqq j \leqq n}$ is an orthonormal basis of $T_a(M)$, and therefore there exists an orthogonal transformation S of $T_a(M)$ such that $\boldsymbol{u}_j(1) = S \cdot \boldsymbol{u}_j(0)$ for $1 \leqq j \leqq n$. We assert that S is a *rotation*. To prove this, it is enough to show that the n-vectors $\boldsymbol{u}_1(0) \wedge \cdots \wedge \boldsymbol{u}_n(0)$ and

$$\boldsymbol{u}_1(1) \wedge \cdots \wedge \boldsymbol{u}_n(1)$$

are equal. Now if υ is a volume-form on M, the function

$$t \mapsto \langle \upsilon(v(t)), \mathbf{u}_1(t) \wedge \cdots \wedge \mathbf{u}_n(t) \rangle$$

is continuous on [0, 1], and never zero because $\upsilon(x) \neq 0$ for all $x \in \mathrm{M}$ and the $\mathbf{u}_j(t)$ form a basis of $\mathrm{T}_{v(t)}(\mathrm{M})$ for all $t \in [0, 1]$; consequently the numbers $\langle \upsilon(a), \mathbf{u}_1(0) \wedge \cdots \wedge \mathbf{u}_n(0) \rangle$ and $\langle \upsilon(a), \mathbf{u}_1(1) \wedge \cdots \wedge \mathbf{u}_n(1) \rangle$ have the same sign and are therefore equal. This proves our assertion.

We remark next that $\mathbf{u}_n(1) = \mathbf{u}_n(0)$, and hence the rotation S stabilizes the hyperplane H in $\mathrm{T}_a(\mathrm{M})$ spanned by $\mathbf{u}_1(0), \ldots, \mathbf{u}_{n-1}(0)$. The restriction S' of S to H is therefore again a rotation, and since H is *odd*-dimensional, there exists a vector $\mathbf{k}_a \neq 0$ in H which is *invariant under* S. The parallel transport $\mathbf{w}$ along v such that $\mathbf{w}(0) = \mathbf{k}_a$ therefore satisfies $\mathbf{w}(1) = \mathbf{k}_a$. Now put $f(t, \xi) = \exp_{v(t)}(\xi \mathbf{w}(t))$, as in (20.20.5.1), so that $f'_\xi(t, 0) = \mathbf{w}(t)$ and $(\nabla_\xi \cdot f'_\xi)(t, \xi) = 0$. With the notation of (20.20.2) we have again $\mathrm{E}'(0) = 0$ by virtue of the periodicity of $\mathbf{w}$, and $\frac{1}{2}\mathrm{E}''(0) = \mathrm{I}(\mathbf{w}, \mathbf{w})$; but because $\mathbf{w}$ is a parallel transport, $\nabla_t \cdot \mathbf{w} = 0$, and hence

$$\mathrm{I}(\mathbf{w}, \mathbf{w}) = -\, \|v' \wedge \mathbf{w}\|^2 \int_0^1 \mathrm{A}(v', \mathbf{w})\, dt,$$

because $\|v' \wedge \mathbf{w}\|$ is constant along v (20.9.5.4). By construction, we have $v' \wedge \mathbf{w} \neq 0$; hence $\mathrm{A}(v', \mathbf{w})$ is a continuous function which by hypothesis is >0 at all points of [0, 1], and therefore $\mathrm{I}(\mathbf{w}, \mathbf{w}) < 0$ (8.5.3). Taylor's formula now shows that for sufficiently small $\xi_0 > 0$, the length of the path $t \mapsto f(t, \xi_0)$ is $< l$; but it is clear that $(t, \xi) \mapsto f(t, \xi)$ $(0 \leqq t \leqq 1,\ 0 \leqq \xi \leqq \xi_0)$ is a loop-homotopy of v to $f(.,\, \xi_0)$. We have therefore obtained a contradiction and the proof is complete.

The example of an odd-dimensional real projective space $\mathbf{P}_{2n+1}(\mathbf{R})$, which is orientable (16.21.11), for which $\mathrm{A}(\mathrm{P}_x) = 1$ (20.21.2), but which is not simply-connected, shows that the hypothesis of *even*-dimensionality in (20.22.6) cannot be dispensed with.

PROBLEMS

1. Let M be a Riemannian manifold, x a point of M, $\mathrm{B}(\mathbf{0}_x; r_0)$ an open ball contained in $\Omega \cap \mathrm{T}_x(\mathrm{M})$, and P_x a plane in $\mathrm{T}_x(\mathrm{M})$. For $0 < r < r_0$, let $\mathrm{C}(\mathrm{P}_x; r)$ be the image under $\exp_x$ of the circle $\|\mathbf{h}_x\| = r$ in P_x. Show that

$$\mathrm{L}(\mathrm{C}(\mathrm{P}_x; r)) = 2\pi r - \tfrac{1}{3}\pi r^3 \mathrm{A}(\mathrm{P}_x) + o_3(r),$$

where $o_3(r)/r^3 \to 0$ as $r \to 0$. (If $\mathbf{h}_1$, $\mathbf{h}_2$ form an orthonormal basis of P_x, consider the family of geodesics

$$f(t, \alpha) = \exp_x(t(\cos \alpha \cdot \mathbf{h}_1 + \sin \alpha \cdot \mathbf{h}_2)).$$

Argue as in Section 20.21, Problem 3(b), and use Section 20.16, Section 7.)

2. Let M be a Riemannian manifold of dimension $n \geqq 2$, such that $\mathrm{Ric}(\mathbf{h}_x) \geqq c\|\mathbf{h}_x\|^2$ for all $\mathbf{h}_x \in T(M)$, where c is a constant >0. Show that on every geodesic, the length of an arc not containing conjugate points is $\leqq \pi((n-1)/c)^{1/2}$. (Argue as in (20.22.3).)

3. Let M_1 and M_2 be two Riemannian manifolds of dimension n, and let v_1 (resp. v_2) be a geodesic in M_1 (resp. M_2) whose interval of definition contains $[a, b]$. Let $\mathbf{w}_1$ (resp. $\mathbf{w}_2$) be a Jacobi field along v_1 (resp. v_2) orthogonal to v_1' (resp. v_2'). Assume that: (1) $\mathbf{w}_1(a) = \mathbf{w}_2(a) = \mathbf{0}$; (2) $\|(\nabla_t \cdot \mathbf{w}_1)(a)\| = \|(\nabla_t \cdot \mathbf{w}_2)(a)\|$; (3) $v_i(t)$ is not conjugate to $v_i(a)$ for $t \in [a, b]$ and $i = 1, 2$; (4) for all $t \in [a, b]$ we have $A(P_{v_1(t)}) \geqq A(Q_{v_2(t)})$ for all planes $P_{v_1(t)}$ in $T_{v_1(t)}(M_1)$ *containing* $v_1'(t)$, and all planes $Q_{v_2(t)}$ in $T_{v_2(t)}(M_2)$ *containing* $v_2'(t)$. Then $\|\mathbf{w}_1(t)\| \leqq \|\mathbf{w}_2(t)\|$ for $a \leqq t \leqq b$ (*Rauch's comparison theorem*). (Put $u_1(t) = \|\mathbf{w}_1(t)\|^2$, $u_2(t) = \|\mathbf{w}_2(t)\|^2$, and $f_i(t) = I_{a,t}(\mathbf{w}_i, \mathbf{w}_i)/u_i(t)$ (notation of 20.20.5) for $i = 1, 2$. Using the formula (20.20.3.7), show that

$$u_1(t)/u_2(t) = \lim_{\varepsilon \to 0} \exp\left(2\int_{a+\varepsilon}^{t} (f_1(s) - f_2(s))\, ds\right)$$

and hence that it is enough to prove that $f_1(t) \leqq f_2(t)$ for $t \in [a, b]$. Fix $t_0 \in [a, b]$ and put $\mathbf{z}_i(t) = \mathbf{w}_i(t)/\|\mathbf{w}_i(t_0)\|$ for $i = 1, 2$. Show that there exists a lifting $\mathbf{s}$ of v_1 such that $\|\mathbf{s}(t)\| = \|\mathbf{z}_2(t)\|$ and $\|(\nabla_t \cdot \mathbf{s})(t)\| = \|(\nabla_t \cdot \mathbf{z}_2)(t)\|$ for $t \in [a, b]$ (use parallel transports from t_0 to t along v_1 and v_2). Then use (20.20.3) to show that

$$I_{a,t_0}(\mathbf{z}_1, \mathbf{z}_1) \leqq I_{a,t_0}(\mathbf{s}, \mathbf{s}) \leqq I_{a,t_0}(\mathbf{z}_2, \mathbf{z}_2),$$

and deduce that $f_1(t_0) \leqq f_2(t_0)$.)

4. With the notation of Problem 3, suppose that $v_1(t)$ is not conjugate to $v_1(a)$ for $a < t \leqq b$, and that $A(P_{v_1(t)}) \geqq A(Q_{v_2(t)})$ for $a \leqq t \leqq b$, where $P_{v_1(t)}$ is any plane containing $v_1'(t)$, and $Q_{v_2(t)}$ any plane containing $v_2'(t)$. Then $v_2(t)$ is not conjugate to $v_2(a)$ for $a < t \leqq b$. (Argue by contradiction, using Problem 3.)

5. Let M be a complete Riemannian manifold such that $A(P_x) \leqq 0$ for all $x \in M$ and all planes $P_x \subset T_x(M)$. With the notation of (20.16.3.1), show that

$$\|T_{\mathbf{h}_x}(\exp_x) \cdot \mathbf{k}_x'\| \geqq \|\mathbf{k}_x\|.$$

(Use Section 20.16, Problem 7.) Deduce that if M is simply-connected, we have

$$d(\exp_x(\mathbf{h}), \exp_x(\mathbf{k})) \geqq \|\mathbf{k} - \mathbf{h}\|$$

for any two vectors $\mathbf{h}$, $\mathbf{k}$ in $T_x(M)$.

6. (a) Let M be a complete, simply-connected Riemannian manifold, such that $A(P_x) \leqq 0$ for all $x \in M$ and all planes $P_x \subset T_x(M)$. Let Z be a compact metric space

and let μ be a positive measure $\neq 0$ on Z; finally, let $f\colon Z \to M$ be a continuous mapping. For each $x \in M$, put

$$h(x) = \int_Z d(x, f(z))^2 \, d\mu(z),$$

where d is the Riemannian distance. Show that h attains its minimum at exactly one point of M. (To show the existence of such a point, observe that $f(Z)$ is contained in a closed ball in M and that such a ball is compact. To show uniqueness, consider a point x_0 at which h attains its minimum. Since $\exp_{x_0}$ is a diffeomorphism of $T_{x_0}(M)$ onto M (20.22.2), we may write $f = \exp_{x_0} \circ \mathbf{f}_0$, where $\mathbf{f}_0$ is a continuous mapping of Z into $T_{x_0}(M)$. If we put

$$h_0(\mathbf{u}) = \int_Z \|\mathbf{u} - \mathbf{f}_0(z)\|^2 \, d\mu(z)$$

for $\mathbf{u} \in T_{x_0}(M)$, show that $h(x_0) = h_0(\mathbf{0}) < h_0(\mathbf{u}) \leq h(\exp_{x_0}(\mathbf{u}))$ for all $\mathbf{u} \neq 0$ in $T_{x_0}(M)$, by using Problem 5.)

(b) With the same assumptions on M, let G be a compact subgroup of the group of isometries I(M). Show that G has a fixed point in M. (Apply (a) to the function $s \mapsto s \cdot x_0$ on G, with μ a Haar measure on G.)

(c) With the same assumptions on M, let G be a closed subgroup of I(M) which acts transitively on M, so that M may be identified with G/K, where K is the stabilizer of a point of M. The subgroup K is compact (Section **20.16**, Problem 11). Show that every compact subgroup of G is conjugate in G to a subgroup of K (use (b)).

7. Let M be a complete connected Riemannian manifold, such that $A(P_x) \leq 0$ for all $x \in M$ and all planes $P_x \subset T_x(M)$. Show that every element (other than the identity element) of the fundamental group $\pi_1(M)$ has infinite order. (Observe that if $\tilde{M}$ is the Riemannian universal covering of M, the elements of $\pi_1(M)$ may be identified with isometries of $\tilde{M}$, and use Problem 6(b).)

8. Let M be a compact Riemannian manifold such that $A(P_x) \geq 0$ for all $x \in M$ and all planes $P_x \subset T_x(M)$.

(a) Show that if M is even-dimensional and nonorientable, then $\pi_1(M)$ is of order 2.
(b) Show that if M is odd-dimensional, then M is orientable. (Argue by contradiction, as in (**20.22.6**).)

9. Let M be a compact submanifold of $\mathbf{R}^n$. Suppose that for each $x \in M$ there exists in $T_x(M)$ a vector subspace E_x of dimension m such that $A(P_x) \leq 0$ for all planes $P_x \subset E_x$. Show that $n \geq m + \dim(M)$. In particular, if $A(P_x) \leq 0$ for all $x \in M$ and all planes $P_x \subset T_x(M)$, then $n \geq 2 \dim(M)$. (Observe that there exists at least one point $x \in M$ such that $\boldsymbol{l}(x) \cdot (\mathbf{h}_x \otimes \mathbf{h}_x) \neq 0$ for all $\mathbf{h}_x \neq 0$ in $T_x(M)$ (Section **20.14**, Problem 3(c)). On the other hand, deduce from Problem 5(a) of Section **20.14** that

$$(\boldsymbol{l}(x) \cdot (\mathbf{h}_x \otimes \mathbf{h}_x) \mid \boldsymbol{l}(x) \cdot (\mathbf{k}_x \otimes \mathbf{k}_x)) \leq \| \boldsymbol{l}(x) \cdot (\mathbf{h}_x \otimes \mathbf{k}_x) \|^2$$

for all $\mathbf{h}_x, \mathbf{k}_x \in E_x$, and use the algebraic lemma of Section **20.14**, Problem 5(c) to obtain a contradiction.)

10. Let M be a Riemannian manifold such that $A(P_x) \leqq 0$ for all $x \in M$ and all planes $P_x \subset T_x(M)$. Let U be a strictly geodesically convex open set in M. For each pair of points x, y in U and each $t \in [0, 1]$, let $u(x, y, t)$ be the point on the unique geodesic arc with endpoints x, y contained in U, such that $d(x, u(x, y, t)) = t \cdot d(x, y)$. Show that for each $a \in U$ we have

$$d(u(a, x, t), u(a, y, t)) \leqq t \cdot d(x, y).$$

(Let $l = d(x, y)$. For $0 \leqq \xi \leqq l$, consider the function $f(t, \xi) = u(a, u(x, y, \xi/l), t)$, and observe that $t \mapsto f(t, \xi)$ is a geodesic. Use Section 20.16, Problem 7, to show that $\|f'_\xi(t, \xi)\| \leqq t \|f'_\xi(1, \xi)\|$.)

11. With the hypotheses and notation of Section 20.16, Problem 3, consider a basis of $T_x(M)$ consisting of $\mathbf{h}$ and any $n-1$ vectors $\mathbf{k}_2, \ldots, \mathbf{k}_n$ orthogonal to $\mathbf{h}$, and let $\mathbf{z}_j$ denote the Jacobi field along $v : t \mapsto \exp_x(t\mathbf{h})$ such that $\mathbf{z}_j(0) = 0$ and $(\nabla_t \cdot \mathbf{z}_j)(0) = \mathbf{k}_j$ $(2 \leqq j \leqq n)$. If we put $f(t) = F(v(t))$, then we have

$$f(t) = \|\mathbf{z}_2(t) \wedge \cdots \wedge \mathbf{z}_n(t)\| / ct^{n-1}$$

for $0 \leqq t < r$, where $c = \|\mathbf{k}_2 \wedge \cdots \wedge \mathbf{k}_n\|$.

(a) Suppose that the $\mathbf{k}_j$ have been chosen so that when $t = r_1$, the $\mathbf{z}_j(t)$ and $v'(t)$ form an orthonormal basis of $T_{v(t)}(M)$. Show that we have

$$\begin{aligned}\frac{f'(r_1)}{f(r_1)} &= \sum_{j=2}^{n} (\mathbf{z}_j(r_1) \mid (\nabla_t \cdot \mathbf{z}_j)(r_1)) - \frac{n-1}{r_1}\\ &= -\frac{n-1}{r_1} + \sum_{j=2}^{n} I(\mathbf{z}_j, \mathbf{z}_j),\end{aligned}$$

where the index form is calculated for liftings of v in $[0, r_1]$. (Use (20.22.1.1).)

(b) Suppose that the Ricci curvature of M satisfies the inequality

$$\operatorname{Ric}(\mathbf{h}_x) \geqq (n-1)a^2 \|\mathbf{h}_x\|^2$$

for all $\mathbf{h}_x \in T(M)$. Show that if g is a piecewise-C^1 function defined on $[0, r_1]$, such that $g(0) = 0$ and $g(r_1) = 1$, then

$$\frac{f'(r_1)}{f(r_1)} \leqq -\frac{n-1}{r_1} + (n-1)\int_0^{r_1} ((g'(t))^2 - a^2(g(t))^2)\, dt.$$

(Consider the parallel transport of $\mathbf{z}_j(r_1)$ along v for $2 \leqq j \leqq n$ and use (20.20.3.9).)

(c) Deduce from (b) that the function $f(t)(at/\sin at)^{n-1}$ is decreasing in $[0, r[$. (Choose the function g suitably in (b).) Deduce that the function $S(x, t)(a/\sin at)^{n-1}$ is decreasing in $[0, r[$.

(d) Suppose that the sectional curvature of M satisfies $A(P_x) \leqq b^2$ for all $x \in M$ and all planes $P_x \subset T_x(M)$. Show that the function $f(t)(bt/\sin bt)^{n-1}$ is increasing in $[0, r[$. (Follow the proof of Rauch's comparison theorem (Problem 3), by taking as comparison manifold a sphere of radius $1/b$.) Deduce that the function $S(x, t)(b/\sin bt)^{n-1}$ is increasing in $[0, r[$.

23. RIEMANNIAN MANIFOLDS OF CONSTANT CURVATURE

(20.23.1) *Let* M *be a Riemannian manifold of dimension* $n \geqq 3$. *At a point* $x \in \mathrm{M}$, *the following conditions are equivalent*:

(a) *The sectional curvature* $\mathrm{A}(\mathrm{P}_x)$ *is the same for all planes* $\mathrm{P}_x \subset \mathrm{T}_x(\mathrm{M})$.

(b) *Relative to an orthonormal moving frame* $(\mathbf{e}_1, \ldots, \mathbf{e}_n)$ *defined in a neighborhood of* x, *the curvature forms satisfy the relations*

$$\Omega_{ij}(x) = \mathrm{A}(x)\sigma_i(x) \wedge \sigma_j(x) \qquad (1 \leqq i, j \leqq n), \tag{20.23.1.1}$$

where $\mathrm{A}(x)$ *is a constant independent of* i, j.

When this condition is satisfied, the relations (20.23.1.1) *are true for all orthonormal moving frames, with the same constant* $\mathrm{A}(x)$.

First of all, it is immediately verified that the relations (20.23.1.1) signify that the vector-valued 2-form $\mathbf{\Omega}^{(\mathbf{R})}(x)$, which is a mapping of $\bigwedge^2 \mathrm{T}_x(\mathrm{M})$ into $\mathrm{End}(\mathrm{T}_x(\mathrm{M}))$, identified with $\mathrm{T}_x(\mathrm{M})^* \otimes \mathrm{T}_x(\mathrm{M})$, is the linear mapping defined by

$$\mathbf{\Omega}^{(\mathbf{R})}(x) \cdot (\mathbf{h}_x \wedge \mathbf{k}_x) = \mathrm{A}(x)((G_x \cdot \mathbf{h}_x) \otimes \mathbf{k}_x - (G_x \cdot \mathbf{k}_x) \otimes \mathbf{h}_x),$$

where the linear mapping $G_x : \mathrm{T}_x(\mathrm{M}) \to \mathrm{T}_x(\mathrm{M})^*$ is that which is canonically defined by $\mathbf{g}(x)$ (20.8.3). Since this mapping does not depend on a choice of basis in $\mathrm{T}_x(\mathrm{M})$, the last assertion of the proposition is proved. This also shows that (b) implies (a), since we may always choose a frame $\boldsymbol{R}$ in which the vectors $\mathbf{e}_i(x)$ and $\mathbf{e}_j(x)$ span a given plane P_x. Let us show conversely that (a) implies (b). If $\mathbf{h}_x = \sum_i \xi^i \mathbf{e}_i(x)$, $\mathbf{k}_x = \sum_i \eta^i \mathbf{e}_i(x)$, we have

$$\langle \mathbf{K}(x), \mathbf{h}_x \otimes \mathbf{k}_x \otimes \mathbf{h}_x \otimes \mathbf{k}_x \rangle = \sum_{h,i,j,k} \mathrm{K}_{hijk}(x)\xi^h \eta^i \xi^j \eta^k,$$

and this is equal to

$$\mathrm{A}(x)\|\mathbf{h}_x \wedge \mathbf{k}_x\|^2 = \mathrm{A}(x) \sum_{i<j} ((\xi^i)^2(\eta^j)^2 + (\xi^j)^2(\eta^i)^2 - 2\xi^i\xi^j\eta^i\eta^j);$$

hence it is clear that the numbers $\mathrm{K}_{hijk}(x)$ are zero except for the $\mathrm{K}_{ijij}(x) = -\mathrm{K}_{ijji}(x)$ $(i \neq j)$, which are all equal to $\mathrm{A}(x)$. In view of (20.10.4.5) and the fact that the $\sigma_i(x) \wedge \sigma_j(x)$ form the basis of $\bigwedge^2 \mathrm{T}_x(\mathrm{M})^*$ dual to the basis $(\mathbf{e}_i(x) \wedge \mathbf{e}_j(x))$ of $\bigwedge^2 \mathrm{T}_x(\mathrm{M})$, this result is equivalent to (20.23.1.1).

(20.23.2) (F. Schur's theorem) *Let* M *be a connected Riemannian manifold of dimension* $n \geqq 3$. *If the equivalent conditions of* (20.23.1) *are satisfied at each point* $x \in$ M, *then the function* A(x) *is constant on* M.

For an orthonormal moving frame **R**, the relations (20.12.3.4) give here, by virtue of (20.23.1.1),

$$d\omega_{ij} = -\sum_k \omega_{ik} \wedge \omega_{kj} + \mathrm{A}\sigma_i \wedge \sigma_j \qquad (1 \leqq i, j \leqq n).$$

Hence, taking the exterior derivative and using also (20.12.3.3), we have

$$\begin{aligned} 0 &= -\sum_k (d\omega_{ik} \wedge \omega_{kj} - \omega_{ik} \wedge d\omega_{kj}) + \mathrm{A}(d\sigma_i \wedge \sigma_j - \sigma_i \wedge d\sigma_j) + d\mathrm{A} \wedge \sigma_i \wedge \sigma_j \\ &= \sum_{h,k} (\omega_{ih} \wedge \omega_{hk} \wedge \omega_{kj} - \omega_{ik} \wedge \omega_{kh} \wedge \omega_{hj}) \\ &\quad - \mathrm{A} \sum_k (\sigma_i \wedge \sigma_k \wedge \omega_{kj} - \omega_{ik} \wedge \sigma_k \wedge \sigma_j) \\ &\quad - \mathrm{A} \sum_k (\omega_{ik} \wedge \sigma_k \wedge \sigma_j - \sigma_i \wedge \omega_{jk} \wedge \sigma_k) + d\mathrm{A} \wedge \sigma_i \wedge \sigma_j, \end{aligned}$$

and since $\omega_{hk} = -\omega_{kh}$, this reduces to

(20.23.2.1) $$d\mathrm{A} \wedge \sigma_i \wedge \sigma_j = 0 \qquad (i \neq j).$$

Put $d\mathrm{A} = \sum_j a_j \sigma_j$. Since $n \geqq 3$, for each index k we can find two indices i, j such that i, j, k are all distinct; hence the formula (20.23.2.1) gives $a_k = 0$ for each k, and therefore $d\mathrm{A} = 0$, which proves the theorem.

(20.23.3) A Riemannian manifold M of dimension $\geqq 2$ is called a *manifold of constant curvature* if the sectional curvature A(P_x) is a constant A independent of $x \in$ M and $\mathrm{P}_x \subset \mathrm{T}_x(\mathrm{M})$. The number A is then called simply the *curvature* of M.

Let M be a Riemannian manifold of constant curvature A, let a be a point of M, and $\mathrm{B}(\mathbf{0}_a; r)$ an open ball contained in $\Omega \cap \mathrm{T}_a(\mathrm{M})$, on which $\exp_a$ is a diffeomorphism onto $\mathrm{B}(a; r)$ (20.16.4). We shall show how to calculate *explicitly* the inverse image $\mathbf{g}_1 = {}^t\exp_a(\mathbf{g})$ of the metric tensor of M. We shall identify $\mathrm{T}(\mathrm{T}_a(\mathrm{M}))$ with $\mathrm{T}_a(\mathrm{M}) \times \mathrm{T}_a(\mathrm{M})$ by means of the canonical trivialization (16.15.5), under which the tangent vector $\tau_{\mathbf{h}_a}^{-1}(\mathbf{k}_a)$ at the point $\mathbf{h}_a$ is identified with $(\mathbf{h}_a, \mathbf{k}_a)$. Suppose that $\mathbf{k}_a$ is *orthogonal* to $\mathbf{h}_a$, and put $v(t) = \exp_a(t\mathbf{h}_a)$. If $\mathbf{w}$ is the Jacobi field along v such that $\mathbf{w}(0) = \mathbf{0}_a$ and $(\nabla_t \cdot \mathbf{w})(0) = \mathbf{k}_a$, then the vector $\mathbf{w}(1)$ is equal to $\mathrm{T}_{\mathbf{h}_a}(\exp_a) \cdot (\mathbf{h}_a, \mathbf{k}_a)$ (18.7.9); and this can be explicitly computed, thanks to the fact that a Jacobi field along v is *proportional to a parallel transport* along v. Let $\mathbf{u}$ be the parallel transport along v such that $\mathbf{u}(0) = \mathbf{k}_a$, so that $\|\mathbf{u}(t)\|^2 = \|\mathbf{k}_a\|^2$ for all t (20.9.5.4); we shall show that there exists a scalar function $\rho(t)$ such that $\mathbf{w} = \rho\mathbf{u}$. We have $\nabla_t \cdot (\rho\mathbf{u}) = \rho'\mathbf{u}$ and $\nabla_t \cdot (\nabla_t \cdot (\rho\mathbf{u})) = \rho''\mathbf{u}$, because $\mathbf{u}$ is a parallel transport (18.6.3); on the other hand, if we complete $\mathbf{h}_a$ and $\mathbf{k}_a$ to an orthogonal basis of $\mathrm{T}_a(\mathrm{M})$ by adjoining vectors $\mathbf{k}_3, \ldots, \mathbf{k}_n$, and if $\mathbf{u}_j$ is the parallel transport of $\mathbf{k}_j$ along v, then it follows from (20.23.1.1) that $((\mathbf{r} \cdot (v' \wedge \mathbf{u})) \cdot v' \,|\, \mathbf{u}_j) = 0$ for $j \geqq 3$. Since in any case we have $((\mathbf{r} \cdot (v' \wedge \mathbf{u})) \cdot v' \,|\, v') = 0$, and since

$$((\mathbf{r} \cdot (v' \wedge \mathbf{u})) \cdot v' \,|\, \mathbf{u}) = -\mathrm{A}\|v' \wedge \mathbf{u}\|^2 = -\mathrm{A}\|\mathbf{h}_a \wedge \mathbf{k}_a\|^2 = -\mathrm{A}\|\mathbf{h}_a\|^2\|\mathbf{k}_a\|^2,$$

it follows that we may write $(\mathbf{r} \cdot (v' \wedge \rho\mathbf{u})) \cdot v' = -\mathrm{A}\|\mathbf{h}_a\|^2\rho\mathbf{u}$. The condition for $\rho\mathbf{u}$ to be a Jacobi field is therefore

(20.23.3.1) $$\rho'' = -\mathrm{A}\|\mathbf{h}_a\|^2\rho,$$

and $\rho\mathbf{u}$ will be equal to $\mathbf{w}$ if we have $\rho(0) = 0$ and $\rho'(0) = 1$ (18.7.5.1), which determines $\rho(t)$ completely: if $\mathrm{A} = 0$, then $\rho(t) = t$; if $\mathrm{A} < 0$, then

(20.23.3.2) $$\rho(t) = \frac{\sinh(t\|\mathbf{h}_a\|\sqrt{-\mathrm{A}})}{\|\mathbf{h}_a\|\sqrt{-\mathrm{A}}};$$

and if $\mathrm{A} > 0$, we have

(20.23.3.3) $$\rho(t) = \frac{\sin(t\|\mathbf{h}_a\|\sqrt{\mathrm{A}})}{\|\mathbf{h}_a\|\sqrt{\mathrm{A}}}.$$

From this calculation we derive an explicit expression for the scalar product in $\mathrm{T}_{\mathbf{h}_a}(\mathrm{T}_a(\mathrm{M}))$ defined by the metric tensor $\mathbf{g}_1$. It follows already from Gauss's lemma (20.16.3.1) that $((\mathbf{h}_a, \mathbf{h}_a) \,|\, (\mathbf{h}_a, \mathbf{h}_a))_{\mathbf{g}_1} = \|\mathbf{h}_a\|^2_{\mathbf{g}}$, and that $((\mathbf{h}_a, \mathbf{h}_a) \,|\, (\mathbf{h}_a, \mathbf{k}_a))_{\mathbf{g}_1} = 0$ if $\mathbf{h}_a$ is orthogonal to $\mathbf{k}_a$ (relative to $\mathbf{g}(a)$); finally, from the preceding work, if $\mathbf{k}'_a$ and $\mathbf{k}''_a$ are orthogonal to $\mathbf{h}_a$ (relative to $\mathbf{g}(a)$), then we have

(20.23.3.4) $$((\mathbf{h}_a, \mathbf{k}'_a) \,|\, (\mathbf{h}_a, \mathbf{k}''_a))_{\mathbf{g}_1} = \rho(1)^2(\mathbf{k}'_a \,|\, \mathbf{k}''_a)_{\mathbf{g}}.$$

The following proposition is an immediate consequence of this:

(20.23.4) *Let* M, M′ *be two Riemannian manifolds of equal constant curvature, and* $\mathbf{g}$, $\mathbf{g}'$ *their respective metric tensors. Let* a (resp. a') *be a point of* M (resp. M′), *and let* F *be a linear isometry of* $\mathrm{T}_a(\mathrm{M})$ (*endowed with the scalar product* $\mathbf{g}(a)$) *onto* $\mathrm{T}_{a'}(\mathrm{M}')$ (*endowed with* $\mathbf{g}'(a')$). *Then, if r is sufficiently small,* $f = \exp_{a'} \circ \mathrm{F} \circ (\exp_a)^{-1}$ *is the unique isometry of the open ball with center a and radius r in* M *onto the open ball with center a′ and radius r in* M′ *such that* $\mathrm{T}_a(f) = \mathrm{F}$.

If the metric tensor of a Riemannian manifold is multiplied by a constant $c > 0$, the Levi–Civita connection remains unchanged (this follows from (20.7.6) and (20.9.4), since the group G is unaltered); the Riemann–Christoffel tensor is multiplied by c, and the sectional curvature $A(P_x)$ is multiplied by c^{-1}. Hence in the study of manifolds of constant curvature we may restrict our considerations to the three cases $A = 1$, $A = -1$, and $A = 0$.

(20.23.5) *A simply-connected, complete Riemannian manifold* M *of constant curvature* $A = 0$ (resp. $A = 1$, $A = -1$) *is isometric to* $\mathbf{R}^n$ (resp. S_n, resp. *hyperbolic space* $\mathbf{Y}_n$ (20.11.7)).

If $A = 0$ or $A = -1$, and if we take $M' = \mathbf{R}^n$ or $M' = \mathbf{Y}_n$, respectively, then for $a \in M$ and $a' \in M'$ the mappings $\exp_a$ and $\exp_{a'}$ are respectively diffeomorphisms of $T_a(M)$ onto M and of $T_{a'}(M')$ onto M′, by virtue of (20.22.1) and (20.20.9); if F is an isometry of $T_a(M)$ onto $T_{a'}(M')$, then it follows from (20.23.4) that $\exp_{a'} \circ F \circ (\exp_a)^{-1}$ is an isometry of M onto M′.

Consider now the case $A = 1$. Then the formulas (20.23.3.3) and (20.23.3.4) show that $\exp_a$ is injective on $B(\mathbf{0}_a; \pi)$, hence is a diffeomorphism onto $B(a; \pi)$ because M is complete (20.20.8), and similarly for $M' = S_n$. Now let a', b' be two points of M′ such that $d(a', b') < \pi$, and let F be a linear isometry of $T_{a'}(M')$ onto $T_a(M)$; then, by (20.23.4), the mapping

$$f = \exp_a \circ F \circ (\exp_{a'})^{-1}$$

is an isometry of $B(a'; \pi)$ onto $B(a; \pi)$. By hypothesis we have $b' \in B(a'; \pi)$; put $b = f(b')$ and $G = T_{b'}(f)$, so that G is a linear isometry of $T_{b'}(M')$ onto $T_b(M)$; then the mapping $g = \exp_b \circ G \circ (\exp_{b'})^{-1}$ is likewise an isometry of $B(b'; \pi)$ onto $B(b; \pi)$. Now the mappings f and g *coincide* in the intersection $U = B(a'; \pi) \cap B(b'; \pi)$. For if C is the great circle arc of length $< \pi$ and end-points a'', b'', the antipodes of a', b' on S_n, then every point x' of $U - C$ is joined to a' (resp. b') by a unique geodesic arc of length $d(a', x')$ (resp. $d(b', x')$) contained in $U - C$. It follows therefore from the choice of G that f and g coincide on $U - C$, and hence by continuity on the whole of U. We have $B(a'; \pi) \cup B(b'; \pi) = M'$; let h be the mapping of M′ into M which is equal to f on $B(a'; \pi)$ and to g on $B(b'; \pi)$, then it is clear that h is a local isometry, and it follows from (20.18.8) that h is an isometry of M′ onto M because M is simply-connected.

The theorem just proved therefore reduces the problem of the determination of all complete connected Riemannian manifolds of constant curvature to that of determining all manifolds whose universal covering is $\mathbf{R}^n$, S_n, or $\mathbf{Y}_n$ (see [73]).

(20.23.6) (E. Cartan) *Let* M *be a connected Riemannian manifold of dimension* $n \geqq 3$. *Then* M *has constant curvature if and only if, for each* $x \in \mathrm{M}$ *and each plane* $\mathrm{P}_x \subset \mathrm{T}_x(\mathrm{M})$, *the surface* $\mathrm{S}(\mathrm{P}_x)$ *generated by the geodesics with origin* x *and tangent vector at* x *belonging to* P_x *is totally geodesic* (20.13.7) *in a neighborhood of* x.

Since the condition is local, to prove its necessity we need consider only the cases of $\mathbf{R}^n$, $\mathbf{S}_n$, and $\mathbf{Y}_n$; but then one sees immediately that there exists a group of isometries of M leaving $\mathrm{S}(\mathrm{P}_x)$ invariant and acting transitively on $\mathrm{S}(\mathrm{P}_x)$, whence the result follows. To prove that the condition is sufficient, consider a point $x \in \mathrm{M}$ and any orthonormal basis $(\mathbf{c}_i)_{1 \leqq i \leqq n}$ of $\mathrm{T}_x(\mathrm{M})$, and take P_x to be the plane spanned by $\mathbf{c}_1$ and $\mathbf{c}_2$. Construct as in (20.12.2) an orthonormal moving frame $\mathbf{R} = (\mathbf{e}_1, \ldots, \mathbf{e}_n)$ in a neighborhood of x, such that $\mathbf{e}_i(x) = \mathbf{c}_i$ for $1 \leqq i \leqq n$, and such that $(\mathbf{e}_1, \mathbf{e}_2)$ restricted to $\mathrm{S}(\mathrm{P}_x)$ is a moving frame for this surface. The hypothesis that $\mathrm{S}(\mathrm{P}_x)$ is totally geodesic signifies that the $n-2$ second fundamental forms $\mathbf{l}_\alpha$ are *identically zero* in $\mathrm{S}(\mathrm{P}_x)$; or, in the notation of (20.12.3), that $\omega'_{\alpha i} = 0$ for $i = 1, 2$ and $\alpha \geqq 3$. Now, by virtue of (20.12.5.1), we have in $\mathrm{S}(\mathrm{P}_x)$

$$d\omega'_{\alpha i} = -\sum_{k=1}^{n} \omega'_{\alpha k} \wedge \omega'_{ki} + \tilde{\Omega}_{\alpha i} \qquad (i = 1, 2, \alpha \geqq 3).$$

It follows that $\tilde{\Omega}_{\alpha i} = 0$ in $\mathrm{S}(\mathrm{P}_x)$ and, in particular, at the point x, that

$$\langle \Omega_{\alpha i}(x), \mathbf{c}_1 \wedge \mathbf{c}_2 \rangle = 0$$

for $i = 1, 2$ and $\alpha \geqq 3$. Since we may replace $\mathbf{c}_1$ and $\mathbf{c}_2$ by any pair $\mathbf{c}_h$, $\mathbf{c}_k$ of basis vectors, we have

(20.23.6.1) $$\langle \Omega_{ij}(x), \mathbf{c}_h \wedge \mathbf{c}_k \rangle = 0,$$

provided i or j is distinct from h and k; or, equivalently (A.14.1.3),

(20.23.6.2) $$\Omega_{ij}(x) = \lambda_{ij}(x)\sigma_i(x) \wedge \sigma_j(x),$$

where $\lambda_{ij}(x)$ is a scalar. If now i, j, k are any three distinct indices and if we replace the vectors $\mathbf{c}_j, \mathbf{c}_k$ by

$$\mathbf{c}'_j = \mathbf{c}_j \cos\alpha + \mathbf{c}_k \sin\alpha, \qquad \mathbf{c}'_k = -\mathbf{c}_j \sin\alpha + \mathbf{c}_k \cos\alpha,$$

then the canonical forms σ'_j, σ'_k for this new frame are given by

$$\sigma'_j(x) = \sigma_j(x)\cos\alpha + \sigma_k(x)\sin\alpha,$$
$$\sigma'_k(x) = -\sigma_j(x)\sin\alpha + \sigma_k(x)\cos\alpha,$$

and the curvature form Ω'_{ij} by

$$\Omega'_{ij}(x) = \Omega_{ij}(x) \cos \alpha + \Omega_{ik}(x) \sin \alpha;$$

the formulas (20.23.6.2) now give

$$\Omega'_{ij}(x) = (\lambda_{ij}(x) \cos^2 \alpha + \lambda_{ik}(x) \sin^2 \alpha)\sigma'_i(x) \wedge \sigma'_j(x)$$
$$-((\lambda_{ij}(x) - \lambda_{ik}(x)) \sin \alpha \cos \alpha)\sigma'_i(x) \wedge \sigma'_k(x),$$

which implies that $\lambda_{ij}(x) = \lambda_{ik}(x)$. Since $\lambda_{ij}(x) = \lambda_{ji}(x)$, it follows that all the $\lambda_{ij}(x)$ are equal to the the same scalar $A(x)$. We have now only to apply (20.23.1) and (20.23.2) to complete the proof.

PROBLEMS

1. Let M be a Riemannian manifold with the property that there exist two constants c_0, c_1 such that $0 < c_0 \leqq A(P_x) \leqq c_1$ for all $x \in M$ and all planes $P_x \subset T_x(M)$. Let $v : [a, b] \to M$ be a geodesic path. If $v(t)$ is not conjugate to $v(a)$ for $a < t < b$, show that

$$\pi c_1^{-1/2} \leqq b - a \leqq \pi c_0^{-1/2}$$

(*O. Bonnet's theorem*). (Use Rauch's comparison theorem (Section 20.22, Problem 3) to compare M with a sphere of appropriate radius.)

2. Let M be a complete connected Riemannian manifold. If there exists a homometry u of M (Section 20.9, Problem 5) of ratio $c \neq 1$, then M has constant zero curvature. (We may assume that $0 < c < 1$. Show first that the sequence $u^n(x)$ of iterated images converges to a fixed point $x_0 \in M$ independent of x. Then show that for each plane $P_x \subset T_x(M)$ we have $A(T(u^n) \cdot P_x) = c^{-2n}A(P_x)$, and deduce that $A(P_x) = 0$.)

3. Let M be a connected Riemannian manifold of dimension n. Show that if the Lie algebra $\mathfrak{i}(M)$ of infinitesimal isometries of M (Section 20.9, Problem 7) has dimension $\frac{1}{2}n(n+1)$, then M is a manifold of constant curvature. (If $\mathbf{\Omega}$ is the curvature form for the Levi-Civita principal connection, show that for every infinitesimal isometry $X \in \mathfrak{i}(M)$ we have $\theta_{\tilde{X}} \cdot (\mathbf{\Omega} \cdot (H_{\mathbf{a}} \wedge H_{\mathbf{b}})) = 0$ (in the notation of Section 20.6, Problem 7). Deduce that, on the one hand, we have $\mathbf{\Omega} \cdot (H_{\mathbf{a}} \wedge H_{\mathbf{b}}) = \mathbf{\Omega} \cdot (H_{s \cdot \mathbf{a}} \wedge H_{s \cdot \mathbf{b}})$ for all $s \in \mathbf{SO}(n, \mathbf{R})$, and that, on the other hand, for fixed $\mathbf{a}$, $\mathbf{b}$ in $\mathbf{R}^n$, the function $((\mathbf{\Omega} \cdot (H_{\mathbf{a}} \wedge H_{\mathbf{b}})) \cdot \mathbf{b} \mid \mathbf{a})$ is locally constant.)

4. Let A be a real number. On an open set U in $\mathbf{R}^n$, define a symmetric covariant tensor field $\mathbf{g}$ by the formula

$$\mathbf{g}(x) = \frac{dx_1 \otimes dx_1 + \cdots + dx_n \otimes dx_n}{(1 + \frac{1}{4}A(x_1^2 + \cdots + x_n^2))^2}$$

for $x = (x_1, \ldots, x_n) \in \mathrm{U}$. Show that U, equipped with the Riemannian structure defined by $\mathbf{g}$, is a manifold of constant curvature equal to A. (If we put

$$u(x) = -\log(1 + \tfrac{1}{4}\mathrm{A}(x_1^2 + \cdots + x_n^2)),$$

show that the forms

$$\sigma_j = e^u dx_i, \qquad \omega_{ij} = \frac{\partial u}{\partial x_j} dx_i - \frac{\partial u}{\partial x_i} dx_j$$

are the canonical and connection forms of the Levi–Civita connection on U.)

APPENDIX

TENSOR PRODUCTS AND FORMAL POWER SERIES

(The numbering of the sections in this Appendix continues that of the Appendix to Volume III.)

20. TENSOR PRODUCTS OF INFINITE-DIMENSIONAL VECTOR SPACES

(A.20.1) Let E, F be two vector spaces (finite-dimensional or not) over a field K; let E*, F* be the *duals* of E, F, respectively (A.9.1), and $\mathscr{L}_2(E^*, F^*; K)$ the vector space of *bilinear forms* on the product $E^* \times F^*$. For each pair of elements $x \in E$, $y \in F$, we define a bilinear form on $E^* \times F^*$ as follows:

$$(x^*, y^*) \mapsto \langle x, x^* \rangle \langle y, y^* \rangle.$$

We denote this bilinear form by $x \otimes y$ and call it the *tensor product* of x and y. The *vector subspace* of $\mathscr{L}_2(E^*, F^*; K)$ spanned by the tensor products $x \otimes y$, where $x \in E$ and $y \in F$, in called the *tensor product* of E and F, and is denoted by $E \otimes_K F$ or simply $E \otimes F$.

(A.20.2) Let $(e_\alpha)_{\alpha \in A}$, $(f_\beta)_{\beta \in B}$ be bases of E, F, respectively. We shall show that the family $(e_\alpha \otimes f_\beta)_{(\alpha, \beta) \in A \times B}$ is a *basis* of $E \otimes_K F$. Since evidently the mapping $(x, y) \mapsto x \otimes y$ of $E \times F$ into $E \otimes_K F$ is *bilinear*, the $e_\alpha \otimes f_\beta$ span $E \otimes_K F$ by definition, and therefore it is enough to show that these elements form a free family. Let $\sum_{\alpha, \beta} \lambda_{\alpha\beta} e_\alpha \otimes f_\beta = 0$ be a linear relation between these elements, with coefficients in K. Then for all $x^* \in E^*$ and $y^* \in F^*$ we have

(A.20.2.1) $$\sum_{\alpha, \beta} \lambda_{\alpha\beta} \langle e_\alpha, x^* \rangle \langle f_\beta, y^* \rangle = 0.$$

For each $\alpha \in A$ (resp. $\beta \in B$), define a linear form e_α^* on E (resp. a linear form f_β^* on F) by the rule $\langle \sum_\mu \xi_\mu e_\mu, e_\alpha^* \rangle = \xi_\alpha$ (resp. $\langle \sum_\nu \eta_\nu f_\nu, f_\beta^* \rangle = \eta_\beta$). Replacing x^* by e_α^* and y^* by f_β^* in (A.20.2.1), we obtain $\lambda_{\alpha\beta} = 0$ for all $\alpha \in A$ and $\beta \in B$, which proves our assertion.

(A.20.3) When E and F are finite-dimensional, the definition given above agress with that of (A.10.3), and in this case $E \otimes_K F$ is identical with

$$\mathscr{L}_2(E^*, F^*; K).$$

If on the other hand at least one of E, F is of infinite dimension, then $E \otimes_K F$ is distinct from $\mathscr{L}_2(E^*, F^*; K)$. The tensor product of any finite number of K-vector spaces is defined in the same way, and the generalization of the properties proved in (A.10.3) and (A.10.4) for finite-dimensional vector spaces is immediate. The isomorphism (A.10.5.1) continues to hold; on the other hand, the mappings (A.10.5.3), (A.10.5.4), and (A.10.5.5) are no longer bijective, but only *injective*, when the vector spaces involved are infinite-dimensional.

(A.20.4) If E and F are two (associative) algebras over K, the tensor product $E \otimes_K F$ possesses a unique structure of K-algebra, such that

(A.20.4.1) $$(x_1 \otimes y_1)(x_2 \otimes y_2) = (x_1 x_2) \otimes (y_1 y_2)$$

for all $x_1, x_2 \in E$, $y_1, y_2 \in F$. For if $(e_\alpha)_{\alpha \in A}$, $(f_\beta)_{\beta \in B}$ are bases of E and F, respectively, we can define an algebra structure (which *a priori* might not be associative) on $E \otimes_K F$ by taking, as multiplication table for the basis $(e_\alpha \otimes f_\beta)$,

$$(e_\alpha \otimes f_\beta)(e_\lambda \otimes f_\mu) = (e_\alpha e_\lambda) \otimes (f_\beta f_\mu);$$

it is clear that this structure is the only one which satisfies (A.20.4.1). The associativity of this algebra then follows immediately from the associativity of E and of F, and (A.20.4.1). If each of E, F has an identity element 1, then $1 \otimes 1$ is the identity element of $E \otimes_K F$; if E, F are commutative, so is $E \otimes_K F$. The algebra $E \otimes_K F$ is called the *tensor product of the algebras* E *and* F.

An important example of a tensor product of commutative algebras is the tensor product $K[X_1, \ldots, X_m] \otimes K[X_{m+1}, \ldots, X_{m+n}]$ of two polynomial algebras, which is isomorphic to $K[X_1, \ldots, X_{m+n}]$.

21. ALGEBRAS OF FORMAL POWER SERIES

(A.21.1) Let K be a field, n an integer >0. Consider the K-vector space $E = K^{\mathbf{N}^n}$, whose elements are all families $(c_\alpha)_{\alpha \in \mathbf{N}^n}$ with $c_\alpha \in K$. We define an *algebra* structure on E by the rule

(A.21.1.1) $$(a_\alpha)(b_\alpha) = (c_\alpha), \qquad \text{where} \quad c_\alpha = \sum_{\lambda+\mu=\alpha} a_\lambda b_\mu .$$

To justify this definition it is enough to verify that the number of pairs of multi-indices (λ, μ) such that $\lambda + \mu = \alpha$ is *finite*; now if $\alpha = (\alpha_i)$, $\lambda = (\lambda_i)$, $\mu = (\mu_i)$, then this relation signifies that $\lambda_i + \mu_i = \alpha_i$ for $1 \leqq i \leqq n$, and hence implies that $0 \leqq \lambda_i \leqq \alpha_i$ and $0 \leqq \mu_i \leqq \alpha_i$ for $1 \leqq i \leqq n$, which proves our assertion.

The algebra so defined is clearly commutative; it is also associative, for if (u_α), (v_α), (w_α) are any three elements of E, we have

$$\sum_{\rho+\nu=\alpha} \Big(\sum_{\lambda+\mu=\rho} u_\lambda v_\mu \Big) w_\nu = \sum_{\lambda+\mu+\nu=\alpha} u_\lambda v_\mu w_\nu = \sum_{\lambda+\sigma=\alpha} u_\lambda \Big(\sum_{\mu+\nu=\sigma} v_\mu w_\nu \Big).$$

Finally, E has an identity element, namely, the element (c_α) such that $c_\alpha = 1$ for $\alpha = (0, \ldots, 0)$ and $c_\alpha = 0$ for all other α.

(A.21.2) For $1 \leqq i \leqq n$ let X_i denote the element (u_α) of E all of whose coordinates are zero except the coordinate with index $\alpha = \varepsilon_i = (\delta_{ij})_{1 \leqq j \leqq n}$ (Kronecker delta), which is equal to 1. It follows from the definition (A.21.1.1) that, for each multi-index α, the element (u_β) of E such that $u_\beta = 0$ for $\beta \neq \alpha$, and $u_\alpha = 1$, may be written as a product $X^\alpha = X_1^{\alpha_1} X_2^{\alpha_1} \cdots X_n^{\alpha_n}$, with the usual notation for multi-indices (9.1). This being so, it is customary to write an element $(c_\alpha)_{\alpha \in \mathbf{N}^n}$ of E in the form $\sum_{\alpha \in \mathbf{N}^n} c_\alpha X^\alpha$; it is to be understood that the summation sign does not in general represent a sum in the usual sense (in algebra, infinite sums are undefined unless all but a finite number of the terms are zero, and this will not in general be the case here). With this notation, the definition of the product in E takes the form

(A.21.2.1) $$\Big(\sum_\alpha a_\alpha X^\alpha\Big)\Big(\sum_\alpha b_\alpha X^\alpha\Big) = \sum_\alpha \Big(\sum_{\lambda+\mu=\alpha} a_\lambda b_\mu \Big) X^\alpha;$$

in other words, the formula for the product of two elements of E has the same form as that for the product of two absolutely convergent power series ((5.5.3) and (9.2.1)). The elements of E are called *formal power-series in* n *indeterminates* X_i $(1 \leqq i \leqq n)$, *with coefficients in* K, and the algebra E is denoted by $K[[X_1, \ldots, X_n]]$. We remark that the polynomial algebra

$K[X_1, \ldots, X_n]$ is the subalgebra of $K[[X_1, \ldots, X_n]]$ consisting of the (c_α) such that $c_\alpha = 0$ for all but a finite number of indices $\alpha \in \mathbf{N}^n$.

If $f = \sum_\alpha c_\alpha X^\alpha$ is a formal power series, the monomials $c_\alpha X^\alpha$ are called the *terms* of the series; the term $c_\alpha X^\alpha$ is said to be of *multidegree* α and (*total*) *degree* $|\alpha| = \alpha_1 + \cdots + \alpha_n$. There are only finitely many terms of f with a given total degree m, and their sum $\sum_{|\alpha|=m} c_\alpha X^\alpha$ is a *homogeneous polynomial* of degree m, called the *homogeneous part of degree m of f*. The homogeneous part of degree 0 of f is c_0, which is also called the *constant term* of f. If $c_0 = 0$, we say that f is *without constant term.*

(A.21.3) Let f be a formal power series belonging to $K[[X_1, \ldots, X_n]]$, and let $g_1, \ldots, g_n$ be n formal power series *without constant terms*, belonging to $K[[X_1, \ldots, X_m]]$. We shall define a formal power series

$$f(g_1, \ldots, g_n) \in K[[X_1, \ldots, X_m]],$$

which coincides with the formal power series denoted by the same symbol when f is a polynomial. For this purpose, let P_r (resp. Q_{jr}) be the polynomial which is the sum of all terms of f (resp. g_j) of total degree $\leqq r$, and consider the polynomial $P_r(Q_{1r}, \ldots, Q_{nr}) \in K[X_1, \ldots, X_m]$. We assert that if $r \leqq s$, *the terms of degree* $\leqq r$ in $P_s(Q_{1s}, \ldots, Q_{ns})$ are *the same* as in $P_r(Q_{1r}, \ldots, Q_{nr})$. Let $f = \sum_\alpha c_\alpha X^\alpha$; then the sum of the terms of degree $\leqq r$ in $P_s(Q_{1s}, \ldots, Q_{ns})$ is obtained by taking the terms of degree $\leqq r$ in each polynomial $c_\alpha Q_{1s}^{\alpha_1} \cdots Q_{ns}^{\alpha_n}$ for which $|\alpha| \leqq s$, and then taking the sum of all these terms. Now, the sum of all the terms of degree $\leqq r$ in $Q_{1s}^{\alpha_1} \cdots Q_{ns}^{\alpha_n}$ is the sum of all the products $\prod_{j=1}^{n} \left(\prod_{k=1}^{\alpha_j} t_{jk} \right)$, where, for $1 \leqq j \leqq n$ and $1 \leqq k \leqq \alpha_j$, t_{jk} is a term of degree v_{jk} in Q_{js}, and the v_{jk} satisfy the inequality

(A.21.3.1)
$$\sum_{j=1}^{n} \left(\sum_{k=1}^{\alpha_j} v_{jk} \right) \leqq r.$$

Since the g_j are without constant terms, we have $v_{jk} \geqq 1$ for all j, k, and therefore $\sum_{j=1}^{n} \left(\sum_{k=1}^{\alpha_j} v_{jk} \right) \geqq |\alpha|$, so that there is no term of degree $\leqq r$ in $c_\alpha Q_{1s}^{\alpha_1} \cdots Q_{ns}^{\alpha_n}$ if $|\alpha| > r$; on the other hand, if $\alpha_j \neq 0$ the relation (A.21.3.1) implies that $v_{jk} \leqq r$ for $1 \leqq k \leqq \alpha_j$, which establishes our assertion.

If R_r is the polynomial which is the sum of the terms of degree $\leqq r$ in $P_r(Q_{1r}, \ldots, Q_{nr})$, we may therefore, for each multi-index $\alpha \in \mathbf{N}^m$, define the monomial $a_\alpha X^\alpha$ to be the monomial of multidegree α in *all* R_r such that $r \geqq |\alpha|$. The formal power series $\sum_\alpha a_\alpha X^\alpha$ is denoted by $f(g_1, \ldots, g_n)$ and is said to be obtained by *substituting* g_j *for* X_j *in f for* $1 \leqq j \leqq n$.

This operation has a property of associativity, corresponding to associativity for the composition of mappings (1.7):

(A.21.4) *Let f be a formal power series belonging to* $K[[X_1, \dots, X_n]]$, *let* g_j $(1 \leqq j \leqq n)$ *be n formal power series without constant terms belonging to*

$$K[[X_1, \dots, X_m]],$$

and let h_k $(1 \leqq k \leqq m)$ *be m formal power series without constant terms belonging to* $K[[X_1, \dots, X_p]]$. *Then, putting* $u = f(g_1, \dots, g_n)$, *we have*

(A.21.4.1)
$$u(h_1, \dots, h_m) = f(g_1(h_1, \dots, h_m), \dots, g_n(h_1, \dots, h_m)).$$

Let P_r (resp. Q_{jr}, R_{kr}) be the sum of the terms of degree $\leqq r$ in f (resp. g_j, h_k) and let U_r be the sum of the terms of degree $\leqq r$ in $f(g_1, \dots, g_n)$. The reasoning in (A.21.3) shows that the sum of the terms of degree $\leqq r$ in $g_j(h_1, \dots, h_m)$ is the same as in $Q_{jr}(R_{1r}, \dots, R_{mr})$, and that the sum of the terms of degree $\leqq r$ in $f(g_1(h_1, \dots, h_m), \dots, g_n(h_1, \dots, h_m))$ is the same as in the polynomial $P_r(Q_{1r}(R_{1r}, \dots, R_{mr}), \dots, Q_{nr}(R_{1r}, \dots, R_{mr}))$; on the other hand, the sum U_r of the terms of degree $\leqq r$ in u is the same as in $P_r(Q_{1r}, \dots, Q_{nr})$ and, repeating the same argument, the sum of the terms of degree $\leqq r$ in $U_r(R_{1r}, \dots, R_{mr})$ is the same as in the polynomial

$$P_r(Q_{1r}(R_{1r}, \dots, R_{mr}), \dots, Q_{nr}(R_{1r}, \dots, R_{mr})).$$

This proves the result.

(A.21.5) If $f_1, f_2 \in K[[X_1, \dots, X_n]]$ and if g_j $(1 \leqq j \leqq n)$ are power series without constant terms, belonging to $K[[X_1, \dots, X_m]]$, and if $f = f_1 + f_2$, then it is clear that

$$f(g_1, \dots, g_n) = f_1(g_1, \dots, g_n) + f_2(g_1, \dots, g_n).$$

Likewise, putting $F = f_1 f_2$, we have

$$F(g_1, \dots, g_n) = f_1(g_1, \dots, g_n) f_2(g_1, \dots, g_n).$$

This may be proved directly as in (A.21.4), or deduced from (A.21.4) by remarking that it is a particular case of this result if f_1, f_2 are without constant terms; and that in general we may write $f_1 = a + F_1$, $f_2 = b + F_2$, where $a, b \in K$ and F_1, F_2 are without constant terms, and

$$f_1 f_2 = ab + aF_2 + bF_1 + F_1 F_2.$$

REFERENCES

VOLUME I

[1] Ahlfors, L., "Complex Analysis," McGraw-Hill, New York, 1953.

[2] Bachmann, H., "Transfinite Zahlen" (Ergebnisse der Math., Neue Folge, Heft 1). Springer, Berlin, 1955.

[3] Bourbaki, N., "Eléments de Mathématique," Livre I, "Théorie des ensembles" (Actual. Scient. Ind., Chaps. I, II, No. 1212; Chap. III, No. 1243). Hermann, Paris, 1954–1956.

[4] Bourbaki, N., "Eléments de Mathématique," Livre II, "Algèbre" (Actual Scient. Ind., Chap. II, Nos. 1032, 1236, 3rd ed.). Hermann, Paris, 1962.

[5] Bourbaki, N., "Eléments de Mathématique," Livre III, "Topologie générale" (Actual. Scient. Ind., Chaps. I, II, Nos. 858, 1142, 4th ed.; Chap. IX, No. 1045, 2nd ed.; Chap. X, No. 1084, 2nd ed.). Hermann, Paris, 1958–1961.

[6] Bourbaki, N., "Eléments de Mathématique," Livre, V, "Espaces vectoriels topologiques" (Actual. Scient. Ind., Chap. I, II, No. 1189, 2nd ed.; Chaps. III–V, No. 1229). Hermann, Paris, 1953–1955.

[7] Cartan, H., Séminaire de l'Ecole Normale Supérieure, 1951–1952: "Fonctions analytiques et faisceaux analytiques."

[8] Cartan, H., "Théorie Élémentaire des Fonctions Analytiques." Hermann, Paris, 1961.

[9] Coddington, E., and Levinson, N., "Theory of Ordinary Differential Equations." McGraw-Hill, New York, 1955.

[10] Courant, R., and Hilbert, D., "Methoden der mathematischen Physik," Vol. I, 2nd ed. Springer, Berlin, 1931.

[11] Halmos, P., "Finite Dimensional Vector Spaces," 2nd ed. Van Nostrand-Reinhold, Princeton, New Jersey, 1958.

[12] Ince, E., "Ordinary Differential Equations," Dover, New York, 1949.

[13] Jacobson, N., "Lectures in Abstract Algebra," Vol. II, "Linear algebra." Van Nostrand-Reinhold, Princeton, New Jersey, 1953.

[14] Kamke, E., "Differentialgleichungen reeller Funktionen." Akad. Verlag, Leipzig, 1930.

[15] Kelley, J., "General Topology." Van Nostrand-Reinhold, Princeton, New Jersey, 1955.

[16] Landau, E., "Foundations of Analysis." Chelsea, New York, 1951.

[17] Springer, G., "Introduction to Riemann Surfaces." Addison-Wesley, Reading, Massachusetts, 1957.
[18] Weil, A., "Introduction à l'Étude des Variétés Kählériennes" (Actual. Scient. Ind., No. 1267). Hermann, Paris, 1958.
[19] Weyl, H., "Die Idee der Riemannschen Fläche," 3rd ed. Teubner, Stuttgart, 1955.

VOLUME II

[20] Akhiezer, N., "The Classical Moment Problem." Oliver and Boyd, Edinburgh–London, 1965.
[21] Arnold, V. and Avez, A., "Théorie Ergodique des Systèmes Dynamiques." Gauthier-Villars, Paris, 1967.
[22] Bourbaki, N., "Eléments de Mathématique," Livre VI, "Intégration " (Actual. Scient. Ind., Chap. I–IV, No. 1175, 2nd ed., Chap. V, No. 1244, 2nd ed., Chap. VII–VIII, No. 1306). Hermann, Paris, 1963–67.
[23] Bourbaki, N., "Eléments de Mathématique: Théories Spectrales" (Actual. Scient. Ind., Chap. I, II, No. 1332). Hermann, Paris, 1967.
[24] Dixmier, J., "Les Algèbres d'Opérateurs dans l'Espace Hilbertien." Gauthier-Villars, Paris, 1957.
[25] Dixmier, J., "Les C*-Algèbres et leurs Représentations." Gauthier-Villars, Paris, 1964.
[26] Dunford, N. and Schwartz, J., "Linear Operators. Part II: Spectral Theory." Wiley (Interscience), New York, 1963.
[27] Hadwiger, H., "Vorlesungen über Inhalt, Oberfläche und Isoperimetrie." Springer, Berlin, 1957.
[28] Halmos, P., "Lectures on Ergodic Theory." Math. Soc. of Japan, 1956.
[29] Hoffman, K., "Banach Spaces of Analytic Functions." New York, 1962.
[30] Jacobs, K., "Neuere Methoden und Ergebnisse der Ergodentheorie" (Ergebnisse der Math., Neue Folge, Heft 29). Springer, Berlin, 1960.
[31] Kaczmarz, S. and Steinhaus, H., "Theorie der Orthogonalreihen." New York, 1951.
[32] Kato, T., "Perturbation Theory for Linear Operators." Springer, Berlin, 1966.
[33] Montgomery, D. and Zippin, L., "Topological Transformation Groups." Wiley (Interscience), New York, 1955.
[34] Naimark, M., "Normal Rings." P. Nordhoff, Groningen, 1959.
[35] Rickart, C., "General Theory of Banach Algebras." Van Nostrand-Reinhold, New York, 1960.
[36] Weil, A., "Adeles and Algebraic Groups." The Institute for Advanced Study, Princeton, New Jersey, 1961.

VOLUME III

[37] Abraham, R., "Foundations of Mechanics." Benjamin, New York, 1967.
[38] Cartan, H., Séminaire de l'École Normale Supérieure, 1949–50: "Homotopie: espaces fibrés."
[39] Chern, S. S., "Complex Manifolds" (Textos de matematica, No. 5). Univ. do Recife, Brazil, 1959.

[40] Gelfand, I. M. and Shilov, G. E., "Les Distributions," Vols. 1 and 2. Dunod, Paris, 1962.

[41] Gunning, R., "Lectures on Riemann Surfaces." Princeton Univ. Press, Princeton, New Jersey, 1966.

[42] Gunning, R., "Lectures on Vector Bundles over Riemann Surfaces." Princeton Univ. Press, Princeton, New Jersey, 1967.

[43] Hu, S. T., "Homotopy Theory." Academic Press, New York, 1969.

[44] Husemoller, D., "Fiber Bundles." McGraw-Hill, New York, 1966.

[45] Kobayashi, S., and Nomizu, K., "Foundations of Differential Geometry," Vols. 1 and 2. Wiley (Interscience), New York, 1963 and 1969.

[46] Lang, S., "Introduction to Differentiable Manifolds." Wiley (Interscience), New York, 1962.

[47] Porteous, I. R., "Topological Geometry." Van Nostrand-Reinhold, Princeton, New Jersey, 1969.

[48] Schwartz, L., "Théorie des Distributions," New ed. Hermann, Paris, 1966.

[49] Steenrod, N., "The Topology of Fiber Bundles." Princeton Univ. Press, Princeton, New Jersey, 1951.

[50] Sternberg, S., "Lectures on Differential Geometry." Prentice-Hall, Englewood Cliffs, New Jersey, 1964.

VOLUME IV

[51] Abraham, R. and Robbin, J., "Transversal Mappings and Flows." Benjamin, New York, 1967.

[52] Berger, M., "Lectures on Geodesics in Riemannian Geometry." Tata Institute of Fundamental Research, Bombay, 1965.

[53] Carathéodory, C., "Calculus of Variations and Partial Differential Equations of the First Order," Vols. 1 and 2. Holden-Day, San Francisco, 1965.

[54] Cartan, E., "Oeuvres Complètes," Vols. 1_{I} to 3_{II}. Gauthier-Villars, Paris, 1952–1955.

[55] Cartan, E., "Leçons sur la Théorie des Espaces à Connexion Projective." Gauthier-Villars, Paris, 1937.

[56] Cartan, E., "La Théorie des Groupes Finis et Continus et la Géométrie Différentielle traitées par la Méthode du Repère Mobile." Gauthier-Villars, Paris, 1937.

[57] Cartan, E., "Les Systèmes Différentiels Extérieurs et leurs Applications Géométriques." Hermann, Paris, 1945.

[58] Gelfand, I. and Fomin, S., "Calculus of Variations." Prentice Hall, Englewood Cliffs, New Jersey, 1963.

[59] Godbillon, C., "Géométrie Différentielle et Mécanique Analytique." Hermann, Paris, 1969.

[60] Gromoll, D., Klingenberg, W. and Meyer, W., "Riemannsche Geometrie im Grossen," Lecture Notes in Mathematics No. 55. Springer, Berlin, 1968.

[61] Guggenheimer, H., "Differential Geometry." McGraw-Hill, New York, 1963.

[62] Helgason, S., "Differential Geometry and Symmetric Spaces." Academic Press, New York, 1962.

[63] Hermann, R., "Differential Geometry and the Calculus of Variations." Academic Press, New York, 1968.

[64] Hochschild, G., "The Structure of Lie Groups." Holden-Day, San Francisco, 1965.

[65] Klötzler, R., "Mehrdimensionale Variationsrechnung." Birkhäuser, Basle, 1970.
[66] Loos, O., "Symmetric Spaces," Vols. 1 and 2. Benjamin, New York, 1969.
[67] Milnor, J., "Morse Theory," Princeton University Press, Princeton, New Jersey, 1963.
[68] Morrey, C., "Multiple Integrals in the Calculus of Variations." Springer, Berlin, 1966.
[69] Reeb, G., "Sur les Variétés Feuilletées." Hermann, Paris, 1952.
[70] Rund, H., "The Differential Geometry of Finsler Spaces." Springer, Berlin, 1959.
[71] Schirokow, P. and Schirokow, A., "Affine Differentialgeometrie." Teubner, Leipzig, 1962.
[72] Serre, J. P., "Lie Algebras and Lie Groups." Benjamin, New York, 1965.
[73] Wolf, J., "Spaces of Constant Curvature." McGraw-Hill, New York, 1967.

INDEX

In the following index the first reference number refers to the chapter in which the subject may be found and the second to the section within that chapter.

D

E

F

G

H

M

N

O

P

R

S

T

U

Pure and Applied Mathematics

A Series of Monographs and Textbooks

Editors **Samuel Eilenberg and Hyman Bass**

Columbia University, New York

RECENT TITLES

Ernst August Behrens. Ring Theory

Morris Newman. Integral Matrices

Glen E. Bredon. Introduction to Compact Transformation Groups

Werner Greub, Stephen Halperin, and Ray Vanstone. Connections, Curvature, and Cohomology: Volume I, De Rham Cohomology of Manifolds and Vector Bundles. Volume II, Lie Groups, Principal Bundles, and Characteristic Classes. *In preparation:* Volume III, Cohomology of Principal Bundles and Homogeneous Spaces

Xia Dao-xing. Measure and Integration Theory of Infinite-Dimensional Spaces: Abstract Harmonic Analysis

Ronald G. Douglas. Banach Algebra Techniques in Operator Theory

Willard Miller, Jr. Symmetry Groups and Their Applications

Arthur A. Sagle and Ralph E. Walde. Introduction to Lie Groups and Lie Algebras

T. Benny Rushing. Topological Embeddings

James W. Vick. Homology Theory: An Introduction to Algebraic Topology

E. R. Kolchin. Differential Algebra and Algebraic Groups

Gerald J. Janusz. Algebraic Number Fields

A. S. B. Holland. Introduction to the Theory of Entire Functions

Wayne Roberts and Dale Varberg. Convex Functions

A. M. Ostrowski. Solution of Equations in Euclidean and Banach Spaces, Third Edition of Solution of Equations and Systems of Equations

H. M. Edwards. Riemann's Zeta Function

Samuel Eilenberg. Automata, Languages, and Machines: Volume A. *In preparation:* Volume B

Morris Hirsch and Stephen Smale. Differential Equations, Dynamical Systems, and Linear Algebra

Wilhelm Magnus. Noneuclidean Tesselations and Their Groups

J. Dieudonné. Treatise on Analysis, Volume IV

In preparation

Brayton Gray. Homotopy Theory: An Introduction to Algebraic Topology

Francois Treves. Basic Linear Partial Differential Equations

4 5 6 7 8 9 0 1 2 3

